OXFORD MATHEMATICS

# Higher GCSE for AQA

## Linear Specification



**Course Editors**
Peter McGuire Ken Smith

**Course Consultants**
Keith Gordon
*Head of Mathematics at Wath Comprehensive School, Rotherham;* ***Chief Examiner***

Trevor Senior
*Head of Mathematics at Wales High School, Rotherham;* ***Chief Examiner***

*Simple geometrical shapes occur commonly in architecture.*

*You can find out more about shapes in section 3.*

OXFORD
UNIVERSITY PRESS

**OXFORD**
UNIVERSITY PRESS
Great Clarendon Street, Oxford OX2 6DP

Oxford University Press is a department of the University of Oxford.
It furthers the University's objective of excellence in research, scholarship,
and education by publishing worldwide in

Oxford New York
Auckland Bangkok Buenos Aires Cape Town Chennai
Dar es Salaam Delhi Hong Kong Istanbul Karachi Kolkata
Kuala Lumpur Madrid Melbourne Mexico City Mumbai Nairobi
São Paulo Shanghai Singapore Taipei Tokyo Toronto
with an associated company in Berlin

Oxford is a registered trade mark of Oxford University Press
in the UK and in certain other countries



10 9 8 7 6 5 4 3 2



Many thanks to the original authors of this series for the use of their material:
Sue Briggs Peter McGuire Derek Philpott Susan Shilton Ken Smith

The authors would like to thank Paul Metcalf for his authoritative coursework guidance.



First published 2001



British Cataloguing in Publication Data

Data available

ISBN 0 19 914813 9

The publishers would like to thank AQA for their kind permission to reproduce past paper questions. AQA accept no responsibility for the answers to the past paper questions which are the sole responsibility of the publishers.

The publishers and authors are grateful to the following:

| | |
|---|---|
| Illustrators | Gecko, Oxford Illustrators, Nick Hawken, Pat Moffett, Ian Dicks, Philip Reeve |
| Photographers | Mike Dudley, Martin Sookias, Andrew Ward |
| Photographic Libraries | Mary Evans Picture Library |
| Suppliers | Eurostar |
| Cover artwork | Pictor Photo Library |

Typeset in Great Britain by Mathematical Composition Setters Ltd, Hilltop Business Park, Devizes Road, Salisbury, Wiltshire.
Printed and bound in pain by Edelvives, Zaragoza

# About this book

This book is designed to help you achieve your best possible grade in the **AQA Specification A** Higher GCSE examination. The book is divided into two parts: a full-colour part containing the GCSE content that you need to learn, and a black-and-white part containing plenty of exam practice and answers.

This is more than just a book of questions: it is a learning package that will help you to make the most of your mathematical talents and expertise. You can be confident that you will be well prepared for your AQA examination.

**Wordfinder**

As well as a detailed **Contents** list and an **index**, the book contains a **Wordfinder** on pages 10 and 11. This provides an alphabetical list of the mathematical terms used in the book, and tells you where to find them.

**Starting points**

The content of the book is divided into sections, each covering a particular topic. Each section begins with **Starting points**, which lists the facts and techniques that you should already know. There are some questions for you to try out before starting the section.

**Sections**

Colour is used in the sections in the following ways:

- Yellow panels contain explanations and worked examples.
- The more difficult questions in an exercise are numbered in blue – for example, on page 17:
  4 Copy and complete: a) $\frac{\square}{x} - \frac{5}{3x} = \frac{19}{3x}$
- Blue text is used at times to stress important points – for example, on page 97, (factor of $ax$ + factor of $c$)(factor of $ax$ + factor of $c$) highlights the key terms in factorising quadratics.
- Words in the margin that link with the main text are coloured red. For example, on page 235: supplementary is defined in the margin and referred to in the main text.

  There are also plenty of exercises to consolidate your learning.
  A **non-calculator icon** is used to identify exercises where you should not use a calculator.

**End points**

Each section finishes with **Endpoints**, where the main work of the section is summarized. These are excellent for revision purposes.

**Skills break**

At certain points throughout the book there are **Skills breaks**. Each one provides a variety of questions all linked to the same data, and allows you to draw on the skills and techniques that you have learned.

**In focus**

Towards the end of the book are **In focus** pages which relate to a particular section, and you can use these for in-depth revision.

**Coursework**

There is a separate **Coursework Guidance** section, which details what you need to do for your investigative task and your statistical task. It contains sample tasks, with **Moderator comments** to help you get better marks.

**Exam questions**

The **Exam questions** section contains hundreds of past paper questions from AQA to help you prepare for your exam. There are also two **practice exam papers**, one calculator and one non-calculator, so that you can gain familiarity and confidence with the new AQA examination style.

**Answers**

Numerical **answers** are given at the end of the book.

**Good luck in your exams!**

# CONTENTS

**Note for students taking the modular GCSE course:**

If you are following **AQA Specification B (modular)**, these symbols will tell you which module the topic is assessed in.

■ Module 1
■ ■ Module 3
■ ■ ■ Module 5

The coursework component (comprising Modules 2 and 4) is covered in the Coursework Guidance section.

■ Means that the Topic goes slightly beyond the Higher level.

## A note on accuracy

Make sure your answer is given to any degree of accuracy stated in the question, for example 2 dp or 1 sf. Where it is not stated, choose a sensible degree of accuracy for your answer, and make sure you work to a greater degree of accuracy through the problem. For example, if you choose to give an answer to 3 sf, work to at least 4 sf through the problem, then round your final answer to 3 sf.

Examination groups differ in their approach to accuracy. Some say that you should not give your final answer to a greater degree of accuracy than that used for the data in the question, but others state answers should be given to 3 sf.

If you are in any doubt, check with your examination group.

## Metric and imperial units

| | Metric | Imperial | Some approximate conversions |
|---|---|---|---|
| **Length** | millimetres (mm)<br>centimetres (cm)<br>metres (m)<br>kilometres (km)<br><br>1 cm = 10 mm<br>1 m = 100 cm<br>1 km = 1000 m | inches (in)<br>feet (ft)<br>yards (yd)<br>miles<br><br>1 ft = 12 in<br>1 yd = 3 ft<br>1 mile = 1760 yd | 1 inch = 2.54 cm<br>1 foot ≈ 30.5cm<br>1 metre ≈ 39.4 in<br>1 mile ≈ 1.61 km |
| **Mass** | grams (g)<br>kilograms (kg)<br>tonnes<br><br>1 kg = 1000 g<br>1 tonne = 1000 kg | ounces (oz)<br>pounds (lb)<br>stones<br><br>1 lb = 16 oz<br>1 stone = 14 lb | 1 pound ≈ 454 g<br>1 kilogram ≈ 2.2 lb |
| **Capacity** | millilitres (ml)<br>centilitres (cl)<br>litres<br><br>1 cl = 10 ml<br>1 litre = 100 cl<br>= 1000 ml | pints (pt)<br>gallons<br><br>1 gallon = 8 pints | 1 gallon ≈ 4.55 litres<br>1 litre ≈ 1.76 pints<br>≈ 0.22 gallons |

## Starting points

You need to know about ... | ... so try these questions

### A Multiples, factors and prime numbers

- The **common multiples** of two numbers are those that are multiples of both.

  **Example** Multiples of 4 are: 4, 8, 12, 16, 20, 24, ... .
  Multiples of 6 are: 6, 12, 18, 24, 30, 36, ... .

  Common multiples of 4 and 6 are: 12, 24, 36, ... .
  The **least common multiple** of 6 and 4 is 12.

- The **common factors** of two numbers are those that are factors of both.

  **Example** Factors of 4 are: 1, 2 and 4.
  Factors of 6 are: 1, 2, 3 and 6.

  Common factors of 4 and 6 are: 1 and 2.
  The **highest common factor** of 4 and 6 is 2.

- A number is **prime** if it has exactly two different factors.

  **Example** 5 is a prime number: its factors are 1 and 5.

### B Writing a number as a product of primes

- The factors of a number that are prime are called **prime factors**.
- A multiplication of prime factors is called a **product of primes**.
- One way to write a number as its product of primes is to break it down into factor pairs until a product of primes is reached.

  **Example** Write 63 as a product of primes.

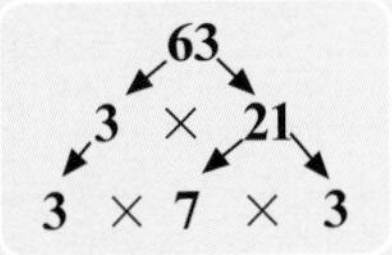


  As a product of primes:
  $63 = 3 \times 3 \times 7$
  $= 3^2 \times 7$ in index notation.

### C Equivalent fractions

- Two fractions equal in value are called **equivalent** fractions, e.g. $\frac{1}{2}$ and $\frac{5}{10}$ are equivalent fractions.
- To find equivalent fractions, multiply or divide the numerator ('top') and denominator ('bottom') by the same number.

  **Examples**

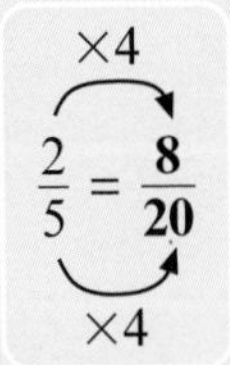


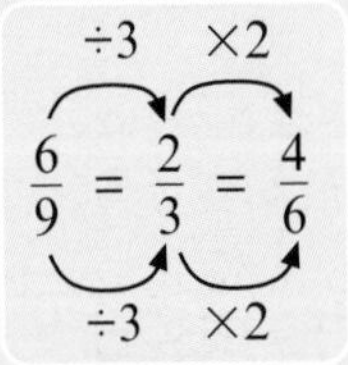


- A fraction in its lowest terms is an equivalent fraction where the numerator and denominator have no common factors except 1.

  **Example** In its lowest terms, $\frac{6}{9}$ is written as $\frac{2}{3}$.

---

**A1** Give three common multiples of:
**a** 5 and 8 **b** 9 and 12

**A2** Find the least common multiple of:
**a** 2 and 7 **b** 3 and 5
**c** 6 and 8 **d** 2 and 4

**A3** List the common factors of:
**a** 24 and 30 **b** 25 and 30

**A4** What is the highest common factor of:
**a** 2 and 7 **b** 6 and 8
**c** 4 and 8 **d** 36 and 54?

**A5** Explain why 1 is not prime.

**A6** List all the prime numbers between 30 and 40.

**A7** Explain why 2 is the only even prime number.

**B1** Write each of these as a product of primes in index notation.
**a** 20 **b** 36
**c** 385 **d** 504

**B2** Evaluate each product of primes:
**a** $2^5 \times 3^2 \times 7$
**b** $3^3 \times 5 \times 11^2$

**C1** Which of these fractions is equivalent to $\frac{4}{10}$?

$\frac{8}{20}, \frac{7}{13}, \frac{12}{30}, \frac{6}{15}, \frac{2}{5}$

**C2** Write three equivalent fractions for:
**a** $\frac{6}{8}$ **b** $\frac{2}{9}$

**C3** Write each fraction in its lowest terms.
**a** $\frac{15}{25}$ **b** $\frac{6}{36}$

## D Mixed numbers and improper fractions

- An improper fraction is one where the numerator is larger than the denominator ('top-heavy'), e.g $\frac{11}{4}$
- An improper fraction is greater than 1, so it can be written as a **mixed number**, a mixture of whole numbers and fractions.

  Example $\frac{11}{4} = \frac{8}{4} + \frac{3}{4} = 2\frac{3}{4}$
- Any whole number can be written as an improper fraction.

  Example $5 = \frac{5}{1}$

**D1** Write as a mixed number:

a $\frac{5}{4}$  b $\frac{8}{3}$

c $\frac{40}{12}$  d $\frac{20}{4}$

**D2** Write as an improper fraction:

a $2\frac{4}{7}$  b $1\frac{1}{2}$

c $4\frac{5}{8}$  d 3

## E Calculating with fractions

- To **add** or **subtract** fractions:
  - find a common multiple of the denominators.
  - find equivalent fractions with a common multiple as the new denominator
  - add or subtract the fractions.

  Example $\frac{3}{4} - \frac{1}{6} = \frac{9}{12} - \frac{2}{12} = \frac{7}{12}$

- To **multiply** fractions:
  - multiply the numerators
  - multiply the denominators.

  Example $\frac{2}{3} \times \frac{1}{5} = \frac{2}{15}$

- Two numbers that multiply together to give 1 are **reciprocals** of each other.

  Example The reciprocal of $\frac{3}{4}$ is $\frac{4}{3}$.

  The reciprocal of 1.5 is $\frac{2}{3}$.

  Note The reciprocal of zero is undefined.

- Dividing by a fraction has the same effect as multiplying by its reciprocal so, to **divide** fractions:
  - write the division as a multiplication
  - multiply the fractions.

  Example $\frac{1}{4} \div \frac{2}{3} = \frac{1}{4} \times \frac{3}{2} = \frac{3}{8}$

**E1** Find the reciprocal of:

a $\frac{5}{4}$  b 3

c $2\frac{1}{2}$  d $1\frac{2}{3}$

e $1\frac{3}{4}$  f $\frac{5}{8}$

g 1.4  h 2.6

i 3.2  j 0

**E2** Give each answer in its lowest terms.

a $\frac{5}{7} - \frac{3}{7}$  b $\frac{1}{3} + \frac{2}{5}$

c $\frac{2}{3} \times \frac{3}{7}$  d $\frac{3}{8} \times \frac{4}{3}$

e $\frac{3}{4} \div \frac{4}{5}$  f $\frac{5}{6} \div 10$

g $1 \div \frac{4}{5}$  h $3 \div \frac{5}{3}$

**E3** What is the value of $a$, in fractional form, when $\frac{1}{a} = \frac{7}{4}$?

# Prime products, multiples and factors

Prime products can be used to find highest common factors and least common multiples.

**Example 1** Find **a** the highest common factor of 72 and 60
**b** least common multiple of 72 and 60

As prime products, **72** = $2 \times 2 \times 2 \times 3 \times 3$
**60** = $2 \times 2 \times 3 \times 5$.

**a** 2 appears at least twice
3 appears at least once in each prime product:
$2 \times 2 \times 2 \times 3 \times 3$ and $2 \times 2 \times 3 \times 5$.

So the **highest common factor** is $2 \times 2 \times 3 = 12$.

**b** 2 appears at most three times in a prime product ($2 \times 2 \times 2 \times 3 \times 3$)
3 appears at most twice in a prime product ($2 \times 2 \times 2 \times 3 \times 3$)
5 appears at most once in a prime product ($2 \times 2 \times 3 \times 5$).

So the **least common multiple** is $2 \times 2 \times 2 \times 3 \times 3 \times 5 =$ **360**.

**Exercise 1.1**
**Multiples, factors and primes**

**1** **a** Find the highest common factor of **i** 240 and 168 **ii** 260 and 180.
**b** Find the least common multiple of **i** 45 and 60 **ii** 450 and 360.

**2** **a** Choose two prime numbers from this list 7, 13, 19, 23, 29.
**b** Find their highest common factor and least common multiple.
**c** Investigate for all different pairs of prime numbers from the list.

A product is the result of a multiplication.

For example, the product of 3 and 6 is 18.

**3** For any pair of numbers, Sue thinks there could be a rule that links the product, the highest common factor and the least common multiple.

She makes a table.

| Numbers | Product | HCF | LCM |
|---|---|---|---|
| 12, 70 | 840 | 2 | 420 |
| 7, 24 | | | |
| 4, 36 | | | |
| 70, 110 | | | |

HCF is shorthand for highest common factor.

LCM is shorthand for least common multiple.

**a** Copy and complete the table.
**b** For any pair of numbers, find a rule that links the product, the highest common factor and the least common multiple.
**c** Use your rule to find the least common multiple of 24 and 36.

**4** Two numbers have a highest common factor of 8 and a least common multiple of 160. Find a pair of numbers that fit this description.

**5** Four flatmates wash their hair regularly: Di every 2 days, Pete every 3 days, Alison every 4 days, and Jake every 5 days.
All four people wash their hair on 1 June.
What is the next date when they will all wash their hair?

**6** Each edge of a cuboid-shaped box measures a whole number of centimetres. The areas of the three different faces are $120\,cm^2$, $96\,cm^2$ and $80\,cm^2$.
Find the volume of the box.

It may help to find a rule that gives the number of factors a number has from its prime product.

**7** **a** Write 84 as a product of primes.
**b** List the factors of 84 as products of primes.
**c** How many factors has 84?
**d** For whole numbers less than 1000, find the number with the most factors.

# Fractions

- One way to add or subtract fractions with different denominators is to use the **least common multiple** of the denominators as the new denominator.

  Example $\frac{5}{6} - \frac{1}{4} = \frac{10}{12} - \frac{3}{12} = \frac{7}{12}$

- One way to deal with mixed numbers is to convert to **improper fractions** before doing any calculation.

  Example $1\frac{1}{4} \times \frac{1}{5} = \frac{5}{4} \times \frac{1}{5} = \frac{5}{20} = \frac{1}{4}$

**Exercise 1.2**
Fractions

**1** Give each answer as a fraction in its lowest terms.

**a** $\frac{3}{4} + \frac{1}{10}$ **b** $1\frac{2}{3} \times 2\frac{1}{4}$ **c** $3\frac{1}{8} - 1\frac{3}{16}$ **d** $1\frac{1}{5} \div \frac{3}{10}$

**2** Give each answer in its lowest terms.

**a** $2\frac{3}{5} - 1\frac{1}{3}$ **b** $1\frac{1}{4} - 1\frac{5}{8}$ **c** $\frac{3}{5} \times 1\frac{1}{4}$ **d** $2\frac{1}{2} \div \frac{3}{5}$

**3** Give each answer in its lowest terms.

**a** $2\frac{2}{3} \div 1\frac{1}{2}$ **b** $\frac{3}{4} \times 1\frac{3}{5}$ **c** $2\frac{1}{2} - 1\frac{5}{9}$ **d** $1\frac{1}{4} + 2\frac{1}{2} + 1\frac{3}{5}$

**4** If $\frac{1}{a} = \frac{1}{b} - \frac{1}{c}$

what is the value of $a$, in fractional form, when $b = 2$ and $c = 5$?

> A fraction with a numerator of 1 is called a unit fraction.
> $\frac{1}{2}$, $\frac{1}{13}$ and $\frac{1}{21}$ are examples of unit fractions.
> Do not include $\frac{1}{1}$ as a unit fraction.

**5** What is the value of $k$ if $1 - \frac{k}{12} = \frac{k}{24}$?

**6** Explain why it is not possible to find two unit fractions that add to give a number greater than 1.

**7** Joe claims that, when $n$ is a positive integer, the value of the expression

$\frac{n^5}{5} + \frac{n^3}{3} + \frac{7n}{15}$ is a positive integer.

Show that Joe's claim is true for values of $n$ from 1 to 5.

> Joe's claim is true for **all** positive integers.

**8** These are the first three lines in a sequence of sums:

| |
|---|
| $\frac{1}{2} + \frac{1}{4} = \frac{3}{4}$ |
| $\frac{1}{2} + \frac{1}{4} + \frac{1}{8} = \frac{7}{8}$ |
| $\frac{1}{2} + \frac{1}{4} + \frac{1}{8} + \frac{1}{16} =$ |

**a** What is the sum of the fractions on:
  **i** the 3rd line
  **ii** the 8th line

**b** Why can the sum of the fractions on any line never be greater than 1?

> Questions **9** – **12** may take some time using trial and improvement.
> Try to find a simpler strategy to solve each problem.

**9** Find two different unit fractions that add to give $\frac{2}{7}$.

**10** Find three different unit fractions that add to give 1.

**11** **a** Show that the sum and product of 3 and $1\frac{1}{2}$ are equal.

**b** Find another pair of fractions whose sum and product are equal.

**12** We can use each digit from 1 to 9 once to make a fraction equivalent to $\frac{1}{4}$:

$$\frac{3942}{15768} = \frac{1}{4}$$

Use each digit from 1 to 9 once to find a fraction equivalent to $\frac{1}{2}$.

We can use these rules for fractions when we work with algebraic fractions.

- To find equivalent algebraic fractions, multiply or divide the numerator and denominator by the same number or expression.

**Example** $\frac{n}{n^2} = \frac{1}{n}$ (numerator $\div n$, denominator $\div n$)

- To add or subtract algebraic fractions with the same denominator, add or subtract the numerators.

**Example** $\frac{5}{x+1} - \frac{2}{x+1} = \frac{3}{x+1}$

- To multiply algebraic fractions, multiply the numerators and multiply the denominators.

**Example** $\frac{3}{x} \times \frac{2}{y} = \frac{6}{xy}$

- To divide algebraic fractions, use the rule that dividing by a fraction has the same effect as multiplying by its reciprocal.

**Example** $\frac{5}{x+1} \div \frac{1}{x-1} = \frac{5}{x+1} \times \frac{x-1}{1} = \frac{5(x-1)}{x+1}$

**Exercise 1.3**
**Fractions and algebra**

**Example**
As a single fraction,
$\frac{3x}{4} \times \frac{2}{y}$ is $\frac{6x}{4y}$.
In its simplest form,
$\frac{6x}{4y}$ is $\frac{3x}{2y}$.

**1** Which of these is equivalent to $\frac{x+y}{8}$?

(K) $\frac{x}{4} \times \frac{y}{2}$ (L) $\frac{x}{8} + \frac{y}{8}$ (M) $\frac{x}{4} + \frac{y}{4}$ (N) $\frac{x}{8} \div \frac{1}{y}$

**2** Sort these into four pairs of equivalent expressions.

(P) $\frac{2}{3}$ (Q) $\frac{2}{x} \div \frac{x}{3}$ (R) $\frac{4}{x} + \frac{2}{x}$ (S) $\frac{2x}{3x}$

(T) $\frac{8}{x}$ (U) $\frac{6}{x^2}$ (V) $\frac{4x}{3} \times \frac{6}{x^2}$ (W) $\frac{6}{x}$

**3** Write each of these as a single fraction. Give each answer in its simplest form.

**a** $\frac{n}{3} \times \frac{1}{n^2}$ **b** $\frac{3}{x} + \frac{5}{x}$ **c** $\frac{a}{10} - \frac{b}{10}$ **d** $\frac{p}{8} + \frac{3p}{8}$

**e** $\frac{m}{4} \div \frac{n}{3}$ **f** $\frac{7v}{8} - \frac{3v}{4}$ **g** $\frac{z}{5y} \times \frac{y}{3}$ **h** $\frac{c}{6} \div \frac{c}{12}$

**4** Copy and complete:

**a** $\frac{x}{5} - \frac{2}{5} = \frac{x-2}{\square}$ **b** $\frac{2}{m} \times \frac{n}{\square} = \frac{n}{3m}$ **c** $\frac{y}{x} + \frac{z}{\square} = \frac{y+z}{x}$

**d** $\frac{6}{\square} \div \frac{1}{n^2} = 6n$ **e** $\frac{3t}{10} + \frac{\square}{10} = \frac{t}{2}$ **f** $\frac{2}{b+3} \div \frac{1}{\square} = \frac{2(b-2)}{b+3}$

**5** Sort these into four pairs of equivalent expressions.

(A) $\frac{x}{3}$ (B) $\frac{1}{x+1}$ (C) $\frac{3}{3(x+1)}$ (D) $\frac{x}{x(x-1)}$

(E) $\frac{3}{x+1}$ (F) $\frac{1}{x-1}$ (G) $\frac{3x}{9}$ (H) $\frac{3(x-1)}{(x+1)(x-1)}$

To add or subtract algebraic fractions with **different** denominators, find equivalent fractions with the same denominator.

**Examples**

◆ $\frac{y}{3} + \frac{y}{4}$

$\frac{y}{3} = \frac{4y}{12}$ (×4) and $\frac{y}{4} = \frac{3y}{12}$ (×3)

So $\frac{y}{3} + \frac{y}{4} = \frac{4y}{12} + \frac{3y}{12}$

$= \frac{4y + 3y}{12} = \frac{7y}{12}$

◆ $\frac{2}{a} - \frac{3}{b}$

$\frac{2}{a} = \frac{2b}{ab}$ (×$b$) and $\frac{3}{b} = \frac{3a}{ab}$ (×$a$)

So $\frac{2}{a} - \frac{3}{b} = \frac{2b}{ab} - \frac{3a}{ab} = \frac{2b - 3a}{ab}$

◆ $\frac{1}{p+1} + \frac{1}{p-2}$

$\frac{1}{p+1} = \frac{(p-2)}{(p+1)(p-2)}$ (×$(p-2)$) and $\frac{1}{p-2} = \frac{(p+1)}{(p+1)(p-2)}$ (×$(p+1)$)

So $\frac{1}{p+1} + \frac{1}{p-2} = \frac{(p-2)}{(p+1)(p-2)} + \frac{(p+1)}{(p+1)(p-2)}$

$= \frac{(p-2)+(p+1)}{(p+1)(p-2)} = \frac{2p-1}{(p+1)(p-2)}$

**Exercise 1.4**
Adding and subtracting algebraic fractions

1 a Which of these is not equivalent to $\frac{5}{c}$?

(A) $\frac{15}{3c}$ (B) $\frac{5(c+1)}{c(c+1)}$ (C) $\frac{5c^3}{c^2}$ (D) $\frac{5a}{ac}$

b Write as a single fraction $\frac{5}{c} + \frac{7}{3c}$.

2 Sort these into four pairs of equivalent expressions.

(E) $\frac{10x}{21}$ (F) $\frac{x}{6} + \frac{x}{30}$ (G) $\frac{x}{5}$ (H) $\frac{x}{10}$

(I) $\frac{3x}{8} - \frac{x}{3}$ (J) $\frac{3x}{10} - \frac{x}{5}$ (K) $\frac{x}{24}$ (L) $\frac{x}{3} + \frac{x}{7}$

3 Write each of these as a single fraction in its simplest form.

a $\frac{2m}{3} + \frac{m}{12}$ b $\frac{3p}{4} - \frac{p}{8}$ c $\frac{q}{2} + \frac{r}{5}$ d $\frac{s}{t} - \frac{s}{2}$

e $\frac{4}{x} - \frac{5}{y}$ f $\frac{6}{f} + \frac{5}{2f}$ g $\frac{1}{p+3} + \frac{1}{p-1}$ h $\frac{2}{z} - \frac{1}{z+6}$

4 Copy and complete:

a $\frac{\square}{x} - \frac{5}{3x} = \frac{19}{3x}$ b $\frac{\square}{x} + \frac{2}{x+5} = \frac{5(x+3)}{x(x+5)}$

c $\frac{\square}{x+1} + \frac{3}{x} = \frac{3}{x(x+1)}$ d $\frac{1}{x+1} + \frac{\square}{x-2} = \frac{3x}{(x+1)(x-2)}$

**Thinking ahead to ...**
using algebraic fractions

**A** This is a sequence of pairs of calculations.

| | |
|---|---|
| $\frac{2}{3} - \frac{2}{5} = \frac{4}{15}$ | $\frac{2}{3} \times \frac{2}{5} =$ |
| $\frac{2}{4} - \frac{2}{6} =$ | $\frac{2}{4} \times \frac{2}{6} =$ |
| $\frac{2}{5} - \frac{2}{7} =$ | $\frac{2}{5} \times \frac{2}{7} =$ |
| ... | ... |

**a** Copy and complete these calculations.
**b** Write down some more pairs of calculations in this sequence.
**c** Comment on the results of these calculations.

## Using algebraic fractions

This explanation is a type of proof.

You can see that it will be true for any value of $n$.

- For the above calculations, the numerator of each fraction is 2 and the denominator of the second fraction is always 2 more than the first.
- So if the first fraction is $\frac{2}{n}$, then the second fraction is $\frac{2}{n+2}$.
- Subtract: $\frac{2}{n} - \frac{2}{n+2} = \frac{2(n+2)}{n(n+2)} - \frac{2n}{n(n+2)} = \frac{2n+4}{n(n+2)} - \frac{2n}{n(n+2)}$

$$= \frac{2n+4-2n}{n(n+2)} = \frac{4}{n(n+2)}$$

- Multiply: $\frac{2}{n} \times \frac{2}{n+2} = \frac{4}{n(n+2)}$
- Both calculations give the result $\frac{4}{n(n+2)}$.

So we have shown that for any pair of fractions of the form $\frac{2}{n}$ and $\frac{2}{n+2}$ the results of subtracting them and multiplying them are equal.

**Exercise 1.5**
Using algebraic fractions

**1** This is a sequence of pairs of calculations.

| | |
|---|---|
| $\frac{3}{4} - \frac{3}{7} = \frac{9}{28}$ | $\frac{3}{4} \times \frac{3}{7} =$ |
| $\frac{3}{5} - \frac{3}{8} =$ | $\frac{3}{5} \times \frac{3}{8} =$ |
| $\frac{3}{6} - \frac{3}{9} =$ | $\frac{3}{6} \times \frac{3}{9} =$ |
| ... | ... |

**a** Complete the calculations above to show that the results are equal for each of these three pairs.
**b** Explain why we can write any pair of fractions in this sequence in the form $\frac{3}{n}$ and $\frac{3}{n+3}$.
**c** Use algebra to show that, for **any** pair of calculations in this sequence the results will be equal.

**2** **a** Find some pairs of **unit** fractions that subtract and multiply to give equal results.
**b** Describe a rule that you think is true for these pairs of unit fractions.
**c** Use algebra to show that your rule is true.

# End points

You should be able to ... | ... so try these questions

**A** Use multiples and factors to solve problems

**A1** Find the highest common factor and least common multiple for each pair of numbers.

**a** 45 and 60 **b** 75 and 325
**c** 120 and 275 **d** 54 and 144
**e** 108 and 56 **f** 81 and 350
**g** 125 and 320 **h** 65 and 196

**A2** **a** Find the highest common factor of 105 and 126.
**b** Find the least common multiple of 105 and 126.

**A3** Two numbers less than 30 have 3 as their highest common factor and 216 as their least common multiple.
What are the two numbers?

**B** Calculate with fractions

**B1** When $x = \frac{1}{3}$, $y = \frac{5}{6}$ and $z = 1\frac{2}{3}$, calculate, in fractional form, the value of:
**a** $2xy$ **b** $y - xz$ **c** $\dfrac{x+y}{z}$

**B2** **a** $1\frac{3}{5} + 2\frac{3}{4}$ **b** $2\frac{5}{8} \div 1\frac{1}{2}$ **c** $3\frac{1}{4} - 1\frac{3}{8}$ **d** $2\frac{2}{3} \times 1\frac{5}{8}$

**B3** What fraction of ABCD is shaded?

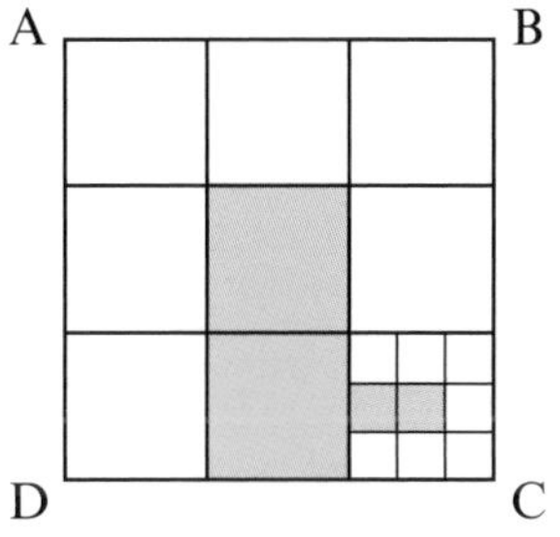


**C** Use algebraic fractions

**C1** Write each of these as a single fraction in its simplest form.
**a** $\dfrac{2x}{3} - \dfrac{x}{6}$ **b** $\dfrac{y}{4} \times \dfrac{3}{y}$ **c** $\dfrac{z}{3} \div \dfrac{1}{9}$
**d** $\dfrac{3}{a} - \dfrac{7}{b}$ **e** $\dfrac{5}{c} - \dfrac{4}{c+1}$ **f** $\dfrac{1}{d+6} + \dfrac{1}{d-5}$

## Some points to remember

- Least common multiple = Product ÷ Highest common factor.
- To add and subtract fractions (including algebraic), first find equivalent fractions with the same denominator.
- Dividing by a fraction has the same effect as multiplying by its reciprocal.

## Starting points
You need to know about ...

... so try these questions

### A Naming angles and triangles

- Any angle less than 90° is an **acute angle**.
- Any angle equal to 90° is a **right angle**.
- Any angle between 90° and 180° is an **obtuse angle**.
- Any angle between 180° and 360° is a **reflex angle**.

- Any triangle which has:
  - three sides of equal length
  - three equal angles (60°)

  is an **equilateral triangle**.

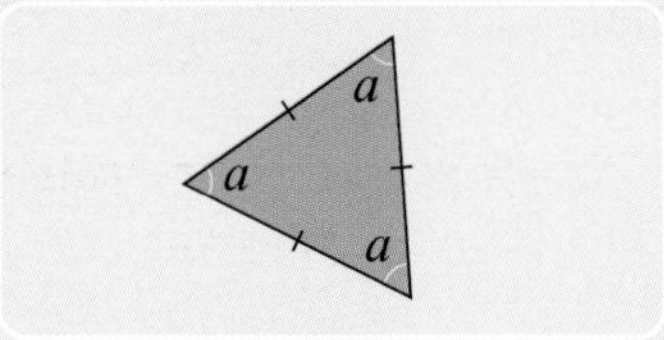

- Any triangle which has:
  - two sides of equal length
  - two equal angles

  is an **isosceles triangle**.

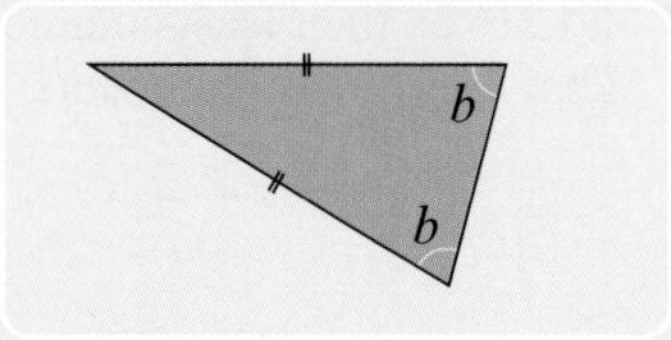

 - Any triangle which has no sides of equal length and no equal angles is a **scalene triangle**.
- Any triangle which has one right angle is a **right-angled triangle**.

### B Angle sums

- Angles at a point on a straight line add up to 180°.

$$a + b = 180°$$

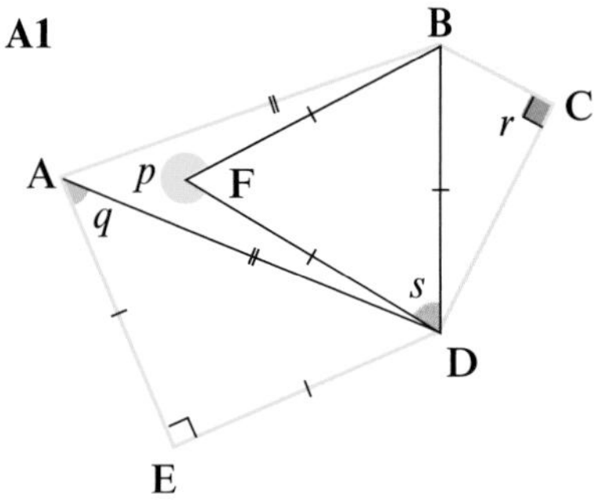

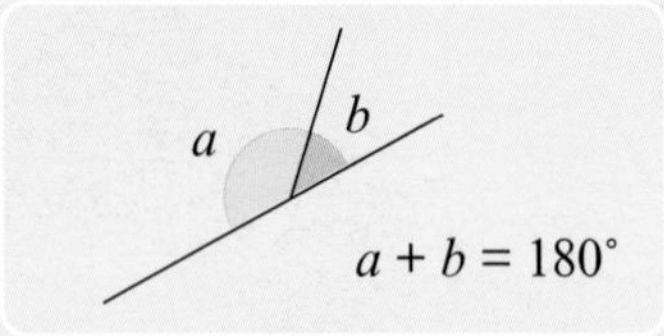

- Angles round a point add up to 360°.

$$c + d + e = 360°$$

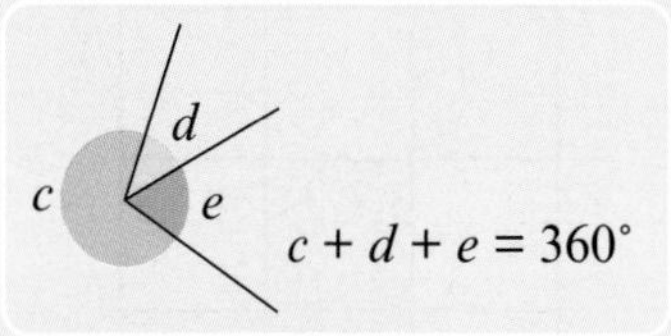

- Vertically opposite angles are equal.

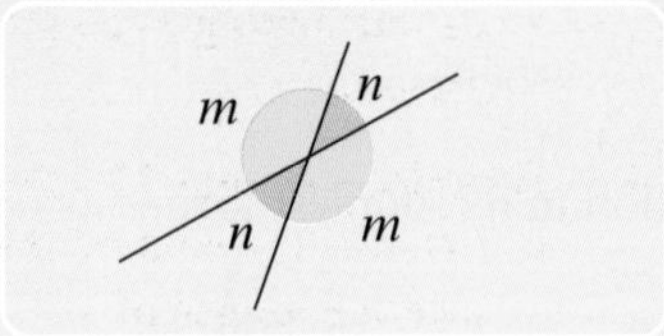

- Angles in a triangle add up to 180°.

$$x + y + z = 180°$$

### C Parallel lines

At each point where a straight line crosses a set of parallel lines there are two pairs of vertically opposite angles.

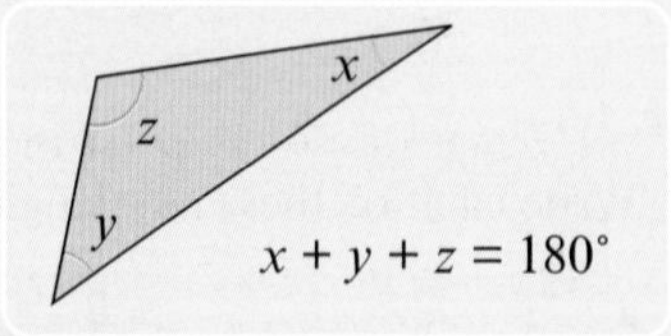

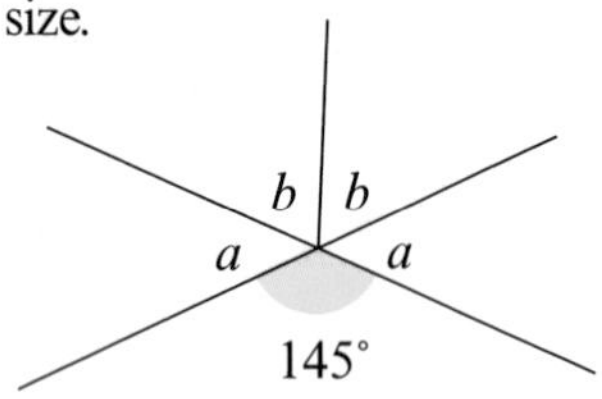

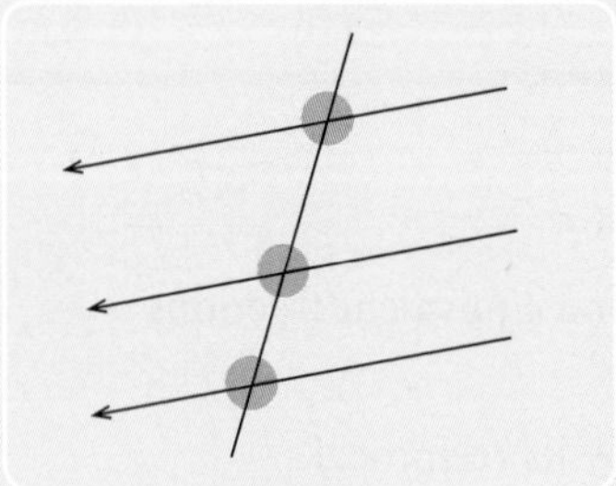

Parallel lines are marked with arrows.

Equal angles are marked with the same colour.

---

**A1**

**a** What type of angle is:
  **i** $p$   **ii** $q$   **iii** $r$?
**b** What type of triangle is:
  **i** BFD   **ii** BCD?
**c** Which triangles are isosceles?

**B1** In the diagram above calculate:
**a** the size of angle $s$
**b** angle $p$
**c** angle $A\hat{D}E$.

**B2** On this diagram, angles marked with the same letter are equal in size.

**a** Work out angles $a$ and $b$.
**b** Explain why a triangle can only have one obtuse angle.

**C1**

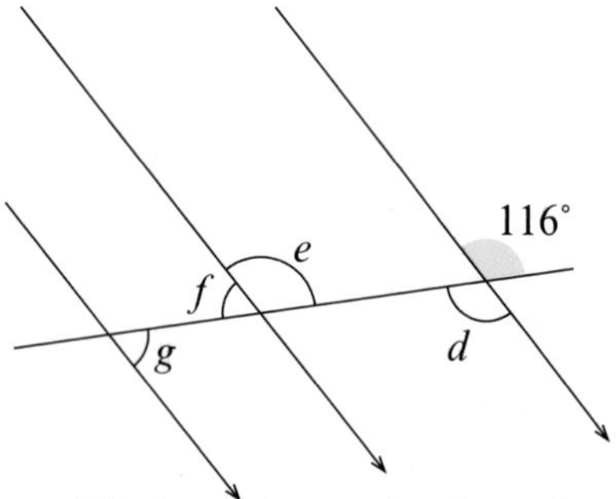

Work out the angles $d$ to $g$ in this diagram.

## D Quadrilaterals

/ = line of symmetry

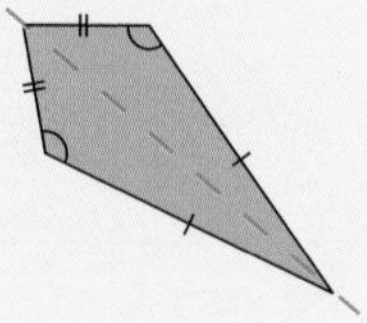

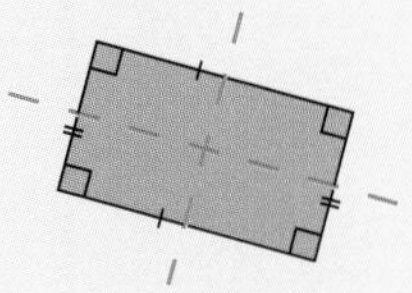

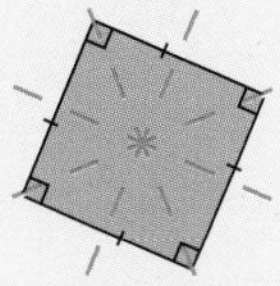

| | Kite | Rectangle | Square |
|---|---|---|---|
| Rotational symmetry of order: | 1 | 2 | 4 |

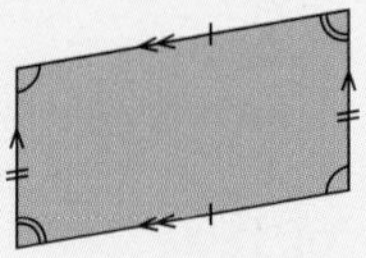

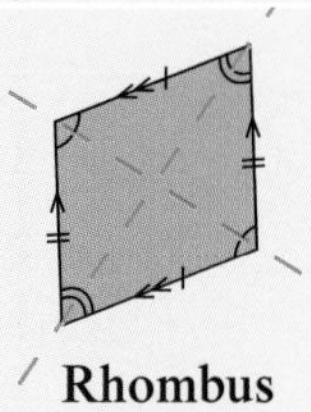

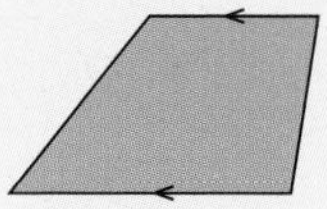

| | Parallelogram | Rhombus | Trapezium |
|---|---|---|---|
| Rotational symmetry of order: | 2 | 2 | 1 |

## E Polygons

- In ABCDE:
  - the **interior angles** are marked in red
  - the angles marked in blue are not interior angles.

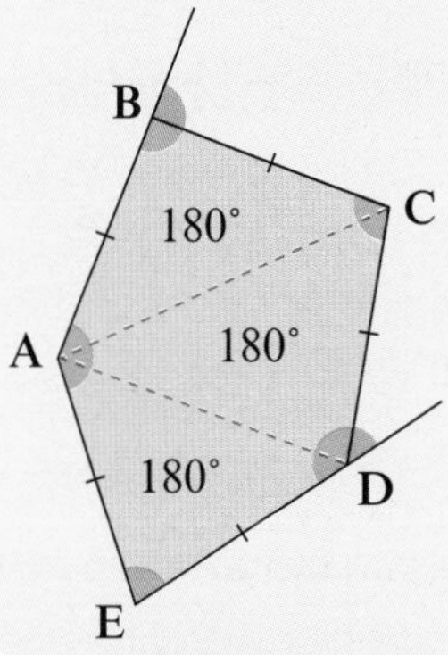


- The **sum of the interior angles** of a polygon with $n$ sides is: $(n-2) \times 180°$
  So for ABCDE the sum of interior angles is: $(5-2) \times 180°$
  $= 3 \times 180°$
  $= 540°$

- In a **regular polygon** all the sides are equal and the interior angles are equal.
- ABCDE is an **irregular polygon**. The sides are all equal but the interior angles are not.

| Name of polygon | Number of sides | Sum of interior angles | Interior angle of a regular polygon |
|---|---|---|---|
| Triangle | 3 | 180° —÷3→ | 60° |
| Quadrilateral | 4 | 360° —÷4→ | 90° |
| Pentagon | 5 | 540° —÷5→ | 108° |
| Hexagon | 6 | 720° —÷6→ | 120° |
| Heptagon | 7 | | |
| Octagon | 8 | | |
| Nonagon | 9 | | |
| Decagon | 10 | | |

Another expression for the **sum of the interior angles** of a polygon with $n$ sides is: $(180° \times n) - 360°$

**D1** Name all the quadrilaterals that fit each of these labels.

(1) There is only one pair of parallel lines.

(2) All angles are equal.

(3) All sides are of equal length.

(4) Opposite angles are equal.

**D2** Draw a trapezium with one line of symmetry.

**E1** What is the sum of the interior angles of an octagon?

**E2** Calculate the angle $a$ in this pentagon.

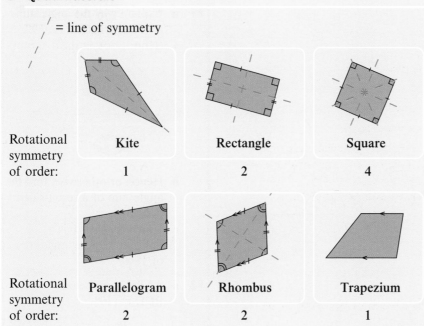

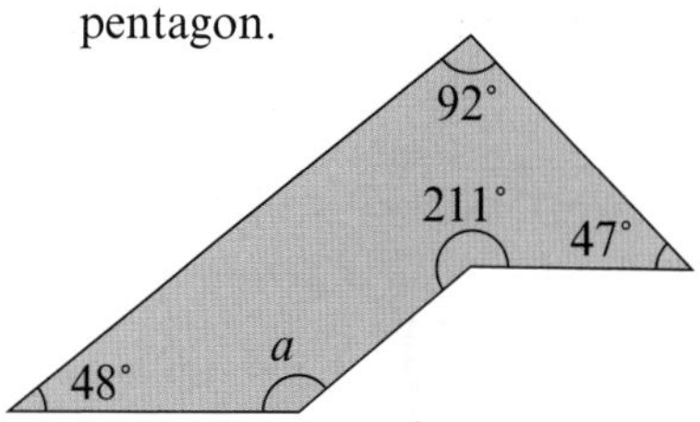


**E3** Calculate the interior angle of a regular heptagon to the nearest degree.

**E4** A dodecagon has 12 sides.
- **a** What is the sum of the interior angles of a dodecagon?
- **b** Calculate the interior angle of a regular dodecagon.

## F Triangles and proof

The angle sum of a triangle is 180°.

How can you prove this for any triangle?

- You could
  - draw a triangle
  - cut off the 3 vertical angles
  - arrange them on a straight line .

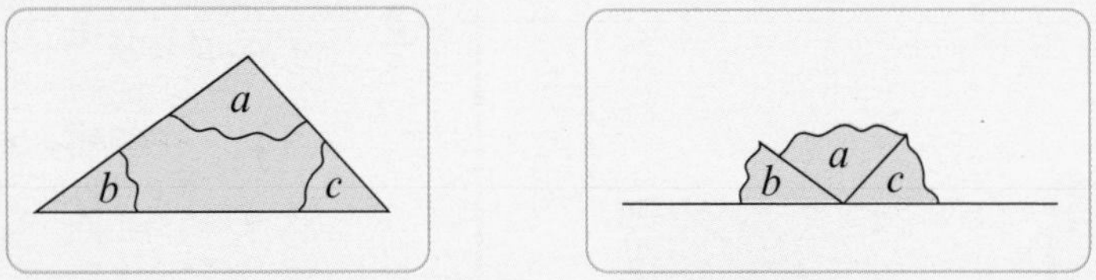


Angles of a straight line = 180°.

So $a + b + c = 180°$

ie the angle sum of a triangle is 180°.

- You could
  - Draw a triangle
  - Provide an exact copy of the triangle drawn
  - Arrange the two triangles to make a quadrilateral

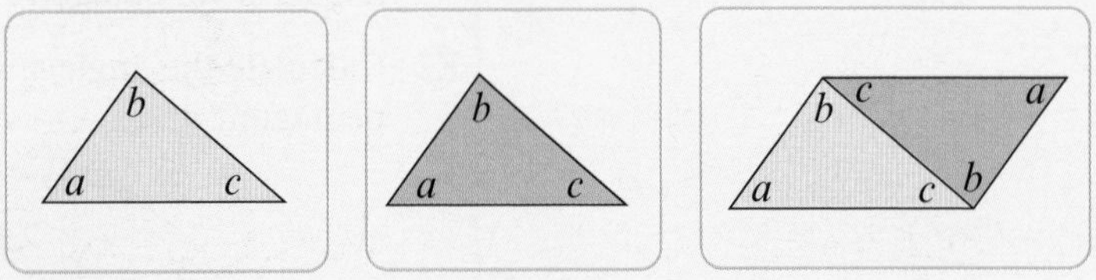


The angle sum of a quadrilateral is 360°.

So $a + b + c + a + b + c = 360°$

ie $2a + 2b + 2c = 360°$

ie $a + b + c = 180°$

ie the angle sum of a triangle is 180°.

The external angle of a triangle is equal to the sum of the two opposite interior angles.

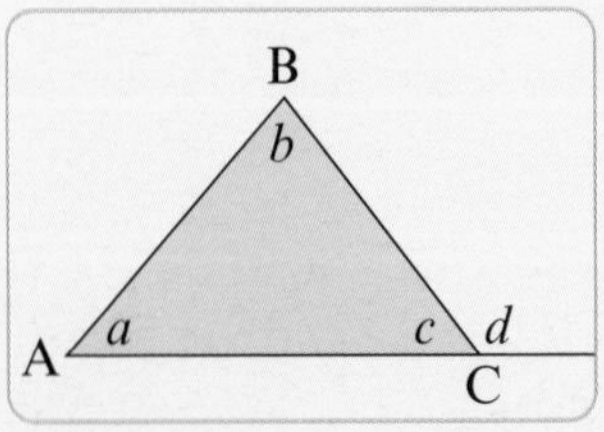


$d$ is the external angle at $c$

now

$c + d = 180°$ [angles on a straight line]

ie $c = 180° - d$

and $a + b + c = 180°$ [angles sum of a triangle]

making $c$ the subject $c = 180° - (a + b)$

So $180° - d = 180° - (a + b)$

ie $d = a + b$

**F1** **a** Explain why the angle sum of a quadrilateral is 360°. (Hint: Look at the diagram).

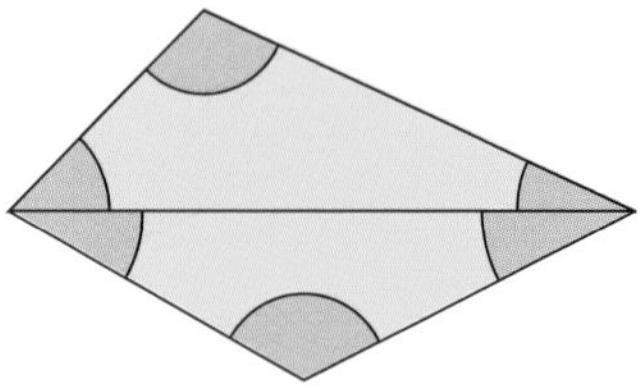

**b** Hence, or otherwise, find the angle sum of a seven-sided polygon.

**c** How may sides has a polygon with an angle sum of 1800°?

**F2** **a** $\Delta ABC$ is isosceles. $B\hat{C}D = 110°$

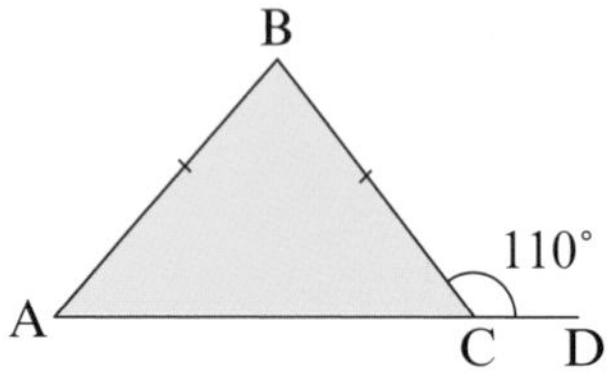


Find $A\hat{B}C$.

**b** $\Delta ABC$ is isosceles. $B\hat{C}D = x°$

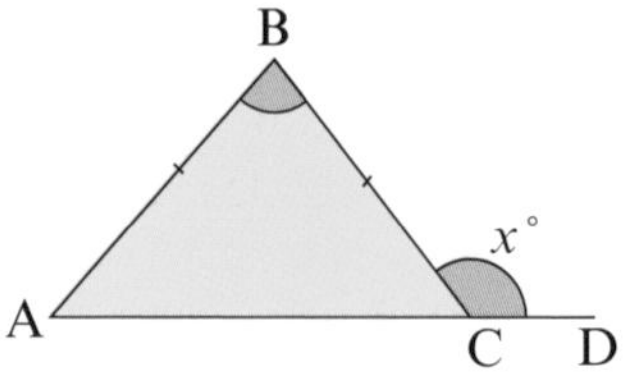


Use the fact that the external angle of a triangle is equal to the sum of the two opposite interior angles to prove that $A\hat{B}C = 2x - 180$.

# Angles in triangles

To calculate an angle you may need to work out some other angles first.

**Example**

Calculate the angle $E\hat{G}F$.

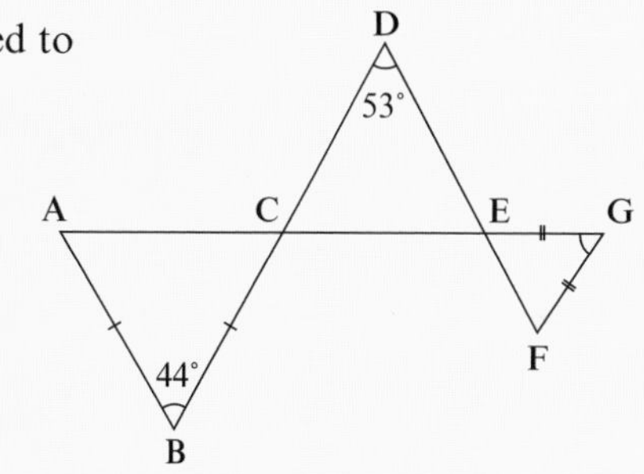


- You should sketch a diagram and label each angle that you calculate.

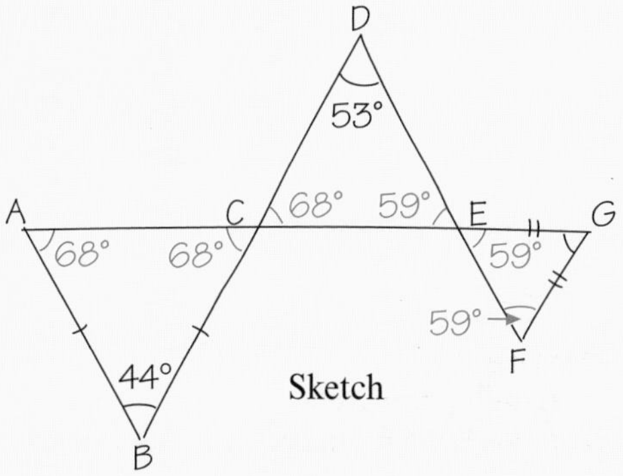


Sketch

Δ stands for triangle

An angle can be written in different ways.

For example:
$D\hat{C}E$ is the same angle as $D\hat{C}G$ and $E\hat{C}D$
$D\hat{E}C$ is the same angle as $D\hat{E}A$ and $C\hat{E}D$.

To calculate $E\hat{G}F$

| *Calculation ...* | *... Reason* |
|---|---|
| $A\hat{C}B = C\hat{A}B$... $= (180° - 44°) \div 2$ $= 68°$ | ...ABC is an isosceles Δ |
| $D\hat{C}E = 68°$... | ...Vertically opposite ACB |
| $D\hat{E}C = 180° - (68° + 53°)$... $= 59°$ | ...Angle sum of Δ |
| $F\hat{E}G = 59°$... | ...Vertically opposite DEC |
| $G\hat{E}F = E\hat{F}G$... | ...EFG is an isosceles Δ |
| $E\hat{G}F = 180° - (59° \times 2)$... | ...Angle sum of Δ |

**So the angle $E\hat{G}F = 62°$**

You may not need to calculate all the intermediate angles.

**Exercise 2.1**
**Angles in triangles**

You will need to work out some other angles first.

**1**

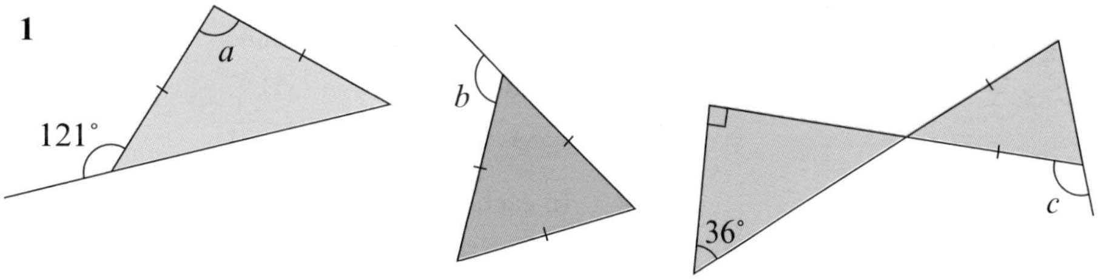


Calculate the angles $a$, $b$ and $c$ in these diagrams.
Give a reason for each calculation.

**2** **a** Which is the easiest angle to calculate in this diagram?
**b** Calculate the angles $a$ to $f$ in this diagram.
**c** In what order did you calculate the angles? Explain why.

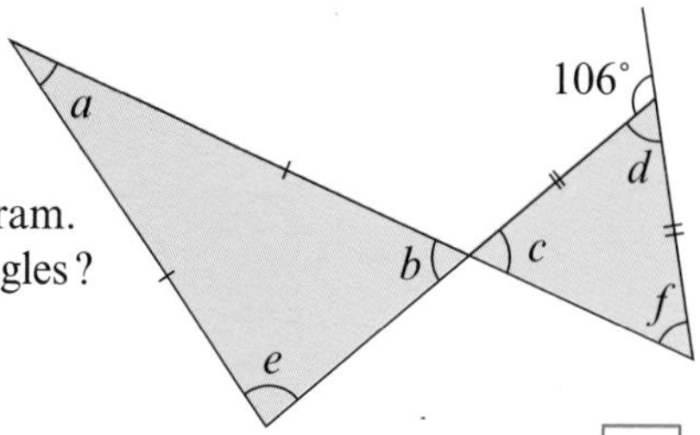

# Parallel lines

In each of these diagrams a straight line crosses two parallel lines.

- In each diagram a pair of **corresponding angles** is labelled. Corresponding angles are equal.

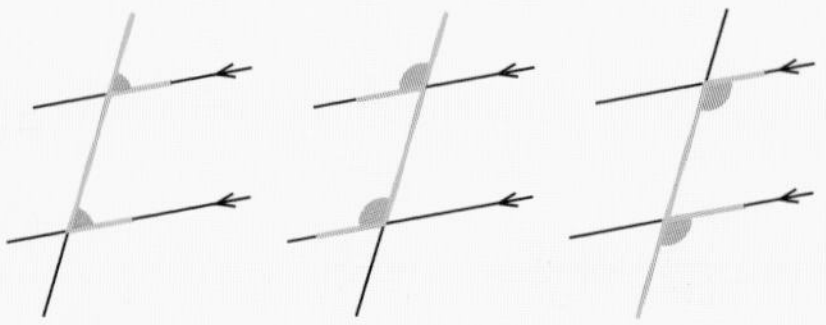

- In each diagram a pair of **alternate angles** is labelled. Alternate angles are equal.

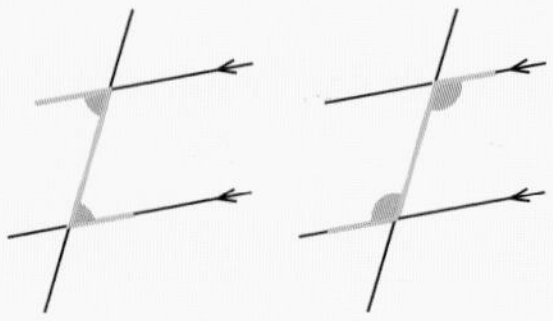

## Exercise 2.2
### Angles in parallel lines

To find corresponding angles in a diagram you could look for an F shape which may be upside down and/or back to front.

To find alternate angles in a diagram you could look for a Z shape which may be back to front.

1 **a** List five pairs of corresponding angles in this diagram.
**b** List three pairs of alternate angles.

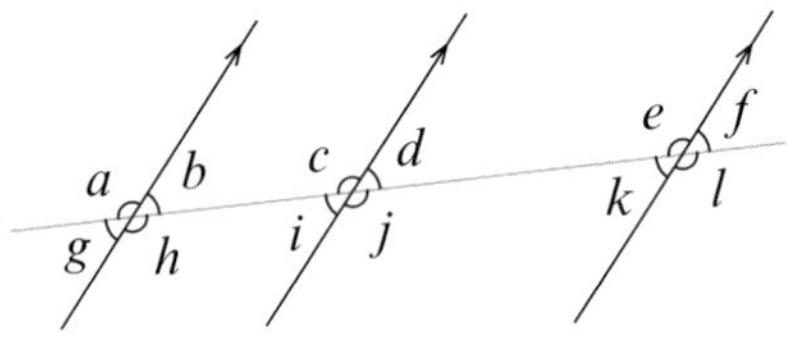


2 Sketch these diagrams.
Work out the angles $a$, $b$ and $c$.
You may need to calculate some other angles first.

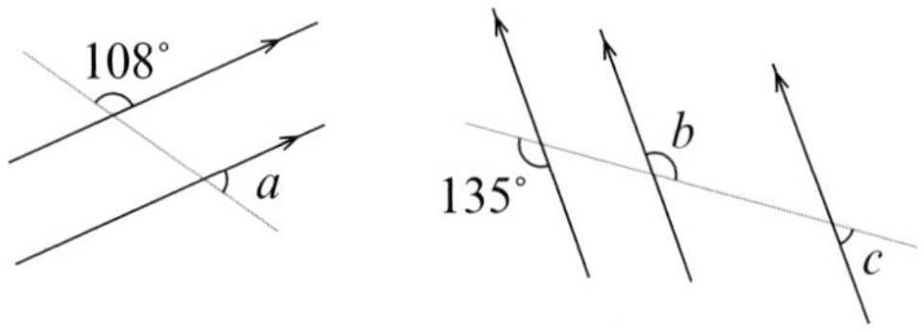


3 In this diagram AS, BR and NQ intersect to make the triangle DPO.
CE, FI, JM and NQ are parallel.

**a** List three pairs of corresponding angles along the line:
**i** AS **ii** BR

**b** Explain why $D\hat{G}H$ and $D\hat{H}I$ are not corresponding angles.

**c** List three pairs of alternate angles in this diagram.

**d** Calculate each of these angles.

**i** $H\hat{D}G$ **ii** $A\hat{D}C$
**iii** $G\hat{K}L$ **iv** $H\hat{L}K$
**v** $Q\hat{P}L$ **vi** $N\hat{O}R$

Give a reason for each calculation that you do.

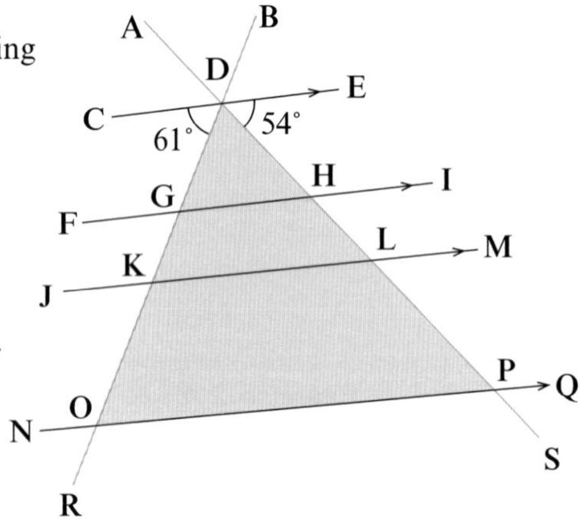


4 In each of these diagrams there is one pair of parallel lines.

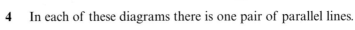

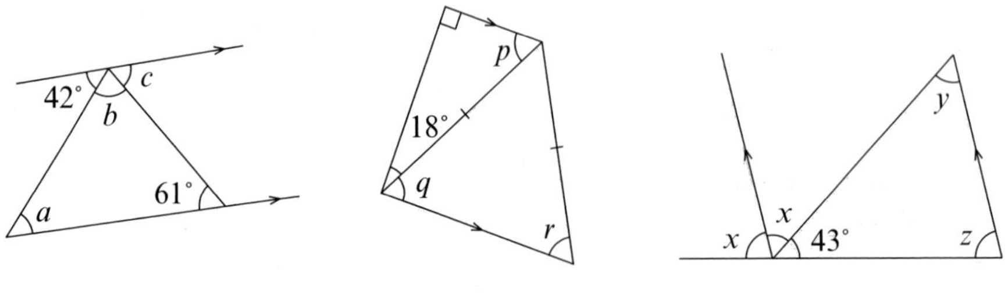


Sketch these diagrams and calculate each of the angles marked with a letter.

5 **a** Work out each of these angles in ABCD.

**i** $A\hat{B}D$ **ii** $A\hat{D}B$

**iii** $B\hat{D}C$ **iv** $B\hat{C}D$

**b** Which two lines are parallel?

**c** What is the mathematical name for ABCD?

**d** Explain why ABCD is not a rhombus.

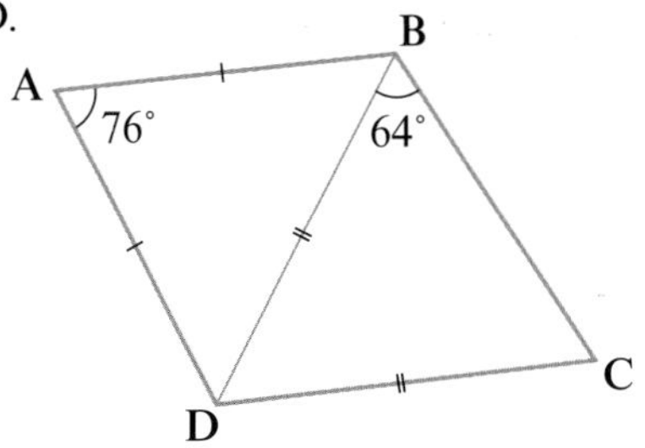


## Properties of polygons

**Exercise 2.3**
Properties of quadrilaterals

1 AB and CD are plastic strips joined by red and yellow elastic bands.
ABCD is a parallelogram; its diagonals intersect at M.
You can stretch the parallelogram if you fix AB and pull CD sideways.

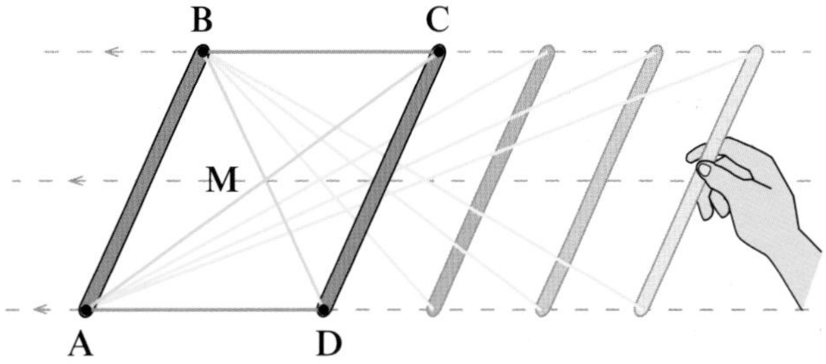


Below are some diagrams of the parallelogram as it is stretched.
ABC′D′ is a rhombus.

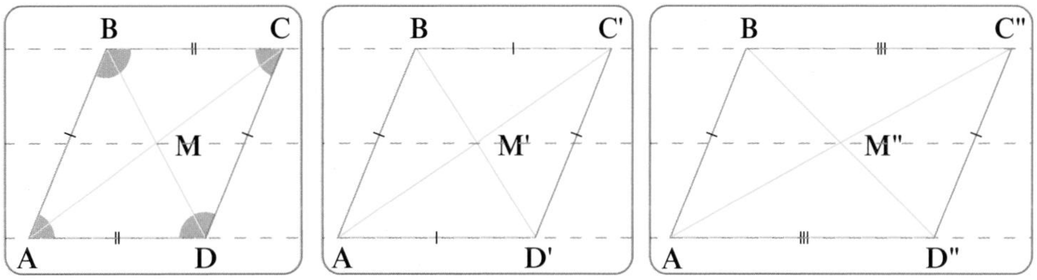


**a** Trace the angles $A\hat{B}C$, $A\hat{B}D$ and $C\hat{B}D$.

**b** Use your tracings to find out what happens to each of these angles as the parallelogram is stretched.

**i** $A\hat{B}C$ **ii** $A\hat{B}D$ **iii** $C\hat{B}D$

**c** What happens to $A\hat{M}B$ as the parallelogram is stretched?

**d** What type of angle is each of these?

**i** $A\hat{M}B$ **ii** $A\hat{M}'B$ **iii** $A\hat{M}''B$

The interior angles of ABCD are marked in green.

2

(1) The diagonals bisect the interior angles.

(2) The diagonals intersect at right angles.

(3) The diagonals bisect each other.

Which of these statements do you think is always true:

**a** for a parallelogram **b** for a rhombus?

If you bisect a line or angle, you cut it into 2 equal parts.

If two lines bisect each other, they are both cut into two equal lengths.

# Angles in polygons

- At each vertex of a polygon the angle between an extended side and the adjacent side is called **an exterior angle**.

  In ABCDE:

  - the exterior angles are marked in orange
  - the interior angles are marked in green.

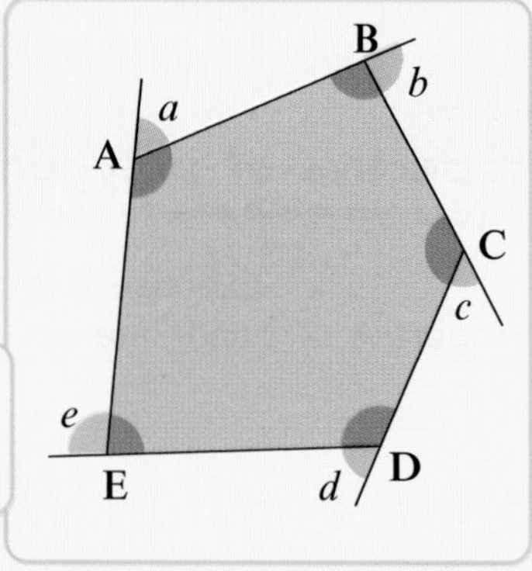


- The sum of the exterior angles of any polygon is 360°.

  In ABCDE:
  $a + b + c + d + e = 360°$

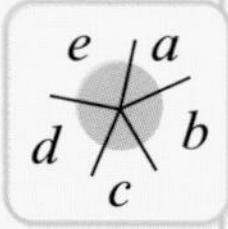


  You can show this by tracing the angles and fitting them together round a point.

- At each vertex the sum of the interior angle and exterior angle is 180°.

**Exercise 2.4**
Exterior angles of polygons

1 For this polygon:
  a calculate each exterior angle
  b check that the total of the exterior angles is 360°.

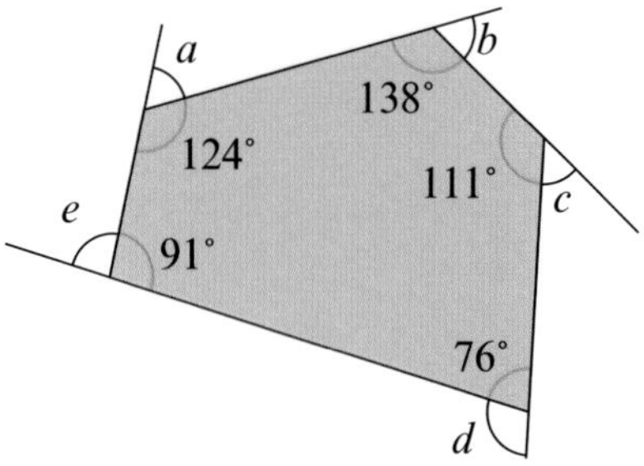


2 A dodecagon has 12 sides.
In this regular dodecagon one side is extended to form the angle *p*.
  a Explain why the exterior angles of a regular dodecagon are all equal to 360° ÷ 12.
  b Calculate the angle *p*.
  c Calculate the interior angle of a regular dodecagon.

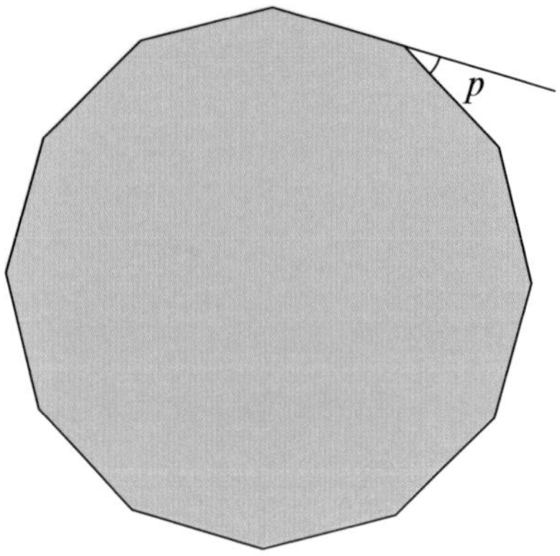


3 This is part of regular nonagon drawn inside a circle with centre C.
One side of the nonagon is extended to form the angle *e*.
Calculate the angles *e* to *h*.

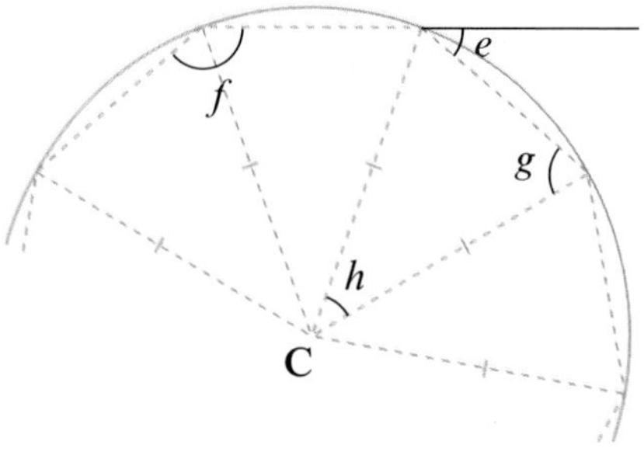

## Exercise 2.5
### Triangles investigation

You can mark eight points that are equally spaced on the circumference of a circle if you:

- draw a circle on square grid paper
- mark in lines that are vertical, horizontal and at 45° to the horizontal.

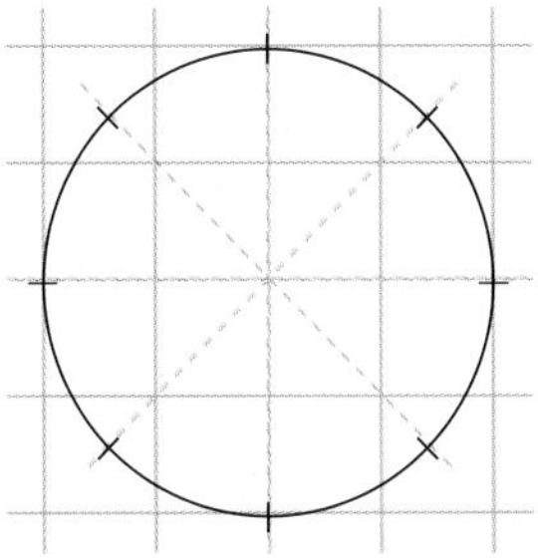

**1** This eight-point circle has the points A to H equally spaced on the circumference.
Δ ABD is drawn by joining three of the points.

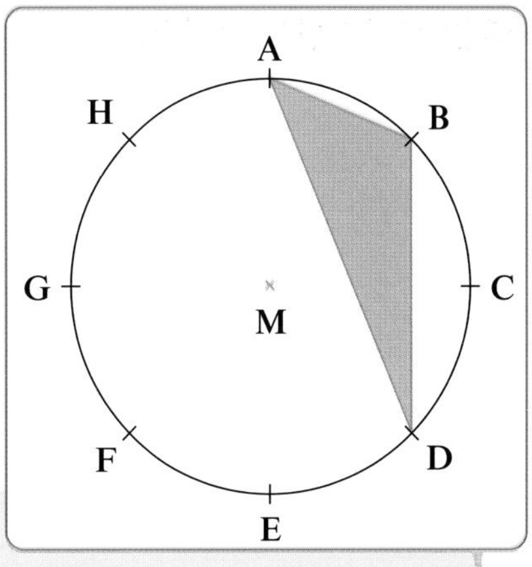


This is how a student calculated the exterior angle at A for Δ ABD.

To find the exterior angle at A

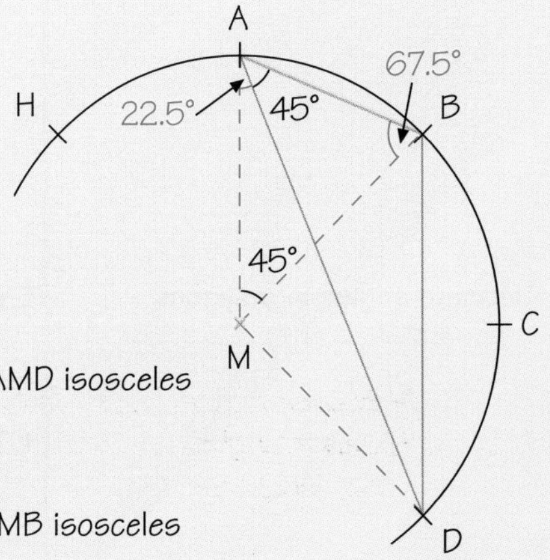


$A\hat{M}D = 135°$
$M\hat{A}D = (180° - 135°) \div 2$   Δ AMD isosceles
$= 22.5°$

$A\hat{M}B = 45°$
$M\hat{A}B = (180° - 45°) \div 2$   Δ AMB isosceles
$= 67.5°$

$D\hat{A}B = M\hat{A}B - M\hat{A}D$
$= 67.5° - 22.5°$
$= 45°$

So the exterior angle at A is **135°**.

**a** Explain why $A\hat{M}D$ is 135°.
**b** For triangle ABD:
  **i** calculate the exterior angles at B and D
  **ii** check that the total of the exterior angles is 360°.

**2** Triangle ACF is also drawn on an eight-point circle.

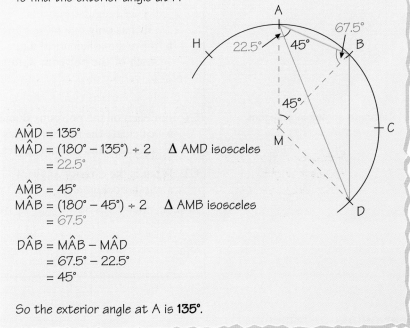

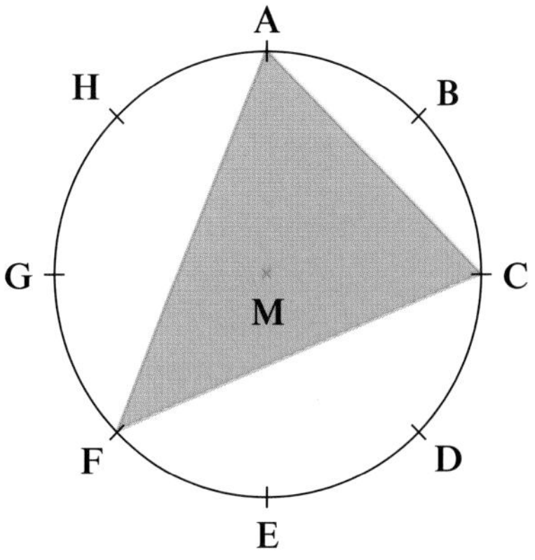


**a** For triangle ACF:
  **i** calculate each interior angle
  **ii** calculate each exterior angle.
**b** How many different triangles is it possible to draw in an eight-point circle?
**c** What different exterior angles are possible for triangles drawn on an eight-point circle?

Do not count any that are reflections or rotations of another polygon.

## End points

**You should be able to ...** | **... so try these questions**

**A** Calculate angles in parallel lines

**A1** Calculate the angles *a*, *b* and *c* in this diagram.

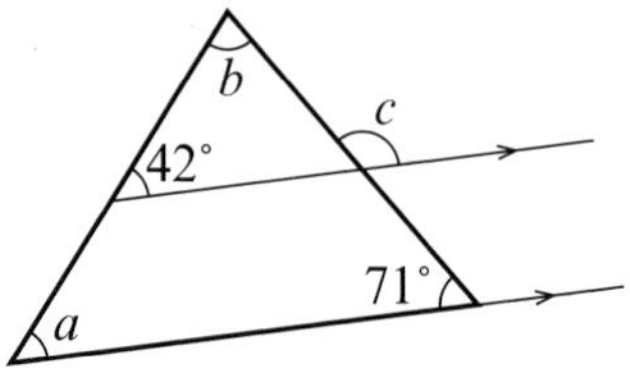


**B** Use the properties of polygons

**B1** Polygons A to E are drawn on an equilateral grid.

**a** Which of these polygons:
- **i** is a regular polygon
- **ii** has only one obtuse angle?

**b** Give a mathematical name for each of the polygons A to E.

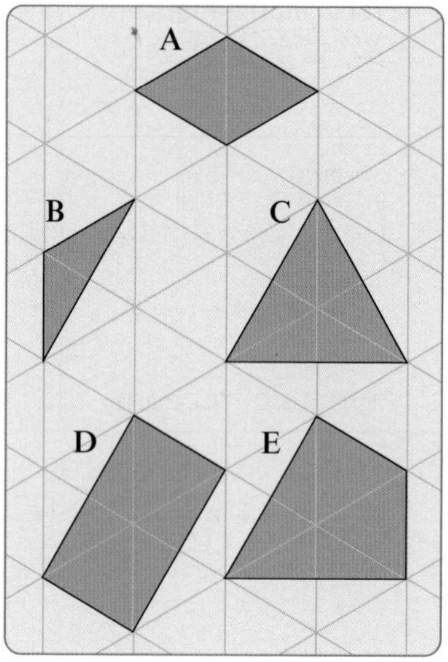


**C** Calculate angles in polygons

**C1** For each of the polygons B and E:
- **a** calculate the interior angles
- **b** calculate the exterior angles.

**C2** What is the exterior angle of a regular octagon?

## Some points to remember

- For a polygon with *n* sides:
  - the sum of the interior angles is $(n - 2) \times 180°$
  - the sum of the exterior angles is 360°
  - each exterior angle of a regular polygon is $360° \div n$.
- Examples of quadrilaterals

| | 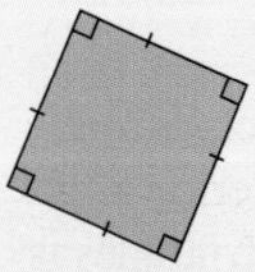 **Square** | 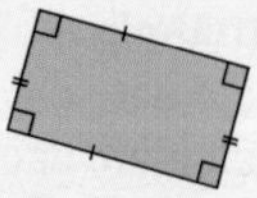 **Rectangle** | 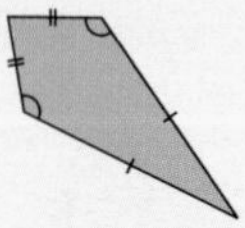 **Kite** | 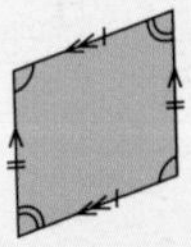 **Rhombus** | 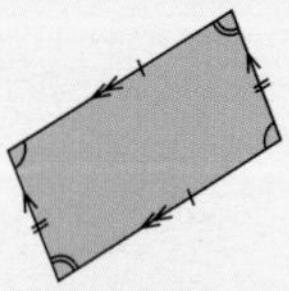 **Parallelogram** |
|---|---|---|---|---|---|
| The diagonals:<br>◆ bisect the interior angles<br>◆ bisect each other<br>◆ intersect at 90°. | ✓<br>✓<br>✓ | ✗<br>✓<br>✗ | ✗<br>✗<br>✓ | ✓<br>✓<br>✓ | ✗<br>✓<br>✗ |
| Number of lines of symmetry | 4 | 2 | 1 | 2 | 0 |
| Order of rotational symmetry | 4 | 2 | 1 | 2 | 2 |

## Starting points

You need to know about ...

### A Multiplying out brackets

- To multiply out brackets, multiply every term inside the bracket by the term outside.

**Example 1**

$$2(3a^2 + 4a - 6) = (2 \times 3a^2) + (2 \times 4a) - (2 \times 6)$$
$$= 6a^2 + 8a - 12$$

We say that $2(3a^2 + 4a - 6)$ and $6a^2 + 8a - 12$ are **equivalent expressions** because:

$2(3a^2 + 4a - 6) = 6a^2 + 8a - 12$ for any value of $a$.

**Example 2**

$$2(3a - 4b - 2) + 4(2a - 6) = 6a - 8b - 4 + 8a - 24$$
$$= 14a - 8b - 28$$

So $14a - 8b - 28$ is equivalent to $2(3a - 4b - 2) + 4(2a - 6)$

### B Algebraic fractions

- To find equivalent algebraic fractions, multiply or divide the numerator and denominator by the same number or expression.

**Example**

$$\frac{n^2}{3n} = \frac{n}{3} \quad (\div n \text{ top and bottom})$$

- To add or subtract algebraic fractions with different denominators:
  - find equivalent fractions with the same denominator
  - add or subtract the numerators.

**Example 1**

$$\frac{2}{p+3} + \frac{3}{p}$$
$$= \frac{2p}{(p+3)p} + \frac{3(p+3)}{(p+3)p}$$
$$= \frac{2p + 3(p+3)}{(p+3)p}$$
$$= \frac{5p+9}{(p+3)p}$$

**Example 2**

$$\frac{2}{a} - \frac{5}{3a}$$
$$= \frac{6}{3a} - \frac{5}{3a}$$
$$= \frac{1}{3a}$$

- To multiply algebraic fractions, multiply the numerators and multiply the denominators.

**Example**

$$\frac{x}{2} \times \frac{5}{x-1} = \frac{5x}{2(x-1)}$$

- To divide algebraic fractions, use the rule that dividing by a fraction has the same effect as multiplying by its reciprocal.

**Example**

$$\frac{3}{x+1} \div \frac{2}{y} = \frac{3}{x+1} \times \frac{y}{2} = \frac{3y}{2(x+1)}$$

... so try these questions

**A1** Which of these expressions is equivalent to $5a + 7$?

A $2(3a + 5)$

B $3(a + 2) + 2(a + 2)$

C $4(2a + 1) + 3(1 - a)$

D $5(2a + 1) + 2(1 - a)$

**A2** Simplify the expression

**a** $3(a - 5) + 5(2a - 3)$
**b** $4(x - 2) + 3(4 - 3x)$
**c** $2(x - 1) + 3(2x - 5) + 4(x - 3)$
**d** $3(x - 5) - 4(2x - 1)$
**e** $3(4a - 2b) + 2(3b - 6a + 2)$

**B1** Write each of these as a single fraction, in its simplest form.

**a** $\frac{2}{p+2} + \frac{4}{p}$ **b** $\frac{8p}{9} \times \frac{3}{2p}$

**c** $\frac{1}{p} - \frac{1}{(p+5)}$ **d** $\frac{5p}{8} \div \frac{3p}{16}$

**e** $\frac{3}{(p+1)} + \frac{2}{p-1}$

**f** $\frac{4}{3p+2} + \frac{1}{p-1}$

**g** $\frac{p+3}{4} + \frac{p+2}{5}$

**h** $\frac{p+3}{4} \times \frac{5}{p-2}$

**i** $\frac{2p-3}{3} \times \frac{4}{p+5}$

**j** $\frac{4p+5}{3} \div \frac{2p-1}{4}$

# Using brackets

- You can use brackets to write an expression for the shaded area in each of these rectangles.

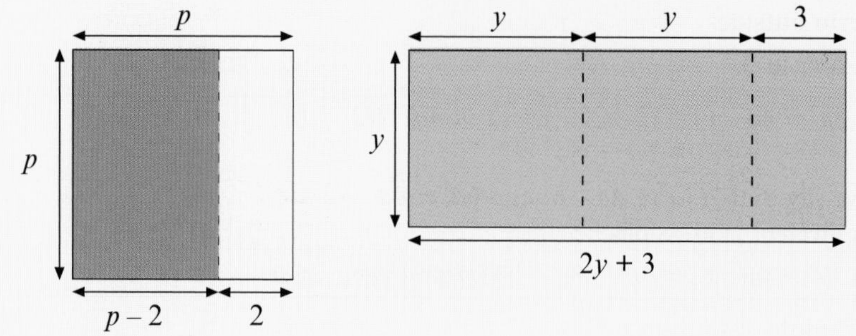


Shaded area $= p(p - 2)$
$= p^2 - 2p$

Shaded area $= y(2y + 3)$
$= 2y^2 + 3y$

**Exercise 3.1**
**Using brackets**

For some of the shaded rectangles you will need to find an expression for a missing dimension first.

**1** For each of these rectangles write an expression for the shaded area:
**a** with brackets **b** without brackets.

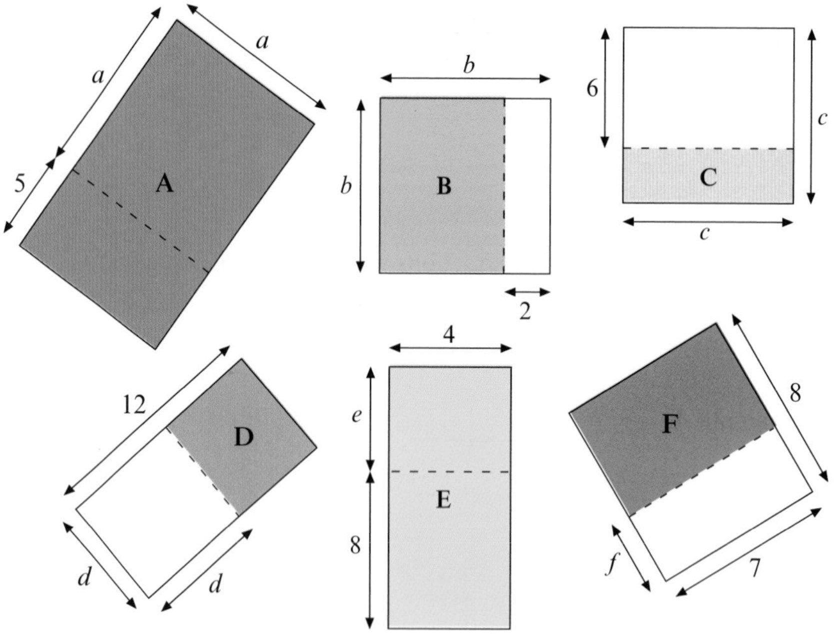


**2** For each of P, Q and R write an expression for the shaded area:
**a** with brackets **b** without brackets.

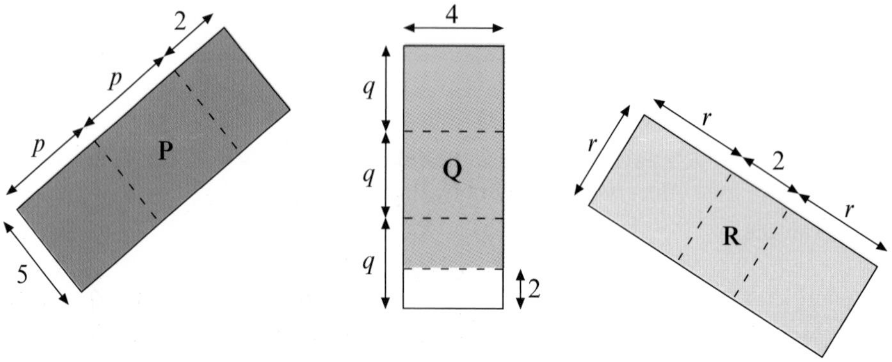

3 Write an expression for the width of rectangles A to E.

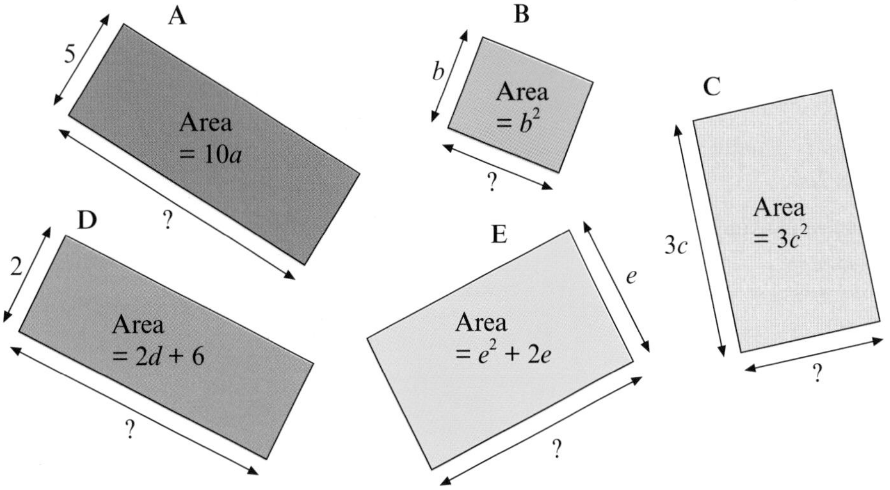


4 Copy and complete these.

a $4x + 12 = 4(\square + \square)$  b $6p - 4 = 2(\square - \square)$
c $2a + 8 = \square(a + \square)$  d $10 - 5q = \square(2 - \square)$
e $3t^2 - 4t = \square(\square - \square)$  f $6s^2 + 7s = \square(\square + 7)$

## Multiplying terms and simplifying

- When you subtract a bracket each term must have the correct sign.

**Example 1**
$30 - (5 + 2)$
$= 30 - 5 - 2$
$= 23$

**Example 2**
$30 - (5 - 2)$
$= 30 - 5 + 2$
$= 27$

**Example 3**
$30 - 2(3 + 5)$
$= 30 - (6 + 10)$
$= 30 - 6 - 10$
$= 14$

**Example 4**
$5x - (x + 3)$
$= 5x - x - 3$
$= 4x - 3$

**Example 5**
$5x - (4 - x)$
$= 5x - 4 + x$
$= 6x - 4$

**Example 6**
$5x - 3(y - 4)$
$= 5x - (3y - 12)$
$= 5x - 3y + 12$

**Exercise 3.2**
Subtracting brackets

1 Multiply out and simplify:

a $3b - 5(2 - b)$  b $12 - 3(2x - 4)$  c $6c - 3(c + 4)$
d $6y - 5(3 - y)$  e $6 - 2(a + 3)$  f $5a - 2(7 - b)$

2 Which of these expressions is equivalent to $2 - x$?

A $2(x + 4) - 3(x + 2)$  B $2(x + 3) - (3x - 4)$
C $3(x - 2) - 2(x - 4)$  D $5(x - 2) - 3(2x - 4)$

3 Simplify each of these expressions.

a $2(3a + 3) - (a + 1)$  b $3(x + 4) - (2x - 3)$
c $2(2a + 3) - 3(4 - a)$  d $2(x + 4) - 3(2x - 1)$
e $5(x - 3) - 2(4 + x)$  f $5(3x + 2) - 2(4x + 3)$
g $2(x - 3) - (x + 3)$  h $3(3x + 2y) - 2(y - 4x)$

In any term the letters are usually written in alphabetical order.

For example:

$2ba$ is usually written as $2ab$

$4n^2m$ is usually written as $4mn^2$.

- You can **multiply terms** by grouping numbers, and each of the letters.

| Example 1 | Example 2 | Example 3 |
|---|---|---|
| $2m \times 3n$ | $2ab \times a$ | $2p \times 3p^2$ |
| $= 2 \times m \times 3 \times n$ | $= 2 \times a \times b \times a$ | $= 2 \times p \times 3 \times p \times p$ |
| $= 2 \times 3 \times m \times n$ | $= 2 \times a \times a \times b$ | $= 2 \times 3 \times p \times p \times p$ |
| $= 6mn$ | $= 2a^2b$ | $= 6p^3$ |

The letters in each term are usually written in alphabetical order.

- **Like terms** must have exactly the same letters in them.

| Example 1 | Example 2 |
|---|---|
| $2p^2r = 2 \times p \times p \times r$ | $3pr^2 = 3 \times p \times r \times r$ |
| $8p^2r = 8 \times p \times p \times r$ | $2p^2r = 2 \times p \times p \times r$ |
| So $2p^2r$ and $8p^2r$ are like terms. | So $2p^2r$ and $3pr^2$ are **not** like terms. |

- To **simplify an expression** collect together any **like terms**.

These **can be simplified** by collecting like terms.
$2a^2b + 3ab^2 + 4a^2b = 6a^2b + 3ab^2$
$2x^2 + 2x + 3x^2 - x + 4 = 5x^2 + x + 4$

These **will not simplify** as there are no like terms.
$2a^2b + 3ab^2$
$2x^2 + 4x + 3$

## Exercise 3.3 Multiplying terms and simplifying

1 Multiply these terms.

**a** $3a \times 2b$ **b** $p \times 3q$ **c** $4y \times 5x$ **d** $5q \times 6p$
**e** $x \times 2x$ **f** $ab \times a$ **g** $2xy \times y$ **h** $2ab \times 3a$
**i** $2p^2 \times 3q$ **j** $a^3 \times a^2$ **k** $2b^2 \times 3b$ **l** $5c^3 \times 2b$

2 Find four pairs of equivalent terms.

A $(2b^2)^3$ B $6b^5$ C $3b^2 \times 2b^4$ D $5b^5$ E $8b^6$
F $3b^2 \times 2b^3$ G $6(b^4)^2$ H $6b^6$ I $6b^8$

3 Multiply out these brackets.

**a** $a(b + 4)$ **b** $m(2n + 3p)$ **c** $2x(3y + 2z)$
**d** $c(a + c)$ **e** $p(p - q)$ **f** $4a(a - b)$
**g** $a(2b - 4c)$ **h** $2a(3a + 4b)$ **i** $4p(2q - 3p)$
**j** $2pq(3p + 2q)$ **k** $4xy(x - 2y)$ **l** $3xy(x^2 - y^2)$

4 Simplify these where possible.

**a** $4x^2 + x - 2x^2$ **b** $5a + 2ab - a + 3ab$
**c** $x^2 + x^3 - 2x$ **d** $4a - 3b + 7a + 5b$
**e** $4p^2 - pq + 6q + pq$ **f** $ab + 2a - ab + 4a$

5 Which of these expressions is equivalent to $2a^2b - 3ab - 5ab^2$?

A $2a(ab + b) + 5b(ab - a)$ B $ab(2a - 2) - a(b - 5b^2)$
C $a(b + 2ab) - ab(4 + 5b)$

6 Multiply out and simplify:

**a** $2(2a + 3b) + 5(a + 4b)$ **b** $2(2x + 4y) + 3(2x - y)$
**c** $x(x - 3) - x(x + 4)$ **d** $2x(3x + 2y) + x(4x - y)$
**e** $ab(a + b) + ab(a - b)$ **f** $3a(ab + b) - 2b(ab - b)$

**7** Simplify these expressions.

**a** $2a(3a - b) + 4b(2a + 3b)$  **b** $5xy(2x + 4y) + 3x(2xy - 2y)$

**c** $pq(2p - 3q) - 2p(3pq - 2q^2)$  **d** $2mn(3m - 4n) - 4m^2(2n - 3m)$

**8** Write each of these as a single fraction in its simplest form.

**a** $\frac{3}{x+1} + \frac{2}{2x+1}$  **b** $\frac{4}{x+2} - \frac{3}{x+1}$

**c** $\frac{5}{x-3} - \frac{2}{x+4}$  **d** $\frac{3x}{x+2} - \frac{2}{x+4}$

**Thinking ahead to ...**
common factors

**A** In this triangle puzzle:

- the numbers in squares on each side are multiplied to give the numbers in circles
- the numbers in the circles are added to give the total.

Copy and complete triangles 1 and 2.

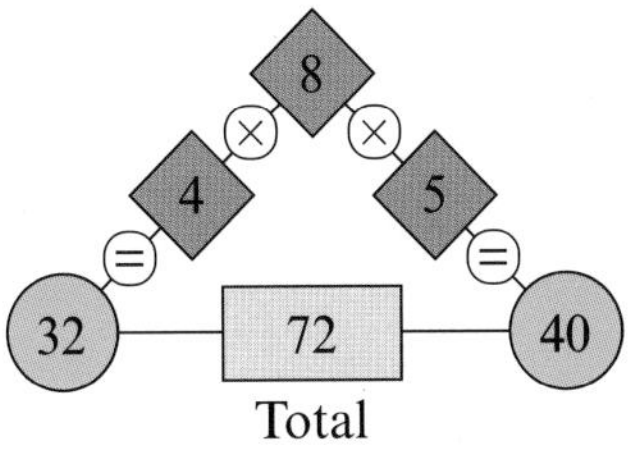


Triangle 1

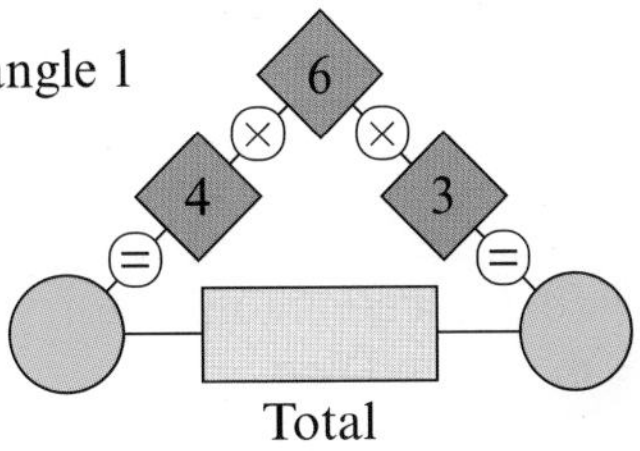


Triangle 2

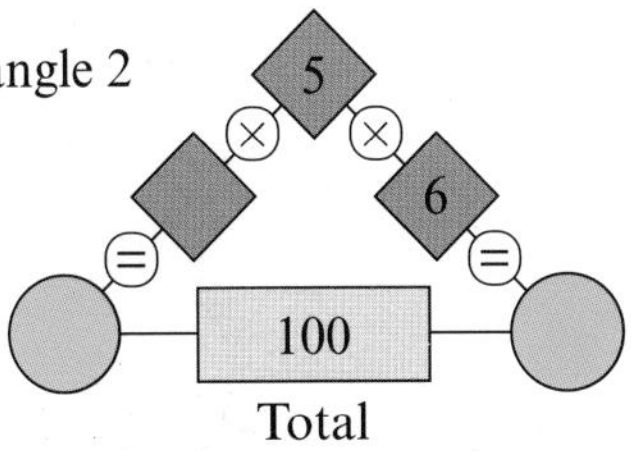


**B**

Triangle 3

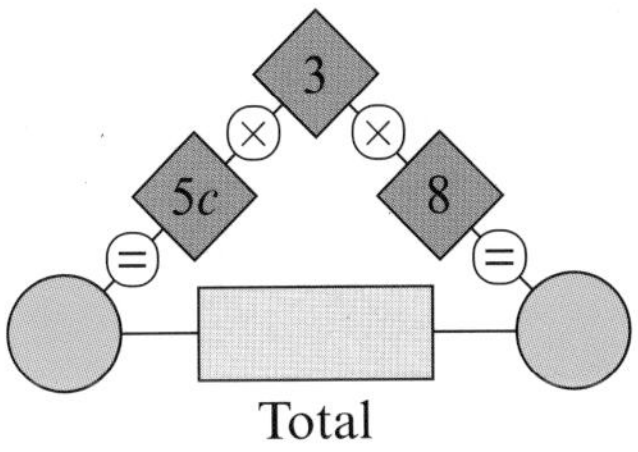


Triangle 4

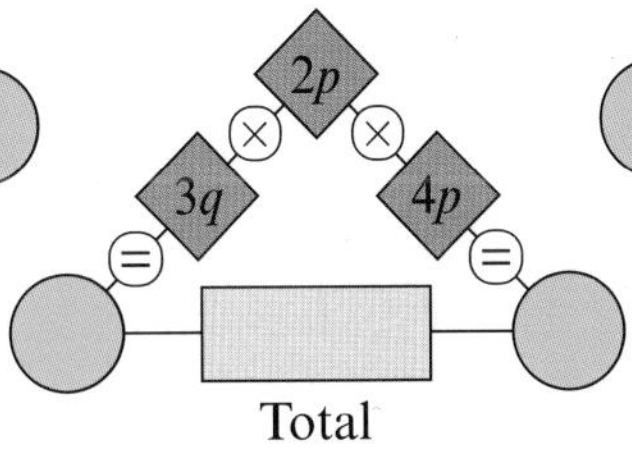


Triangle 5

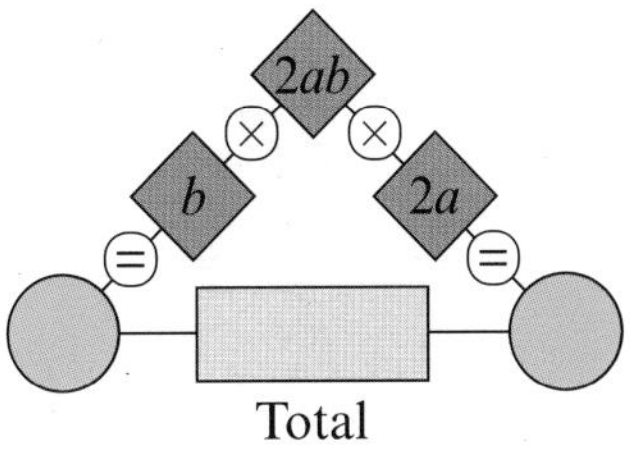


For each of the triangles 3, 4 and 5, write an expression for the total.

**C** Copy and complete these triangle puzzles.

Triangle 6

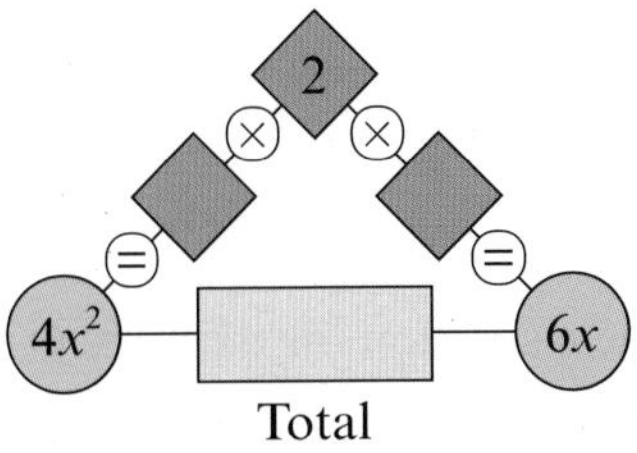


Triangle 7

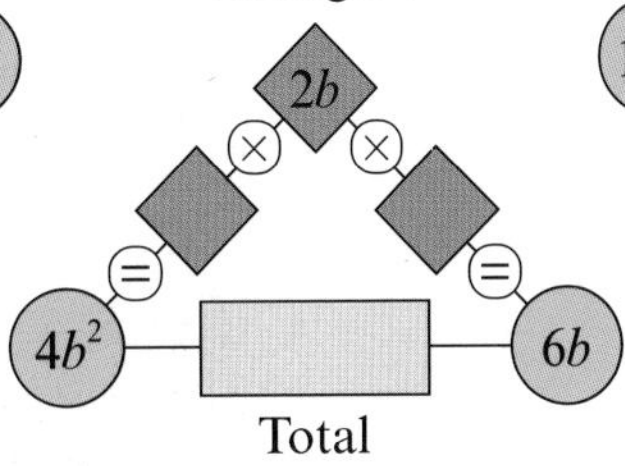


Triangle 8

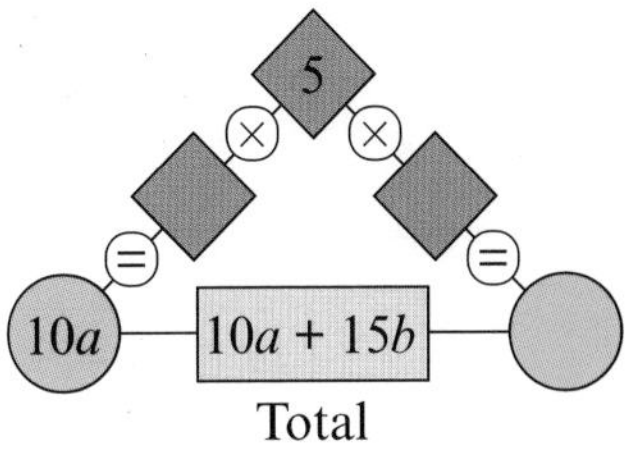

## Common factors

The factors of a number are all **whole numbers**.

- A whole number can be written as a product of factors.

**Examples** $48 = 12 \times 4$ $\quad 72 = 12 \times 6$
$48 = 2 \times 24$ $\quad 72 = 2 \times 36$
… …
12 and 2 are common factors of 48 and 72.

- A term can be written as a product of factors.

**Examples** $3a^2 = 3 \times a^2$ $\quad 6ab = 3 \times 2ab$
$3a^2 = a \times 3a$ $\quad 6ab = a \times 6b$
… …
3 and $a$ are common factors of $3a^2$ and $6ab$.

- To **factorise an expression** look for a common factor of the terms and write the expression using brackets.

**Example**

$$3a^2 + 6ab = 3(a^2 + 2ab) = 3a(a + 2b)$$

$3 \times a^2$ $\quad 3 \times 2ab$ $\quad a \times a$ $\quad a \times 2b$

So 3 is a common factor of $3a^2$ and $6ab$

So $a$ is a common factor of $a^2$ and $2ab$

**So $3a^2 + 6ab = 3a(a + 2b)$**

This is **factorised fully** as there are no other common factors.

**Exercise 3.4**
**Common factors**

When you factorise an expression, check that the two expressions are equivalent by multiplying out the bracket.

**Example**

Factorise $3ab^2 + 6a^2$

$3ab^2 + 6a^2 = 3a(b^2 + 2a)$

Check

| $\times$ | $b^2$ | + | $2a$ |
|---|---|---|---|
| $3a$ | $3ab^2$ | + | $6a^2$ |

$3a \times b^2 = 3ab^2$

$3a \times 2a = 6a^2$

**1** Factorise these fully.

| | | | |
|---|---|---|---|
| **a** $2x + 14$ | **b** $8x - 10y$ | **c** $5x + xy$ | **d** $pq + 7p$ |
| **e** $6d + 4de$ | **f** $6ab + 9a$ | **g** $2a - 8ab$ | **h** $3a^2 + 12a$ |
| **i** $15xy + 20yz$ | **j** $a^2b + ab^2$ | **k** $12c^2d - 15cd$ | **l** $25x^2y + 15yz$ |

**2** Draw a complete solution for triangles L and M.

Triangle L

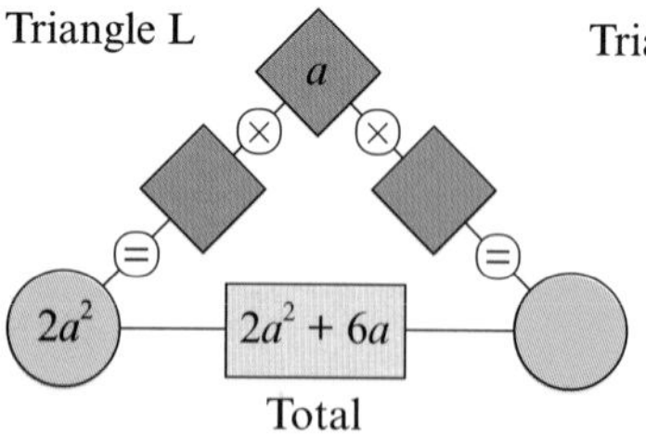


Triangle M

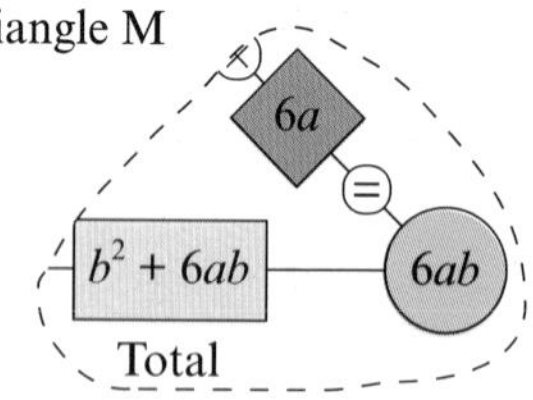


**3** **a** In triangle N what term, other than 1, can be in the top square?
**b** Draw a complete solution for triangle N.

Triangle N

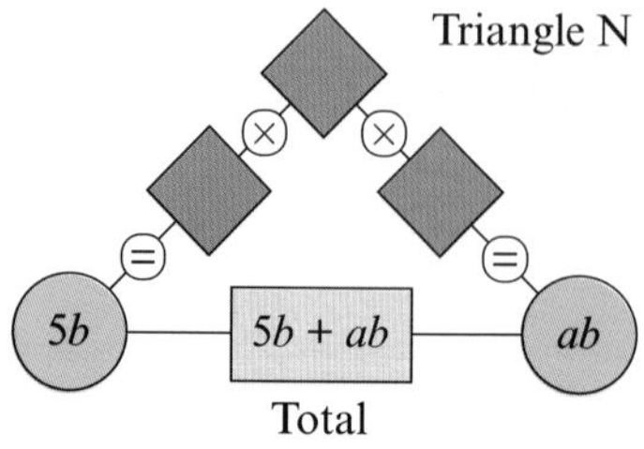


**4** Draw a complete solution for triangles O and P.

Triangle O

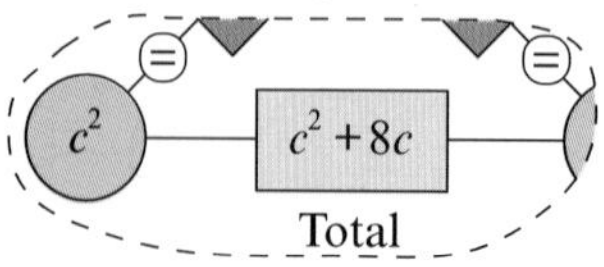


Triangle P

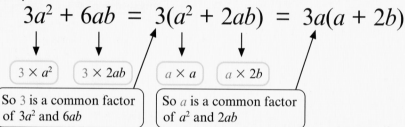

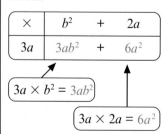

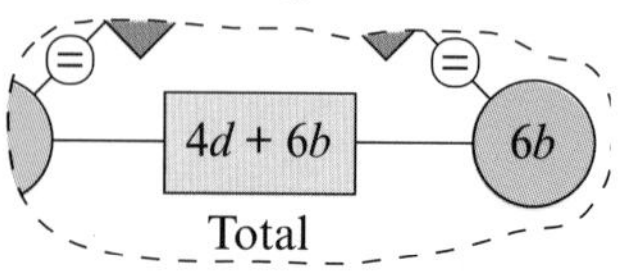

In these puzzles do not use the number 1 in the top square.

**5** **a** Draw two different solutions for triangle Q.
**b** How many different solutions are there for triangle Q?

Triangle Q

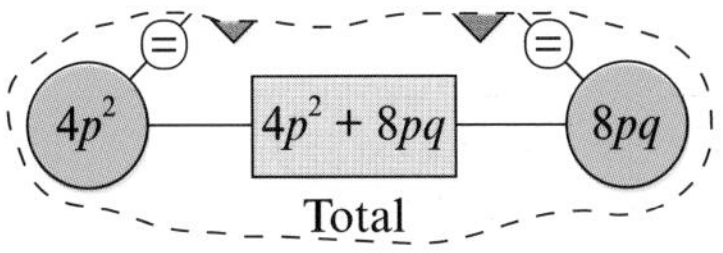


**6** Draw a complete solution for triangle R.

Triangle R

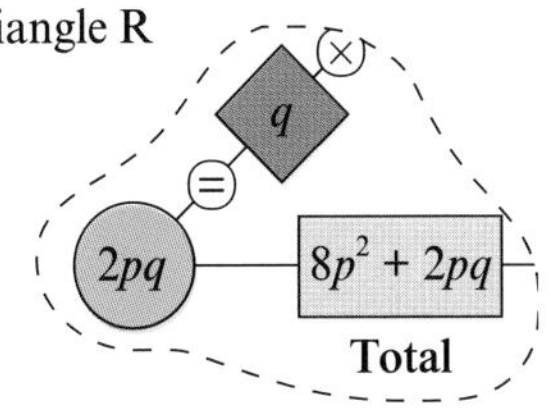


**7** Factorise these fully.

**a** $5x + 10y + 20$ **b** $6a - 9b + 12c$ **c** $2x^2 + 3x + xy$
**d** $14x - 28y + 21z$ **e** $2x^2 + 8xy + 6x$ **f** $6ab^2 + 2a^2b + 5ab$

**Thinking ahead to ... multiplying two brackets**

The total area of this shape is $43 \times 28\,\text{cm}^2$.

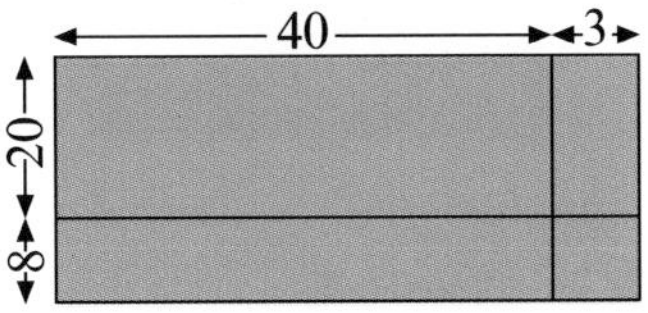


One way to find the area is:

- split the shape into 4 rectangles
- find the area of each rectangle in $\text{cm}^2$
  $20 \times 40 = 800$
  $20 \times 3 = 60$
  $8 \times 40 = 320$
  $8 \times 3 = 24$
- add the areas.
  The total area is $1204\,\text{cm}^2$.

**A** Using brackets the area of this shape can be written as $(x + 3)(x + 2)$.
This shape is split into four rectangles.
Find an expression for:

**a** the area of each rectangle
**b** the total area of the shape.

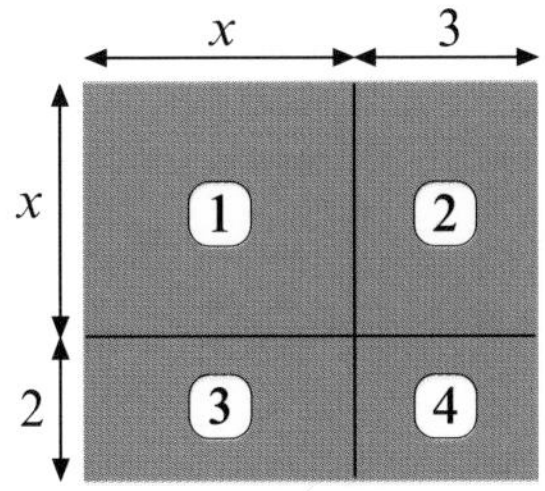


**B** Shapes P and Q are split into rectangles.

P

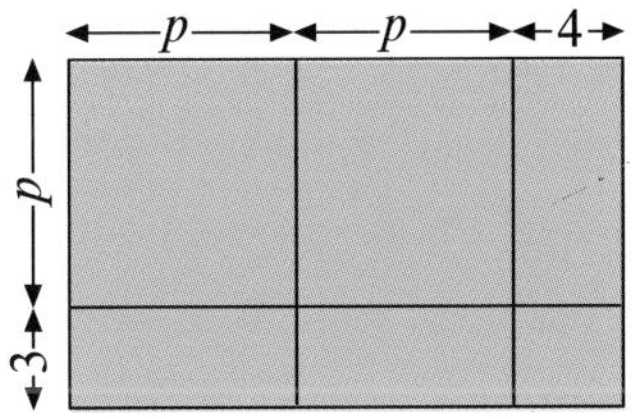


Q

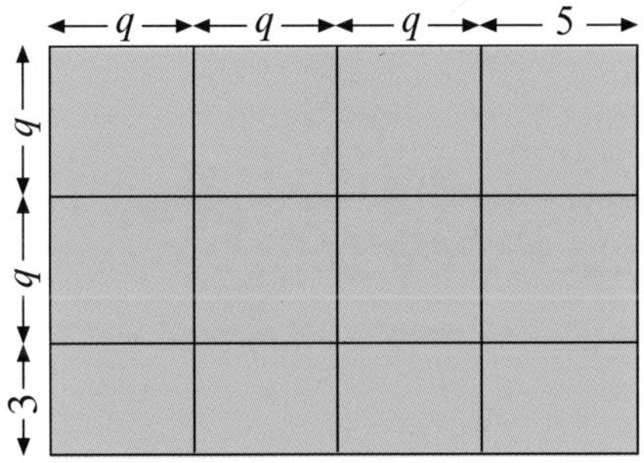


For each of the shapes P and Q write an expression for the total area:

**a** with brackets **b** without brackets in its simplest form.

**C** Shape R is split into four rectangles.

**a** Which of these is an expression for the shaded area?

There is more than one correct expression.

| | | | |
|---|---|---|---|
| A | $n(n + 2) - 4n$ | B | $(n + 2)(n - 4)$ |
| C | $n(n + 2) - 4(n + 2)$ | D | $(n - 2)(n - 4)$ |
| E | $n(n - 4) + 2(n - 4)$ | F | $n(n + 2) - 8$ |

R

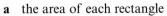

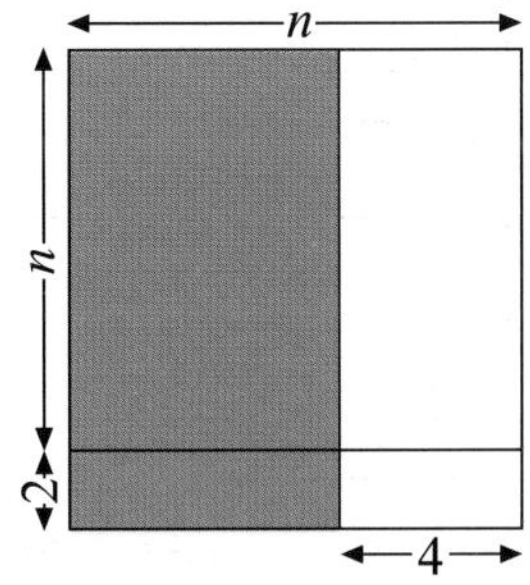


**b** Write and simplify an expression for the shaded area without brackets.

# Multiplying two brackets

- You can multiply out brackets from $(2n + 3)(3n + 4)$ like this:
  - $(2n + 3)(3n + 4) = (2n + 3) \times (3n + 4)$
  - Work in a table:

| $\times$ | $2n$ | $+3$ |
|---|---|---|
| $3n$ | $6n^2$ | $9n$ |
| $+4$ | $8n$ | $12$ |

  - The total is: $6n^2 + 9n + 8n + 12$
    which simplifies to: $6n^2 + 17n + 12$
  - So, $(2n + 3)(3n + 4) = 6n^2 + 17n + 12$ for any value of $n$.
  - $(2n + 3)(3n + 4)$ and $6n^2 + 17n + 12$ are equivalent expressions.

**Exercise 3.5**
Multiplying out brackets

1 Which of these expressions is equivalent to $(n + 1)(n + 8)$?

A $n^2 + 8n + 8$ B $2n + 9$ C $n^2 + 9n + 8$ D $2n + 8$

2 **a** Which of these expressions gives the area of this rectangle?

A $n^2 + 10n + 7$ B $n^2 + 10n + 10$

C $n^2 + 7n + 7$ D $n^2 + 7n + 10$

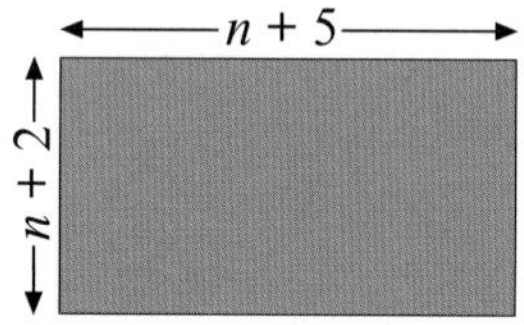


**b** Explain your answer.

3 For each of these, multiply out the brackets and simplify.

**a** $(a + 1)(a + 3)$ **b** $(d + 5)(d + 9)$ **c** $(3e + 1)(e + 7)$
**d** $(2a + 1)(5a + 4)$ **e** $(2b + 3)(3b + 5)$ **f** $(2c + 7)(3c + 2)$

4 For each of these, multiply out the brackets and simplify.

$(x + 1)^2 = (x + 1)(x + 1)$

**a** $(x + 1)^2$ **b** $(x + 2)^2$ **c** $(x + 3)^2$

Comment on any pattern you see in your results.

5 Copy and complete:

| $\times$ | $x$ | $-4$ |
|---|---|---|
| $x$ | ☐ | ☐ |
| $-2$ | ☐ | ☐ |

So $(x - 4)(x - 2) =$ ☐

6 **a** This shaded square is surrounded by four rectangles, each $x$ cm by 3 cm, where $x > 3$.
Write an expression for the area of the shaded square:
**i** with brackets
**ii** without brackets.

**b** Write an expression for the shaded square if the rectangles are:
**i** $x$ cm by 4 cm, where $x > 4$
**ii** $x$ cm by 5 cm, where $x > 5$
**iii** $x$ cm by $y$ cm, where $x > y$.
Write each answer without brackets and simplify.

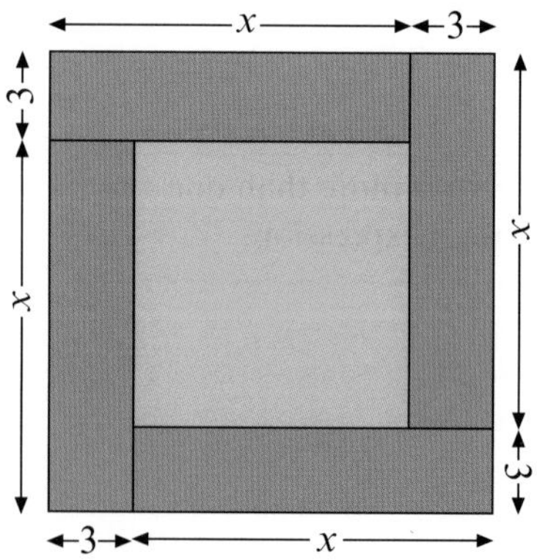

**7** Copy and complete:

| × | $2x$ | $-5$ |
|---|---|---|
| $3x$ | ☐ | $-15x$ |
| $+1$ | ☐ | ☐ |

So $(2x - 5)(3x + 1) =$ ☐

**8** For each of these, multiply out the brackets and simplify.

**a** $(z + t)(2z + t)$ **b** $(3k + m)(2k + 5m)$ **c** $(2c + 3d)(c + 7d)$
**d** $(3p - q)(2p + 5q)$ **e** $(f - g)(5f - 3g)$ **f** $(3w - v)(7w + v)$

**9** Find four pairs of equivalent expressions.

A $(2a + 3)(a - 2)$ B $(2a - 3)(a + 2)$ C $(2a - 1)(a + 6)$
D $(2a - 1)(a - 6)$ E $2a^2 - 13a - 6$ F $2a^2 + a - 6$
G $2a^2 - 13a + 6$ H $2a^2 + 11a - 6$ I $2a^2 - a - 6$

**10** For each of these, multiply out the brackets and simplify.

**a** $(z + t)(z - t)$ **b** $(2k + m)(2k - m)$ **c** $(2c - 3d)(2c + 3d)$
**d** $(3p - q)(3p + q)$ **e** $(5f + 3g)(5f - 3g)$ **f** $(7w - v)(7w + v)$

Comment on any pattern you see in your results.

There is more than one correct expression.

**11** Which of these expressions J to N gives the area shaded in this square?

J $(a - 3b)^2$ K $a^2 - 6ab + 9b^2$
L $6ab + a^2$ M $a^2 - 9b^2$
N $(a - 3b)(a + 3b)$

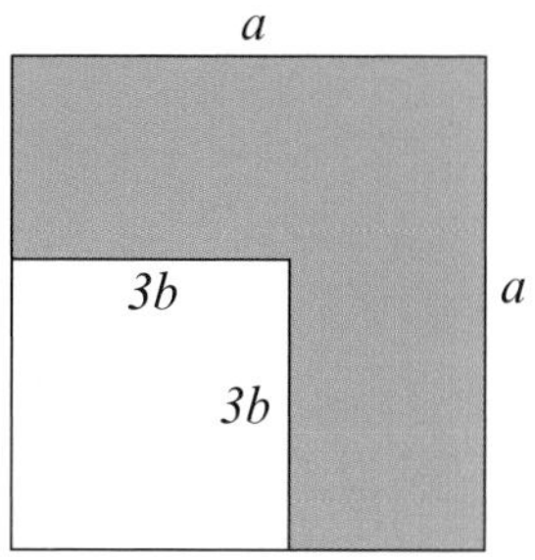


**12** **a** Multiply out and simplify:
i $(x + 1)^2$ ii $(x + 1)^3$ iii $(x + 1)^4$
**b** Look at the pattern in the coefficients.
Use it to write down $(x + 1)^6$ without brackets (in its simplest from).

**13** Write each of these without brackets and simplify.

**a** $a(a - 2)(a + 5)$ **b** $(p - 3)(p - 4)(p + 2)$
**c** $2q(2q + 3)(q - 7)$ **d** $(x + 5)(2x - 1)(x + 2)$

**Thinking ahead to ... algebraic proofs**

**A** **a** Copy and complete these questions which follow a pattern.

$(3 \times 4) - (1 \times 6) =$ ☐
$(4 \times 5) - (2 \times 7) =$ ☐
$(5 \times 6) - (3 \times 8) =$ ☐
⋮
$(☐ \times ☐) - (100 \times ☐) =$ ☐

**b** Describe what happens to the answers in the pattern.
**c** Write an expression for the $n$th question in this pattern.

# Algebraic proof

- Algebra can be used to prove that some statements are true.

**Example** Prove that the answer to every line in this pattern is 6.

$3 \times 4 - 1 \times 6$
$4 \times 5 - 2 \times 7$
$5 \times 6 - 3 \times 8$
…

- For any value of $n$, the $n$th line in this pattern can be written as: $(n + 2)(n + 3) - n(n + 5)$
- This can be simplified:

$$\begin{aligned}(n + 2)(n + 3) - n(n + 5) &= (n^2 + 5n + 6) - (n^2 + 5n) \\ &= n^2 + 5n + 6 - n^2 - 5n \\ &= 6\end{aligned}$$

This proves that for any value of $n$ the answer will be 6.

## Exercise 3.6 Algebraic proof

**1** For each of these patterns:

**a** copy and complete the examples

A
$2 \times 3 - 1 \times 4 =$
$3 \times 4 - 2 \times 5 =$
$4 \times 5 - 3 \times 6 =$
⋮

B
$2 \times 2 - 1 \times 3 =$
$3 \times 3 - 2 \times 4 =$
$4 \times 4 - 3 \times 5 =$
⋮

C
$4 \times 5 - 1 \times 8 =$
$5 \times 6 - 2 \times 9 =$
$6 \times 7 - 3 \times 10 =$
⋮

**b** write an expression for the $n$th line

**c** prove that every line of the pattern gives the same answer and state the answer.

**2** Each of these is an expression for the $n$th line of a pattern.

D $(n + 1)(n + 6) - n(n + \square) =$

E $(n + 2)(n + \square) - n(n + 5) =$

F $(n + \square)(n + 4) - n(n + 9) =$

**a** For each of the expressions D, E and F:
**i** find the missing number
**ii** calculate the answer for the pattern.

**b** In expression G below, $a$, $b$ and $c$ are constants.

G $(n + a)(n + b) - n(n + c) =$

**i** Write a relationship that connects $a$, $b$ and $c$.
**ii** What is the answer for this pattern?

**3** For each of these patterns:

**a** copy and complete the examples

A
$3 \times 4 - 1 \times 8 =$
$4 \times 5 - 2 \times 9 =$
$5 \times 6 - 3 \times 10 =$
⋮

B
$3 \times 5 - 1 \times 8 =$
$4 \times 6 - 2 \times 9 =$
$5 \times 7 - 3 \times 10 =$
⋮

C
$1 \times 6 - 3 \times 3 =$
$2 \times 7 - 4 \times 4 =$
$3 \times 8 - 5 \times 5 =$
⋮

**b** write an expression for the $n$th line

**c** describe what happens to the answers in each pattern.

**4** Write some lines in a pattern for which the answer to the $n$th line is:

**a** $n + 4$  **b** $n - 2$  **c** $10 - n$

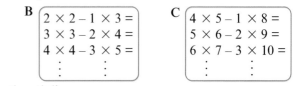

# Factorising

To factorise an expression is to write it as a **multiplication** of its factors.
For example:
$x^2 + 5x + 4$ factorises to give $(x + 4)(x + 1)$
or $(x + 1)(x + 4)$

**Example** The expression $n^2 + 2n - 8$ has no common factor but it can be factorised. This is one way to factorise $n^2 + 2n - 8$.

- Think about a table that gives a total of $n^2 + 2n - 8$ and fill in the parts you are sure about.
- Try values that give $-8$ in the correct position.
- So $n^2 + 2n - 8 = (n + 4)(n - 2)$ or $(n - 2)(n + 4)$

| $\times$ | $n$ | |
|---|---|---|
| $n$ | $n^2$ | |
| | | $-8$ |

- The total for this table is $n^2 - 2n - 8$ which **is not** what you want.

| $\times$ | $n$ | $-4$ |
|---|---|---|
| $n$ | $n^2$ | $-4n$ |
| $+2$ | $2n$ | $-8$ |

- The total for this table is $n^2 + 2n - 8$ which **is** what you want.

| $\times$ | $n$ | $+4$ |
|---|---|---|
| $n$ | $n^2$ | $4n$ |
| $-2$ | $-2n$ | $-8$ |

## Exercise 3.7 Factorising

**1** Factorise these expressions.

**a** $x^2 + 8x + 7$ **b** $p^2 + 9p + 8$ **c** $t^2 + 10t + 9$
**d** $m^2 + 8m + 12$ **e** $n^2 + 6n + 9$ **f** $a^2 + 13a + 36$

**2** Copy and complete these statements.

**a** $x^2 + 6x + \square = (x + 4)(x + 2)$ **b** $q^2 + 6q + \square = (q + 5)(q + 1)$
**c** $r^2 + \square r + 10 = (r + 5)(r + 2)$ **d** $y^2 - \square y + 10 = (y - 1)(y - 10)$
**e** $t^2 + 7t + 12 = (t + 4)(\square\square\square)$ **f** $p^2 - 7p - 8 = (p + 1)(\square\square\square)$

**3** Which pair of expressions multiply to give $x^2 + 3x - 4$?

$x + 1$ $x - 1$ $x + 2$ $x - 2$ $x + 4$ $x - 4$

**4** Which pair of expressions multiply to give $x^2 - 6x + 8$?

$x + 4$ $x - 1$ $x + 2$ $x - 2$ $x - 8$ $x - 4$

**5** Factorise these expressions.

**a** $p^2 + 2p - 3$ **b** $m^2 + 3m - 10$ **c** $t^2 - 4t - 5$
**d** $r^2 - 3r - 28$ **e** $x^2 - 8x + 15$ **f** $n^2 - 7n + 6$
**g** $t^2 - t - 12$ **h** $b^2 - 4b - 12$ **i** $x^2 + 4x - 12$

**6** Which pair of expressions multiply to give $x^2 - 16$?

$x - 16$ $x - 8$ $x - 4$ $x + 1$ $x + 4$ $x + 2$

**7** Factorise these expressions.

**a** $t^2 - 9$ **b** $r^2 - 25$ **c** $16p^2 - 1$
**d** $9x^2 - 4$ **e** $x^2 - y^2$ **f** $x^2 - 9p^2$

**8** Which of these cannot be factorised as in the Example above?

A $x^2 + x + 15$ B $x^2 + 9x + 20$ C $x^2 - 6x + 12$

- You can use a table to factorise expressions like $2n^2 - 7n + 3$, where the **coefficient** of $n^2$ is not equal to 1.
  - Think about a table that gives a total of $2n^2 - 7n + 3$ and fill in the parts you are sure about.
  - Try values that give +3 in the correct position, and give a negative coefficient of $n$.

| × | $n$ | |
|---|---|---|
| $2n$ | $2n^2$ | |
| | | +3 |

| × | $n$ | $-1$ |
|---|---|---|
| $2n$ | $2n^2$ | $-2n$ |
| $-3$ | $-3n$ | +3 |

  - The total for this table is $2n^2 - 5n + 3$ which **is not** what you want.

| × | $n$ | $-3$ |
|---|---|---|
| $2n$ | $2n^2$ | $-6n$ |
| $-1$ | $-n$ | +3 |

  - The total for this table is $2n^2 - 7n + 3$ which **is** what you want.
  - So $2n^2 - 7n + 3 = (n - 3)(2n - 1)$ or $(2n - 1)(n - 3)$

## Exercise 3.8 Factorising

**1** Which pair of expressions multiply to give $2x^2 + 7x + 6$?

$2x + 6$ | $2x + 3$ | $2x + 2$ | $2x + 1$ | $x + 1$ | $x + 2$ | $x + 3$ | $x + 6$

**2** Factorise these expressions.

**a** $3x^2 - 10x + 3$ **b** $4y^2 + y - 3$ **c** $4p^2 - 11p - 3$
**d** $4x^2 - x - 3$ **e** $3x^2 + 8x - 3$ **f** $3a^2 + 10a + 3$
**g** $5n^2 - 11n + 2$ **h** $3x^2 - x - 4$ **i** $3x^2 - 4x + 1$

**3** This puzzle has been completed but most clues are missing.

**Algebra Puzzle**

| | | | | | | |
|---|---|---|---|---|---|---|
| | 1 $+ 9x^2$ | | 2 $+ 1$ | $+ 4x$ | 3 $+ 4x^2$ | |
| 4 $+ x^2$ | $- 1$ | | $+ x^2$ | | $- 1$ | |
| $- 4x$ | | 5 $+ 3x^2$ | $- 2x$ | 6 $- 5$ | | |
| 7 $+ 3$ | $+ 4x^2$ | $+ 8x$ | | 8 $- 4x$ | $- 1$ | 9 $+ 5x^2$ |
| | | 10 $+ 4$ | 11 $- 5x$ | $+ x^2$ | | $+ x$ |
| | 12 $- 9$ | | $+ x^2$ | | 13 $+ x^2$ | $- 4$ |
| | 14 $+ 4x^2$ | $- x$ | $- 14$ | | $- 49$ | |

**Clues**

**Across**
**2** $(2x + 1)(2x + 1)$
**4** $(x - 1)(x$
**5** $(x + 1)($
**7** $(2x + 1)$
**8** $(x - 1)$
**10**
**13**
**14**

**Down**
**1** $(3x - 1)(3x + 1)$
**2**
**3**
**4**
**5**
**6**
**9**
**11**
**12**
**13**

For every clue find a pair of brackets that multiply to give the correct expression.

**a** Write a full set of clues for this puzzle.
**b** Make up your own puzzle like this.

4 For each of these find the numerical value of $a$ and $b$ such that for all values of $x$:

a $x^2 - 12x + a = (x - b)^2$ b $x^2 - ax + 9 = (x - b)^2$
c $9x^2 - ax + 1 = (3x - b)^2$ d $x^2 - ax + a = (x - b)^2$

If there is a common factor in all the terms, take that out first then factorise.

Example

$2x^2 + 10x + 8$
$= 2(x^2 + 5x + 4)$
$= 2(x + 1)(x + 4)$
This is factorised fully.

5 Find a common factor and factorise each of these expressions fully.

a $3x^2 - 9x + 6$ b $6x^2 + 21x + 15$ c $6x^2 + 16x - 6$
d $14x^2 - 21x - 14$ e $3x^2 - 6x + 3$ f $4x^2 - 4$

6 a Factorise fully $n^3 - n$.
b Show that $n^3 - n$ is even for all integer values of $n$.

7 Factorise these fully.

a $x^2 + 2xy + y^2$ b $4x^2 + 8xy + 4y^2$ c $9x^2 - 16y^2$

## Using factorisation

**Exercise 3.9**
**Using factorisation**

1 Three consecutive numbers are added together.

a If the smallest number is $p$, write each other number in terms of $p$.
b i Write an expression for the sum of the three numbers.
ii Simplify the expression.
iii Explain why this shows that the sum of three consecutive numbers is always a multiple of 3.

2 a Evaluate the expression $n^2 + n - 6$ for:
i $n = 1$ ii $n = 2$ iii $n = 3$ iv $n = 4$
b Factorise $n^2 + n - 6$.
c Explain why this shows that the value of $n^2 + n - 6$ is even for any value of $n$.

3 Show that $n^2 - 6n + 9 \geqslant 0$ for all values of $n$.

4 Kim puts this cross anywhere on this size 8 grid. She multiplies the corner numbers in pairs and finds the difference.

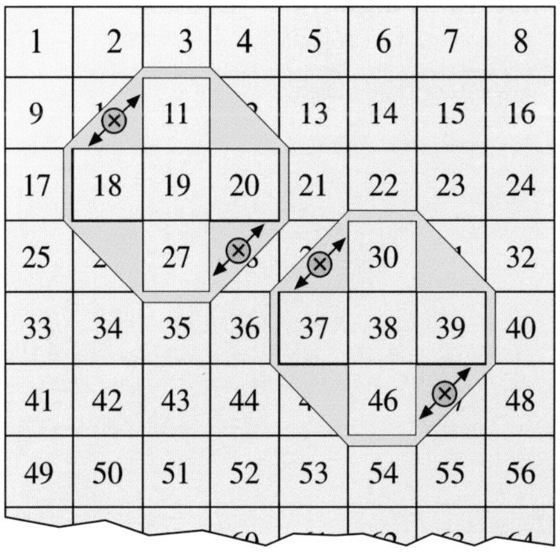


Example

$20 \times 27 - 11 \times 18 = 342$
$39 \times 46 - 30 \times 37 = 684$

a i If the number in the centre of the cross is $y$, write each of the corner numbers in terms of $y$.
ii Write an expression for the difference of the products.
iii Show that the difference will be $18y$ for all possible values of $y$.
b Kim uses the same cross on a size $n$ grid.
i Write the corner numbers in terms of $n$ and $y$.
ii Write an expression for the difference of the products.
iii What would be the difference for a cross with $y = 16$ on a size 9 grid?

Because the cross cannot be placed on the edge of the grid some values of $y$ are not possible.

Example

On the 8 grid these values of $y$ are not possible:
1, 2, 3, 4, 5, 6, 7, 8, 9, 16, 17, 24, 25, … .

5 Show that the product of three consecutive numbers must be a multiple of 6.

**Exercise 3.10**
**Using the difference of two squares**

1 The expression $a^2 - b^2$ is called the difference of two squares.

a Factorise $a^2 - b^2$.

b Use this to find the value of:
i $98^2 - 2^2$ ii $47^2 - 3^2$

2 a Copy and complete this sequence of lines.

b Describe any patterns that you notice.

c Write an expression for the $n$th line in this sequence and simplify it.

d Write down the exact value of:
i $1234^2 - 1233^2$
ii $123123123123^2 - 123123123122^2$

$2^2 - 1^2 =$
$3^2 - 2^2 =$
$4^2 - 3^2 =$
$5^2 - 4^2 =$
$6^2 - 5^2 =$
⋮ ⋮

3 a Use algebra to explain the pattern in these answers.

b Use the pattern to write down the value of:
i $128^2 - 123^2$
ii $307^2 - 302^2$

$6^2 - 1^2 = 35$
$7^2 - 2^2 = 45$
$8^2 - 3^2 = 55$
$9^2 - 4^2 = 65$
⋮ ⋮ ⋮

- Some algebraic fractions can be simplified by factorising.

**Example**

$$\frac{x^2 + 5x + 6}{x^2 + 2x - 3} = \frac{(x + 3)(x + 2)}{(x + 3)(x - 1)} = \frac{(x + 2)}{(x - 1)}$$

(numerator $\div (x + 3)$, denominator $\div (x + 3)$)

**Exercise 3.11**
**Factors in algebraic fractions**

1 a Find the value, in fractional form, of:

$$\frac{n - 6}{n^2 - 5n - 6}$$

when:
i $n = 1$ ii $n = 2$ iii $n = 3$

b Describe any patterns that you notice.

c i Factorise and simplify the fraction.
ii Use this to explain any patterns you found.

2 a Simplify each of these fractions as far as possible.

i $\dfrac{2n + 6}{2n^2 + 4n - 6}$ ii $\dfrac{n^2 - n - 6}{n^2 - 3n}$

iii $\dfrac{n^2 + 5n}{2n + 10}$ iv $\dfrac{2n^2 - 4n}{n^2 - 5n + 6}$

b Give the value of each fraction when $n = 50$.

3 a Express $\dfrac{x^2 + 3}{3} + \dfrac{x(x + 9)}{6}$ as a single fraction and simplify.

b Show that the value of this is an integer for all integer values of $x$.

# End points

You should be able to ... | ... so try these questions

**A** Multiply out brackets

**A1** Simplify these expressions.
- **a** $2(2b - 4) + 6(4 + 3b)$
- **b** $4(2b - 4) + 6(5 + 3b)$
- **c** $3ab(b - a) + 7ab(b + a)$
- **d** $3xy(x - 3) - 7xy(y + 4)$

**A2** Which of these expressions is equivalent to $(t + 5)(t - 3)$?

A $t^2 + 8t + 15$ B $t^2 + 8t - 15$ C $t^2 + 2t - 15$

**A3** For each of these, multiply out the brackets and simplify.
- **a** $(a + 1)(a + 8)$
- **b** $(b + 4)^2$
- **c** $(5c + 2)(c + 3)$
- **d** $(d + 2)(d - 1)$
- **e** $(2e + 5)(3e - 10)$
- **f** $(f - 7)(3f - 2)$

**A4** These questions follow a pattern.

$(2 \times 7) - (1 \times 5)$
$(3 \times 8) - (2 \times 6)$
$(4 \times 9) - (3 \times 7)$
...

- **a** Write the next three lines of this pattern.
- **b** What is the answer to the 20th line of this pattern?
- **c** Write an expression for the $n$th line of this pattern.
- **d** Prove that the answer to any line of this pattern will be a multiple of 3.

**B** Factorise expressions

**B1** Factorise these fully.
- **a** $8p + 4q$
- **b** $6a - 12b$
- **c** $a^2 + ab$
- **d** $7m + 3mn$
- **e** $8xy + 10y$
- **f** $4ab + 6a^2$
- **g** $3xy^2 - 5x^2y$
- **h** $4pq - 6p^2q$
- **i** $2g^2h - 5h^2$

**B2** A $k + 2$ B $k - 7$ C $k - 2$ D $k + 14$ E $k + 7$ F $k - 1$

Which pair of expressions multiply to give:
- **a** $k^2 + 5k - 14$
- **b** $k^2 - 9k + 14$
- **c** $k^2 + 13k - 14$?

**B3** Factorise these expressions.
- **a** $x^2 + 12x + 11$
- **b** $x^2 + 9x + 14$
- **c** $x^2 + 4x - 5$
- **d** $x^2 + x - 6$
- **e** $x^2 - x - 20$
- **f** $x^2 - 5x + 6$
- **g** $x^2 - 100$
- **h** $2x^2 - 11x + 15$
- **i** $9x^2 - 16$
- **j** $5x^2 + 4x - 1$
- **k** $x^2 - 4y^2$
- **l** $4x^2 - 4xy + y^2$

**B4** Find two pairs of equivalent expressions.

A $\frac{1}{a+1} + \frac{2}{a}$ B $\frac{3a+6}{a^2+2a}$ C $\frac{3}{a-1} - \frac{3}{a+1}$

D $\frac{2a^2}{5a^3} \times \frac{15a}{2a}$ E $\frac{15a}{a^2-1} \times \frac{2}{5a}$ F $\frac{2}{a-1} \div \frac{a-1}{3}$

**B5** **a** Simplify $\frac{x^2 - 8x + 16}{3x - 12}$ as far as possible.

**b** For what values of $x$ is $\frac{x^2 - 8x + 16}{3x - 12}$ positive?

## Starting points

You need to know about ...

### A Types of data

Data may be described as discrete or continuous.

- Discrete data is data which can only be of certain definite value. Data on shoe sizes would be discrete data.
- Continuous data is data which can have any value within a range. Data on the width of hand-spans would be continuous.

### B Finding averages and the range

- For data in categories like colour, you can only find one average:
  - the **mode** (the **modal** category is the most common).

Red Blue Yellow Blue Green Black Red Blue

The modal colour is **Blue**.

- For numerical data, you can find several averages and the range:
  - the **mode** is the most common value (or values)
  - the **range** is the difference between the highest and lowest values

87 43 101 56 87 67

Mode = **87**
Range = **58** (101 – 43)

  - the **median** is the middle value when the data is in order (for an **even** number of values, take the median as halfway between the middle pair of values)

43 56 67 ¦ 87 87 101

Median = **77**

  - the **mean** is the total of all the values divided by the number of values.

$$\frac{87 + 43 + 101 + 56 + 87 + 67}{6}$$ Mean = **73.5**

### C Deciding which average to use

- The **mean** is the most widely used average. It is not a sensible average to use for data with **extreme values**, values which are much smaller or much greater than the others.

| G.T. Small & Son – Monthly Salaries (£) | | | | | |
|---|---|---|---|---|---|
| 870 | 870 | 870 | 870 | 1050 | 1050 |
| 1050 | 1210 | 1210 | 1210 | 2080 | 2330 |

Mean = £1222.50
Median = £1050
Mode = £870

10 of the 12 salaries are smaller than the mean, so the **median** is the more sensible average to use.

- The **mode** can also be a poor choice of average. For this data, it is a poor choice because it is the lowest salary.

... so try these questions

**A1** State whether the following data is continuous or discrete:
- **a** the number of fish in an aquarium
- **b** the weight of apples in an orchard
- **c** the quoted age of each person in a cinema
- **d** the number of chairs in each classroom of a school.

**B1**

X: 15 42 33 37 84 42 50 81 29 26 67 15 19 55

Y: 38 25 106 78 44 62 13 90 25 31 25

For each set of data, find:
- **a** the mode
- **b** the median
- **c** the mean
- **d** the range.

**B2**

Mode 3 and 7
Median 6
Mean 6

These are averages for a set of data with 8 values. List what the values might be.

**C1**

*S. Fry & Partners*
**Weekly Wages (£)**
112 285 285 340 340 340 372 372 388

Find:
- **a** the modal wage
- **b** the median wage
- **c** the mean wage
- **d** For each average wage:
  - **i** decide whether it is sensible to use it or not
  - **ii** if you think it is not sensible, explain why.

## D Finding averages from a frequency table

♦ For data in categories:

❖ the **mode** is the category with the highest frequency.

| Make of car | Ford | Rover | Vaux. | Others |
|---|---|---|---|---|
| Frequency | 9 | 6 | 2 | 7 |

The modal make of car is **Ford**.

♦ For numerical data:

❖ the **mode** is the value (or values) with the highest frequency
❖ the **range** is the difference between the highest and lowest values

| No. of people in car | 1 | 2 | 3 | 4 | 5 |
|---|---|---|---|---|---|
| Frequency | 3 | 10 | 6 | 3 | 2 |

Mode = **2** people
Range = **4** people
(5 – 1)

❖ the **median** is the middle value when the data is in order

1 1 1 2 2 2 2 2 2 2 2 2 3 3 3 3 3 3 4 4 4 5 5

Median = **2** people

❖ the **mean** is the total of all the values divided by the total frequency.

| Number of people in car | Frequency | | Total number of people |
|---|---|---|---|
| 1 | 3 | 1 × 3 | 3 |
| 2 | 10 | 2 × 10 | 20 |
| 3 | 6 | 3 × 6 | 18 |
| 4 | 3 | 4 × 3 | 12 |
| 5 | 2 | 5 × 2 | 10 |
| | 24 | | 63 |

Mean = $\frac{63}{24}$ = **2.6** people (to 1 dp)

## E Constructing pie charts

| Make of car | Ford | Rover | Vaux. | Others | Total |
|---|---|---|---|---|---|
| Frequency | 9 | 6 | 2 | 7 | 24 |

♦ For a percentage pie chart, calculate the percentage in each category:
❖ there are 2 Vauxhall cars out of 24, so
100 ÷ 24 × 2 = 8.3% (to 1 dp)

♦ For an angle pie chart, calculate the angle for each category:
❖ there are 9 Ford cars out of 24, so
360 ÷ 24 × 9 = 135°.

|  | Freq | Sector | Size |
|---|---|---|---|
| Ford | 9 | 37.5% | 135° |
| Rover | 6 | 25% | 90° |
| Vauxhall | 2 | 8.3% | 30° |
| Others | 7 | 29.2% | 105° |
| Totals | 24 | 100% | 360° |

Ford
Rover
Vauxhall
Others

---

**NUMBER OF CHILDREN IN CAR**

1 0 1 2 1 1 0 0
4 1 0 1 0 0 2 0
2 0 3 0 0 0 2 0 1

**COLOUR OF CAR**

| Red | Blue | Green | White |
|---|---|---|---|
| Silver | Red | White | Black |
| Black | Green | Red | Blue |
| Black | Blue | Blue | Silver |
| Red | White | Blue | Brown |
| Blue | Black | Brown | Red |

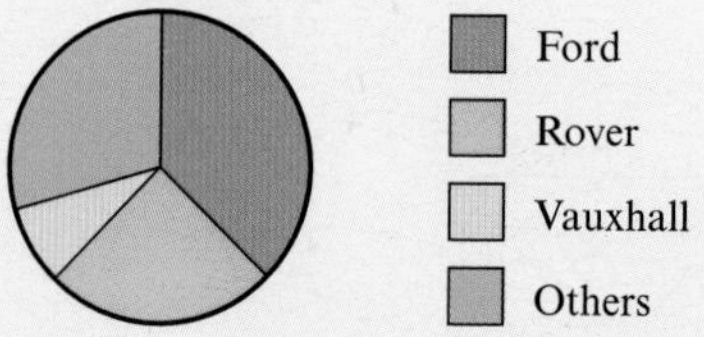

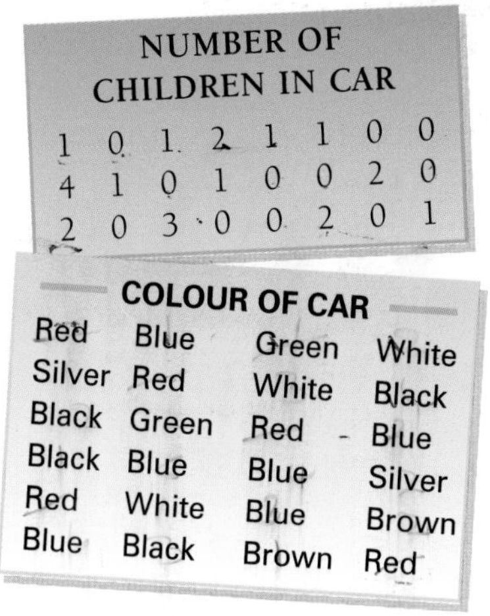

**D1** Present the colour-of-car data in a frequency table.

**D2** Find the modal colour of car.

**D3** Present the number-of-children-in-car data in a frequency table.

**D4** Find:
**a** the modal number of children in a car
**b** the range of the number of children in a car.

**D5** Find:
**a** the median number of children in a car
**b** the mean number of children in a car.

**E1** Use your frequency table from Question **D1** to calculate:
**a** the percentage for each colour
**b** the angle for each colour.

**E2** Draw a pie chart to show the colour of car data.

# F Diagrams that present data

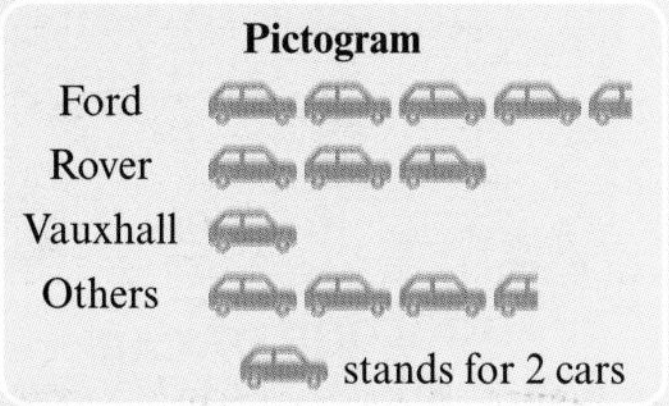


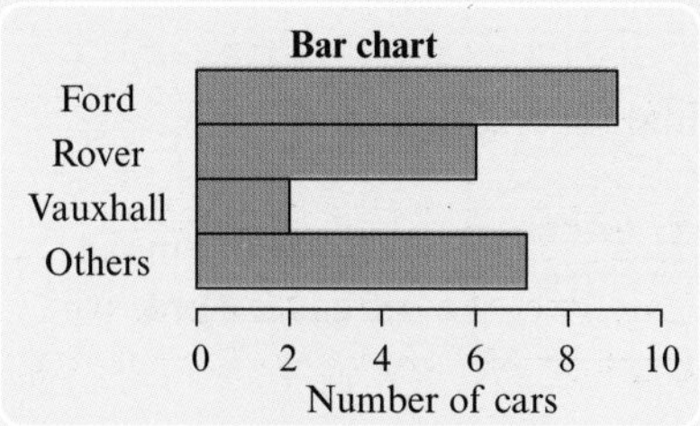


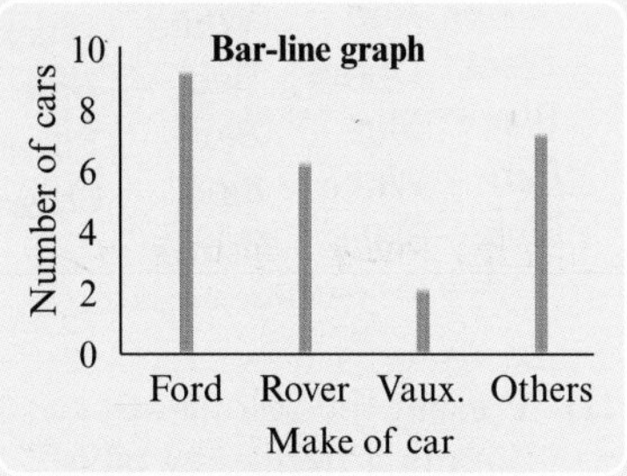


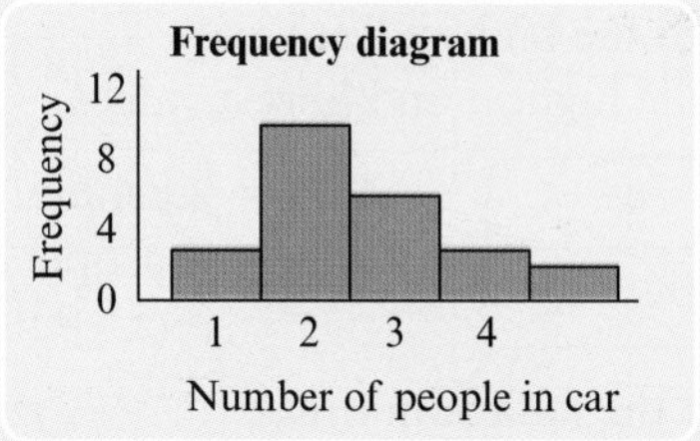


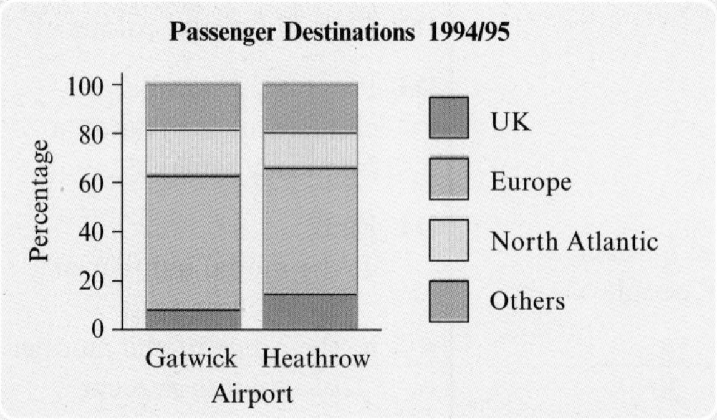


**Split bar chart or Component bar chart**

**Line graph**

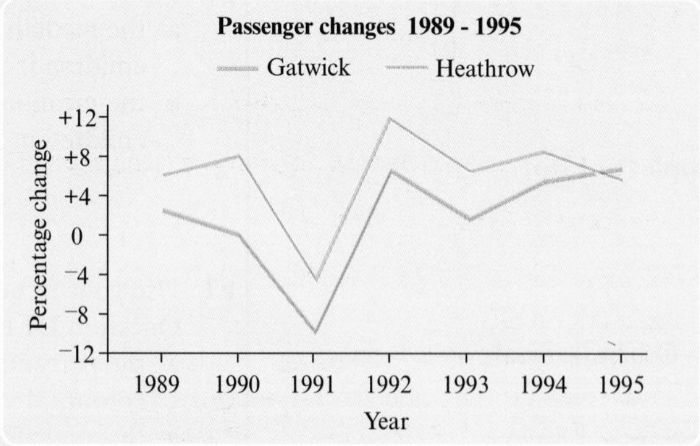


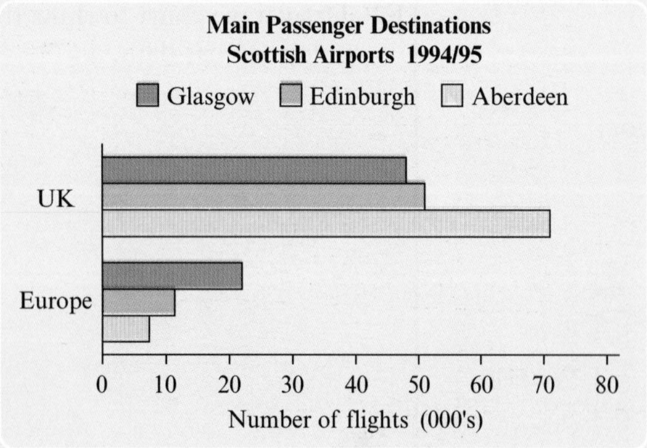


**Comparative bar chart**

Use your frequency tables from Questions **D1** and **D3** to answer Questions **F1** to **F4**.

**F1** Draw a bar-line graph to show the colour-of-car data.

**F2** Draw a frequency diagram to show the number-of-children-in-car data.

**F3** Draw a pictogram to show the colour-of-car data.

**F4** Draw a bar chart to show the colour-of-car data.

**F5**

**Southampton to Channel Islands *1994/95***

| | Number of passengers (000's) | Number of flights (000's) |
|---|---|---|
| Jersey | 156 | 4.0 |
| Guernsey | 100 | 3.4 |
| Alderney | 31 | 2.9 |
| Totals | 287 | 10.3 |

Calculate the percentage of passengers going to:
**a** Jersey **b** Guernsey
**c** Alderney

**F6** Calculate the percentage of flights going to each of the three islands.

**F7** Draw a split bar chart to show the two sets of data.

**F8**

**% change in no. of passengers**

| | 1990 | 1991 | 1992 | 1993 | 1994 |
|---|---|---|---|---|---|
| **Glasgow** | 11.0 | -3.1 | 12.4 | 7.4 | 8.8 |
| **Edinburgh** | 5.3 | -6.1 | 8.4 | 7.2 | 10.3 |

Draw a line graph to show this data for Glasgow and Edinburgh.

**F9**

**NUMBER OF FLIGHTS 1994/95 (000's)**

| | Heathrow | Gatwick |
|---|---|---|
| UK | 75 | 33 |
| Europe | 252 | 111 |
| North Atlantic | 40 | 19 |
| Others | 47 | 21 |

Draw a comparative bar chart to show this data.

**Thinking ahead to ... calculating the interquartile range**

10-ring target

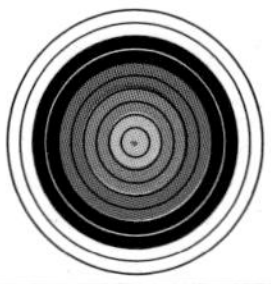

**A** These are Viv's scores in 12 arrows at a 10-ring target.

| Ring score | 1 | 2 | 3 | 4 | 5 | 6 | 7 | 8 | 9 | Total |
|---|---|---|---|---|---|---|---|---|---|---|
| Frequency | 1 | 0 | 0 | 2 | 1 | 3 | 4 | 0 | 1 | 12 |

**a** Calculate the range of the scores.
**b** Explain why the range is not a sensible measure of spread for Viv's scores.

## Calculating the interquartile range

Because the spread of the middle 50% of the data is measured, the interquartile range is not affected by any extreme values in the data.

- The **interquartile range** of a set of data measures how spread out the middle 50% of the data is.
- To calculate the interquartile range:

| Ring score | 1 | 2 | 3 | 4 | 5 | 6 | 7 | 8 | 9 | Total |
|---|---|---|---|---|---|---|---|---|---|---|
| Frequency | 1 | 0 | 0 | 2 | 1 | 3 | 4 | 0 | 1 | 12 |

- list the data in order

1 4 4 5 6 6 6 7 7 7 7 9

- divide the data into four quarters

1 4 4 ¦ 5 6 6 ¦ 6 7 7 ¦ 7 7 9

- find the end values of the middle 50% of the data

1 4 4 ¦ 5 6 6 ¦ 6 7 7 ¦ 7 7 9

**4.5** **Lower quartile** **7** **Upper quartile**

- calculate the difference between the upper and lower quartiles.

Interquartile range = 7 – 4.5
= 2.5

You can divide the data into four quarters by first dividing it into two halves.

The end values are the medians of the two halves.

**Exercise 4.1**
**Interquartile range**

**1** These are the scores for three archers.

William

| Ring score | 2 | 3 | 4 | 5 | 6 | 7 | 8 | Total |
|---|---|---|---|---|---|---|---|---|
| Frequency | 1 | 0 | 3 | 2 | 5 | 3 | 2 | 16 |

Bryony

| Ring score | 4 | 5 | 6 | 7 | 8 | 9 | Total |
|---|---|---|---|---|---|---|---|
| Frequency | 2 | 4 | 3 | 3 | 0 | 2 | 14 |

Daniel

| Ring score | 2 | 3 | 4 | 5 | 6 | 7 | 8 | 9 | 10 | Total |
|---|---|---|---|---|---|---|---|---|---|---|
| Frequency | 1 | 1 | 2 | 3 | 5 | 3 | 0 | 3 | 2 | 20 |

**a** **i** For each archer list the data.
**ii** Calculate the interquartile range of their scores.

**b** Which archer is the most consistent?
Explain your answer.

A set of data presented in a frequency table is called a **frequency distribution**.

# Drawing and using a stem and leaf plot

- The stem and leaf plot is a frequency diagram.
- The actual data is displayed together with its frequency.
- Part of the value of each piece of data is used to fix the data class ie the **stem**.

  Part of the value is listed in the diagram ie the **leaves**

### Example

Show this set of data with a stem and leaf plot

6, 13, 23, 42, 30, 32, 36, 27, 24, 32, 40, 45, 8

As the data is listed and shown in order on a stem and leaf plot the median value can be easily found. Here the median value is 6.

| stem | leaves |
|---|---|
| 4 | 0 2 5 |
| 3 | 0 2 2 6 |
| 2 | 3 4 7 |
| 1 | 3 |
| 0 | 6 8 |

**Key** 1|3 = 13

**Exercise 4.2**
Stem and leaf plots

1 Draw a stem and leaf plot for each data set.

**a** 23, 12, 14, 26, 38, 46, 33, 32, 19, 7, 24, 12, 40, 20
**b** 15, 22, 24, 20, 10, 12, 31, 30, 24, 26, 26, 24, 32, 40
**c** 42, 16, 9, 7, 6, 18, 25, 6, 24, 30, 36, 25, 30, 41, 16, 44
**d** 38, 14, 22, 56, 18, 24, 28, 37, 40, 30, 32, 28, 46, 40, 28

2 The data gives the number of minutes waited by people to fly on the 'London Eye'.

44, 56, 50, 52, 42, 40, 38, 44, 52, 54, 53, 38, 30
41, 40, 51, 59, 62, 34, 40, 47, 50, 54, 50, 57

**a** Draw a stem and leaf plot to show this data.
**b** Find the median value for the data set.

You can use a stem and leaf plot to help you to compare distributions.

### Example

This data shows the points scored by members of two teams over the past six weeks.

**Team A**
12, 23, 34, 18, 7, 0, 9, 8, 12, 0, 7, 0, 6, 0, 22, 15, 0, 9

**Team B**
8, 14, 6, 22, 31, 14, 0, 0, 9, 12, 0, 17, 0, 18, 6, 0, 4, 16

Use a stem and leaf plot and compare the data sets.

| Team A | | Team B |
|---:|:---:|:---|
| | 4 | |
| 4 | 3 | 3 |
| 3 2 | 2 | 2 |
| 8 6 5 2 | 1 | 2 4 4 6 7 8 |
| 9 9 8 7 7 6 0 0 0 0 0 0 | 0 | 0 0 0 0 0 4 6 6 8 9 |

**Key** |1| 2 = 12
3 |2| = 23

The median for each data set is between the $9^{th}$ and the $10^{th}$ values. You can easily find this from the stem and leaf plot.

**Median**
Both teams have the same median points start value of 8.5.

You may choose to use the interquartile range as part of your comparison.

**Range**
Team A has a range of 34 [34 – 0]

Team B of has a range of 31 [31 – 0]

Overall the performance of the teams is similar, but team B is slightly more consistent.

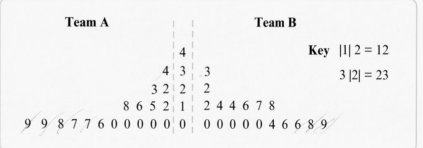

**Exercise 4.3**
**Comparing data using stem and leaf plots**

1 These data sets show the length of time (to the nearest minute) spent by customers in two different supermarkets on the same day.

**Supermarket A**

31, 14, 12, 27, 8, 38, 15, 42, 46, 32, 28
9, 28, 30, 34, 42, 21, 20, 12, 8, 4, 6
24, 33, 52, 46, 38, 35, 30, 18, 22, 34

**Supermarket B**

26, 15, 19, 34, 36, 28, 24, 14, 22, 37
40, 29, 22, 20, 28, 33, 30, 16, 41, 22
17, 26, 30, 41, 44, 43, 30, 32, 7, 12, 26, 44

**a** Draw a stem and leaf plot to show both data sets.
**b** Give the median value for the data in each set.
**c** Give the range for each data set.
**d** Compare the two sets of data.
What general statements do you feel able to make?

2 This system and leaf plot shows the distribution of marks in the same test by two groups of students.

| Group A | | Group B |
|---:|:---:|:---|
| 5 5 5 | 5 | 1 3 4 |
| 2 1 1 0 | 4 | 0 3 5 8 8 |
| 7 6 6 5 5 5 4 0 | 3 | 2 5 6 6 6 8 9 9 |
| 8 7 4 4 4 4 3 1 | 2 | 0 3 4 4 6 6 6 9 |
| 8 6 6 4 0 | 1 | 6 8 9 9 9 |
| 7 5 | 0 | 4 |

**Key** |1| 6 = 16

4 |3| = 34

**a** Give the median test score for each group
**b** What can you say about the range for each data set?
**c** Which group do you think performed better in the test?
Give reasons for your answer.

3 This data gives the length of time (seconds) taken for telephone calls by two different call centres.

**Call Centre A**

24, 38, 46, 30, 27, 46, 54, 18, 20, 37, 35, 62
58, 64, 70, 32, 27, 20, 18, 26, 35, 44, 56, 68
75, 60, 36, 45, 19, 27, 35, 50, 63, 74, 55, 40

**Call Centre B**

53, 42, 47, 16, 22, 12, 20, 30, 35, 38, 47, 76
66, 54, 46, 19, 30, 32, 38, 45, 52, 64, 72, 75
11, 20, 36, 42, 53, 38, 24, 19, 23, 28, 24, 20

**a** Draw a stem and leaf plot to show both data sets.
**b** Which call centre is more efficient ?
Give reasons for your choice.
**c** Calculate the mean time for a call from each centre.

**Thinking ahead to ...**
**calculating the standard deviation**

**A** In a competition, archers fire ten arrows in each round.
Kate and Bob each have a mean round score of 61 in eight rounds.

Kate: 35 50 74 41 56 79 73 80

Bob: 36 54 35 65 63 82 83 70

Whose round scores do you think are the more consistent?
Explain your answer.

## Calculating the standard deviation

- The most widely used measure of dispersion is the **standard deviation**.

  To calculate the standard deviation:
  - calculate the mean of the values
  - calculate the **squared deviation** for each value
    i.e. the square of the difference between the value and the mean
  - calculate the square root of the mean squared deviation.

**Example** Calculate the standard deviation of Kate's scores.

| $x$ | | Squared Deviation $(x - \bar{x})^2$ |
|---|---|---|
| 35 | $(35 - 61)^2$ | 676 |
| 41 | $(41 - 61)^2$ | 400 |
| 50 | $(50 - 61)^2$ | 121 |
| 56 | $(56 - 61)^2$ | 25 |
| 73 | $(73 - 61)^2$ | 144 |
| 74 | $(74 - 61)^2$ | 169 |
| 79 | $(79 - 61)^2$ | 324 |
| 80 | $(80 - 61)^2$ | 361 |
| 488 | | 2220 |
| $\frac{488}{8} = 61$ Mean score ($\bar{x}$) | | $\sqrt{\frac{2220}{8}} =$ **16.7** (to 3 sf) Standard deviation |

The standard deviation, $\sigma$, can be written as a formula:

$$\sigma = \sqrt{\frac{\Sigma f(x - \bar{x})^2}{n}}$$

where $\sigma$ is the Greek letter 'sigma'
$\Sigma(x - \bar{x})^2$ is the total of the squared deviations
and $n$ is the number of values

**Exercise 4.4**
**Standard deviation**

**1** Use the data given above to calculate the standard deviation of Bob's scores.

**2** **a** Calculate the mean of Paula's round scores.
**b** **i** Calculate their standard deviation.
**ii** Check your answer using the alternative formula.

Paula: 35 45 50 60 75

Standard deviation will not be tested in a written exam but may be useful for the Ma4 coursework task.

An alternative formula for the standard deviation is:

$$\sigma = \sqrt{\frac{\Sigma x^2}{n} - \bar{x}^2}$$

where $\Sigma x^2$ is the total of the squares of each value

**Thinking ahead to ...**
calculating the standard deviation of a frequency distribution

**A** These are Bob's scores from his two best rounds in the competition.

Round 6

| Ring score | 5 | 6 | 7 | 8 | 9 | 10 |
|---|---|---|---|---|---|---|
| Frequency | 1 | 0 | 1 | 3 | 4 | 1 |

Round 7

| Ring score | 6 | 7 | 8 | 9 | 10 |
|---|---|---|---|---|---|
| Frequency | 2 | 2 | 0 | 3 | 3 |

For each round:

**a** list Bob's scores
**b** calculate the standard deviation of the scores.

## Calculating the standard deviation of a frequency distribution

- To calculate the standard deviation of a frequency distribution:
  - calculate the mean of all the values
  - calculate the squared deviation for each different value
  - multiply each squared deviation by its frequency
  - calculate the square root of the mean squared deviation.

**Example** Calculate the standard deviation of Bob's Round 6 scores.

| Score $x$ | Frequency $f$ | Total score $fx$ | Squared deviations $(x - \bar{x})^2$ | $f(x - \bar{x})^2$ |
|---|---|---|---|---|
| 5 | 1 | 5 | 10.24 | 10.24 |
| 6 | 0 | 0 | 4.84 | 0 |
| 7 | 1 | 7 | 1.44 | 1.44 |
| 8 | 3 | 24 | 0.04 | 0.12 |
| 9 | 4 | 36 | 0.64 | 2.56 |
| 10 | 1 | 10 | 3.24 | 3.24 |
| | 10 | 82 | | 17.60 |

$\frac{82}{10} = 8.2$ Mean score ($\bar{x}$)

$\sqrt{\frac{17.60}{10}} = \mathbf{1.33}$ (to 3 sf) Standard deviation

The standard deviation, σ, can be written as a formula:

$$\sigma = \sqrt{\frac{\Sigma f(x - \bar{x})^2}{\Sigma f}}$$

where $\Sigma f(x - \bar{x})^2$ is the total of the squared deviations
and $\Sigma f$ is the total frequency.

**Exercise 4.5**
Standard deviation of a frequency distribution

**1** **a** Use this method to calculate the standard deviation of Bob's Round 7 scores from Question **A** above.
**b** In which round do Bob's scores show less variation? Explain.

**2**

Paula

| Ring score | 2 | 3 | 4 | 5 | 6 | 7 | 8 | 9 | Total |
|---|---|---|---|---|---|---|---|---|---|
| Frequency | 11 | 8 | 5 | 3 | 3 | 4 | 6 | 10 | 50 |

Zoe

| Ring score | 2 | 3 | 4 | 5 | 6 | 7 | 8 | 9 | 10 | Total |
|---|---|---|---|---|---|---|---|---|---|---|
| Frequency | 4 | 5 | 7 | 11 | 13 | 10 | 6 | 3 | 1 | 60 |

Matt

| Ring score | 2 | 3 | 4 | 5 | 6 | 7 | 8 | 9 | 10 | Total |
|---|---|---|---|---|---|---|---|---|---|---|
| Frequency | 6 | 10 | 12 | 9 | 6 | 9 | 11 | 10 | 7 | 80 |

**a** Draw a frequency diagram to show:
**i** Paula's scores **ii** Zoe's scores **iii** Matt's scores.
**b** Calculate the standard deviation of each archer's scores.
**c** Comment on the shape of your frequency diagrams and your answers for the standard deviations.

# Comparing sets of data

◆ One way to compare sets of data is to compare two types of statistic:

**a** an average
**b** a measure of dispersion.

**Example** Compare Zoe's and Matt's round scores using the mean and the standard deviation.

Zöe: 56 61 67 49 58 48

Matt: 65 58 72 71 54 68 41 53

Use the statistical function on your calculator to check these values for the mean and the standard deviation.

| | Zoe | Matt |
|---|---|---|
| Mean | 56.5 | 60.25 |
| Standard deviation (to 3 sf) | 6.60 | 10.0 |

**a** Matt's round scores are higher on average.
**b** Zoe is more consistent because her scores show less variation than Matt's scores.

## Exercise 4.6 Comparing sets of data

**1**

Zeta: 42 61 54 53 67 72 78

Carlo: 48 44 52 61 45 79 56 75

**a** Calculate the mean round score for each archer.
**b** Calculate the standard deviation of each set of scores.
**c** Use the mean and the standard deviation to compare the sets of data.

**2**

Dean

| Ring score | 4 | 5 | 6 | 7 | 8 | Total |
|---|---|---|---|---|---|---|
| Frequency | 5 | 5 | 7 | 6 | 2 | 25 |

Leah

| Ring score | 3 | 4 | 5 | 6 | 7 | 8 | Total |
|---|---|---|---|---|---|---|---|
| Frequency | 1 | 3 | 7 | 6 | 2 | 1 | 20 |

Jack

| Ring score | 3 | 4 | 5 | 6 | 7 | Total |
|---|---|---|---|---|---|---|
| Frequency | 7 | 9 | 4 | 3 | 7 | 30 |

Compare these distributions using the mean and the standard deviation.

**3**

Eliza

| Ring score | 3 | 4 | 5 | 6 | 7 | 8 | 9 | 10 | Total |
|---|---|---|---|---|---|---|---|---|---|
| Frequency | 3 | 7 | 4 | 1 | 1 | 0 | 0 | 2 | 18 |

Sam

| Ring score | 2 | 3 | 4 | 5 | 6 | 7 | 8 | Total |
|---|---|---|---|---|---|---|---|---|
| Frequency | 1 | 4 | 3 | 4 | 2 | 1 | 1 | 16 |

When you compare data which has extreme values, using the median and the interquartile range can give a better comparison.

**a** **i** For each archer list the scores.
**ii** Find the median score
**iii** Calculate the interquartile range of the scores.

**b** Use the median and interquartile range to compare the two distributions.

**Thinking ahead to ...**
using cumulative frequencies to find the median

**A**

| Ring score | 5 | 6 | 7 | 8 | 9 | 10 | Total |
|---|---|---|---|---|---|---|---|
| Frequency | 3 | 2 | 4 | 0 | 3 | 3 | 15 |

Salim

| Ring score | 5 | 6 | 7 | 8 | 9 | Total |
|---|---|---|---|---|---|---|
| Frequency | 7 | 8 | 13 | 18 | 14 | 60 |

Emily

**a** **i** List all Salim's scores.
**ii** Find Salim's median score.

**b** **i** List all Emily's scores.
**ii** Find Emily's median score.

## Using cumulative frequencies to find the median

- The median can be found without listing all the data.
- To find the median using cumulative frequencies:
  - calculate the cumulative frequencies
  - use the total frequency to decide where the median is
  - find the middle value.

The cumulative frequencies are the running totals of the frequencies. For example: 7 + 8 = 15.

**Example** Find the median of Emily's scores.

| Ring score | Frequency | Cumulative Frequency |
|---|---|---|
| 5 | 7 | 7 |
| 6 | 8 | 15 |
| 7 | 13 | 28 |
| 8 | 18 | 46 |
| 9 | 14 | 60 |

The total frequency is 60, so the median score is halfway between the 30th largest and the 31st largest.

The 28th largest score is 7.

All the scores between the 29th and 46th largest are 8.

The final cumulative frequency here is 60: check it is equal to the total frequency.

The 30th and 31st largest scores are both 8, so
Median score = **8**

**Exercise 4.7**
Finding the median using cumulative frequencies

**1**

Lee

| Ring score | 3 | 4 | 5 | 6 | 7 | 8 | Total |
|---|---|---|---|---|---|---|---|
| Frequency | 7 | 10 | 5 | 1 | 6 | 11 | 40 |

Faith

| Ring score | 4 | 5 | 6 | 7 | 8 | Total |
|---|---|---|---|---|---|---|
| Frequency | 11 | 7 | 8 | 9 | 10 | 45 |

Aqib

| Ring score | 1 | 2 | 3 | 4 | 5 | 6 | 7 | 8 | 9 | Total |
|---|---|---|---|---|---|---|---|---|---|---|
| Frequency | 2 | 4 | 2 | 3 | 1 | 4 | 7 | 9 | 7 | 39 |

Tegan

| Ring score | 1 | 2 | 3 | 4 | 5 | 6 | 7 | 8 | 9 | 10 | Total |
|---|---|---|---|---|---|---|---|---|---|---|---|
| Frequency | 6 | 9 | 5 | 4 | 2 | 5 | 6 | 4 | 3 | 4 | 48 |

Find each archer's median score.

**2**

Jake

| Ring score | 4 | 5 | 6 | 7 | 8 | 9 | 10 |
|---|---|---|---|---|---|---|---|
| Cumulative Frequency | 7 | 18 | 31 | 40 | 48 | 62 | 80 |

**a** Explain why Jake's median score is 7.5.
**b** Explain why the interquartile range of this distribution is 3.

**Thinking ahead to ...**
box-and-whisker plots

**A** A survey of three fast food restaurants recorded the number of chips in a serving. A sample of 80 servings was taken from each restaurant.

| | Number of chips | | | | | | | | | | | | | |
|---|---|---|---|---|---|---|---|---|---|---|---|---|---|---|
| Restaurant | 32 | 33 | 34 | 35 | 36 | 37 | 38 | 39 | 40 | 41 | 42 | 43 | 44 | Total |
| X | 1 | 2 | 3 | 5 | 9 | 13 | 14 | 12 | 8 | 5 | 4 | 3 | 1 | 80 |
| Y | 4 | 6 | 11 | 14 | 13 | 11 | 8 | 6 | 4 | 3 | 0 | 0 | 0 | 80 |
| Z | 1 | 1 | 2 | 6 | 9 | 13 | 15 | 13 | 10 | 6 | 2 | 1 | 1 | 80 |

**a** **i** Which restaurant do you think gives more chips in a serving?
**ii** Explain your answer.

**b** **i** Find the median number of chips for each restaurant.
**ii** Calculate the interquartile range for each distribution.

## Drawing box-and-whisker plots

- A **box-and-whisker plot** shows a frequency distribution by using:
  - the lowest and highest values
  - the median
  - the lower and upper quartiles.

**Example**

Draw a box-and-whisker plot for these chips from restaurant W.

| Number of chips | 34 | 35 | 36 | 37 | 38 | 39 | 40 | 41 | 42 | 43 |
|---|---|---|---|---|---|---|---|---|---|---|
| Frequency | 1 | 2 | 4 | 7 | 12 | 14 | 16 | 15 | 7 | 2 |
| Cumulative frequency | 1 | 3 | 7 | 14 | 26 | 40 | 56 | 71 | 78 | 80 |

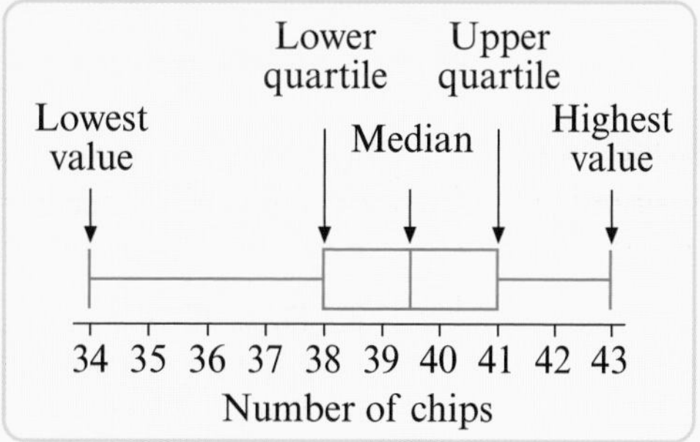


**Exercise 4.8**
Drawing box-and-whisker plots

**1** Copy this box-and-whisker diagram on squared paper. Use the data from Question **A** to draw a box-and-whisker plot for each restaurant.

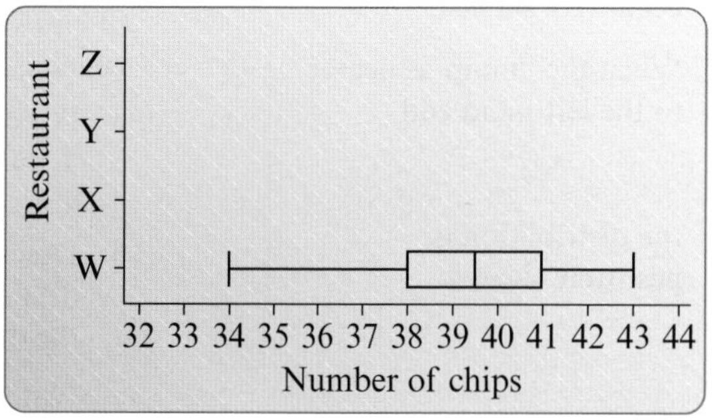


**2** Write a short report that compares the number of chips per serving from the four fast food restaurants.

## Comparing frequency distributions

A **frequency polygon** is a line graph which shows the frequencies that make up a distribution.

You can compare two or more distributions on the same diagram by plotting the frequency polygons.

- You can compare frequency distributions by considering their shape:
  - this distribution can be described as **skewed** (the larger frequencies are nearer to one end than the other)

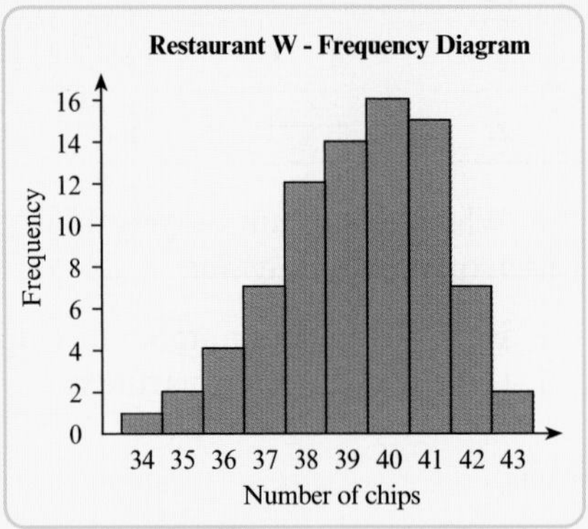


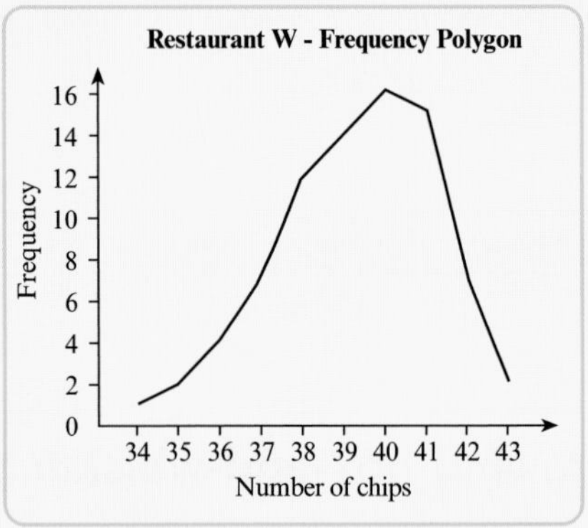


  - this distribution can be described as **bell-shaped** (the frequencies are larger in the middle, and tail off towards each end).

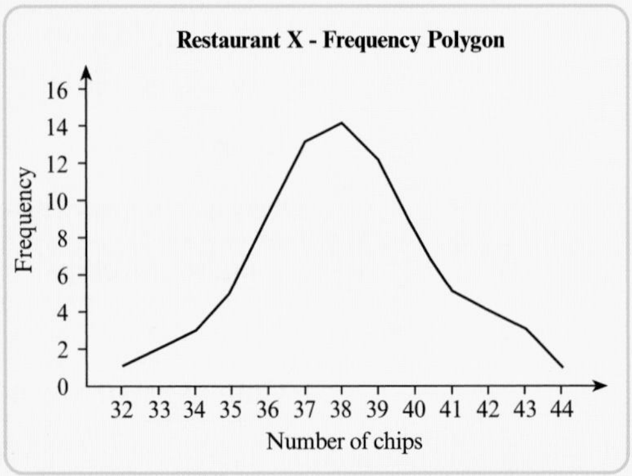


### Exercise 4.9
### Comparing distributions

A skewed distribution that has the 'hump' nearer to

the right-hand end is **negatively skewed**.

When the 'hump' is nearer to the left-hand end,

the distribution is **positively skewed**.

**1** **a** Draw the frequency polygon for restaurant X on squared paper.
**b** Draw a frequency polygon for restaurant Z on the same diagram.
**c** Compare the two distributions.

**2** **a** Draw the skewed frequency polygon for restaurant W, starting the scale on the horizontal axis at 32 chips.
**b** Draw a frequency polygon for restaurant Y on the same diagram.
**c** Compare the two distributions.

**3** For this distribution, the mean number of chips is less than the median number, which is less than the modal number.

**a** For each of the other 3 restaurants:
**i** calculate the mean number of chips
**ii** draw a similar diagram.

**b** Describe the link between the shape of a distribution and the mean, the median, and the mode.

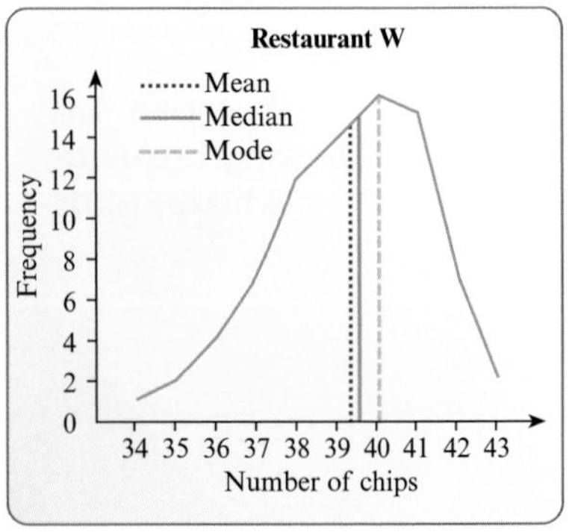

## Misleading diagrams

- Statistical diagrams can be misleading in several ways, including:
  - when the vertical axis does not start at 0

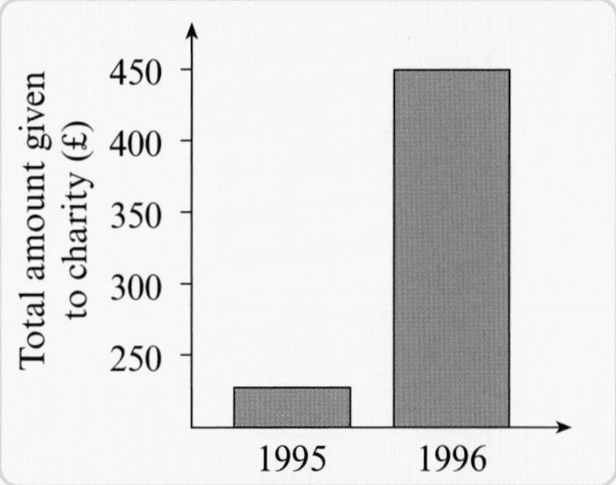


The vertical axis starts at £200.

This gives the impression that the amount given in 1996 was 10 times the amount given in 1995, not 2 times.

  - when enlargements of a shape are used.

The 1996 note is 2 times the height *and* 2 times the width of the 1995 note.

This gives the impression that the amount given in 1996 was 4 times the amount given in 1995.

**Exercise 4.10**
Misleading diagrams

1 Draw a misleading diagram to show this revenue data by using:

**a** a vertical scale which does not start at 0
**b** enlargements of a shape.

*WD Wilton Dale Films*

| | 1994 | 1996 |
|---|---|---|
| Revenue (£m) | 6.1 | 18.3 |

2 Draw a misleading diagram to show this data for number of visitors.

*WD Wilton Dale Theme Parks*

| | 1994 | 1995 | 1996 |
|---|---|---|---|
| Number of visitors (000's) | 24 | 36 | 72 |

3

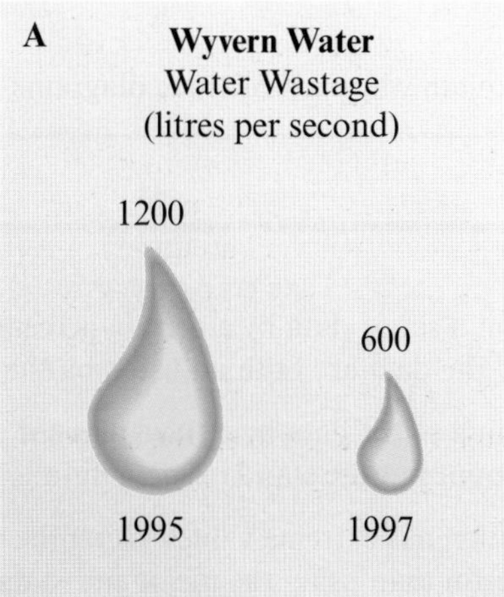


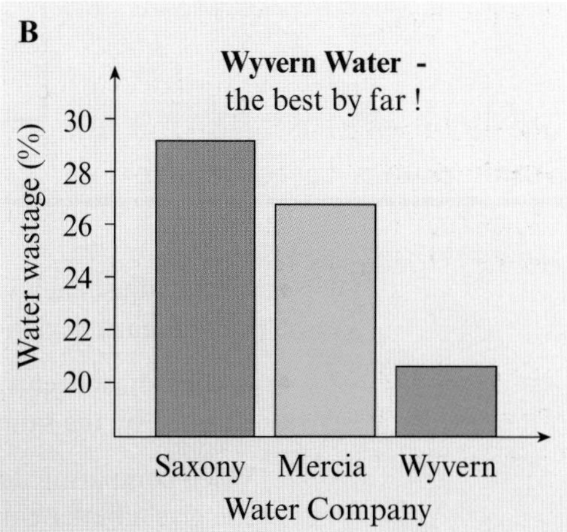


Explain why each of these diagrams is misleading.

## End points

You should be able to ...    ... so try these questions

**A** Calculate the mean and the standard deviation

**A1** For Mel's round scores, calculate:
**a** the mean
**b** the standard deviation.

| | | | | |
|---|---|---|---|---|
| 43 | 64 | 67 | 58 | 62 |
| 69 | 83 | 78 | | |

Mel

**B** Understand the effect of adjusting a set of data on the mean and standard deviation

**B1** Jason scores 5 less than Mel in each round.
Give the mean and standard deviation of Jason's round scores.

**C** Calculate the standard deviation of a frequency distribution

**C1**

| Ring score | 5 | 6 | 7 | 8 | 9 | 10 | Total |
|---|---|---|---|---|---|---|---|
| Frequency | 6 | 9 | 13 | 14 | 11 | 7 | 60 |

Robin

Calculate the mean and standard deviation of Robin's scores.

**D** Compare sets of data

**D1** **a** Copy and complete this table using your answers to Question **C1**.
**b** Use the statistics in your table to compare Robin and Tina.

| | Robin | Tina |
|---|---|---|
| Mean | | 7.42 |
| Standard deviation | | 1.30 |

**E** Find the median and the interquartile range of a frequency distribution

**E1** For Robin's scores, find:
**a** the median
**b** the interquartile range.

**F** Draw a box-and-whisker plot

**F1** Draw a box-and-whisker plot for Robin's scores.

**G** Recognise when diagrams used to present data are misleading

**G1**

**X**

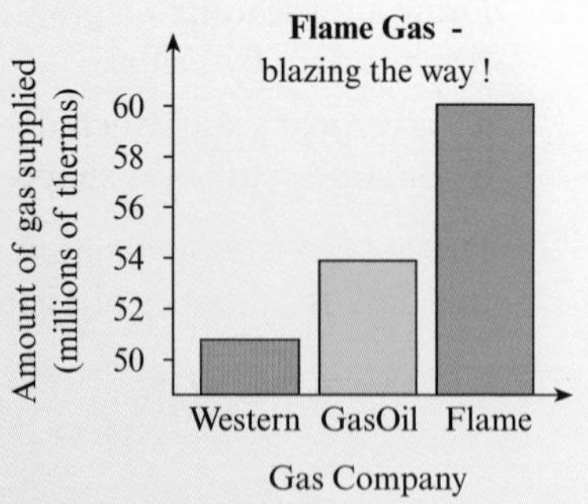


**Y**

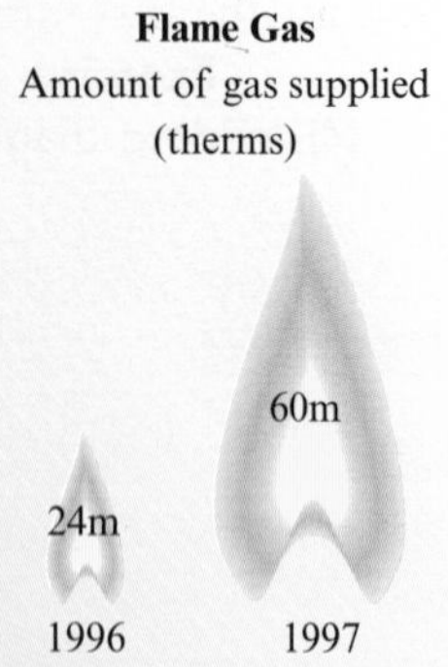


Explain why each of these diagrams is misleading.

### Some points to remember

- If you adjust each value in a set of data by a **certain percentage**, the mean and the standard deviation of the data are both adjusted by the same percentage.
- If you adjust each value in a set of data by a **fixed amount**, the mean of the data is adjusted by the same amount but the standard deviation is left unchanged.
- When comparing data which has extreme values, using the median and interquartile range can give a better comparison than using the mean and standard deviation.

## Starting points

You need to know about ... | ... so try these questions

### A Continuing a sequence

- A **sequence** of numbers usually follows a pattern or rule.
- Each number in a sequence is called a **term**.

  For example, in the sequence: 3, 5, 7, 9, 11, ... , the 3rd term is 7.
- A sequence can often be continued by finding a pattern in the **differences**.

Sequence P 3, 9, 15, 21, 27, ...
+6 +6 +6 +6

❖ In Sequence P, the **first difference** is 6 each time.

Sequence Q 5, 6, 9, 14, 21, ...
+1 +3 +5 +7
+2 +2 +2

❖ In Sequence Q, the **second difference** is 2 each time.
So continue the sequence by adding 9, then 11, and so on.

**A1** Find the 6th and 7th term in:
**a** sequence P
**b** sequence Q.

**A2** What is the 10th term in this sequence?

4, 7, 10, 13, 16, ...

**A3** For each sequence, find the next three terms.
**a** 2, 8, 14, 20, 26, ...
**b** 2, 5, 11, 20, 32, ...
**c** 2, 3, 9, 20, 36, ...

**A4** **a** What are the second differences in this sequence?

7, 16, 31, 52, ...

**b** What is the 5th term?

### B A rule for a sequence

These are the first three matchstick patterns in a sequence.

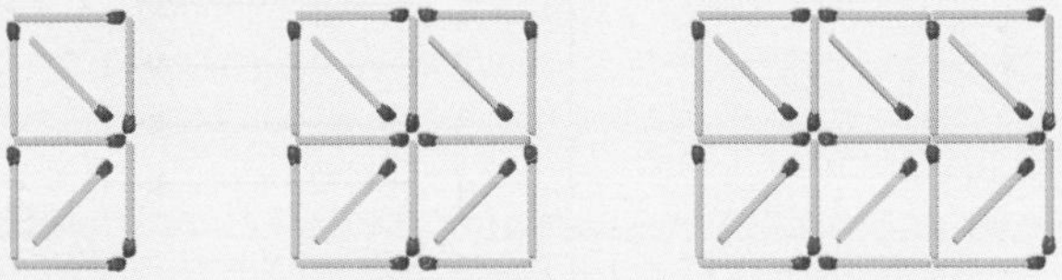

Pattern 1 Pattern 2 Pattern 3

- The number of matches in each pattern can be shown in a table.

| Pattern number ($n$) | Number of matches ($m$) |
|---|---|
| 1 | 9 |
| +1 → 2 | 16 (+7) |
| +1 → 3 | 23 (+7) |
| +1 → 4 | 30 (+7) |

❖ The pattern number goes up by 1 each time.

❖ The number of matches goes up by 7 each time.

❖ So a rule that links the number of matches ($m$) with the pattern number ($n$) begins $m = 7n$ ...

❖ A rule that fits all the results in the table is $\boldsymbol{m = 7n + 2}$

- This rule can be used to calculate the number of matches in **any** pattern,
  for example: in Pattern 10 there are $(7 \times 10) + 2 = 72$ matches.

**B1** These are the first three patterns in a sequence.

Pattern 1

Pattern 2

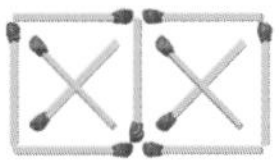

Pattern 3

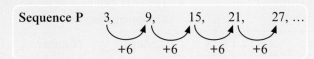

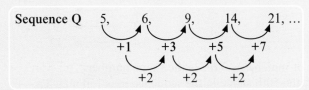

**a** Draw pattern 5.

**b** How many matches are in pattern 9?

**c** Make a table for the first five patterns in this sequence.

**d** Which of these rules fits the results in your table?

$m = 4n + 2$ $m = 5n + 1$

$m = 6n - 1$

# Sequences and mappings

These equilateral triangles are the first three in a sequence of shapes.

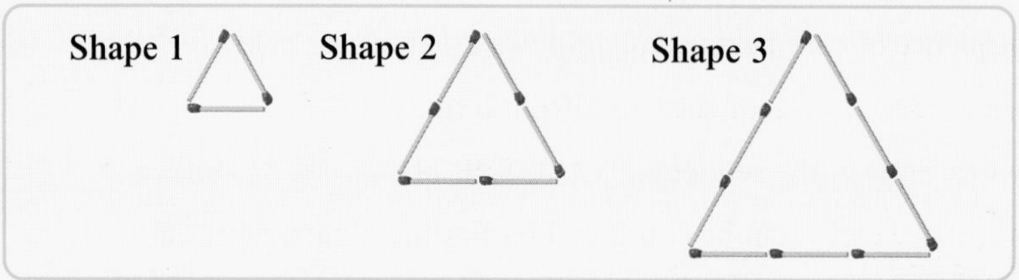


- Data for this sequence can be shown in a table.

| Shape number | 1 | 2 | 3 | 4 |
|---|---|---|---|---|
| Number of matches | 3 | 6 | 9 | 12 |

- The data can also be shown in a **mapping diagram** like this.

| Shape number | | Number of matches |
|---|---|---|
| 1 | ⟶ | 3 |
| 2 | ⟶ | 6 |
| 3 | ⟶ | 9 |
| 4 | ⟶ | 12 ... |

- The number of matches is 3 times the shape number.
  For example, the 50th shape in the sequence uses 150 matches.

  Using $n$ to stand for the shape number,
  the rule for the mapping diagram can be written: $n \longrightarrow 3n$

**Exercise 5.1**
Sequences and mappings

**1** These are the first three patterns in a sequence.

Copy and complete the mapping diagram for the sequence.

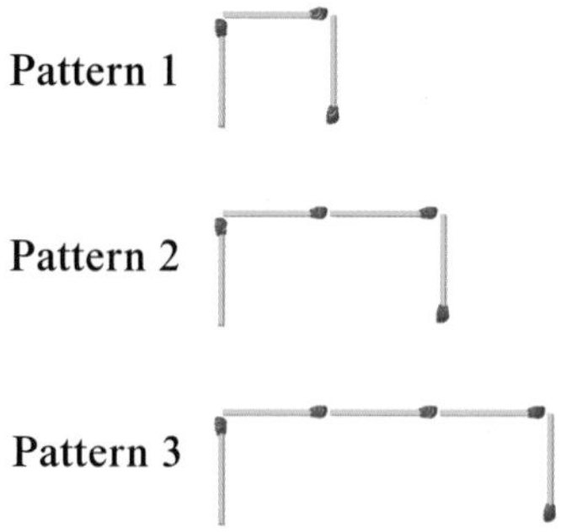


| Pattern number | | Number of matches |
|---|---|---|
| 1 | ⟶ | 3 |
| 2 | ⟶ | ☐ |
| 3 | ⟶ | ☐ |
| 4 | ⟶ | ☐ |
| ⋮ | | ⋮ |
| 20 | ⟶ | ☐ |
| ⋮ | | ⋮ |
| $n$ | ⟶ | ☐ |

**2** These patterns of touching squares are the first four in a sequence.

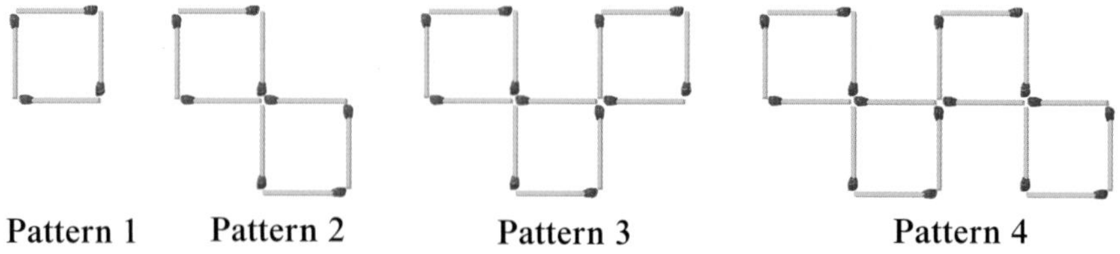


**a** Draw a mapping diagram for the first six patterns of touching squares.
**b** Find a rule for the sequence in the form $n \longrightarrow$ .... , where $n$ is the pattern number.
**c** Use your rule to calculate the number of matches in the 100th pattern.
**d** Which pattern would use 620 matches?
**e** Each touching square now has one extra match added as a diagonal. Which pattern will now use 620 matches?

**Thinking ahead to ... finding rules**

**A** These are the first three patterns in sequence A.

Sequence A

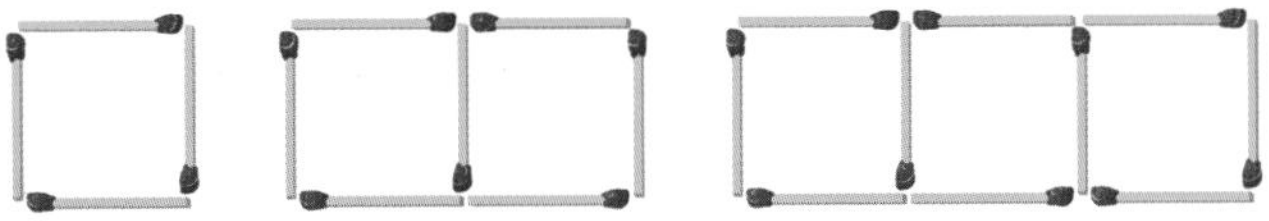

Pattern 1 Pattern 2 Pattern 3

How many matchsticks are in the 100th pattern?
Explain how you worked it out.

## Finding rules

**Example** **Find a rule for the number of matches ($m$) in the $n$th pattern in sequence A above.**

◆ **Method 1** Look at how the patterns are made.

Pattern 1 Pattern 2 Pattern 3

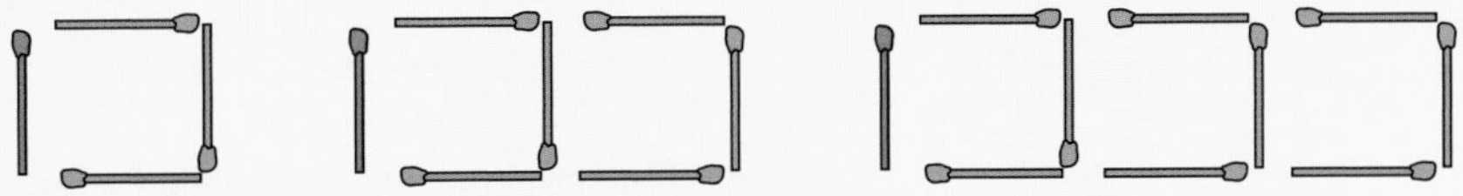

$1 + (3 \times 1)$ $1 + (3 \times 2)$ $1 + (3 \times 3)$

❖ So a rule for the number of matches ($m$) in the $n$th pattern is $\boldsymbol{m = 1 + 3n}$

◆ **Method 2** Look at differences.

| Pattern number ($n$) | $3n$ | Number of matches ($m$) | |
|---|---|---|---|
| 1 | 3 | 4 | |
| | | | +3 |
| 2 | 6 | 7 | |
| | | | +3 |
| 3 | 9 | 10 | |
| | | | +3 |
| 4 | 12 | 13 | |
| | | | +3 |
| 5 | 15 | 16 | |

❖ The pattern number goes up by 1 each time.

❖ The number of matches goes up by 3 each time so there is a linear rule that begins $m = 3n$ ... .

❖ Compare $3n$ with the number of matches.

❖ The number of matches is 1 more than $3n$ each time.

❖ So a rule for the number of matches ($m$) in the $n$th pattern is $\boldsymbol{m = 3n + 1}$.

Examples of linear rules are:

$m = 4n + 3$
$y = 2 - 5x$
$a = 3b - 1$

Rules such as $m = n^2 + 1$ and $y = \frac{5}{x} - 6$ are non-linear.

**Exercise 5.2**
**Finding rules**

**1** These triangle patterns are the first three in sequence B.

Sequence B

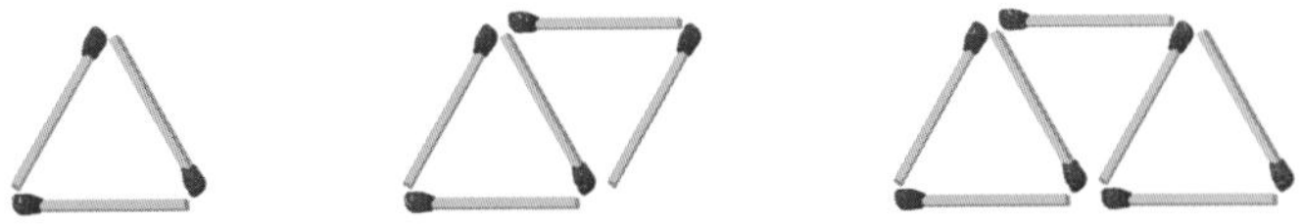

Pattern 1 Pattern 2 Pattern 3

**a** Find a rule for the number of matches ($m$) in the $n$th triangle pattern. Explain your method.
**b** Use your rule to find the number of matches in the 8th pattern.
**c** Check your answer by drawing the 8th pattern and counting the matches.

**2** **Sequence C**

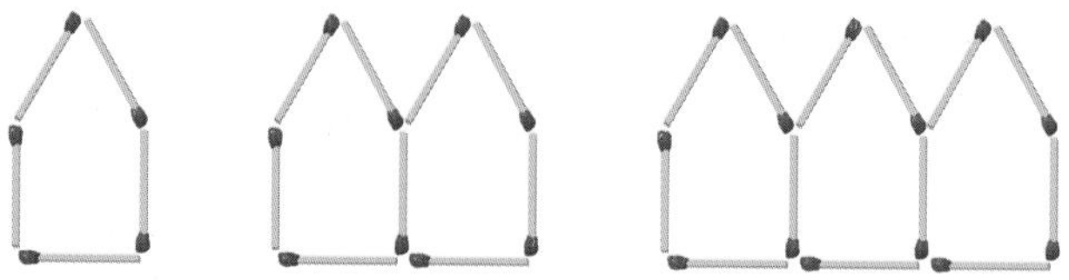

**Pattern 1** **Pattern 2** **Pattern 3**

**a** For sequence C, find a rule for the number of matches ($m$) in the $n$th pattern.
Explain your method.

**b** Calculate the number of matches in the 40th pattern.

**c** Which pattern uses exactly 129 matches?

**3** **Sequence D**

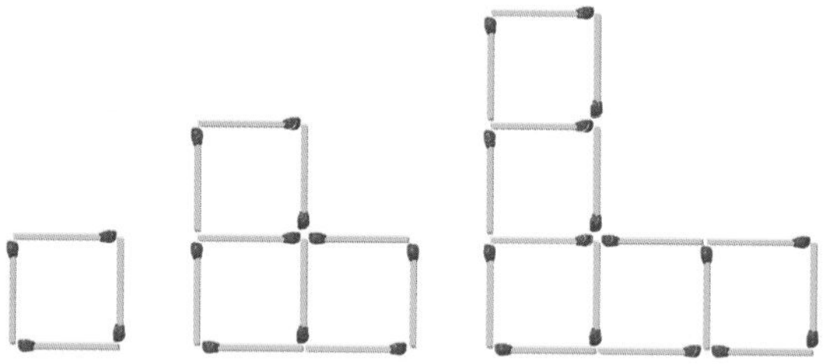

**Pattern 1** **Pattern 2** **Pattern 3**

**a** For sequence D, find a rule for the number of matches ($m$) in the $n$th pattern.
Explain your method.

**b** How many matches are in the 100th pattern?

**4** This mapping diagram fits a sequence of matchstick patterns.

| Pattern number ($n$) | | Number of matches ($m$) |
|---|---|---|
| 1 | → | 6 |
| 2 | → | 10 |
| 3 | → | 14 |
| 4 | → | 18 |
| 5 | → | 22 |

**a** Draw a sequence of matchstick patterns that fits this mapping diagram.

**b** Find a rule for the number of matches ($m$) in the $n$th pattern.

**5** Copy and complete each mapping diagram.

**a**

| | | |
|---|---|---|
| 1 | → | 4 |
| 2 | → | 9 |
| 3 | → | 14 |
| 4 | → | 19 |
| ⋮ | | ⋮ |
| 50 | → | □ |
| ⋮ | | ⋮ |
| $n$ | → | □ |

**b**

| | | |
|---|---|---|
| 1 | → | 26 |
| 2 | → | □ |
| 3 | → | □ |
| 4 | → | □ |
| ⋮ | | ⋮ |
| □ | → | $^-14$ |
| ⋮ | | ⋮ |
| $s$ | → | $30 - 4s$ |

**c**

| | | |
|---|---|---|
| 1 | → | 6 |
| 2 | → | 3 |
| 3 | → | 0 |
| 4 | → | $^-3$ |
| ⋮ | | ⋮ |
| □ | → | $^-75$ |
| ⋮ | | ⋮ |
| $p$ | → | □ |

Any letter can stand for the shape number.

For example, the rule
$n = \longrightarrow 3n + 2$

can also be written as:
$p = \longrightarrow 3p + 2$

or
$s = \longrightarrow 3s + 2$

and so on.

**Thinking ahead to ...**
**finding the $n$th term**

A

7, 9, 11, 13, 15, 17, ...

**a** Write the next two terms in this sequence.
**b** Find the 12th term.
**c** What is the 50th term?

## The *n*th term of a sequence

**Example** Find an expression for the $n$th term in the sequence
7, 9, 11, 13, 15, ...

- The sequence can be displayed in a table like this:

| Term number ($n$) | 1 | 2 | 3 | 4 | 5 |
|---|---|---|---|---|---|
| Term | 7 | 9 | 11 | 13 | 15 |

The difference between terms in a sequence is often called the **list difference**.

- The difference for the terms is 2 each time.
  So there is a linear expression for the $n$th term that begins $\mathbf{2n}$ ... .
- Comparing the sequence with the sequence $n \longrightarrow 2n$

| Term number ($n$) | 1 | 2 | 3 | 4 | 5 |
|---|---|---|---|---|---|
| $2n$ | 2 | 4 | 6 | 8 | 10 |

shows that each term is 5 more than $2n$.
So an expression for the $n$th term of the sequence
7, 9, 11, 13, 15, ... is $\mathbf{2n + 5}$.

**Exercise 5.3**
**Finding the $n$th term**

**1** A 6, 9, 12, 15, 18, ... B 1, 6, 11, 16, 21, ...
C 13, 23, 33, 43, 53, ... D 2, 10, 18, 26, 34, ...

For each of the sequences A to D:

**a** find an expression for the $n$th term
**b** use your expression to calculate the 50th term.

**2** A student has tried to find the $n$th term of this sequence.

5, 8, 11, 14, 17, ...
$n$th term is $n + 3$ ✗

**a** Explain the mistake you think has been made.
**b** Find a correct expression for the $n$th term of this sequence.

**3** The 2nd term of a sequence is 7.
Which of these could not be an expression for the $n$th term?

$3n + 1$ $11 - 2n$ $n + 5$ $n + 7$ $5n - 3$

**4** List the first 5 terms of the sequence whose $n$th term is $4 - 3n$.

**5** Find an expression for the $n$th term of each of these sequences.

**a** 20, 18, 16, 14, 12, ...
**b** 22, 12, 2, $^{-}8$, $^{-}18$, ...
**c** $^{-}1$, $^{-}3$, $^{-}5$, $^{-}7$, $^{-}9$, ...
**d** 10, 5, 0, $^{-}5$, $^{-}10$, ...

# Extending number patterns

**Exercise 5.4**
**Extending patterns**

1 Morag finds an expression for the $n$th number in this sequence.

4, 10, 18, 28, 40, ...

This is her working:

| | | |
|---|---|---|
| 1st number | 4 | $= 1 \times 4$ |
| 2nd number | 10 | $= 2 \times 5$ |
| 3rd number | 18 | $= 3 \times 6$ |
| 4th number | 28 | $= 4 \times 7$ ... |
| So $n$th number | | $= n \times (n + 3)$ |

**a** Show Morag's line of working for the 5th number.
**b** Find the 10th number in this sequence.
**c** Explain how Morag's working helps to find an expression for the $n$th number in this sequence.

2 These are the first five triangle numbers:

1, 3, 6, 10, 15, ...

The triangle numbers follow this pattern:

1st triangle number $1 = \dfrac{1 \times 2}{2}$

2nd triangle number $3 = \dfrac{2 \times 3}{2}$

3rd triangle number $6 = \dfrac{3 \times 4}{2}$

4th triangle number $10 = \dfrac{4 \times 5}{2}$

**a** What is the next line in this pattern?
**b** Use the pattern to find the 12th triangle number.
**c** Find an expression for the $n$th triangle number.

3 These are the first five powers of 2:

2, 4, 8, 16, 32, ...

Powers of 2 follow this pattern:

| | | | |
|---|---|---|---|
| 1st power | 2 | | $= 2^1$ |
| 2nd power | 4 | $= 2 \times 2$ | $= 2^2$ |
| 3rd power | 8 | $= 2 \times 2 \times 2$ | $= 2^3$ |
| 4th power | 16 | $= 2 \times 2 \times 2 \times 2$ | $= 2^4$ |

**a** What is the next line in this pattern?
**b** Use the pattern to find the 8th power of 2.
**c** Write an expression for the $n$th power of 2.

## Writing rules in different ways

**Exercise 5.5**
**Different ways to write rules**

**1** These hollow square tile designs are the first three in a sequence.

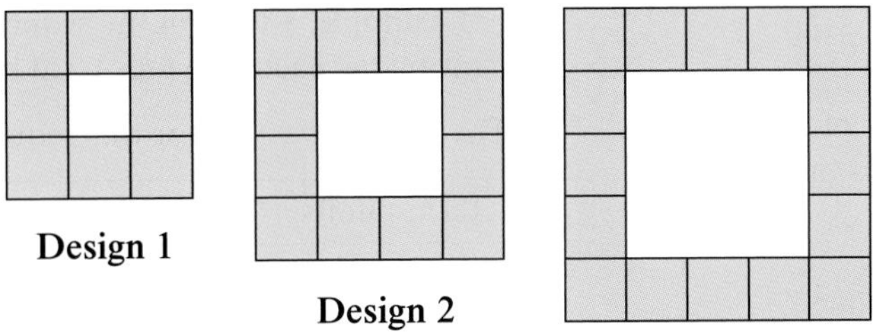

Design 1 Design 2 Design 3

Three students find a rule for the number of tiles in the $n$th design.
All three students are correct.

| Andrew | Aisha | Fiona |
|---|---|---|
| Number of tiles $= 4(n + 1)$ | Number of tiles $= 4n + 4$ | Number of tiles $= 2(n + 2) + 2n$ |

The way each student wrote their rule shows how they found it.

Andrew drew this diagram to show how he found his rule.

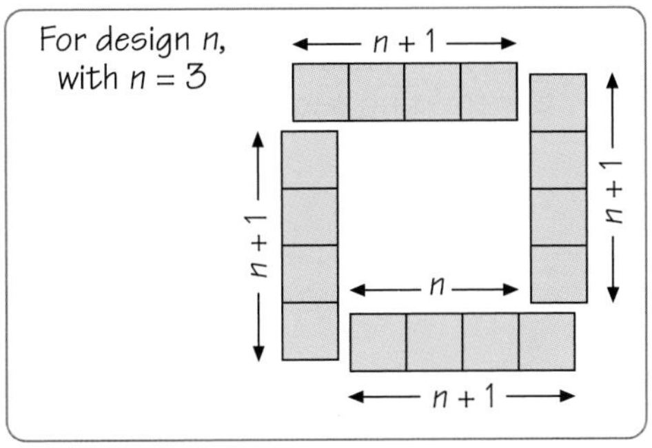


Fiona's rule is not as easy to use as the other two rules.

Using Fiona's rule to find the number of tiles in the 10th design gives:

Number of tiles
$= 2 \times (10 + 2) + (2 \times 10)$
$= 24 + 20$
$= 44$

**a** Draw a diagram for Aisha's rule.
**b** Draw a diagram for Fiona's rule.
**c** Show how you can calculate the number of tiles in the 50th design using:
**i** Andrew's rule **ii** Aisha's rule **iii** Fiona's rule
**d** Which design has 324 tiles?
Explain how you found your result.

**2** These designs are the first three in a sequence.

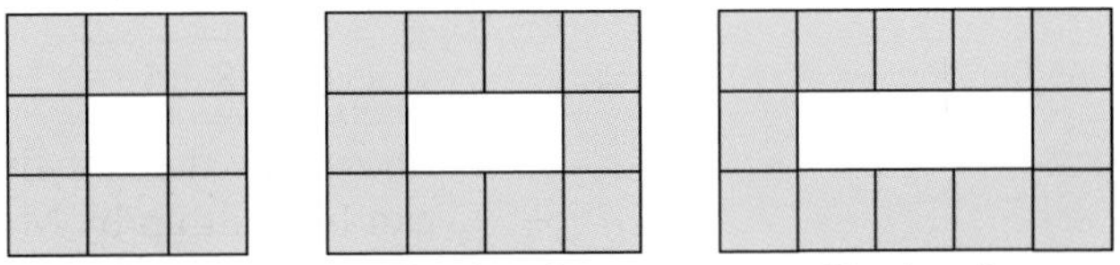

Design 1 Design 2 Design 3

**a** Find two different ways of writing a rule for the number of tiles in the $n$th design.
**b** Explain how you found your results.

## Generalising other sequences

The general terms of a sequence is the $n$th value. It is often written $u_n$.
$u_1$ is the first value
$u_2$ is the second value...

Any number to the power of 0 is 1, so $3^0 = 1$

- Some sequences are generated by multiplying by one number each time. For example, a term in the sequence 3, 9, 27, 81, ... is obtained by multiplying each previous term by 3.

  The general term $u_n$ for this sequence is $3^n$.

| Term number ($n$) | 1 | 2 | 3 | 4 | 5 ... | $n$ |
|---|---|---|---|---|---|---|
| Term | $3^1$ | $3^2$ | $3^3$ | $3^4$ | $3^5$ | $3^n$ |

  Sequences like this where subsequent terms are multiplied by the same value are known as **geometric sequences.**

- Other sequences can be based on the one above. For instance, the sequence 1, 3, 9, 27, ... has the general term $u_n$ as $3^{n-1}$. In this case the multiplier between terms is still 3.

- Where a sequence is given only in fractions treat it as two separate sequences, one for the numerator and another for the denominator.

  For example, the sequence $\frac{2}{5}, \frac{3}{11}, \frac{4}{29}, \frac{5}{83}, \frac{6}{245}, \ldots$

  has the general term $\frac{\boldsymbol{n+1}}{\boldsymbol{3^n+2}}$.

**Exercise 5.6**
**Generalising sequencies**

1 Find the general term for each of these geometric sequences.

a 2, 4, 8, 16, 32, ...
b 1, 3, 7, 15, 31, ...
c 1, 2, 4, 8, 16, ...
d $\frac{1}{4}, \frac{1}{2}, 1, 2, 4, \ldots$
e $1, \frac{1}{3}, \frac{1}{9}, \frac{1}{27}, \frac{1}{81}, \ldots$
f $\frac{1}{4}, \frac{1}{16}, \frac{1}{64}, \frac{1}{256}, \ldots$

2 Find the general term for these fractional sequences:

a $\frac{1}{99}, \frac{2}{98}, \frac{3}{97}, \frac{4}{96}$
b $\frac{1}{4}, \frac{4}{8}, \frac{9}{16}, \frac{16}{32}, \ldots$
c $\frac{3}{1}, \frac{6}{8}, \frac{9}{27}, \frac{12}{64}$
d $\frac{2}{3}, \frac{5}{5}, \frac{10}{9}, \frac{17}{17}, \ldots$

3 The general term for the sequence 4, 16, 64, 256, ... is $4^n$.
Use this to find the general term for the sequences:

a 196, 184, 136, $^-56$, ...
b 256, 64, 16, 4, 1, ...

4 Morag is finding the general term for the sequence 4, 10, 18, 28, 40, ... . This is her working. (see Exercise 5.4 question 1)

1st number 4 = 1 x 4
2nd number 10 = 2 x 5
3rd number 18 = 3 x 6
4th number 28 = 4 x 7

So $u_n = n \times (n + 3)$
$= n^2 + 3n$

The general term for the sequence can be found in a similar way.

$^-1$, 0, 3, 8, 15, ...

a Show how the first five terms can be built up by Morag's method.

b Write the general term for the sequence in the form $u_n = \ldots$ .

# End points

## You should be able to ... ... so try these questions

**A** Find a rule that fits a sequence of patterns

**A1** These are the first three matchstick patterns in a sequence.

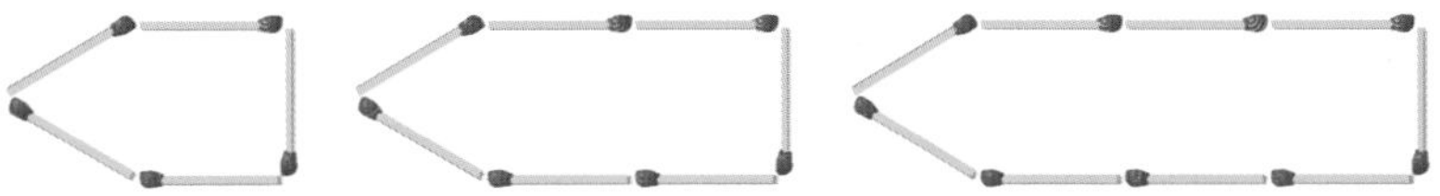

Pattern 1 Pattern 2 Pattern 3

**a** Find a rule for the number of matches ($m$) in the $n$th pattern. Explain your method.

**b** Use your rule to find the number of matches in the pattern 246.

**B** Find a general rule for a sequence of numbers or fractions

**B1** A 5, 9, 13, 17, ... B 6, 7, 8, 9, 10, ...
C 9, 12, 15, 18, 21, ... D 5, 20, 45, 80, 125, ...
E $^{-}3$, $^{-}8$, $^{-}13$, $^{-}18$, ... F $^{-}49$, $^{-}41$, $^{-}33$, $^{-}25$, $^{-}17$, ...
G $^{-}1$, 11, 23, 35, 47, ... H 4, 13, 22, 31, 40, ...

For each of the sequences A to F:

**a** find a general term, $u_n$

**b** calculate $u_{20}$.

**B2** Find the general term for each of these geometric sequences.

**a** 27, 81, 243, 729, ... **b** 8, 16, 32, 64, ...
**c** 4, 6, 10, 18, ... **d** 1, 13, 61, 253, ...

**B3** Find the general term for each of these fractional sequences.

**a** $\frac{2}{2}$ $\frac{4}{5}$ $\frac{8}{10}$ $\frac{16}{17}$ $\frac{32}{26}$ **b** $\frac{3}{3}$ $\frac{5}{12}$ $\frac{7}{27}$ $\frac{9}{48}$ $\frac{11}{75}$

## Some points to remember

- It is often possible to find a rule for the $n$th pattern in a sequence of patterns by looking at how each pattern can be made.
- In a sequence:
  - if the first differences are $k$ each time, there is a simple linear expression for the $n$th term that begins $kn$ ... .
  - if the terms are fractions treat the numerators and denominators as separate sequences.

# Decorum Design

DDC is the favourite shop for trade and private buyers who want a new look for bathrooms, kitchens and bedrooms. Here are some of our items but come to the shop to see our full range.

## WALLPAPER - rolls

width 53 cm, length 10 metres.

FLORAL DESIGN **£5.49** per roll.

**£4.97** each for 12 rolls or more

ANTIQUE EMBOSSED **£7.99** per roll

WALLPAPER PASTE **£4.99** - covers 10 sq metres

TRY OUR ANTIQUE EMBOSSED PAPERS TO COVER THAT TATTY WALL

ALL PAINT PRICES REDUCED BY 20% FOR NEXT THREE WEEKS

## EMULSION PAINTS - Top quality own brand

| | | |
|---|---|---|
| BRILLIANT WHITE | 1 litre | **£3.42** |
| | 2 ($\frac{1}{2}$) litre | **£8.45** |
| | 5 litre | **£16.99** |
| PASTEL SHADES | 2 ($\frac{1}{2}$) litre | **£11.99** |
| | 5 litre | **£19.49** |

A litre tin will cover about 8 $m^2$ with a single coat.

Two coats needed over very dark surfaces.

When calculating how much paint to order, do not subtract the area of doors and windows.

## WALL TILES - imported Italian and French

JARDIN RANGE - Box of 10 tiles **£1.79**

ASSISI RANGE - Box of 10 tiles **£1.99**

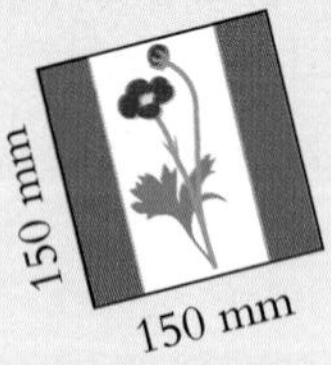


JARDIN

ASSISI

For those who have not caught up with metric units yet.
1 foot = 0.3048 metres 1 inch = 25.4 millimetres

All prices include VAT at 17.5%

TILE-FIX CEMENT - **£7.99** a tub, covers 4 sq. metres

FORRET - Box of 10 wall tiles **£2.14**

BENETIA - Box of 10 tiles **£2.14**

FORRET

BENETIA

## FLOOR TILES

QUARRY TILES - terracotta, 120mm x 120 mm, 31p each

CERAMIC REGULAR SHAPED TILES - choice of patterns

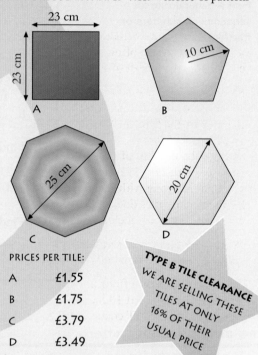

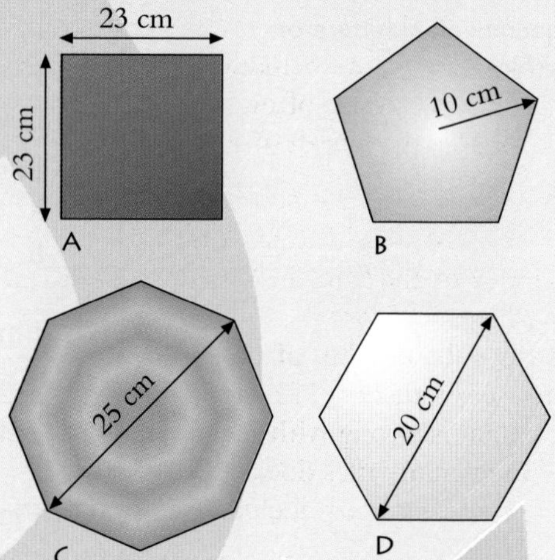


PRICES PER TILE:

| | |
|---|---|
| A | **£1.55** |
| B | **£1.75** |
| C | **£3.79** |
| D | **£3.49** |

TYPE B TILE CLEARANCE
WE ARE SELLING THESE TILES AT ONLY 16% OF THEIR USUAL PRICE

We also sell small square tiles to fit with our type C tile. Pack of ten **£5.75**

TILE CEMENT - 12 kg bag - **£17.89** - enough for 10 square metres of floor.

COLOURED GROUT **£7.69** per tub - enough for 15 square metres of floor.

When calculating what to order, allow one complete tile for every part of a tile you need.

## DESIGN SERVICE

We offer a free design service for bathrooms and kitchens. Just give us your plans to a scale of 1:50 and we will calculate how much paint, wallpaper or tiling you need.

1 Spencer decides to tile the wall above a bath with Jardin style tiles.
He wants to cover an area 2.5 m by 65 cm.
 **a** How many tiles will he need to buy?
 **b** What will this cost him (including Tile-Fix)?
 **c** What is the total cost ex-VAT?

2 **a** Describe the shape of a type B floor tile and give the size of one interior angle.
 **b** What is the reduced price of one type B tile?

3 Mike uses Decorum's design service to calculate the number of quarry tiles he should buy for a floor which is 3.82 metres by 4.15 metres.
 **a** Make a scale drawing of the floor to the correct scale.
 **b** Calculate the number of tiles he needs.
 **c** Decorum allow for 5 % of the tiles breaking. How many should Mike buy?

4 For type D tiles:
 **a** What is the mathematical name of the shape?
 **b** What is the size of an internal angle?
 **c** What is the length of a side?

5 Steve's floor is 3.95 metres by 2.37 metres, and he wants to tile it with type A tiles. He works out that the area of the floor is 9.36 $m^2$ and that the area of one tile is 0.0529 $m^2$.
He says the number of tiles he needs is: $9.36 \div 0.0529 = 177$ tiles
 **a** What is wrong with Steve's method?
 **b** How many tiles does he really need?
 **c** What is the cost including grout and cement?

6 Which size tin of brilliant white paint is the best value for money? Explain your answer.

7 The dimensions of a floor are given as 13 ft 6 inches by 11 ft 10 inches.
Calculate the ex-VAT cost of tile cement and grout for this floor.

8 Decorum are thinking of selling smaller bags of tile cement. They have chosen a 5 kg size as most useful. If they price it at the same unit cost as the 12 kg bag, at what price should they sell it?
Give your answer to the nearest penny.

9 **a** What is the mathematical name of the shape of the type C floor tile?
 **b** Calculate the size of an internal angle of this shape.
 **c** Calculate the discounted price of 50 type B tiles.

10 The $2\frac{1}{2}$ litre size can of paint is 18 cm high.
Show that the diameter of this can, to the nearest millimetre is 133 mm.

11 The contents of a tub of Tile-Fix cement weigh 3.5 kg.
New regulations state that:
"Wall tile cement must be applied at a rate of **at least** 885 gm/$m^2$".
 **a** Does the information given by Decorum Design meet the new regulations?
 Explain your answer.
 **b** The £7.99 tub of Tile-Fix is discontinued.
 Decorum Design sell a new tub to cover 5 $m^2$ at a price of £10.49.
  **i** Is this an increase or decrease in price?
  **ii** Give any change in price as a percentage of the £7.99 price, to the nearest whole number.

12 Type A floor tiles are 6 millimetres thick.
The contents of a box of 25 tiles weighs 17.8 kg.
Calculate the weight of 1 $cm^3$ of tile to 1 dp.

13 Sally is asked to find the area of one type C floor tile. She measures a side to be 96 mm.
This is her working:

> Tile shape made up of 8 triangles.
> Area of one triangle is given by:
> $0.5 \times 9.6 \times 12.5 = 60$
> Area of tile is given by:
> $60 \times 8 = 480$
> Area of tile is 480 $cm^2$.

 **a** Explain why Sally's calculation must be wrong.
 **b** Calculate the area of a type C tile.
 Give your answer correct to the nearest $cm^2$.

14 For a display, type D floor tiles are to be cut so that each tile gives two rhombus-shaped parts and two parts in the shape of an equilateral triangle.
 **a** Sketch a type D tile and show how it can be cut to give the shapes for the display.
 **b** Explain why each rhombus shape is more than 25% of the tile.
 **c** Give the area of one equilateral triangle shape as a percentage of a tile (to 1 dp).

## Starting points
You need to know about ...

... so try these questions

### A Some mathematical terms

- Lines AB and CD are **perpendicular** to line XY because they would meet XY at right angles.

XY is also **perpendicular** to AB and CD.

AB and CD are **parallel**.

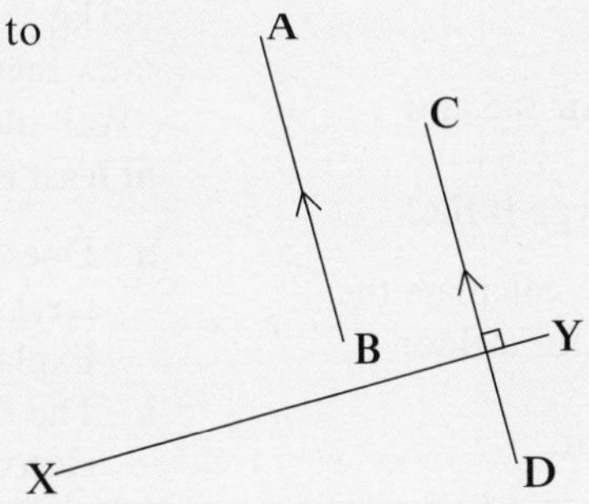

- These are some different types of triangle.

**Equilateral**
All sides equal
All angles equal

**Isosceles**
Two sides equal
Two angles equal

**Right-angled**
One right angle

**Scalene**
No sides equal
No angles equal

### B Congruent shapes

- Shapes are said to be congruent if they have the same shape or size or if they are reflections of each other.
  **Example**   Triangles A and B are **congruent** to each other.

### C Constructing a triangle, given all three sides

**Example**
Construct a triangle ABC where
AB = 4.4 cm, BC = 3.8 cm and CA = 3.2 cm.

- Make a rough sketch first.
- Draw a base line (say BC).
- Draw the locus of points 3.2 cm from C (purple arc).
- Draw the locus of points 4.4 cm from B (red arc).
- Where the arcs cross is point A
- Draw the triangle ABC.

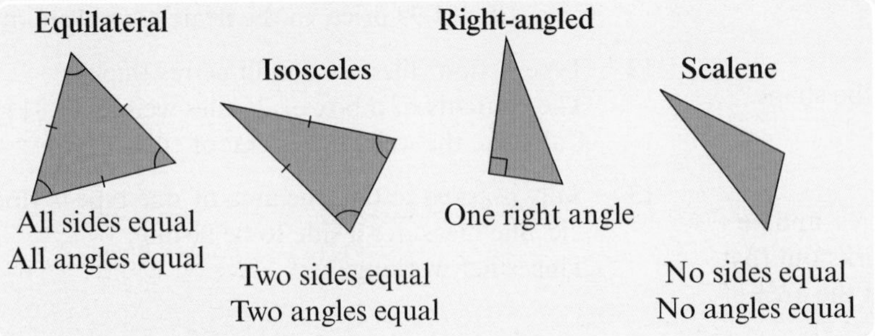

**A1 a** Draw a straight line PQ which is at an angle of 30° to the horizontal.
**b** Now draw another straight line RS which you **estimate** to be perpendicular to PQ.
**c i** At what angle is your RS to the horizontal?
**ii** What angle do you think this should be?

**A2** Can a triangle be both:
**a** isosceles and scalene
**b** right-angled and scalene
**c** isosceles and right-angled
**d** equilateral and right-angled?

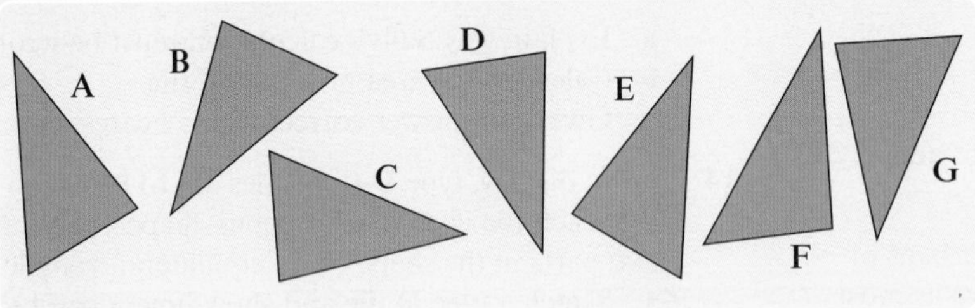

**B1** Which of the triangles C to G is not congruent to triangle A?

**B2** Draw another triangle which is congruent to triangle A.

**B3** If two quadrilaterals both have sides of 12 cm, 13 cm, 14 cm and 15 cm are they necessarily congruent?
Explain your answer.

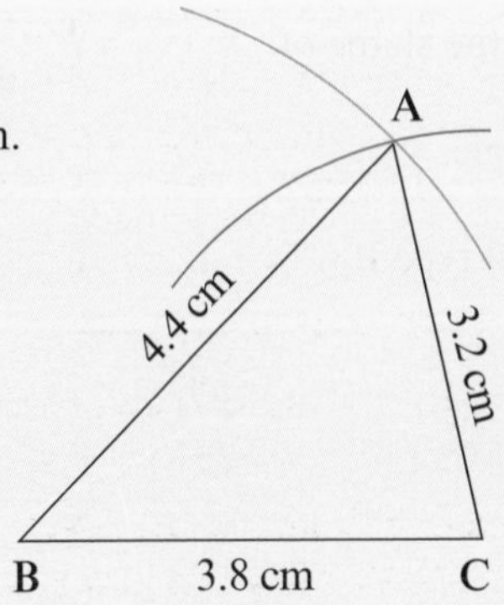

**C1** Draw a triangle PQR, where PR = 8 cm, PQ = 6 cm and RQ = 9 cm.

**C2** Draw a triangle RST, where RS = 7.5 cm, ST = 7.5 cm and TR = 12.2 cm.
What type of triangle is this?

**C3** A triangular field is 350 metres by 840 metres by 760 metres. Use a scale of 1 : 10 000 to construct a scale drawing of the field.

## Constructing triangles from other data

$\Delta$ABC means Triangle ABC.

$\angle$B means angle B.

- You can construct a triangle when you are not given the length of all three sides. You might be given **one side and two angles.**

**Example** Draw $\Delta$ABC, where AB = 5 cm, $\angle$B = 55°, and $\angle$A = 43°.

**Stage** 1 Make a rough sketch.
2 Make the side you know (AB) the base and draw it.
3 At A, draw an angle of 43° with a protractor.
4 At B, draw an angle of 55°.

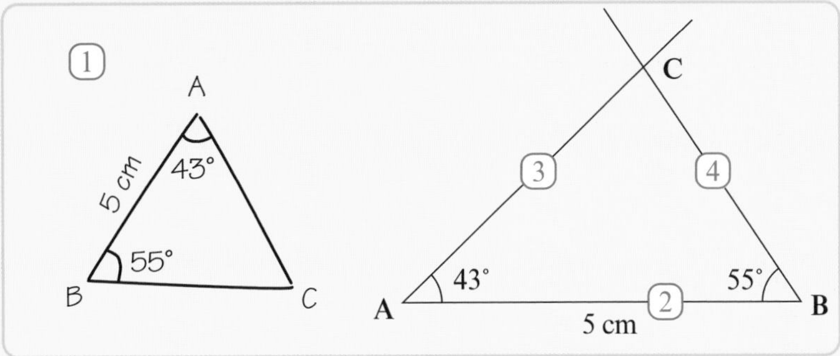


- You might be given **two sides and one angle.**

**Example** Draw $\Delta$RST, where RT = 6.5 cm, RS = 4.5 cm and $\angle$R = 46°.

**Stage** 1 Make a rough sketch.
2 Make the long side (RT) the base, and draw it.
3 At R, draw an angle of 46° and mark S, 4.5 cm from R.
4 Draw the last side TS.

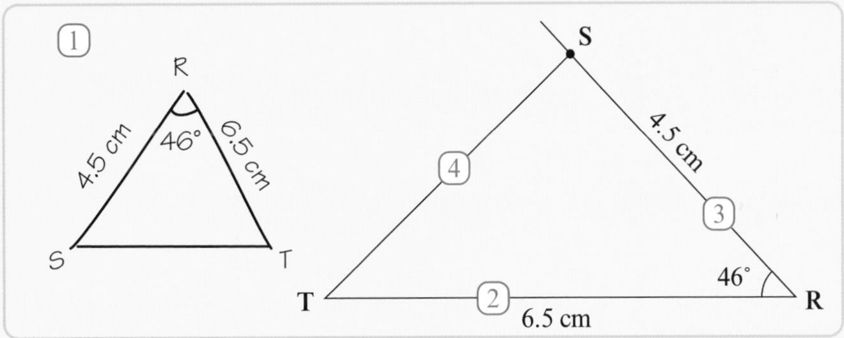


**Exercise 6.1**
**Constructing triangles**

1 Construct these triangles:

a $\Delta$DEF, where DF = 6 cm, EF = 5 cm and $\angle$F = 69°
b $\Delta$GHI, where GI = 7.2 cm, $\angle$G = 53° and $\angle$I = 42°
c $\Delta$JKL, where JK = 8 cm, $\angle$J = 25° and $\angle$K = 125°
d $\Delta$MNP, where NP = 6.3 cm, MP = 5.2 cm and $\angle$P = 131°.

2 Construct $\Delta$QRS, where RQ = 8.3 cm, $\angle$R = 35° and $\angle$S = 68°. You will need to calculate another angle first.

Two of the triangles in Question **3** are almost identical but they cannot be referred to as congruent unless they **are** identical.

3 Construct these triangles to decide which two look identical:

a $\Delta$IJK, where JK = 5.5 cm, IJ = 8 cm and $\angle$J = 50°
b $\Delta$LMN, where LM = 8 cm, $\angle$L = 60° and $\angle$M = 50°
c $\Delta$PQR, where PR = 6.5 cm, QR = 7.5 cm and $\angle$R = 70°.

**Thinking ahead to ... congruent triangles**

**A** **a** Construct a triangle BCD where CD = 8.5 cm, BD = 9.4 cm and ∠B = 54°.

**b** What is the length BC on your triangle?

**c** Construct a different triangle which still meets all the conditions in Part **a** but has a different length for side BC.

## Congruent triangles

SSS stands for the data which is equal for both triangles: Side, Side, Side

The order of the letters is important, so SAS shows that the angle is between the two known sides (i.e. the included angle):

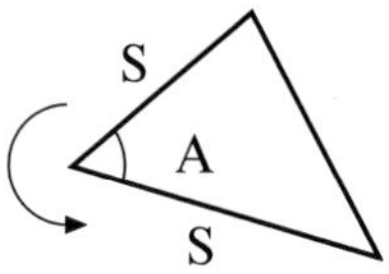


- There are four sets of conditions which can help you decide if two triangles are congruent.

  1 The **three sides** of one are equal to the three sides of the other. This can be referred to as SSS.

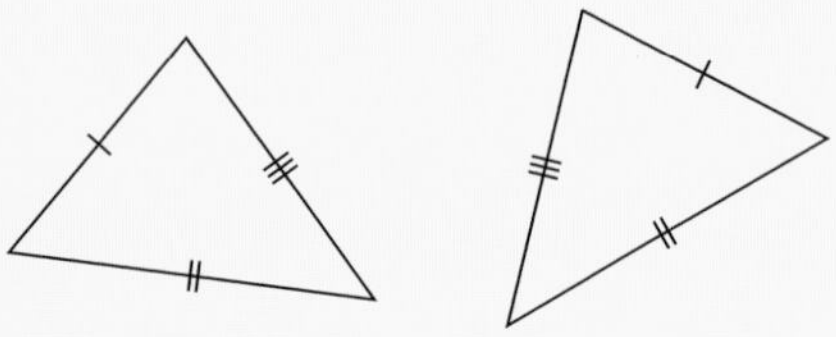

  2 **Two angles and a side** of one are equal to two angles and corresponding side of the other.

  This is known as **ASA** (or **AAS**).

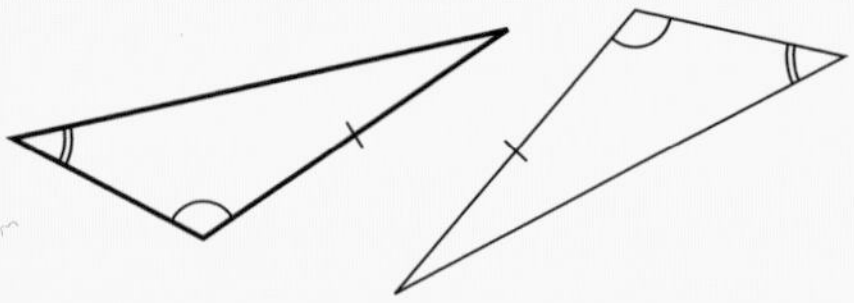

  3 **Two sides and the included angle** of one are equal to two sides and the included angle of the other.

  This is known as SAS.

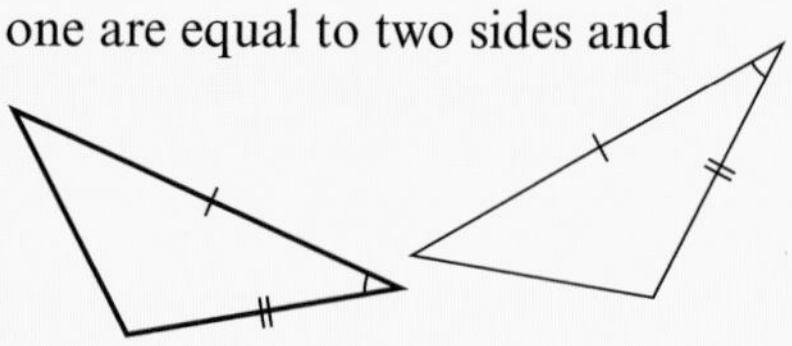

  4 They are both **right-angled** and have the **hypotenuse and one other side** equal.

  This is known as **RHS**.

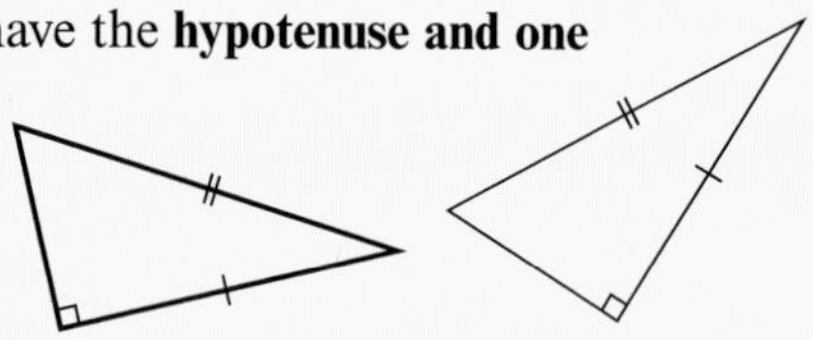

- The cases SSS, SAS, ASA (or AAS) and RHS can be stated as reasons why two triangles are congruent.

**Example** Name a triangle which is congruent to Δ BAD and state the case for congruence.

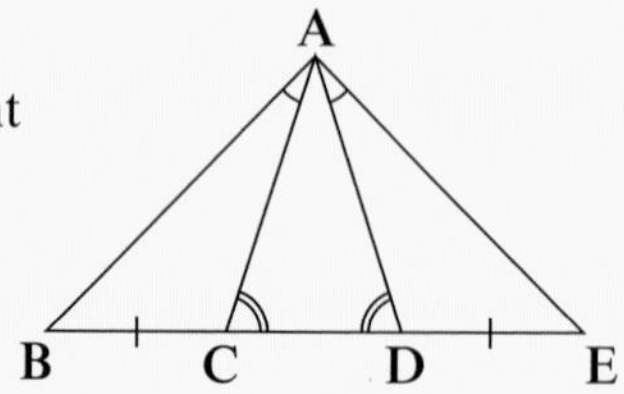


∠BAD = ∠EAC since ∠CAD is common to both angles.
BD = CE, since CD is common to both lengths.

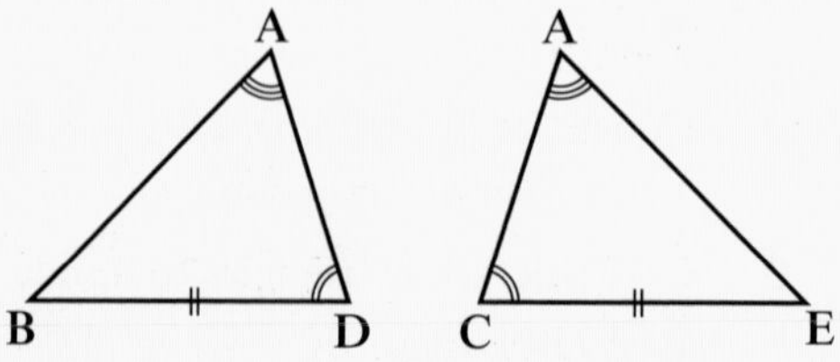


Therefore Δ BAD is congruent to Δ EAC. The case for congruence is AAS.

Note that when you name two shapes which are congruent you must state the letters in the corresponding order.

i.e. Δ BAD is congruent to Δ EAC, but Δ BAD is not congruent to Δ AEC.

**Exercise 6.2**
**Congruent triangles**

1 In each of the following diagrams name a triangle which is congruent to $\Delta$ABC and state the case for congruence.

**a**

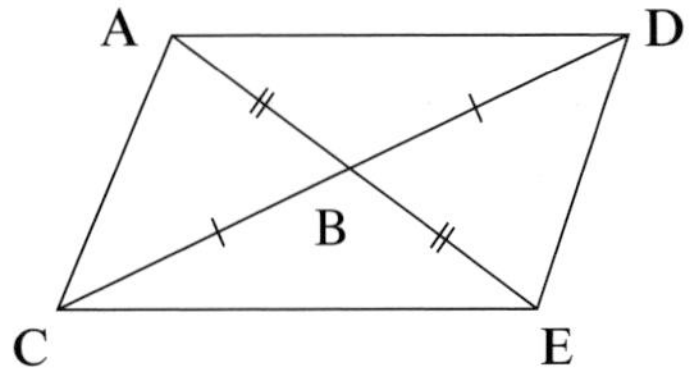


**b**

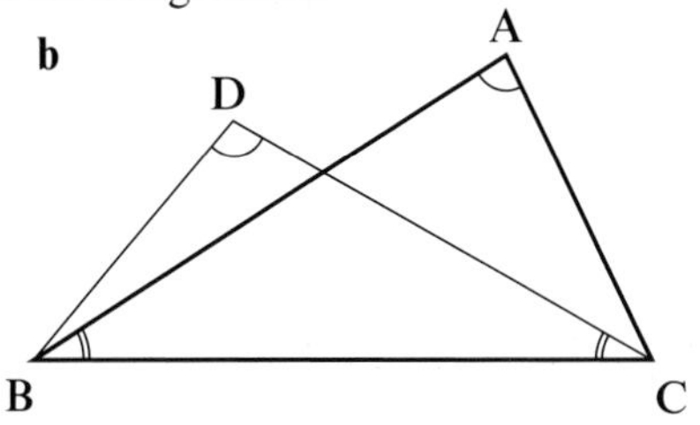


**c**

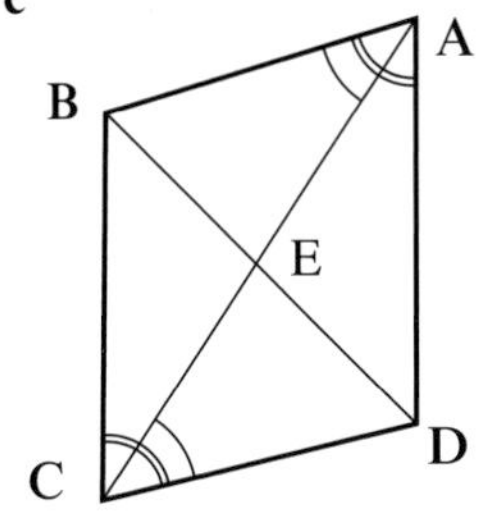


**d**

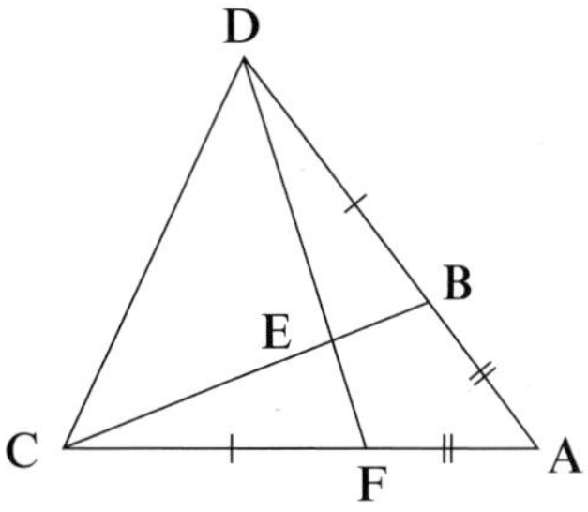


**e**

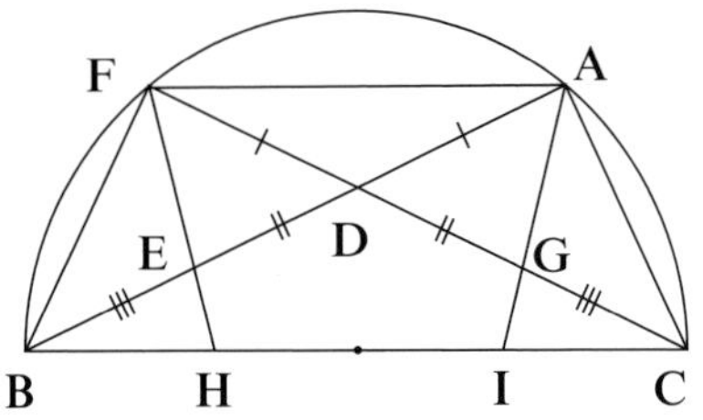


**f**

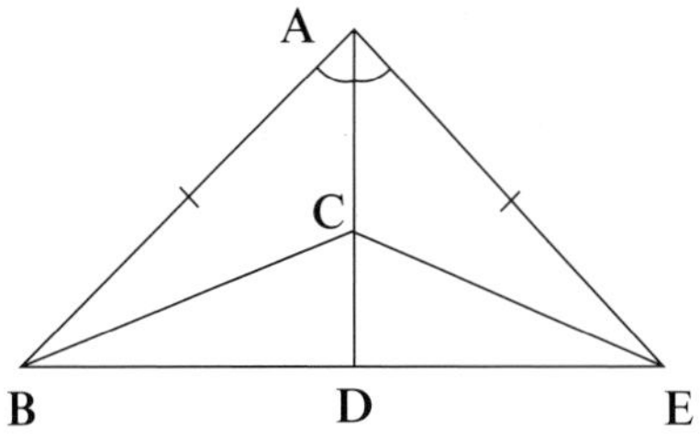


2 Simon said that the following two triangles were definitely congruent. He said there was a new case for congruence, **ASS**, which fitted them.

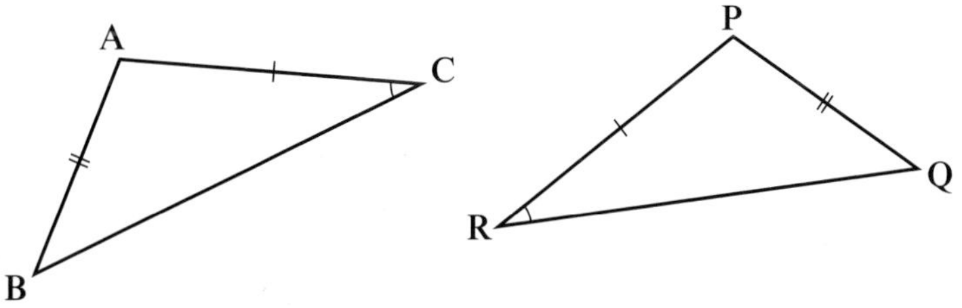


**a** Although the triangles might be congruent, explain why they need not necessarily be so.
**b** Construct a pair of triangles which match ASS but which are not congruent.

3 Sketch a pair triangles for each of these, state if the two are not necessarily congruent or certainly congruent and, if they are, the case for congruence.

**a** In $\Delta$DEF and $\Delta$GHI: DF = GI, DE = GH, EF = HI.
**b** In $\Delta$ABC and $\Delta$EFD: AB = DE, BC = EF, $\angle$ABC = $\angle$DEF.
**c** In $\Delta$LMN and $\Delta$RST: LN = RT, $\angle$LMN = $\angle$RTS, $\angle$LNM = $\angle$TSR.
**d** In $\Delta$PQR and $\Delta$STU: PR = ST, QR = TU, $\angle$QPR = $\angle$TSU = 90°.
**e** In $\Delta$BCD and $\Delta$EFG: BC = EG, BD = FG, $\angle$BCD = $\angle$FEG.
**f** In $\Delta$RST and $\Delta$UVW: $\angle$RTS = $\angle$VUW, VW = SR, $\angle$RST = $\angle$UVW.

**Thinking ahead to ... bisecting an angle**

A pedestrian area is edged by two buildings which meet at an angle of 36°.
The plans say trees must be planted an equal distance from both buildings.

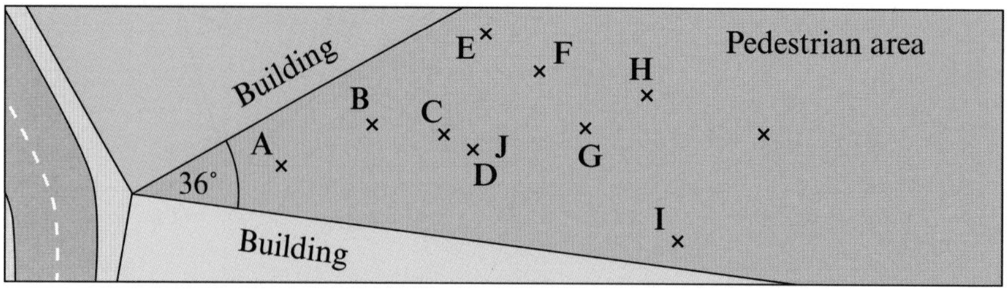
Building
Building
Pedestrian area
36°
A
B
C
D
E
F
G
H
I
J

'Equidistant from' means 'the same distance from'.

**A** Which crosses on the diagram mark where trees could be planted?

**B** Draw two lines which meet at an angle of 36°.
Mark the locus of all points which are equidistant from both lines.

## Bisecting an angle

To bisect an angle means to draw a line which cuts it in two equal parts from its vertex.

It is useful to be able to bisect an angle without measuring it.
You can do this using a pair of compasses.

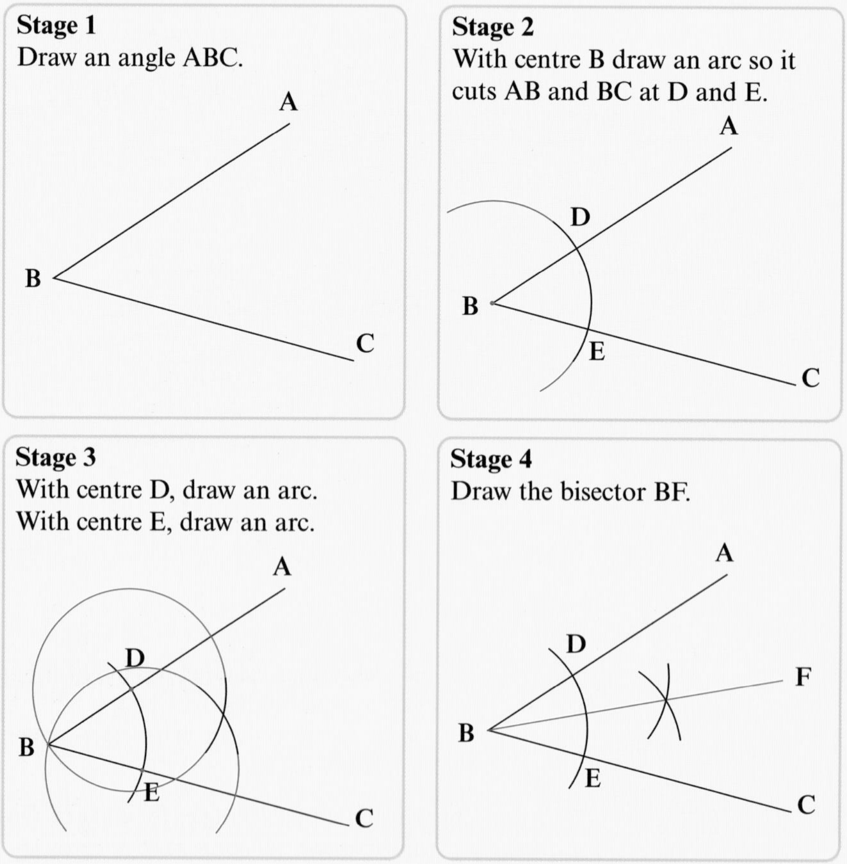
Stage 1
Draw an angle ABC.
A
B
C
Stage 2
With centre B draw an arc so it cuts AB and BC at D and E.
A
D
B
E
C
Stage 3
With centre D, draw an arc.
With centre E, draw an arc.
A
D
B
E
C
Stage 4
Draw the bisector BF.
A
D
F
B
E
C

**Exercise 6.3**
**Bisecting an angle**

**1** Use a protractor to draw each angle then bisect it using compasses.
**a** 44° **b** 100° **c** 146° **d** 90°

**2** Use compasses to draw an equilateral triangle with sides of 8 cm.
What size is each angle?
Bisect one of the angles. What angle have you made?

**Thinking ahead to ...**
**perpendicular bisectors**

When two lines cross they are said to intersect.

**A** Draw a straight line and mark two points, A and B, 6 cm apart.

**B** Draw a circle at A and another with the same radius at B.

**C** Draw larger circles with equal radii at A and B. If they intersect, mark the points of intersection.

**D** Continue by drawing larger circles and marking the points of intersection.

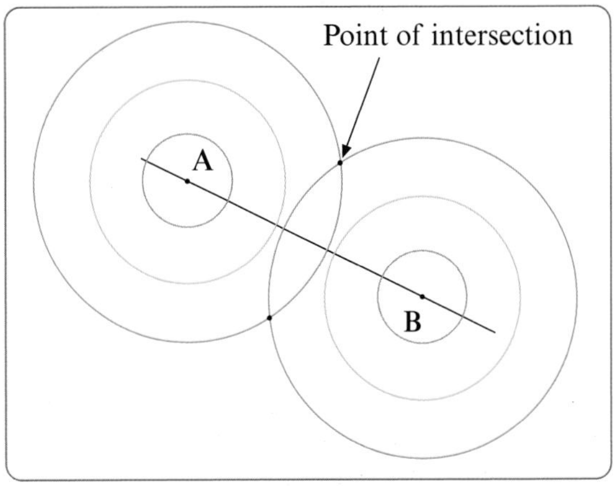


**E** **a** Draw the locus of all points of intersection of equal-sized circles.
**b** Describe the link between this locus and the line AB.

## Perpendicular bisectors

A line which cuts a straight line exactly in half at right angles is called a **perpendicular bisector**.

Example

To construct the perpendicular bisector of the line EF.

- Draw the line EF.
- With centre E draw an arc with a radius greater than half of EF.
- With centre F draw another arc with the same radius.
- Join the two points of intersection.

N marks the midpoint of EF.

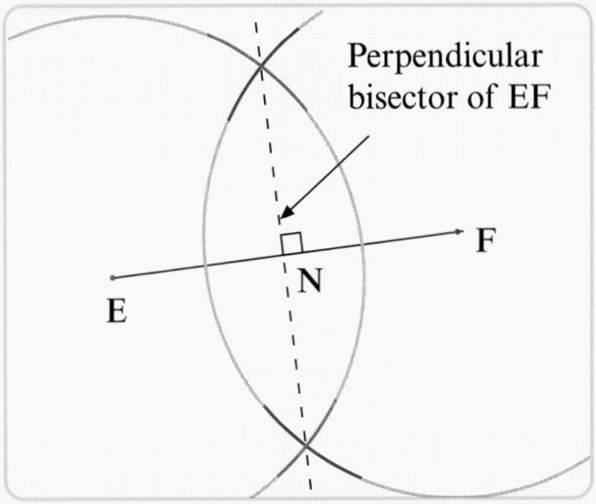


**Exercise 6.4**
**Perpendicular bisectors**

**1** **a** Draw a line AB, 6 cm long. Construct the perpendicular bisector of AB.
**b** Mark any point on the bisector and measure its distance to A and to B.
**c** What can you say about the distance of any point on the bisector from A and from B?

**2** Draw the perpendicular bisectors of lines with these lengths:

**a** 10 cm **b** 7.7 cm **c** 4.6 cm

Check that both sides are equal in length and that you have right angles.

**3** **a** Draw this triangle.
**b** Construct the perpendicular bisector of each side.
**c** Where do all three bisectors intersect?
**d** Does this happen for other triangles?

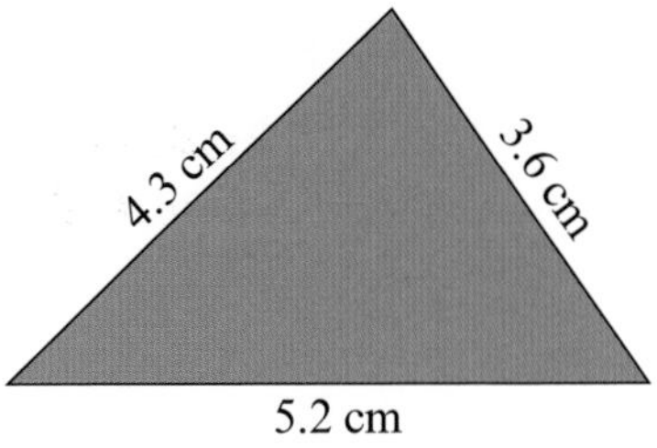

4 a Draw a circle of radius 6 cm and mark its centre C.
b Draw a chord (AB).
c Construct the perpendicular bisector of the chord.
d Draw two more chords and bisect them.
e What do you notice about where the bisectors intersect?

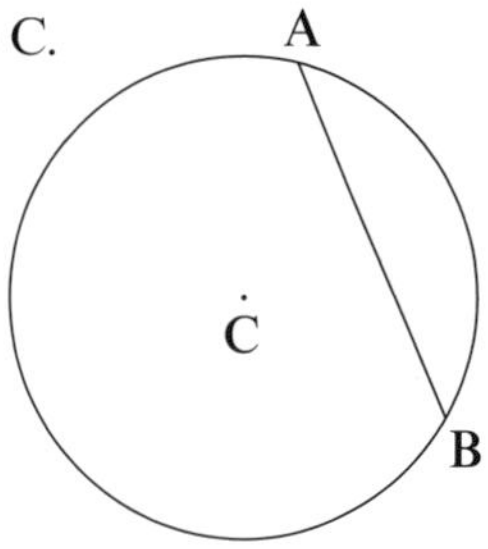


## Meeting conditions

Rules that have to be met such as 'the tap must be the same distance from A and B' are known as conditions.

Constructions such as bisecting angles or lines can be used for making scale drawings where conditions have to be met.

**Example**

A water tap is to be put in a large garden but it must meet these conditions:
1 It must be the same distance from the two greenhouses, A and B.
2 It must be the same distance from the grape vine wires as from the hedge.

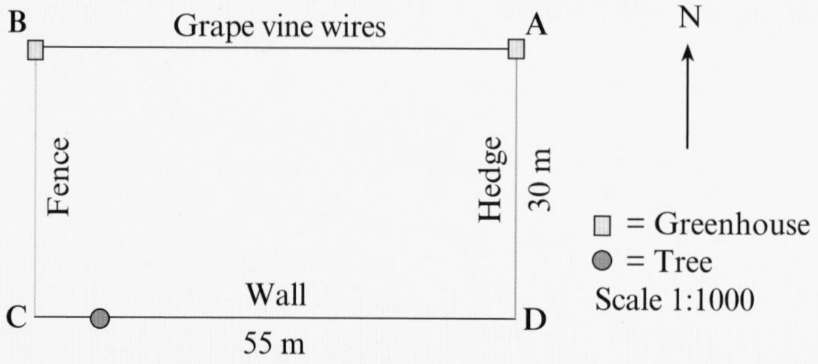


Where must the tap be placed? How far is it from the tree?

To meet condition **1** you draw the perpendicular bisector of BA.
This line is the locus of all points equidistant from B and A.
To meet condition **2** you bisect angle BAD.
This line is the locus of all points equidistant from line BA and line AD.

Where the two loci intersect both conditions are met, so the tap must be at this point.

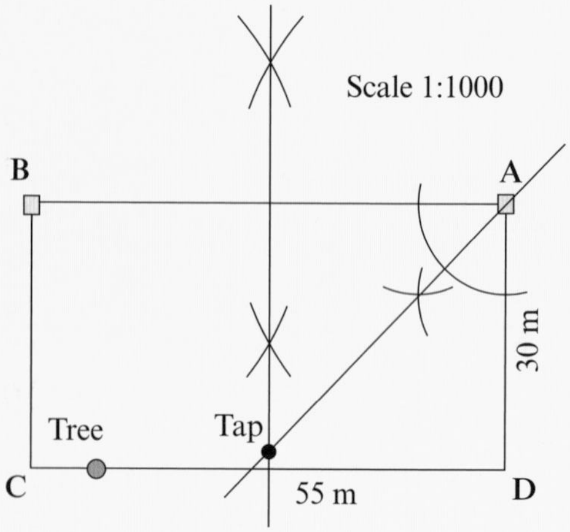


**Exercise 6.5**
Meeting conditions

1 a On the scale diagram above measure the distance in centimetres between the tap and the tree.
b What is this actual distance in the garden?

2 Make a scale drawing to show where the tap will be if:
condition **1** stays the same
condition **2** says the tap must be equidistant from the wall and the hedge.

3 An Olympic javelin field has lines which make an angle of 29° to each other.
A thrower aims the javelin so that it flies equidistant from both lines.
The thrower hopes to reach the club record of 88 metres.

This diagram only approximates to how a true javelin field is marked out. The throwing point actually lies on an arc about 2 metres wide which comes at the end of a 36 metre run-up.

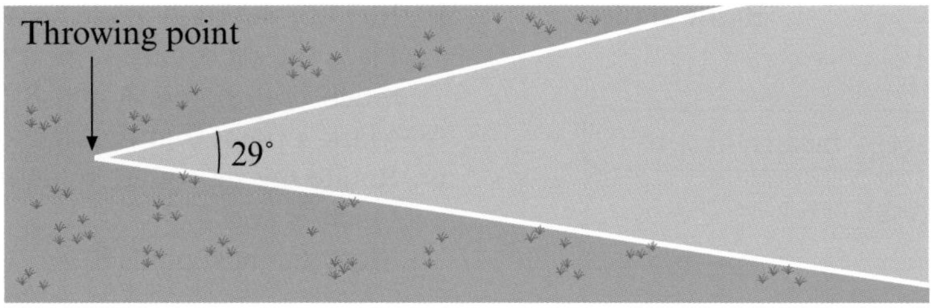


**a** Make a scale drawing of the field for throws up to 100 metres. Use a scale of 1:1000.
**b** Mark the locus of all points 88 metres from the throwing point.
**c** Construct the locus of points equidistant from the sidelines.
**d** Mark where the thrower hopes the javelin will land.

4 Two lighthouses are 3.6 miles apart on a straight coastline.
A ferry sails into port by keeping the same distance from both lighthouses.
A fishing boat sails so that it is always 3 miles from the coast.

**a** Make a scale drawing of the coast to show the position of the lighthouses. Use a scale of 1 cm to 0.5 miles.
**b** Show and label the course taken by the fishing boat.
**c** Construct and label the course taken by the ferry.
**d** Mark the point where there is the greatest risk of a collision.

To construct an angle of 60° without a protractor you can construct an equilateral triangle.

To construct an angle of 30°, first construct a 60° angle, as above, then bisect it.

5 Coastguard A is 14 km due west of coastguard B along a stretch of straight coastline.
A ship S can be seen on a bearing of 030° from A and on a bearing of 300° from B.

**a** Use a ruler and compasses only to make a scale drawing of the situation. Use a scale of 1 cm to represent 2 km.
**b** From your drawing find the distance of the ship from coastguard B.

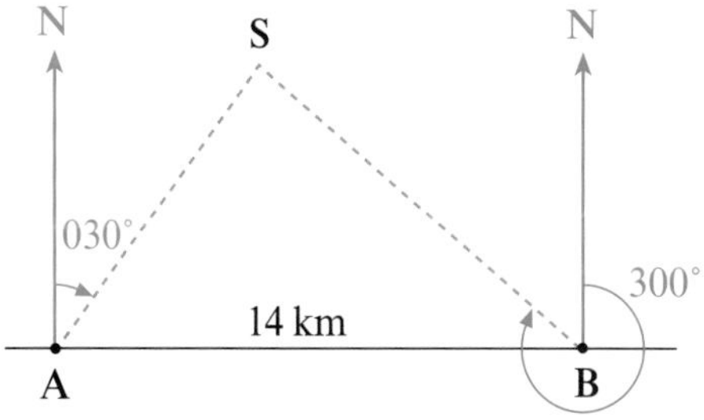


**c** The ship sails into shore so that its path bisects the angle ASB. Construct this path and find its distance from A when it docks.

6 **a** Construct a quadrilateral PQRS where RSP = 90°, QR = 8.4 cm, RS = 9 cm, SP = 6 cm and PQ = 5.4 cm.
**b** Bisect angles RSP and SPQ, and let these bisectors meet at point T.
**c** Construct the perpendicular from T to meet PS at N.
**d** Draw a circle with radius TN and centre T. Which sides of the quadrilateral does your circle touch ?

7 A buoy is moored 1000 metres from a long straight harbour wall.
A small boat sails so that it is always the same distance from the buoy as from the wall. Investigate the path of the boat with the help of a scale drawing.

To draw a perpendicular from a point X to a line AB:
- draw an arc from X to intersect AB in two places
- from each intersection draw equal arcs on the other side of AB
- join the intersection of these arcs to X.

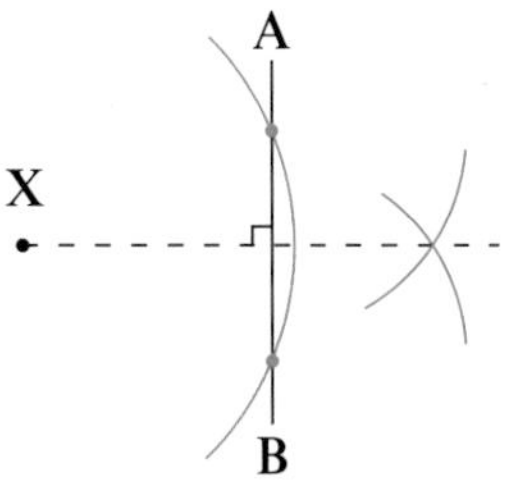

# The locus of a right angle

**Exercise 6.6**
Right angles

'Adjacent' means 'next to'.

1 **a** Draw a line AB which is 8 centimetres long.
**b** Identify the right angle on a set square.
Place the set square so that the sides adjacent to the right angle touch both points A and B.
Mark a point P at the right-angled vertex.
**c** Rotate the set square to another position so adjacent sides touch A and B.
Mark the new point P.

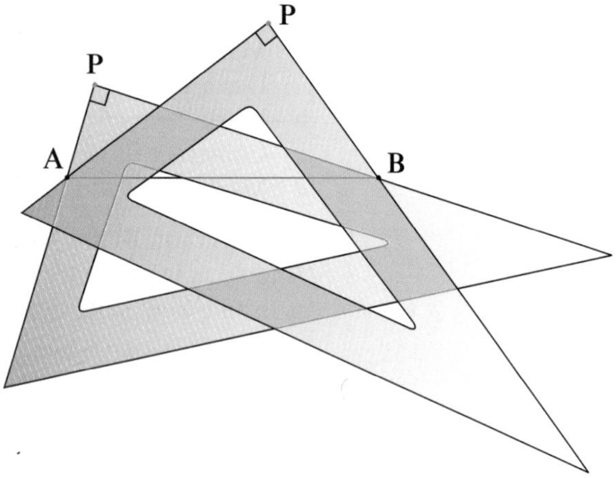


**d** Mark the locus of all positions of P as you rotate the set square 360°.
What shape is this locus?

2 **a** Draw a circle of any size.
**b** Draw in a diameter and label it RS.
**c** Join point R to any point, C, on the circumference of the circle.
**d** Join C to S.
**e** Measure angle RCS. What do you notice?
**f** For any diameter RS and any point on the circumference C, what can you say about the triangle RCS?

3 In this circle point H is the centre and lines AG and GC are equal in length.
Name six angles in the diagram which must be right angles.

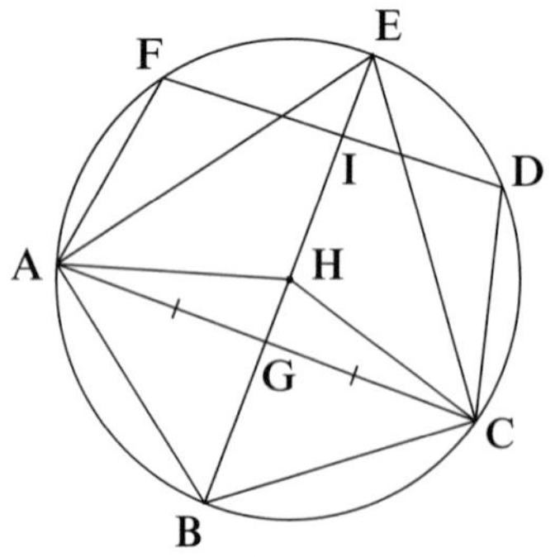


4 The cross-section of a roof is semicircular in shape.

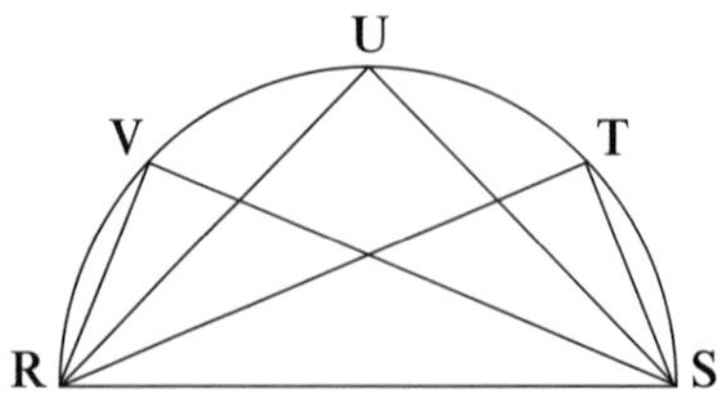


Beam RV is 12 metres long and beam VS is 26 metres.
Beam RU is equal in length to US.

**a** Use what you know about the angles in a semicircle to calculate the diameter RS to 1 dp.
**b** Calculate the length of RU to 1 dp.

## End points

You should be able to … … so try these questions

**A** Construct and draw triangles when you are given their sides or angles

**A1** Construct $\Delta ABC$ where $AB = 7$ cm, $AC = 6$ cm and $BC = 6$ cm.

**A2** Draw $\Delta DEF$, where $\angle EDF = 92°$, $\angle DFE = 47°$ and $DF = 4.5$ cm.

**A3** Draw $\Delta PQR$, where $PQ = 7.4$ cm, $QR = 4.3$ cm and $\angle Q = 123°$.

**A4** Draw two different triangles STU which each have $TU = 6$ cm, $\angle SUT = 40°$ and $ST = 4.3$ cm.

**B** Construct and use perpendicular bisectors

**B1**

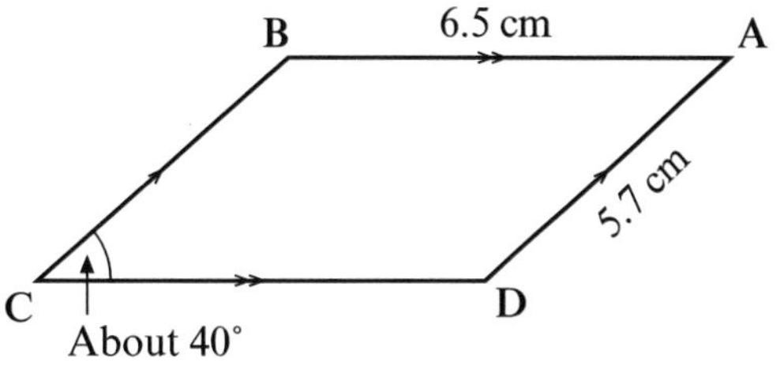


**a** Draw the parallelogram ABCD with $B\hat{C}D$ about 40°.
**b** Construct perpendicular bisectors for the sides BC and AD.
**c** What can you say about the gradients of the two bisectors?

**B2**

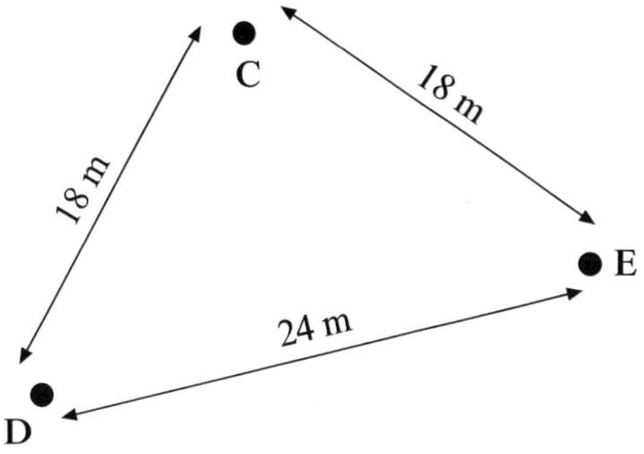


Two dogs are tied at points D and E, 24 metres apart.
A cat at point C wishes to pass between the dogs so that it is always the same distance from each one.
Make a scale drawing and construct the path the cat must take.
Use a scale of 1 cm to 3 metres.

**C** Construct and use the bisectors of an angle

**C1** Draw an angle of 110° with a protractor.
Construct the bisector of the angle.

**C2** Construct $\Delta$ ABC where $AB = 4.3$ cm, $BC = 6.8$ cm and $AC = 5.9$ cm.
Bisect each of the angles.
What do you notice about where the bisectors intersect?

**D** Recognise and state the conditions for congruence

**D1** For each pair of triangles, state if they are congruent or not necessarily congruent. For those that are congruent state their case for congruence.

**a** 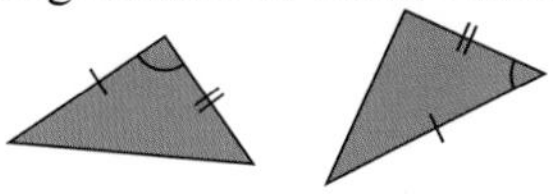

**b** 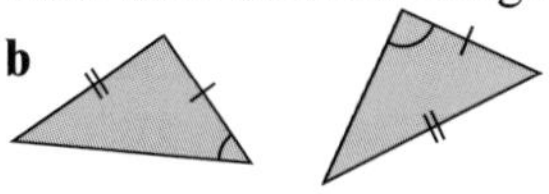

**c** 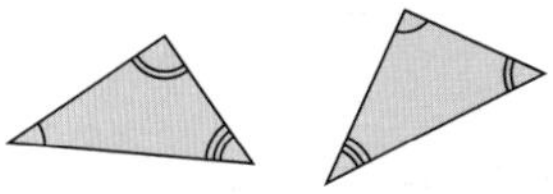

**d** 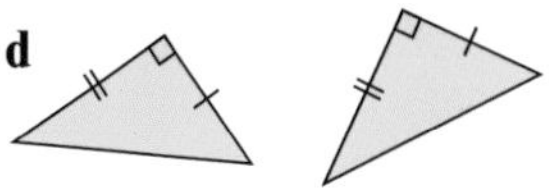

**e** 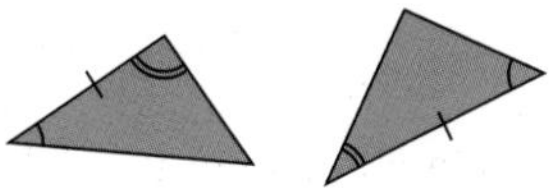

**f** 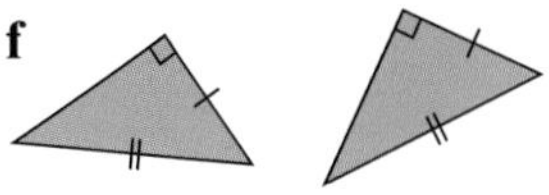

**E** Draw loci from some conditions given

**E1** This garden is the shape of a trapezium with the dimensions given.

Trees are at the midpoints of sides AD and DC.

A new bush is to be planted:

- equidistant from each fence
- equidistant from each tree.

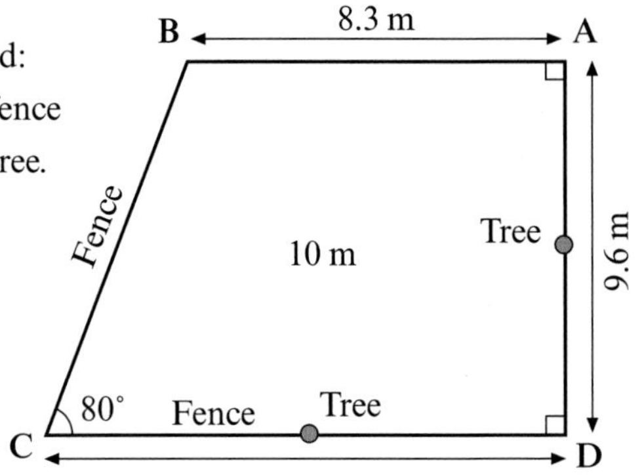


Make a scale drawing of the garden and show by construction the position of the bush. Use a scale of 1 : 100.

## Some points to remember

- To construct an angle of 90°, draw a line and construct its perpendicular bisector.
- When asked to construct something do not rub out the construction lines when you finish.
- To construct an angle of 60°, construct an equilateral triangle.
- The perpendicular bisector of any chord will pass through the centre of the circle.

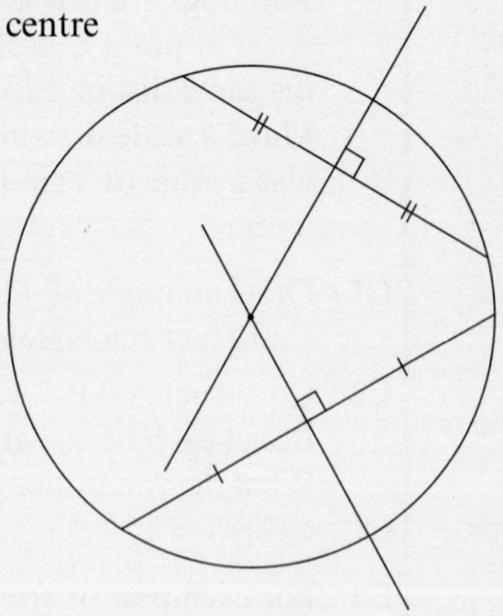

- All angles in a semicircle which are made by the diameter and a point on the circumference are right angles.

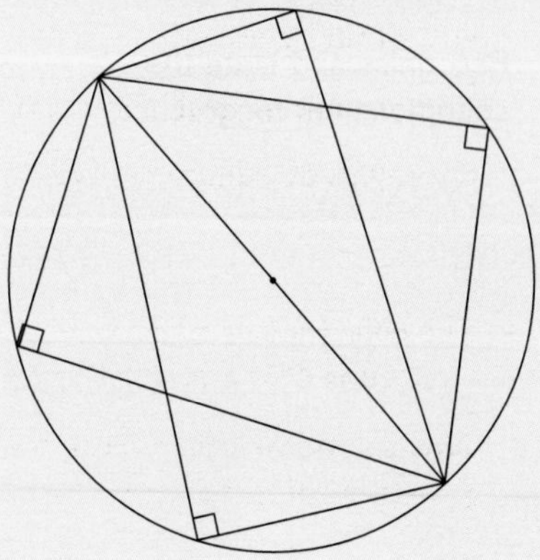

# SECTION 7 POWERS AND SURDS

## Starting points

You need to know about … … so try these questions

### A Fractions and decimals

- In recurring decimal notation, dots are placed over the first and last of the set of recurring digits.

  For example: $0.236666666\ldots = 0.23\dot{6}$
  $0.236363636\ldots = 0.2\dot{3}\dot{6}$
  $0.236236236\ldots = 0.\dot{2}3\dot{6}$

- We can multiply recurring decimals by powers of 10.

  For example: if $n = 0.3232323232\ldots$
  then $10n = 3.2323232323\ldots$
  $100n = 32.3232323232\ldots$ and so on.

- Any fraction is equivalent to a terminating or recurring decimal.

  For example: $\frac{3}{4} = 3 \div 4 = 0.75$
  $\frac{11}{30} = 11 \div 30 = 0.366666\ldots = 0.3\dot{6}$

**A1** Use recurring decimal notation to write these recurring decimals.
**a** 0.777777 …
**b** 0.12323232323 …
**c** 4.56666666 …

**A2** If $n = 0.565656\ldots$, what is:
**a** $10n$ **b** $100n$ **c** $1000n$?

**A3** Convert these fractions to decimals, using recurring decimal notation where necessary.
**a** $\frac{1}{4}$ **b** $\frac{1}{3}$ **c** $\frac{6}{11}$
**d** $\frac{17}{30}$ **e** $\frac{3}{8}$

### B Index notation

- We say $5^4$ is **5 to the power 4** or **5 raised to the power 4**.

  $5^4 = 5 \times 5 \times 5 \times 5 = 625$ (4 is the **index**)

- Numbers raised to a **negative power** are defined so that:

  $x^{-y} = \frac{1}{x^y}$ e.g. $5^{-2} = \frac{1}{5^2} = \frac{1}{25}$ (= 0.04)

  So, the division pattern in this table is continued:

| $5^3$ | $5^2$ | $5^1$ | $5^0$ | $5^{-1}$ | $5^{-2}$ | $5^{-3}$ | $5^{-4}$ |
|---|---|---|---|---|---|---|---|
| 125 | 25 | 5 | 1 | $\frac{1}{5}$ | $\frac{1}{25}$ | $\frac{1}{125}$ | $\frac{1}{625}$ |
| | ÷5 | ÷5 | ÷5 | ÷5 | ÷5 | ÷5 | ÷5 |

**B1** Write these in index notation.
**a** 3 raised to the power 5
**b** 4 to the power $^{-}3$

**B2** Evaluate as an integer or in fractional form:
**a** $3^4$ **b** $3^{-2}$ **c** $4^0$
**d** $5^{-1}$ **e** $2^9$ **f** $(^{-}6)^2$
**g** $(\frac{2}{3})^2$ **h** $(^{-}\frac{1}{4})^2$

### C Square roots and cube roots

- The **square roots** of 9 are 3 and $^{-}3$ because,
  3 squared is $3^2 = 3 \times 3 = 9$ and $(^{-}3)$ squared is $(^{-}3)^2 = {}^{-}3 \times {}^{-}3 = 9$

  Where a square root can be positive or negative, you should give the positive root unless stated otherwise.

  'The square root of 9 is 3' can be written $\sqrt{9} = 3$ or $\sqrt[2]{9} = 3$.

  Examples of square roots are:
  - $\sqrt{1.44} = 1.2$
  - $\sqrt{1\frac{7}{9}} = \sqrt{\frac{16}{9}} = \frac{4}{3}$

- The **cube root** of 64 is 4, because
  4 cubed is $4^3 = 4 \times 4 \times 4 = 64$

  This can be written $\sqrt[3]{64} = 4$.
  Examples of cube roots are:
  - $\sqrt[3]{729} = 9$
  - $\sqrt[3]{(^{-}8)} = {}^{-}2$

**C1** What is the value of:
**a** the square root of 169
**b** the cube root of 1000
**c** the square root of 6.76
**d** the cube root of 9.261?

**C2** Evaluate correct to one decimal place:
**a** $\sqrt{12}$ **b** $\sqrt{105}$ **c** $\sqrt{6.7}$

**C3** Evaluate:
**a** $\sqrt{49}$ **b** $\sqrt[3]{3.375}$
**c** $\sqrt[3]{(^{-}1000)}$

**C4** Evaluate in fractional form:
**a** $\sqrt{\frac{1}{25}}$ **b** $\sqrt{\frac{49}{100}}$ **c** $\sqrt{1\frac{11}{25}}$

**C5** Explain why any number has no more than one cube root.

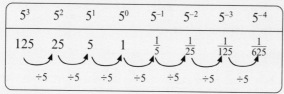

## Roots

It is also true that

$(^{-}1.2)^4 = 2.0736$ so a fourth root of 2.0736 is $^{-}1.2$.

If a root can be positive or negative, give the positive root as your answer (unless stated otherwise).

- As well as square and cube roots, there are fourth, fifth, sixth, ... roots.

  **Examples**

  - $3^5 = 3 \times 3 \times 3 \times 3 \times 3 = 243$ so a fifth root of 243 is 3.
    This can be written: $\sqrt[5]{243} = 3$.
  - $1.2^4 = 1.2 \times 1.2 \times 1.2 \times 1.2 = 2.0736$ so a fourth root of 2.0736 is 1.2
    This can be written: $\sqrt[4]{2.0736} = 1.2$
  - $\left(\frac{1}{2}\right)^6 = \frac{1}{2} \times \frac{1}{2} \times \frac{1}{2} \times \frac{1}{2} \times \frac{1}{2} \times \frac{1}{2} = \frac{1}{64}$ so a sixth root of $\frac{1}{64}$ is $\frac{1}{2}$.
    This can be written: $\sqrt[6]{\frac{1}{64}} = \frac{1}{2}$.

- Sometimes a root cannot be written exactly as a decimal or fraction.

  **Example** Find $\sqrt[5]{20}$ correct to 1 dp using trial and improvement.
  $2^5 = 32$ ($> 20$ so 2 is too big)
  $1.5^5 = 7.593\,75$ ($< 20$ so 1.5 is too small)
  $1.9^5 = 24.760\,99$ ($> 20$ so 1.9 is too big)
  $1.8^5 = 18.895\,68$ ($< 20$ so 1.8 is too small)
  $1.85^5 \approx 21.669\,99$ ($> 20$ so 1.85 is too big)
  The solution lies between 1.8 and 1.85 so $\sqrt[5]{20} = 1.8$ (to 1 dp).

**Exercise 7.1**
**Roots**

**1** Calculate:
**a** $\sqrt[4]{16}$ **b** $\sqrt[6]{11.390\,625}$ **c** $\sqrt[5]{^{-}1}$

**2** Which of these are integers?
**a** $\sqrt[3]{8}$ **b** $\sqrt[4]{12}$ **c** $\sqrt[6]{600}$ **d** $\sqrt[5]{^{-}32}$

**3** Evaluate these, giving each answer as a fraction.
**a** $\sqrt[3]{\frac{27}{64}}$ **b** $\sqrt[5]{\frac{32}{243}}$ **c** $\sqrt[7]{\frac{1}{78125}}$

**4** Evaluate $\sqrt[4]{10}$ correct to 1 decimal place.

**5** Find the value of $x$ when:
**a** $\sqrt[4]{x} = 3.5$ **b** $\sqrt[x]{1.728} = 1.2$ **c** $\sqrt[3]{x} = {}^{-}4.5$

**6** Copy and complete:
**a** $\sqrt[2]{6^4} = 6^{\square}$ **b** $\sqrt[3]{2^6} = 2^{\square}$ **c** $\sqrt[4]{5^8} = 5^{\square}$

**7** **a** Solve:
**i** $\sqrt{x^x} = 16$ **ii** $\sqrt{x^x} = 216$
**b** Make up two more problems like this that give integer values for $x$.

## Using positive and negative indices

To multiply powers of the same number, **add** the indices.

**Example 1**

$3^2 \times 3^4$
$= 3^{2+4}$
$= 3^6$

$3^2 \times 3^4$
$= (3 \times 3) \times (3 \times 3 \times 3 \times 3)$
$= 3^6$

**Example 2**

$3^{-2} \times 3^4$
$= 3^{-2+4}$
$= 3^2$

$3^{-2} \times 3^4 = \frac{1}{3^2} \times 3^4$
$= \frac{3 \times 3 \times 3 \times 3}{3 \times 3}$
$= 3^2$

To divide powers of the same number, **subtract** the indices.

**Example 1**

$$\frac{4^5}{4^2} = 4^{5-2} = 4^3$$

$$\frac{4^5}{4^2} = \frac{4 \times 4 \times 4 \times 4 \times 4}{4 \times 4} = 4^3$$

**Example 2**

$$\frac{2^3}{2^7} = 2^{3-7} = 2^{-4}$$

$$\frac{2^3}{2^7} = \frac{2 \times 2 \times 2}{2 \times 2 \times 2 \times 2 \times 2 \times 2 \times 2} = \frac{1}{2 \times 2 \times 2 \times 2} = \frac{1}{2^4} = 2^{-4}$$

To raise a power of a number to another power, **multiply** the indices.

**Example 3**

$$(3^2)^3 = 3^{2 \times 3} = 3^6$$

$$(3^2)^3 = (3 \times 3) \times (3 \times 3) \times (3 \times 3) = 3^6$$

## Exercise 7.2
### Multiplying and dividing

Algebraic expressions involving indices can be simplified using the rules for indices.
For example:

$$x^2 \times x^3 = x^{3+2} = x^5$$

$$\frac{a^7}{a^2} = a^{7-2} = a^5$$

**1** Give the answer to each of these in index notation.

**a** $6^2 \times 6^3$ **b** $5^5 \times 5^{-1}$ **c** $(7^2)^4$ **d** $\frac{3^5}{3^4}$ **e** $\frac{2^8}{2^{-3}}$ **f** $(5^3)^{-2}$

**2** Copy and complete these calculations.

**a** $7^3 \times 7^{\square} = 7^8$ **b** $4^{\square} \times 4^5 = 4^5$ **c** $3^{-4} \times 3^{\square} = 3^2$

**d** $\frac{5^7}{5^{\square}} = 5^{-3}$ **e** $\frac{2^{\square}}{2^4} = 2$ **f** $(7^3)^{\square} = 7^{15}$

**3** Simplify:

**a** $x^5 \times x^3$ **b** $x^8 \div x^2$ **c** $(x^2)^5$ **d** $\frac{x^4}{x^2}$ **e** $\frac{x^5}{x}$ **f** $\frac{x^7}{x^9}$

**4** Copy and complete:

**a** $(t^{\square})^4 = t^{12}$ **b** $(q^5)^{\square} = q^{-5}$ **c** $(p^2)^{\square} = \frac{1}{p^8}$

**5** To what power must $x^2$ be raised to give:

**a** $x^{10}$ **b** $x^{-4}$ **c** $\frac{1}{x^2}$?

**6** A student has made mistakes in trying to simplify some expressions.

**a** $a^2 \times a^4 = a^8$ ✗ **b** $a^{10} \div a^2 = a^5$ ✗

**c** $a^3 \div a^{12} = a^9$ ✗ **d** $(a^4)^3 = a^7$ ✗

Explain the mistakes, and simplify each expression correctly.

**7** Solve these equations.

**a** $2^x \times 2^x = 2^{16}$ **b** $(7^x)^x = 7^{36}$ **c** $\frac{3^x}{3^{4x}} = \frac{1}{27}$

**8** Simplify these expressions as far as you can.

**a** $\frac{4p^3 \times p^2}{2p^4}$ **b** $\frac{2q \times 3q^5}{5q^2}$ **c** $\frac{4r^5 \times 6r^2}{12r^3}$

**Thinking ahead to ... fractional indices**

According to the rules for multiplying powers:

- $25^{\frac{1}{2}} \times 25^{\frac{1}{2}} = 25^{\frac{1}{2} + \frac{1}{2}} = 25^1 = 25$
- $27^{\frac{2}{3}} \times 27^{\frac{2}{3}} \times 27^{\frac{2}{3}} = 27^{\frac{2}{3} + \frac{2}{3} + \frac{2}{3}} = 27^2 = 729$

**A** What do you think could be the value of:

**a** $25^{\frac{1}{2}}$ **b** $27^{\frac{2}{3}}$?

**B** What value does your calculator give for $16^{\frac{1}{4}}$?

## Fractional indices

Fractional powers are defined so that rules for calculating with integer indices can also be used with fractional indices.

**Example 1**

- $9^{\frac{1}{2}} \times 9^{\frac{1}{2}} = 9^{\frac{1}{2} + \frac{1}{2}} = 9^1 = 9$, so $9^{\frac{1}{2}} = \sqrt[2]{9} = 3$
- $8^{\frac{1}{3}} \times 8^{\frac{1}{3}} \times 8^{\frac{1}{3}} = 8^{\frac{1}{3} + \frac{1}{3} + \frac{1}{3}} = 8^1 = 8$, so $8^{\frac{1}{3}} = \sqrt[3]{8} = 2$

In general, $x^{\frac{1}{n}} = \sqrt[n]{x}$

**Example 2**

- $8^{\frac{2}{3}} \times 8^{\frac{2}{3}} \times 8^{\frac{2}{3}} = 8^{\frac{2}{3} + \frac{2}{3} + \frac{2}{3}} = 8^2 = 64$

  so, $8^{\frac{2}{3}} = \sqrt[3]{8^2} = \sqrt[3]{64} = 4$
- $16^{\frac{3}{4}} \times 16^{\frac{3}{4}} \times 16^{\frac{3}{4}} \times 16^{\frac{3}{4}} = 16^{\frac{3}{4} + \frac{3}{4} + \frac{3}{4} + \frac{3}{4}} = 16^3 = 4096$

  so $16^{\frac{3}{4}} = \sqrt[4]{16^3} = \sqrt[4]{4096} = 8$

In general, $x^{\frac{m}{n}} = \sqrt[n]{x^m}$

**Example 3**

- $9^{-\frac{1}{2}} \times 9^{\frac{1}{2}} = 9^{-\frac{1}{2} + \frac{1}{2}} = 9^0 = 1$, so $9^{-\frac{1}{2}} = \dfrac{1}{9^{\frac{1}{2}}} = \dfrac{1}{\sqrt{9}} = \dfrac{1}{3}$

In general, $x^{-\frac{m}{n}} = \dfrac{1}{x^{\frac{m}{n}}} = \dfrac{1}{\sqrt[n]{x^m}}$

**Exercise 7.3 Fractional indices**

**1** Find four matching pairs of equal value.

**a** $16^{\frac{1}{2}}$ **b** $^{-}2$ **c** $\frac{1}{2}$ **d** $512^{\frac{1}{3}}$

**e** $4^{-\frac{1}{2}}$ **f** $8$ **g** $4$ **h** $(^{-}32)^{\frac{1}{5}}$

**2** Evaluate these as integers or in fractional form.

**a** $36^{\frac{1}{2}}$ **b** $125^{\frac{1}{3}}$ **c** $81^{\frac{1}{4}}$

**d** $49^{-\frac{1}{2}}$ **e** $8^{-\frac{1}{3}}$ **f** $1^{-\frac{1}{5}}$

**3** From the list of fractions, $\frac{2}{9}, \frac{3}{2}, \frac{1}{3}, \frac{2}{3}, \frac{1}{9}$, which one is equivalent to:

**a** $(\frac{4}{9})^{\frac{1}{2}}$ **b** $(\frac{1}{27})^{\frac{1}{3}}$?

**4** Find four pairs of equivalent expressions.

a $x^{-\frac{5}{3}}$ b $\frac{1}{\sqrt[5]{x^3}}$ c $x^{\frac{3}{5}}$ d $\sqrt[3]{x^5}$

e $\frac{1}{\sqrt[3]{x^5}}$ f $\sqrt[5]{x^3}$ g $x^{-\frac{3}{5}}$ h $x^{\frac{5}{3}}$

**5** Sort these into four pairs of equal value.

a $64^{\frac{1}{2}}$ b 512 c 32 d $4^{\frac{5}{2}}$

e $216^{\frac{2}{3}}$ f 8 g 36 h $16^{\frac{9}{4}}$

**6** Which of these fractions is equivalent to $32^{-\frac{2}{5}}$?

a $\frac{1}{4}$ b $-\frac{64}{5}$ c $\frac{5}{64}$ d $-\frac{1}{4}$

**7** Evaluate these as integers or in fractional form.

a $125^{\frac{2}{3}}$ b $81^{\frac{3}{4}}$ c $243^{\frac{2}{5}}$

d $64^{\frac{7}{6}}$ e $4^{-\frac{3}{2}}$ f $27^{-\frac{4}{3}}$

**8** A $x^{\frac{1}{3}}$ B $9^x$ C $x^2$ D $x^{\frac{1}{2}}$ E $x^{\frac{3}{4}}$ F $x^{\frac{2}{3}}$

**a** Which of these expressions gives an integer value when:
**i** $x = 9$ **ii** $x = 64$ **iii** $x = 16$?

**b** Find a value for $x$ that gives an integer value for all six expressions.

**9** What is the value of $p$ when $(p^{\frac{2}{3}})^{\frac{3}{2}} = 243$?

**10** A number $y$ is such that $y^{\frac{1}{2}} = \frac{1}{3}$. Find the value of:

a $y^{\frac{3}{2}}$ b $y^{-1}$

**Exercise 7.4**
**Solving equations**

**1** Solve:

a $64^{\frac{1}{x}} = 4$ b $7^{2y} = \frac{1}{49}$ c $1024^{\frac{1}{5}} = 2^z$

d $32^a = 2$ e $125^b = \frac{1}{5}$ f $2^{2c+1} = \frac{1}{128}$

g $3 \times 4^p = 96$ h $\frac{8^q}{4} = 4$ i $8 \times 16^r = 1$

**2** In this puzzle, the solution to each clue is the solution of an equation. Copy and complete the puzzle.

**Across**

**2** $3721^{-\frac{1}{2}} = \frac{1}{x}$

**3** $y^{\frac{2}{5}} = 16$

**6** $z^{\frac{2}{3}} = 121$

**8** $v^{-\frac{1}{3}} = \frac{1}{3}$

**Down**

**1** $w^{-2} = \frac{1}{100}$

**2** $a^{\frac{1}{6}} = 2$

**3** $\left(\frac{1}{b}\right)^{\frac{1}{2}} = \frac{1}{4}$

**4** $c^{\frac{3}{5}} = 27$

**5** $k^{-\frac{3}{4}} = \frac{1}{27}$

**6** $m^{-1} = \frac{1}{17}$

**7** $n^{-\frac{4}{5}} = \left(\frac{1}{2}\right)^4$

**Thinking ahead to ... square roots**

**A** Use your calculator to decide which of these are false.

(A) $\sqrt{20} = \sqrt{10} \times \sqrt{2}$

(B) $\sqrt{20} = \sqrt{17} + \sqrt{3}$

(C) $\sqrt{20} = 2 \times \sqrt{5}$

(D) $\sqrt{20} = \dfrac{\sqrt{40}}{\sqrt{2}}$

(E) $\sqrt{20} = \sqrt{25} - \sqrt{5}$

## Simplifying square roots

The rules for multiplying and dividing square roots also apply to other roots.

For example:
$a^{\frac{1}{3}} \times b^{\frac{1}{3}} = (a \times b)^{\frac{1}{3}}$

Rules for multiplying and dividing square roots are:

◆ $\sqrt{a} \times \sqrt{b} = \sqrt{ab}$

**Example**

$\sqrt{9} \times \sqrt{16} = 3 \times 4$
$= 12$

$\sqrt{9 \times 16} = \sqrt{3 \times 3 \times 4 \times 4}$
$= \sqrt{(3 \times 4) \times (3 \times 4)}$
$= 3 \times 4$
$= 12$

So $\sqrt{9} \times \sqrt{16} = \sqrt{(9 \times 16)}$
$= \sqrt{144}$
$= 12$

◆ $\dfrac{\sqrt{a}}{\sqrt{b}} = \sqrt{\dfrac{a}{b}}$

**Example**

$\dfrac{\sqrt{100}}{\sqrt{4}} = \dfrac{10}{2} = 5$

$\sqrt{\dfrac{100}{4}} = \sqrt{\dfrac{10}{2} \times \dfrac{10}{2}}$

$= \dfrac{10}{2} = 5$

So $\dfrac{\sqrt{100}}{\sqrt{4}} = \sqrt{\dfrac{100}{4}} = \sqrt{25}$

$a\sqrt{b} = a \times \sqrt{b}$

For example:
$2\sqrt{11} = 2 \times \sqrt{11}$

So some expressions involving square roots can be simplified.

**Example**

◆ $(2\sqrt{11})^2 = (2 \times \sqrt{11}) \times (2 \times \sqrt{11})$
$= 2 \times 2 \times \sqrt{11} \times \sqrt{11}$
$= 4 \times 11$
$= 44$

◆ $\sqrt{20} = \sqrt{4} \times \sqrt{5}$
$= 2\sqrt{5}$

◆ $(\sqrt{2} + \sqrt{3})^2 = 2 + \sqrt{6} + \sqrt{6} + 3$
$= 5 + 2\sqrt{6}$

| × | $\sqrt{2}$ + | $\sqrt{3}$ |
|---|---|---|
| $\sqrt{2}$ | 2 | $\sqrt{6}$ |
| $+\sqrt{3}$ | $\sqrt{6}$ | 3 |

Expressions such as $5 + 2\sqrt{6}$ are said to be in **surd form**. They include roots that are not equivalent to integers or fractions.

Greek mathematicians thought of numbers such as $\sqrt{2}$ and $\sqrt{3}$ as absurd.

**Exercise 7.5**
**Simplifying square roots**

In this exercise, do not use a calculator unless stated otherwise.

**1** **a** Which of these is not equivalent to $\sqrt{24}$ ?

(A) $2\sqrt{6}$

(B) $4\sqrt{6}$

(C) $\sqrt{8} \times \sqrt{3}$

(D) $\sqrt{4} \times \sqrt{6}$

(E) $\dfrac{\sqrt{72}}{\sqrt{3}}$

**b** Check your results with a calculator.

**2**

(A) $(\sqrt{5} + \sqrt{2})^2$

(B) $(\sqrt{5} \times \sqrt{2})^2$

(C) $\dfrac{\sqrt{54}}{\sqrt{6}}$

(D) $\dfrac{\sqrt{100}}{\sqrt{5}}$

(E) $\sqrt{6} \times \sqrt{2}$

(F) $\sqrt{8} \times \sqrt{2}$

(G) $(\sqrt{12} - \sqrt{3})^2$

(H) $(4\sqrt{5})^2$

For each of these expressions:

**a** decide if it gives an integer value

**b** explain how you made your decision.

3 Sort these into five pairs of equal value.

(A) $\sqrt{3} \times \sqrt{5}$ (B) $\sqrt{8}$ (C) $\sqrt{2} \times \sqrt{2}$ (D) $2\sqrt{2}$ (E) $(\sqrt{2} + \sqrt{8})^2$

(F) $2$ (G) $(\sqrt{3} - 1)^2$ (H) $\sqrt{15}$ (I) $4 - 2\sqrt{3}$ (J) $18$

When simplifying surds, write square roots in the form $a\sqrt{b}$ so that $a$ and $b$ are integers and $b$ is as small as possible.

4 Simplify as far as you can, giving answers in surd form where necessary.

**a** $\sqrt{40}$ **b** $\sqrt{7} \times \sqrt{3}$ **c** $\dfrac{\sqrt{90}}{\sqrt{10}}$

**d** $\sqrt{2} \times \sqrt{32}$ **e** $\sqrt{3} \times \sqrt{15}$ **f** $(3\sqrt{5})^2$

**g** $(2 + \sqrt{7})^2$ **h** $(\sqrt{6} + \sqrt{12})^2$ **i** $(3 + \sqrt{5})(1 - \sqrt{2})$

**j** $(\sqrt{8} - \sqrt{2})^2$ **k** $(3\sqrt{2} + \sqrt{15})^2$ **l** $(2\sqrt{3})^4$

5 Give the area of each shape and leave your answer in simplified surd form.

**a**

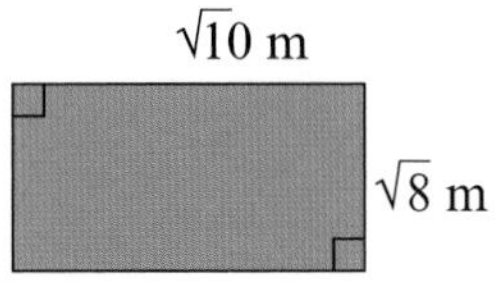


**b**

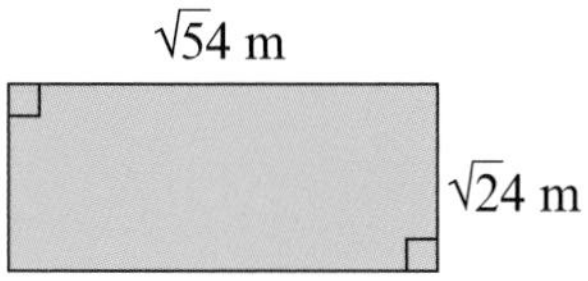


**c**

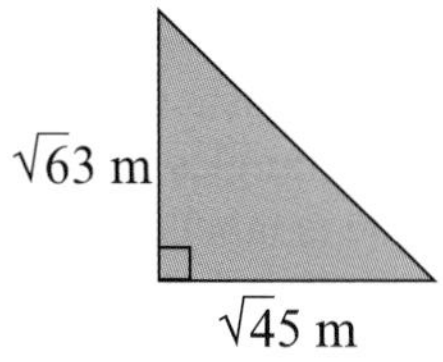


**d**

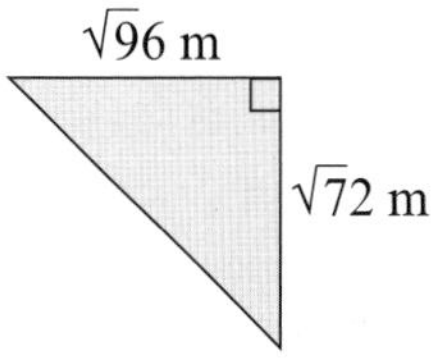


**e**

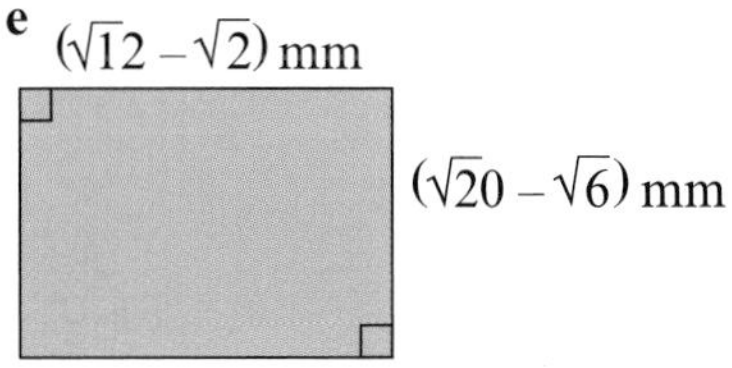


**f**

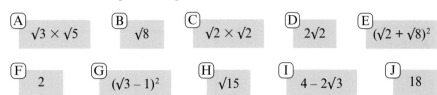

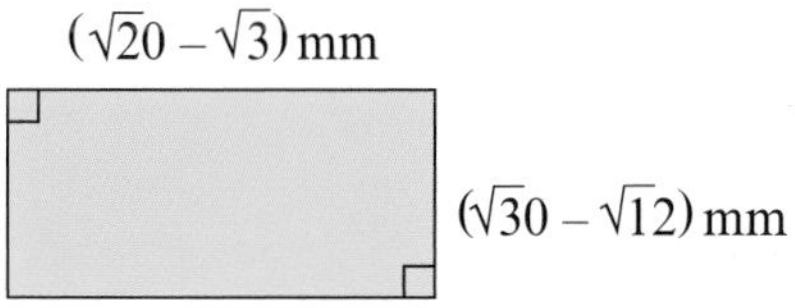


6 Explain why $\dfrac{\sqrt{54}}{\sqrt{216}}$ can be written as $\frac{1}{2}$.

Use surd form in your explanation.

7 A rectangle is $4\sqrt{5}$ cm wide.
The area of the rectangle is given as $48\sqrt{10}$
Find the length of the long side of the rectangle.
Give your answer in simplified surd form.

8 Find the value of:

**a** $(8 \times 10^{21})^{\frac{1}{3}}$ **b** $2^{\frac{1}{4}} \times 8^{\frac{1}{4}}$ **c** $\dfrac{54^{\frac{1}{3}}}{2^{\frac{1}{3}}}$

## End points

You should be able to ... | ... so try these questions

**A** Evaluate roots

**A1** Evaluate:
**a** $\sqrt[5]{243}$ **b** $\sqrt[3]{216}$ **c** the fourth root of 16

**B** Use rules for multiplying and dividing powers

**B1** Copy and complete:
**a** $2^4 \times 2^{\square} = 2^{12}$ **b** $\frac{3^4}{3^{\square}} = \frac{1}{3^2}$ **c** $(7^2)^{\square} = 7^6$

**B2** Simplify:
**a** $2x^3 \times x^4$ **b** $\frac{x^6}{x^2}$ **c** $(3x^3)^4$

**C** Work with indices

**C1** Evaluate these as integers or in fractional form.
**a** $100^{\frac{1}{2}}$ **b** $64^{\frac{2}{3}}$ **c** $32^{-\frac{3}{5}}$

**C2** Solve these equations.
**a** $27^x = 3$ **b** $6^x = \frac{1}{36}$ **c** $81^x = \frac{1}{3}$

**D** Simplify square roots

**D1** Simplify these as far as you can.
**a** $\sqrt{2} \times \sqrt{5}$ **b** $\sqrt{32}$ **c** $(3\sqrt{7})^2$ **d** $(\sqrt{3} - \sqrt{2})^2$

**E** Work with surds

**E1** The length and width of a rectangle are given as $3\sqrt{45}$ and $4\sqrt{30}$. Give the area of the rectangle in simplified surd form.

**E2** This pattern consists of a set of six right-angled triangles as shown. Each triangle is labelled. In surd form, how long is the hypotenuse in:
**a** triangle 1
**b** triangle 2?

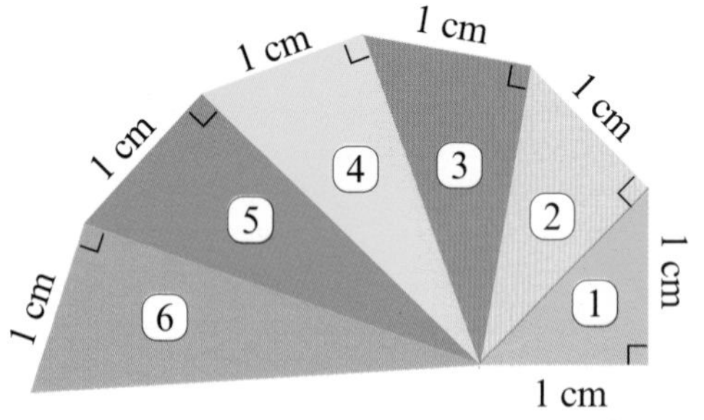


### Some points to remember

Rules for indices:
- $a^m \times a^n = a^{m+n}$
- $a^{-m} = \frac{1}{a^m}$
- $a^m \div a^n = a^{m-n}$
- $a^{\frac{m}{n}} = \sqrt[n]{a^m}$
- $(a^m)^n = a^{mn}$

Rules for square roots
- $\sqrt{a} \times \sqrt{b} = \sqrt{ab}$
- $\sqrt{\frac{a}{b}} = \frac{\sqrt{a}}{\sqrt{b}}$

## Starting points

You need to know about ... / ... so try these questions

### A Decimal places for rounding

- Numbers can be approximated to a set number of decimal places.

Examples

32.3743 to 2 decimal places (2 dp) is **32.37**.
32.3753 to 2 dp is **32.38** (5, or greater, in this position rounds the 7 up).

34.496 to 2 dp is **34.50** (the 6 makes 49 one larger, i.e. 50).

102.499 81 to 3 dp is **102.500** (3 places after the point must be shown).

**A1** Round each number to 2 dp.
**a** 23.4671 **b** 4.1627
**c** 142.855 **d** 12.2962

**A2** Round each number to 1 dp.
**a** 56.48932 **b** 0.33928
**c** 5.9642 **d** 599.999

### B Significant figures for rounding

- Rounding to a set number of significant figures is another form of approximating. This is the most natural way to round.

Examples

32.3743 to 2 significant figures (2 sf) is **32**.
32.5743 to 2 sf is **33** (5, or greater, in this position rounds the 2 up).

34.496 to 3 sf is **34.5**.

0.000 2756 to 2 sf is **0.000 28**.

346 534.7945 to 3 sf is **347 000** and to 7 sf is **346 534.8**.

**B1** Round each number to 2 sf.
**a** 146 **b** 15.634
**c** 3425.7 **d** 6.2521
**e** 0.357 26 **f** 1097 423

**B2** Round each number to 1 sf.
**a** 32.87 **b** 167.2
**c** 0.049 14 **d** 473 524
**e** 96.65 **f** 0.000 086

### C Calculating with indices

- When two index numbers are multiplied their powers are added.

Examples $6^5 \times 6^3 = 6^8$
$10^8 \times 10^{-3} = 10^5$
$4^{-6} \times 4^{-3} = 4^{-9}$

- When two index numbers are divided their powers are subtracted.

Examples $6^7 \div 6^2 = 6^5$
$10^{-5} \div 10^2 = 10^{-7}$

- When an index numbers is itself raised to a further power the powers are multiplied.

Example $(10^3)^4 = 10^{12}$
$(8^{-3})^2 = 8^{-6}$

**C1** Calculate the following leaving your answers in index notation.
**a** $8^7 \times 8^3$ **b** $6^6 \div 6^4$
**c** $10^6 \div 10^8$ **d** $10^{-4} \times 10^2$
**e** $4^{-4} \times 4^{-6}$ **f** $10^{-6} \div 10^{-8}$
**g** $(5^2)^4$ **h** $(6^4)^{-3}$

### D Standard form

- Very large numbers or very small ones are often expressed in standard form. The first part is a number between 1 and 10 and the second part is a power of ten.

Examples

| Number | In standard form |
|---|---|
| 15 000 000 | $1.5 \times 10^7$ |
| 0.000 000 002 54 | $2.54 \times 10^{-9}$ |
| 523 800 | $5.238 \times 10^5$ |
| 0.04 | $4 \times 10^{-2}$ |
| 146.356 | $1.46356 \times 10^2$ |

**D1** Write each number in standard form.
**a** 345 000 **b** 0.004 17
**c** 42.97 **d** 4 million
**e** 0.000 0023 **f** 6413.234

**D2** Convert these standard form numbers to normal numbers.
**a** $6.74 \times 10^3$
**b** $1.5 \times 10^{-4}$
**c** $6.5241 \times 10^2$
**d** $2 \times 10^{-7}$

## Limits of accuracy

The tolerance is the amount by which a value may vary.

The sign ± means **plus or minus**. So D is 74 miles, plus or minus 3 miles.

- Any measurement is only approximate because it is only as accurate as the instrument used to make it.
- Measurements are often given as having a certain tolerance.
  You may measure the distance between Bristol and Oxford as 74 miles but mean it is 74 miles with a tolerance of 3 miles because your mileometer can have an error of 3 miles (i.e. the distance is between 71 and 77 miles).

  This can be written as $\boldsymbol{D = 74 \pm 3}$ or as $\boldsymbol{71 \leqslant D \leqslant 77}$ where $D$ is the distance in miles.

  The numbers 71 and 77 are known as the **bounds** between which the distance can lie.
- Because of the problem of accuracy, measured values are rounded to a certain number of decimal places or significant figures.

  **Example** A pencil measured as 12.6 cm to the nearest millimetre can have a true length anywhere between 12.55 cm and 12.65 cm.

  The value 12.55 is known as the **lower bound** and 12.65 as the **upper bound**.

### Exercise 8.1 Limits of accuracy

1 Steve buys a piece of material of length 3 metres, correct to the nearest centimetre. What is the minimum length of the material?

2 Alison has a piece of wood of length 15.6 cm, correct to the nearest millimetre. What is:

**a** the maximum length it could be
**b** its minimum length?

3 Australia has a land area of 7 700 000 km$^2$ to the nearest 100 000 km$^2$.

**a** What is the upper bound for the land area?
**b** State the lower bound.
**c** Write an inequality in $L$ for the range of values for the land area.

4 The thickness $t$ of a sheet of steel is given as 0.17 mm ± 0.01 mm. Write an inequality in $t$ for the range of values the thickness can have.

Errors or approximations for measured values are often increased when these measurements are used in a calculation.

**Example 1** A machine can cut card to the nearest millimetre. A card is to be cut 12.2 cm wide by 16.5 cm long. What are the maximum and minimum areas the card can have?

The width can lie between 12.15 cm and 12.25 cm and the length between 16.45 and 16.55 cm.

So **the maximum area of card = 12.25 × 16.55 = 202.7375 cm$^2$**
and **the minimum area of card = 12.15 × 16.45 = 199.8675 cm$^2$.**

In this case, the maximum and minimum areas only appear similar when they are both given to 2 sf (i.e. area = 200 cm$^2$ to 2 sf).

In Example 1 the length and width are both given correct to 3 sf but, surprisingly, the answer is only accurate to 2 sf.

A volume would be even more inaccurate as three lengths are multiplied – but it is interesting to consider how an answer can be less accurate than 1 sf!

## Example 2

A cylinder has a volume of $640\,cm^3$ correct to 2 sf.
Ezra measures its height $h$ to the nearest millimetre as 9.3 cm.
Between what bounds does the radius $r$ of the cylinder lie?

The volume of the cylinder is given by $V = \pi r^2 h$

Rearranging gives $r^2 = \dfrac{V}{\pi h}$

| **For upper bound** | **For lower bound** |
|---|---|
| $r^2 = \dfrac{645}{\pi \times 9.25}$ | $r^2 = \dfrac{635}{\pi \times 9.35}$ |
| $r^2 = 22.195\,662\,33\ldots$ | $r^2 = 21.617\,837\,19\ldots$ |
| $\mathbf{r = 4.711\,227\,264\ldots\,cm}$ | $\mathbf{r = 4.649\,498\,596\ldots}$ |

Note that for the upper bound, the greatest value of $V$ is taken but the smallest value of $h$ is used – because dividing by a smaller number gives a larger answer.

Here, the difference between the upper and lower bounds for $r$ is about 6 hundredths of a centimetre.

### Exercise 8.2
### Accuracy in calculations

1 A box of cornflakes weighs 805 grams. The empty box weighs 55 grams. Both weights are only accurate to the nearest 5 grams.
Calculate:

**a** the maximum weight of cornflakes in the box
**b** the minimum weight of cornflakes.

2 Twelve volumes of an encyclopedia are placed on a shelf. Volumes 1 to 6 are each 3.5 cm wide and volumes 7 to 12 are each 4.3 cm wide, all measurements being to the nearest millimetre.

**a** What is the lower bound for the width of all twelve volumes?
**b** Calculate the difference between the upper and lower bounds.

3 A triangular field has the measurements shown. Each side has been measured to the nearest metre.

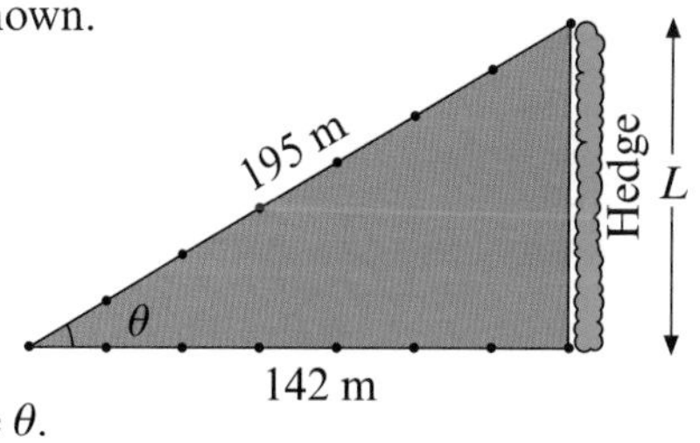


**a** Calculate the range of possible values for the length $L$ of the hedge to the nearest metre.
Express this as an inequality in $L$.
**b** Calculate the lower bound for the angle $\theta$.

4 A rectangular badminton court measures 13.40 metres by 6.10 metres where measurements are taken correct to two decimal places.

Calculate the upper and lower bounds for:
**a** the area of the court
**b** the perimeter of the court.

5 The area of a trapezium is given by $A = \dfrac{h(a + b)}{2}$

where $h$ is the perpendicular height and $a$ and $b$ are the lengths of the parallel sides.

Calculate the range of values for the area of this trapezium if all the measurements are given correct to 1 dp.

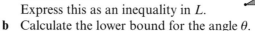

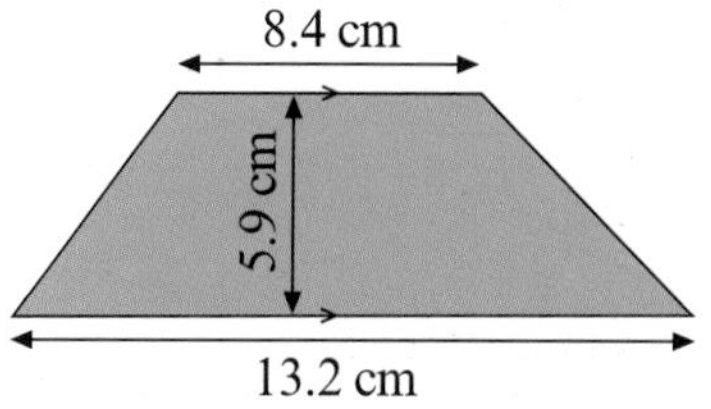

# Using significant figures to estimate answers

One way to check if a calculation gives an answer of about the right size is to round each of the numbers to 1 sf.

**Example**

Here are some answers given by four students when they had to calculate the value of **43 183.5** × **184.23** without using a calculator.

**a** 795 569.6 **b** 79 556 962 **c** 7 955 696 **d** 79 557

Which answer is likely to be most accurate?

To 1 sf these numbers become 40 000 × 200
which is 40 000 × 2 × 100
= 80 000 × 100
= **8 000 000**

Answer **c** (7 955 696) is about the same order of magnitude as 8 000 000 so it is likely to be most accurate.

A value which is of the correct order of magnitude is about the right size.

**Exercise 8.3**
**Estimating answers**

1 Work out estimates of the answers to each of these.
Show all the stages you use.

**a** 31.2 × 241.45 **b** 5677 × 3.764
**c** 54 856 × 83.42 **d** 542 × 52
**e** 56 234 ÷ 82.5 **f** 62 381.23 ÷ 578.23
**g** 452 ÷ 2.34 **h** 28 536 ÷ 0.9623

2 A theatre sells 562 tickets at £28.50 each.

**a** Roughly what is their income from ticket sales?
**b** Why does rounding both numbers to 1 sf give too large an estimate?

3 France has a land area of 549 619 km$^2$.
In 1990 the population was 56 304 000.
Estimate the population density in people per km$^2$.

Population density means the average (mean) number of people to each square kilometre.

4 When the numbers in 341.2 × 14.25 are rounded to 1 sf and then multiplied the estimate is much smaller than the true answer.
For the problem 156 ÷ 34.7 the estimate is much larger.
Explain why.

5 For each of these problems, say if rounding all numbers to 1 sf makes estimates too large, too small, or about the right size.

**a** 56.5 × 1763.2 **b** 184 ÷ 19.6
**c** 491.432 × 2061.4 **d** 445 × 84 632
**e** 2265 ÷ 27.7 **f** 6834 ÷ 14.23
**g** 453 782 + 242 565 **h** 7452.3 – 2837.324

6 For the problem 342 561 + 453, why is rounding to 1 sf not helpful?

7 In problems which only use addition and subtraction, when is rounding to 1 sf useful for finding an estimate?

## Working with numbers less than 1

**Example 1** Estimate the value of **342 × 0.052**.

Approximating to 1 sf, this becomes 300 × 0.05.
The answer to this estimate will be smaller than 300.
One way to work out the value is to look for patterns.

$300 \times 5 = 1500$

$300 \times 0.5 = 150$

$300 \times 0.05 = 15$ **So the estimate is 15.**

**Example 2** Estimate for the value of **26 ÷ 0.0056**.

To 1 sf this is: 30 ÷ 0.006.

$30 \div 6 = 5$

$30 \div 0.6 = 50$

$30 \div 0.06 = 500$

$30 \div 0.006 = 5000$ **So the estimate is 5000.**

**Exercise 8.4**
Calculating and estimating answers

1 **a** Calculate 342 × 52 without a calculator.
**b** Use the example above to help decide what 342 × 0.052 is.

2 Estimate the value of 45 × 0.0023. Show all the stages you use.
Calculate the exact answer and check it with your estimate.

3 Estimate then calculate the values of these. Which are best and worst estimates?

**a** $346.3 \div 0.04$ **b** $26.23 \times 0.67$ **c** $876.2 \times 0.000\,23$

**d** $2.448 \div 0.0018$ **e** $\dfrac{2567.245 \times 0.032}{54.23}$ **f** $\dfrac{78.4567 - 7.64}{0.0421}$

4 Estimate the volumes of these cuboids. Show the rounding you do.
Do not use a calculator.

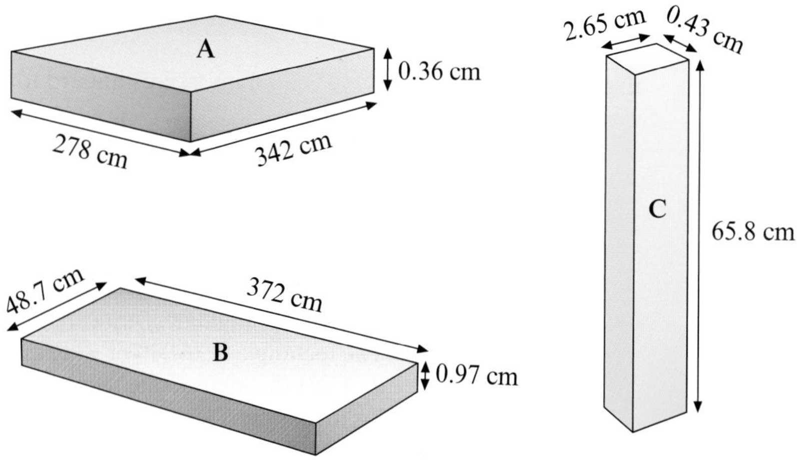


5 A cuboid is 113.6 cm long, 0.42 cm deep and has a volume of 18.42 cm$^3$.
Estimate its width.

## Calculating with numbers in standard form

Calculations with very large numbers can be estimated by approximating each part to a number of significant figures and then using standard form.

**Example 1** The land area of Australia is 7 682 300 km² and its population is 16 260 400. Singapore has a land area of 580 km² and a population of 2 645 400.
How do the population densities compare?

For Australia:
Rounding to 2 sf gives population as 16 000 000 and area as 7 700 000 km².

$$\text{Population density} = \frac{\text{Population}}{\text{Land area}} = \frac{1.6 \times 10^7}{7.7 \times 10^6} = 0.21 \times 10 \text{ (to 2 sf)}$$

= 2.1 people per kilometre².

For Singapore:
Rounding to 2 sf gives population as 2 600 000 and area as 580 km².

$$\text{Population density} = \frac{2.6 \times 10^6}{5.8 \times 10^2} = 0.45 \times 10^4 \text{ (to 2 sf)}$$

= 4500 people per kilometre².

Singapore therefore has a population density over 2000 times as great as that of Australia.

When powers of ten are divided their indices are subtracted. For example:

$10^7 \div 10^6 = 10^1$ or 10

$10^6 \div 10^2 = 10^4$

**Example 2** The mass $M$ of Jupiter is $1.25 \times 10^{27}$ kg and it has a radius of 71 942 km. Use this formula to find its density.

$$\text{Density} = \frac{M}{\frac{4}{3}\pi r^3}$$

$$\text{Density} = \frac{1.25 \times 10^{27}}{\frac{4}{3} \times \pi \times (7.2 \times 10^4)^3} = \frac{3 \times 1.25 \times 10^{27}}{4 \times \pi \times (7.2)^3 \times 10^{12}}$$

$= 0.000\,799\,51 \times 10^{15}$.

$= \mathbf{8 \times 10^{11}}$ **kg km⁻³ (2 sf).**

When powers of ten are raised to a further power the indices are multiplied. For example:

$(10^4)^3 = 10^{12}$

**Exercise 8.5**
**Standard form**

1 Write each number in standard form.

**a** 3647.5 **b** 102 072.62 **c** 576 174.87
**d** 0.056 **e** 0.000 756 6 **f** 0.000 006 25

2 Give each value, not in standard form.

**a** $3.746 \times 10^5$ **b** $5.808 \times 10^3$ **c** $3.42 \times 10^7$
**d** $5.675 \times 10^{-2}$ **e** $6.388 \times 10^{-5}$ **f** $9.02 \times 10^{-9}$

3 A large rectangular of moorland is chosen as a part of a survey.
The rectangle is 36 040 m wide and 38 220 m long.
**a** Give the area of the rectangle in standard form.
**b** In standard form give the perimeter of the rectangle.

**Exercise 8.6**
**Using standard form**

1 When lightning strikes the Earth there is a lead stroke which travels down at about $1.6 \times 10^2$ km per second and a brighter return stroke which flashes up at about $1.4 \times 10^5$ km per second.
How many times faster is the return stroke than the lead stroke?

**2** The Earth is approximately $1.5 \times 10^8$ km from the Sun. Light travels at about $3 \times 10^5$ km per second.

How long does it take light from the Sun to reach the Earth?

**3** The Earth can be considered to have a circular orbit around the Sun of radius 149 600 000 km. Earth travels through space at 107 244 km per hour.

**a** Approximate the speed and radius to 3 sf and write each in standard form.

**b** Use your figures to find roughly how long in hours the Earth takes to make one orbit. What do we call this length of time?

**4** A ream of paper is a stack of 500 sheets. Each sheet of paper is 0.000 096 metres thick.

**a** Write 500 and 0.000 096 in standard form.

**b** Use your answers to part **a** to calculate the height of a ream of paper.

**5** The mass of one atom of carbon is about $2 \times 10^{-23}$ grams and the mass of an oxygen atom is $2.66 \times 10^{-23}$ grams. A molecule of carbon dioxide gas ($CO_2$) has two atoms of oxygen and one atom of carbon.

**a** Calculate the mass of one carbon dioxide molecule.

**b** Calculate the number of molecules in 1 gram of carbon dioxide gas.

**6** The Amazon river discharges $1.05 \times 10^{13}$ litres of water per minute into the Atlantic. There are estimated to be about $5.74 \times 10^9$ people alive today.

How long would it take the Amazon to fill a bath of 80 litres for every person on Earth?

**7** The largest gasholder in the world is at Simmering near Vienna. It is shaped like a cylinder, stands 84 metres tall, and holds $3 \times 10^5$ cubic metres of gas. Calculate the diameter of the gasholder in metres giving your answer in standard form.

The volume $V$ of a cylinder is given by:

$$V = \pi r^2 h$$

where $h$ is its height and $r$ is its base radius.

**8** $4 \times 10^4$ tonnes of cosmic dust are estimated to fall on the Earth each year. The Earth has a diameter of about $1.3 \times 10^7$ metres.

**a** Taking the Earth as a sphere, estimate its surface area giving your answer in standard form to 2 sf.

**b** How much cosmic dust falls on each square metre of the Earth's surface each year?

The surface area $A$ of a sphere is given by:

$$A = 4\pi r^2$$

where $r$ is its radius.

## End points

You should be able to ... ... so try these questions

**A** Understand upper and lower bounds

**A1** A cylinder has a diameter of 6.7 cm and a height of 9.8 cm, both measurements given to the nearest millimetre.

**a** What is the upper bound of the diameter?

**b** What is the lower bound of the cylinders height?

**c** Calculate the range of values for the volume of the cylinder.

**d** Calculate the upper bound for the curved surface area of the cylinder. (Hint: the curved surface area of a cylinder of radius $r$ and height $h$ is $2\pi rh$).

**B** Use significant figures to estimate answers

**B1** By rounding numbers to 1 significant figure estimate the value of each of these:

**a** $6453 \times 48.734$ **b** $865.27 \div 0.000\,342$

**c** $\dfrac{4.645 \times 0.000\,32}{34.82}$ **d** $\dfrac{0.259 \times 0.422}{0.000\,772}$

**C** Calculate with numbers in standard form

**C1** Give each number in a standard form.

**a** 360 000 **b** 100 700 **c** 0.000 007 57

**C2** Calculate $346\,572.8 \times 0.000\,02$ giving the answer in standard form.

**C3** The Strahov sports stadium in Prague can hold $2.4 \times 10^5$ spectators. If the average mass of a spectator is $5.4 \times 10^{-2}$ tonnes calculate the total mass of the spectators when the stadium is full.

## Some points to remember

- When a number is written as say 8.6 correct to 1 decimal place its true value can actually lie anywhere between 8.55 and 8.65.
8.55 is known as the **lower bound** and 8.65 as the **upper bound**.
- One method to calculate an estimate of the answer to a calculation is to round every number to 1 sf at the start. Beware: it is not sensible to give these answers to more than 1 sf.

## Starting points

You need to know about ...

... so try these questions

### A Expanding brackets

Expressions are often written including brackets.
By multiplying the terms you can remove, or expand, the brackets.

To remove the brackets from the expression $(y + 4)(y - 5)$ you can:

- Use a table

| $\times$ | $y$ | $+4$ |
|---|---|---|
| $y$ | $y^2$ | $+4y$ |
| $-5$ | $-5y$ | $-20$ |

The table gives: $y^2 + 4y - 5y - 20$
Simplified gives: $\mathbf{y^2 - y - 20}$
So removing the brackets we have: $(y + 4)(y - 5) = y^2 - y - 20$

- Multiply each term in the second bracket by each term in the first bracket.

This gives: $(y + 4)(y - 5) = y(y - 5) + 4(y - 5)$
$= y^2 - 5y + 4y - 20$
$= \mathbf{y^2 - y - 20}$

With a systematic approach you can deal with more than two brackets.
Remove the brackets from the expression $(n + 2)(n - 3)(2n + 1)$:

$$(n + 2)(n - 3)(2n + 1) = (n + 2)(2n^2 - 5n - 3)$$
$$= 2n^3 - 5n^2 - 3n + 4n^2 - 10n - 6$$
$$= 2n^3 - n^2 - 13n - 6$$

**A1** Expand the brackets and simplify these expressions.
**a** $(y + 3)(y + 5)$
**b** $(k - 4)(k + 3)$
**c** $(n + 3)(n - 9)$
**d** $(w - 5)(w - 4)$
**e** $(v - 8)(v + 8)$

**A2** Remove the brackets and simplify these expressions.
**a** $(2n + 3)(n - 4)$
**b** $(w - 8)(3w + 2)$
**c** $(3y - 4)(4y - 3)$
**d** $(2k - 3)^2$
**e** $(3v + 1)(v^2 - 3)$

**A3** Remove the brackets and simplify these expressions.
**a** $2y(y - 3)(y + 2)$
**b** $3x(x + 1)^2$
**c** $(x + 1)(x - 3)(x + 5)$
**d** $(2n - 1)(2n - 3)(n + 1)$
**e** $(x + 2)^3$
**f** $(k - 3)(4 - 2k)(k + 2)$

### B Factorising quadratic expressions

To factorise an expression is to find two, or more, other expressions which when multiplied together produce the original expression.

- As $(y + 4)(y - 5) = y^2 - y - 20$
we say that $(y + 4)$ and $(y - 5)$ are factors of $y^2 - y - 20$
Factorising $y^2 - y - 20$ gives: $y^2 - y - 20 = (y + 4)(y - 5)$

- You can factorise expressions like $3n^2 + 2n - 8$ where the coefficient of $n^2$ is greater than 1.
Factorising $3n^2 + 2n - 8$ gives: $3n^2 + 2n - 8 = (3n - 4)(n + 2)$

- In general terms if you think of a quadratic expression as:

$$ax^2 + bx + c$$

Factorising gives:

$$(\text{factor of } ax + \text{factor of } c)(\text{factor of } ax + \text{factor of } c)$$

We must ensure that the factors of $a$ and the factors of $c$ are chosen so that they can be combined to give the value of $b$.

Factorising the expression $5x^2 - 8x - 4$ gives:

$$5x^2 - 8x - 4 = (5x + 2)(x - 2)$$

The factors of 5 are $5 \times 1$, the factors of 4 are $4 \times 1$, or $2 \times 2$
A total of $^-8$ cannot be made combining 5 and 1 with 4 and 1.
Combining 5 and 1 with 2 and 2 we have:

$$5 \times {}^-2 = {}^-10,\ 1 \times 2 = 2 \text{ and } {}^-10 + 2 = {}^-8$$

**B1** Factorise these expressions.
**a** $y^2 + 9y + 14$
**b** $k^2 - 3k - 40$
**c** $x^2 + 4x - 12$
**d** $w^2 + 5w - 36$
**e** $x^2 - 11x + 24$
**f** $y^2 - 9y + 8$
**g** $p^2 + 7p - 60$
**h** $b^2 + b - 72$
**i** $x^2 - 16x + 48$
**j** $n^2 - 25$
**k** $y^2 - 100$
**l** $a^2 - b^2$
**m** $4y^2 - 36$
**n** $16x^2 - 36y^2$
**o** $100a^2 - b^4$

## C Solving linear equations

Linear equations can be solved by manipulating the terms of the equation so that the unknown is equated to a value.

Manipulation can be thought of as keeping the equation balanced. This can be seen when the equation $5x - 8 = 3x + 5$ is solved.

| | |
|---|---|
| | $5x - 8 = 3x + 5$ |
| [+8] | $5x = 3x + 13$ |
| [$-3x$] | $2x = 13$ |
| [÷2] | $\mathbf{x = 6.5}$ |

Manipulation can also involve dealing with brackets or fractions.

Dealing with brackets.

| | |
|---|---|
| Solve the equation | $3(4x + 1) = 5(x - 3)$ |
| [Remove the brackets] | $12x + 3 = 5x - 15$ |
| [$-3$] | $12x = 5x - 18$ |
| [$-5x$] | $7x = {}^{-}18$ |
| [÷7] | $\mathbf{x = {}^{-}2.57}$ (2 dp) or $x = \frac{^{-}18}{7}$ |

Dealing with fractions.

| | |
|---|---|
| Solve the equation | $\frac{3}{4}(x - 2) = \frac{1}{2}(3x - 4)$ |
| [Multiply each side by 4] | $3(x - 2) = 2(3x - 4)$ |
| [Remove the brackets] | $3x - 6 = 6x - 8$ |
| [+6] | $3x = 6x - 2$ |
| [$-6x$] | ${}^{-}3x = {}^{-}2$ |
| [÷$^{-}3$] | $\mathbf{x = 0.67}$ (2 dp) or $x = \frac{2}{3}$ |

## D Forming linear equations from problems

A problem given in words can often be solved by creating and solving a linear equation. This will be possible if one unknown is chosen and any other unknowns expressed in terms of the one unknown.

This can be seen when this problem is solved.

Lisa is 4 years younger than her sister Emma and 19 years older than her daughter Nicole. Together their ages total 96. How old is each of them?

Let Lisa be $n$ years old.
Emma is 4 years older than Lisa i.e. $n + 4$
Nicole is 19 years younger than Lisa i.e. $n - 19$

Their ages total 96, so we have:

| | |
|---|---|
| | $n + n + 4 + n - 19 = 96$ |
| | $3n - 15 = 96$ |
| [+15] | $3n = 111$ |
| [÷3] | $n = 37$ |

This gives:

| | |
|---|---|
| **Lisa ($n$) is** | **37 years old** |
| **Emma ($n + 4$) is** | **41 years old** |
| **Nicole ($n - 19$) is** | **18 years old** |

You can check your answer to this sort of problem, as follows.

The total of the three ages is:

$$37 + 41 + 18 = 96$$

and 96 is the total for the ages given in the problem.

**C1** Solve these linear equations.
- **a** $9y + 5 = 5y - 13$
- **b** $y - 8 = 11y - 42$
- **c** $3(4x - 5) = 9$
- **d** $28 = 7(3x - 5)$
- **e** $19 - 4x = 3(2x - 7)$

**C2** Solve these linear equations.
- **a** $5(4x + 2) = 3(2x - 6)$
- **b** $8(2 - 3n) = 4(2n + 1)$
- **c** $7(w + 3) = 12(w - 7)$
- **d** $2(3k + 3) = 9(5k - 8)$
- **e** $6(9v - 2) = 5(6v + 8)$

**C3** Solve these linear equations.
- **a** $\frac{2}{3}(3y + 1) = 18$
- **b** $\frac{3}{5}(3u - 4) = 12$
- **c** $\frac{5}{8}(2k + 4) = 15$
- **d** $\frac{1}{4}(8h - 12) = 21$
- **e** $\frac{7}{10}(4a - 5) = 2.1$

**C4** Solve these linear equations.
- **a** $\frac{3}{5}(2w - 1) = \frac{1}{3}(3w - 6)$
- **b** $\frac{2}{3}(k + 1) = \frac{3}{4}(2k - 3)$
- **c** $\frac{1}{2}(8n - 4) = \frac{1}{3}(4n - 5)$

**D1** A rectangle is 6 centimetres longer than it is wide.
The rectangle has a perimeter of 48 cm.

Write and solve an equation to find the length and width of the rectangle.

**D2** In a game a player was asked to think of a number and add seven, then to multiply the answer by three.
The player answered 81.

Write and solve an equation to find the number the player first thought of.

**D3** Three buses took fans to a match. Bus A carried 9 more fans than bus B. Bus C carried 3 fewer fans than bus A.

The three buses carried a total of 147 fans.

Write and solve an equation to find the number of fans on each bus.

**Thinking ahead to ...**
solving simultaneous linear equations

**A** In a cafe, two teas and 4 coffees cost £4.60.
From this information:

- **a** Calculate the cost of one tea and two coffees. Explain your method.
- **b** Calculate the cost of three teas and six coffees. Explain your method.
- **c** Explain why the cost of six teas and eight coffees cannot easily be calculated.

**B** In a shop:
1 cola and 3 bags of crisps cost £1.52
2 colas and 4 bags of crisps cost £2.48.
Find the cost of:

- **a** 3 colas and 7 bags of crisps
- **b** 1 cola and 1 bag of crisps
- **c** 1 bag of crisps
- **d** 1 cola.

## Solving simultaneous linear equations

When a linear relationship between two variables can be expressed by two or more different equations, you can find the value of each variable by solving the equations simultaneously.

One way to solve these equations is algebraically, the most common being a combination of elimination and substitution.

**Example**

Find the values of $a$ and $b$ that satisfy the equations:

$$2a + b = 19$$
$$3a + 4b = 26$$

- Label the equations (1) and (2):

$$2a + b = 19 \text{ ... (1)}$$
$$3a + 4b = 26 \text{ ... (2)}$$

- Eliminate one variable by making the coefficients of $a$ or $b$ the same in both equations.
In this case multiply equation (1) by 4 (to eliminate $b$)

$$8a + 4b = 76 \text{ ... (3)}$$
$$3a + 4b = 26 \text{ ... (2)}$$

The coefficients of $b$ are now 4 in each of the equations (3) and (2).

- Subtract (2) from (3) to eliminate $b$

$$5a = 50$$

- Find the value of $a$:

$$\mathbf{a = 10}$$

- Substitute the value of $a$ in one equation

$$2a + b = 19$$
$$2(10) + b = 19$$
$$20 + b = 19$$
$$\mathbf{b = {}^{-}1}$$

- Check your solution in one other equation
When $a = 10$ and $b = {}^{-}1$

$$3a + 4b = 26 \text{ ... (2)}$$

LHS gives $3(10) + 4({}^{-}1) = 26$ = RHS

LHS means left-hand side. RHS means right-hand side.

**Exercise 9.1**
Solving simultaneous equations

**1** Solve each pair of equations simultaneously to find values for $x$ and $y$.

| | | |
|---|---|---|
| **a** $x + 4y = 42$<br>$2x + 5y = 57$ | **b** $11x + 3y = 91$<br>$3x + y = 25$ | **c** $5x + 7y = 44$<br>$x + 3y = 12$ |
| **d** $5x + 3y = 7$<br>$4x + y = 7$ | **e** $12x + 5y = {}^{-}9$<br>$5x + y = {}^{-}7$ | **f** $8x + y = 20$<br>$11x + 4y = 17$ |

In these equations the only way to make the coefficients of one variable the same, is to multiply each equation by a different value.

This means that you create two new equations which you can label (3) and (4).

**2** To find a value of $m$ and $n$ that satisfies these equations:

$$2m + 3n = 28 \ldots (1)$$
$$3m + 4n = 37 \ldots (2)$$

- **a** Multiply equation (1) by 3.
- **b** Multiply equation (2) by 2.
- **c** Subtract and find a value for $n$ that satisfies both equations.
- **d** Substitute for $n$ in one equation to find a value for $m$.
- **e** Check your solutions are correct.

**3** Solve each pair of equations.

| | | |
|---|---|---|
| **a** $2v + 3w = 40$<br>$5v + 2w = 34$ | **b** $3v + 2w = 3$<br>$6v + 10w = 24$ | **c** $4v + 2w = 8$<br>$3v + 7w = {}^{-}5$ |
| **d** $3v + 5w = 19$<br>$2v + 4w = 14$ | **e** $2v + 3w = 3$<br>$3v + 5w = 4$ | **f** $8v + 4w = 16$<br>$3v + 3w = 3$ |
| **g** $6v + 3w = 9$<br>$4v + 5w = 3$ | **h** $11v + 2w = 17$<br>$2v + 3w = 11$ | **i** $7v + 4w = 1$<br>$2v + 3w = 4$ |

Sometimes to eliminate one variable when the coefficients are the same you have to add two equations.

**Example**

Find the values of $x$ and $y$ that satisfy the equations:

$$6x - 2y = 18 \ldots (1)$$
$$5x + 3y = 1 \ldots (2)$$

- Make the coefficients of $y$ the same by multiplying: (1) × 3 and (2) × 2 — $18x - 6y = 54 \ldots (3)$, $10x + 6y = 2 \ldots (4)$
- Eliminate $y$ by adding: (3) + (4) — $28x = 56$
- Find $x$ — $\mathbf{x = 2}$
- Substitute $x = 2$ in (1) — $6(2) - 2y = 18$, $12 - 2y = 18$, $-2y = 6$, $\mathbf{y = {}^{-}3}$
- Check $x = 2$ and $y = {}^{-}3$ in equation (2) — $5x + 3y = 1 \ldots (2)$

$$\text{LHS} = 5(2) + 3({}^{-}3) = 1 = \text{RHS}$$

**Exercise 9.2**
**Solving simultaneous equations**

**1** Solve each pair of equations.

| | | |
|---|---|---|
| **a** $a - b = 8$<br>$4a + b = 42$ | **b** $5b + 2a = 29$<br>$b - 2a = 1$ | **c** $3a - b = 15$<br>$4a + 2b = 10$ |
| **d** $b + 3a = 2$<br>$3b - a = 26$ | **e** $5a + 2b = 17$<br>$2a - 3b = 3$ | **f** $7a + 5b = 22$<br>$3a - 2b = 26$ |
| **g** $3a + 2b = 4$<br>$2a - 3b = 7$ | **h** $5a - 2b = 24$<br>$2a - 7b = 22$ | **i** $7a + 3b = 15$<br>$5a - 4b = 23$ |
| **j** $4a + 5b = 13$<br>$3a + 2b = 8$ | **k** $3a - 4b = 18$<br>$2a - 5b = 19$ | **l** $5b + 3a = 17$<br>$7b - 4a = 32$ |

**2** Find values for $k$ and $n$ that satisfy $5k - n = 15$ and $3k - n = 5$.

In some cases you will need to rearrange one or more of the equations before you can begin to solve them simultaneously.

3 Find values for $m$ and $n$ that satisfy the equation $5m - 2n = 28$, and also the equation $7m - 37 = 5n$.

4 Solve each pair of equations.

| | | |
|---|---|---|
| **a** $5x = 16 + 2y$<br>$2x = 3y + 2$ | **b** $4y = x + 17$<br>$3y - 4x = 3$ | **c** $5x = 7 + 3y$<br>$2x = 4y$ |
| **d** $5y = 7 - 3x$<br>$7x = 3 - 5y$ | **e** $3y = 17 - 4x$<br>$5x = 7y + 32$ | **f** $3x = 13 - 2y$<br>$59 + 6y = 5x$ |
| **g** $7y = 3x - 5$<br>$5y - 2 = 4x$ | **h** $y + 31 = 9x$<br>$5x - 27 = 3y$ | **i** $12x = 13 - 11y$<br>$5x + 8y = 2$ |
| **j** $x = y + 8$<br>$29 + 4y = 3x$ | **k** $4y = 5 - 3x$<br>$12 = 5y - 2x$ | **l** $7y = 7 - 7x$<br>$11 - 2x = 5y$ |

## Solving problems with simultaneous equations

From some problems you can create two equations with the variables. Solving the equations simultaneously will produce values that satisfy the problem.

**Example**

Two groups of friends met in a cafe. One group bought three teas and four coffees, this cost them a total of £5.35. The other group paid a total of £5.80 for five teas and three coffees.
Find the price charged in the cafe for coffee and tea.

For ease of working you may prefer to work with integer values.

Here the 535 is the total for the first group in pence.

- Decide on the variables: let the price of coffee be $c$ (pence) and the price of tea be $t$ (pence)
- Create two different equations.

for the first group $3t + 4c = 535$
the second group $5t + 3c = 580$

- Solve the equations:

$3t + 4c = 535$ ... (1)
$5t + 3c = 580$ ... (4)

Multiply (1) by 3 and (2) by 4 to eliminate $c$:

$9t + 12c = 1605$ ... (3)
$20t + 12c = 2320$ ... (4)

Subtract: (4) – (3) $\quad 11t = 715$

Find $t$ $\quad$ $\mathbf{t = 65}$

Substitute $t = 65$ in (1)

$3t + 4c = 535$ ... (1)
$3(65) + 4c = 535$
$195 + 4c = 535$

Find $c$

$4c = 340$
$\mathbf{c = 85}$

- Check in equation (2) $\quad 5t + 3c = 580$
LHS = 5(65) + 3(85) = 325 + 255 = 580 = RHS

**The charge for coffee was 85 pence, and for tea was 65 pence.**

**Exercise 9.3**
**Solving problems**

1 Two groups visited Waterworld. The first group of four adults and three children paid a total of £38 for their tickets. The second group of five adults and two children £40.50 for their tickets.
What are the charges for adult and child tickets at Waterworld?

When you check your answers it is always best to use the original word problem.

This is because you may have created an incorrect equation. This will not be shown up if you substitute your solutions in anything other than the original word problem.

2 Crispers is a snack food sold in bags of two sizes: regular and jumbo. Three regular bags and four jumbo bags weigh a total of 264 g. Five of each size weigh a total of 350 g.

Find the weight of each size of Crispers.

3 A bag contains 89 marbles. Some are large; some are small. Each small marble weighs 2 g, and each large marble weighs 5 g. The total weight of the marbles in the bag is 256 g.

**a** Write two different equations that describe this situation.
**b** Solve your equations to find the number of each marble type in the bag.

4 Mina pays her gas bill with a mixture of £5 and £10 notes. Her gas bill is for £155, and she pays with a total of 22 notes.

How many of each type of note does she use to pay the bill?

5 A parking meter takes 10 p and 20 p coins. The meter was emptied and found to contain 380 coins with a total of £63.70.

How many of each type of coin were in the meter?

6 Two numbers are such that when you add three times one number to five times the other number a total of 46 is produced. The difference between the two numbers is 2.

What are the two numbers?

7 The line $y = mx + c$ passes through the points (3,10) and (5, 18). Find the values of $m$ and $c$.

8 A promoter sold 28 500 tickets for a pop concert. A standing ticket cost £8, and a ticket for a seat cost £12.50. All the tickets were sold and ticket sales produced a total of £283 350.

How many of each type of ticket were sold?

9 The line $mx + y = c$ passes through the points (2, 2) and ($^-1$, 11).

**a** Find values of $m$ and $c$.
**b** Give the equation of the line and one other point that lies on the line.

10 The curve $y = ax^2 + bx + c$ passes through the points (3, 13), (1, $^-1$) and (0, $^-5$).

**a** Find values for $a$, $b$, and $c$
**b** Give the equation of the curve.

11 The curve y= $ax^2 + bx + c$ passes through the points (1, 0), (0, 1) and ($^-1$, 6).

**a** Find values for $a$, $b$, and $c$.
**b** Give the equation of the curve.

12 The curve $y - bx = ax^2 + c$ passes through the points (2, 6), ($^-1$, 0),($^-2$, 10).

**a** Find values for $a$, $b$, and $c$.
**b** Give the equation of the curve.

13 Print Express print T-shirts. They have two printing machines: machine A prints 64 T-shirts an hour, and machine B prints 44 T-shirts an hour. Print Express have an order for 1430 T-shirts and only a total of 25 hours to complete the printing.

For how many hours should they print T-shirts on each machine?

14 Two numbers are such that a half the smaller number added to two thirds of the larger number gives a total of sixteen. What are the two numbers if they have a difference of ten?

## Solving quadratic equations by inspection (factors)

Quadratic equations must have a term like $x^2$. Often they will have a term like $3x$ and a term which is a numerical value.
In general, quadratic equations have two different solutions or roots.

**Example**

- Solve the equation $x^2 + 5x - 14 = 0$
- Factorise $(x + 7)(x - 2) = 0$

Here we have a situation where two brackets are multiplied to give an answer of zero.
This is only possible if either, or both, of the brackets is equal to zero.

In this case if $(x + 7)(x - 2) = 0$
then: either $(x + 7) = 0$ or $(x - 2) = 0$
which means $x = {}^-7$ or $x = 2$

- Solving the equation $x^2 + 5x - 14 = 0$
gives $\mathbf{x = {}^-7}$ **or** $\mathbf{x = 2}$

**Exercise 9.4**
Solving quadratic equations by inspection

1 Solve these quadratic equations.

| | | |
|---|---|---|
| **a** $y^2 + 5y + 4 = 0$ | **b** $y^2 + 9y + 18 = 0$ | **c** $y^2 + 5y - 24 = 0$ |
| **d** $x^2 + 6x - 7 = 0$ | **e** $x^2 + 5x - 36 = 0$ | **f** $x^2 + 2x - 63 = 0$ |
| **g** $w^2 - 7w - 44 = 0$ | **h** $w^2 - 8w - 105 = 0$ | **i** $w^2 - 9w - 112 = 0$ |
| **j** $a^2 - 8a + 15 = 0$ | **k** $a^2 - 12a + 35 = 0$ | **l** $a^2 - 11a + 30 = 0$ |
| **m** $b^2 - 7b + 10 = 0$ | **n** $b^2 + 31b + 84 = 0$ | **o** $b^2 - 13b - 30 = 0$ |
| **p** $c^2 + 7c - 60 = 0$ | **q** $c^2 - 14c + 45 = 0$ | **r** $c^2 + 12c - 45 = 0$ |
| **s** $h^2 + h - 210 = 0$ | **t** $h^2 - 2h - 399 = 0$ | **u** $h^2 - 17h + 42 = 0$ |
| **v** $n^2 + 3n - 180 = 0$ | **w** $n^2 - 15n + 56 = 0$ | **x** $n^2 - 9n - 136 = 0$ |

The basis of solving quadratics in this way is being able to factorise the quadratic equation that is given.
Remember that the equation might not be in the form: $ax^2 + bx + c = 0$

**Example**

Solve the equation $x^2 + 9x = 0$

- Factorise $x(x + 9) = 0$

which means $\mathbf{x = 0}$ **or** $\mathbf{x = {}^-9}$

**Example**

- Solve the equation $x^2 - 49 = 0$
- Factorise

$x^2 - 49$ is known as a difference of two squares the squares being $x^2$ and 49 (i.e. $7^2$)

$x^2 - 49 = 0$
gives $(x + 7)(x - 7) = 0$
which means $\mathbf{x = {}^-7}$ **or** $\mathbf{x = 7}$

Factorising a difference of two squares always produces a result in the same format.

Factorising $a^2 - b^2$ gives $(a + b)(a - b)$

**Exercise 9.5**
Solving quadratic equations

1 Solve these quadratic equations.

| | | |
|---|---|---|
| **a** $y^2 + 8y = 0$ | **b** $y^2 - 5y = 0$ | **c** $y^2 - 18y = 0$ |
| **d** $h^2 - 100 = 0$ | **e** $h^2 - 25 = 0$ | **f** $h^2 - 81 = 0$ |

**2** Solve these quadratic equations.

**a** $x^2 - 12x = 0$ **b** $x^2 + 79x = 0$ **c** $3x^2 + x = 0$
**d** $4x^2 - x = 0$ **e** $x^2 - 169 = 0$ **f** $x^2 - x = 0$
**g** $5x^2 - 2x = 0$ **h** $4x^2 + 5x = 0$ **i** $7x^2 + 2x = 0$

Quadratic equations, where the coefficient of (say) $x^2$ is greater than 1, can also be solved by factorising.

**Example**

Solve the equation $6x^2 - 17x + 12 = 0$

◆ Factorise $(3x - 4)(2x - 3) = 0$

which means $\mathbf{x = \frac{4}{3}}$ **or** $\mathbf{x = \frac{3}{2}}$

**Exercise 9.6**
**Solving quadratic equations**

**1** Solve these quadratic equations.

**a** $12x^2 - 5x - 2 = 0$ **b** $6x^2 - 13x - 5 = 0$ **c** $4x^2 - 4x - 3 = 0$
**d** $10y^2 - 3y - 1 = 0$ **e** $9y^2 + 9y - 4 = 0$ **f** $6y^2 - 7y + 2 = 0$
**g** $9h^2 + 3h - 2 = 0$ **h** $14a^2 - 12a - 2 = 0$ **i** $12a^2 + 2a - 4 = 0$
**j** $5c^2 - 11c + 2 = 0$ **k** $6c^2 + 16c - 6 = 0$ **l** $3c^2 - 16c + 16 = 0$
**m** $4w^2 - 4w - 24 = 0$ **n** $6w^2 - 16w + 8 = 0$ **o** $12w^2 - 22w - 4 = 0$

**2** Solve these equations.

**a** $12x^2 - 6x = 0$ **b** $4x^2 - 12x = 0$ **c** $56x^2 + 14x = 0$
**d** $5y^2 = 8y$ **e** $16y^2 = 8y$ **f** $25y^2 - 1 = 0$
**g** $k^2 - \frac{1}{16} = 0$ **h** $3k^2 - 27k = 0$ **i** $5k = k^2$
**j** $w^2 = w$ **k** $w^2 - \frac{1}{9} = 0$ **l** $3w^2 - 4w = 0$
**m** $c^2 = 31c$ **n** $42c^2 - 18c = 0$ **o** $16c^2 - 9 = 0$

## Solving quadratic equations using the formula

In the formula:
$a$ is the coefficient of the term in $x^2$

$b$ is the coefficient of the term in $x$

$c$ is the numerical or (constant) value.

When you use the formula you need to check that:
$b^2 - 4ac$
gives a positive result.

If $b^2 - 4ac$ gives a negative answer then √(negative) is not real. We say the equation has no real roots, so no solution is found.

If a quadratic equation cannot be solved by factorising, a formula can be used to provide the solutions.

To solve the equation $ax^2 + bx + c = 0$ for $x$ this formula is used:

$$x = \frac{^-b \pm \sqrt{(b^2 - 4ac)}}{2a}$$

**Example**

Solve the equation $3x^2 + 4x - 1 = 0$

◆ Using the formula:
matching coefficients we have $a = 3, b = 4, c = {}^-1$

The formula gives $x = \frac{^-4 \pm \sqrt{16 - (4 \times 3 \times {}^-1)}}{(2 \times 3)}$

$$x = \frac{^-4 \pm \sqrt{16 - ({}^-12)}}{6} = \frac{^-4 \pm \sqrt{28}}{6}$$

$$x = \frac{^-4 \pm 5.292}{6}$$

So $x = \frac{^-4 + 5.292}{6} = \mathbf{0.22}$ (2 dp) or $x = \frac{^-4 - 5.292}{6} = \mathbf{{}^-1.55}$ (2 dp)

**Exercise 9.7**
**Solving quadratic equations using the formula**

1 Use the formula to solve these quadratic equations.
Give your answers correct to two decimal places (2 dp).

**a** $2x^2 + 3x - 4 = 0$ **b** $4x^2 + 3x - 5 = 0$ **c** $x^2 + 3x - 5 = 0$
**d** $y^2 + 5y + 3 = 0$ **e** $2y^2 + 7y + 1 = 0$ **f** $5y^2 + 8y + 1 = 0$
**g** $3p^2 - 5p + 1 = 0$ **h** $4p^2 - 8p + 2 = 0$ **i** $3p^2 - 3p - 3 = 0$
**j** $2w^2 + w - 5 = 0$ **k** $3w^2 - w - 8 = 0$ **l** $5w^2 - 5w + 1 = 0$

2 Solve these equations. Give your answers in surd form.

**a** $2a^2 + 2a - 3 = 0$ **b** $6a^2 + a - 6 = 0$ **c** $4a^2 - a - 4 = 0$
**d** $8c^2 + 8c - 1 = 0$ **e** $5c^2 + 8c + 1 = 0$ **f** $7c^2 + 9c + 1 = 0$
**g** $4 - 3n - 2n^2 = 0$ **h** $3 + 5n - n^2 = 0$ **i** $12 + 6n - 2n^2 = 0$

3 Solve these equations giving your answers correct to 2 dp.

**a** $x(2x + 3) = 7$ **b** $1 = x(7 - 3x)$ **c** $x(4x + 5) = 4$

## Solving quadratic equations by completing the square

Another way to solve a quadratic equation that cannot be factorised is by **completing the square**. This might be quicker than using the formula.

### Example

Solve the equation $x^2 + 6x - 1 = 0$

Check that $b^2 \geqslant 4ac$ for the equation.
If $b^2 < 4ac$ then:
$(b^2 - 4ac)$ is negative and the equation will have no solution.

- The equation cannot be factorised, but $b^2 \geqslant 4ac$ so it can be solved.
- If we look for a complete square which includes $x^2 + 6x$ we must start with $(x + 3)^2$.

The diagram shows that $(x + 3)^2$ gives:
$x^2 + 6x + 9$

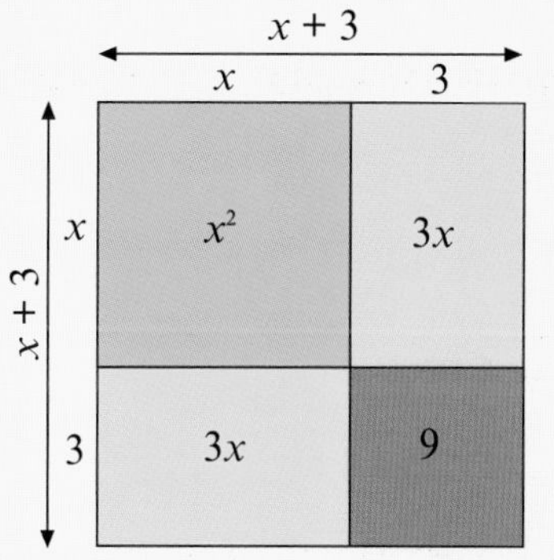


We have:

$$x^2 + 6x + 9 = (x + 3)^2$$
$$x^2 + 6x = (x + 3)^2 - 9$$

When we replace $x^2 + 6x$ the original equation becomes:

$$(x + 3)^2 - 9 - 1 = 0$$
$$(x + 3)^2 - 10 = 0$$
$$(x + 3)^2 = 10$$

[√ each side] $x + 3 = \pm\sqrt{10}$

$$x = \pm\sqrt{10} - 3$$
$$x = \pm 3.162 - 3$$

$x = -$ **6.16** (2dp) or $x =$ **0.16** (2dp)

**Exercise 9.8**
**Solving quadratic equations by completing the square**

**Accuracy**
Give your answers correct to 2 dp, unless surd form is asked for.

1 **a** Show that $x^2 + 10x - 5$ can be solved using the square of $(x + 5)$.
**b** Solve $x^2 + 10x - 5 = 0$ by completing the square.

2 Solve $x^2 + 8x - 3 = 0$ by completing the square.

3 Explain why $x^2 + 5x + 9 = 0$ cannot be solved.

4 By completing the square, where possible, solve these equations.

**a** $x^2 + 12x - 2 = 0$ **b** $x^2 + 6x - 8 = 0$ **c** $x^2 + 14x - 1 = 0$
**d** $x^2 + 4x + 5 = 0$ **e** $x^2 + 6x + 27 = 0$ **f** $x^2 + 8x + 6 = 0$

5 Complete the square to solve these. Give your answer in surd form.

**a** $x^2 + 5x - 4 = 0$ **b** $x^2 + 7x - 2 = 0$ **c** $x^2 + 3x + 1 = 0$

## Rearranging quadratic equations before solving them

A quadratic equation might be written in such a way that it does not look like your expectation of a quadratic equation.

**Example**

Solve the equation $(2x + 1)^2 = 3x(x - 1) + 9$

- Remove the brackets

$$(2x + 1)(2x + 1) = 3x(x - 1) + 9$$
$$4x^2 + 4x + 1 = 3x^2 - 3x + 9$$
$$x^2 + 7x - 8 = 0$$

- The quadratic equation can then solved by an appropriate method. $\mathbf{x^2 + 7x - 8 = 0}$

**Example**

Solve the equation $x + 6 = \dfrac{27}{x}$

- Remove the fraction by multiplying each term by $x$ $\qquad x^2 + 6x = 27$
- The quadratic equation can then be solved by an appropriate method. $\mathbf{x^2 + 6x - 27 = 0}$

**Example**

Solve the equation $\dfrac{x + 3}{4} = \dfrac{3}{x - 1}$

- Remove the fractions $\qquad (x + 3)(x - 1) = 12$
- Remove the brackets $\qquad x^2 + 2x - 3 = 12$
- The quadratic equation can then be solved by an appropriate method. $\mathbf{x^2 + 2x - 15 = 0}$

**Example**

Solve the equation $2x + 3 = \sqrt{(x^2 - 1)}$

- Remove the √ by squaring $\qquad (2x + 3)^2 = x^2 - 1$
- Remove the bracket

$$(2x + 3)(2x + 3) = x^2 - 1$$
$$4x^2 + 12x + 9 = x^2 - 1$$
$$3x^2 + 12x + 10 = 0$$

- The quadratic equation can then be solved by an appropriate method. $\mathbf{3x^2 + 12x + 10 = 0}$

These are just some forms of manipulation you might use when faced with an equation you need to rearrange.

**Exercise 9.9**
Rearranging and solving quadratic equations

**1** Rearrange and solve these equations. Give solutions correct to 2 dp.

**a** $x - 5 = \dfrac{36}{x}$ **b** $x(x - 7) = 60$ **c** $(x - 1)^2 = 64$

**d** $y = \dfrac{27}{(y + 6)}$ **e** $2y + 5 = \sqrt{(y^2 - 2)}$ **f** $y + 5 = \dfrac{6}{y}$

**g** $\dfrac{w - 2}{2} = \dfrac{7}{w - 3}$ **h** $\dfrac{w + 4}{2} = \dfrac{1}{2w - 1}$ **i** $36 = \dfrac{1}{w^2}$

**j** $\dfrac{4}{v^2} = 100$ **k** $3v^2 = 5v$ **l** $3v + 4 = \dfrac{5}{v} - 3$

**m** $\dfrac{3}{a} - 5 = 4a + 1$ **n** $\dfrac{2}{a - 1} + \dfrac{3}{a + 2} = 4$ **o** $100a = \dfrac{1}{a}$

# Creating and solving quadratic equations from problems

Some problems enable you to create a quadratic equation which you can solve to find an answer. While a quadratic equation will have two solutions it is possible that only one of them will be appropriate as an answer.

**Example**

A rectangle is 5 metres longer than it is wide. The rectangle has an area of 456 $m^2$. Calculate the length of a diagonal of the rectangle.

- Show the situation on a diagram.

  Let the width of the rectangle be $w$.
  The length of the rectangle is $w + 5$
  AB is a diagonal of the rectangle.

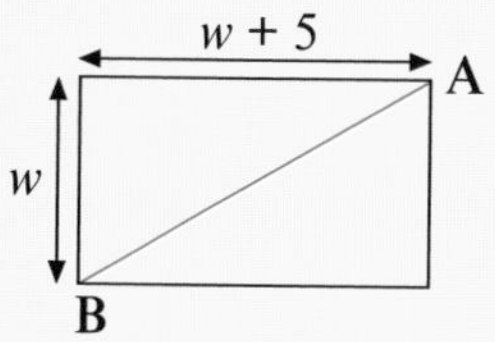


- Create and solve equation in $w$.

  The area of the rectangle is 456 $m^2$

  i.e.
  $$w(w + 5) = 456$$
  $$w^2 + 5w = 456$$
  $$w^2 + 5w - 456 = 0$$
  $$(w - 19)(w + 24) = 0$$
  $$w = 19 \text{ or } w = {}^{-}24$$

  **The width of the rectangle is 19 m** (the width cannot be $^{-}24$ m).
  **The length of the rectangle is 24 m** (i.e. 19 + 5).

- To find the length of the diagonal AB.
  By Pythagoras' rule:
  $$AB^2 = 19^2 + 24^2$$
  $$AB^2 = 937$$
  $$AB = \sqrt{937} = 30.6 \text{ (1 dp)}$$

  **The length of a diagonal of the rectangle is 30.6 m (1 dp).**

**Exercise 9.10**
Solving problems using quadratic equations

1 The width of a rectangle is 12 cm less than its length. The rectangle has an area of 448 $cm^2$.

**a** Write and solve an equation to find the dimensions of the rectangle.
**b** Calculate the length, to 2 dp, of a diagonal of the rectangle.

2 Two positive numbers multiplied together give 432.
One number is greater than the other number by six.
Find the two numbers.

3 A rectangle has a perimeter of 126 cm. The length of a diagonal of the rectangle is 45 cm.

Write and solve an equation to find:

**a** the length of the rectangle
**b** the area of the rectangle.

4 In triangle ABC the base is 4 cm longer than its height. ABC has an area of 117 $cm^2$.

Calculate the size of $A\hat{B}C$ to 2 dp.

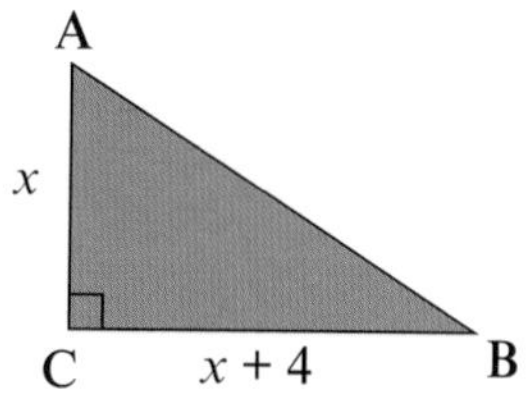

**5** Use the formula $p = vt + 0.5gt^2$ to find values for $t$ to 2 dp when $p = 20$, $v = 6$ and $g = 1$.

**6** The base of a right-angled triangle is 9 cm longer than the height of the triangle. The area of the triangle is 486 cm$^2$.

**a** Calculate the length of the base and the height of the triangle.
**b** Calculate the perimeter of the triangle.

**7** The diagram shows a trapezium ABCD which has a height $w$. The lengths of AB and CD are given in terms of $w$. The area of ABCD is 93.5 cm$^2$.

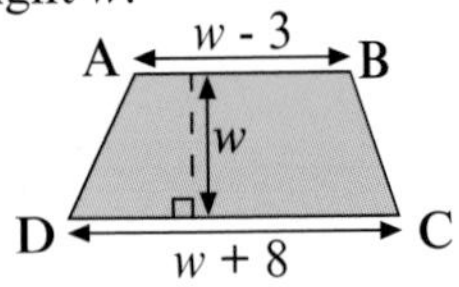


**a** Write and solve an equation to find a value for $w$.
**b** The trapezium ABCD has the same area as triangle RST.
Calculate the length of RT to 2 dp.

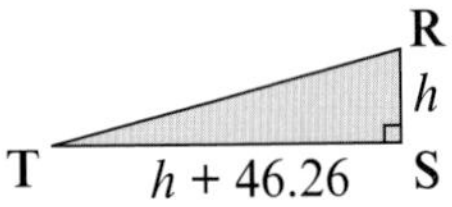


**c** Calculate the size of RTS to 2 dp.

**8** Two positive numbers differ by 4, and when multiplied give 2752.25
Write and solve an equation to find the two numbers.

**9** In triangle TVM all the sides are given in terms of $m$.

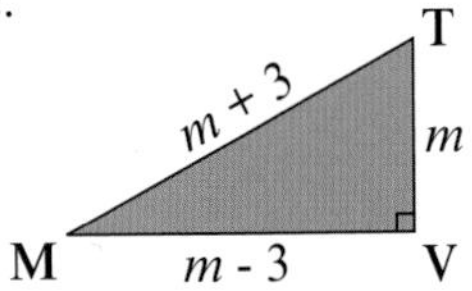


**a** Show that $m^2 = 12m$
**b** Calculate the area of TVM.
**c** Show that, to the nearest degree, $\angle MTV = 37°$.

**10** A concrete beam used to construct a bridge is in the shape of a cuboid. The beam is 2.8 metres longer than it is wide, and is 2.5 metres high. The volume of the beam is given as 52.7 $m^3$.

**a** Show that the beam is more than 6 metres long.
**b** If the width of the beam is $y$ metres show that the total surface area of the beam ($S$) is given by the equation:

$$S = 2y^2 + 15.6y + 14$$

**11** This cuboid has a width of $x$ cm. The height of the cuboid is 4 cm longer than its width. The depth of the cuboid is 6 cm longer than the width,

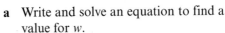

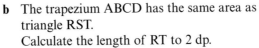

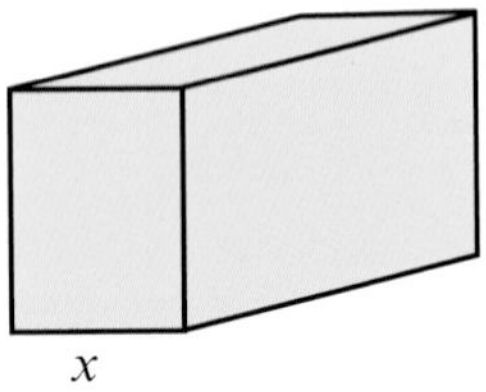


**a** Write an expression in terms of $x$ for the surface area of the cuboid.
**b** Write your expression as a quadratic.

The surface area of the cuboid is 398 cm$^2$.

**c** Write and show an equation to find a value for $x$.
**d** Calculate the volume of the cuboid.

## End points

You should be able to ... ... so try these questions

**A** Solve simultaneous linear equations

**A1** Solve these equations for $x$ and $y$.

**a** $5x - 3y = 21$
$2x + 5y = {}^-4$

**b** $2x - 3y = {}^-3$
$7x - 2y = {}^-19$

**B** Solve problems with simultaneous equations

**B1** Two groups of people bought tickets on the Cliff Railway. The first group of five adults and two children paid a total of £29.30. The second group of two adults and three children paid a total of £17.55.

Find the charges for adult and child tickets for the Cliff Railway.

**C** Solve quadratic equations by inspection

**C1** Solve these quadratic equations.

**a** $x^2 + 5x - 24 = 0$
**b** $x^2 - 8x - 65 = 0$
**c** $x^2 - 15x + 36 = 0$
**d** $2x^2 - 2x - 12 = 0$

**D** Solve quadratic equations using the formula

**D1** Solve these quadratic equations to 2 dp using the formula.

**a** $x^2 + 6x - 2 = 0$ **b** $x^2 - 9x + 3 = 0$
**c** $2x^2 + 9x + 3 = 0$ **d** $3x^2 - 5x + 6 = 0$

**D2** Solve these equations leaving your answer in surd form.

**a** $x^2 + 8x - 4 = 0$ **b** $x^2 - 6x - 8 = 0$

**E** Solve quadratic equations by completing the square

**E1** By completing the square solve the equation $x^2 + 5x - 3 = 0$.

**E2** Complete the square to solve the equation $x^2 + 8x - 5 = 0$.
Give your answer in surd form.

**F** Create and solve quadratic equations for problems

**F1** A rectangle has an area of 570 cm². The rectangle is 23 cm longer than it is wide.
Write and solve an equation to find the dimensions of the rectangle.

## Some points to remember

- When you are solving simultaneous equations you may have to rearrange one or both of them at the start.
- Check when starting to solve a quadratic equation that $b^2 \geqslant 4ac$.
- When creating equations to solve word problems do not introduce more unknowns than you need. Always try to express one unknown in terms of another unknown.
- You can always check a solution to an equation by substitution.

## Starting points

You need to know about ... ... so try these questions

### A The equation of a straight line

- You can use the coordinates of points on a line to find the equation of the line.

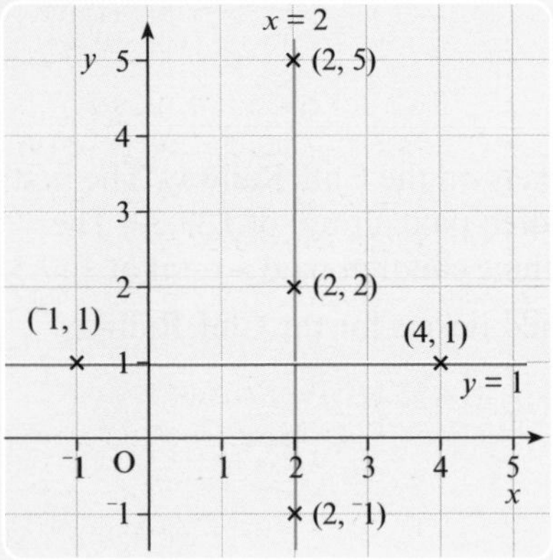


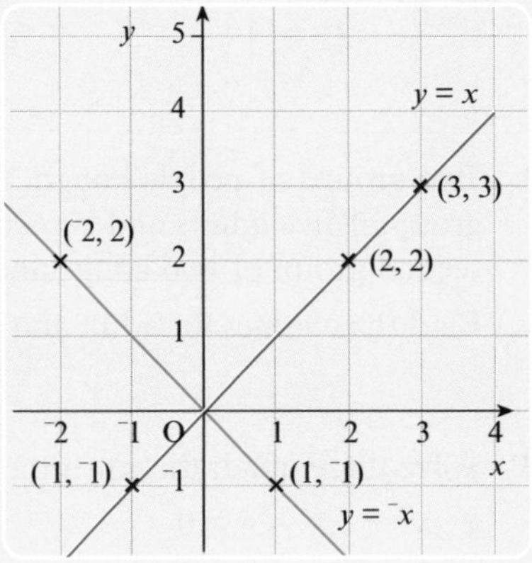


**A1**

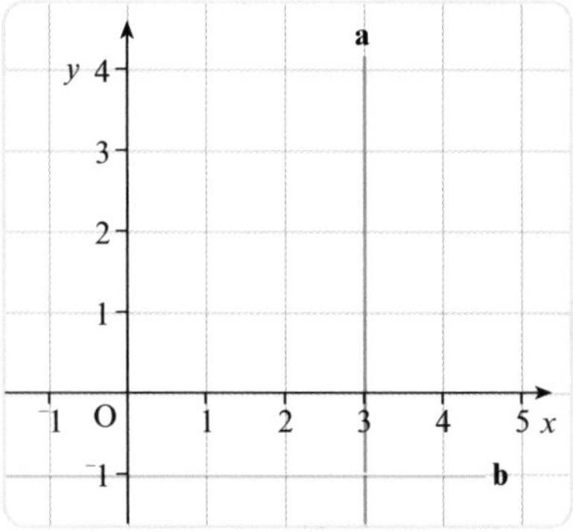


Write down the equation of the lines **a** and **b**.

### B Reflections

- To describe a reflection you must give the mirror line. On a grid you can give the equation of the mirror line.

**Example**

B is the image of A after a reflection in $y = {}^-x$.

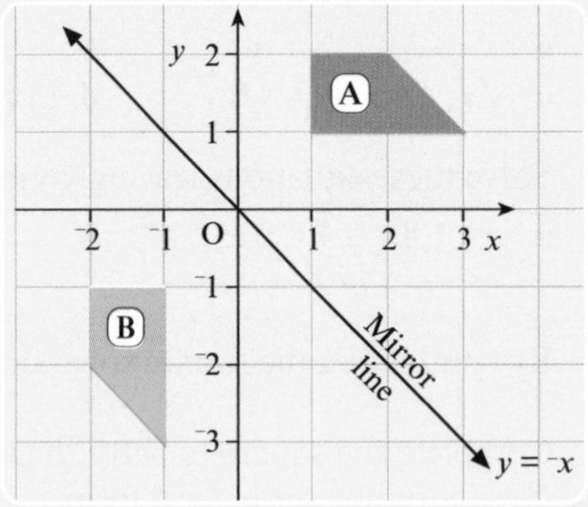


**B1** **a** Copy this diagram.

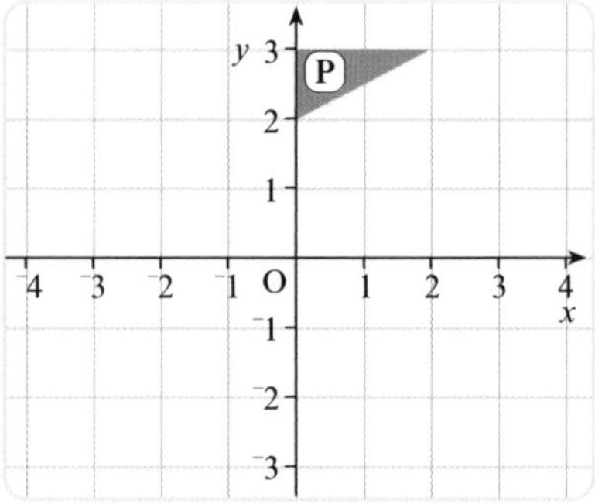


**b** Draw the image of P after a reflection in:
**i** $y = x$ **ii** $x = {}^-1$

### C Rotations

- To describe a rotation you must give:
  - the angle and direction of rotation
  - the centre of rotation.
- On a grid you can give the coordinates of the centre.

**Example**

- C is the image of A after a rotation of ⁺90° (90° anticlockwise) about (4, 1).
- D is the image of A after a rotation of ⁻90° (90° clockwise) about (⁻2, 2).

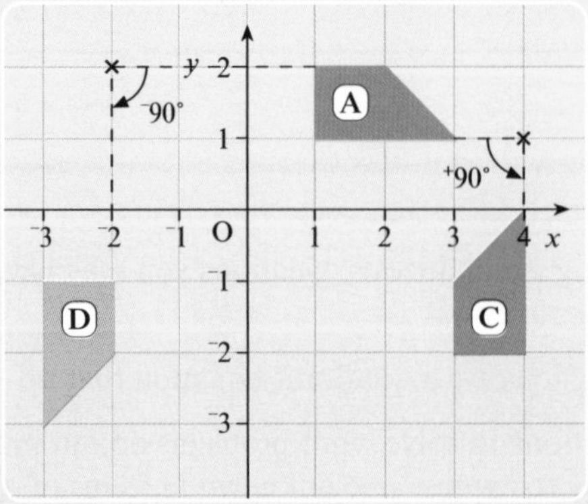


**C1**

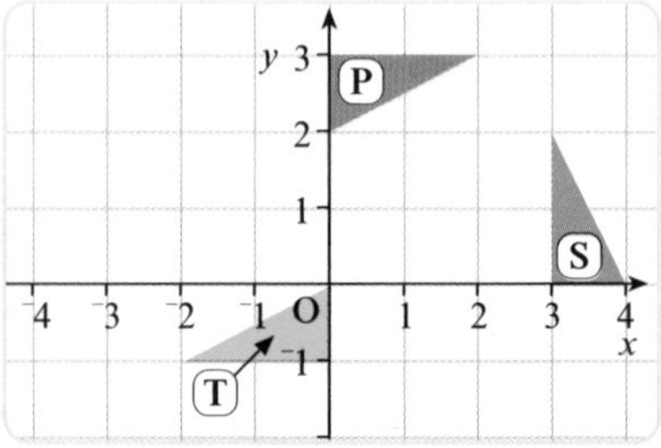


**a** Triangle S is the image of P after a rotation about (3, 3). What is the angle of rotation?

**b** Triangle T is the image of P after a rotation of 180°. Give the coordinates of the centre of rotation.

## D Translations

- A translation only changes the position of a shape.
- On a grid you can use a vector to describe a translation.

**Example**

- A is translated on to E by the vector $\begin{pmatrix} 1 \\ -2 \end{pmatrix}$.
- A is translated on to F by the vector $\begin{pmatrix} -3 \\ -1 \end{pmatrix}$.

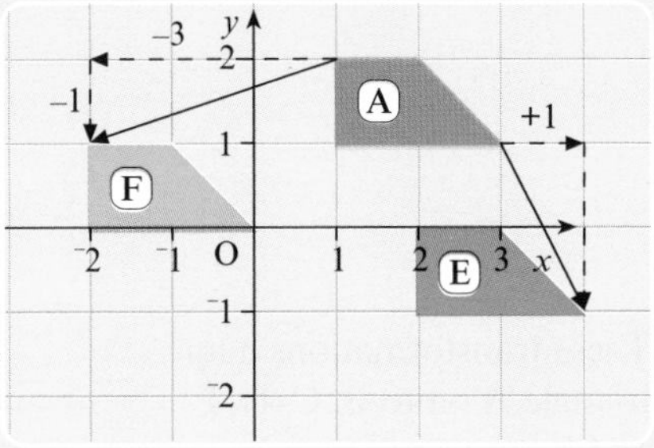


## E Enlargements

- To describe an enlargement you must give:
  - the centre of enlargement
  - the scale factor (SF).

**Example**

G is an enlargement of A with centre (4, $-1$) and SF 2.

H is an enlargement of A with centre ($-1$, 0) and SF $\frac{1}{2}$.

I is an enlargement of A with centre (3, 3) and SF $-1$.

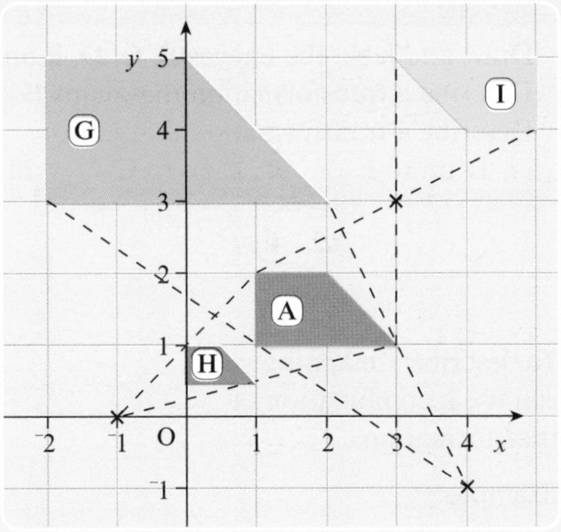


- If the SF is **greater than 1**, or **less than $-1$**, the image is **larger** than the object.
- If the SF is **between 0 and 1**, or **between $-1$ and 0**, the image is **smaller** than the object. (Although the image is smaller, it is still called an enlargement!)
- If the scale factor is **negative**, the image is on the opposite side of the centre of enlargement and upside down.

## F Congruence and similarity

- Of the four main types of transformation, enlargement is the only one that changes the **size** of the object (unless the SF is $-1$ or 1).
  - Under either reflection, rotation, or translation: the object and image are always **congruent**, i.e. they have the same shape and size.
  - Under enlargement: the object and image are only **similar**, i.e. they have the same shape but are different in size.

**D1** These translations map P on to U and V.

| Object | Translation | Image |
|---|---|---|
| P | $\begin{pmatrix} 2 \\ 0 \end{pmatrix}$ | U |
| P | $\begin{pmatrix} -1 \\ -3 \end{pmatrix}$ | V |

On your diagram from Question **B1**, draw and label the image of P after each of these translations.

**E1** Copy this diagram and enlarge triangle P with:

**a** SF 2 and centre ($-1$, 4)
**b** SF $\frac{1}{2}$ and centre (0, 5)
**c** SF $-2$ and centre (2, 2.5).

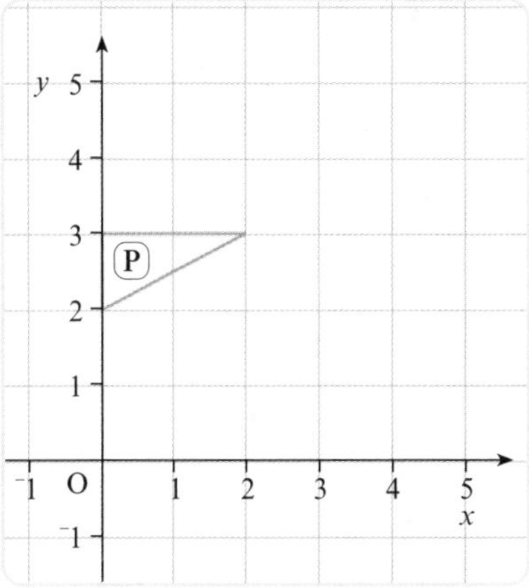


**E2** On a pair of axes draw the shape with vertices A(1, 3) B(1, 5) C(2, 4) D(2, 3)
Enlarge the shape with:

**a** SF $-1.5$ centre (0, 2)
**b** SF $-2$ centre (2, 2)
**c** SF $-3$ centre (1, 2).

**F1** **a** In this diagram which shapes are congruent to U?

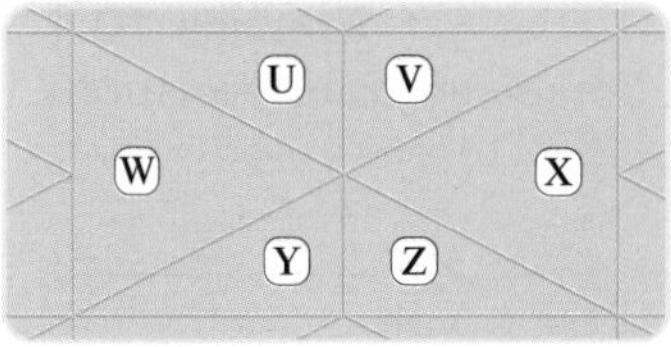


**b** Explain why the triangle made by W, Y, and Z is similar to triangle U.

**Thinking ahead to ...**
combined transformations

**A** **a** Draw triangle A on axes with $^{-}5 \leqslant x \leqslant 5$ and $^{-}5 \leqslant y \leqslant 5$.

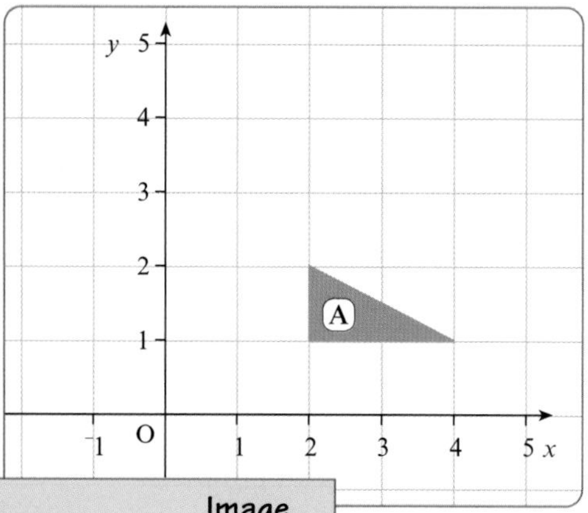


**b** These transformations map triangle A on to B, C, D, E and F.

| Object | Transformation | Image |
|---|---|---|
| A | Reflection in the line $y = x$ | B |
| A | Reflection in the line $x = 1$ | C |
| A | Rotation of 180° about the origin | D |
| A | Rotation of $^{-}90°$ about the point (1, 0) | E |
| A | Translation $\begin{pmatrix} ^{-}4 \\ 2 \end{pmatrix}$ | F |

Draw and label the images B, C, D, E and F.

**c** Describe a transformation that maps B on to C.

**d** Describe a transformation that maps:
**i** D on to E **ii** E on to C **iii** F on to E.

## Combined transformations

- To describe a mapping you can use a combination of transformations.

**Example**

To map F on to G you can use:

a rotation of $^{-}90°$ about (1, 0)

followed by

a translation $\begin{pmatrix} ^{-}4 \\ 2 \end{pmatrix}$

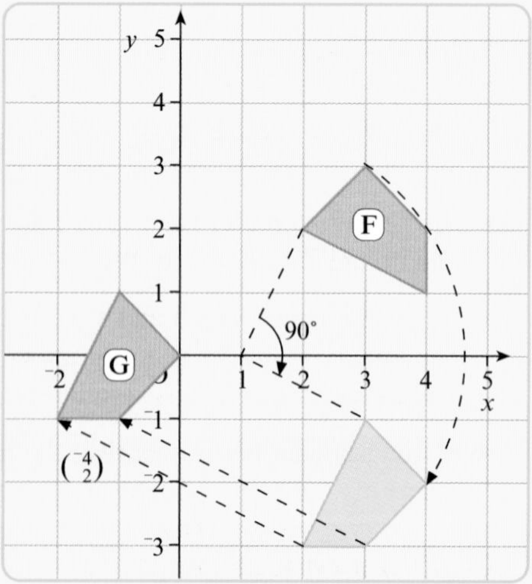


**Exercise 10.1**
Combined transformations

Use axes with:
$^{-}5 \leqslant x \leqslant 6$ and
$^{-}5 \leqslant y \leqslant 5$

**1** H is the image of F after:

a translation $\begin{pmatrix} ^{-}4 \\ 2 \end{pmatrix}$

followed by

a rotation of $^{-}90°$ about (1, 0)

**a** Draw and label F and its image H.

**b** Compare G and H and comment on any differences.

**2**  These transformations map the quadrilateral F on to J and K.

| Transformations | | | |
|---|---|---|---|
| Object | First | Second | Image |
| F | Rotate 180° about (2, 1) | Rotate ⁺90° about (2, 1) | J |
| F | Rotate ⁺90° about (2, 1) | Rotate 180° about (2, 1) | K |

> Use the same axes as you drew for Question **1**.

**a**  Draw and label the images J and K.
**b**  Compare J and K and comment on any differences.

**3**

| ① | a reflection in $y = x$ |
|---|---|

| ② | a rotation of 180° about (0, 0) |
|---|---|

| ③ | a reflection in $y = 0$ |
|---|---|

| ④ | a translation $\begin{pmatrix} ^-2 \\ ^-2 \end{pmatrix}$ |
|---|---|

Sort the transformations 1 to 4 into two pairs so that:

> in each pair the transformations are commutative.

> A pair of transformations are commutative if the image of an object is the same whichever transformation you use first.

Draw diagrams to explain your answers.

◆ Combined transformations can be equivalent to a single transformation.

**Example**

A rotation of ⁻90° about (0, 3) maps F on to G.
So

| a rotation of ⁻90° about (1, 0) |
|---|

followed by

| a translation $\begin{pmatrix} ^-4 \\ 2 \end{pmatrix}$ |
|---|

is **equivalent to the single transformation**

| a rotation of ⁻90° about (0, 3) |
|---|



---

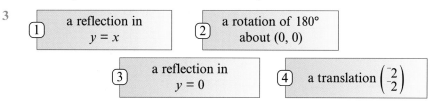

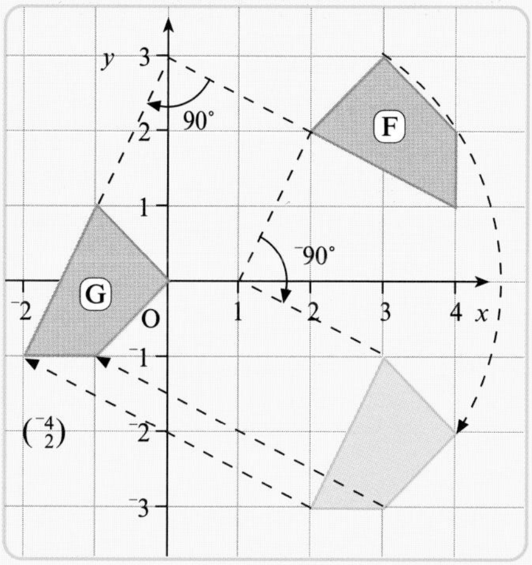

**Exercise 10.2**
Equivalent transformations

**1**  These transformations map the quadrilateral F on to L, M and N.

| Transformations | | | |
|---|---|---|---|
| Object | First | Second | Image |
| F | Reflect in y = 1 | Reflect in y = x | L |
| F | Rotate 180° about (1, 1) | Translate $\begin{pmatrix} 0 \\ 2 \end{pmatrix}$ | M |
| F | Reflect in y = ⁻1 | Rotate ⁻90° about (1, ⁻1) | N |

> Use axes with:
> ⁻5 ≤ x ≤ 5 and
> ⁻5 ≤ y ≤ 5

**a**  On a new diagram draw and label the images L, M and N.
**b**  What single transformation maps F on to:
  **i** L  **ii** M  **iii** N?

**Exercise 10.3**
**Transformations**

Each tile is a transformation of tile P.

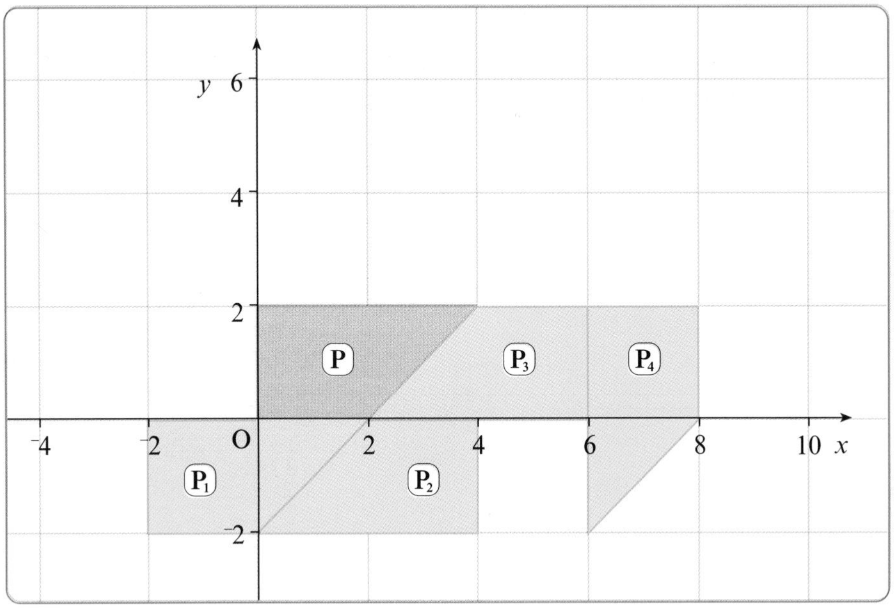


**1** **a** Draw these tiles on axes with $^-4 \leqslant x \leqslant 10$ and $^-4 \leqslant y \leqslant 8$.
**b** Describe a single transformation that maps P on to:
**i** $P_1$ **ii** $P_2$ **iii** $P_3$

**2** In this table each pair of transformations maps tile P on to $P_4$.
**a** For each mapping describe the second transformation fully.

| | Object | Image | Transformations First | Second |
|---|---|---|---|---|
| **i** | P | $P_4$ | Rotate ⁻90° about (2, 0) | |
| **ii** | P | $P_4$ | Rotate 180° about (1, 4) | |
| **iii** | P | $P_4$ | Reflect in y = x | |

**b** Describe another pair of transformations that maps P on to $P_4$.

**3** These transformations map tile P on to $P_5$, $P_6$ and $P_7$.

| Object | Transformations First | Second | Image |
|---|---|---|---|
| P | Rotate ⁻90° about (2, ⁻2) | Reflect in x = 5 | $P_5$ |
| P | Reflect in y = x | Translate $\begin{pmatrix} ^-2 \\ 0 \end{pmatrix}$ | $P_6$ |
| P | Rotate ⁺90° about (0, 2) | Reflect in x = 1 | $P_7$ |

**a** On your diagram draw the images $P_5$, $P_6$ and $P_7$.
**b** Describe a pair of transformations that map $P_5$ on to P.

## Enlargements

**Exercise 10.4**
Enlargements

**1** These tiles are arranged in the shape of a Greek cross.

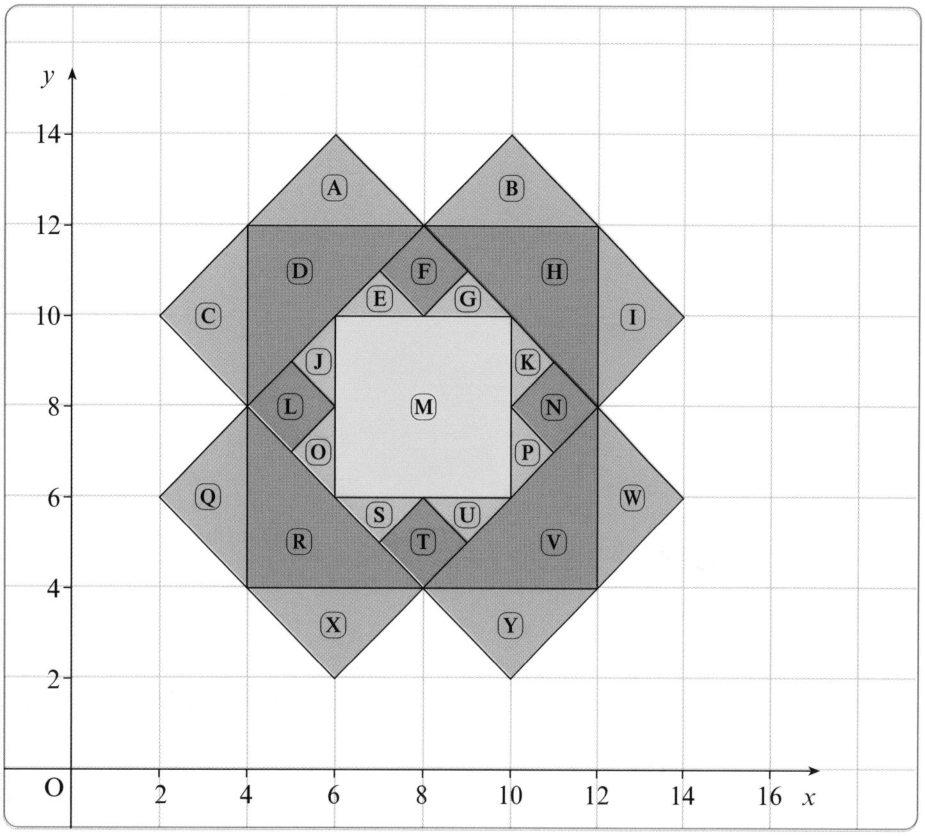


You could put tracing paper over your drawing for each question.

If you draw lines through corresponding points on the object and image, they will meet at the centre of enlargement.

**a** Copy the diagram and label the tiles A to Y.

**b** What is the image of E after an enlargement with centre (4, 8) and SF 2?

**c** Find the scale factor and centre for each of these enlargements.

| Object | Image | SF | Centre |
|---|---|---|---|
| S | Y | | |
| Y | U | | |
| E | A | | |
| Q | W | | |

**d** Explain why tile T will not map on to M with just an enlargement.

For an enlargement give the scale factor and the coordinates of the centre.

For a translation give the vector.

For a reflection give the equation of the mirror line.

For a rotation give direction and angle of rotation and the coordinates of the centre.

**e** This table gives a pair of transformations for three mappings. Describe fully the second transformation for each of these mappings.

| | Object | Image | Transformations First | Transformations Second |
|---|---|---|---|---|
| **i** | G | W | Enlarge SF 2 with centre (8, 8) | |
| **ii** | Q | E | Rotate $^-90°$ about (6, 6) | |
| **iii** | A | S | Enlarge SF $\frac{1}{2}$ with centre (4, 8) | |

## Exercise 10.5
## Drawing enlargements

1 Draw the triangle A on axes with:
$^{-}6 \leqslant x \leqslant 10$
$^{-}6 \leqslant y \leqslant 10$.

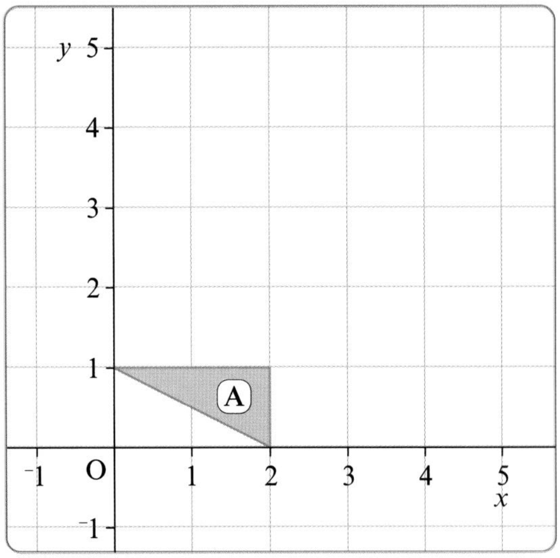


For an enlargement: the distance from the centre to a point on the object multiplied by the scale factor gives the distance from the centre to the corresponding point on the image.

2 These transformations map the triangle A on to $B_1$, $B_2$, $B_3$ and $B_4$.

| Object | First transformation | Second transformation | Image |
|---|---|---|---|
| A | Enlarge SF 2 with centre (0, 0) | Translate $\begin{pmatrix}1\\2\end{pmatrix}$ | $B_1$ |
| A | Rotate $^{-}90°$ about (2, 0) | Enlarge SF 2 with centre ($^{-}1$, 0) | $B_2$ |
| A | Enlarge SF 2 with centre ($^{-}1$, 0) | Translate $\begin{pmatrix}4\\4\end{pmatrix}$ | $B_3$ |
| A | Rotate $^{+}90°$ about (0, 1) | Enlarge SF 2 with centre ($^{-}3$, $^{-}2$) | $B_4$ |

**a** On your diagram draw and label the images $B_1$, $B_2$, $B_3$ and $B_4$.
**b** Describe a single transformation that will map $B_1$:
**i** on to $B_2$ **ii** on to $B_3$ **iii** on to $B_4$.
**c** What single transformation maps A:
**i** on to $B_1$ **ii** on to $B_3$.

3 In this table each pair of transformations maps $B_2$ on to A. For each mapping describe the second transformation fully.

| | Object | Image | First transformation | Second transformation |
|---|---|---|---|---|
| a | $B_2$ | A | Rotate $^{+}90°$ about (5, 0) | |
| b | $B_2$ | A | Rotate $^{+}90°$ about (5, 2) | |

**4** Each tile in this pattern is also a transformation of A.

**a** Copy this pattern on axes with:
$^{-}6 \leq x \leq 10$
$^{-}6 \leq y \leq 10$

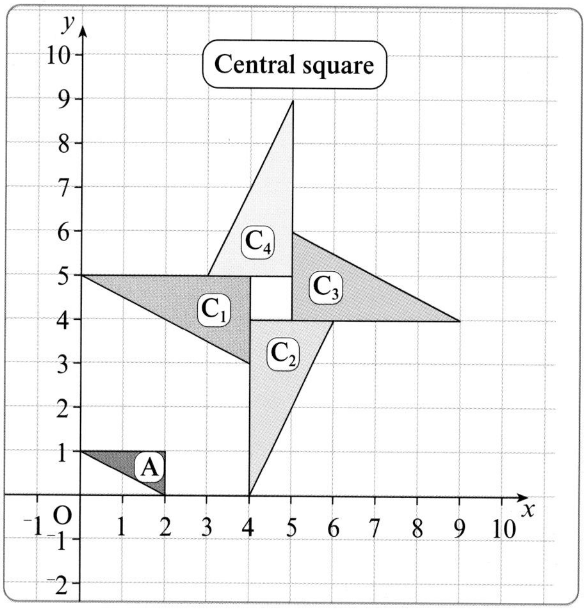


You do not need more than two transformations for each mapping.

**b** In a table show what transformations map A on to each triangle in this pattern.

**c** **i** Make another pattern using four triangles.
**ii** In a table to show what transformations map A on to each triangle in your pattern.
**iii** Pass your table to a partner and ask them to draw your pattern.

**5**

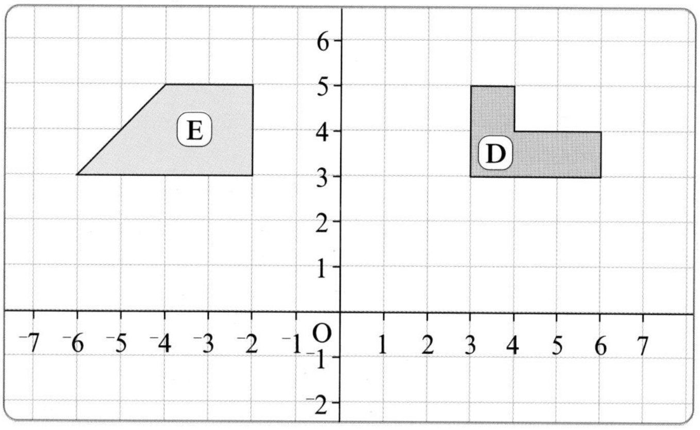


**a** Copy these shapes on axes with: $^{-}12 \leq x \leq 12$
$^{-}7 \leq y \leq 10$

**b** On your diagram, draw and label the images $D_1$ and $E_1$.

| Object | Transformations First | Transformations Second | Image |
|---|---|---|---|
| D | Enlarge SF 2 with centre (0, 1) | Rotate 180° about (0, 1) | $D_1$ |
| E | Enlarge SF $^{-}1.5$ with centre (0, 1) | Enlarge SF $^{-}1.5$ with centre (0, 1) | $E_1$ |

**c** Describe a single transformation that maps:
**i** D on to $D_1$ **ii** E on to $E_1$.

**d** Describe a single transformation that maps:
**i** $D_1$ on to D **ii** $E_1$ on to E.

**Thinking ahead to ...**
inverse transformations

**A a** Draw hexagon R on axes with:
$^{-}7 \leqslant x \leqslant 7$ and $^{-}7 \leqslant y \leqslant 7$.

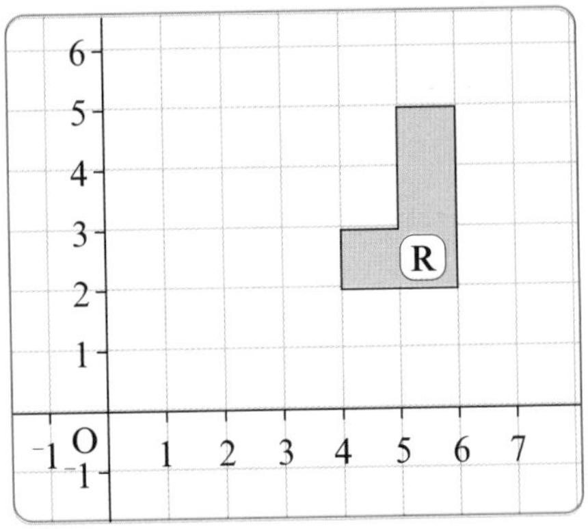


**b** Describe the effect on R of each of these pairs of transformations.

| Object | Transformations: First | Transformations: Second |
|---|---|---|
| R | Enlarge SF $^{-}2$ with centre (2, 1) | Enlarge SF $^{-}0.5$ with centre (2, 1) |
| R | Rotate $^{-}90°$ about (1, 0) | Rotate $^{-}270°$ about (1, 0) |

## Inverse transformations

- If a combination of two transformations maps the object on to itself then each transformation is the **inverse** of the other.

**Exercise 10.6**
Inverse transformations

1 Describe the inverse transformation of:

a Translation $\begin{pmatrix} 1 \\ ^{-}2 \end{pmatrix}$

**b** Rotation $^{-}90°$ about (1, 0)

**c** Enlargement SF 2 with centre ($^{-}1$, 2)

2 a Describe the inverse transformation of: Reflection in $y = {}^{-}x$

b What is special about the inverse transformation of any reflection?

3 a Draw the triangle S on axes with:
$^{-}1 \leqslant x \leqslant 5$
$^{-}5 \leqslant y \leqslant 5$.

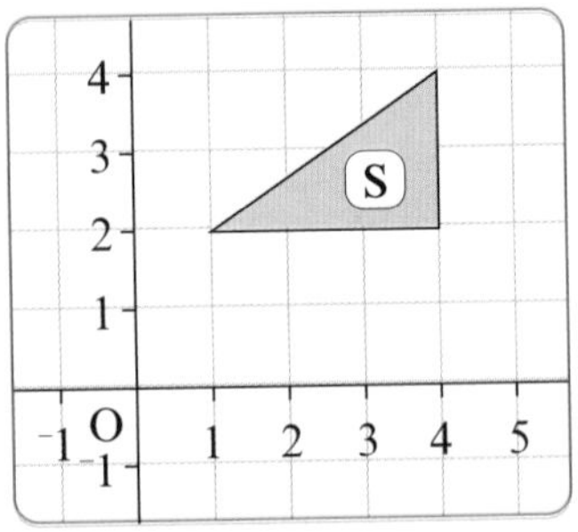


b On your diagram, draw and label the images $S_1$ and $S_2$.

$S_1$ Rotation 180° about (3, 0)

$S_2$ Enlargement SF $^{-}1$ with centre (3, 0)

c Describe two inverse transformations of Rotation 180° about $(x, y)$

# End points

You should be able to ...     ... so try these questions

**A** Use combined transformations

**A1** **a** Draw shape E on axes with: $^{-}12 \leqslant x \leqslant 6$ and $^{-}6 \leqslant y \leqslant 6$.

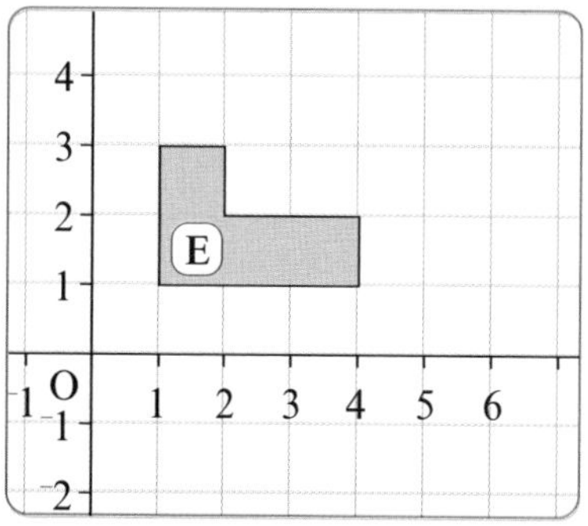


**b** On your diagram, draw and label the images K, L and M.

| | Transformations | | |
|---|---|---|---|
| Object | First | Second | Image |
| E | Enlarge SF $^{-}2$ with centre ($^{-}1$, 0) | Enlarge SF 0.5 with centre (3, $^{-}4$) | K |
| E | Reflect in $y = {}^{-}x$ | Reflect in $y = 0$ | L |
| E | Rotate 180° about (0, 0) | Rotate 180° about (0, $^{-}2$) | M |

**c** What single transformation maps E on to:
**i** K    **ii** L    **iii** M?

**d** These transformations map K on to L, M and E.
Describe the second transformation.

| | | | Transformations | |
|---|---|---|---|---|
| | Object | Image | First | Second |
| **i** | K | L | Rotate $^{-}90°$ about ($^{-}1$, $^{-}3$) | |
| **ii** | K | M | Reflect in $x = 0$ | |
| **iii** | K | E | Translate $\begin{pmatrix} 2 \\ 2 \end{pmatrix}$ | |

**e** **i** What single transformation maps K on to L?
**ii** Describe two single transformations that each maps K on to M.

**B** Identify inverse transformations

**B1** Describe the inverse transformation of each of these.

**a** Enlargement SF 0.25 with centre (3, $^{-}1$)

**b** Rotation $^{-}270°$ about (0, 2)

**c** Translation $\begin{pmatrix} 0 \\ ^{-}3 \end{pmatrix}$

**B2** Describe the inverse transformation of: Reflection in $y = x + 1$

# Toujours Paris

## Ile de la Cité

This boat-shaped island in the River Seine is where Paris was first inhabited by Celtic tribes over 2000 years ago. It is where Notre Dame is situated. This cathedral is a superb example of French medieval architecture and is particularly known for its wonderful stained glass rose windows.

The width of this South Window is 13 metres.

In the Ile de la Cité you will also find Point Zéro. This is a geometer's mark from which all distances in France are measured.

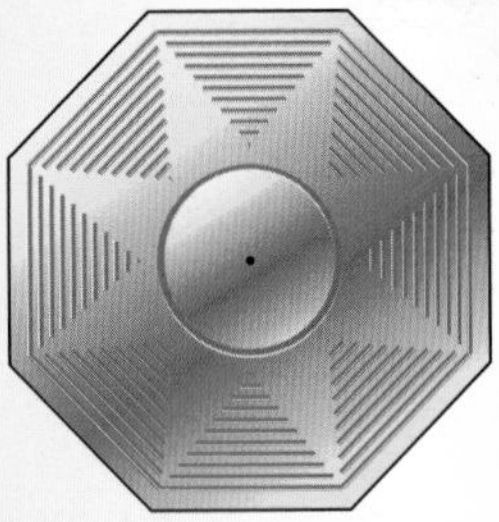

Each side of the shape is 12 cm long.

### Paris au Quotidien

**Arrondissements**: Il faut le savoir, Paris est divisé en 20 arrondissements se déroulant en spiraleà partir du 1er (le quartier du Louvre).
**Banques**: Ouvertes en général du lundi au vendredi de 9h à 16h30, quelques (rares) agences le samedi. Les Caisses d'Epargne ouvrent plus souvent le samedi et ferment le lundi.
**Change**: On ne peut pas tout prévoir à l'avance; vous pourrez changer vos devises dans les gares, les aéroports, les grandes agences de banque, les points change (ouverts tard le soir), ainsi qu'à notre bureau d'accueil des Champs-Elysées.

### Daily Life in Paris

**Districts**: You should know that Paris is divided into 20 districts numbered in a circular direction, starting with the 1st district (the Louvre area).
**Banks**: They are generally open from Monday to Friday from 9 am to 4.30 pm, some (rare) branches on Saturday. The savings banks are more often open on Saturday and closed on Monday.
**Exchange**: You cannot foresee everything; you can therefore change your foreign currency in railway stations, airports, major bank branches, exchange offices (open late in the evening), as well as in the visitors' office on the Champs-Elysées.

## Eiffel Tower Factfile – 1996

- Built in 1889 for the Universal Exhibition
- Built by Gustave Eiffel (1832–1923)
- Built from pig iron girders
- Total height 320 metres
- The world's tallest building until 1931
- Height to 3rd level is 899 feet
- There are 1652 steps to the third level
- Two and a half million rivets were used
- The tower is 15 cm higher on a hot day.
- Its total weight is 10100 tonnes
- 40 tons of paint are used every 4 years
- On a clear day it is possible to see Chartres Cathedral, 72 km away to the South West.
- The tower is visited by about $5\frac{1}{2}$ million people every year

*Admission charge 56 Francs*

### DAY TRIPS BY EUROSTAR

**Waterloo Station (London) to Paris**

Celebrate that special occasion in style with a day trip to Paris!

**ADULT** £
**CHILD (4–11 YRS)** £

### *Entry fees in Paris – 1996*

| | |
|---|---|
| Eiffel Tower | 56 FF |
| Louvre | 45 FF |
| Pompidou Centre | 35 FF |
| Picasso Museum | 28 FF |
| Museum of Modern Art | 27 FF |
| Versailles Palace | 45 FF |
| Parc de la Villette | 45 FF |
| Cluny Museum | 28 FF |

### Datafile

Population of Paris (in 1982) 2 188 918
In 1996, £1 sterling was equivalent to 7.54 French Francs.
1 metre = 3.281 feet
1 kilometre = 0.62 miles

### Paris Lucky dip – Superb value !

In the hat are four tickets to the: Eiffel Tower, Picasso Museum, Versailles Palace and the Museum of Modern Art. Pick two tickets at random from the four.
Entry fee – only 83 Francs.

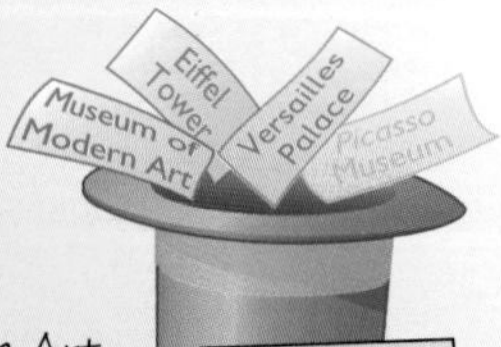


**83 FF**

Give answers to 3 significant figures where appropraite.

**1** **a** How many lines of symmetry has the South Window of Notre Dame?
**b** What is the order of rotational symmetry of the South Window?
**c** Draw a window shape where the number of lines of symmetry is not equal to the order of rotational symmetry.
**d** Calculate the area of the South Window inside the frame.

The window is surrounded by a frame with the cross-section below.

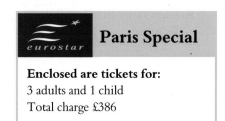

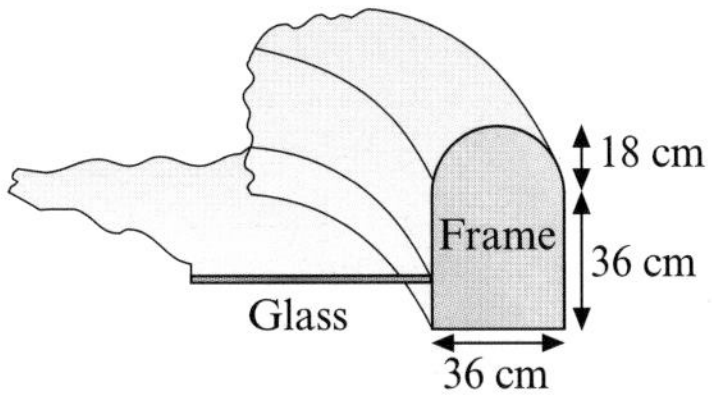


**e** Calculate the area of the frame's cross-section.
**f** Calculate the approximate volume of the frame in cubic metres.

**2** For Point Zéro in the Ile de la Cité, the outer polygon is a regular octagon.
**a** Calculate the size of an exterior angle.
**b** Calculate the size of an interior angle.
**c** Calculate the distance across the octagon from one flat face to the opposite one.
**d** Calculate the total area of Point Zéro.

**3** **a** How many tons of paint will have been used on the Eiffel Tower from when it was built up to the year 2000?
**b** How much higher than the third level is the total height of the tower?
Give your answer to the nearest metre.
**c** From a point 2000 metres away what is the angle of elevation to the topmost point on the Eiffel Tower, to the nearest degree?
**d** The spire of Notre Dame reaches 90 metres from the ground and is 4.3 km as the crow flies from the Eiffel Tower.
What is the angle of depression from the third level of the Eiffel Tower to the tip of Notre Dame's spire? Give your answer to 2 sf.
**e** The ratio of steps to the third level, to steps to the first level is about 9 to 2.
About how many steps are there to the first level?
**f** Calculate the approximate percentage that the tower grows on a hot day.

**4** What is the approximate bearing of Paris from Chartres?

**5** **a** In 1996, what was the Eiffel Tower entry fee equivalent to in £ Sterling?
**b** By 1997 one French Franc had changed in value to 10.6 pence. What effect did this change have on the entry fee in £ sterling?

**6** **a** Compare the French and English text in the extract on Daily Life in Paris.
What is the relative frequency of a vowel in each language? (Vowels are a, e, i, o, and u.)
**b** Compare the French and English extracts. Which language uses the longest words? Explain how you decided.

**7** For the Paris Lucky Dip use the following shorthand:

*E* – Eiffel Tower  *P* – Picasso Museum
*V* – Versailles Palace  *M* – Museum of Modern Art

**a** What is the probability that a pair of tickets is picked which includes the Versailles Palace?
**b** What is the probability of picking P and M?
**c** Give the probability of picking a pair of tickets
**i** worth more than the entry fee
**ii** worth less than the entry fee.
**d** Is the seller likely to make a profit or loss on every hundred entries? Give your reasons.

**8** These two bills were for Eurostar day trips to Paris.

eurostar **Paris Special**

**Enclosed are tickets for:**
2 adults and 3 children
Total charge £465

eurostar **Paris Special**

**Enclosed are tickets for:**
3 adults and 1 child
Total charge £386

Calculate:
**a** the cost of an adult's ticket **and** a child's ticket
**b** the total cost for 1 adult and 2 children.

## Starting points

You need to know about ...

### A Solving equations in one unknown

- where the unknown is on both sides of the equation

| | |
|---|---|
| | $5c - 4 = 3c - 8$ |
| [$-3c$] | $2c - 4 = {}^-8$ |
| [$+4$] | $2c = {}^-4$ |
| [$\div 2$] | $c = {}^-2$ |

- where brackets are involved

| | |
|---|---|
| | $3(4p + 5) = 5(3p - 4)$ |
| | $12p + 15 = 15p - 20$ |
| [$-15$] | $12p = 15p - 35$ |
| [$-15p$] | ${}^-3p = {}^-35$ |
| [$\div {}^-3$] | $p = \frac{35}{3} = 11.66\ldots$ |

- where fractions are involved

| | |
|---|---|
| | $\frac{2}{3}(w + 3) = 4w - 3$ |
| [$\times$ both sides by 3] | $2(w + 3) = 12w - 9$ |
| | $2w + 6 = 12w - 9$ |
| [$+9$] | $2w + 15 = 12w$ |
| [$-2w$] | $15 = 10w$ |
| | $\mathbf{w = 1.5}$ |

### B Factorising expressions

- when a numerical factor can be identified
  factorise $12a - 9x + 18y$
  i.e. $\mathbf{3(4a - 3x + 6y)}$
- when an algebraic factor can be identified
  factorise $a^2b + ab^2 - ac$
  i.e. $\mathbf{a(ab + b^2 - c)}$
- when more than one factor can be identified
  factorise $15xy - 20x^3 + 5xy^2 + 10x^2y^3$
  i.e. $\mathbf{5x(3y - 4x^2 + y^2 + 2xy^3)}$

### C Dealing with powers and roots

- when powers are involved

| | |
|---|---|
| | $x^2 = 3a^2b - 4c$ |
| [$\sqrt{}$ both sides] | $\mathbf{x = \sqrt{(3a^2b - 4c)}}$ |

- when roots are involved

| | |
|---|---|
| | $\sqrt{c} = 3xy + b$ |
| [square both sides] | $\mathbf{c = (3xy + b)^2}$ |

... so try these questions

**A1** Solve these equations:

a $7d - 8 = 3d + 14$
b $3y + 9 = 2y - 5$
c $5(2x + 1) = 41 + x$
d $3(3p + 4) = 2(p + 27)$
e $5(4 - x) = 3(2x + 1)$
f $3(k - 1) = 2(1.5 - 3k)$
g $8(n + 5) = 4(3n - 8)$
h $15(g - 3) = 10(2g + 1)$

**A2** Solve these equations:

a $\frac{2}{3}c - 1 = 4$
b $\frac{3}{4}x + 1 = 2x$
c $3 - \frac{1}{2}x = x - 15$
d $\frac{3}{5}(v + 4) = 3v$
e $\frac{1}{2}(w - 3) = 2(2w + 1)$
f $\frac{1}{3}(4a + 2) = 2(a - 6)$
g $\frac{2}{3}(x - 1) = 2(2x - 6)$
h $\frac{3}{8}y + 2 = \frac{1}{3}y - 3$

**B1** Factorise these expressions:

a $6bc - 15ax + 3by$
b $18a^2 + 24b^2c$
c $3a^2x - 5ay^2 + 6ab - ac$
d $axy - a^2x^2 + ax^2y - abx$
e $6c^2d - 15bcd$
f $8axy^2 + 12abx - 20a^2x + 16ax$

**C1** Write each of these as an equation in $x$:

a $x^2 - 3 = 4a$
b $3x^2 = 5c + 1$
c $5 + \sqrt{x} = 3a - 2$
d $\sqrt{(ax)} = 7$
e $5\sqrt{x} = 4a$

**Thinking ahead to ...**
changing the subject
of formulas

**A**  A cook book gives this formula to find
the time to cook a piece of lamb.

> Allow 30 minutes per pound
> and an extra 20 minutes.

**a**  How long would it take to
cook 6 pounds of lamb?

**b**  What weight of lamb would be cooked in 1 hour and 50 minutes?

## Changing the subject of formulas

The subject of a formula is
the variable that is to be
calculated.

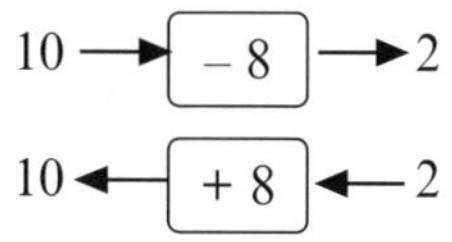

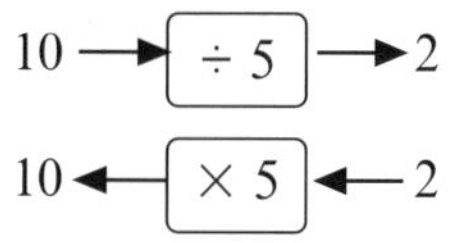

The formula that links cooking time ($T$) with weight ($W$) can be written as:

$$T = 30W + 20$$    $T$ is the subject of this formula

♦ With the formula you to calculate a value for $T$ for a value of $W$.

♦ To calculate $W$ for a value $T$, we need to make $W$ the subject of the formula.

Addition 'undoes'
subtraction and vice versa.
For example:

Flowchart method for changing the subject of a formula

♦ Draw a flowchart
for the formula.

$$W \longrightarrow \boxed{\times 30} \xrightarrow{30W} \boxed{+20} \xrightarrow{30W + 20} T$$

♦ Reverse the flow
chart to rearrange
the formula.

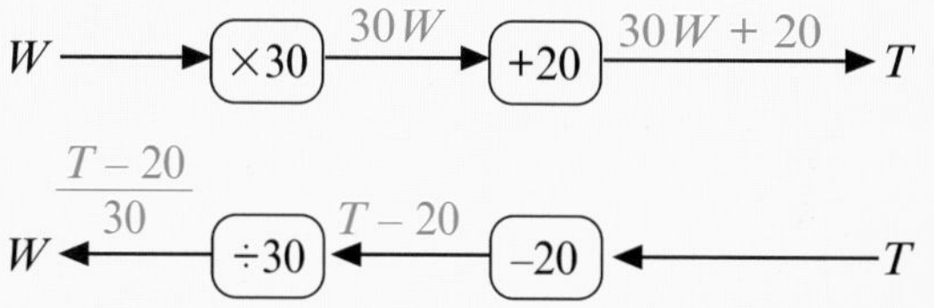

$$W \xleftarrow{\frac{T-20}{30}} \boxed{\div 30} \xleftarrow{T-20} \boxed{-20} \longleftarrow T$$

$$W = \frac{T - 20}{30}$$

Multiplication 'undoes'
division and vice versa.
For example:

Algebraic method for changing the subject of a formula

♦ $T$ is the subject

$$-20 \left( \begin{array}{c} T = 30W + 20 \\ T - 20 = 30W \\ \frac{T - 20}{30} = W \end{array} \right) \begin{array}{l} -20 \\ \div 30 \end{array}$$

♦ $W$ is the subject

$$W = \frac{T - 20}{30}$$

**Exercise 11.1**
Changing the subject
of formulas

**1**  Make $y$ the subject of each of these formulas:

**a**  $5x = 3y - 4$       **b**  $9x + 2y = 8$       **c**  $5x - 4 = 4y + 5$
**d**  $3x - 4 = 5y - 8$   **e**  $3xy + 4 = 2a$     **f**  $2x + 2xy = 3a$
**g**  $4x = 3(2y + 1)$    **h**  $3(4y - 2) = 5x$   **i**  $2(3 - 4y) = 5x$
**j**  $2(3x - y) = 5x - 4$  **k**  $5a^2 + 3 = 4 - 2y$  **l**  $4(3y - 1) = x^2 + 2$
**m**  $2(3a - 5y) = a - 9x$  **n**  $2(3 - 4y) = x^2$   **o**  $2x(3 - 4y) = 5$
**p**  $a^2(b + 2y) = 5$   **q**  $x^2 - 4 = 3a(y - a)$  **r**  $3\pi + 2ay = a^2r$

**2**  The formula that gives the cost in pence ($c$) of placing an advertisement in a
local paper, where $n$ is the number of words is:

$$c = 15n + 50$$

**a**  Calculate the cost of a 65 word advert.
**b**  **i**  Make $n$ the subject of the formula.
  **ii**  How many words are in an advert that costs £19.85?
**c**  For £10, what is the maximum number of words in an advert?

3 A formula for the sum ($S$) of the interior angles of a polygon is:

$S = 180(n - 2)$ where $n$ is the number of sides

**a** Calculate the sum of the interior angles of a heptagon.

**b** **i** Make $n$ the subject of the formula.

**ii** The sum of the interior angles of a polygon is 3240°. How many sides does the polygon have?

**c** Explain why it is not possible to draw a polygon where the sum of the interior angles is 600°.

4 When a metal rail of length $w$ metres is increased in temperature by $t$° C it increases in length. The new length $x$ metres is given by the formula:

$x = w(1 + kt)$, where $k$ has a value of 0.0008

**a** The temperature of a rail 55 metres in length is increased by 21° C.

**i** Calculate the length of the rail after the temperature increase.

**ii** With an increase in temperature of 21° does the length of the rail increase by more that 15%. Explain your answer.

**b** **i** Make $t$ the subject of the formula.

**ii** What increase in temperature increases the rail length by 1.342 metres?

5 The surface area ($S$) of a cylindrical axle is given by the formula:

$S = 2\pi r(r + h)$ where $r$ is the radius, and $h$ the length.

**a** Calculate the value of $S$, to 4 sf, when $r = 22.5$ cm and $h = 2.2$ metres.

**b** **i** Make $h$ the subject of the formula.

**ii** An axle has a radius of 15 cm, and a surface area of 3016 cm$^2$. Calculate the length of the axle to 2 sf.

## Changing the subject of formulas that involve fractions

### Example 1

- An approximate formula to convert kilometres to miles is: $m = \frac{5}{8}k$
  - Making $k$ the subject gives a formula for converting miles to kilometres.

$$m = \frac{5}{8}k$$

[× 8] $8m = 5k$

[÷ 5] $\frac{8m}{5} = k$

  - So $k = \frac{8m}{5}$

### Example 2

- A formula that links $t$ and $a$ is given as: $\frac{3t}{4} = \frac{5}{a-1}$
  - for subject $a$: [× both sides by 4] (×4, ×4) $3t = \frac{4 \times 5}{a-1}$

[× both sides by $(a - 1)$] $3t(a - 1) = 20$

$3at - 3 = 20$

[+3] $3at = 23$

[÷3$t$] $a = 23 \div 3t$

$$\mathbf{a = \frac{23}{3t}}$$

This is just one way to rearrange the above formula so that it reads $a = \ldots$. Different rearrangements may have more, or fewer, steps.

Multiplying both sides by 4, and then $(a - 1)$ produces an equation with no fractions. In this way manipulation is easier.

## Exercise 11.2 Rearranging formulas with fractions

For these questions, round each answer to 2 dp.

Remember manipulation is made easier if you can produce an equation with no fractions involved.

**1** A formula for converting kilograms ($k$) to pounds ($p$) is: $p = \frac{11}{5}k$

**a** How many pounds are equivalent to 8 kilograms?
**b** Make $k$ the subject of the formula.
**c** Convert 3 pounds to kilograms.

**2** The formula for the area ($A$) of a triangle with base length $b$ and height $h$ is:

$$A = \frac{1}{2}bh$$

**a** Make $h$ the subject of the formula.
**b** Find the height of a triangle with area 100 cm$^2$ and base length 2.5 cm.

**3** Temperature in °F can be converted to °C using this formula:

$$C = \frac{5(F - 32)}{9}$$

**a** Convert 68°F to °C.
**b** Rearrange the formula to give a formula to convert °C to °F.
**c** Convert 24° C to °F.

**4** Make $y$ the subject of each formula:

**a** $z = \frac{3}{4}y$ **b** $m = \frac{1}{3}xy$ **c** $d = \frac{1}{2}(y + 4)$ **d** $k = \frac{4}{7}(y - h)$

**5** When $d$ metres are travelled in $t$ seconds,
a formula for average speed ($s$) in km/h is: $s = \frac{18d}{5t}$

**a** In 1992, Linford Christie ran 100 metres in 9.96 seconds.
What was his average speed in km h$^{-1}$?
**b** Make $d$ the subject of the formula.
**c** Make $t$ the subject.

**6** When the temperature at ground level is $G$°C and the height above the ground in metres is $h$, the approximate temperature $T$(° $C$) is given by:

$$T = G - \frac{h}{300}$$

**a** The temperature at ground level is 26° C. Calculate the temperature outside a jet flying at 15 000 metres above ground.
**b** Make $h$ the subject of the formula.
**c** Calculate $h$ when $G = 20$ and $T = {}^{-}30$.

**7** A formula for the area ($A$) of a trapezium is given as: $A = \frac{h(b + c)}{2}$.

Calculate the value of $c$ when:
$A = 250.25$, $b = 12.40$, and $h = 17.50$.

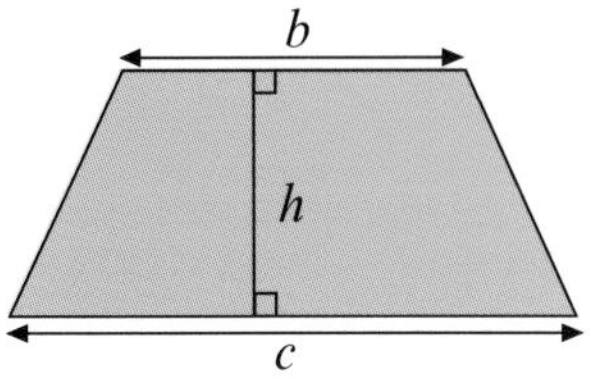


**8** Make $w$ the subject of each formula:

**a** $\frac{5}{3w} = 2x$ **b** $\sin 32° = \frac{3}{4w}$ **c** $\frac{w}{3} - 5a = 4b$

**d** $t = \frac{a}{w - b}$ **e** $\frac{a^2}{2w} = 0.5b$ **f** $\frac{c^2 - b}{2w} = 2ab$

**g** $\frac{3x}{2w + 3} = 4$ **h** $\frac{3x}{w - 1} = 2$ **i** $\frac{\tan 50°}{2w + 1} = 3$

**j** $\frac{2a}{w - 3} = 5$ **k** $\frac{w + 4}{5} = 2a$ **l** $\frac{3w - 2}{4} = l$

**m** $\frac{2w + 3}{2} = 3a - 1$ **n** $\frac{2}{w} = \frac{3}{a}$ **o** $\frac{3w}{x} = \frac{4a}{5}$

**p** $\frac{2a}{w - 3} = 4$ **q** $\frac{2}{3}(w - 3) = 5a + 1$ **r** $3a + 1 = \frac{5a}{w + 2}$

## Rearranging formulas when factorising is involved

**Example**

- A formula that links $b$, $c$, and $w$ is: $2b + 3 = \dfrac{b + c}{w}$

  - Make $b$ the subject:

| | |
|---|---|
| [× both sides by $w$] | $w(2b + 3) = b + c$ |
| | $2bw + 3w = b + c$ |
| | $2bw - b = c - 3w$ |
| [Factorise] | $b(2w - 1) = c - 3w$ |
| [÷ both sides by $(2w - 1)$] | $\boldsymbol{b = \dfrac{c - 3w}{2w - 1}}$ |

**Exercise 11.3**
**Using factorising**

**1** A formula for the surface area ($S$) of this cuboid is: $S = 2bw + 2ab + 2aw$

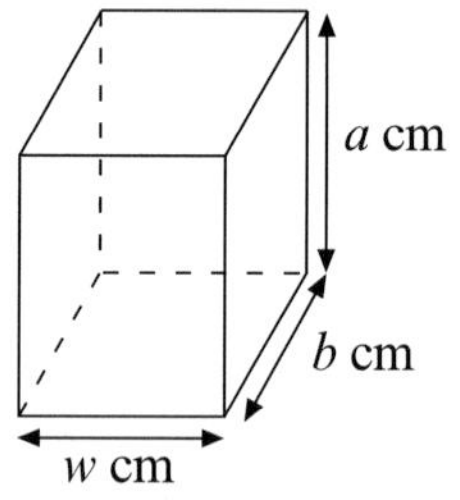


**a** Make $w$ the subject of the formula.
**b** Make $a$ the subject of the formula.
**c** When $w = 4.50$ cm and $a = 6.40$ cm the surface area is $136.08\text{ cm}^2$.
Calculate the value of $b$ in this cuboid.

**2** The diagram shows a circular play area centre O and radius $w$ metres. Part of the area is to be fenced, shown in the diagram as a red line.
A formula for the length of fencing ($P$) is:

$$P = 2w + \frac{\pi w}{2}$$

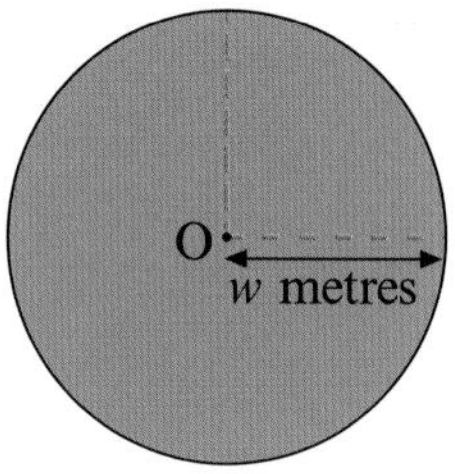


**a** Rearrange the formula so that $w$ is the subject.
**b** The length of fencing used was 89.27 metres.
Calculate, to 2 sf, the diameter of the play area.

**3** This is a block of packaging material with a rectangular slot cut through it.
A formula for the volume $V$ of packaging material in the block is: $V = abc - 2abd$

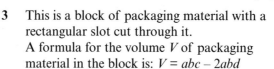

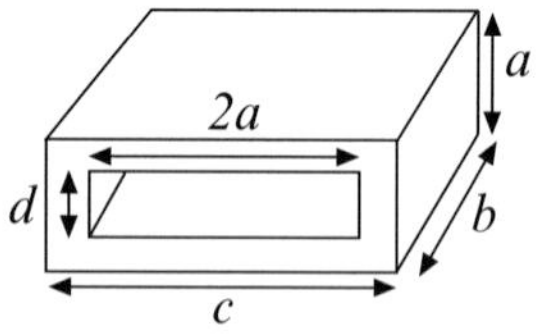


**a** **i** Rearrange the formula to make $b$ the subject.
**ii** Calculate the value of $b$ when:
$V = 1602.25\text{ cm}^3$, $a = 6.50$ cm, $c = 25.00$ cm, and $d = 4.00$ cm.
**b** Calculate the value of $a$ when:
$V = 4212.00\text{ cm}^3$, $b = 18.00$ cm, $c = 32.50$ cm, and $d = 6.50$ cm.

**4** Rearrange each of these formulas to make $h$ the subject:

**a** $3ac = 4ah - hy$ **b** $3h = 4x + 2ah$ **c** $3xh + 5 = 2x + ah$
**d** $ah + b^2 - c^2h = 4$ **e** $3h - 5 + hx = 2$ **f** $ah + bh - 4 = 5$
**g** $3ah + y + hx = 2$ **h** $3ah + 4 = h - 1$ **i** $4 - 3ah = 5bh + 1$
**j** $3 - 4ah + by = h + 1$ **k** $3 - hy + hx = 5$ **l** $3(2ah + 1) - hx = 2$
**m** $3(ah + 1) = 2(bh - 2)$ **n** $\dfrac{ah}{4} = 5x$ **o** $\dfrac{3}{h} + 4 = a$
**p** $2(3ah - 4) = \frac{1}{3}$ **q** $\frac{2}{3}a = 4(xh + b)$

# Rearranging formulas with squares and roots

**Example 1**

Four square photographs, of side $x$, are placed in a square mount of side $c$ as shown in the diagram.

**a** Write a formula for the area ($A$) of mount that is still visible.
**b** Rearrange the formula to make $x$ the subject.

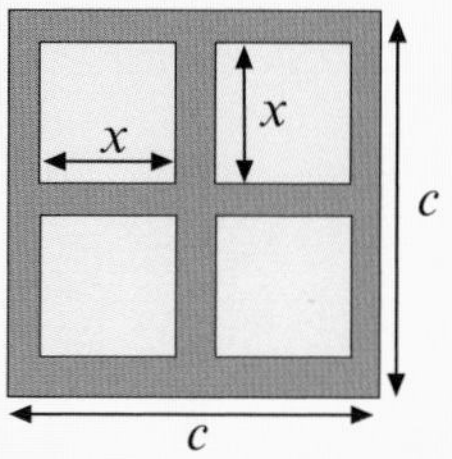


**a** A formula for the area ($A$) is: $A = c^2 - 4x^2$

**b** The area is: $A = c^2 - 4x^2$

[+ $4x^2$] $4x^2 + A = c^2$

[$-A$] $4x^2 = c^2 - A$

[÷ both sides by 4] $x^2 = \frac{(c^2 - A)}{4}$

[√both sides] $x = \sqrt{\frac{(c^2 - A)}{4}} = \frac{\sqrt{(c^2 - A)}}{\sqrt{4}} = \mathbf{\frac{\sqrt{(c^2 - A)}}{2}}$

**Example 2**

Make $y$ the subject of the formula: $\sqrt{(y + 3a)} = a - 2$

[square both sides] $y + 3a = (a - 2)^2$

$y + 3a = a^2 - 4a + 4$

[$-3a$] $\mathbf{y = a^2 - 7a + 4}$

> To square both sides, you square the *expression* on each side. This is not simply squaring each *term* on both sides.

**Exercise 11.4**
**Dealing with squares and roots**

1 When an object is dropped, a formula for the approximate distance ($d$ metres) travelled in a time of $t$ seconds is:

$$d = (\tfrac{49}{10})t^2$$

The Sears tower in Chicago is 443 metres high.
About how long will it take an apple to fall from the top to the ground?

2 A formula for velocity ($v$) is : $v^2 = u^2 + 2as$

**a** Show why a formula for $u$ is: $u = \sqrt{(v^2 - 2as)}$
**b** Calculate the value of $u$, to 1 sf, when: $v = 14$, $a = 3.5$, and $s = 18$

3 The formula for the volume ($V$) of a right cone of radius $r$, and height $h$ is: $V = (\frac{1}{3})\pi r^2 h$.
Rearrange the formula to make $r$ the subject.

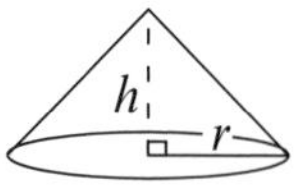


4 A formula for the volume ($V$) of a sphere of radius $r$ is : $V = (\frac{4}{3})\pi r^3$.

**a** Make $r$ the subject of the formula.
**b** Calculate the radius of a sphere which has a volume of 100 cm³.

> $p^3 = p \times p \times p$
>
> If $y = p^3$ then the **cube root** of $y$ is $p$.
>
> $\sqrt[3]{y} = p$

5 A formula for the area ($A$) of a triangle is: $A = \sqrt{s(s-a)(s-b)(s-c)}$

Rearrange the formula to make $b$ the subject.

6 Rearrange each formula to make $b$ the subject:

| | | |
|---|---|---|
| a $b^2 + 3a - 4 = y^2$ | b $3y = x - 2ab^2$ | c $b^2x = y - 1$ |
| d $b^2 - 3x = 4 - 2x$ | e $4x = 2(b^2 - 5)$ | f $3x + 4a = 2(1 - b^3)$ |
| g $b^2 = \frac{x - 1}{b}$ | h $b^2 = \frac{b^2 - c}{2a}$ | i $\sqrt{(3b - 1)} = a$ |
| j $2a = \sqrt{(b - 2)}$ | k $w = \sqrt{(ab^2 + 3)}$ | l $bc = \sqrt{(b^2 + 1)}$ |
| m $2\sqrt{(a^2 - b^2)} = xy$ | n $\pi\sqrt{(a + b)} = r$ | o $k = 2\pi\sqrt{\frac{b}{y}}$ |
| p $\sqrt{(2b - x)} = \sqrt{(a + c)}$ | q $c - 3 = \sqrt{(2b + 1)}$ | r $2\sqrt{(b - x)} = x - 2$ |

# Evaluating non-linear formulas

- For the formula, $a = (b + 5)(b - 1)$, find the value of a when $b = 3$.

$$\begin{aligned} a &= (3 + 5) \times (3 - 1) \\ &= 8 \times 2 \\ &= 16 \end{aligned}$$

- For the formula, $y = x^2 + 4x + 1$, find the value of $y$ when $x = 7$.

$$\begin{aligned} y &= 7^2 + (4 \times 7) + 1 \\ &= 49 + 28 + 1 \\ &= 78 \end{aligned}$$

**Exercise 11.5**
**Evaluating non-linear formulas**

**Accuracy**
For these questions, give answers correct to 2 dp when no other degree of accuracy is stated.

1 Stopping distance is the distance a vehicle travels as it brakes to a halt. A formula that gives the stopping distance of a car on a dry road is:

$$d = \frac{v^2}{200} + \frac{v}{5},$$

where $d$ is the stopping distance (metres) and $v$ the speed (km h$^{-1}$).
Calculate the stopping distance of a car travelling at 110 km h$^{-1}$ on a dry road.

2 The net for a square-based box is made by cutting four square corners from a 12 cm square piece of card as shown.
A formula for the volume ($V$ cm$^3$) of the box is:

$$V = 4x(x^2 - 12x + 36)$$

where $x$ is the length shown on the diagram.

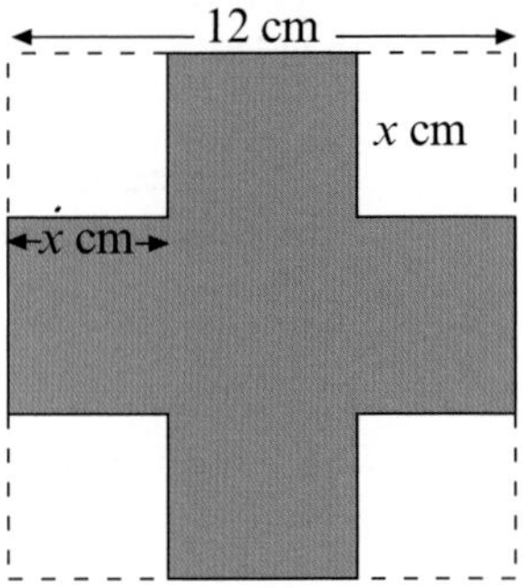


**a** Find the volume of the box when $x = 1.5$.
**b** Calculate $V$ when $x = 4$.
**c** What value for $x$ gives the box with the greatest volume?

To calculate the greatest volume think about using a graph, or a spreadsheet.

3 A $y = x^2 + 5x - 8$ B $y = x^2 + 3x + 2$ C $y = x^3 - 3x^2 - x + 2$
D $y = (x + 1)(x + 2)$ E $y = (x - 2)(x + 9)$ F $y = (x + 3)(x + 1)(x - 3)$

For each formula find the value of $y$ when:
**a** $x = 5$ **b** $x = {}^{-}5$ **c** $x = 2$ **d** $x = {}^{-}2$ **e** $x = 2.5$ **f** $x = {}^{-}2.5$
Comment on any patterns in your results.

4 The formula $v = \sqrt{(u^2 + 2as)}$ gives:
the velocity of an object ($v$) in m/s, for an initial velocity of $u$ m/s, an acceleration ($a$) in m s$^{-2}$, and a distance ($s$) travelled in metres.

**a** Calculate to 1 dp the value of $v$ when:
**i** $u = 30$, $a = 5$ and $s = 6.8$ **ii** $u = 20$, $a = 9.8$ and $s = 2.4$
**b** Calculate to 2 sf the value of $u$ when:
**i** $s = 15.40$, $v = 120.00$ and $a = 3.60$
**ii** $v = 100$, $a = {}^{-}14.50$, $s = 250$
**iii** $a = 3\frac{1}{2}$, $v = 18\frac{1}{4}$, and $s = 6\frac{1}{3}$

5 The volume ($V$) of a sphere of radius $r$ is given by the formula $V = \frac{4}{3}\pi r^3$.
The Earth is an approximate sphere of radius $6.4 \times 10^6$ metres.

**a** Calculate the volume of the Earth, correct to 4 sf.
**b** Rearrange the formula to give $r$ in terms of $V$.

## End points

You should be able to ... ... so try these questions

**A** Rearrange a formula

**A1** Rearrange each formula to make $x$ the subject:
- **a** $3x - a = 5$
- **b** $a - 4x = 2x + 3$
- **c** $2(3x + y) = 5$
- **d** $3(a - xy) = b + 2xy$
- **e** $3(2 - 3x) = 3(x + bc)$

**B** Handle formulas where fractions are involved

**B1** Make $v$ the subject of each formula:
- **a** $\frac{3}{4}v = a - x$
- **b** $2v + 1 = \frac{3}{5}(a - 3)$
- **c** $\frac{2a + 3}{v} = 3a^2$
- **d** $\frac{2v + 5}{4} = ax + 2y$
- **e** $\frac{3v + 2a}{3} = \frac{a + v}{5}$
- **f** $\frac{3}{b - 2v} = c^2$
- **g** $\frac{1}{w} + \frac{2}{v} = \frac{3}{x}$

**C** Recognise when you need to factorise when rearranging a formula

**C1** Rearrange each formula to give $c$ in terms of the other variables:
- **a** $ax^2 + cx = 3$
- **b** $3x + 2ac = ac - 5x + cy$
- **c** $2a(a + 3c) = a(1 + c)$
- **d** $a^2b - a^2c = 3 + 2a$
- **e** $(w - cx^2) = \frac{2}{3}$
- **f** $2 + a^2c + cx = ax + 1$

**D** Deal with powers and roots in formulas

**D1** In each formula rearrange the terms to make $y$ the subject:
- **a** $\sqrt{xy} = 5$
- **b** $(a + x) = \sqrt{(1 + y)}$
- **c** $\frac{2}{\sqrt{y}} = 2x - 1$
- **d** $3\sqrt{\frac{y}{a}} = b$
- **e** $5 = 2\sqrt{(a^2 - y)}$
- **f** $\sqrt{(1 - ay + by)} = 3$
- **g** $3 - y^2 = 4a + x$
- **h** $2(3y^2 - 4) = 5x$
- **i** $3y^2 = 6ax - 9$

**E** Evaluate formulas

**E1** A formula for the surface area ($S$) of a cylinder is:
$S = 2\pi rh + 2\pi r^2$, where $r$ is the radius and $h$ the height.
- **a** Calculate the value of $S$ when: $r = 5.7$ cm and $h = 12.5$ cm. Give your answer to 3 sf.
- **b** Calculate the value of h when: $r = 5$ cm and $S = 424\,\text{cm}^2$. Give your answer to 1 dp.

**E2** Given the formula $p = \frac{a + b^2}{3a}$,

calculate to 2 sf the value of $b$ when: $a = 3.5$, and $p = 5.5$.

## Starting points

You need to know about ... / ... so try these questions

### A Writing ratios

- A **ratio** compares the size of two or more quantities.

  3 : 4  1 : 4 : 2  $3\frac{1}{2} : \frac{1}{2}$

- In the ratio 3 : 4, the **number of parts** are 3 and 4.

**A1** Roughcast is used to cover walls. It is a mix of 3 parts mortar to 1 part gravel.
Write this mix as a ratio of:
**a** mortar to gravel
**b** gravel to mortar.

### B Equivalent ratios

- The ratios 3 : 4 and 6 : 8 are **equivalent ratios**.

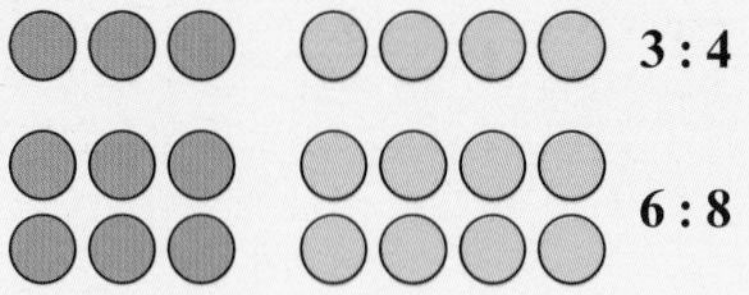


- To write an equivalent ratio:
  - multiply or divide each number of parts by the same number.

  6 : 8 (÷2) → **3 : 4**  2 : 3 : 7 (×5) → **10 : 15 : 35**

- A ratio in its **simplest terms**, or **lowest terms**, is written with the smallest whole numbers possible.

  $1 : \frac{1}{4} : 3$ (×4) → **4 : 1 : 12**  15 : 24 (÷3) → **5 : 8**

**B1** Copy and complete these sets of equivalent ratios.
**a** 4 : 1, 12 : ?  **b** 2 : 5 : 4, 1 : ? : 2  **c** 9 : 12, ? : 4, 6 : ?

**B2**

| | | |
|---|---|---|
| 2 : 5 | 1 : 3 : 2 | 10 : 20 |
| 10 : 25 | $2\frac{1}{2} : 5$ | 1 : 2 : 3 |
| 3 : 9 : 6 | 2 : 1 | $1 : 2\frac{1}{2}$ |

List the sets of equivalent ratios.

**B3** Write each of these ratios in its simplest terms.
**a** 9 : 6  **b** 14 : 21
**c** 4 : 8 : 2  **d** $5 : \frac{1}{2}$
**e** $2 : 1\frac{1}{2} : 3$

### C Unitary ratios

- Two or more ratios can be compared by writing each one as a **unitary ratio** in the form ***n*** **: 1** or **1 :** ***n***.

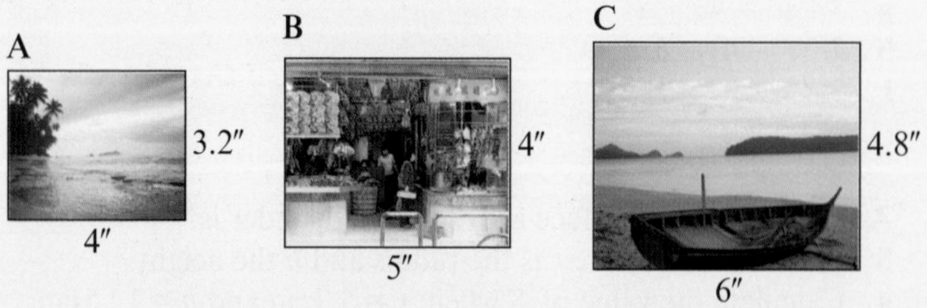


The ratio of **width to height** for each of these photographs is:

| A | B | C |
|---|---|---|
| 4 : 3.2 | 5 : 4 | 6 : 4.8 |
| **1.25 : 1** | **1.25 : 1** | **1.25 : 1** |
| or **1 : 0.8** | or **1 : 0.8** | or **1 : 0.8** |

The ratios are equivalent so the photographs are **in proportion**.

**C1**

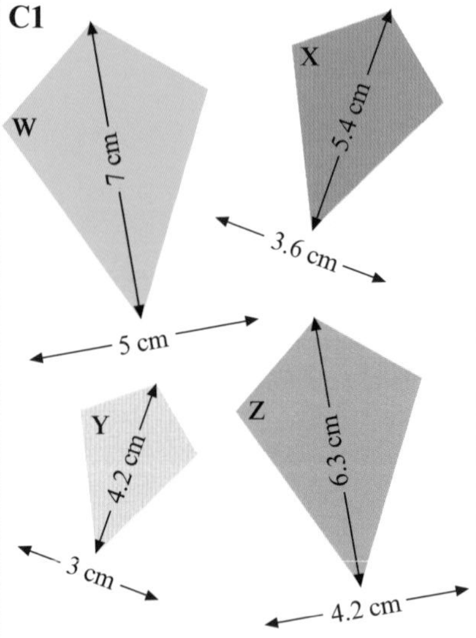


Which of these kites are in proportion?

## D Writing ratios as fractions, decimals and percentages

- A ratio that compares **two** quantities can be written as a fraction, as a decimal, and as a percentage.

4″

5″

The **width-to-height ratio** is 5 : 4 or:

$$\frac{\textbf{width}}{\textbf{height}} = \frac{5}{4} = \mathbf{1.25 = 125\%}$$

- the width is $\frac{5}{4}$ of the height
- the width is 1.25 times the height
- the width is 125% of the height.

## E Sharing in a given ratio

- An amount can be shared in a given ratio.

**Example** Share £140 in the ratio 2 : 1 : 4

- find the total number of parts
  2 + 1 + 4 = 7
- find the amount for one part
  £140 ÷ 7 = £20
- multiply each number of parts by the amount for one part.

| | | | | | | Total |
|---|---|---|---|---|---|---|
| 2 | : | 1 | : | 4 | | 7 |
| ×£20 | | ×£20 | | ×£20 | | ×£20 |
| **£40** | : | **£20** | : | **£80** | | £140 |

## F Similar triangles

- Triangles that have the same three angles are similar.

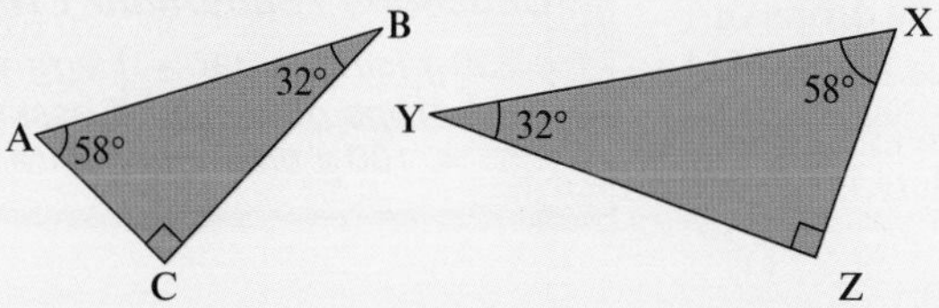


- $\hat{A}$ and $\hat{X}$ are **corresponding angles** because they are equal.
- AC and XZ are **corresponding sides** because they are opposite corresponding angles.

- Triangles that are enlargements of each other are similar.

## G Reciprocals

- Two numbers which multiply together to equal 1 are **reciprocals** of each other.

| | |
|---|---|
| **0.8** is the reciprocal of 1.25 | $\frac{4}{5}$ is the reciprocal of $\frac{5}{4}$ |
| **1.25** is the reciprocal of 0.8 | $\frac{5}{4}$ is the reciprocal of $\frac{4}{5}$ |

- The reciprocal of a number $n$ is $\frac{1}{n}$ or $1 \div n$.

$$\frac{1}{0.8} = \mathbf{1.25} \qquad \frac{1}{1.25} = \mathbf{0.8}$$

**D1** Write each of these ratios as:
**i** a fraction
**ii** a decimal
**iii** a percentage.

**a** 5 : 8 **b** 9 : 5
**c** 7 : 12 **d** 15 : 11
**e** 11 : 18 **f** 17 : 24

**E1** Amy, Ben and Zoe are left £150 by their grandmother. They share the money in the ratio of their ages: 5, 3, and 2. Calculate how much each gets.

**E2** Liz trains for the 100 m hurdles. She splits her time 3 : 5 between speed work and technique. Calculate how long Liz spends on:
**a** technique in 2 hours training
**b** speed work in 4 hours training

**F1** EFG and JKL are similar.

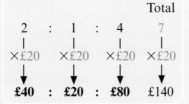

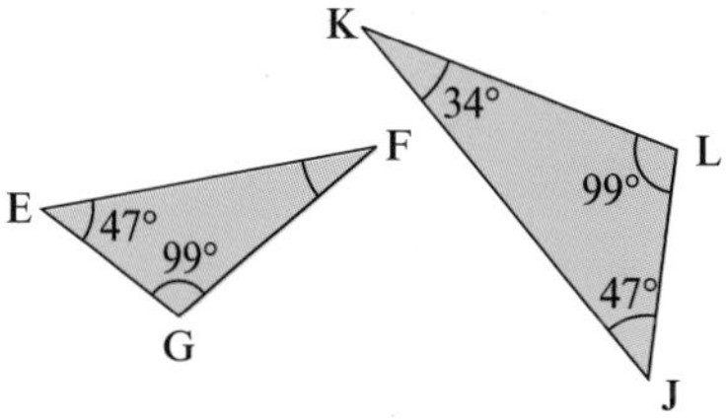


**a** Explain why $\hat{F}$ is equal to 34°.
**b** Give the corresponding angle to $\hat{G}$.
**c** Give the corresponding side to:
**i** EG **ii** KL

**G1** Give the reciprocal of:
**a** 2 **b** 0.25
**c** 3 **d** $1\frac{1}{3}$

# Keeping in proportion

- When adjusting a recipe, you need to keep the ingredients in proportion.

**Example** This recipe for Swiss Hot Chocolate serves 2 people.
How much of each ingredient is needed for 5 people?

**Swiss Hot Chocolate**
- *600 ml milk*
- *140 g drinking chocolate*
- *80 ml whipped cream*

| Milk | | Chocolate | | Cream | Serves |
|---|---|---|---|---|---|
| 600 ml | : | 140 g | : | 80 ml | 2 |
| ×2.5 ↓ | | ×2.5 ↓ | | ×2.5 ↓ | ×2.5 ↓ |
| **1500 ml** | : | **350 g** | : | **200 ml** | 5 |

You can keep the multiplier in your calculator memory.

- find the **multiplier:** $5 \div 2 = 2.5$
- multiply each amount by the multiplier.

**Exercise 12.1**
**Recipes**

1 This recipe serves 6 people.
How much of each ingredient do you need for 9 people?

**Chilled Chocolate Drink**
*230 g caster sugar • 300 ml water*
*50 g cocoa • 1200 ml chilled milk*

2 This recipe for chocolate fudge makes 36 pieces.

**a** How much cocoa is needed to make 72 pieces?
**b** How much milk is needed to make 18 pieces?
**c** Calculate how much of each ingredient you need to make 24 pieces.

**Chocolate Fudge**
- *450 g white sugar*
- *150 ml milk*
- *150 ml water*
- *75 g butter*
- *30 g cocoa*

3 This recipe makes 24 sweets.

**a** Calculate how many drops of peppermint essence are needed to make 42 sweets.
**b** How much chocolate is needed for 42 sweets?

**Chocolate Peppermint Creams**
*• 230 g icing sugar • 1 egg white*
*• 4 drops peppermint essence*
*• 100 g plain chocolate*

- When you use an equivalent ratio to solve a problem, you are keeping the number of parts in proportion.

**Example**

An orange dye is a mix of red and yellow in the ratio 2 : 3.

**a** What amount of red is mixed with 480 ml of yellow?
**b** How much orange dye is produced?

| | Red | | Yellow | Total |
|---|---|---|---|---|
| Ratio | 2 | : | 3 | |
| Amounts | ? | : | 480 | ? |

- find the multiplier, $480\text{ ml} \div 3 = 160\text{ ml}$
- multiply the other number of parts by the multiplier
- find the total of the amounts.

| Red | | Yellow | Total |
|---|---|---|---|
| 2 | : | 3 | |
| ×160 ml ↓ | | ×160 ml ↓ | |
| 320 ml | : | 480 ml | 800 ml |

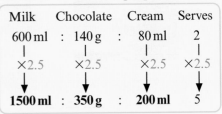

**Exercise 12.2**
**Using equivalent ratios**

1 Calculate:

**a** the amount of yellow dye to mix with 210 ml of red dye
**b** how much orange dye is produced.

2 Walls can be covered with mortar to protect them from the weather.

**a** For a clay wall how much cement is mixed with 1500 kg of sand?
**b** For a concrete wall:
  **i** write the mortar mix as a ratio in its simplest terms
  **ii** calculate how much concrete is produced with 400 kg of cement.

| Wall surface | Mortar mix Cement : Lime : Sand |
|---|---|
| Clay | 1 : 1 : 6 |
| Concrete | $1 : \frac{1}{2} : 4\frac{1}{2}$ |

3 On a map, 4.8 cm stands for a distance of 240 m.

**a** What length on the map represents a distance of 315 m?
**b** Give the map scale as a unitary ratio.
**c** Explain why the total number of parts has no meaning in this ratio.

◆ When one shape is an enlargement of another, each dimension has been multiplied by the same number to keep it in proportion.

**Example** I′J′K′L′ is an enlargement of IJKL. Find the height K′L′.

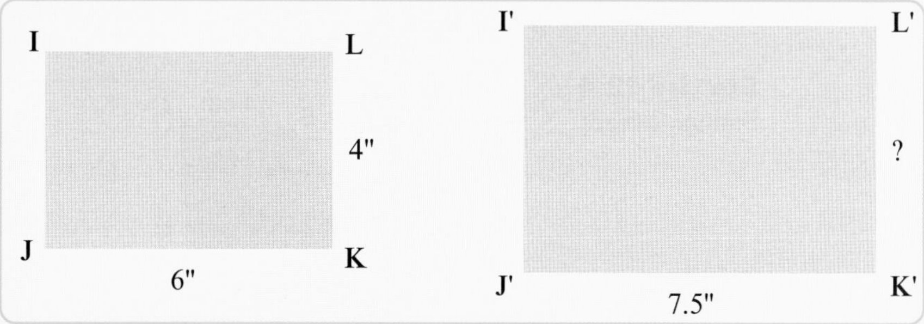


The multiplier is the scale factor of the enlargement.

❖ find the multiplier,
$7.5 \div 6 = 1.25$

❖ multiply the height KL by the multiplier.

The shapes have the same width-to-height ratio, so they are in proportion.

| Width | | Height | | $\frac{\text{Width}}{\text{Height}}$ |
|---|---|---|---|---|
| 6″ | : | 4″ | IJKL | $\frac{6}{4} = 1.5$ |
| ×1.25 ↓ | | ×1.25 ↓ | | |
| **7.5″** | : | **5″** | **I′J′K′L′** | $\frac{7.5}{5} = 1.5$ |

**Exercise 12.3**
**Enlargement**

1

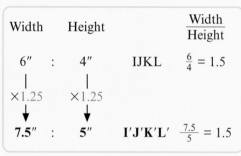

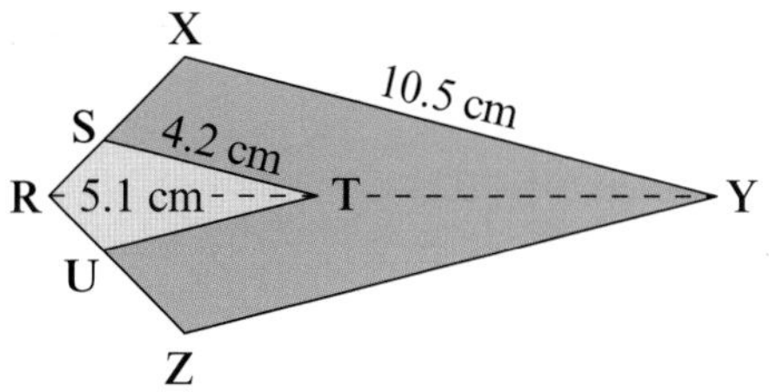


Kite RXYZ is an enlargement of kite RSTU.

**a** Calculate the scale factor of the enlargement.
**b** Find the length RY.
**c** Calculate the ratio:
  **i** $\frac{RT}{ST}$ **ii** $\frac{RY}{XY}$
**d** Explain why the ratios are equal.

## Similar shapes

- Two shapes are **similar** if:
  - all corresponding angles are equal, **and**
  - all ratios of lengths of corresponding sides are equal.

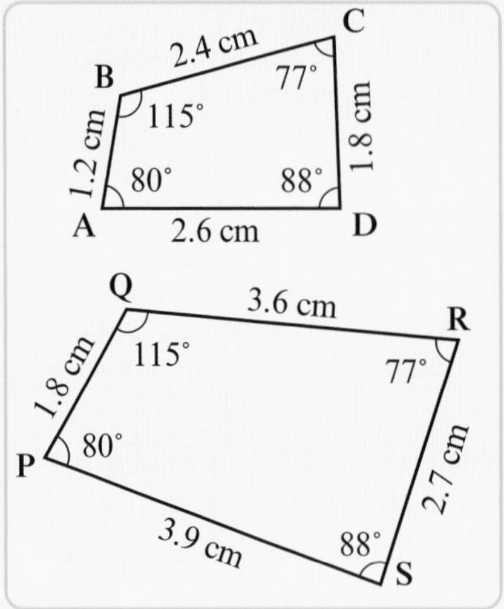


For shapes ABCD and PQRS:

- corresponding angles are equal:
  $\hat{A} = \hat{P}, \quad \hat{B} = \hat{Q}, \quad \hat{C} = \hat{R}, \quad \hat{D} = \hat{S}$
- ratios of lengths of corresponding sides are equal:

$$\frac{PQ}{AB} = \frac{1.8}{1.2} = 1.5 \qquad \frac{RS}{CD} = \frac{2.7}{1.8} = 1.5$$

$$\frac{QR}{BC} = \frac{3.6}{2.4} = 1.5 \qquad \frac{PS}{AD} = \frac{3.9}{2.6} = 1.5$$

So ABCD and PQRS are similar.

These ratios are greater than 1 because lengths in the larger shape have been divided by lengths in the smaller shape.

A ratio is less than 1 if you divide lengths in the smaller shape by lengths in the larger shape.

$$\frac{AB}{PQ} = \frac{BC}{QR} = \frac{CD}{RS} = \frac{AD}{PS} = 0.\dot{6}$$

**Exercise 12.4**
Similar shapes

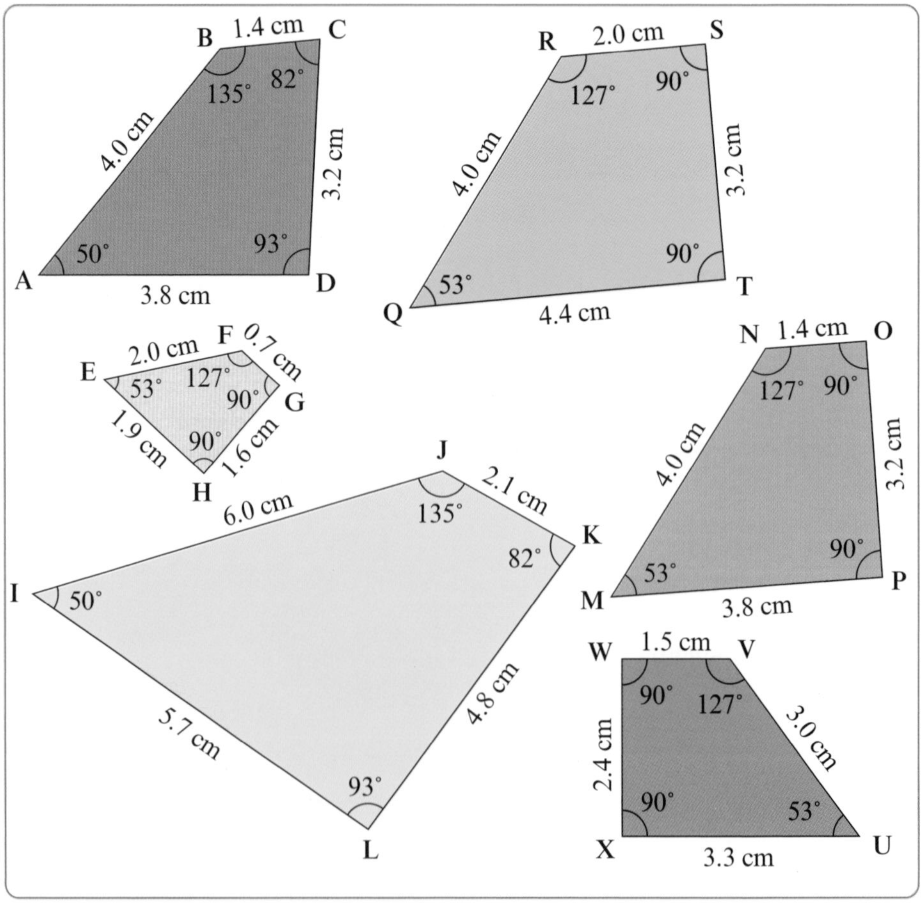


1 **a** Which shape is similar to EFGH?
  **b** Explain why.

2 **a** List all the pairs of similar shapes.
  **b** Give a ratio of the lengths of corresponding sides for each pair.

## Finding lengths in similar shapes

- You can use a ratio to find a length in the **larger** of two similar shapes.

  **Example** ABC and PQR are similar triangles.
  Find the length PQ.

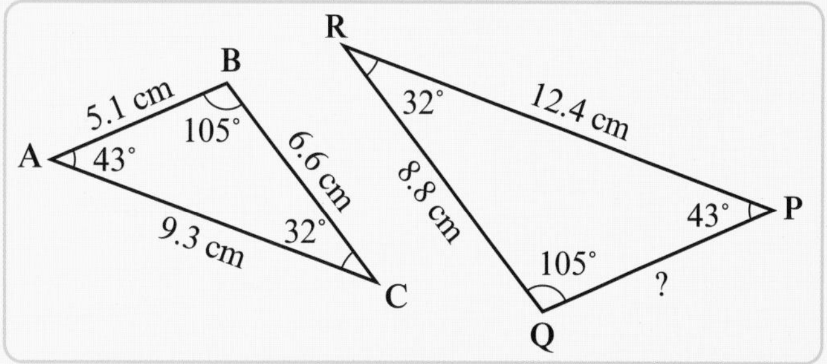


  - Calculate the ratio greater than 1 for any two corresponding lengths. $\frac{PR}{AC} = \frac{12.4}{9.3} = 1.\dot{3}$
  - Multiply AB, the corresponding length to PQ, by the ratio. AB 5.1 cm $\xrightarrow{\times 1.\dot{3}}$ PQ **6.8 cm**

- To find a length in the **smaller** of two similar shapes:
  - use the same method with the ratio less than 1.

Dividing a length on the larger shape by a length on the smaller shape gives the ratio greater than 1:

$$\frac{PR}{AC} = \frac{12.4}{9.3} = 1.\dot{3}$$

The ratio less than 1 is:

$$\frac{AC}{PR} = \frac{9.3}{12.4} = 0.75$$

**Exercise 12.5**
**Finding lengths in similar shapes**

**1**

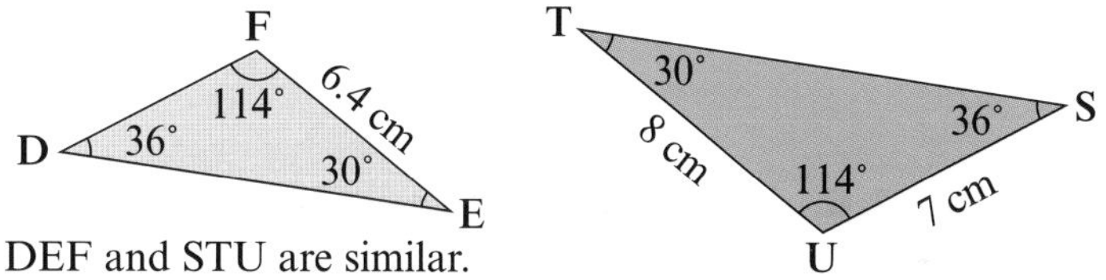


DEF and STU are similar.
Explain why this calculation is wrong.

To find DF: $\frac{TU}{FE} = \frac{8}{6.4} = 1.25$

So DF = US × 1.25 = 7 cm × 1.25
= 8.75 cm ✗

**2**

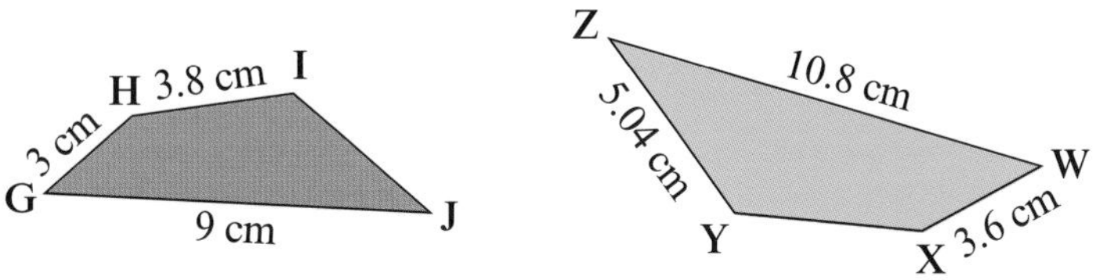


GHIJ and WXYZ are similar.
Find the length:

**a** YX **b** IJ

**3** LMN and LJK are similar triangles.

**a** Find the length JM.
**b** Explain why JK and MN are parallel.

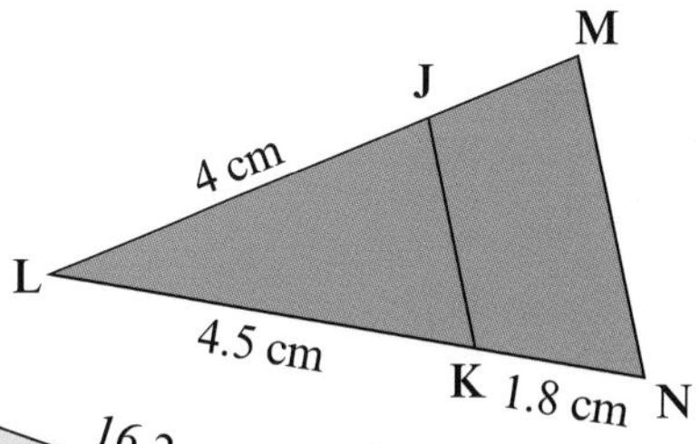


**4** KLNP and QMNO are similar.
Find the length LN.

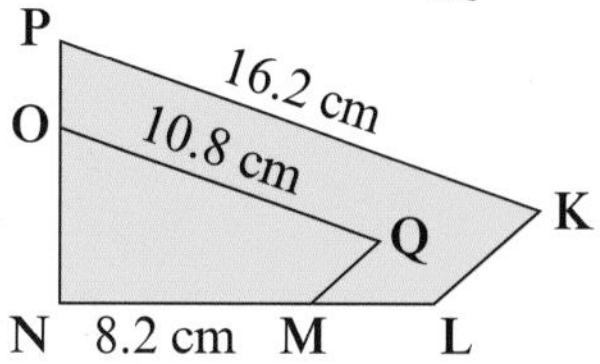

# Similar triangles

If two triangles have the same three angles then they must be similar.

All ratios of corresponding lengths, therefore, are equal.

- You need to recognise when triangles are similar.

**Example** In this diagram, PQ is parallel to ST.

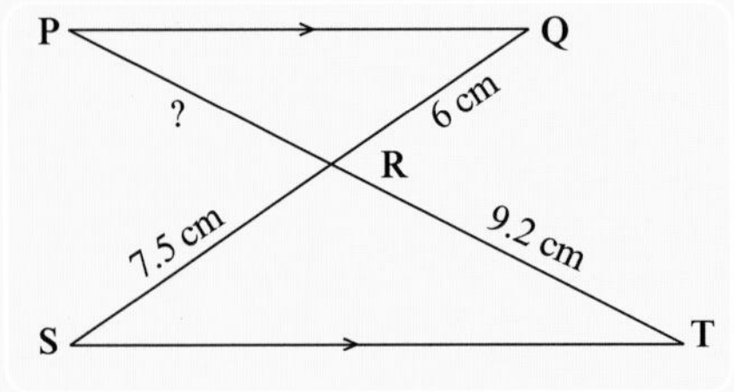


**a** Explain why triangles PQR and TSR are similar.
**b** Find the length PR.

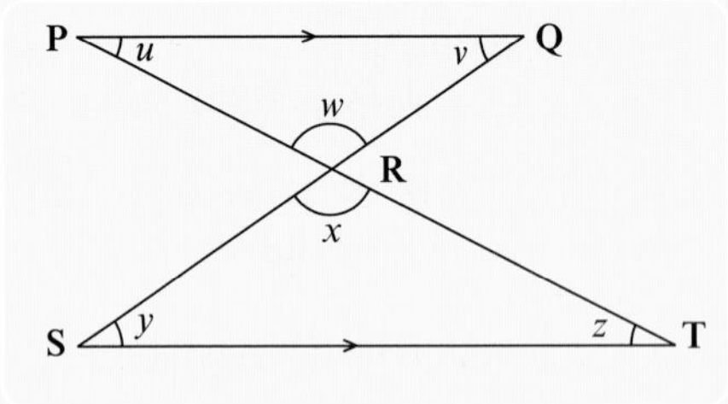


$u$ and $z$ are alternate angles,
$v$ and $y$ are alternate angles,
$w$ and $x$ are opposite angles, so
$u = z,\ v = y,\ w = x$

The triangles have the same three angles, so they are similar.

**b** The ratio is: $\frac{QR}{RS} = \frac{6}{7.5} = 0.8$

**PR** = RT × 0.8
= 9.2 cm × 0.8
= **7.36 cm**

## Exercise 12.6 Similar triangles

**1** **a** Explain why ABD and CAD are similar triangles.
**b** Give the corresponding side to:
**i** AC **ii** AD
**c** Find the length:
**i** AB **ii** BC

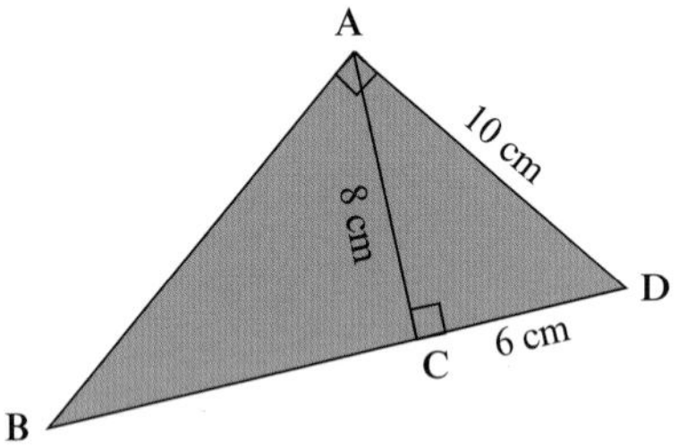


**2** **a** Calculate the ratio:
**i** $\frac{KM}{NM}$ **ii** $\frac{OM}{LM}$
**b** Explain why KMO and NML are similar triangles.

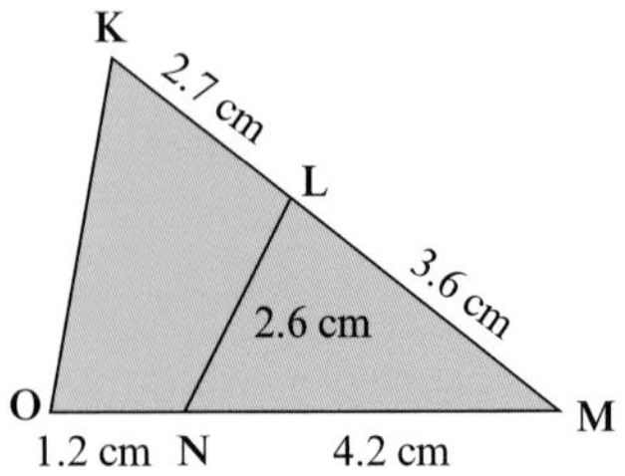


**3** **a** Explain why DEF and GHF are similar triangles.
**b** Find the length GH.

**4** Explain why enlarging GHF, using centre of enlargement F, does **not** give DEF.

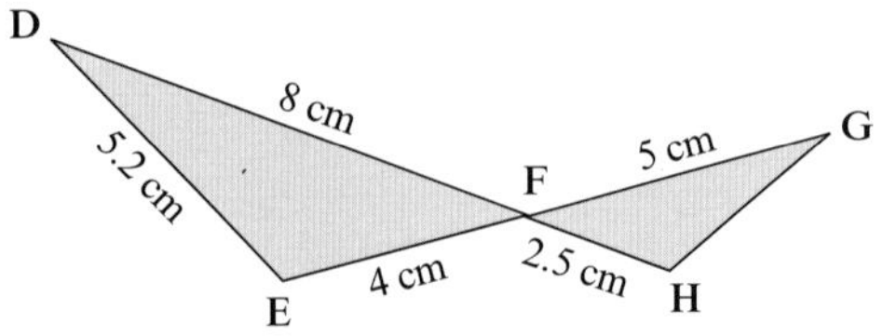

# Proportionality

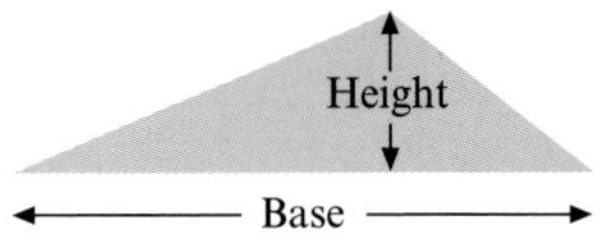


- All the shapes in a set of similar shapes must be in proportion.

**Example** This logo is made up of four similar triangles. The dimensions of the triangles, in centimetres, are:

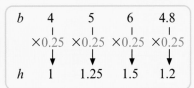

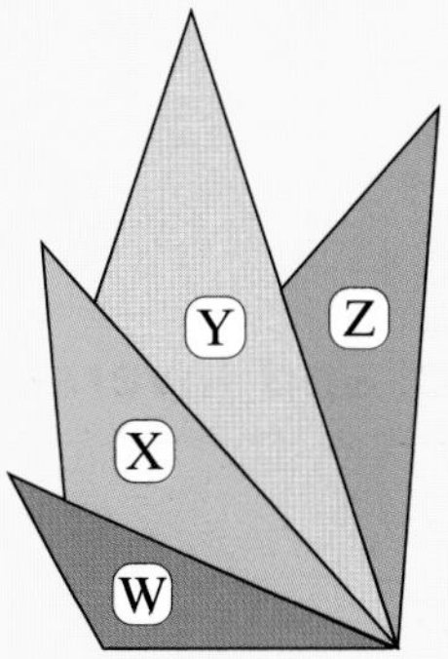


| | W | X | Y | Z |
|---|---|---|---|---|
| Base ($b$) | 4 | 5 | 6 | 4.8 |
| Height ($h$) | 1 | 1.25 | 1.5 | 1.2 |

| | | | | |
|---|---|---|---|---|
| $b$ | 4 | 5 | 6 | 4.8 |
| | ×0.25 ↓ | ×0.25 ↓ | ×0.25 ↓ | ×0.25 ↓ |
| $h$ | 1 | 1.25 | 1.5 | 1.2 |

This ratio is a **constant** because it is the same for each pair of values.

All the ratios $\frac{h}{b}$ are the same, 0.25, so $h$ is **proportional to** $b$.

The symbol $\propto$ means 'is proportional to'.

- If one variable, $y$, is proportional to another variable, $x$, you can write:

$$y \propto x$$

**Exercise 12.7**
**Proportionality**

1 **a** Draw a graph of $h$ against $b$.
The four points lie on a straight line.
**b** Explain why the line, when extended, passes through (0, 0).
**c** What is the link between the constant ratio and the gradient of the line ?
**d** Give an equation connecting $b$ and $h$ in the form $h = \ldots$ .

2 **a** For each of the four similar triangles, calculate the ratio $\frac{b}{h}$.
**b** Give an equation connecting $b$ and $h$ in the form $b = \ldots$ .

This angle is the angle between the base of each triangle and the horizontal.

3 **a** Draw a graph of $a$ against $b$.
**b** Is $a$ proportional to $b$?
Explain your answer.

| | W | X | Y | Z |
|---|---|---|---|---|
| Base ($b$) | 4 | 5 | 6 | 4.8 |
| Angle ($a°$) | 24° | 48° | 72° | 84° |

4 **a** Copy and complete this table.
**b** Draw a graph of $A$ against $h$.
**c** Is $A \propto h$ ?
Explain your answer.

| | W | X | Y | Z |
|---|---|---|---|---|
| Height ($h$) | 1 | 1.25 | 1.5 | 1.2 |
| Area ($A$) | | | | |

5 **a** Copy and complete this table.
**b** Draw a graph of $A$ against $h^2$.
**c** Is $A \propto h^2$?
Explain your answer.

| | W | X | Y | Z |
|---|---|---|---|---|
| $h^2$ | | | | |
| $A$ | | | | |

6 These are the dimensions of another four triangles.
**a** Draw a graph of $h$ against $b$.
**b** Describe the shape of the graph.
**c** Calculate ratios to show that $h$ is not proportional to $b$.

| | A | B | C | D |
|---|---|---|---|---|
| $b$ | 4 | 4.8 | 5 | 6 |
| $h$ | 2.5 | 2.7 | 2.75 | 3 |

**7** **a** Each of these five triangles has an area of 3 cm².
**i** Draw a graph of $h$ against $b$.
**ii** Is $h \propto b$?
Explain your answer.

| | P | Q | R | S | T |
|---|---|---|---|---|---|
| Base ($b$) | 3 | 4 | 5 | 6 | 8 |
| Height ($h$) | 2 | 1.5 | 1.2 | 1 | 0.75 |

**b** **i** Copy and complete this table.
**ii** Draw a graph of $h$ against $\frac{1}{b}$.
**iii** Is $h \propto \frac{1}{b}$?
Explain your answer.

| | P | Q | R | S | T |
|---|---|---|---|---|---|
| $\frac{1}{b}$ | | | | | |
| $h$ | 2 | 1.5 | 1.2 | 1 | 0.75 |

## Direct and inverse proportion

When two variables are directly proportional, an **increase** in one matches an **increase** in the other.

- If two variables, $x$ and $y$, are proportional to each other, $y \propto x$, then they are in **direct** proportion.

 The relationship between the variables can be given as: $y = kx$, where $k$ is a constant.

 **Example**

| $b$ | 4 | 5 | 6 | 4.8 |
|---|---|---|---|---|
| | ×0.25 ↓ | ×0.25 ↓ | ×0.25 ↓ | ×0.25 ↓ |
| $h$ | 1 | 1.25 | 1.5 | 1.2 |

 $h = 0.25\,b$

When two variables are inversely proportional, an **increase** in one matches a **decrease** in the other.

- Two variables, $x$ and $y$, are in **inverse** proportion when one of the variables is in direct proportion to the reciprocal of the other: $y \propto \frac{1}{x}$

 The relationship between the variables can be given as: $y = \frac{k}{x}$ or $xy = k$, where $k$ is a constant.

 **Example**

| $b$ | 3 | 4 | 5 | 6 | 8 |
|---|---|---|---|---|---|
| $h$ | 2 | 1.5 | 1.2 | 1 | 0.75 |
| $bh$ | 6 | 6 | 6 | 6 | 6 |

 $h = \frac{6}{b}$ or $bh = 6$

**Exercise 12.8 Direct and inverse proportion**

**1** The dimensions of these metric paper sizes, given in millimetres, are in direct proportion.

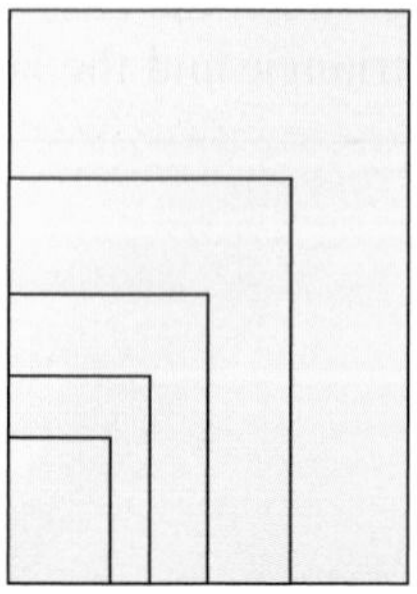

| | A4 | A3 | A2 | A1 | A0 |
|---|---|---|---|---|---|
| Width ($w$) | 210 | 297 | 420 | 594 | 841 |
| Height ($h$) | 297 | 420 | 594 | 841 | 1189 |

**a** Calculate the height to width ratio, $\frac{h}{w}$, for each size.
**b** Explain why these ratios are not exactly equal to √2.
**c** Give an equation connecting $w$ and $h$.

**2** The dimensions of the sticky labels in this packet, given in centimetres, are inversely proportional.

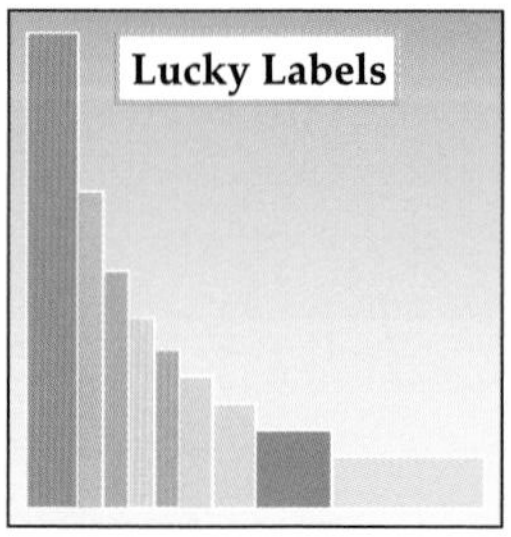


| Width ($w$) | 18 | 12 | 9 | 7.2 | 6 |
|---|---|---|---|---|---|
| Height ($h$) | 2 | 3 | 4 | 5 | 6 |

**a** What is the area of each of the labels?
**b** Give an equation connecting $w$ and $h$ in the form:
**i** $h = \ldots$ **ii** $wh = \ldots$ **iii** $w = \ldots$ .

**3** The mass of glaze, $g$ grams, needed to glaze a plate is in direct proportion to the square of the plate's diameter, $d$.

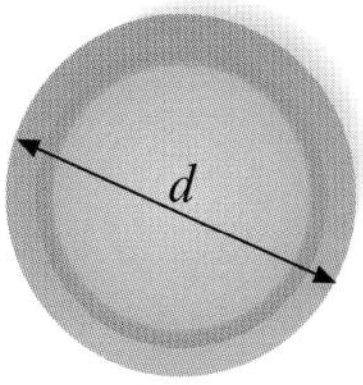


**a** Copy and complete this table.

| $d$ | 5 | 7 | 9 | 10 | 14 |
|---|---|---|---|---|---|
| $d^2$ | | | | | |
| $g$ | 10 | 19.6 | 32.4 | 40 | 78.4 |

**b** Give an equation connecting $d$ and $g$ in the form $g = kd^2$.

## Finding the constant

- To give an equation connecting two variables, you need to find the constant.

**Example** $w$ is directly proportional to the square of $f$, and $w = 45$ when $f = 5$. Give an equation connecting $w$ and $f$.

Write down the relationship. $w \propto f^2$ so $w = kf^2$

Substitute $w = 45$ and $f = 5$ into the formula. $45 = k \times 5^2$

$45 = 25k$

Rearrange the formula to find the constant. $\frac{45}{25} = k$

This gives $k = 1.8$, so $\mathbf{w = 1.8f^2}$

- For two variables in inverse proportion, it is easier to use the formula $xy = k$ to find the constant.

**Example** $d$ is inversely proportional to the positive square root of $z$, and $d = 2.5$ when $z = 36$. Give $d$ in terms of $z$.

$d \propto \frac{1}{\sqrt{z}}$ so $d\sqrt{z} = k$

$2.5 \times \sqrt{36} = k$

$2.5 \times 6 = k$

This gives $k = 15$, so $\mathbf{d = \frac{15}{\sqrt{z}}}$

**Exercise 12.9**
**Finding the constant**

**1** $m$ is inversely proportional to $v$, and $m = 6$ when $v = 9$.

**a** Write down the relationship between $m$ and $v$.
**b** Find the constant.
**c** Give an equation connecting $m$ and $v$.
**d** Calculate $m$ when $v = 4.5$.
**e** Explain what happens to $m$ when $v$ is halved.

**2** $s$ is directly proportional to the positive square root of $q$, and $s = 28$ when $q = 16$.

**a** Write down the relationship between $s$ and $q$.
**b** Give an equation connecting $s$ and $q$.
**c** Calculate $s$ when $q = 36$.

**3** $p$ is inversely proportional to the square of $t$, and $p = 2.5$ when $t = 6$.

**a** Give an equation connecting $p$ and $t$ in the form $p = \dots$ .
**b** Calculate $p$ when $t = 3$.
**c** Explain what happens to $p$ when $t$ is halved.
**d** Rearrange your equation to give $t$ in terms of $p$.
**e** Calculate $t$ when $p = 3.6$.

4 The number of plates, $n$, that can be glazed with one bag of glaze powder is inversely proportional to the square of the diameter, $d$, of the plate.
One bag glazes 800 side plates, diameter 5″.

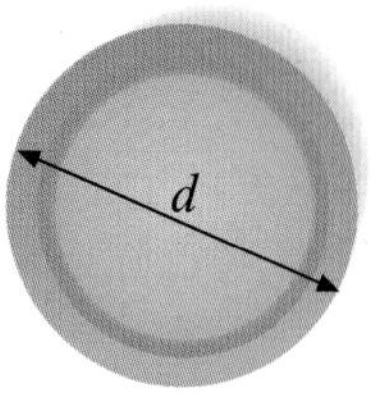


**a** Write down the relationship between $n$ and $d$.
**b** Find the constant.
**c** Give an equation connecting $n$ and $d$.
**d** Calculate how many plates one bag of powder will glaze for:
**i** 10″ dinner plates **ii** 14″ serving plates.

5 The area of cardboard needed to make a box for a basketball, $A\,\text{cm}^2$, is directly proportional to the square of the diameter of the ball, $d$ cm.
A 26 cm basketball box needs 2366 $\text{cm}^2$ of cardboard.

**a** Write down the relationship between $A$ and $d$.
**b** Find the constant.
**c** Give $A$ in terms of $d$.
**d** Calculate how much cardboard is needed for a 19 cm basketball box.

You need to decide whether to work in centimetres or metres.

6 The length of time, $t$ seconds, that a basketball bounces when dropped is directly proportional to the square root of the height, $h$ cm, from which it is dropped.
The ball bounces for 16 seconds when dropped from 1 m.

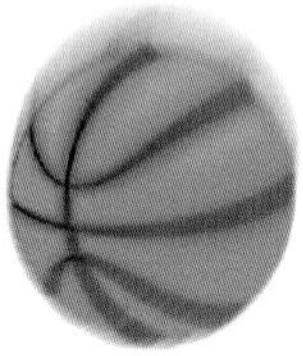

**a** Give an equation connecting $t$ and $h$.
**b** Calculate the bounce time, to the nearest 0.1 seconds, from a height of 50 cm.

The aperture is the adjustable opening in the lens that lets the light in when you take a photograph.

7 The $f$/stop on this camera lens is inversely proportional to the aperture diameter, $a$ mm.
$f8$ has an aperture of 5.25 mm.

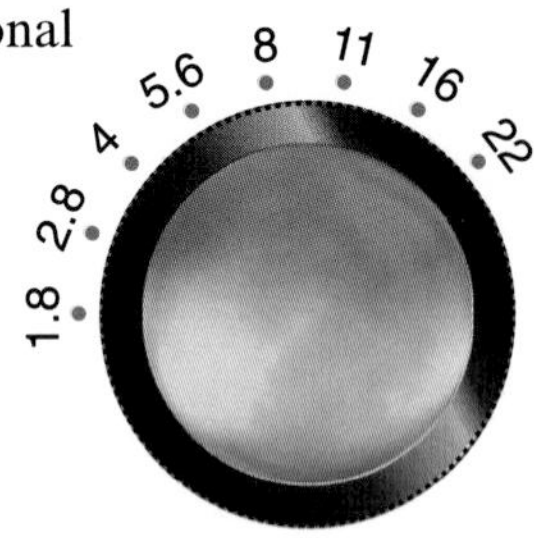


**a** Write down the relationship between $f$ and $a$.
**b** Find the constant.
**c** Give $f$ in terms of $a$.
**d** Calculate the aperture diameter for $f16$.

The brightness ratio compares how much light is let in at two different apertures.
For example: an aperture with a brightness ratio of 0.5 lets twice as much light in as an aperture with a brightness ratio of 0.25.

8 The brightness ratio, $r$, of a camera lens is in inversely proportional to the square of the $f$/stop.
$f4$ has a brightness ratio of 0.25.
**a** Give an equation connecting $r$ and $f$.
**b** Calculate the brightness ratio for $f16$.
**c** Find the $f$/stop that has a brightness ratio of 1.

9 Chocolate Chunks are cubic pieces of chocolate.
The mass of one piece, $m$ grams, is directly proportional to the cube of the width, $w$ cm.
The 1.5 cm chunk has a mass of 10.8 grams.

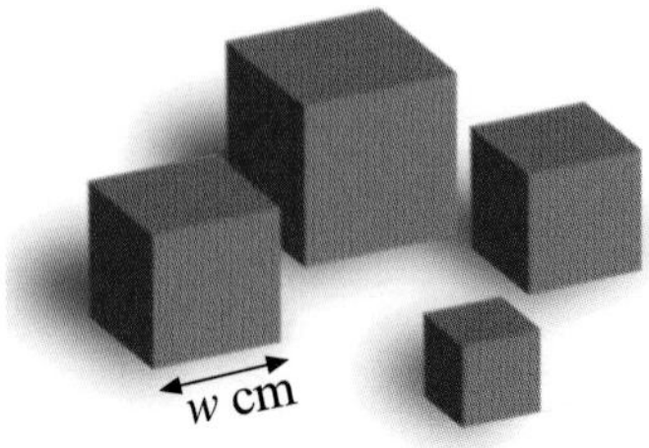


**a** Give $m$ in terms of $w$.
**b** Calculate the mass of a 3 cm chunk.
**c** The width of two chunks is in the ratio 1 : 2.
Give the ratio of the mass of the two chunks.

# End points

You should be able to ... ... so try these questions

**A** Keep amounts in the same proportion

**A1** This ice-cream recipe serves 8 people.

**Chocolate Ice Cream**
- 150 g caster sugar
- 20 g cocoa • 4 eggs
- 410 g can evaporated milk
- 4 tablespoons boiling water

**a** Calculate how much of each of these ingredients you need to serve 20 people.
**i** caster sugar **ii** boiling water

**b** How many eggs are needed to serve 6 people?

**B** Identify similar shapes

**B1**

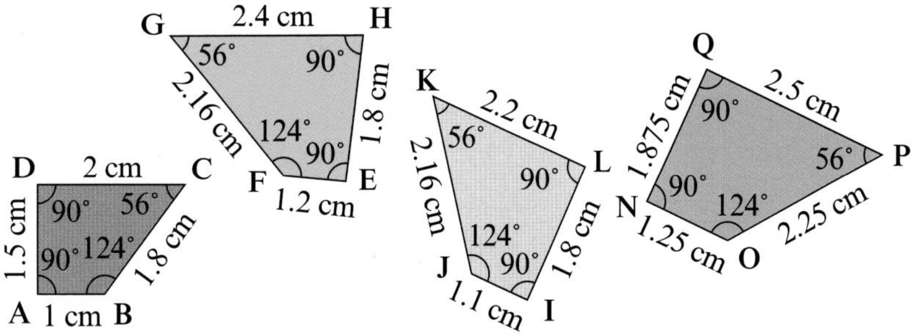


Which of these shapes are similar to ABCD?

**C** Find lengths in similar shapes

**C1**

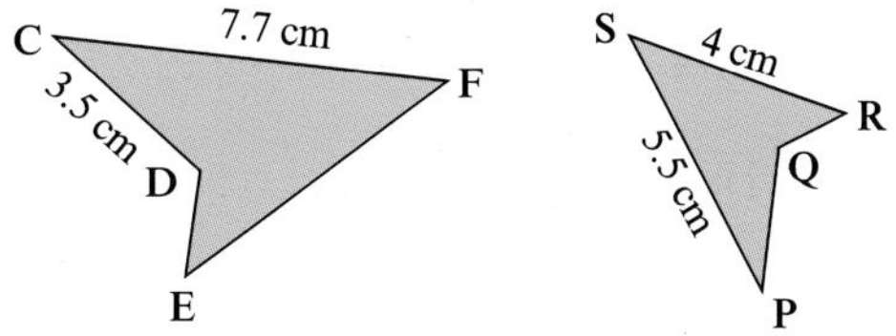


CDEF and PQRS are similar. Find the length:
**a** PQ **b** EF

**D** Recognise and use similar triangles

**D1**

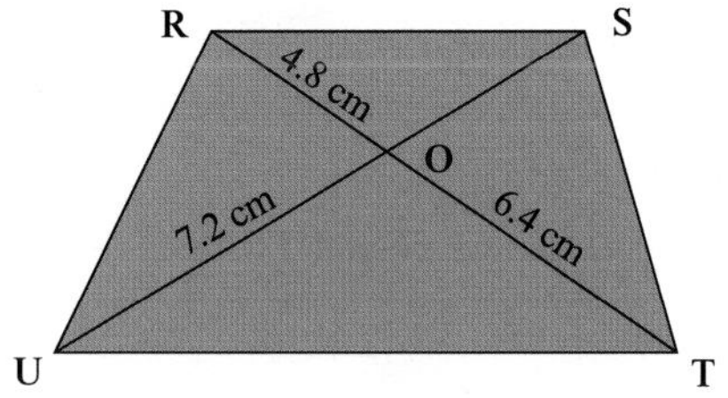


RSTU is a trapezium.
**a** Explain why triangles ROS and TOU are similar.
**b** Find the length OS.

**E** Understand direct and inverse proportion

**E1** The length of time, $s$ seconds, that a tennis ball bounces when dropped is in direct proportion to the square root of the height, $h$ cm, from which it is dropped.
The ball bounces for 18 seconds when dropped from 64 cm.
**a** Give an equation connecting $s$ and $h$.
**b** Calculate the bounce time from a height of 1 m.

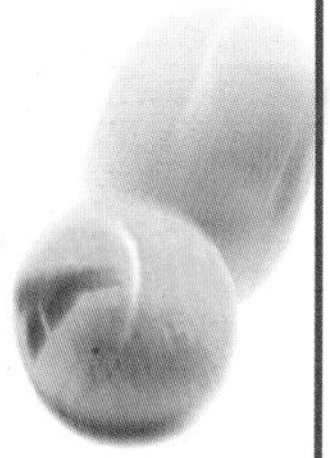

**E2** $w$ varies inversely as $q$, and $w = 1.2$ when $q = 5$.
**a** Give $w$ in terms of $q$.
**b** Calculate $w$ when $q = 10$.
**c** Explain what happens to $w$ when $q$ is doubled.
**d** Calculate $q$ when $w = 7.5$.

## Starting points

You need to know about ... | ... so try these questions

### A Prisms and pyramids

- This solid has a cross-section which is the same all through the solid. This is known as a **uniform cross-section.**

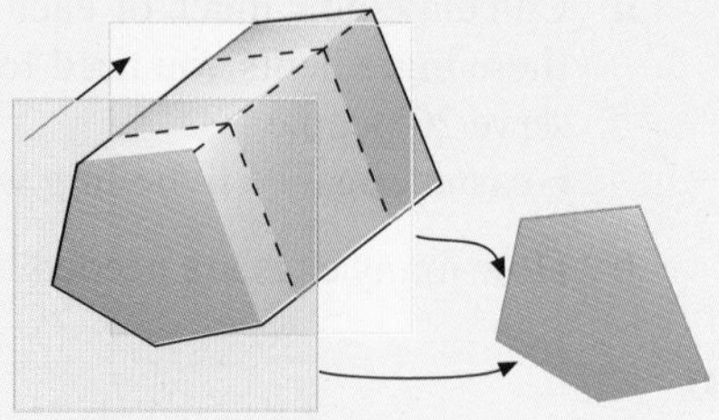

- Solids with a uniform cross section are known as **prisms**.
- These are other examples of prisms.

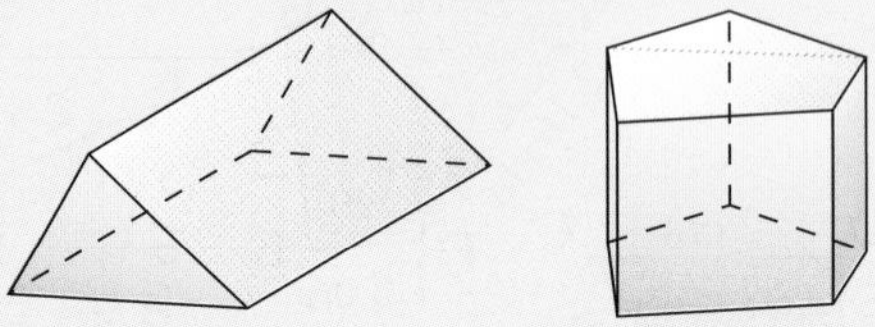

Triangular prism    Pentagonal prism

- A **pyramid** is a solid with a polyhedral base with triangular sides which meet at one vertex.
- These are two examples of pyramids.

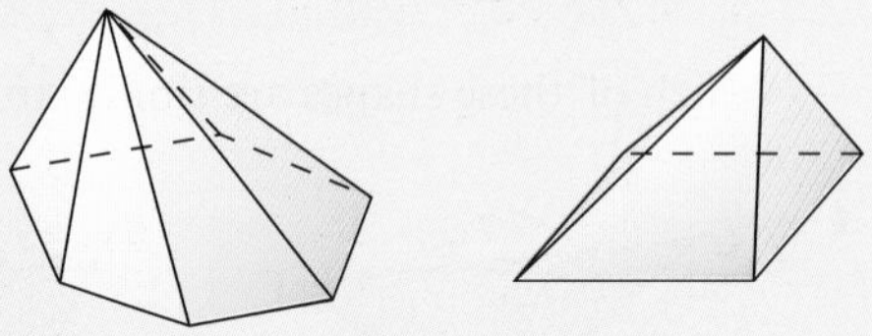

Hexagonal pyramid    Square pyramid

**A1** Each of the solids is either a prism or a pyramid.

**a** List all the prisms.

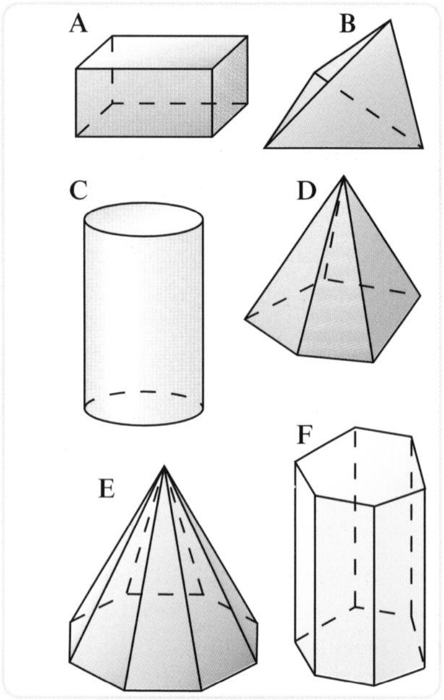


**b** Name each solid.

### B The area of a plane shape

- Area of a triangle $= \frac{1}{2}bh$

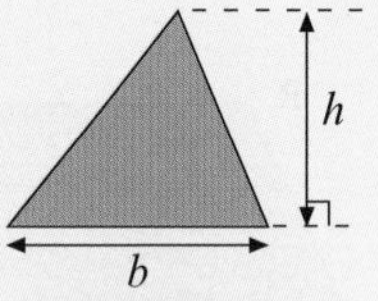


- Area of a parallelogram $= bh$

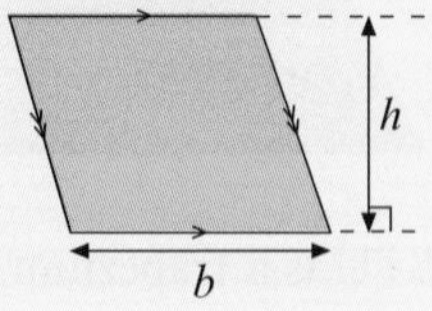


- Area of a trapezium $= \dfrac{h(a+b)}{2}$

  where $h$ is the perpendicular height.

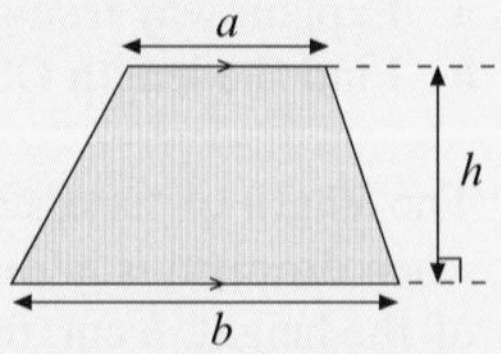


**B1** Calculate the areas of these shapes.

**a**

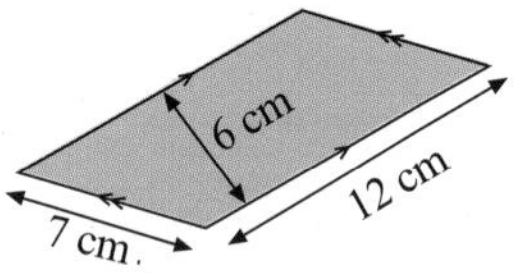


**b**

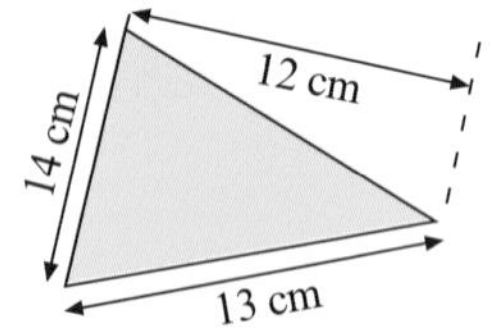


**c**

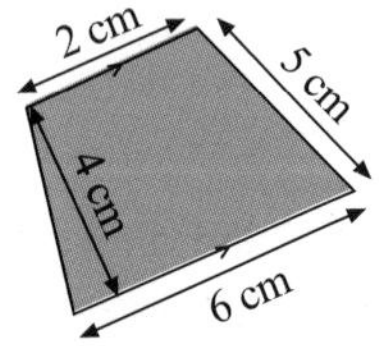


### C The area and circumference of a circle

- Area of a circle $= \pi r^2$
- Circumference of circle $= 2\pi r$ or $\pi d$

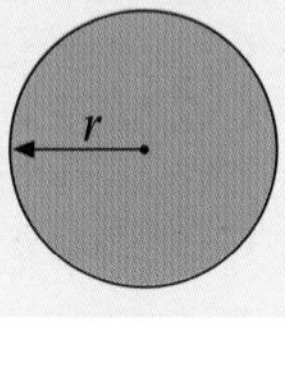


**C1** Calculate the area of a circle with a diameter of 2.4 m.

**C2** A circle has a circumference of 52 cm. What is its radius?

**C3** Write an expression for the area and circumference of a circle with radius $2y$.

## D Volumes and surface areas of prisms

- The volume of any prism = Area of uniform cross-section × Depth

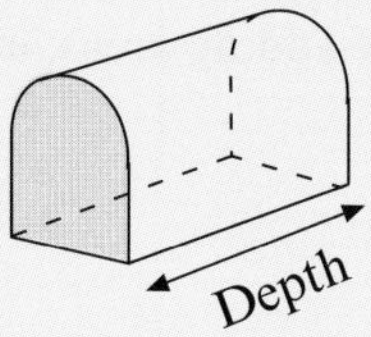


**Example** What are the volume and surface area of this prism?

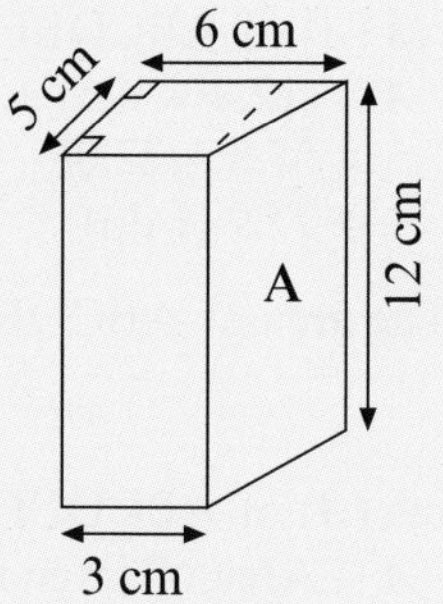


The uniform cross-section is trapezium-shaped.

Area of trapezium $= \dfrac{5 \times (3 + 6)}{2}$

$= 22.5\,\text{cm}^2$

**Volume of prism** $= 22.5 \times 12$

$\mathbf{= 270\,cm^3}$

- The surface area is the total of the areas of **all the exterior faces**.

To find the area of face A you need to know its width PQ.

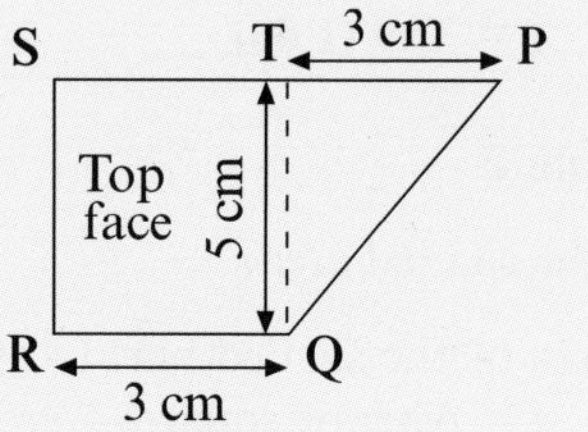


By Pythagoras' rule:

$PQ^2 = PT^2 + TQ^2$

$PQ = \sqrt{3^2 + 5^2} = \sqrt{34}$

$PQ = 5.83$ (2 dp)

Area of face A $= 5.83 \times 12 = 70\,\text{cm}^2$ (2 sf)
Area of B $= 5 \times 12 = 60\,\text{cm}^2$
Area of top C $= 22.5\,\text{cm}^2$
Area of bottom D $= 22.5\,\text{cm}^2$
Area of back E $= 6 \times 12 = 72\,\text{cm}^2$
Area of front F $= 3 \times 12 = 36\,\text{cm}^2$

**Surface area of prism = 283 cm²** (3 sf)

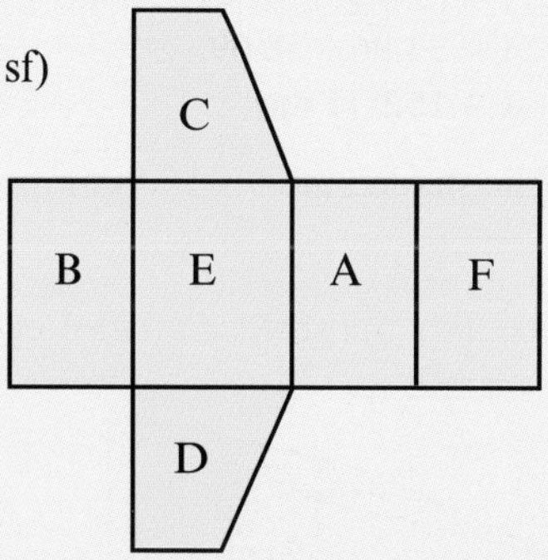


## E The volume and surface area of a cylinder

- Volume of cylinder $= \pi r^2 h$
- Area of curved surface $= 2\pi rh$ or $\pi dh$
- Total surface area $= 2\pi r^2 + 2\pi rh$

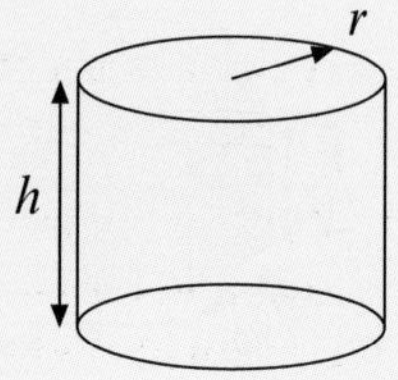


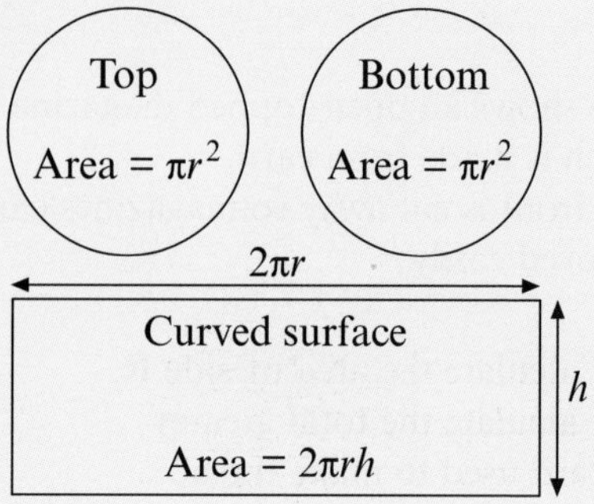


**D1** Calculate the volume and surface area of each of these prisms.

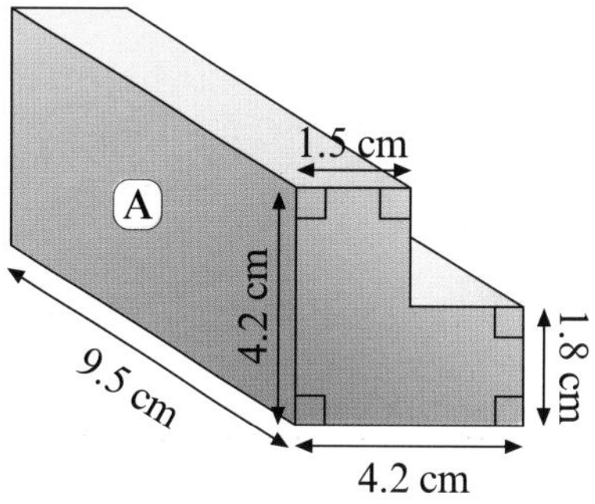


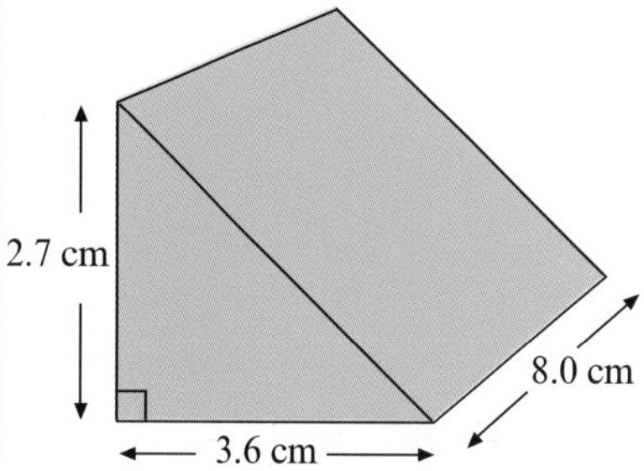


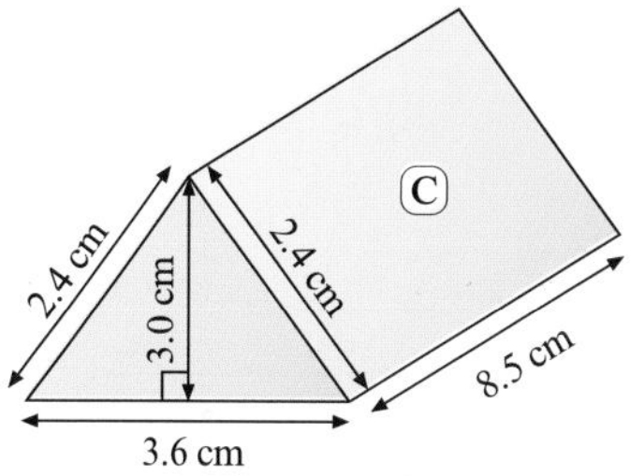


**E1** Calculate the volume and total surface area of this cylinder.

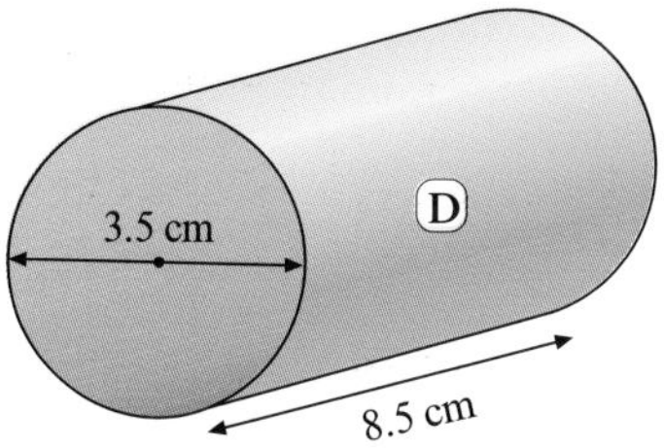


**E2** A cylinder has a radius of 8 cm and a length of 12 cm.
Write an expression in π for:
**a** the volume of the cylinder
**b** the total area of the cylinder.

# The area of composite shapes

Composite shapes are made up from more than one shape.
Sometimes you can find the area by adding areas; at other times it is easier to subtract.

### Example 1

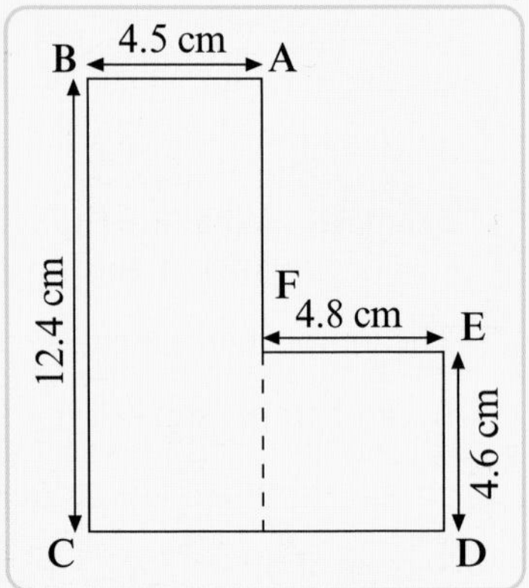


Find the area of ABCDEF.

The area ($A$) of ABCDEF can be given by:

$A = (12.4 \times 4.5) + (4.6 \times 4.8)$
$= 55.8 + 22.08$
$= 77.9$ (1 dp)

The area of ABCDEF is $77.9\,\text{cm}^2$

or

The area ($A$) of ABCDEF is given by:

$A$ = Area of large rectangle – Area cut out for L-shape
$A = (12.4 \times 9.3) - (4.8 \times 7.8)$
$A = 115.32 - 37.44$
$\mathbf{A = 77.9}$ **(1 dp)**

### Example 2

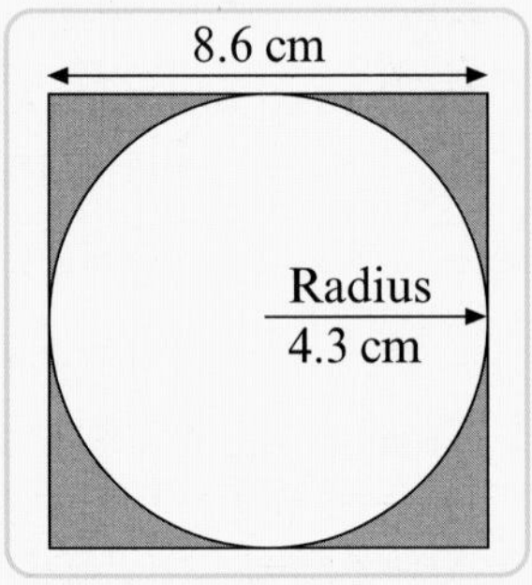


Find the shaded area.

The area of the shaded part ($A$) is given by:

$A$ = Area of square – Area of circle
$A = (8.6 \times 8.6) - (\pi \times 4.3^2)$
$A = 73.96 - 58.09 \ldots$
$\mathbf{A = 15.9}$ **(1 dp)**

The answer is rounded at the end of the calculation.

The shaded area is $15.9\,\text{cm}^2$.

**Exercise 13.1**
**Working with composite shapes**

1 Calculate the surface area of each shape.

**a**

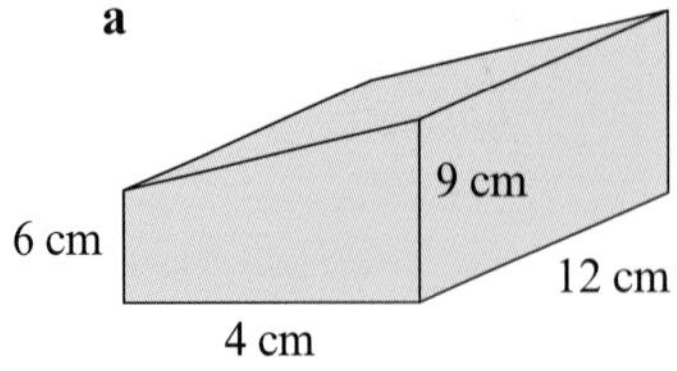


**b**

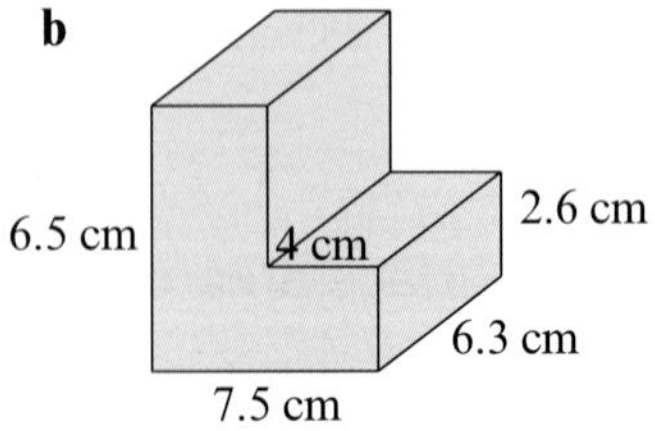


**c**

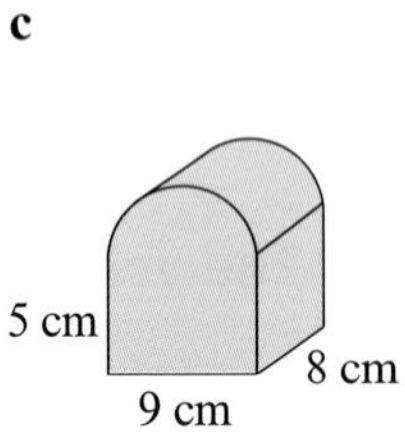


2 This shows an open-topped magazine holder which is made from card.
The front is cut away so magazines can be removed easily.

**a** Calculate the area of side R.
**b** Calculate the total area of card used to make the box.

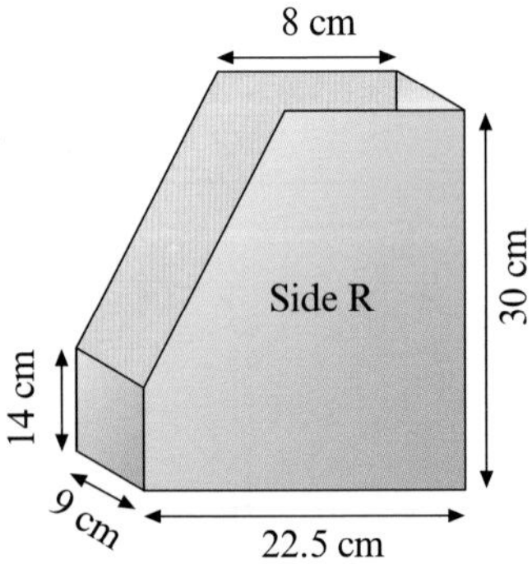

**3** This surround is used to frame photographs. A circle of radius 4 cm is cut from a rectangle of card. Gold lines are printed 1 cm from the edge of the circle and the card as shown.

**a** Calculate the area of card after the cut out.

**b** Is the area of photograph showing more or less than $\frac{1}{4}$ of the area of the surround? Explain your answer.

**c** Calculate the length of gold line on the card.

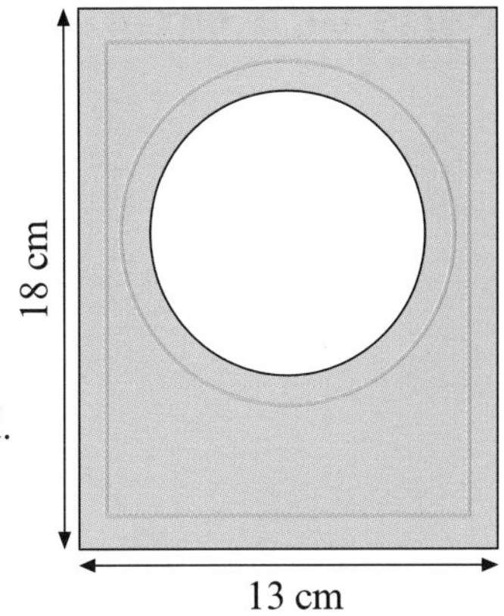


**4** This diagram shows the net of a box for playing cards. The net was cut from a rectangle of card that it just fitted.

**a** What were the original dimensions of the piece of card?

**b** Calculate the area of the glue flap on the right.

**c** The box opens at either end. Find the area of an opening end.

**d** Calculate the total area of the net.

**e** Roughly, what fraction of the card is wasted? Explain how you decided.

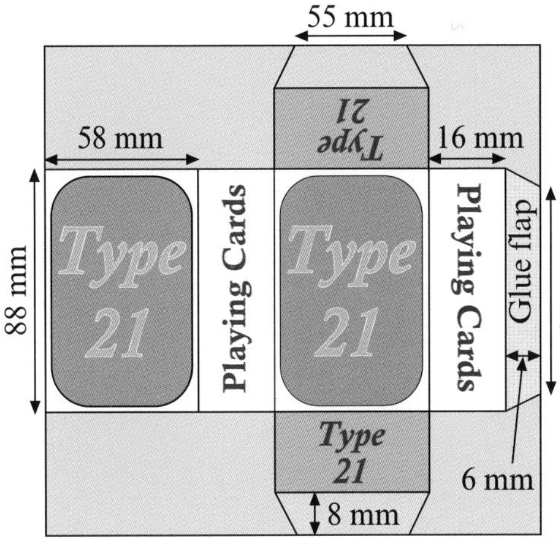


**5** This logo design for a company that makes helmets uses two parts of a circle in a red square.

**a** Calculate the area of the visor in the logo (the blue part).

**b** Find the area of the helmet in the logo (the yellow part).

**c** What percentage of the logo is red?

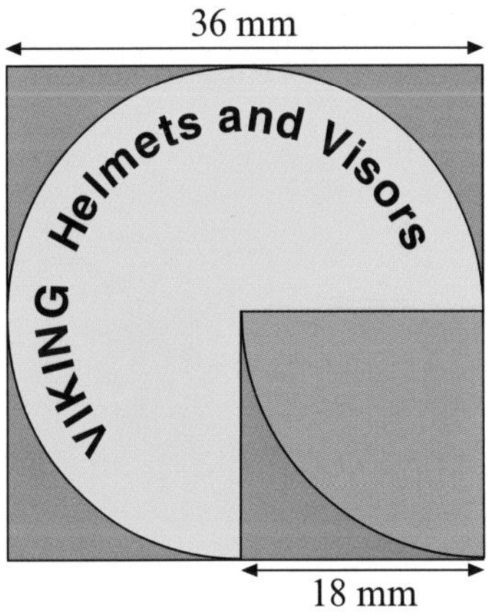


**6** These are diagrams of filters used in spotlights. Blue and yellow filters can give blue, yellow and green light.

Each filter is in a square frame of length $x$.

Give an expression in $x$ for:

**a** the area of the blue filter

**b** the non-coloured area in the yellow filter frame

**c** the area of green where the two filters overlap.

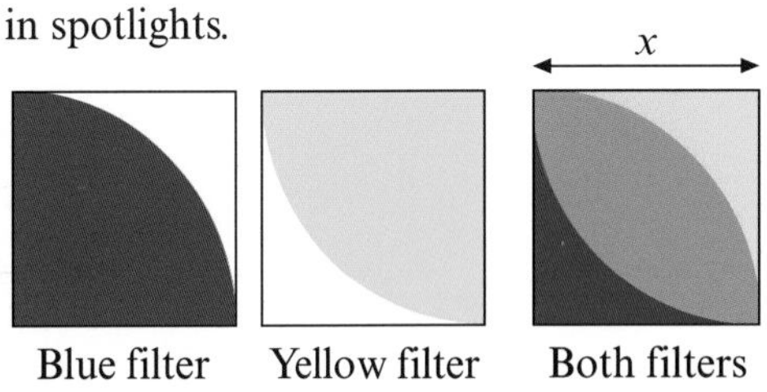

# Parts of a circle

For some problems you need to calculate the area of a sector of a circle or the length of an arc.
For both of these the ratio of the angle $x$ to a full circle (360°) is used.

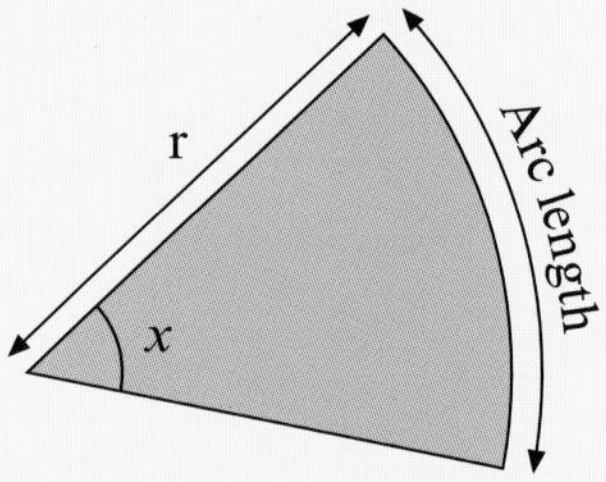


- Area of sector $= \frac{x}{360} \times$ area of circle

$= \frac{x}{360} \pi r^2$

- Arc length $= \frac{x}{360} \times$ circumference

$= \frac{x}{360} \pi d$

**Example** A piece of cheese of angle 56° is cut from a round cheese of radius 15 cm. Find its area and perimeter.

Area of sector $= \frac{56}{360} \times \pi r^2$

$= \frac{56}{360} \times \pi \times 15^2$

$=$ **110.0 cm²** (to 1 dp)

Arc length $= \frac{56}{360} \times \pi d$

$= \frac{56}{360} \times \pi \times 30$

$= 14.7$ cm

Perimeter = 15 + 15 + 14.7 = **44.7 cm**

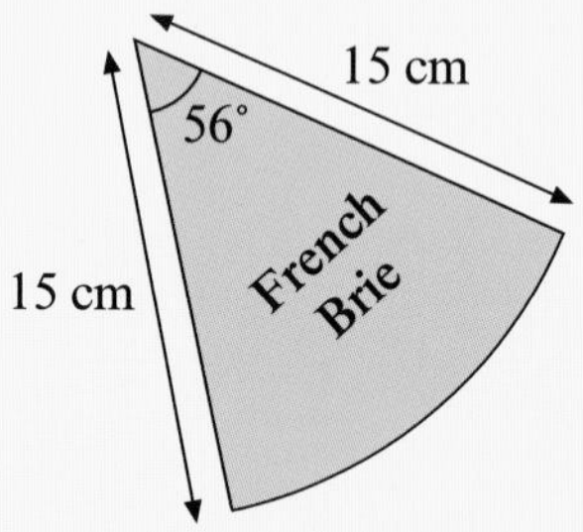


**Exercise 13.2**
Arcs and sectors

1 In each case calculate the area of the sector and its perimeter.

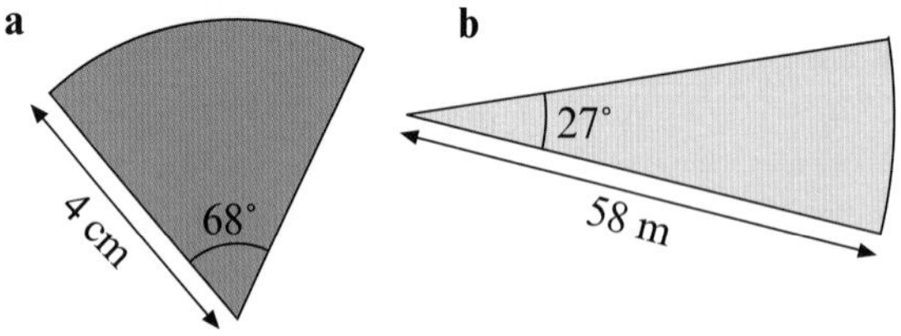


2 In each case calculate the area of the sector and its perimeter. Give each answer in terms of $\pi$.

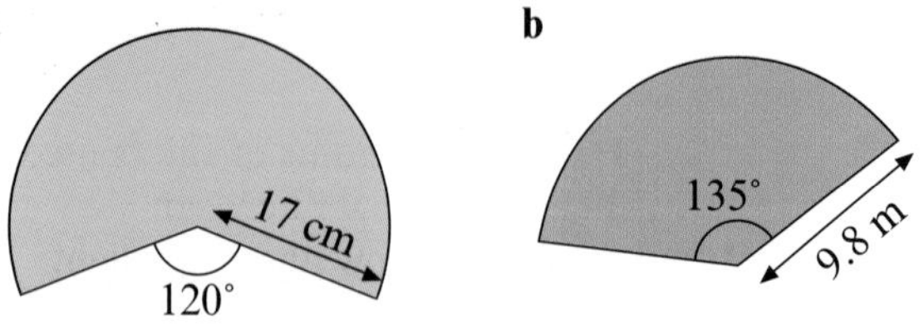


3 A baby's bib is made from towelling and is designed around two arcs of circles with the same centre.

a Calculate the area of towelling in the bib.
b The bib is edged all round with piping. What length of piping is used?

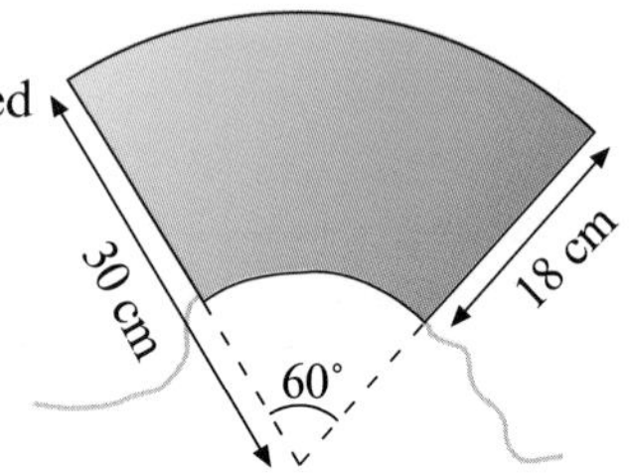

## Calculating volumes and surface areas

**Exercise 13.3**
Volumes and surface area problems

1 Calculate the volume and surface area of each shape.

**a**

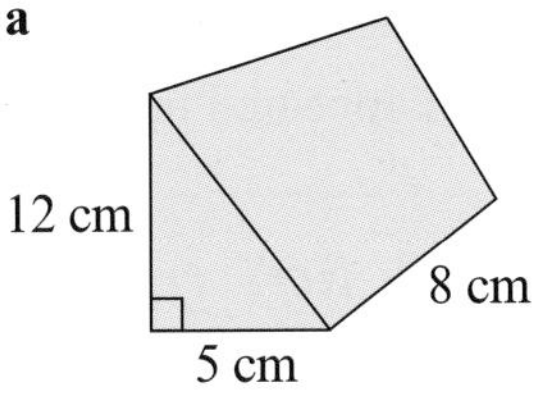


**b**

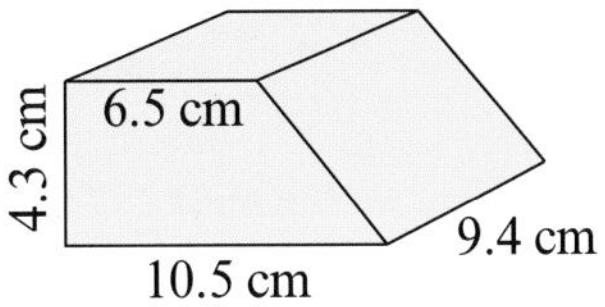


**c**

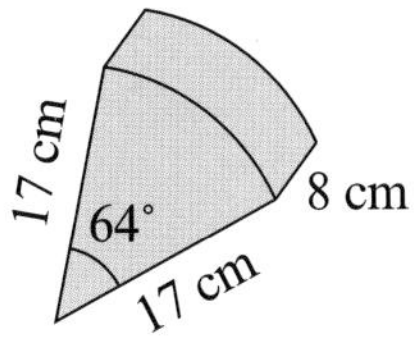


**d** A cylinder of diameter 13.5 cm and height 15.4 cm.

2 This collecting box is in the shape of a cylinder. The circumference of its base is 29.2 cm and its capacity is 1120 cm$^3$.
Calculate the height of the box.

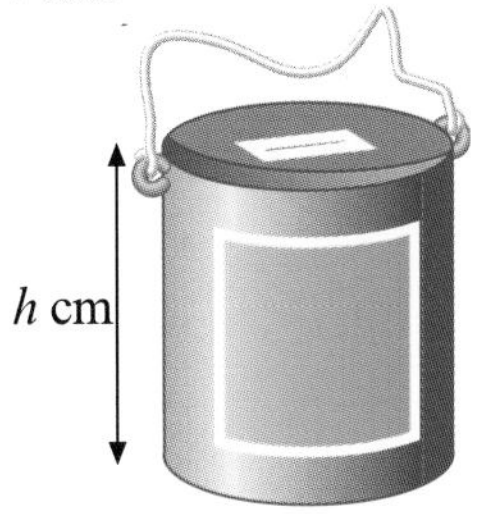


3 In a childrens' play area this space under the swings is dug to a depth of 0.3 m and filled with bark chippings.

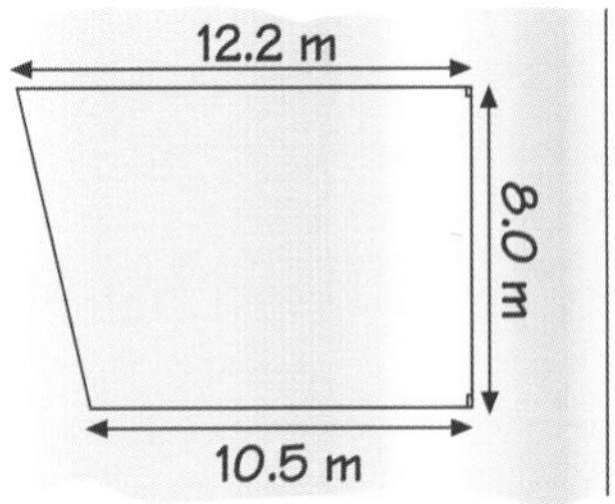


**a** Calculate the volume of bark chippings needed.

**b** Forestry Products sell bark chippings in bags. Each bag costs £5.12 and contains 0.07 m$^3$ of chippings.

**i** How many bags of chippings should be bought?

**ii** What is the total cost of chippings?

4 The diagram shows the uniform cross-section of an open-air swimming pool.

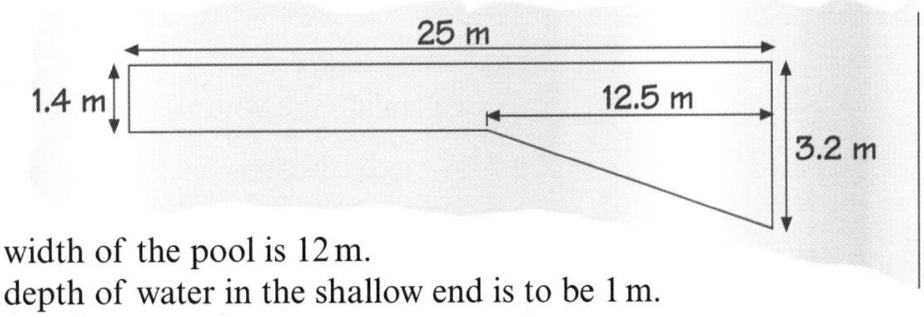


The width of the pool is 12 m.
The depth of water in the shallow end is to be 1 m.

**a** Calculate the area of the cross-section.

**b** What is the total capacity of the swimming pool?

**c** Calculate the volume of water in the pool.

**d** The amount of chlorine added to the water in this pool is 1 millilitre per cubic metre of water.
How much chlorine, in litres, will be added?

The surface area of a solid is the **total** area of all its **exposed** faces, including curved ones.

5 A solid spigot is made from a cylinder of diameter 6 cm and height 4 cm which is fixed on to a cylinder of diameter 12 cm and height 8 cm.
Calculate the surface area of the spigot.

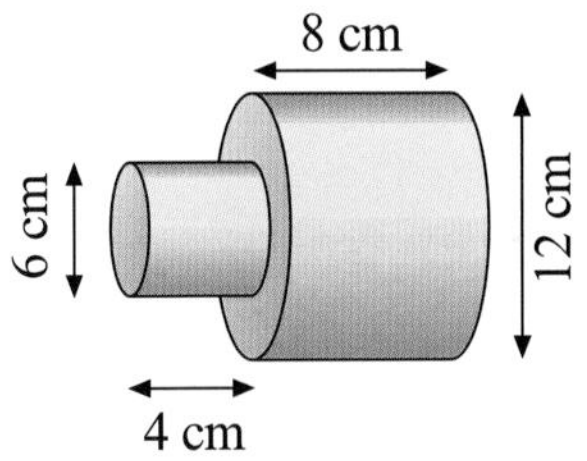

6 Kevin usually makes a fruit cake in a 20 cm square cake tin.
The finished cake has a depth of 8 cm.
Just to be different, he makes a cake of the same volume in a round tin with a diameter 24 cm.
What is the expected depth of this cake?

7 This competition appeared in a puzzle magazine.

> **WIN A BOX OF POUND COINS !!**
>
> - A cuboid-shaped box is full of pound coins.
> - The areas of the three different rectangular faces are 120 cm$^2$, 96 cm$^2$ and 80 cm$^2$.
> - Each edge measures a whole number of centimetres.
>
> **FIND THE VOLUME OF THE BOX AND WIN THE MONEY !**

**a** Find the volume of the box described in the competition.
**b** A pound coin has a diameter of 2.2 cm and a thickness of 0.3 cm.
Estimate how much money is in the box.

8 A concrete post to support a handrail has this symmetrical uniform cross-section. The post is 16 cm thick.

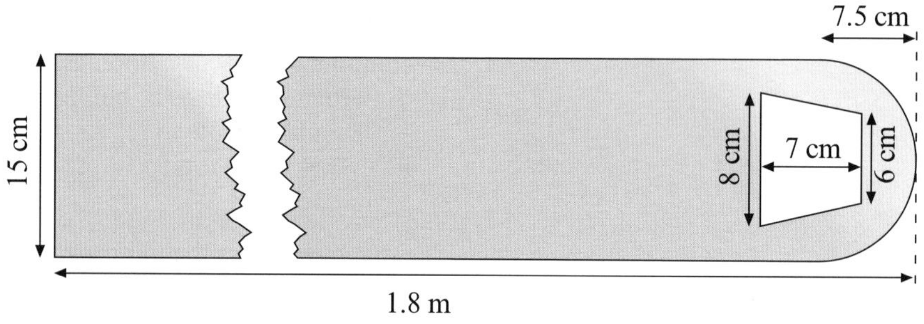


Calculate the volume of concrete in the post.

9 A piece of cheese is cut from a large cylinder of cheddar.
Each cut passes vertically down through the centre of the end circle.
The piece has an angle of 54°.

**a** Calculate the area of one end of the cheese piece.
**b** Calculate the area of the curved surface.
**c** What is the total surface area of the piece of cheese?
**d** If no other pieces have been cut, what is the surface area of the remainder of the cylinder of cheese?

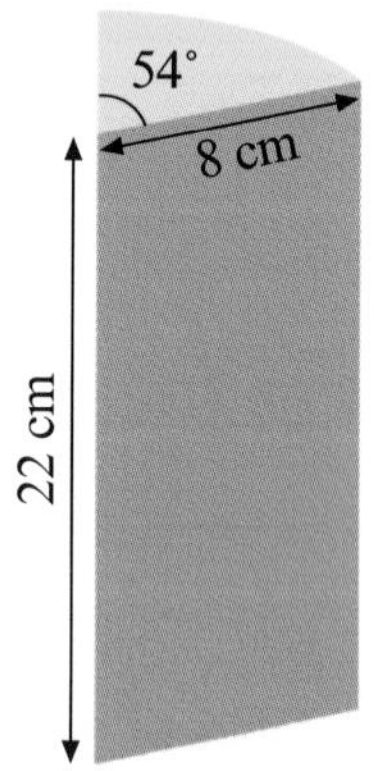


Beware: you will need to calculate a further dimension before you can do part **b**.

**10** A tunnel is cut through a hillside.
The cross-section of the tunnel is shown where the arc AB is a semicircle.
The tunnel is 540 metres long.

**a** Calculate the volume of soil which must be removed.

**b** All the interior surfaces are lined with a skin of concrete.
Calculate the internal surface area of the tunnel.

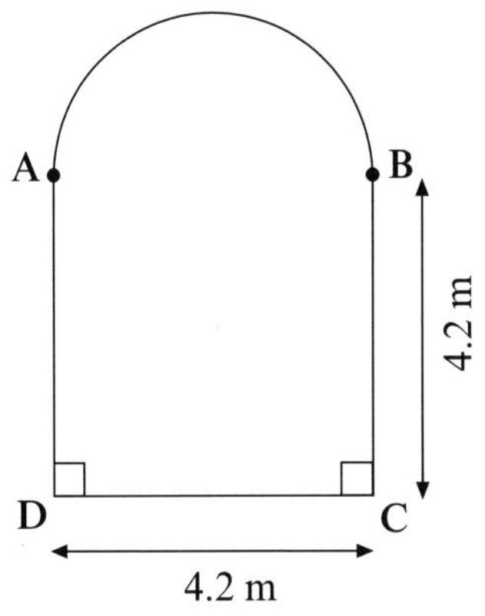


**11** The diagram shows a two kilogram weight.
The weight is made from a hollow cylinder of metal from which a sector of angle 30° is cut out.
This cut-out allows the weight to be slid on to a pole.

**a** Calculate the volume of the weight.

**b** Calculate the mass of metal which is cut out from the hollow cylinder.

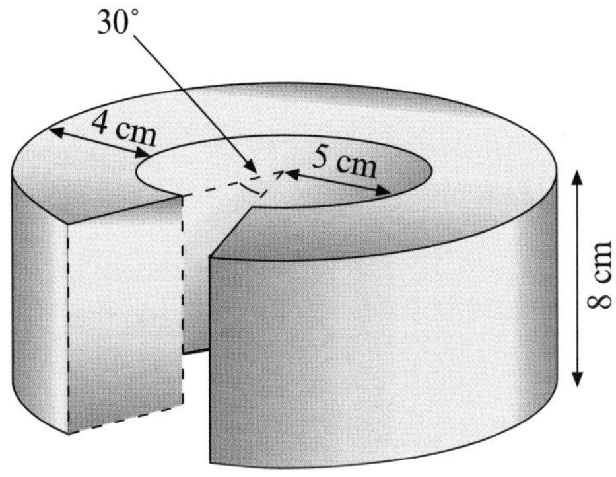


**12** A solid is made from a cuboid with dimensions $x$, $y$ and $z$.
A hole of radius $r$ is drilled right through from one side as shown.

Write and simplify an expression for:

**a** the volume of the solid

**b** the surface area of the solid.

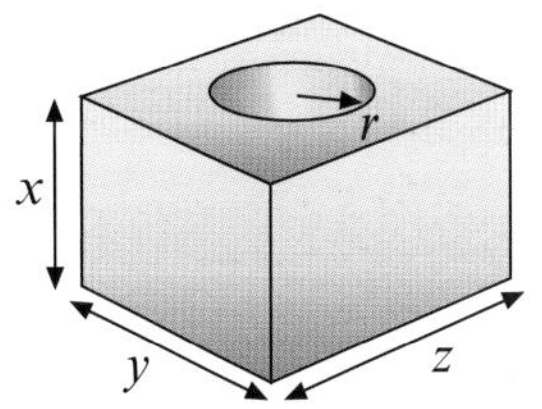


**13** In a solid cylinder the height is half the radius.
Write an expression for the radius $R$ in terms of the volume $V$ and the surface area $S$.

# Capacity, volume and density

- For liquids, volume or capacity is often measured in litres (l) or millilitres (ml).

  1 litre = 1000 millilitres
  1 millilitre is equivalent in volume to 1 $cm^3$.

- Mass (often called weight) is measured in kilograms (kg) or grams (g).

  1 kilogram = 1000 grams
  1 ml of water weighs 1 gram.

- Density can be measured in grams per cubic centimetre ($g/cm^3$).

$$\text{Density} = \frac{\text{mass in grams}}{\text{volume in cm}^3}$$

**Exercise 13.4**
Capacity, volume and density

1 Write down one choice from each bracket to complete the sentence.
**a** The volume of an orange is about (30 $cm^3$, 300 $cm^3$, 3000 $cm^3$).
**b** The capacity of a wine glass is about (75 litres, 7.5 litres, 75 millilitres).
**c** The weight of an apple is about (17 grams, 170 grams, 1.7 kilograms).

2 A dairy plans to sell cartons containing 550 ml of milk.

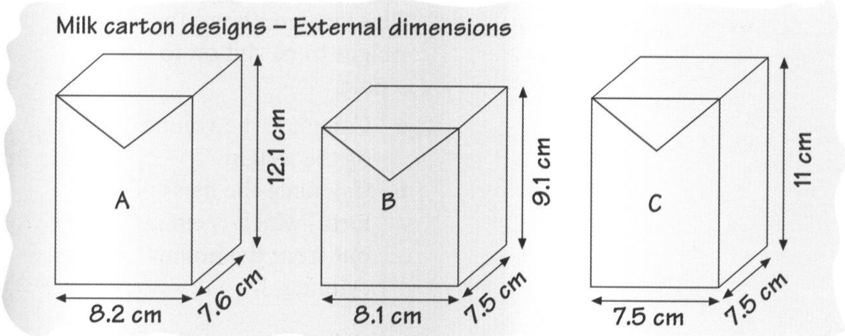


**a** Which of these cartons do you think the dairy should use?
The waxed cardboard used for the cartons is 1 mm thick.
**b** Which carton do you now think they should use?
**c** What is the capacity of carton C?
**d** What percentage of the capacity of carton C would be airspace if 550 ml of milk were added?
**e** What volume of waxed card is used to make carton C? Ignore the flaps.
**f** Find the mass of C if the density of cardboard is 0.86 $g/cm^3$.

Question **2** shows the difference between volume and capacity.
Carton A has a **volume** of $8.2 \times 7.6 \times 12.1$ $cm^3$, but its **capacity** is less than this because of the thickness of the cardboard.

3 A milk tank is a cylinder of radius 1.2 m and length 4.8 m.
Calculate the volume of the tank in:
**a** $cm^3$ **b** litres.

4 Each of these pieces of cheese has a uniform cross-section.

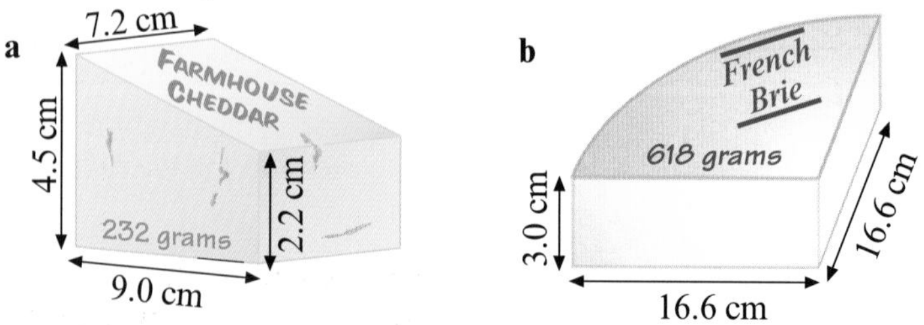


Calculate the volume and density of each piece.

## Planes of symmetry

Some solids have **plane symmetry**.
For example, this solid has plane symmetry.

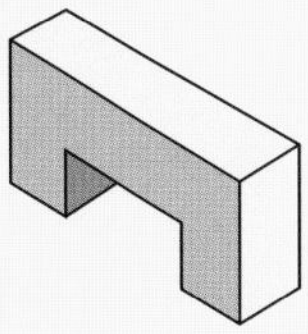

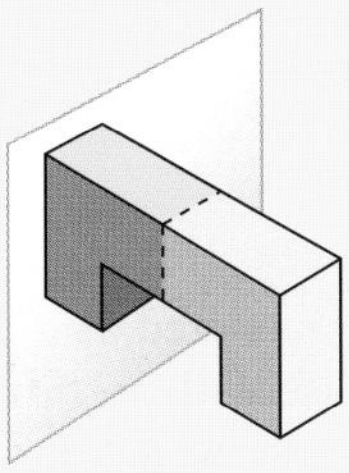

In the mirror, look at half of the solid like this and you see the other half.

So the mirror shows the position of a **plane of symmetry**.

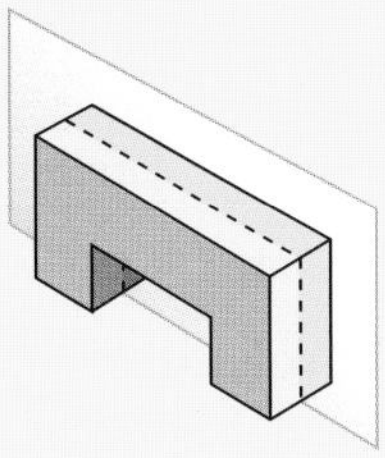

This mirror shows the position of another **plane of symmetry**.

This solid is said to have two **planes of symmetry**.

**Exercise 13.5**
**Properties of solids**

1 Each of these solids is made from four cubes.

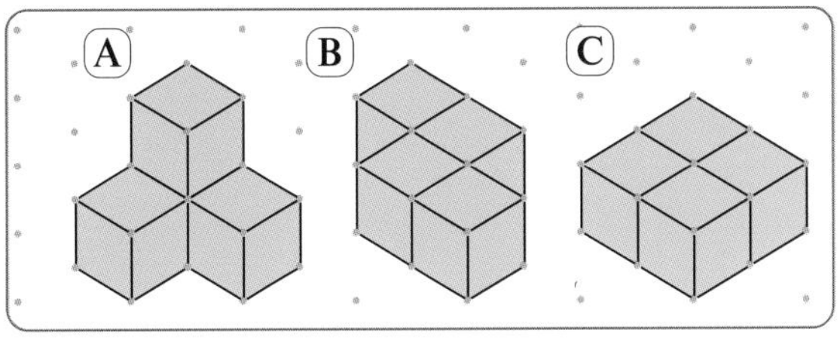


a Which of these solids are prisms?
b How many planes of symmetry has solid B?
c Which solid has exactly 3 planes of symmetry?

2 Draw a prism made from four cubes that has:

a 2 planes of symmetry  b 5 planes of symmetry.

3 Each of these nets A to D is for a prism or a pyramid.

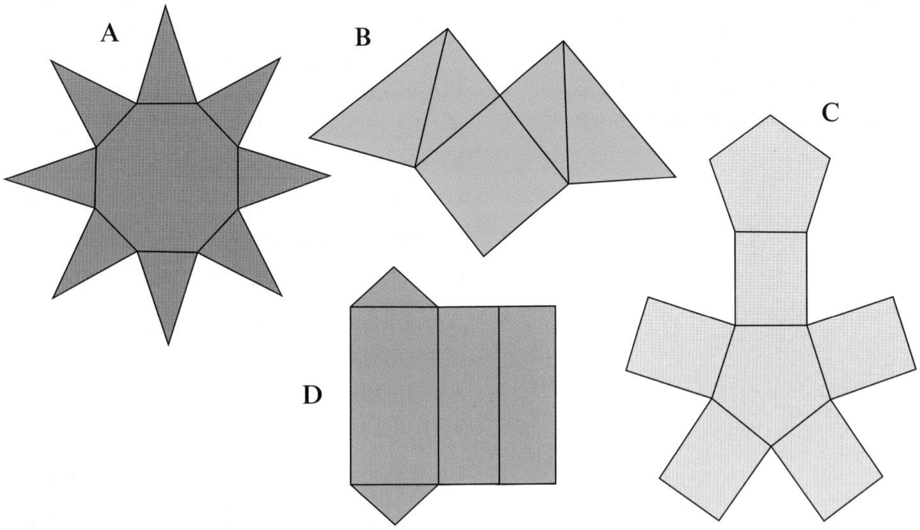


a Which of the nets is for:
i a prism  ii a pyramid?
b How many planes of symmetry would each assembled solid have?

# Pyramids and cones

- For any pyramid:
  - **Volume = $\frac{1}{3}$Base area × Perpendicular height**

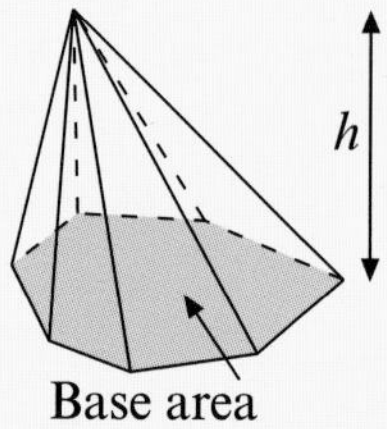


- A cone can be considered as a pyramid but with a circular base (i.e. a polygon base with an infinite number of sides). The base area of a cone of radius $r$ is $\pi r^2$.
- So for a cone:
  - **Volume of a cone = $\frac{1}{3}\pi r^2 h$**
  - **Curved surface area = $\pi \times r \times$ Slant height**

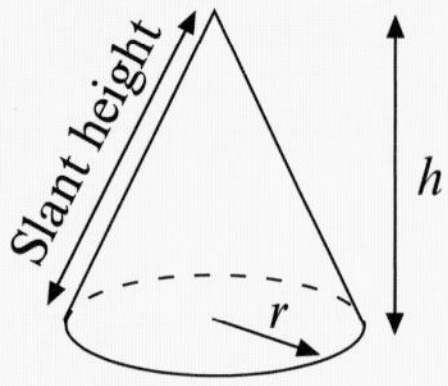


A right cone or right pyramid is one where the vertex lies directly over the centre of the base.

**Example** Find the volume and surface area of a right cone with a base diameter of 12 cm and a perpendicular height of 14 cm.

**Volume of cone** $= \frac{1}{3}\pi r^2 h$

$= \frac{1}{3} \times \pi \times 6^2 \times 14$

$=$ **528 cm³** (nearest integer).

To find the slant height ($s$)

Using Pythagoras' rule: $s = \sqrt{14^2 + 6^2}$

$= 15.23$ cm

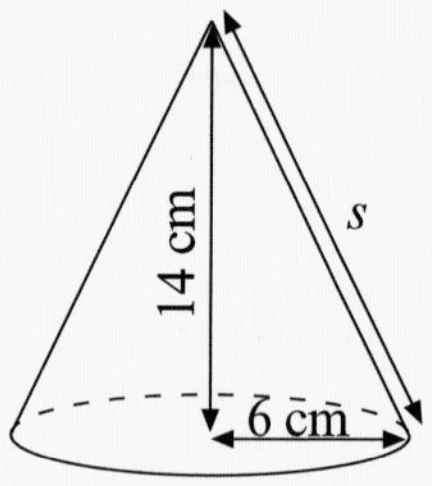


Curved surface area $= \pi \times r \times s$

$= \pi \times 6 \times 15.23$

$= 287$ cm² (nearest integer).

**Surface area** = Curved surface area + Area of base

$= 287 + (\pi \times 6^2)$

$=$ **400 cm²** (nearest integer).

**Exercise 13.6**
**Pyramids and cones**

1 For a right cone of height 10 cm and base radius of 4 cm, calculate:
  **a** the volume
  **b** the curved surface area
  **c** the surface area.

2 A pyramid has a perpendicular height of 2.4 m and a square base of length 0.8 m. Calculate the volume of the pyramid.

3 A right hexagonal pyramid has a base which is a regular hexagon of side 4 cm and a perpendicular height of 12 cm.
  **a** Calculate the area of its base.
  **b** Calculate its volume.

4 Cone A has a base radius of 5 cm and a height of 10 cm.
Cone B has a base radius of 10 cm and a height of 5 cm.
Which cone has the larger volume?

# The sphere

Archimedes, a Greek mathematician and inventor, was famous for his marvellous machines which were used against the invading Roman army. He was killed in 212 BC by the Romans while defending his home town of Syracuse.

What seems obvious now that a curved surface can have an area was first put forward by Archimedes in about 250 BC. He also discovered the formulas for the surface area and volume of a sphere.

- For a sphere of radius r
  - **Volume** $= \frac{4}{3}\pi r^3$
  - **Curved surface area** $= 4\pi r^2$

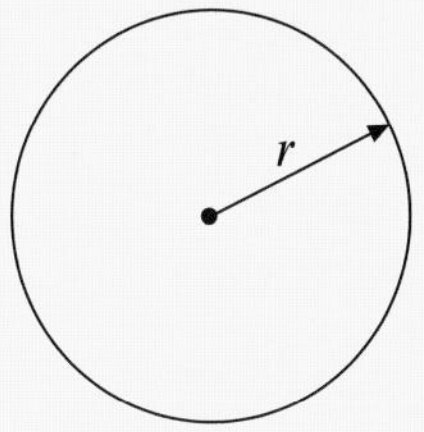


Example Calculate the volume and curved surface area of a basketball which has a diameter of 26 cm.

**Volume of basketball** $= \frac{4}{3}\pi r^3$
$= \frac{4}{3} \times \pi \times 13^3$
$= 9203\text{ cm}^3$ (nearest integer).

**Curved surface area** $= 4\pi r^2$
$= 4 \times \pi \times 13^2$
$= 2124\text{ cm}^2$ (nearest integer)

**Exercise 13.7**
Volumes and surface areas of spheres

1 Ball bearing A has a diameter of 1 cm and ball bearing B one of 2 cm.
 a Calculate the surface area of each ball bearing, leaving your answers in terms of $\pi$.
 b How many times larger is the surface area of B than A?
 c Calculate the volumes of each one, leaving your answers in terms of $\pi$.
 d How many times larger is the volume of B than A?

2 a A ball has a surface area of $3632\text{ cm}^2$.
 What is its diameter to the nearest centimetre?
 b A different ball has a volume of $3632\text{ cm}^3$.
 Find its diameter to the nearest millimetre.

3 Saturn has a diameter of $1.2 \times 10^5$ km.
Write an expression for its surface area in terms of $\pi$.

4 What are the dimensions of a square piece of card which has the same surface area as a sphere of diameter 14 cm?

5 An Eiffel Tower paperweight is shaped like a hemisphere with a radius of 55 mm.
 a What is its volume?
 b What is the external surface area of the clear plastic part?

Archimedes had the diagram in Question **6** carved on his tomb because of its special properties. That's real pride in your work!

6 A sphere of diameter 15 cm fits exactly into a cylinder of height 15 cm and diameter 15 cm.

Calculate:
 a the curved surface area of the cylinder
 b the curved surface area of the sphere.

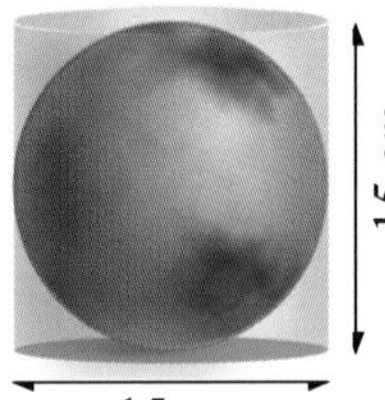


7 A sphere, diameter $d$, fits exactly inside a cylinder of diameter $d$ and height $d$.
Show why both always have the same curved surface area regardless of the value of $d$.

# Parts of solids and composite solids

A cone or pyramid which has had its vertex cut off by a plane parallel to its base is known as a frustum.

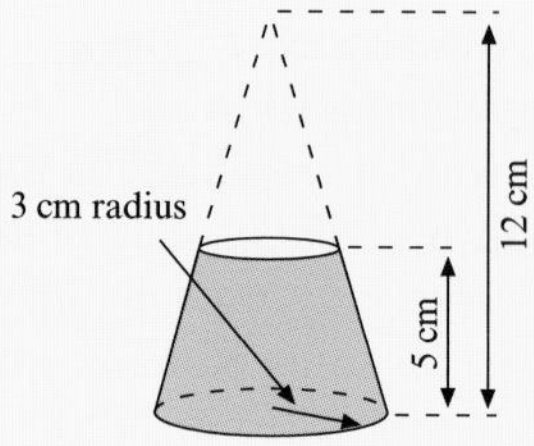


**Example** What is the volume of this frustum of a cone ?

Volume of frustum = Vol. of 12 cm cone – Vol. of 7 cm cone

To calculate the volume of the 7 cm cone you need to know its base radius $b$.
The easiest way to find this is to consider the similar triangles ABE and ACD.

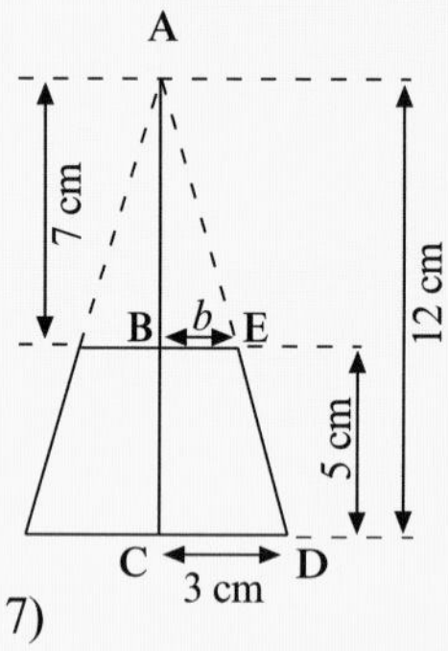


$$\frac{BE}{CD} = \frac{AB}{AC} \quad \text{so} \quad \frac{b}{3} = \frac{7}{12}$$

$$b = \frac{3 \times 7}{12} = 1.75\,\text{cm}$$

**Vol. of frustum** $= (\frac{1}{3}\pi \times 3^2 \times 12) - (\frac{1}{3}\pi \times 1.75^2 \times 7)$
$= (\frac{1}{3}\pi \times 108) - (\frac{1}{3}\pi \times 21.4375)$
$= \frac{1}{3}\pi\,(108 - 21.4375)$
**= 90.6 cm³** (to 1 dp)

- Cones, pyramids and spheres can be parts of composite solids.

**Example** A perfume spray fits tightly inside a clear plastic cube of side 12 cm. The shape of the spray is a hemisphere that sits on a cylinder. Calculate the volume of air between the spray and the cube.

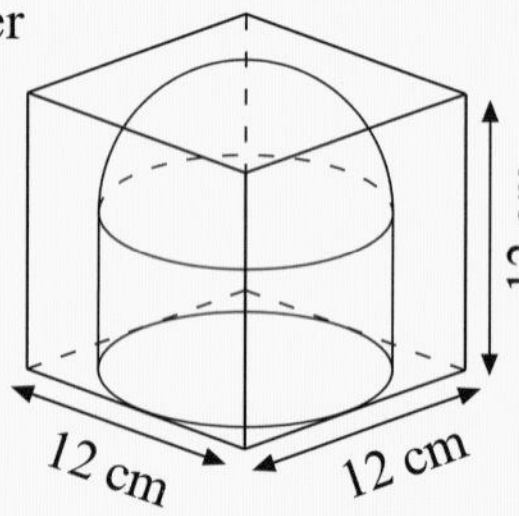


Vol. spray = Vol. hemisphere + Vol. cylinder
$= (\frac{1}{2} \times \frac{4}{3}\pi r^3) + \pi r^2 h$
$= (\frac{4}{6} \times \pi \times 6^3) + (\pi \times 6^2 \times 6)$
$= 144\pi + 216\pi$
$= 360\pi$
$= 1131\ \text{cm}^3$ (nearest integer)

**Vol. airspace** = Vol. cube – Vol. spray
$= 12^3 - 1131$
$= 1728 - 1131$
**= 597 cm³** (nearest integer)

**Exercise 13.8**
Volume of complex solids

1 A cylindrical pencil is 17.6 cm long and has a radius of 3 mm. It has 1.5 cm of its length shapened to a point. Calculate:

**a** its volume **b** its surface area.

2 A hemispherical hollow of radius 2 cm is made in the top of an 8 cm cube.:

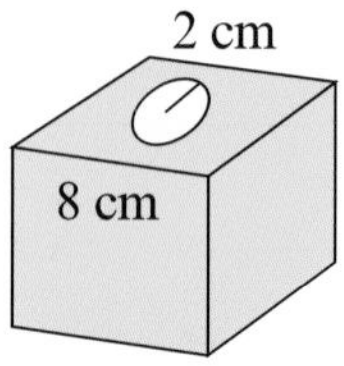


**a** Calculate the volume of the solid
**b** Calculate the total surface area of the shape.

3 Calculate the volume and total surface area of each solid.

a
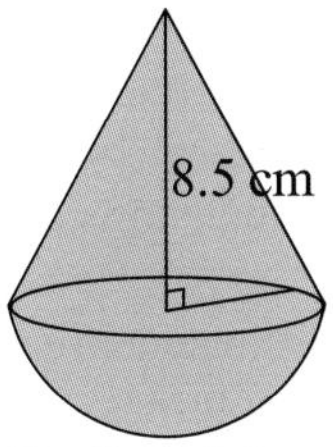


b
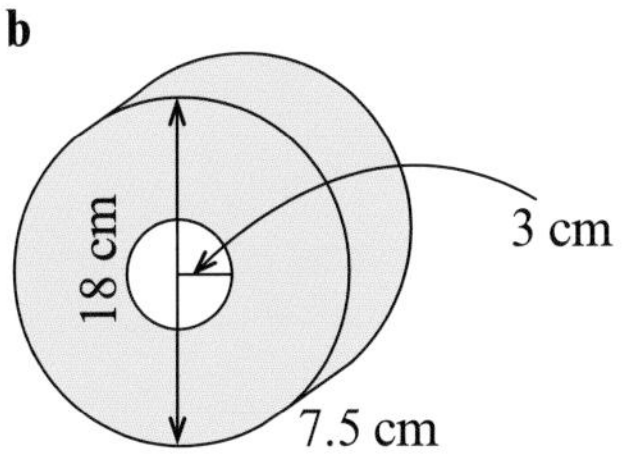


c
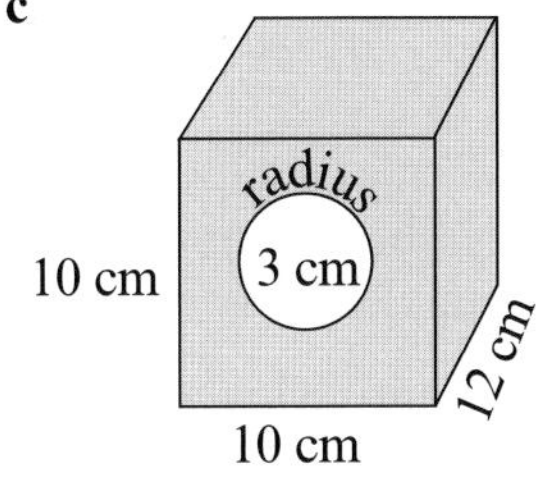


d
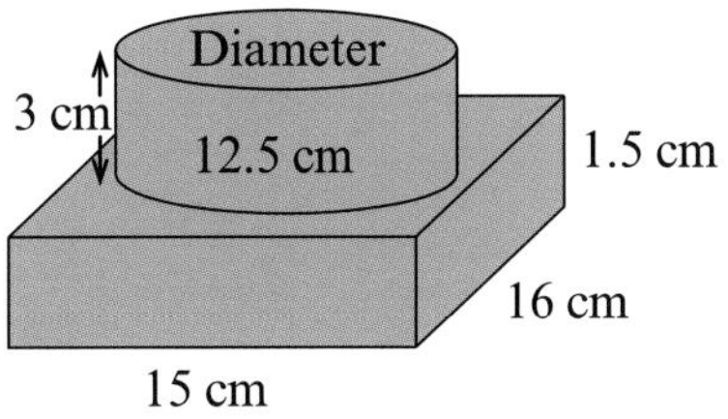


4 An ice lolly mould is shaped like an inverted cone of height 9 cm.

Green juice is poured into the mould until it reaches half the height, then red juice is added to the top.

a What volume of red juice is added?
b What percentage of the volume of the lolly is green?

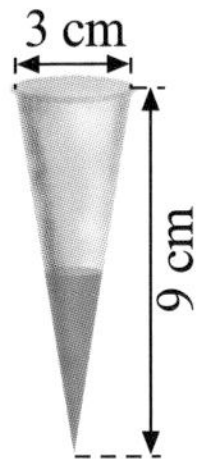


Note that the formula for the volume of a cone still applies to cones which are not right cones, i.e. sloping ones.

5 A model of a clock tower is made from one piece of card.

It is shaped like a square-based pyramid on a square-based prism.

A scale of 1 : 50 is used.

a What is the volume of the clock tower in $m^3$?
b What is the volume of the model in $cm^3$?
c Sketch a possible net for the model.
d Calculate the area of card needed for your net (ignore tabs and other fixings).

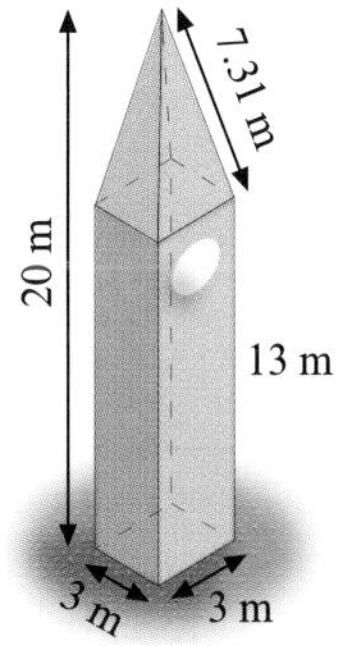


6 Three spheres of diameter 4 cm sit inside a cuboidal tin with internal dimensions of 12 cm, 16 cm and 14 cm.
The space in the tin is filled with water and this water is then poured out into a measuring cylinder of radius 4 cm.
How high up the cylinder does the water reach?

7 This frustum was made by cutting the top off a right cone.
Let $h$ be the height of the original cone in cm.

a Write an expression for the height of the part which was cut off.
b Use similar triangles to show why $5h = 48$, and hence calculate the height of the original cone and the part which was cut off.
c Calculate the volume of the frustum.

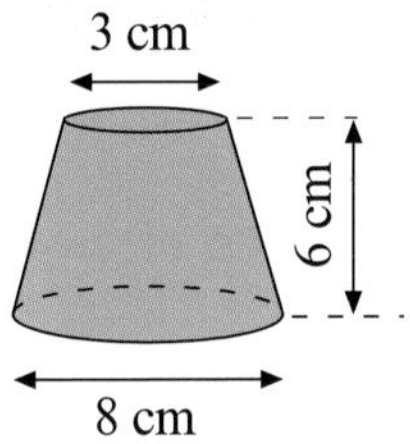

**Thinking ahead to ...**
the ratio of lengths areas and volumes

A Cube A has a side length of 1 cm and cube B has a side length of 2 cm.

a Calculate the surface areas of both cubes.
b Calculate the volumes of both cubes.
c The ratio of the dimensions of the cubes is 1 : 2.
Calculate:
i the ratio of the surface areas
ii the ratio of the volumes.

## The ratio of lengths, areas and volumes

- As an object is enlarged the ratio of its lengths obviously increases. Less obviously, the area and volume will also increase but not in the same ratio as the lengths.
- Areas involve multiplying a length by a length, so if the all corresponding lengths on an object increase $x$ times, **any** area on the object will increase $x \times x$ or $x^2$ times.
- Volumes involve multiplying a length by a length by a length, so if all the corresponding lengths on an object increase $x$ times, **any** volume on the object will increase $x \times x \times x$ or $x^3$ times.
- This may seem obvious as cubes or cuboids are enlarged. It also applies to any **similar** objects such as spheres, cones, or irregular or compound shapes.

**Example** Two similar wine bottles have base areas of 30.8 cm$^2$ and 50.3 cm$^2$.

a Calculate the ratio of the heights of the two bottles in the form $1 : n$.
b The large bottle has a height of 30 cm. Calculate the height of the small bottle.
c If the large bottle is a standard bottle, is it true to call the small bottle a 'half bottle'?

It is best not to round these ratios too much because squaring and cubing numbers tends to exaggerate any errors.

a The ratios of the base areas is:
30.8 : 50.3 = 1 : 1.6331169 (to 7 dp)
so the ratio of corresponding lengths will be $1 : \sqrt{1.6331169}$
$= 1 : 1.2779346$

**The ratio of the heights of the bottles is 1 : 1.2779** (to 4 dp)

b **The height of the small bottle** $= \dfrac{30}{1.2779} =$ **23.5 cm** (to 3 sf)

c The ratio of lengths is 1 : 1.2779346
so the ratio of the volumes will be $1 : (1.2779346)^3$
$= 1 : 2.087$
so the large bottle has roughly twice the capacity of the small one.
**so the small bottle can justifyably be called a 'half bottle'.**

**Exercise 13.9**
Ratio of areas and volumes

1 Two similar cones have heights of 4 cm and 12 cm.

a The volume of the larger cone is 353 cm$^3$.
Calculate the volume of the smaller cone.
b The surface area of the small cone is 13.8 cm$^2$.
Calculate the surface area of the large one.

2

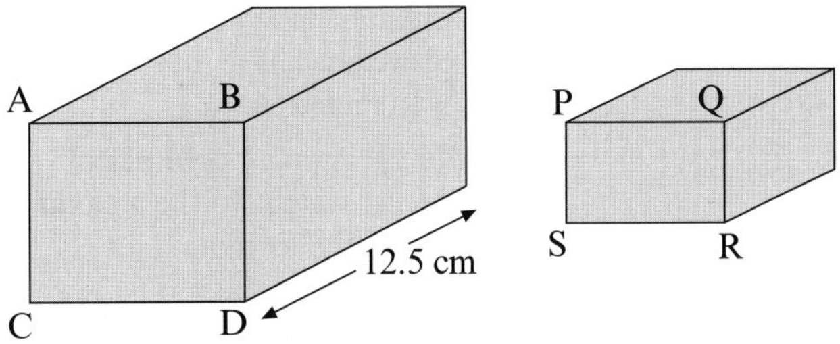


These cuboids are similar.
The area of ABCD is 36.5 $cm^2$, the area of PQRS is 22.5 $cm^2$

**a** Give the ratio of their surface areas in the ratio $1:n$.

**b** Calculate the ratio of the lengths of the cuboid in the form $1:n$.

**c** Calculate the length of the smaller cuboid.

**d** For these cuboids, give the ratio of the volumes in the ratio $1:n$.

3 Two cylinders are similar.
Cylinder A has a length of 8.5 cm.
Cylinder B has a length of 14.5 cm.

**a** Give the ratio of the base area of the cylinders in the ratio $1:n$.

**b** The volume of cylinder A is 153 $cm^3$.
Find the volume of the larger cylinder.

4 Two pyramids are similar.
The ratio of the dimensions of the bases is 2:5.
Give the ratio of the volumes of the pyramids in the form $1:n$.

5 Two cuboidal tanks have volumes in the ratio 1:4.5
Give the ratio of the dimensions of the tanks in the form $1:n$.

6 Two cylinders are similar. Their volumes are in the ratio 1:5.5.
The diameter of the smaller cylinder is 8 cm.
Calculate the diameter of the larger cylinder.

7 A scale model of a object is built to a scale of 1 to 32.

**a** The surface area of the model is 125.4 $cm^2$.
What would you expect the surface area of the real object to be? Explain your answer.

**b** The volume of the model is 238 $cm^3$.
What do you expect to be the volume of the real object? Explain your answer.

**8** Two similarly shaped weights have masses of 5 kg and 2 kg. The 5 kg weight has a base which is 9 cm across.

**a** Write the ratio of volumes of the larger weight to the smaller weight in the form $n:1$.

**b** How wide is the base of the 2 kg weight?

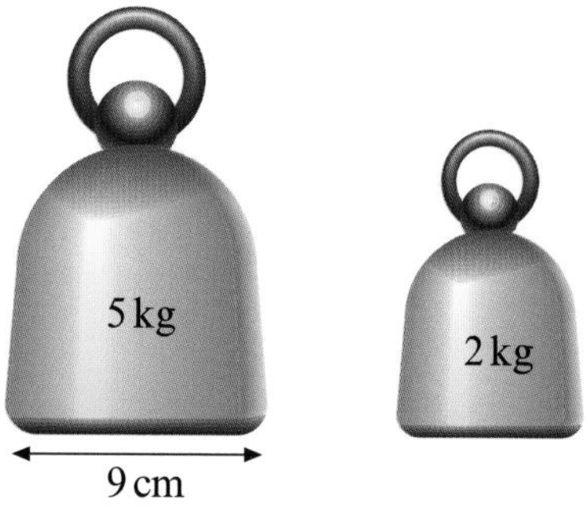


**9** A newspaper displayed this chart as part of a feature on how two companies donated money to charity.

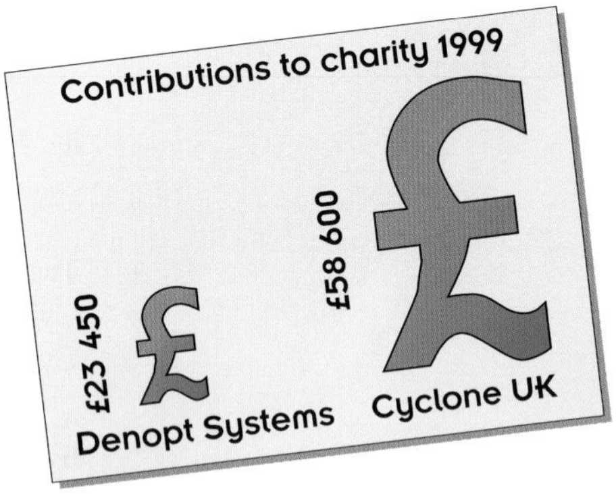


**a** The newspaper believe they have displayed the data accurately, but why is the chart misleading?

**b** The £ sign for Cyclone UK is 36 mm tall.
How should the newspaper change Denopt System's £ sign to accurately reflect the companies' contributions?

**10** This irregular solid of end area 16 cm$^2$ and volume 132 cm$^3$ is enlarged by certain scale factors. Calculate the following with the stated scale factors:

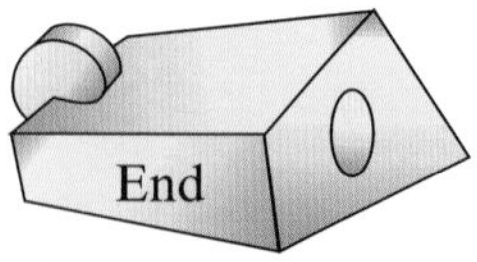


**a** End area, SF 2 **b** Volume, SF 3 **c** End area, SF $\frac{1}{3}$

**d** End area, SF $1\frac{3}{4}$ **e** Volume, SF $\frac{2}{3}$ **f** Volume, SF 2.

**11** A barn with a curved roof has a maximum height of 6 metres and holds 1740 tons of straw.
A similarly shaped but smaller barn holds half as much straw.

Calculate the maximum height of the smaller barn.

**12** Three similar plane shapes, A, B and C, are enlarged.
When A is enlarged by SF 4, B by SF $\frac{1}{2}$ and C by SF 2, the images all have the same area.

Calculate the scale factor which would enlarge:

a shape A on to shape C b shape B on to shape C
c shape A on to shape B d shape B on to shape A
e shape C on to shape A f shape C on to shape B.

# Formulas for length, area and volume

In calculations with length, area and volume:

- a length added to or subtracted from a length gives a length
- an area added to or subtracted from an area gives an area
- a volume added to or subtracted from a volume gives a volume
- a length multiplied by a length gives an area
- a length multiplied by a length multiplied by a length gives a volume
- the square root of an area gives a length
- to add, subtract, multiply or divide by a number that is not a length does not change whether an expression gives a length, area or volume.

For example, if the letters $a$, $b$ and $c$ represent lengths:

$3a + b$ represents a **length**: the lengths $3a$ and $b$ add to give a length

$\frac{1}{2}a(b - c)$ represents an **area**: the expression $(b - c)$ is a length; the lengths $(b - c)$ and $a$ multiply to give an area

$5a(b^2 + 3c^2)$ represents a **volume**: the areas $b^2$ and $3c^2$ add to give an area; the length $a$ and the area $(b^2 + 3c^2)$ multiply to give a volume.

**Exercise 13.10**
Formulas for length, area and volume

1 In the following expressions, $l$ and $w$ each represent a length.
For each expression, decide if it represents a length, area or volume.

**a** $4(l + w)$ **b** $4lw^2$ **c** $lw + w^2$

2 One of these expressions gives the volume of the prism.

$\frac{1}{2}(x + y + z + l)$ $x^2 + yz - l^2$ $2zx + zl$

$\frac{1}{2}lz(x + y)$ $\frac{1}{4}l(x + y + z)$

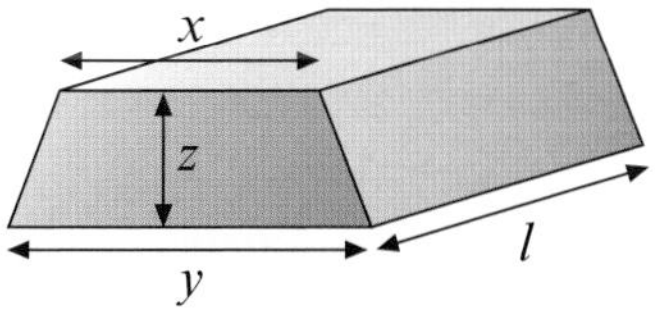


**a** Which is the correct expression for the volume of the prism?
**b** Give reasons for your answer.

π is a number, not a length.

3 One of these expressions gives the perimeter of this shape and one gives the area.

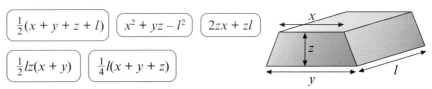

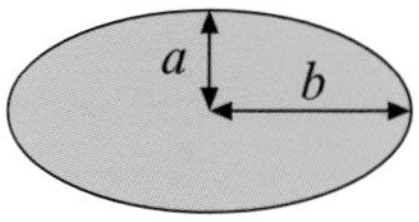


$\pi a^2b$ $\pi ab$ $\pi b^2a^2$ $\pi(a + b)$ $\pi b^2(a + 4)$

Which is the correct expression for:

**a** the perimeter **b** the area?

4 In the following expressions, $r$ and $h$ each represent a length.
For each expression, decide if it represents a length, area or volume.
Give reasons for each answer.

**a** $\frac{1}{4}rh$ **b** $\sqrt{r^2 + h^2}$ **c** $3(r + h) + \pi h$

**d** $r^2(r + h)$ **e** $\frac{4}{3}\pi r^3$ **f** $\pi r(r + h)$

5 The letters $x$, $y$ and $z$ represent lengths.
Explain why $xy + xyz + y(x - z)$ cannot represent a length, area or volume.

## End points

You should be able to ... | ... so try these questions

**A** Calculate areas of composite shapes

**A1** Calculate the area of this shape.

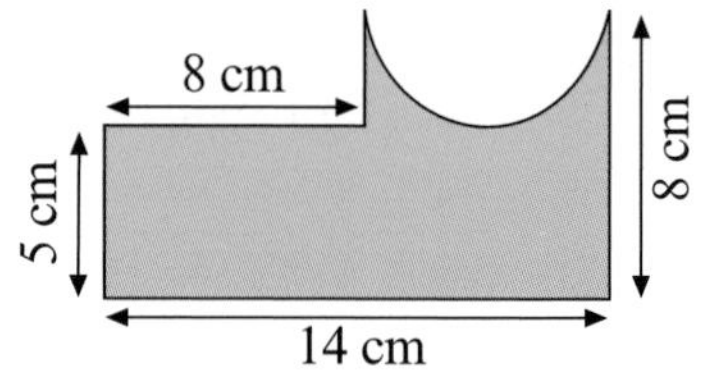


**B** Understand about planes of symmetry

**B1** This solid is made from six cubes.

How many planes of symmetry does it have?

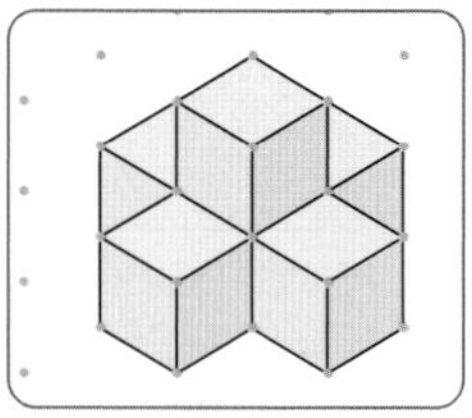

**C** Calculate volumes, capacities and surface areas of solids

**C1** This is a section through a plastic closed tank which holds heating oil. The tank is shaped like a cylinder with a hemisphere at each end. The thickness of the plastic can be ignored.

Calculate:

**a** the capacity of the tank in $m^3$
**b** surface area of the tank in $m^2$
**c** how many litres of oil there are in a full tank.

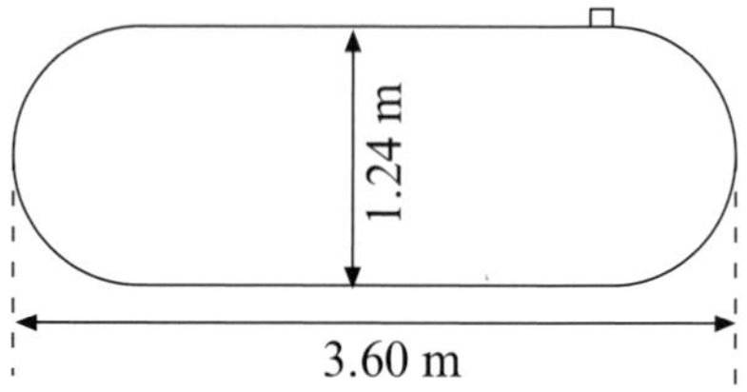


**D** Understand the effect of scale on area and volume

**D1** A large dumb-bell solid has a surface area of $852\,cm^2$. A smaller similar solid has a surface area of $284\,cm^2$.

**a** The length of the small solid is 15 cm. What is the length of the larger one?

**b** The volume of the large solid is $3742\,cm^3$. What is the volume of the smaller one?

**E** Decide if a formula could be for length, area or volume

**E1** Amy knows that some of the following formulas could give the curved surface area ($A$) of a solid but she has forgotten which ones they are. $s$ and $r$ are lengths on the solid.

**a** $A = 4\pi r^2 s$  **b** $A = \pi(3r - s)$  **c** $A = 2\pi(r^2 + s^2)$

**d** $A = 4\pi rs^2$  **e** $A = 3\pi rs$  **f** $A = \dfrac{15\pi r^3}{3s}$

Which formulas cannot be for the surface area?

## Some points to remember

- The volume of a cone with base radius $r$ and height $h$ is given by: $V = \frac{1}{3}\pi r^2 h$
- The curved surface area of a cone of slant height $s$ is given by: $A = \pi rs$
- The volume of a sphere of radius $r$ is given by: $V = \frac{4}{3}\pi r^3$
- The surface area of a sphere of radius $r$ is given by: $A = 4\pi r^2$

## Starting points

You need to know about ... / ... so try these questions

### A Theoretical probability

- Theoretical probability can be calculated just by examining possible outcomes.
- It is usually given as a fraction (it must be less than 1).

**Example 1**

What is the probability that this spinner will stop on B?
There are twelve sections of equal size.
Four sections have B on them.
So the probability that the spinner stops on B is $\frac{4}{12} = \frac{1}{3}$.

This can be written as $P(\mathbf{B}) = \frac{1}{3}$.

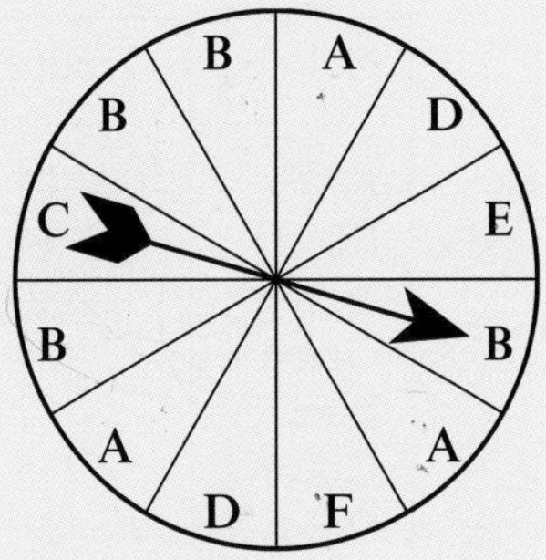


**Example 2**

In the same way, $P(\mathbf{A\ or\ B}) = \frac{7}{12}$.

- Probabilities can be shown on a probability scale from 0 to 1.

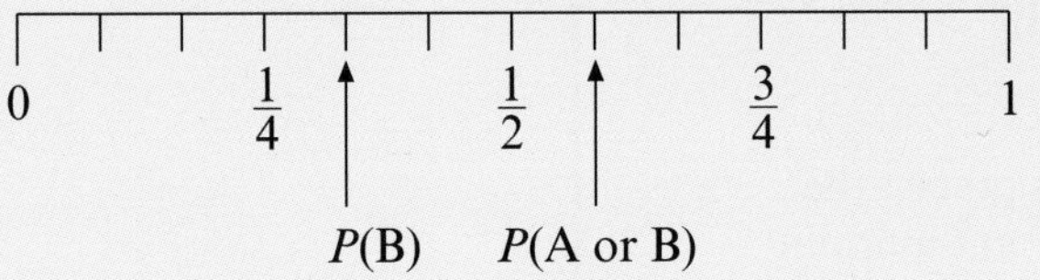


**A1** What is the probability that the wheel will stop on:
**a** A **b** E **c** D?

**A2** What are these probabilities for the wheel?
- **a** $P$(B or F)
- **b** $P$(E or A)
- **c** $P$(A, B or C)
- **d** $P$(a letter after D in the alphabet)
- **e** $P$(a letter of the alphabet)
- **f** $P$(the letter N)

**A3** Draw a probability scale and label the positions of $p$(A), $P$(C), $P$(D), and $P$(A or B or E or F).

**A4** For a 1 to 6 dice what is the probability that for one roll you will get:
- **a** a multiple of 3
- **b** a prime number
- **c** a factor of 12?

### B The probability of a non-event

If you know the probability of something happening, then you can also calculate the probability of it **not** happening.

**Example**

$P(\mathbf{B})$ on the spinner above is $\frac{1}{3}$.

$P(\mathbf{not\ B})$ is $1 - \frac{1}{3} = \frac{2}{3}$.

**B1** The probability of getting a red colour on a spinner is $\frac{4}{5}$.
What is the probability of not getting red?

**B2** In a game $P$(2 points) is $\frac{1}{4}$ and $P$(3 points) is $\frac{1}{8}$.
What is $P$(not 2 nor 3 points)?

### C Multiplication of fractions

In probability, sometimes you need to multiply fractions.

You can think of $\frac{3}{4} \times \frac{2}{5}$ like this:

$3 \times 2 = 6$

$$\frac{3}{4} \times \frac{2}{5} = \frac{6}{20} = \frac{3}{10}$$

$4 \times 5 = 20$

$\frac{2}{5}$ of $\frac{3}{4} = \frac{2}{5} \times \frac{3}{4}$

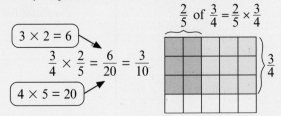

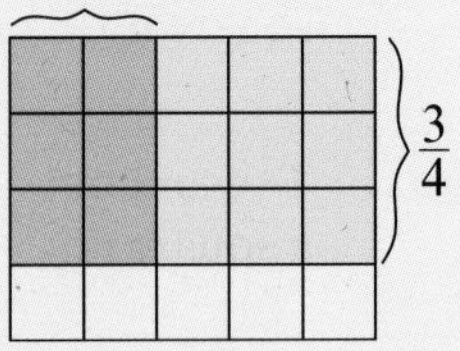


- Multiply the numerators.
- Multiply the denominators.
- Then reduce to the simplest terms if you need to.

**C1** Multiply these fractions. Give each answer in its simplest terms.
- **a** $\frac{3}{4} \times \frac{1}{4}$
- **b** $\frac{5}{8} \times \frac{2}{5}$
- **c** $\frac{2}{3} \times \frac{6}{7}$
- **d** $\frac{1}{9} \times \frac{3}{4}$

## D Relative frequency

- Relative frequency is a way of **estimating** a probability.
- It can be used both in equally likely situations or when outcomes are not equally likely.
- It is found by experiment (**experimental probability**), by survey, or by looking at data already collected (**empirical probability**).
- It is can be given as a decimal, a fraction or a percentage.

### Example

This data shows the colour of cars passing a factory gate one morning.

Estimate the probability that, at a random time, a car passing will be red.

| Colour | Frequency |
|---|---|
| Red | 68 |
| Black | 14 |
| Yellow | 2 |
| Green | 34 |
| Blue | 52 |
| Grey | 35 |
| Other | 23 |
| Total | 228 |

The relative frequency of red cars is $\frac{68}{228} = 0.30$ (to 2 dp).

So the probability of a red car is 0.30 or 30%.

- This estimate of the probability can then be used to predict the number of red cars that will pass the gate on a different morning.

**D1** What is the relative frequency of blue cars in the survey? Give your answer to 2 dp.

**D2** Estimate the probability that, at a random time, a car passing will be:
- **a** not blue
- **b** either yellow or green
- **c** neither green nor red
- **d** either black, grey or blue.

**D3** For which of these are the outcomes not equally likely.

The probability that:
- **a** a dropped matchbox will land face up
- **b** a person will catch measles before they are ten
- **c** the next snowtorm will be in December.

## E Tree diagrams for combined events

- A tree diagram can help to organise data when calculating the probability of several events happening together.
- To find the probability of a particular joint outcome you can multiply the probabilities along the branches that lead to it.

This is a tree diagram for spinning a fair coin and rolling a fair 1 to 6 dice.

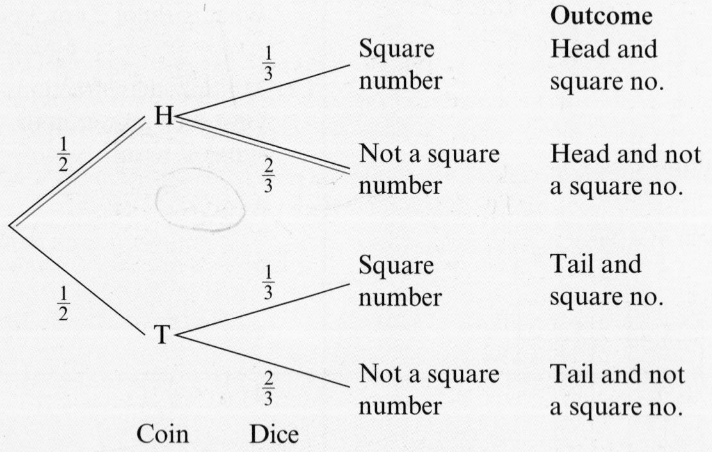


### Example

To find the probability of a head **and** a non-square number you multiply the probabilities along the red branches.

Probability of a head and non-square no. is $\frac{1}{2} \times \frac{2}{3} = \frac{2}{6}$

**E1** What is the probability of a head and a square number?

**E2** What is the probability of a tail and a non-square number?

**E3** A red 1 to 6 dice and a blue 1 to 6 dice are rolled.
- **a** Draw a tree diagram to show the probability of a triangle number on the red dice and an even or odd number on the blue one.
- **b** What is the probability of getting a non-triangle number on the red with an even number on the blue dice?

## Ways of pairing things

**Exercise 14.1**
Pairing

When people come down to breakfast in France the custom is that each person shakes hands with everyone else.
This can mean a large number of handshakes, but exactly how many depends on the number of people. Two people shaking hands counts as one handshake.

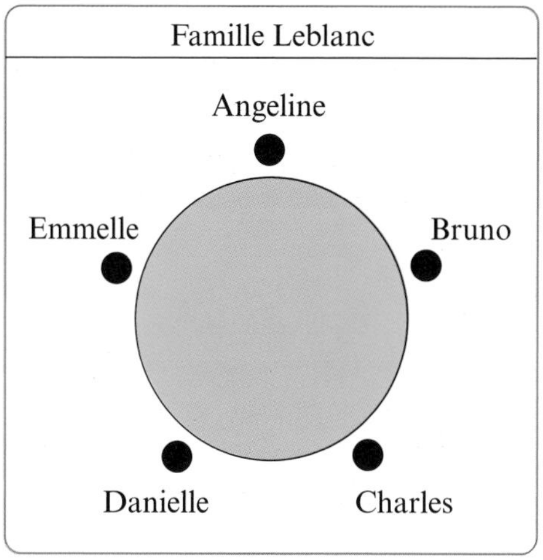


There are five members of the Leblanc family: Angeline, Bruno, Charles, Danielle and Emmelle.
Here are three of the handshakes that are made: AB, AD, BE.

1 **a** List all the handshakes for the Leblanc family above.
**b** Arrange your list so you can be sure you have every handshake.
**c** How many handshakes in total?

2 A guest, Frederic, stays overnight with the Leblancs.
**a** When they all come down to breakfast list the handshakes made.
**b** How many handshakes are made?

3 **a** Draw up a table like this.

| No. of people | 1 | 2 | 3 | 4 | 5 | 6 |
|---|---|---|---|---|---|---|
| No. of handshakes | 0 | 1 | | | | |

**b** Describe any patterns you can see in the table.
**c** Use your pattern to predict how many handshakes there will be for seven people. Check your prediction.

4 Emmelle is going on a trip. She wishes to take two books. Her bookshelf has five books she has not read. She picks two books at random.

**a** List all the different pairs of books she might pick.
**b** What is the probability that she picks two books by Sartre?
**c** What is the probability that she picks two books by the same author?
**d** What is the probability that she picks at least one blue book?
**e** Give the probability of picking two books of different colour.

5 Bruno has eight books on his shelf, four of which are by Sartre.
He picks two books at random.
What is the probability that they are both by Sartre?

## Dependent and independent events

If one cube is pulled from this bag at random, **replaced**, then another cube is chosen the probability of getting a red is $\frac{3}{9}$ on both occasions.

The same applies to the probability of other colours. This can be represented on a tree diagram like this:

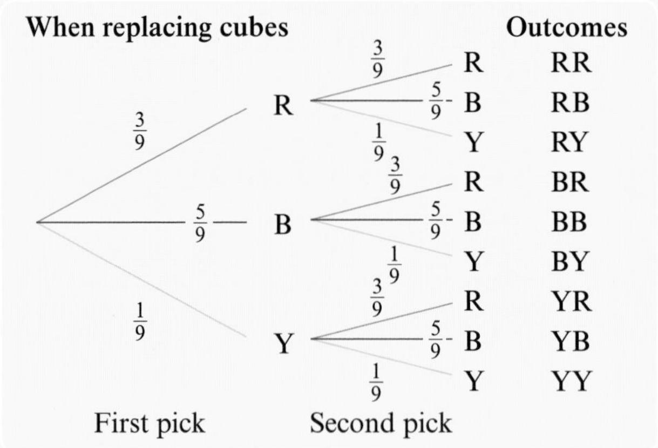


The probability of getting two reds is therefore $\frac{3}{9} \times \frac{3}{9} = \mathbf{\frac{1}{9}}$.

In this case the probability for the second pick **does not depend** on the first. Events of this type are known as **independent events**.

Two or more events or outcomes are independent if the happening of one of them has no effect on the other.

Two events or outcomes are dependent if the happening of one of them directly affects the other.

Suppose that when a cube has been picked it is **not replaced** for the second pick. If a red cube is picked first there will be 8 cubes left and only 2 of them will be red. For the second pick the probability of a red will be $\frac{2}{8}$.

The probabilities on the tree diagram will now look like this.

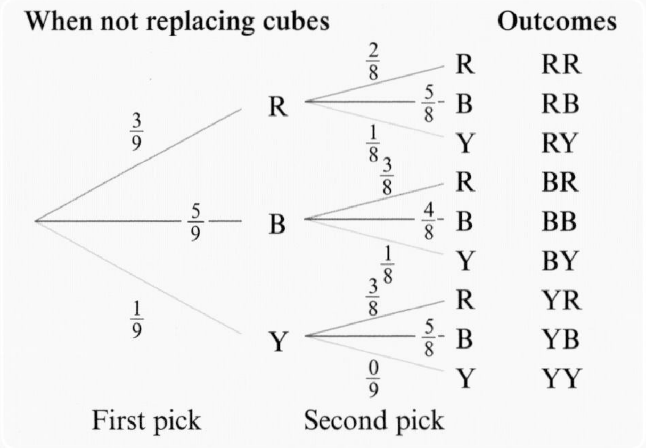


The probability of getting two reds has now changed to $\frac{3}{9} \times \frac{2}{8} = \mathbf{\frac{1}{12}}$.

In this case the probabilities for the second pick **do depend** on the first. Events of this type are known as **dependent events**.

**Exercise 14.2**
**Probabilities for dependent events**

1 For the problem above, what is the probability of getting:

**a** two blues when cubes are replaced
**b** two blues when cubes are not replaced
**c** a yellow and a red when cubes are replaced
**d** a yellow and a red when cubes are not replaced?

2 A washing line has these socks hanging on it.

Andy the thief steals two socks from the line at random.

**a** What is the probability that the first sock he takes has a white stripe?
**b** If the first sock has a white stripe, what is the probability that the second will also have a white stripe?
**c** Draw a tree diagram to show white stripes or no white stripes when two socks are taken from the line.
**d** From your diagram calculate the probability that neither sock the thief takes has a stripe.

Note that these are dependent events. Even if the thief takes both socks at exactly the same instant, in probability terms this is the same as taking one then the other.

Andy has a pang of conscience and puts the socks back on the line. His friend, Lisa, then steals two socks from the twelve, also at random.

**e** What is the probability that Lisa steals:
  **i** two green socks
  **ii** a blue sock and a yellow sock
  **iii** a pair of the same colour (ignore the stripes)?

Even though a tree diagram can get crowded when dealing with four colours, it can still provide a visual way to organise the data.

For instance you could use a reduced tree diagram such as this for question **2e** part **i**.

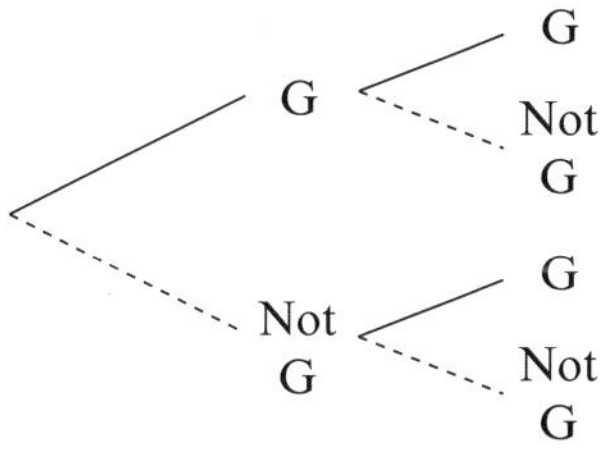


3 A card is picked at random from a pack of nine number cards.
It is not replaced then a second card is picked.
The cards are numbered 2, 3, 4, 5, 6, 7, 8, 9, 10.

**a** Copy and complete the reduced tree diagram for picking cards with prime numbers.
**b** Calculate the probability that both cards show prime numbers.
**c** What is the probability that only one card will have a prime number?

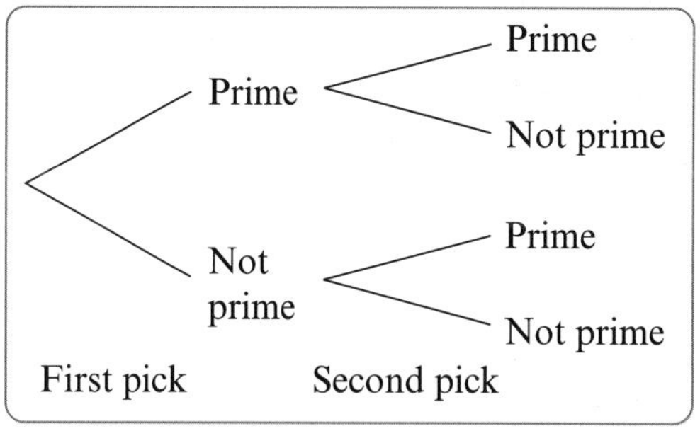


**d** Draw a different tree diagram to help you calculate the probability that the product of the two numbers will be even.

4 Three stable hands decide who mucks out the horses by drawing straws.
There are 3 short straws and 2 long ones.
When a straw has been picked it is not replaced.
Any person picking a short straw must help with mucking out.

What is the probability that:

**a** all three people have to muck out
**b** no one has to muck out
**c** only one person mucks out
**d** exactly two people muck out?

**5** A charity decides to raise funds by using a scratch card with ten hidden numbers. There are always five 1's, three 3's and two 4's but they are in different positions on each card.
You win if you uncover a total of 5. More than 5 and you lose.

**a** List all the winning combinations.

A tree diagram has been started to show the winning combinations.

**b** Why do you think the ends of some of the branches are crossed out and others are ticked?

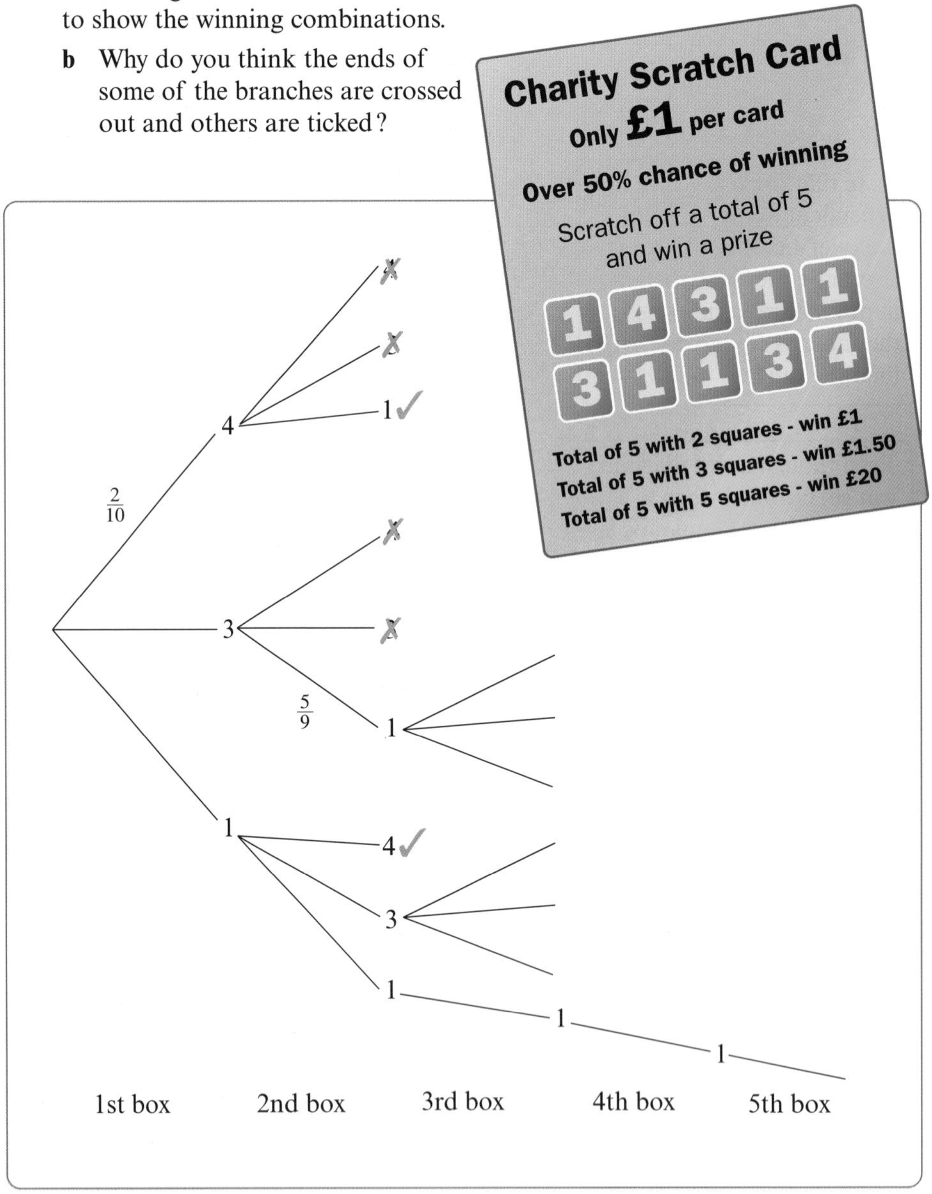
Charity Scratch Card
Only £1 per card
Over 50% chance of winning
Scratch off a total of 5
and win a prize
1 4 3 1 1
3 1 1 3 4
Total of 5 with 2 squares - win £1
Total of 5 with 3 squares - win £1.50
Total of 5 with 5 squares - win £20
$\frac{2}{10}$
$\frac{5}{9}$
4
3
1
1st box
2nd box
3rd box
4th box
5th box

Note that not every branch need always be shown on a tree diagram. You only need to draw those useful to you.

**c** Copy and complete the tree diagram and the probabilities to show winning combinations.

**d** Calculate the probability of getting a total of 5 by uncovering:
**i** 5 squares **ii** 3 squares **iii** 2 squares.

**e** Is it true to say there is 'Over 50% chance of winning' with one card? Explain your answer.

**f** If the charity sells 500 cards, what profit or loss are they likely to make?

# Conditional probability

The probability of something happening may depend on what happened earlier. For instance, the probability that a student will be absent from school on one day may be affected by whether the same student was absent the day before. These probabilities are often estimates based on relative frequencies.

The probability that Jenny goes to the cinema on Friday night is an example of **conditional probability**.

### Example

The probability that Jenny goes to the cinema on Friday is 0.27 based on her usual behaviour, but if she buys a video the probability goes down to 0.12. The probability that she buys a video is 0.3

Draw a tree diagram and calculate the probability that Jenny goes to the cinema.

The probabilities in blue have been calculated from the knowledge that the probabilities at any point must total 1.

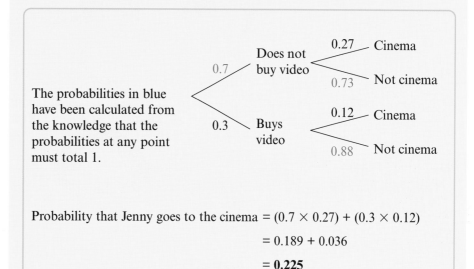

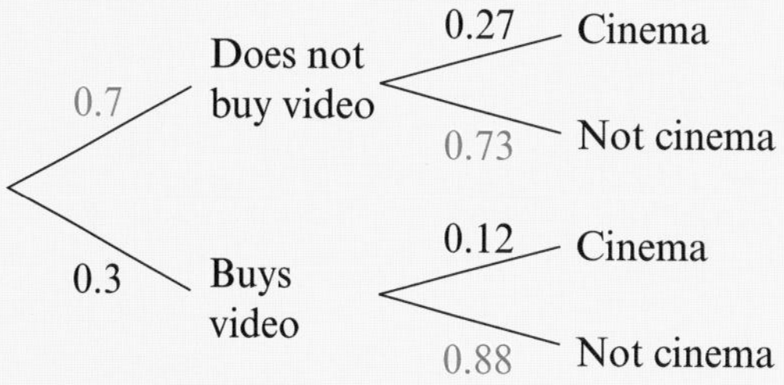


Probability that Jenny goes to the cinema = (0.7 × 0.27) + (0.3 × 0.12)

= 0.189 + 0.036

= **0.225**

**Exercise 14.3**
Conditional probability

1 The probability that a new road needs repair after one year is 0.14.
If the road is not repaired in the first year, the probability that it needs repair in the second year is 0.45.
If it is repaired, the probability drops to 0.33 in the second year.

**a** Draw a tree diagram to show the probabilities of repair over the two years.

**b** Calculate the probability that the road:

**i** does not need repair over the two years

**ii** needs repair in the second year.

2 If it is a sunny day, the probability that the following day is sunny is 0.53.
If it is not sunny, the probability for sun the following day drops to 0.13.
Amy has her holiday on Monday, Tuesday and Wednesday.
It is sunny on Monday.

What is the probability that Amy has:

**a** three days of sun **b** only one day of sun **c** two days of sun?

3 The probability that a person chosen at random is left-handed is 0.08.
If they are left-handed, the probability that they are also left-footed is 0.91.
If they are not left-handed, the probability of being left-footed is 0.01.

Use a tree diagram to calculate the probability that a person chosen at random catches a ball with one hand but kicks it with the other.

## When can you add probabilities?

**Exercise 14.4**
Adding probabilities

This shows the set of males with criminal records in the town of Humbleton.

Give your probabilities as fractions and do not cancel them down.

1 What is the probability that a male criminal in Humbleton has:

**a** a beard  **b** a necklace  **c** a beard or a necklace?

2 Compare your answers for the probabilities in Question **1**.
What link can you find between them?

3 What is the probability that a male criminal in Humbleton has:

**a** an earring  **b** a moustache?

4 **a** Predict the probability of a criminal having either an earring or a moustache.

**b** Use the full set above to find the probability that a criminal has either an earring or a moustache.

**c** Why do you think your prediction may not be the same as the true answer?

5 What is the probability that a criminal, chosen at random, has:

**a** a hat  **b** a beard  **c** a beard or a hat?

Venn diagrams are named after the mathematician John Venn, who invented them in 1881 for his work on logic.

You can use Venn Diagrams to help you see why you can sometimes add probabilities, but at other times you must not.

Look at those criminals who have:
- ◆ dark hair ◆ earrings.

All those with dark hair are in the red loop and all those with an earring are in the blue loop.

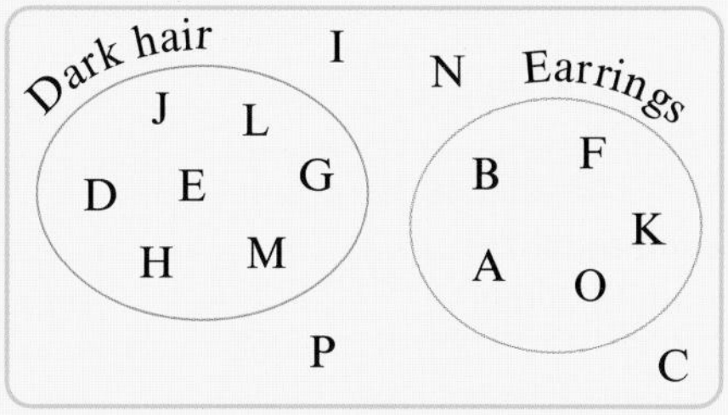


From this you can see that the probability of having dark hair is $\frac{7}{16}$ and of having an earring is $\frac{5}{16}$.

The probability of having either is $\frac{12}{16}$ (which is $\frac{7}{16} + \frac{5}{16}$).

The outcomes 'dark hair' and 'earrings' in Hambleton are said to be mutually exclusive.

Mutually exclusive outcomes are ones that cannot happen together. For instance, if you are born in January in England you cannot also be born in a summer month.

Outcomes which are not mutually exclusive can happen together. For example, if you are born in January you can be a teenager.

But now look at those with:
- ◆ a beard ◆ a moustache.

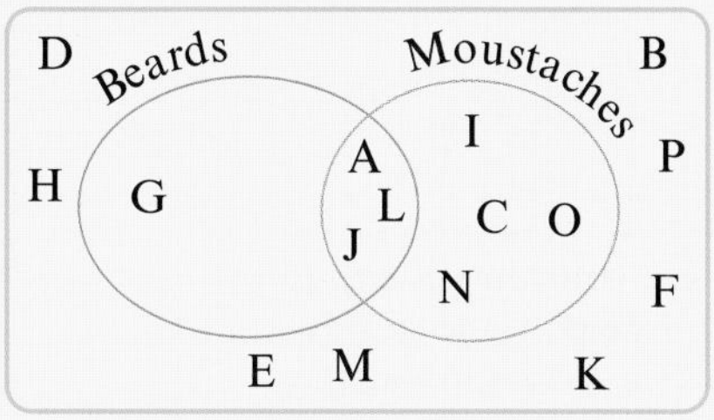


This time the loops overlap because Alf, Liam and Jim have both beards and moustaches.

The probability of having a beard is $\frac{4}{16}$ and of having a moustache is $\frac{7}{16}$.

The probability of having either is $\frac{8}{16}$ (which is **not** $\frac{4}{16} + \frac{7}{16}$).

A beard and a moustache in Hambleton are not mutually exclusive outcomes.

**You can only add probabilities when outcomes are mutually exclusive.**

## Exercise 14.5 Probability and Venn diagrams

1 a Draw a Venn diagram to show the criminals with a necklace and those with fair hair.
  b Can you add the separate probabilities to find the probability that a criminal has either a necklace or fair hair? Explain your answer.

2 Use your diagram to decide the probability that a criminal has:
  a **either** a necklace **or** fair hair
  b **both** a necklace **and** fair hair
  c **neither** a necklace **nor** fair hair
  d a necklace but not fair hair
  e fair hair but not a necklace.

3 Draw Venn diagrams to help you decide on the probabilities of having:
  a either a hat or an earing
  b either a necklace or dark hair
  c either glasses or fair hair
  d neither glasses nor dark hair.

4 For each of these, describe whether the two parts are mutually exclusive – the probability of:
  a a girl playing hockey in a school or a girl playing basketball
  b a boy playing basketball in a school or a girl playing hockey
  c a driver or a passenger
  d a bus driver or a tall woman
  e a dancer or a mechanic.

## Outcomes from a biased dice

A biased dice is one where each outcome is not equally likely.

This table shows the probabilities of getting each number on a biased 0–9 dice.

| Outcome | 0 | 1 | 2 | 3 | 4 | 5 | 6 | 7 | 8 | 9 |
|---|---|---|---|---|---|---|---|---|---|---|
| Probability | 0.13 | 0.21 | 0.05 | 0.10 | 0.01 | 0.25 | 0.13 | 0.07 | 0 | 0.05 |

Other probabilities can be found from these probabilities.

**Example**

What is the probability of getting a number less than 4 in one roll of the dice?

There are four numbers (0, 1, 2, and 3) less than 4.
There is no overlap between them (you can't get a 3 and a 1 in the same roll).
So you can add the probabilities for 0, 1, 2 and 3.

**So the probability of a number less than 4 is 0.13 + 0.21 + 0.05 + 0.1 = 0.49.**

Take care with some probabilities where outcomes are not mutually exclusive.

Here you can get a multiple of 3 **and** an odd number in the same roll of the dice.

**Example** What is the probability of getting either a multiple of 3 or an odd number?

The numbers 3 and 9 come in both sets but must only be counted once.

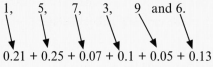

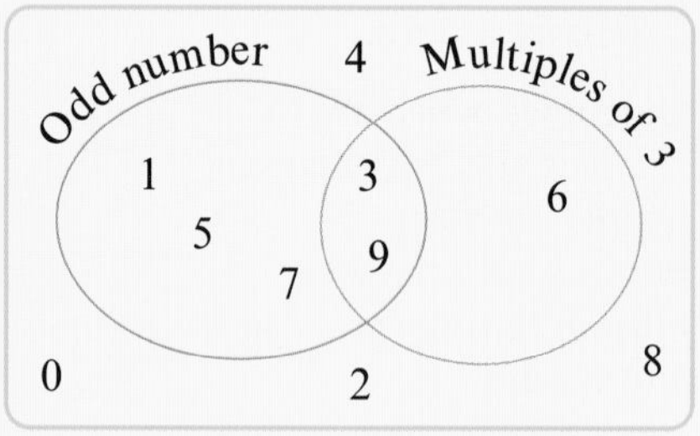


So the probabilities to add are those for:
1, 5, 7, 3, 9 and 6.

0.21 + 0.25 + 0.07 + 0.1 + 0.05 + 0.13

**So the probability of a multiple of three or an odd number is 0.81.**

**Exercise 14.6 Outcomes from a biased dice**

Note that 0 is counted as the first even number.

1 What is the probability, with this dice, of getting:
- **a** a number greater than 6
- **b** an even number
- **c** a prime number
- **d** a number which is not prime
- **e** a number which is a multiple of 3 **and** an even number?

2 Use the Venn diagram to calculate the probability of getting in one roll:
- **a** a multiple of three **and** an odd number
- **b** a number which is **neither** a multiple of 3 **nor** an odd number
- **c** a number which **is not** a multiple of 3 but **is** an odd number.

3 Use a Venn diagram to find the probability of getting a number greater than 5 or an even number.

4
- **a** Why can you add all the probabilities when you find the probability that in one roll you get a number less than 5 or a number greater than 7?
- **b** Calculate this probability.
- **c** What is the probability of **not** getting a number less than 5 **nor** a number greater than 7?

# Using Venn diagrams to find probabilities

Venn diagrams can be very useful when you have to find, for instance, the probability of: (one event **and** another event), (one event **or** another event), (**neither** one event **nor** another) or (one event **but not** another event).

**Example** On Saturday night 20 young people went into town. 12 of these visited Kingston's club, 9 went to Night Owls, and 3 decided not to go to any club.

Calculate:

**a** $P$(Kingston's or Night Owls) **b** $P$(Kingston's and Night Owls)

**c** $P$(Kingston's but not Night Owls).

This problem seems strange at first. You know only 20 people went out but 12 + 9 + 3 appears to make 24 people in town that night.

This must mean 4 people have been counted twice – in other words, they went to both clubs.

The Venn diagram would therefore look like this.

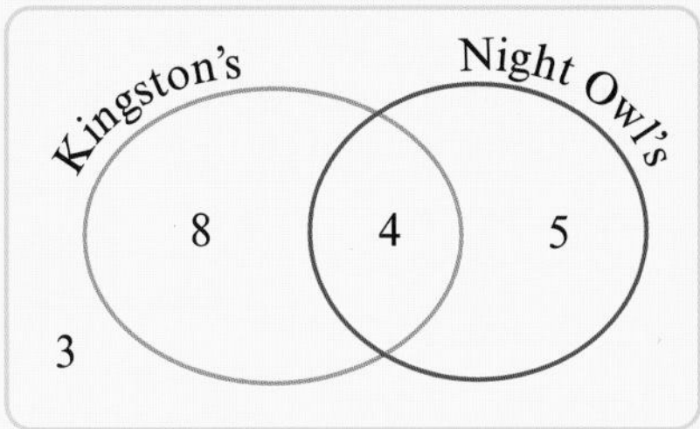


Therefore:

$$P(\text{Kingston's or Night Owls}) = \frac{(8 + 4 + 5)}{20} = \frac{17}{20}$$

$$P(\text{Kingston's and Night Owls}) = \frac{4}{20}$$

$$P(\text{Kingston's but not Night Owls}) = \frac{8}{20}$$

**Exercise 14.7**
Venn diagrams and probabilities

1 Liam carried out a survey into what people had for breakfast.

He found that, of the 44 people asked, 20 had English breakfast(E), 6 of these had both cereal(C) and English breakfast, and 5 people had neither English breakfast nor cereal.

**a** Draw a Venn diagram to show the information.

**b** Calculate the following probabilities:

**i** $P$(cereal but not English breakfast) **ii** $P$(cereal or English breakfast)

**iii** $P$(cereal) **iv** $P$(not cereal nor English breakfast)

**c** Explain why $P(\text{E}) + P(\text{C}) \neq P(\text{E or C})$.

The symbol ≠ means is not equal to …

2 Two hundred TV viewers were asked about the soaps they watched on Monday night. These are some of the findings where E stands for Eastenders, C for Coronation Street and N for Neighbours.

26 watched only N, in total 100 watched E, 8 watched all three, 35 watched E and C, 20 watched E and N but not C, 22 watched C and N, and 29 watched none of the soaps.

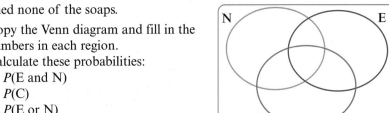

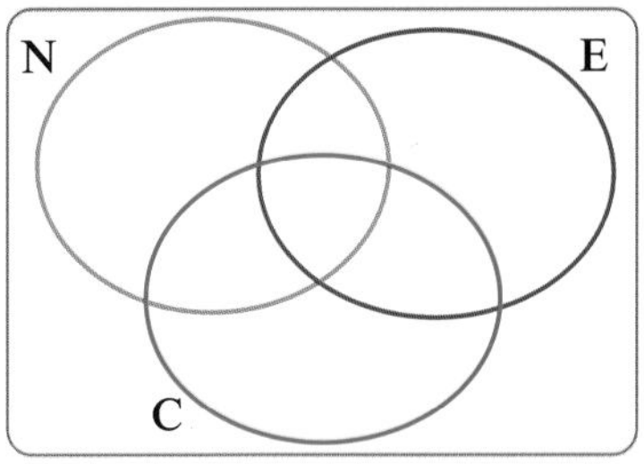


**a** Copy the Venn diagram and fill in the numbers in each region.

**b** Calculate these probabilities:

**i** $P$(E and N)

**ii** $P$(C)

**iii** $P$(E or N)

**iv** $P$(E and C but not N).

## End points

| You should be able to ... | ... so try these questions |
| --- | --- |
| **A** Calculate probabilities for dependent events | **A1** A box contains 7 coach bolts, 4 machine bolts and 2 galvanised bolts. No other bolts are in the box. Two bolts are taken at random from the box at the same time.<br>**a** If one of the bolts taken is a machine bolt, what is the probability that the other one is also a machine bolt?<br>**b** Draw a tree diagram to show the probabilities when two bolts are taken.<br>**c** What is the probability that:<br>**i** both are coach bolts<br>**ii** the two bolts are of different types? |
| **B** Understand the effect of conditional probability | **B1** The probability that Claire gets up before 7 am is 0.26. If she does get up before 7 am, the probability that she has a bath is 0.68. If not, the probability of a bath drops to 0.24.<br>**a** Draw a tree diagram with the probabilities to show the situation.<br>**b** Calculate the probability that Claire has a bath in the morning. |
| **C** Decide when you can find a probability by adding the separate probabilities | **C1** For each of these, describe whether the two parts are mutually exclusive. The probability of:<br>**a** a 1 to 6 dice giving a multiple of 3 or a square number<br>**b** a 1 to 10 spinner giving a square number or a prime number<br>**c** a 1 to 6 dice giving a prime number or an even number<br>**d** a 1 to 6 dice giving a square number or a triangular number. |
| **D** Use Venn diagrams to help find probabilities | **D1** Amongst a group of 34 jugglers, 12 wore green hats and 19 had red trousers. The number of jugglers who had neither green hats nor red trousers was twice the number who had both red hats and green trousers.<br>**a** Draw a Venn diagram to display the information.<br>**b** What was the probability that a juggler, chosen at random, had:<br>**i** a green hat but no red trousers<br>**ii** a green hat or red trousers<br>**iii** a green hat and red trousers? |

### Some points to remember

- With combined probability questions it is often helpful to draw a tree diagram. A reduced tree diagram may be necessary where there are many outcomes at each stage.
- With combined probability questions first establish if the events are dependent or independent.
- Where two objects are picked at the same time it is equivalent to two dependent events where one happens then the other.
- Before you add separate probabilities check that there is no overlap between the sets. A Venn diagram can sometimes help you decide this.

## Starting points

You need to know about ...

### A Finding the median and the interquartile range

- You can find the median and interquartile range of a set of ungrouped data by listing the data in order.
- For a distribution with a large total frequency, using cumulative frequencies is quicker than listing the data.

**1996 Open Golf Championship – Top 40 finishers**
1st & 2nd round scores

| Score | 65 | 66 | 67 | 68 | 69 | 70 | 71 | 72 | 73 | 74 | 75 | 76 | 77 |
|---|---|---|---|---|---|---|---|---|---|---|---|---|---|
| Frequency | 3 | 3 | 14 | 13 | 11 | 15 | 9 | 4 | 5 | 0 | 1 | 1 | 1 |
| Cumulative Frequency | 3 | 6 | 20 | 33 | 44 | 59 | 68 | 72 | 77 | 77 | 78 | 79 | 80 |

Lower quartile: 20th 67 ¦ 21st 68 → 67.5

Upper quartile: 60th 71 ¦ 61st 71 → 71

80 = 20 + 20 + 20 + 20

Interquartile range = Upper quartile – Lower quartile
= 71 – 67.5
= **3.5**

Median score = **69**  (40th) 69 ¦ 69 (41st) → 69

### B Calculating the standard deviation of a distribution

- To calculate the standard deviation of a frequency distribution:

**1996 Open Golf Championship – Top 40 finishers**
2nd round scores

| Score | 65 | 66 | 67 | 68 | 69 | 70 | 71 | 72 | 73 |
|---|---|---|---|---|---|---|---|---|---|
| Frequency | 2 | 3 | 6 | 8 | 6 | 8 | 3 | 3 | 1 |

- calculate the mean of all the values
- calculate the square root of the mean squared deviation.

| Score $x$ | Frequency $f$ | Total score $fx$ | Squared deviation $f(x-\bar{x})^2$ |
|---|---|---|---|
| 65 | 2 | 130 | 28.125 |
| 66 | 3 | 198 | 22.6875 |
| 67 | 6 | 402 | 18.375 |
| 68 | 8 | 544 | 4.5 |
| 69 | 6 | 414 | 0.375 |
| 70 | 8 | 560 | 12.5 |
| 71 | 3 | 213 | 15.1875 |
| 72 | 3 | 216 | 31.6875 |
| 73 | 1 | 73 | 18.0625 |
| | 40 | 2750 | 151.5 |

$\frac{2750}{40} = 68.75$ Mean score ($\bar{x}$)

$\sqrt{\frac{151.5}{40}} = \mathbf{1.95}$ (to 3 sf) Standard deviation

... so try these questions

**1996 Open Golf Championship**

| Score | Frequency 3rd round | Frequency 4th round | Score | Frequency 3rd round | Frequency 4th round |
|---|---|---|---|---|---|
| 64 | 1 | 0 | 70 | 11 | 7 |
| 65 | 0 | 1 | 71 | 3 | 6 |
| 66 | 2 | 1 | 72 | 3 | 6 |
| 67 | 0 | 5 | 73 | 3 | 3 |
| 68 | 5 | 4 | 74 | 3 | 2 |
| 69 | 6 | 2 | 75 | 3 | 3 |

**A1** Find the median and interquartile range of:
**a** the 3rd round scores
**b** the 4th round scores.

**A2** Find the median and interquartile range of the 3rd and 4th round scores combined.

**1996 Open Golf Championship**
Top 40 finishers

| 1st round Score | Frequency | Score | Frequency |
|---|---|---|---|
| 65 | 1 | 72 | 1 |
| 66 | 0 | 73 | 4 |
| 67 | 8 | 74 | 0 |
| 68 | 5 | 75 | 1 |
| 69 | 5 | 76 | 1 |
| 70 | 7 | 77 | 1 |
| 71 | 6 | | |

**B1** **a** Calculate the mean and standard deviation of the 1st round scores.

**b** Copy and complete this table using your answer to part **a**.

| | 1st round | 2nd round |
|---|---|---|
| Mean | | 68.75 |
| Standard deviation | | 1.95 |

## C Comparing sets of data

- You can compare sets of data using two types of statistic: an average, and a measure of dispersion.

| | 3rd round | 4th round |
|---|---|---|
| Mean | 70.55 | 70.4 |
| Standard deviation (to 3 sf) | 2.52 | 2.56 |

The 4th round scores are slightly lower on average.
The amount of variation in the scores is almost identical.

- When you compare data which has extreme values, the median and the interquartile range can give a better comparison.

## D Grouped data

- Data that is collected in groups, or grouped to make it easier to present, is called a **grouped frequency distribution**.

**1996 Open Golf Championship – Top 40 finishers**
1st round scores

| Score | 64 – 66 | 67 – 69 | 70 – 72 | 73 – 75 | 76 – 78 |
|---|---|---|---|---|---|
| Frequency | 1 | 18 | 14 | 5 | 2 |

- Each group of data is a **class**: the size of a class is the **class interval**.
  - the 64-66 class has a class interval of 3.

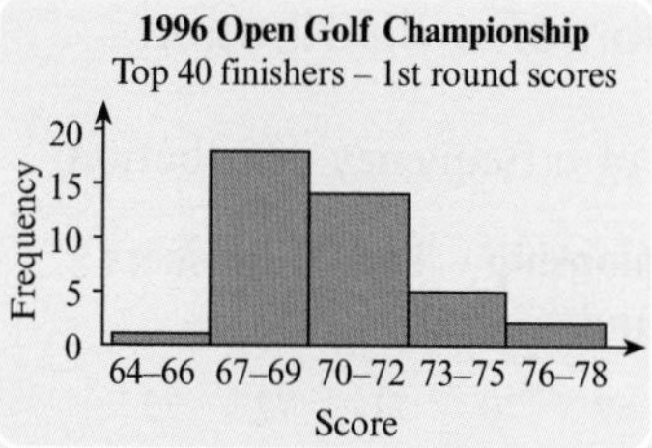


- You can present grouped data on a **grouped frequency diagram.**
- The class with the highest frequency is called the **modal class**:
  - the modal class here is 67–69.

## E Estimating the mean of grouped data

- To calculate an estimate of the mean of a grouped frequency distribution:
  - find the **mid-class value** for each class
  - estimate the total for each class i.e. multiply the mid-class value by the frequency
  - divide the sum of the estimated totals by the total frequency.

| 1st round score | Frequency $f$ | Mid-class value $m$ | Total score estimate $fm$ |
|---|---|---|---|
| 64 –66 | 1 | 65 | 65 |
| 67 – 69 | 18 | 68 | 1224 |
| 70 – 72 | 14 | 71 | 994 |
| 73 – 75 | 5 | 74 | 370 |
| 76 – 78 | 2 | 77 | 154 |
| | 40 | | 2807 |

Estimate of mean score ($\bar{m}$) = $\frac{2807}{40}$ = 70.175

**C1** Use the statistics in your table from Question **B1b** to compare the 1st and 2nd round scores.

**C2** **a** Copy and complete this table using your answer to Question **A2**.

| | 1st & 2nd rounds | 3rd & 4th rounds |
|---|---|---|
| Median | 69 | |
| Interquartile range | 3.5 | |

**b** Compare the two sets of data.

**D1** Use the ungrouped data above Question **A1**, and classes 64–66, 67–69, 70–72, 73–75, to create a grouped frequency distribution for:
**a** the 3rd round scores
**b** the 4th round scores.

**D2** For each distribution:
**a** draw a grouped frequency diagram
**b** give the modal class.

**D3** Use the ungrouped data above Question **A1**, and a class interval of 2 to group the 3rd round scores.

**E1** Use your grouped frequency distribution from Question **D1a** to calculate an estimate of the mean of the 3rd round scores.

**E2**

| 4th round Score | Frequency $f$ | Mid-class value $m$ |
|---|---|---|
| 64 – 65 | 1 | 64.5 |
| 66 – 67 | 6 | 66.5 |
| 68 – 69 | 6 | |
| 70 – 71 | 13 | |
| 72 – 73 | 9 | |
| 74 – 75 | 5 | |
| | 40 | |

Copy and complete this working to calculate an estimate of the mean.

# Histograms

- A grouped frequency diagram can also be represented by a **histogram**.
- Data can be collected, or presented, in groups with **unequal** class intervals.

To draw a histogram for a distribution with unequal class intervals:

- find the class interval for each class
- calculate the **frequency density** for each class
  i.e. the frequency divided by the class interval
- use the frequency density as the height of each bar.

**Example** Draw a histogram for this grouping of the day 2 scores.

| Points, $p$ | Frequency | Class interval | Calculation | Frequency density |
|---|---|---|---|---|
| $400 \leqslant p < 700$ | 9 | 300 | $9 \div 300$ | 0.03 |
| $700 \leqslant p < 800$ | 10 | 100 | $10 \div 100$ | 0.10 |
| $800 \leqslant p < 850$ | 13 | 50 | $13 \div 50$ | 0.26 |
| $850 \leqslant p < 950$ | 8 | 100 | $8 \div 100$ | 0.08 |
| $950 \leqslant p < 1100$ | 10 | 150 | $10 \div 150$ | $0.0\dot{6}$ |

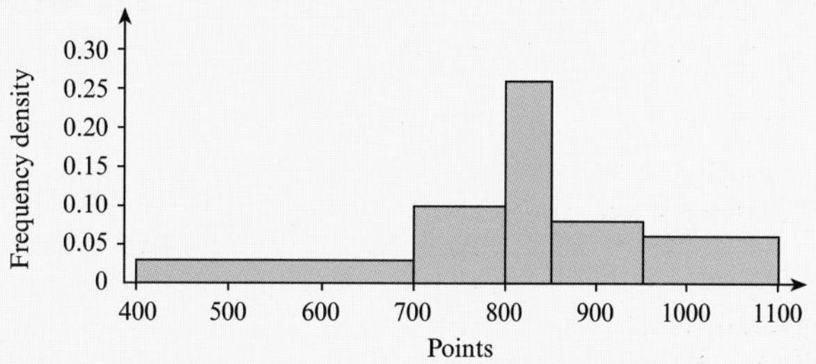


**Exercise 15.1**
**Histograms**

**1** **1996 Olympic Games – Decathlon (Day 1)**
Points scored in each of the 5 events by the top 10 decathletes

| Points | 700 – 799 | 800 – 849 | 850 – 899 | 900 – 949 | 950 – 999 | 1000 – 1099 | Total |
|---|---|---|---|---|---|---|---|
| Frequency | 8 | 10 | 13 | 8 | 8 | 3 | 50 |

**a** Using 2 mm graph paper, draw a histogram for this distribution.
**b** The frequencies for three of the classes are the same.
Explain why the height of the bar for the 700–799 class is only half the height of the bars for classes 900–949 and 950–999.

**2** **1996 Olympic Games – Decathlon**
Total points scored by each decathlete

Alex Kruger (GB) retired on Day 1 of the Decathlon because of a knee injury.

| Points | 6500 – 7499 | 7500 – 7799 | 7800 – 7999 | 8000 – 8099 | 8100 – 8199 | 8200 – 8299 | 8300 – 8999 |
|---|---|---|---|---|---|---|---|
| Frequency | 2 | 3 | 4 | 3 | 6 | 4 | 6 |

Draw a histogram for this distribution.

## Discrete and continuous data

- There are two types of numerical data:
  - **discrete** data can only be certain definite values

**1996 Olympic Games – Decathlon**
Dan O'Brien (US) – Points scored in each event

| | | | | | | | | | |
|---|---|---|---|---|---|---|---|---|---|
| 975 | 952 | 830 | 868 | 967 | 991 | 845 | 910 | 842 | 644 |

This data is discrete because it is not possible to score fractional numbers of points such as 873.5 in a decathlon event.

  - **continuous** data can take any value on the real number line, including fractional values.

Tessa Sanderson was taking part in her sixth Olympic Games in 1996. She won the Javelin Gold Medal in 1984.

**1996 Olympic Games – Women's Javelin**
Tessa Sanderson (GB) – Qualifying throws

58.86 m 56.80 m 56.64 m

This data is continuous because the distance thrown can be any value, but it is measured and rounded down to the nearest centimetre.

- Some data may appear to be continuous, but is actually discrete.

Steve Smith won the Bronze Medal with his jump of 2.35 m.

**1996 Olympic Games – Men's High Jump**
Steve Smith (GB) – Clearances in Final

2.25 m 2.32 m 2.35 m

This is discrete data because the bar is only set at certain heights in a high jump competition. For example, in the men's final it was not possible to set a height of 2.30 m.

### Exercise 15.2
### Discrete and continuous data

Fiona May won the Silver Medal with her jump of 7.02 m.

Great Britain won the Silver Medal in the men's 4 × 400 m relay, and set a new European record time.

**1** A

**1996 Olympic Games – Women's Long Jump**
Fiona May (GB) – Distances jumped in Final

6.68 m 7.02 m 6.78 m 6.73 m 6.76 m 6.88 m

B

**1996 Olympic Games – Pole Vault**
Jean Galfione (GB) – Heights cleared in Final

5.60 m 5.80 m 5.86 m 5.92 m

C

**1996 Olympic Games**
**4 × 400 m Men's Relay Final**
Great Britain
400 m split times (seconds)

| | |
|---|---|
| Iwan Thomas | 44.91 |
| Jamie Baulch | 44.19 |
| Mark Richardson | 43.62 |
| Roger Black | 43.87 |

For each set of data:

**a** decide whether the data are discrete or continuous
**b** explain your decision.

**2** Gail Devers and Merlene Ottey were given the same time in the final of the women's 100 m.
Their times, to 3 dp, could have been any value between 10.930 and 10.940 seconds.

Explain why it is appropriate in athletics events to round times **up** to the nearest unit, and distances **down** to the nearest unit.

**Women's 100 m Final**

| | Time (s) |
|---|---|
| Devers | 10.94 |
| Ottey | 10.94 |
| Torrence | 10.96 |
| Sturrup | 11.00 |

3 a Calculate the mean age of these athletes.

b Explain why the exact mean age could be 0.5 years greater than your answer.

c When age data is given in years, do you think it is discrete or continuous ? Explain your answer.

| **Great Britain's Men's 4 × 400 m Relay Team Ages at 20th July 1996** | |
|---|---|
| Jamie Baulch | 23 |
| Roger Black | 30 |
| Mark Richardson | 23 |
| Iwan Thomas | 22 |

## Using continuous data

- When you group continuous data, you need to decide what happens to values at the ends of the classes, the **class limits**.

$16.5 \leqslant x < 18.0$
$18.0 \leqslant x < 18.5$ ← 18.0 → $16.5 < x \leqslant 18.0$
$18.0 < x \leqslant 18.5$

A distance of 18.0 m goes into different classes in each grouping because of the different ways the classes are defined.

- A common way of showing classes such as:
$16.5 \leqslant x < 18.0$, $18.0 \leqslant x < 18.5$, etc. is 16.5 – , 18.0 – , 18.5 – , etc.
When grouping data using unequal class intervals, you should also show the upper class limit of the final class, e.g. 21.0

**1996 Olympic Games – Women's Shot Final**

| Distance (metres) | 16.5 – | 18.0 – | 18.5 – | 19.0 – | 19.5 – 21.0 |
|---|---|---|---|---|---|
| Frequency | 5 | 13 | 10 | 9 | 4 |

- When age data is given in its usual form, in completed years, you can group it in the same way as discrete data.

**Exercise 15.3**
**Using continuous data**

1 **1996 Olympic Games – Great Britain's Athletics Team**

| Age | 18 – 23 | 24 – 26 | 27 – 29 | 30 – 32 | 33 – 41 |
|---|---|---|---|---|---|
| Frequency | 18 | 18 | 19 | 18 | 9 |

Draw a histogram for this distribution.

The width of the 18 – 23 class is 6 years, and the mid-class value is 21.

2 **1996 Olympic Games – Men's Shot Final**

| Distance (metres) | 19.0 – | 19.5 – | 20.0 – | 20.5 – | 21.0 – 22.0 | Total frequency |
|---|---|---|---|---|---|---|
| Frequency | 5 | 11 | 17 | 8 | 1 | 42 |

a Draw a histogram for this distribution.
b The mid-class value of the 19.0 – class is 19.25
List the mid-class values for the other four classes.
c Calculate an estimate of the mean distance

You can use the same method to estimate the mean as you used before for discrete data.

3 For the data from the women's shot final above:

a draw a histogram
b use the statistical function on your calculator to estimate the mean.

4 Use your answers to Questions **2** and **3** to compare the men and women.

5 This data shows the reaction times of athletes in the sprint events.

The reaction time is the length of time between the starter's gun firing and the rear foot leaving the starting block.

s is the abbreviation for seconds.

Linford Christie (GB), the 100 m Gold Medal winner in 1992, was disqualified after a second false start when his recorded reaction time was 0.086 s.

Scientists believe a reaction time under one-tenth (0.100) of a second is impossible.

**Men's 100 m Final**

| | Reaction time (s) |
|---|---|
| Bailey | 0.174 |
| Fredericks | 0.143 |
| Boldon | 0.164 |
| Mitchell | 0.145 |
| Marsh | 0.147 |
| Ezinwa | 0.157 |
| Green | 0.169 |
| Christie | DISQ |

**Men's 200 m Final**

| | Reaction time (s) |
|---|---|
| Johnson | 0.161 |
| Fredericks | 0.200 |
| Boldon | 0.208 |
| Thompson | 0.202 |
| Williams | 0.182 |
| Garcia | 0.229 |
| Stevens | 0.151 |
| Marsh | 0.167 |

**110 m Hurdles Final**

| | Reaction time (s) |
|---|---|
| Johnson | 0.170 |
| Crear | 0.124 |
| Schwarthoff | 0.164 |
| Jackson | 0.133 |
| Valle | 0.179 |
| Swift | 0.151 |
| Vander-Kuyp | 0.167 |
| Batte | 0.160 |

**Women's 100 m Final**

| | Reaction time (s) |
|---|---|
| Devers | 0.166 |
| Ottey | 0.166 |
| Torrence | 0.151 |
| Sturrup | 0.176 |
| Trandenkova | 0.151 |
| Voronova | 0.133 |
| Onyali | 0.174 |
| Pintusevych | 0.176 |

**Women's 200 m Final**

| | Reaction time (s) |
|---|---|
| Perec | 0.174 |
| Ottey | 0.194 |
| Onyali | 0.231 |
| Miller | 0.172 |
| Malchugina | 0.198 |
| Sturrup | 0.165 |
| Cuthbert | 0.175 |
| Guidry | 0.207 |

**100 m Hurdles Final**

| | Reaction time (s) |
|---|---|
| Engquist | 0.132 |
| Bukovec | 0.164 |
| Girard-Leno | 0.133 |
| Devers | 0.189 |
| Rose | 0.179 |
| Freeman | 0.181 |
| Shekhodanova | 0.175 |
| Goode | 0.160 |

**1996 Olympic Games – Reaction Times**
**Semifinals – 100 m, 200 m, 100 m/110 m Hurdles**

| Reaction time (s) | 0.12– | 0.14– | 0.15– | 0.16– | 0.17– | 0.18– | 0.19– | 0.21–0.27 | Total |
|---|---|---|---|---|---|---|---|---|---|
| Frequency | 7 | 11 | 13 | 16 | 14 | 16 | 9 | 8 | 94 |

**a** Group the data for the six finals in the same way as the semifinals data.
**b** Draw a histogram for:
**i** the semifinals **ii** the finals.

6 **a** Copy your grouped frequency distribution for the finals data, and double each of the frequencies.
**b** Draw a frequency polygon to show this distribution by plotting each frequency against its mid-class value.
**c** On the same diagram, draw a frequency polygon for the semifinalists.
**d** Do you think your frequency polygons give a fair comparison of the reaction times in the finals and semifinals ?
Explain your answer.

There were exactly twice as many semifinalists as finalists.

7 Use your calculator to estimate the mean reaction time in:
**a** the semifinals **b** the finals.

8 'Athletes' reaction times are faster in finals than in semifinals'

**a** Do you think this hypothesis is true or false ?
**b** Give reasons for your answer.

9 Investigate the effects of grouping the finals data in different ways.

## Drawing a cumulative frequency curve

**1996 Olympic Games – Decathlon (Day 2)**
Points scored in each of the 5 events by the top 10 decathletes

| Points | 400 – 499 | 500 – 599 | 600 – 699 | 700 – 799 | 800 – 899 | 900 – 999 | 1000 – 1099 | Total |
|---|---|---|---|---|---|---|---|---|
| Frequency | 1 | 0 | 8 | 10 | 17 | 10 | 4 | 50 |

- To draw a **cumulative frequency curve**:
  - construct a cumulative frequency table

If you include a cumulative frequency of 0 in your table then you have a point to start the curve from: (400, 0).

| Points | <400 | <500 | <600 | <700 | <800 | <900 | <1000 | <1100 |
|---|---|---|---|---|---|---|---|---|
| Cumulative frequency | 0 | 1 | 1 | 9 | 19 | 36 | 46 | 50 |

  - plot the cumulative frequencies on a graph
  - join the points with a smooth curve.

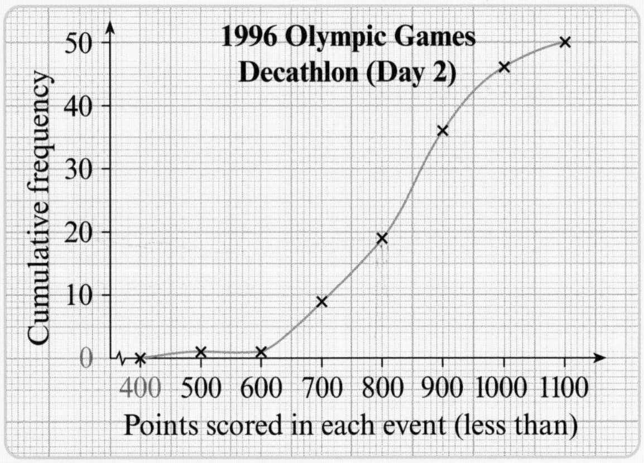


**Exercise 15.4**
Drawing cumulative frequency curves

**1**

**1996 Olympic Games – Decathlon (Day 1)**
Points scored in each of the 5 events by the top 10 decathletes

| Points | 700 – 799 | 800 – 849 | 850 – 899 | 900 – 949 | 950 – 999 | 1000 – 1099 | Total |
|---|---|---|---|---|---|---|---|
| Frequency | 8 | 10 | 13 | 8 | 8 | 3 | 50 |

**a** Copy and complete the cumulative frequency table below.

| Points | <700 | <800 | <850 | <900 | <950 | <1000 | <1100 |
|---|---|---|---|---|---|---|---|
| Cumulative frequency | 0 | | | | | | |

**b** Use your table to draw a cumulative frequency curve.

**2**

Steve Backley (GB) won the javelin silver medal with the very first throw of the final.

**1996 Olympic Games – Men's Javelin Final**

| Distance (metres) | 76 – | 80 – | 81 – | 82 – | 84 – | 87 – 89 |
|---|---|---|---|---|---|---|
| Frequency | 6 | 6 | 8 | 13 | 13 | 2 |

**a** Copy and complete the following cumulative frequency table.

| Distance (metres) | <76 | <80 | <81 | <82 | <84 | <87 | <89 |
|---|---|---|---|---|---|---|---|
| Cumulative frequency | 0 | 6 | | | | | 48 |

**b** Draw a cumulative frequency curve for the Men's Javelin final.

**3** Using your answers to Question **2**, can you think of a way to find the median distance thrown in the final? Explain your method.

# Estimating the median and the interquartile range

There were 48 throws in the final, so the **exact** median distance is halfway between the 24th and 25th longest.

It is impossible to find these distances from the table, so you can only **estimate** the median.

- A cumulative frequency table shows which class the median is in.

**1996 Olympic Games – Men's Javelin Final**

| Distance (metres) | <76 | <80 | <81 | <82 | <84 | <87 | <89 |
|---|---|---|---|---|---|---|---|
| Cumulative Frequency | 0 | 6 | 12 | 20 | 33 | 46 | 48 |

The median distance must be between 82 m and 84 m.

As you are only estimating distances within the classes, you can divide the 48 throws into four quarters using the **12th**, **24th**, and **36th** longest (cumulative frequencies 12, 24, and 36).

- To estimate the median and quartiles from a cumulative frequency curve:
  - divide the cumulative frequency into four quarters
  - go across to the curve, then go down and read off each value.

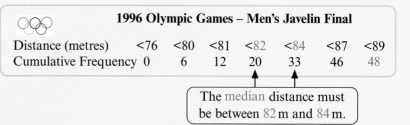

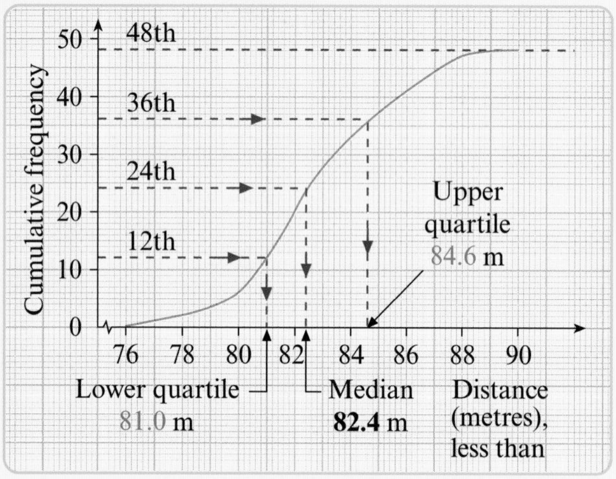


Estimate of median distance = **82.4 m**

Estimate of interquartile range = 84.6 – 81.0
= **3.6 m**

**Exercise 15.5**
**Estimating the median and interquartile range**

1

**1996 Olympic Games – Men's Hammer Final**

| Distance (metres) | 73 – | 75 – | 76 – | 77 – | 78 – | 80 – 82 |
|---|---|---|---|---|---|---|
| Frequency | 6 | 6 | 16 | 7 | 13 | 4 |

a Copy and complete the following cumulative frequency table.

| Distance (metres) | < 73 | | < 80 | < 82 |
|---|---|---|---|---|
| Cumulative frequency | 0 | | | 52 |

b Draw a cumulative frequency curve for the Men's Hammer final.

2 Use your cumulative frequency curve to estimate:

a the median distance thrown  b the interquartile range.

3 Compare the distances thrown in the Men's Hammer final and the men's Javelin final.

4 a From page 179, copy the cumulative frequency table and the cumulative frequency curve for the Decathlon (day 2).

b Estimate the median points score.

c Estimate the interquartile range of the scores.

Use cumulative frequencies 12.5, 25, and 37.5 to divide the data into four quarters.

Use cumulative frequencies 12.5, 25, and 37.5 to divide the data into four quarters.

**5** Use your answers to Exercise 15.4, Question **1** to estimate:

**a** the median points score on day 1 of the decathlon
**b** the interquartile range of the scores.

**6** Compare the points scored on days 1 and 2 of the Decathlon.

**7**

**1996 Olympic Games – Women's Javelin Final**

| Distance (metres) | 56 – | 58 – | 60 – | 62 – | 64 – | 66 – 68 | Total frequency |
|---|---|---|---|---|---|---|---|
| Frequency | 7 | 11 | 10 | 10 | 7 | 1 | 46 |

**a** Make a cumulative frequency table for the Women's Javelin final.
**b** Draw a cumulative frequency curve.
**c** Decide how to divide the data into four quarters.
**d** Estimate the median distance thrown and the interquartile range.

## Estimating cumulative frequencies

◆ You can estimate cumulative frequencies from a cumulative frequency curve.

**Example** Estimate how many Men's Javelin throws were:

**a** less than 81.5 m **b** greater than 85 m

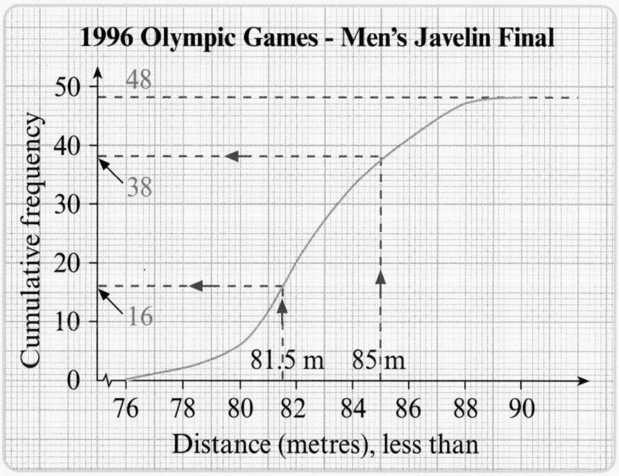


**a** Estimated number of throws less than 81.5 m = **16**

**b** Estimated number of throws greater than 85 m = 48 – 38
= **10**

**Exercise 15.6**
**Estimating cumulative frequencies**

**1** Use the cumulative frequency curve above to estimate how many throws were less than 83 m.

**2** The winning throw in the 1988 Men's Javelin final was 84.28 m. Estimate how many throws were greater than this in 1996.

Jonathan Edwards (GB) won the Silver Medal in the Triple Jump.

**3**

**1996 Olympic Games – Men's Triple Jump Final**

| Distance (metres) | 15.5 – | 16.0 – | 16.5 – | 17.0 – | 17.5 – | 18.0 – 18.5 |
|---|---|---|---|---|---|---|
| Frequency | 1 | 7 | 14 | 6 | 2 | 1 |

**a** Draw a cumulative frequency curve for the Men's Triple Jump final.
**b** Estimate how many jumps were greater than 16.8 m.

# Sampling from a population

- A **population** is the name given to any set of data under investigation. If you are investigating a very large population, it may not be possible to collect all the data.
- A **sample** is a smaller set of data, chosen from a population, that can be analysed and used to make conclusions about the population.

**Exercise 15.7**
**Sampling from a population**

1 The British Team for the 1996 Olympic Games consisted of 500 members. This is the distribution of their ages.

**1996 Olympic Games**
**Ages of Great Britain Team at 20/7/96**

| Age | 16 – | 21 – | 24 – | 27 – | 29 – |
|---|---|---|---|---|---|
| Frequency | 25 | 54 | 68 | 59 | 63 |
| Age | 31 – | 34 – | 37 – | 42 – | 52 – 71 |
| Frequency | 51 | 41 | 47 | 55 | 37 |

Draw a histogram for this age distribution.

In a random sample, each member of the population has the same chance of being chosen.

One way of choosing a random sample is to:

- allocate a number to each member of the population
- use a random number table to select the sample.

If the population is listed at random then you could choose a **systematic** sample, e.g. every 10th member on the list is chosen until the sample is complete.

2 These are three random samples taken from the 500 members of the Great Britain Olympic Team.

**1996 Olympic Games**
**Great Britain Team**

| Age | Frequency |
|---|---|
| 19 – | 5 |
| 24 – | 4 |
| 29 – | 7 |
| 39 – | 4 |
| 49 – 68 | 5 |

**Sample size 25**

**1996 Olympic Games**
**Great Britain Team**

| Age | Frequency |
|---|---|
| 19 – | 8 |
| 24 – | 13 |
| 29 – | 12 |
| 34 – | 3 |
| 39 – | 5 |
| 49 – 68 | 9 |

**Sample size 50**

**1996 Olympic Games**
**Great Britain Team**

| Age | Frequency |
|---|---|
| 19 – | 14 |
| 24 – | 25 |
| 29 – | 28 |
| 34 – | 10 |
| 39 – | 11 |
| 49 – | 8 |
| 59 – 68 | 4 |

**Sample size 100**

**a** Draw a histogram for each random sample.
**b** Compare the shapes of your distributions with your answer to Question **1**.

The estimate of the mean for each sample can be used as an estimate of the mean of the population.

3 For each random sample in Question **2**, use mid-class values of 21.5, 26.5, … etc. to calculate an estimate of the mean.

4 The actual mean age of the Great Britain Olympic Team was 33.4 years. Use your answers to Question **3** to describe the effect of the size of sample on the estimate of the mean.

5 The Great Britain Team had 312 competitors and 188 support staff.

a Do you think the mean age of these two groups is likely to be the same? Explain your answer.

b When choosing a random sample of 50, what do you think the advantage might be of choosing random samples of 31 from the competitors and 19 from the support staff ?

## End points

You should be able to ... ... so try these questions

**1996 Olympic Games – Discus Finals**

| Distance (metres) | | 56 – | 60 – | 62 – | 64 – | 66 –70 | Total |
|---|---|---|---|---|---|---|---|
| Frequency | Women | 7 | 10 | 13 | 13 | 6 | 49 |
| (of throws) | Men | 5 | 8 | 16 | 13 | 3 | 45 |

**A** Use histograms and frequency polygons

**A1** Draw a histogram to show the distances thrown in the:
**a** Women's Discus final **b** Men's Discus final.

**A2** **a** Draw a frequency polygon for the Women's Discus final.
**b** On the same diagram, draw a frequency polygon for the Men's final.
**c** Compare the two distributions.
**d** Explain why the frequency polygons would not have given a fair comparison if the number of throws in the finals had been 59 and 45.

**B** Calculate an estimate of the mean of a grouped frequency distribution

**B1** (Do **not** use the statistical function on your calculator for this question.) For the Men's Discus final, calculate an estimate of the mean distance thrown.

**B2** Use the statistical functions on your calculator to estimate the mean of the distances thrown in the Women's Discus final.

**1996 Olympic Games – Men's Long Jump Final**

| Distance (metres) | 6.4 – | 7.2 – | 7.6 – | 7.8 – | 8.0 – | 8.2 – | 8.4 – 8.6 |
|---|---|---|---|---|---|---|---|
| Frequency | 2 | 3 | 4 | 13 | 13 | 2 | 1 |

**1996 Olympic Games – Women's Long Jump Final**

| Distance (metres) | 6.2 – | 6.4 – | 6.6 – | 6.8 – | 7.0 – 7.2 |
|---|---|---|---|---|---|
| Frequency | 3 | 6 | 9 | 14 | 3 |

**C** Estimate the median and the interquartile range

**C1** **a** Draw a cumulative frequency curve for the Men's Long Jump final.
**b** Use your cumulative frequency curve to estimate:
**i** the median distance jumped **ii** the interquartile range.

**C2** Use cumulative frequencies 9, 18, and 27 to estimate the median and interquartile range for the Women's Long Jump final.

**D** Compare grouped frequency distributions

**D1** Compare the distances jumped in the Women's and Men's finals.

**D2** Use your answers to Questions **B1** and **B2** to compare the distances thrown in the Women's and Men's Discus finals.

**E** Estimate cumulative frequencies

**E1**

**1996 Olympic Games – Women's Triple Jump Final**

| Distance (metres) | 13.5 – | 13.9 – | 14.1 – | 14.4 – | 14.7 – | 15.0 – 15.4 |
|---|---|---|---|---|---|---|
| Frequency | 10 | 9 | 12 | 8 | 7 | 1 |

**a** Draw a cumulative frequency curve for the Women's Triple Jump final.
**b** Use your curve to estimate how many jumps were:
**i** less than 14.0 m **ii** greater than 14.5 m.

### Some points to remember

- When you choose a sample from a population, usually the larger the sample the closer its mean and standard deviation are to those of the whole population.

# WHEEL GARDEN CENTRE

## OUR WATER WHEEL PRODUCES OVER 60% OF THE ELECTRICITY WE USE

**Watering cans - plastic**

| 5 litre | 8 litre | 10 litre | 12 litre |
|---|---|---|---|
| 99p | £1.29 | £1.69 | £1.99 |

**Watering cans - traditional**

| 1 gallon | 1.5 gallon | 2 gallon | 2.5 gallon |
|---|---|---|---|
| £4. 99 | £6.49 | £7.99 | £9.99 |

**Visit our Coffee Shop**

**Rolls of Plastic Sheet**

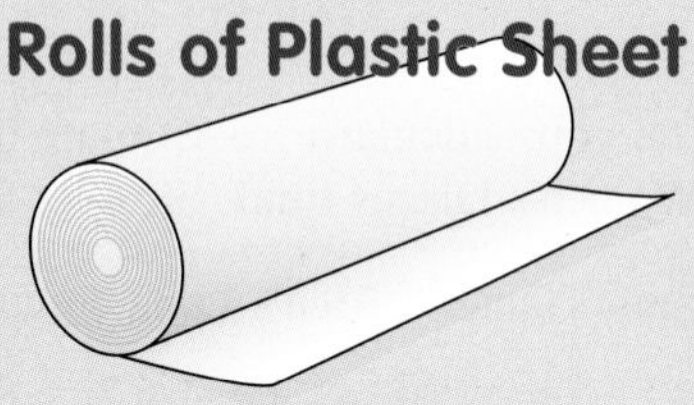

The easy way to stop weeds
15 metres long
and
1650 mm wide

**Only £3.75**
**per roll**
**while stocks last**

*Special Offers on Spring bulbs*

**Traditional Watering Can Roses**
**Now in stock**

**£15.99**

**Cloches**

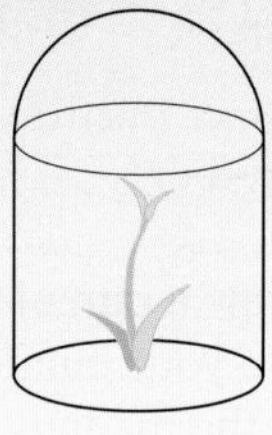

**Victorian Bell Cloches**

**Cloche Frames**

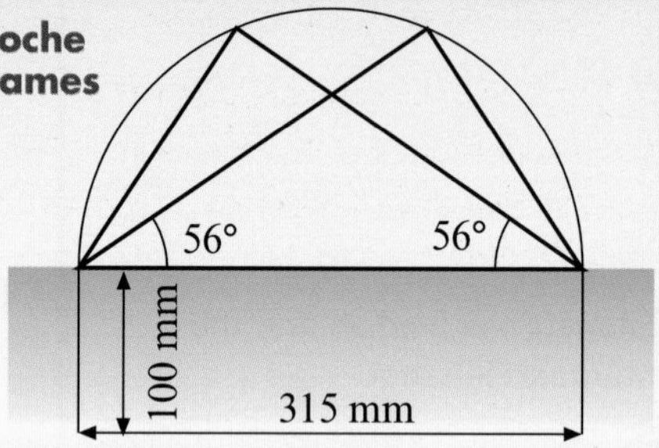


**The Super 9 Garden Store – 100% treated timber**

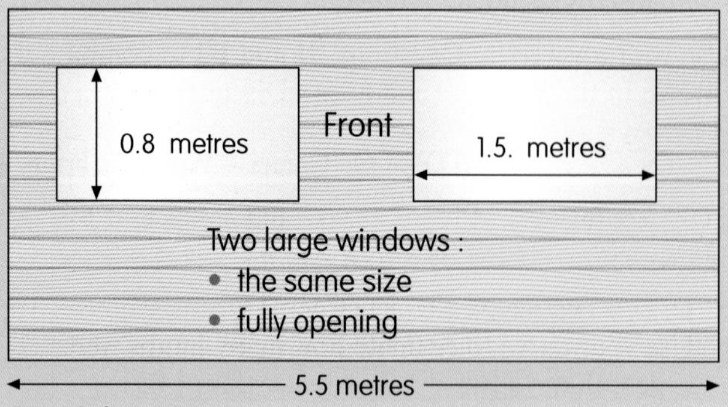


Roof angle gives fast run-off of rainwater

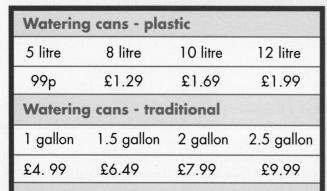

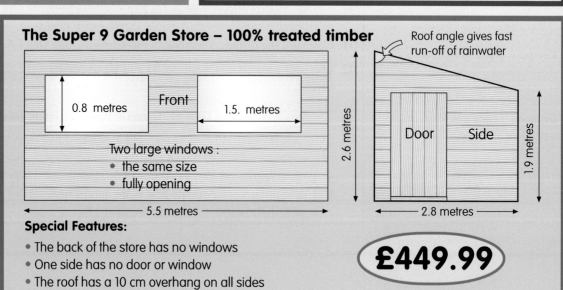

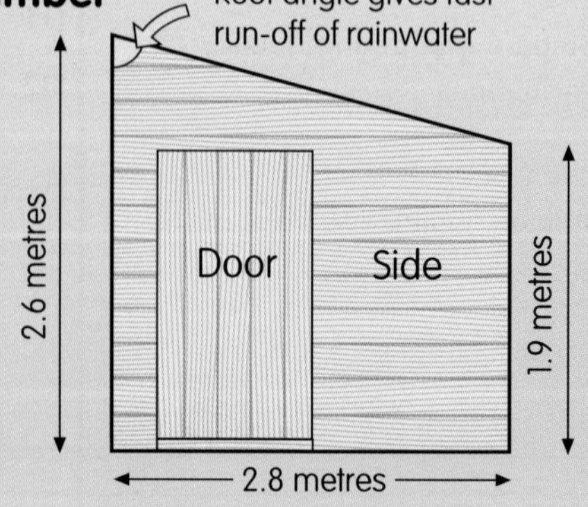


**Special Features:**

- The back of the store has no windows
- One side has no door or window
- The roof has a 10 cm overhang on all sides

**£449.99**

1 The plastic sheet is 120 microns thick.
(One micron is $\frac{1}{10\,000}$th of a centimetre.)

**a** Give the thickness of the sheet in metres as a number in standard form.

**b** Give the volume of the plastic sheet on the roll in $m^3$ in standard form.

2 The plastic sheet is rolled onto a cardboard tube. The inside diameter of the tube is 86 mm and the outside diameter is 94 mm.

**a** Calculate the volume of cardboard used to make the tube. Do not round your answer.

**b** The outside of the tube is coated with adhesive. The adhesive is applied at a rate of 135 ml/$m^2$, and is supplied in 5 litre cans.
How many tubes can be coated with adhesive from one can? (There is no wastage.)

3 Calculate, to the nearest millimetre, the length of wire used to make one cloche frame.

4 A four-frame open-ended cloche 2.5 metres long is made by covering four frames, above ground level, with plastic sheet. At either end, an extra 15 mm of sheet is allowed for a seam.

**a** How many cloches can be made from one roll of plastic sheet?

**b** Calculate the percentage of each roll of sheet that is wasted making these cloches.

5 A Victorian bell-cloche is in the shape of a cylinder surmounted by a hemisphere. The internal dimensions of the cloche are: diameter 28 cm and height 42 cm.

**a** Show that the capacity of the bell-cloche is approximately 23 litres.

**b** The bell-cloche is made of 9 mm thick glass. Calculate the volume of glass used to make the cloche.

**c** If the internal diameter (cm) of a bell-cloche is $2r$, and the internal height (cm) is $h$, show that the capacity $C$ of the cloche is given by:

$$C = \frac{1}{3}\pi r^2(3h - r)$$

**d** Rearrange the formula for capacity $C$ to give a formula for $h$ in terms of $r$ and $C$.

6 Wheel Garden Centre has a policy that price increases will never be more than 6%.
Next month they will increase the price of plastic sheet to £3.99 per roll.

Is this increase in line with their policy? Explain.

7 In metres the height $h$ of a bucket, painted red, on the water wheel above the pool is given by the equation:

$$h = 4 + 3\sin(30t)$$

where $t$ is the time in seconds after the wheel has been set in motion.

**a** Calculate $h$ (to 1dp) for values of $t$ from 0 to 12.

**b** Draw a graph to show how the height of the red bucket varies during the first 12 seconds.

**c** The floor of the Coffee Shop is 3.7 metres above the pool. In the first 12 seconds when was the red bucket level with the floor of the Coffee Shop?

8 In the Coffee Shop Jim paid £3.29 for three coffees and two teas. Ella paid £4.10 for five teas and 2 coffees.

Write and solve two equations to find the prices of tea and coffee in the Coffee Shop.

9 Packs of spring bulbs are made up with daffodils and tulips in the ratio 5 : 3.

If there are $n$ bulbs in a pack, give the number of tulip bulbs in terms of $n$.

10 The outside of the Super 9 store is sprayed with timber preservative (excluding the roof). The preserver is sprayed at a rate of 250 ml/$m^2$.

Will 5 litres of preserver cover one store? Explain your answer.

11 Which traditional watering can is the best value for money in your opinion? Explain your answer.

12 The traditional copper watering can 'rose' is the shape of a truncated right cone.

**a** Calculate, in $cm^3$, to 2 dp the volume of water needed to fill the rose.

**b** Calculate, in $cm^3$, to 1 dp the curved surface area of the rose.

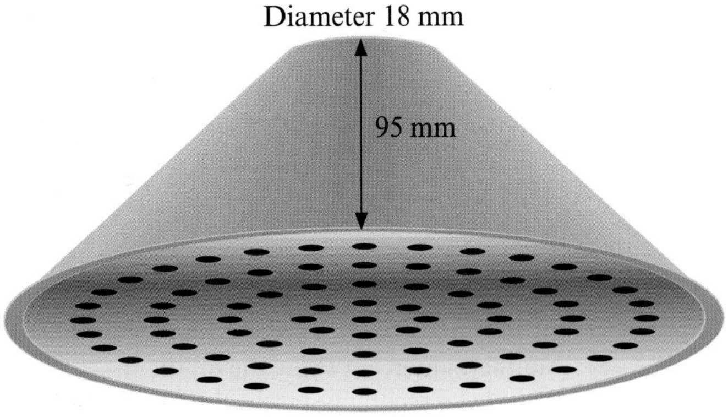

## Starting points

You need to know about ...

### A Linear (straight line) graphs

A linear graph can be identified from its equation.
The equation of all linear graphs is of the form

$y = mx + c$ where $m$ is the gradient and $c$ the intercept with the $y$-axis

Line A has a gradient of $\frac{3}{4}$, and an intercept with the $y$-axis at 1.

The equation of line A is

$\mathbf{y = \frac{3}{4}x + 1}$

Line B has a gradient of $^{-}\frac{5}{2}$, and an intercept with the $y$-axis at $^{-}3$.

The equation of line B is:

$\mathbf{y = {}^{-}\frac{5}{2}x - 3}$

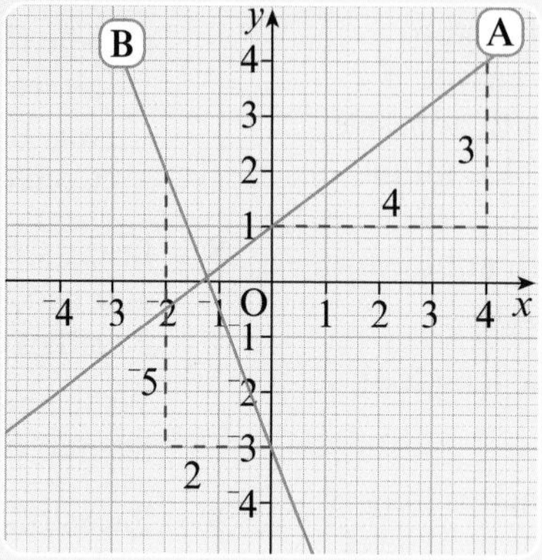


Line A has a **positive gradient** sloping upwards from left to right.
Line B has a **negative gradient** sloping downwards from left to right.

Being able to identify the gradient and intercept with the $y$-axis from a linear equation enables you to sketch or draw a graph.

To sketch a graph of $y = \frac{2}{3}x - 5$ you can identify that:

the gradient $m$ is $\frac{2}{3}$

the $y$-intercept $c$ is $^{-}5$

The sketch shows a graph of

$\mathbf{y = \frac{2}{3}x - 5}$

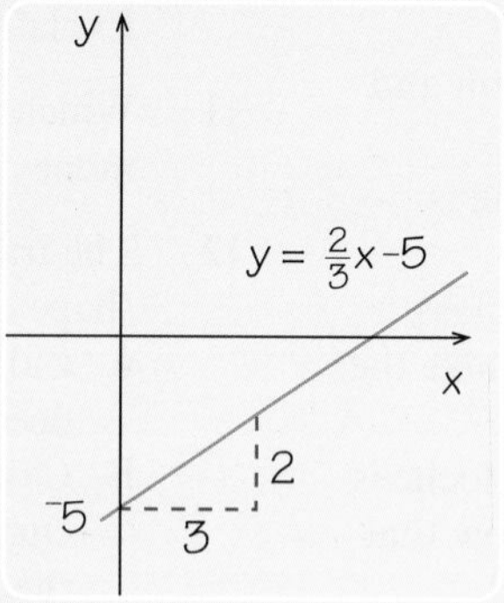


(Note this is a sketch, so no accurate plotting or measuring is needed.)

To draw a graph of $y = \frac{2}{3}x - 5$

you can identify that the gradient is $\frac{2}{3}$, and the $y$-intercept is $^{-}5$.

To draw the graph:

- draw, graduate, and label a pair of axes
- plot the point $(0, {}^{-}5)$, i.e. the $y$-intercept
- from the $y$-intercept show the gradient, i.e. to plot a second point on the graph
- draw a straight line passing through the two points
- label the linear graph with its equation.

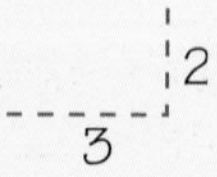


(Note this is drawing a graph, so accuracy is essential.)

... so try these questions

**A1** Identify the gradient and intercept with the $y$-axis for these linear graphs.

**a** $y = 4x + 3$
**b** $y = \frac{1}{5}x - 8$
**c** $y = 0.8x + 1.5$
**d** $y = \frac{7}{4} + \frac{5}{4}x$
**e** $y = 1 - 0.25x$
**f** $y = 6x$
**g** $y = x$
**h** $y = 12$

**A2** A linear graph has an intercept with the $y$-axis at (0, 3) and a gradient of 2.5.

Write the equation of this graph.

**A3** A linear graph has a gradient of $^{-}1$ and an intercept with the $y$-axis at the point $(0, {}^{-}2)$.

Give the equation of this line.

**A4** Sketch each of these linear graphs.

**a** $y = 3x + 2$
**b** $y = 5x - 4$
**c** $y = \frac{3}{5}x + 1$
**d** $y = \frac{5}{8}x$
**e** $y = 3 - 4x$
**f** $y = 0.5 - 1.5x$

**A5** Draw each of these linear graphs.

**a** $y = \frac{1}{2}x + 1.2$
**b** $y = \frac{3}{4} - 2x$
**c** $y = 4 + 3x$
**d** $y = \frac{3}{4}x$
**e** $y = 1.5 - 4x$
**f** $y = x - 2$

## B Rearranging linear equations

Linear equations may have to be rearranged to put them in the form

$$y = mx + c$$

(Note that rearranging an equation is **not** changing the equation; it is producing an equivalent equation.)

- To draw the graph of $3y + 4x = 6$ using the gradient and intercept method, the equation will need to be in the form $y = mx + c$.

**Example** To rearrange $3y + 4x = 6$

| | |
|---|---|
| | $3y + 4x = 6$ |
| [– $4x$ from both sides] | $3y = {}^{-}4x + 6$ |
| [÷ both sides by 3] | $y = {}^{-}\frac{4}{3}x + 2$ |

So $3y + 4x = 6$ and $y = {}^{-}\frac{4}{3}x + 2$ are equivalent equations.

The graph of $3y + 4x = 6$ has a gradient of ${}^{-}\frac{4}{3}$ and a $y$-intercept at (0,2).

- Rearranging the linear equation may involve dealing with brackets.

**Example** To rearrange $4 = 2(3y - 5x)$

| | |
|---|---|
| | $4 = 2(3y - 5x)$ |
| [multiply out bracket] | $4 = 6y - 10x$ |
| [+ $10x$ to both sides] | $4 + 10x = 6y$ |
| [÷ both sides by 6] | $\frac{4}{6} + \frac{10}{6}x = y$ |

So $4 = 2(3y - 5x)$ and $y = \frac{5}{3}x + \frac{2}{3}$ are equivalent equations.

- Rearranging the linear equation may involve dealing with fractions.

**Example** To rearrange $\frac{3}{5}x = \frac{1}{2}y - 1$

| | |
|---|---|
| | $\frac{3}{5}x = \frac{1}{2}y - 1$ |
| [multiply both sides by 10] | $6x = 5y - 10$ |
| [+ 10 to both sides] | $6x + 10 = 5y$ |
| [÷ both sides by 5] | $\frac{6}{5}x + 2 = y$ |

So $\frac{3}{5}x = \frac{1}{2}y - 1$ and $y = \frac{6}{5}x + 2$ are equivalent equations.

- Rearranging the linear equation may involve fractions and brackets.

**Example** To rearrange $\frac{3}{4}(2y - 3x) = 2$

| | |
|---|---|
| | $\frac{3}{4}(2y - 3x) = 2$ |
| [× both sides by 4] | $3(2y - 3x) = 8$ |
| [multiply out the bracket] | $6y - 9x = 8$ |
| [+ $9x$ to both sides] | $6y = 9x + 8$ |
| [÷ both sides by 6] | $y = \frac{3}{2}x + \frac{4}{3}$ |

So $\frac{3}{4}(2y - 3x) = 2$ and $y = \frac{3}{2}x + \frac{4}{3}$ are equivalent equations.

(Note that when you are dividing both sides of an equation by a value it is better to write the coefficient of $x$ as a fraction rather than a decimal. In this way you can work with a gradient as a fraction.)

**B1** Rearrange each of these equations so that they are in the form $y = mx + c$.

**a** $5y = 3x + 15$
**b** $2y + 5x = 8$
**c** $3y - 4x = 3$
**d** $4 - 2y = 3x$
**e** $7x - 2y = 1$
**f** $2(x - 3y) = 6$
**g** $5 = 4(2x + y)$
**h** $2(3 - 2y) = 5x$
**i** $4(2x - 1) = 2y$
**j** $\frac{2}{3}y - 1 = 3x$
**k** $4 - \frac{1}{2}x = 5y$
**l** $y = \frac{3}{5}x - 5$
**m** $3x + \frac{2}{3}y = 4$
**n** $5(x + 2y) = 3$
**o** $\frac{3}{5}x + 2y = 4$
**p** $4(x - 3) = \frac{1}{2}y$
**q** $6 = 3(2x - y)$
**r** $\frac{3}{4} = 3x - y + 1$
**s** $2(x - 3) = 3(4 - 2y)$
**t** $3(x + 1) = 4(3 - y)$
**u** $1 + \frac{1}{3}x = 2y$
**v** $3 - \frac{1}{4}y = \frac{1}{2}x$
**w** $1 - \frac{1}{2}x = \frac{2}{3}y$
**x** $\frac{1}{2}(y + 4) = \frac{2}{3}x$
**y** $\frac{2}{3}(x + 2y) = 1$
**z** $\frac{5}{8}(y - 2x) = 3$

## Solving simultaneous linear equations graphically

When a linear relationship between two variables can be expressed by two or more different equations, you can find the value of each variable by drawing graphs of each equation and finding a point where the coordinates satisfy both equations (i.e. where the line graphs intersect).

**Example**

Find values of $x$ and $y$ that satisfy the equations:

$$2x + 3y = {}^{-}5 \;\ldots\; (1)$$
$$x + y = {}^{-}1 \;\ldots\; (2)$$

Linear graphs are of the form $y = mx + c$. This can be rearranged to give:

$mx = y - c$
$c = y - mx$
$y = c + mx$
$\dfrac{y - c}{m} = x$
...

So, $y = \frac{^{-}5}{3} - \frac{2}{3}x$ is a rearranged form of:

$y = -\frac{2}{3}x - \frac{5}{3}$

- To draw graphs using the gradient and intercept the equations need to be in the form $y = mx + c$

To rearrange Equation (1)

$2x + 3y = {}^{-}5$
[i.e.] $3y = {}^{-}5 - 2x$
[i.e.] $y = \dfrac{^{-}5}{3} - \dfrac{2}{3x}$

To rearrange Equation (2)

$x + y = {}^{-}1$
[i.e.] $y = {}^{-}1 - x$

- Draw graphs of $y = \dfrac{^{-}5}{3} - \dfrac{2}{3x}$ and $y = {}^{-}1 - x$ on the same pair of axes.

When the scales for the axes are not given in the question you must decide on appropriate scales.

Having drawn your graphs you may have to extend your axes to find a point that both graphs pass through (point of intersection).

The graph of

$y = \dfrac{^{-}5}{3} - \dfrac{2}{3x}$ has:

a gradient of $^{-}\frac{2}{3}$

a $y$-intercept at $^{-}\frac{5}{3}$

The graph of
$y = {}^{-}1 - x$ has:
a gradient of $^{-}1$
a $y$-intercept at $^{-}1$

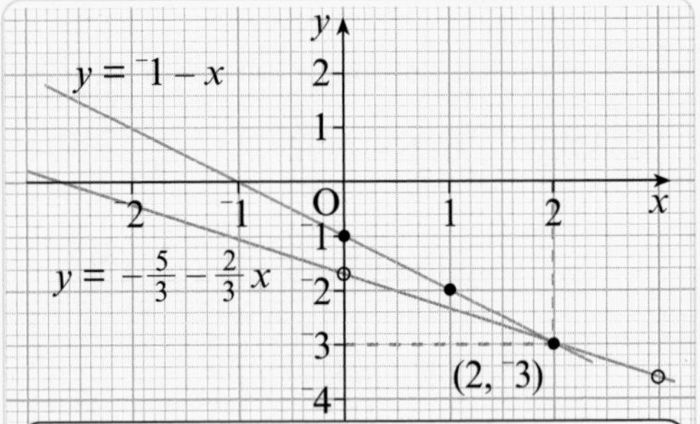


The two linear graphs both pass through the point (2, $^{-}3$). The values of $x$ and $y$ at this point must satisfy both equations, i.e. $x = 2$ $y = {}^{-}3$.

Where the two linear graphs cross, the coordinates give the solution to the equations.
**The solutions are $x = 2$ and $y = {}^{-}3$.**

Check the graphical solution by substituting the values in one equation.

Check in Equation (1) for $x = 2$, $y = {}^{-}3$:
LHS = $2(2) + 3({}^{-}3) = 4 - 9 = {}^{-}5$ = RHS (solution is correct.)

It is always a good idea to check the solution to an equation. Finding an error will allow you to rework your solution and not lose marks.

**Exercise 16.1**
Solving simultaneous equations graphically

1 Solve each pair of equations graphically.

**a** $x + 4y = 7$
$2x + 5y = 8$

**b** $5x - 3y = {}^{-}7$
$3x + y = 7$

**c** $5x + 7y = 4$
$x + 3y = 4$

**d** $3x + 2y = 12$
$4x - y = 5$

**e** $12x + 5y = 9$
$5x - y = 13$

**f** $8x + y = 7$
$11x + 4y = 7$

2 By drawing two linear graphs find values of $x$ and $y$ that satisfy the two equations $3x + 2y = 5$ and $3y - 3x = 15$.

3 **a** On a pair of axes draw a graph of $y = 3x - 5$.
**b** Show that $2(y + 5) = 6x$, and $y = 3x - 5$ are equivalent.
**c** Draw another linear graph on your axes and solve these simultaneous equations.

$$2(y + 5) = 6x \ldots (1)$$
$$3x - 2y = 4 \ldots \quad (2)$$

4 **a** On a pair of axes draw a graph of $3x + y = 9$.
**b** By drawing another linear graph on the axes solve the simultaneous equations:

$$3x + y = 9 \ldots (1)$$
$$x + 2y = 6 \ldots (2)$$

**c** Use graphs to explain why this pair of equations

$$y - 3x = 5 \ldots \quad (1)$$
$$2y - 4 = 6x \ldots (2)$$

cannot be solved simultaneously.

5 A group of male and female students went on a fell walk.
In total, sixty-three students were in the group.
The number of females was nine more than twice the number of males.
Let $x$ be the number of females in the group and $y$ the number of males.

**a** Write two different equations in $x$ and $y$ from the student data.
**b** **i** Draw and label a pair of axes with values: $0 \leqslant x \leqslant 63$ and $0 \leqslant y \leqslant 63$
**ii** By drawing suitable linear graphs find the number of females and the number of males in the group.

6 A blend of olive oil is made by mixing Extra Virgin and Virgin olive oils.
When making 500 litres of the blended oil the amount of Virgin oil used was 25 litres more than four times the amount of Extra Virgin oil used.
Let $x$ be the number of litres of Extra Virgin oil used and $y$ the number of litres of Virgin oil used.

**a** Write two different equations in $x$ and $y$ from the olive oil data.
**b** **i** Draw and label a pair of axes with values: $0 \leqslant x \leqslant 500$ and $0 \leqslant y \leqslant 500$.
**ii** Draw two linear graphs to find the amount of each type of oil used when 500 litres of the blended oil are produced.

The next 500 litres was a different blend, with the amount of Virgin oil being 40 litres more than six times the amount of Extra Virgin oil.

**c** **i** Write a different equation in $x$ and $y$ from the new data.
**ii** Draw a third linear graph on the axes to find the amount of each type of oil used in this next blend.

7 Two variables $x$ and $y$ are such that:

$$y = \tfrac{1}{2}(x + 7)$$
and $$x = \tfrac{1}{3}(29 - 4y)$$

**a** On a pair of axes, draw two linear graphs to show the relationship between $x$ and $y$.
**b** Find a value for $x$, and a value for $y$, that will satisfy both equations.

8 The equation $2y + 2 = 3x$ is solved simultaneously with a second linear equation in $x$ and $y$. The solution of $x = 4$ and $y = 5$ satisfies both equations.
The graph of the second equation has a $y$-intercept at $(0, 3)$.
Draw graphs to find the second linear equation.

# Quadratic graphs

A quadratic expression in one variable (e.g. $n$) is an expression where the highest power of the variable is 2 (e.g. $n^2$).

Quadratic graphs are shown on these axes.

The graph of $y = x^2 + 2x$ can be plotted by calculating values of $y$ for chosen values of $x$.

**Example**

For $\quad y = x^2 + 2x$

| $x$ | $-1$ | 0 | 1 | 2 | 3 |
|---|---|---|---|---|---|
| $x^2$ | 1 | 0 | 1 | 4 | 9 |
| $2x$ | $-2$ | 0 | 2 | 4 | 6 |
| $y$ | $-1$ | 0 | 3 | 8 | 15 |

i.e. the graph of $y = x^2 + 2x$ passes through the points: $(^{-}1, {}^{-}1)$, $(0, 0)$, $(1, 3)$, $(2, 8)$, $(3, 15)$ …

Note that the points should be joined by a smooth curve.

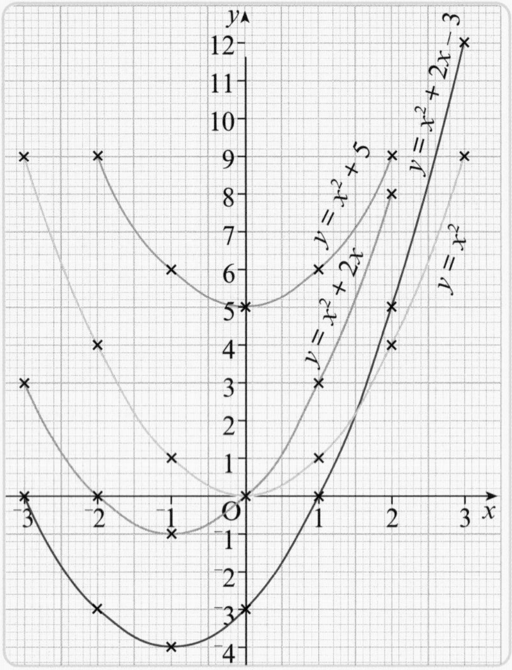


The smooth curve of a quadratic graph is known as a **parabola**.

From the graphs drawn on the axes you can note that:

- quadratic graphs are symmetrical
  graphs such as $y = x^2$, $y = 2x^2$, $y = 0.5x^2$, are symmetrical about the $y$-axis (the line $x = 0$).
- the value of any constant, i.e. not a term in any variable, fixes the $y$-intercept of the graph.
  $y = x^2 + 5$ has a $y$-intercept at $(0, 5)$; $y = x^2 + 3x - 3$ has a $y$-intercept at $(0, {}^{-}3)$.

**Exercise 16.2**
Quadratic graphs

**1** From the graphs above give the line of symmetry of:

**a** the graph of $y = x^2 + 2x - 3$ **b** the graph of $y = x^2 + 2x$.

**2** Draw a pair of axes where $^{-}4 \leqslant x \leqslant 4$, and $^{-}12 \leqslant y \leqslant 28$.

**a** On the axes draw these quadratic graphs.
$y = x^2 \qquad y = x^2 + 6 \qquad y = x^2 - 3x \qquad y = x^2 - 3x - 8$

**b** Give the line of symmetry of the graph $y = x^2 - 3x$.

**c** Give the equation of a quadratic graph that has the same line of symmetry as $y = x^2 - 3x$ but has a $y$-intercept at $(0, {}^{-}8)$.

**3** Draw a pair of axes where $^{-}4 \leqslant x \leqslant 4$, and $^{-}4 \leqslant y \leqslant 32$.

**a** On the axes draw these quadratic graphs and compare them.
$y = x^2 + 2x \qquad y = x^2 + 3x \qquad y = x^2 - 2x \qquad y = x^2 - 4x$

**b** What does the coefficient of $x$ tell you about a quadratic graph?

You might choose to use a graphical calculator when you have to compare a number of graphs.

A sketch graph shows the shape of the graph and known data. This data might show $x$-intercepts and/or $y$-intercepts, or maybe the relative steepness or tightness of a graph.

Sketching a graph is not the same as drawing a graph. Points are not plotted and graduations on axes do not have to be accurate.

**4** Give the line of symmetry and the $y$-intercept for each of these graphs.

**a** $y = x^2 + 4x - 3$ **b** $y = x^2 - x + 2$ **c** $y = x^2 + 5x$
**d** $y = x^2 - 6x + 1$ **e** $y = x^2 + 6x - 2$ **f** $y = x^2 + 8x + 3$
**g** $y = x^2 - 7$ **h** $y = x^2 - 6x + 8$ **i** $y = x^2 - 7x + 5$

**5** This is a sketch graph of $y = x^2$.

**a** Make a copy of this sketch.
**b** On your copy sketch and label these graphs.
  **i** $y = x^2 + 3$
  **ii** $y = x^2 - 5x$
  **iii** $y = x^2 - 4x - 5$

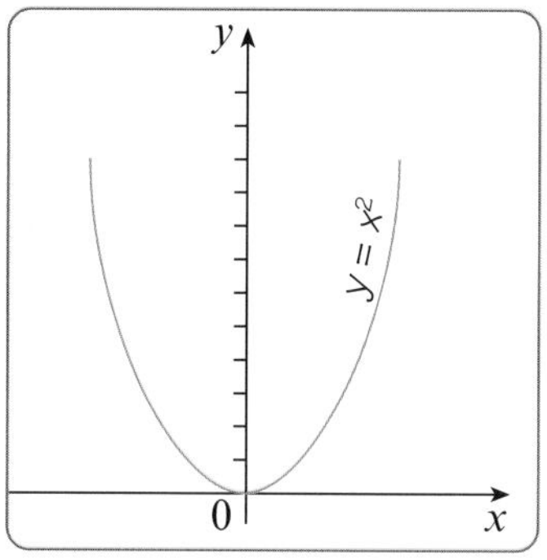


**6** On a pair of axes, sketch these graphs and label them.

**a** $y = x^2 - 5x - 6$
**b** $y = x^2 + x + 4$

We have considered the effect of the coefficient of $x$, and the constant value in a quadratic expression, on its graph.

The coefficient of $x^2$ will also have an effect on a quadratic graph.

**Example**

When the coefficient of $x^2$ is not 1.

For the graph of $y = 2x^2$
when $x = 0$, $y = 0$
$x = 1$, $y = 2$
$x = 2$, $y = 8$
$x = {}^{-}1$, $y = 2$
$x = {}^{-}2$, $y = 8$
…, …

So, on the same axes the graph of $y = 2x^2$ is a tighter curve than the graph of $y = x^2$

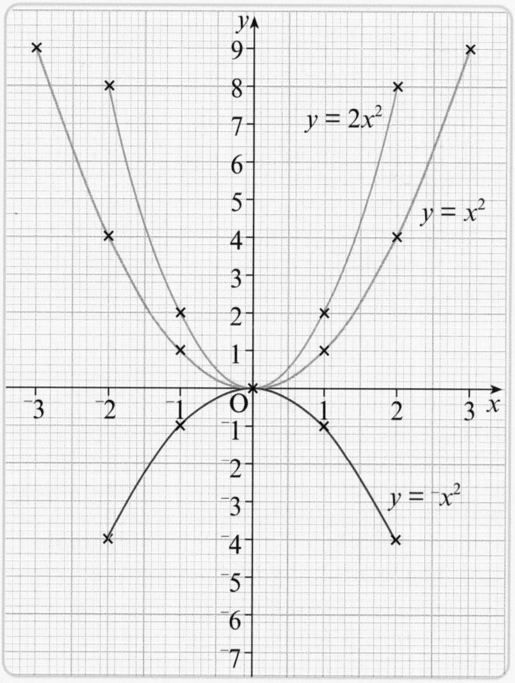


**Example**

When the coefficient of $x^2$ is negative.

For the graph of $y = {}^{-}x^2$
when $x = 0$, $y = 0$
$x = 1$, $y = {}^{-}1$
$x = 2$, $y = {}^{-}4$
$x = {}^{-}1$, $y = {}^{-}1$
$x = {}^{-}2$, $y = {}^{-}4$
…, …

So, the graph of $y = {}^{-}x^2$ is inverted compared with the graph of $y = x^2$.

When we say the graph is inverted, we could say that it is upside down. We might also say that, in this case it is a reflection in the $x$-axis, i.e. the line $y = 0$.

**7** Draw a pair of axes where ${}^{-}3 \leqslant x \leqslant 3$, and $10 \leqslant y \leqslant {}^{-}30$.

On the axes draw these quadratic graphs.

**a** $y = {}^{-}3x^2$ **b** $y = 2 - 4x - x^2$ **c** $y = 3x - 2x^2$ **d** $y = 6x - 8 - 2x^2$

# Solving quadratic equations graphically

This is the graph of $y = x^2 + 5x - 6$, for values of $x$ from ${}^{-}8$ to ${}^{+}6$.

We can use the graph to solve the equation $x^2 + 5x - 6 = 0$.

The graph of $y = x^2 + 5x - 6$ crosses the graph of $y = 0$ where:

$x = {}^{-}6$ and where $x = 1$

At $x = {}^{-}6$, and $x = 1$ the two equations must be equal.
So we can write:

$$x^2 + 5x - 6 = 0$$

From the graphs, the values of $x$ that satisfy the equation are:
$x = {}^{-}6$ or $x = 1$

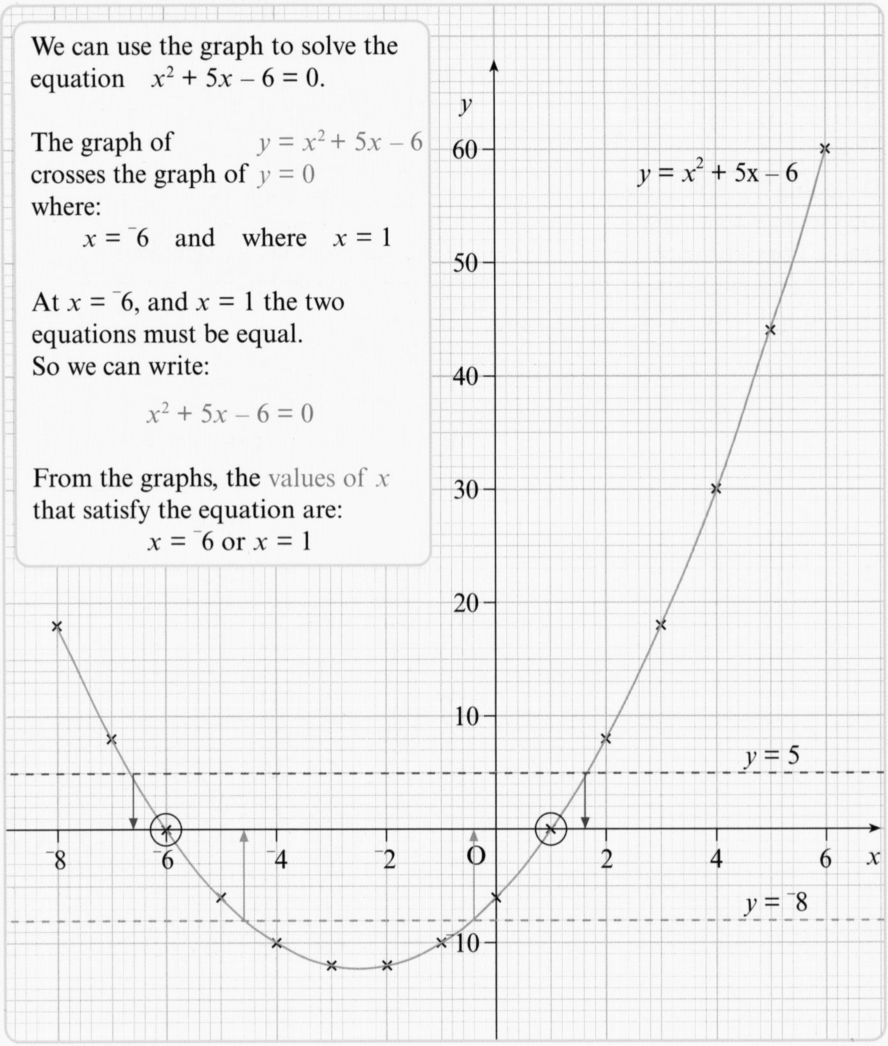


$y = 0$ is the equation of the $x$-axis.

You can be fairly confident of these values of $x$ as solutions because they are both integer values.

If you are in any doubt, substitute in the equation to check each value.

For this graph:

| | | |
|---|---|---|
| when | $x = {}^{-}8$ | $y = 18$ |
| | $x = {}^{-}7$ | $y = 8$ |
| | $x = {}^{-}6$ | $y = 0$ |
| | $x = {}^{-}5$ | $y = {}^{-}6$ |
| | $x = {}^{-}4$ | $y = {}^{-}10$ |
| | $x = {}^{-}3$ | $y = {}^{-}12$ |
| | $x = {}^{-}2$ | $y = {}^{-}12$ |
| | $x = {}^{-}1$ | $y = {}^{-}10$ |
| | $x = 0$ | $y = {}^{-}6$ |
| | $x = 1$ | $y = 0$ |
| | $x = 2$ | $y = 8$ |
| | $x = 3$ | $y = 18$ |
| | $x = 4$ | $y = 30$ |
| | $x = 5$ | $y = 44$ |
| | $x = 6$ | $y = 60$ |

The graph of $y = x^2 + 5x - 6$ can be used to solve many more equations.

**Example** From the graph solve $x^2 + 5x - 6 = 5$.
The graph of $y = x^2 + 5x - 6$ crosses the graph of $y = 5$
so we can say: $x^2 + 5x - 6 = 5$
The two equations are equal where $x \approx {}^{-}6.6$ and where $x \approx 1.6$

**The solutions for $x^2 + 5x - 6 = 5$ are: $x \approx {}^{-}6.6$ and $x \approx 1.6$**

**Example** From the graph solve $x^2 + 5x - 6 = {}^{-}8$.
The graph of $y = x^2 + 5x - 6$ crosses the graph of $y = {}^{-}8$
so we can say: $x^2 + 5x - 6 = {}^{-}8$

**The solutions for $x^2 + 5x - 6 = {}^{-}8$ are: $x \approx {}^{-}4.6$ and $x \approx {}^{-}0.4$**

These values of $x$, from the graph, are not integer values. The values are not exact, they are a good approximation. They answer the question, from the graph.

For more accurate solutions, use the values from the graph as a starting point for the trial-and-improvement method.

## Exercise 16.3
### Solving quadratic equations graphically

1 The graph of $y = x^2 + 5x - 6$ can also be used to solve the equation:

$$x^2 + 5x - 6 = 10$$

**a** What other graph would you draw on the axes to solve $x^2 + 5x - 6 = 10$?
**b** Explain how you would use the two graphs to solve the equation.
**c** From the graph on page 192, solve $x^2 + 5x - 6 = 10$.

2 The graph on page 192 can be used to solve these equations:

$$x^2 + 5x - 6 = 0$$
$$x^2 + 5x - 6 = 5$$
$$x^2 + 5x - 6 = 8$$
$$x^2 + 5x - 6 = 1$$

**a** Give two other equations you think can be solved from the same graph.
**b** Explain how you would solve your equations.

3 Asif used the same graph to solve $x^2 + 5x - 6 = 22$, and only found one solution.

**a** Explain why.
**b** Give the solution you think Asif did find from the graph.
**c** Is the other solution greater or less than $x = {}^-8$? Explain.

4 Draw the graph of $y = x^2 - 5x + 6$ for values of $x$ from $^-2$ to 6.

**a** Use your graph to solve $x^2 - 5x + 6 = 0$.
**b** On your axes draw and label the graph of $y = 4$.
**c** Use your graphs to solve $x^2 - 5x + 6 = 4$.
**d** Use the graph solve $x^2 - 5x + 6 = 10$.

5 Draw the graph of $y = x^2 - 4x - 5$ for values of $x$ such that $^-3 \leqslant x \leqslant 7$.

**a** Use the graph to solve:
**i** $x^2 - 4x - 5 = 0$ **ii** $x^2 - 4x - 5 = 11$
**b** From the graph solve $x^2 - 4x - 5 = {}^-9$.

6 **a** **i** Draw the graph of $y = x^2 - 4x + 3$ for $x$ such that $^-3 \leqslant x \leqslant 5$.
**ii** Use the graph to solve $x^2 - 4x + 3 = 0$.
**b** **i** For values of $x$ such that $^-3 \leqslant x \leqslant 5$ draw the graph of $y = x^2 + 3$.
**ii** On the same axes draw and label a graph of $y = 4x$.
**iii** Show that the solutions to $x^2 + 3 = 4x$ are $x = 1$ and $x = 3$.

For $x$: $^-6 \leqslant x \leqslant 2$ is a notation that can be interpreted in this way:

for values of $x$ such that, $x$ is greater than or equal to $^-6$, and less than or equal to 2.

7 In each of the following draw a graph and solve the equation.

**a** For $x$: $^-6 \leqslant x \leqslant 2$ draw the graph of $y = x^2 + 5x - 4$, and solve:
**i** $x^2 + 5x - 4 = 0$ **ii** $x^2 + 5x - 4 = {}^-6$ **iii** $x^2 + 5x = 4$
**b** For $x$: $^-4 \leqslant x \leqslant 2$ draw the graph of $y = x^2 + 3x - 1$, and solve:
**i** $x^2 + 3x - 1 = 0$ **ii** $x^2 + 3x + 1 = 0$ **iii** $x^2 + 3x - 1 = 0.5$
**c** For $x$: $^-2 \leqslant x \leqslant 5$ draw the graph of $3 + 4x - x^2$, and solve:
**i** $4x - x^2 + 3 = 0$ **ii** $4x + 3 - x^2 = 4$ **iii** $4x - x^2 + 1 = 0$
**d** For $x$: $^-4 \leqslant x \leqslant 2$ draw the graph of $y = 1 - 3x - 2x^2$, and solve:
**i** $2x^2 = 1 - 3x$ **ii** $^-3 = 1 - 3x - 2x^2$
**e** For $x$: $^-2 \leqslant x \leqslant 3$ draw the graph of $y = 5x - 2x^2$, and solve:
**i** $5x - 2x^2 = 0$ **ii** $5x - 2x^2 + 2 = 0$

8 Draw a suitable graph and solve the equation $(x + 2)^2 - 5 = 0$.

9 For $x$: $^-2 \leqslant x \leqslant 3$, solve $3x(x - 2) + 1 = 0$ by drawing a graph.

10 **a** Draw a graph to show that if $x(x + 4) = 6$, then $x \approx {}^-5$ or $x \approx 1$.
**b** Use graphs to explain that the equations $x(x + 4) = 6$ and $x(x + 4) - 6 = 0$ have the same solutions.

In some cases to use a graph to solve an equation is only possible if you can decide on an additional graph to draw on the axes.
Rearranging the equation is one way to decide on additional graphs.

**Example**

Draw the graph of $y = 2x^2 + x - 4$ for $x$: $^-2 \leqslant x \leqslant 2$.
Use the graph to find approximate solutions to:

**a** $2x^2 + x - 4 = 0$ **b** $2x^2 + x = x + 3$ **c** $2x^2 + 1 = x + 2$

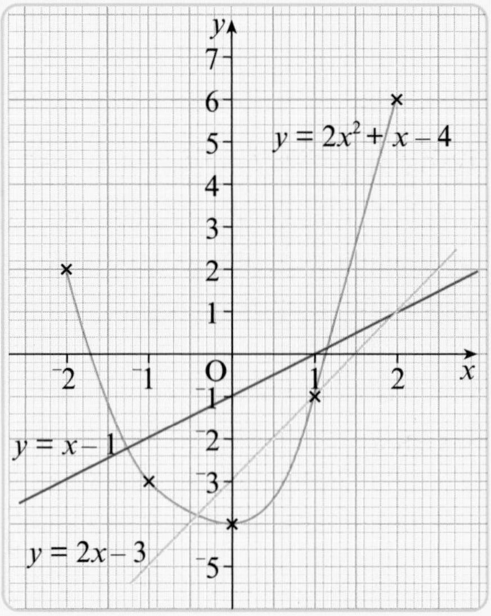


The graph of $y = 2x^2 + x - 4$ is drawn.

**a** To solve $2x^2 + x - 4 = 0$
From the graph:
$x \approx {}^-1.7$, or $x \approx 1.2$

**b** To solve $2x^2 + x = x + 3$
the equation must be rearranged.

The LHS is $2x^2 + x$.
By subtracting 4 from the LHS
we have $2x^2 + x - 4$ (graph drawn)

So, to solve $2x^2 + x = x + 3$
we subtract 4 from both sides.

$2x^2 + x - 4 = x + 3 - 4$
i.e. $2x^2 + x - 4 = x - 1$

On the axes draw the graph of $y = x - 1$.
Where the curve $2x^2 + x - 4$ and the line $x - 1$ intersect is the solution to the equation $2x^2 + x = x + 3$.

**The solution to $2x^2 + x = x + 3$ is $x \approx {}^-1.2$, or $x \approx 1.2$**

**c** To solve $2x^2 + 1 = x + 2$

The LHS is $2x^2 + 1$.
By subtracting 5 and adding $x$ LHS is $2x^2 + 1 - 5 + x = 2x^2 + x - 4$.

So, to solve $2x^2 + 1 = x + 2$, we subtract 5 and add $x$ to both sides.

$2x^2 + 1 - 5 + x = x + 2 - 5 + x$
i.e. $2x^2 + x - 4 = 2x - 3$

On the axes we draw the graph of $y = 2x - 3$.
Where the curve $2x^2 + x - 4$ and the line $2x - 3$ intersect is the solution to the equation $2x^2 + 1 = x + 2$.

**The solution to $2x^2 + 1 = x + 2$ is $x \approx {}^-0.5$, or $x = 1$.**

**11** **a** Draw the graph of $y = x^2 - 4x - 2$, for $x$: $^-3 \leqslant x \leqslant 6$.
**b** Use the graph to find approximate solutions to $x^2 - 5x - 5 = x - 10$.
**c** Find approximate solutions to $x^2 - 2x + 3 = 8 - x$ from the graph.

**12** **a** Draw the graph of $y = 9 - 2x - x^2$, for $x$: $^-5 \leqslant x \leqslant 3$.
**b** Find approximate solutions to $12 - x = 13 - 5x - x^2$.
**c** Find approximate solutions to $7 + 2x - x^2 = 3 + x$.

**13** If you had drawn the graph of $y = 12 - 4x - x^2$, find the equation of each additional line you would draw to solve these.

**a** $10 - x^2 = 5x + 6$ **b** $7 - 6x = 15 - 6x - x^2$

**14** **a** For $x$: $^{-}8 \leqslant x \leqslant 3$ draw the graph of $y = 12 - 4x - x^2$.

**b** Use the graph to find approximate solutions to $1 - x = 8 - 7x - x^2$.

**c** Find approximate solution to $4 - 3x - x^2 = 1 - 2x$ from the graph.

**d** Use the graph to find approximate solutions to $2x + 7 = 4 - x^2 - 2.5x$.

The graph of $y = 12 - 4x - x^2$ was used to solve an equation and the approximate solutions were given as $x \approx 1.7$ or $x \approx {}^{-}4.7$.

**e** Show from the graph that the equation solved was $9 - 4x = 17 - x^2 - 7x$.

**15** **a** For $x$: $^{-}7 \leqslant x \leqslant 3$, draw the graph of $y = x^2 + 5x - 8$.

Use the graph to find approximate solutions to these equations.

**b** $x(x + 5) = 8$ **c** $5(x - 1) + x^2 = 0$ **d** $x^2 + 5x = 10$

**e** $x(x + 7) - 3 = 1 + 7x$ **f** $3(x - 2) = x(x + 11) - 12$

## Graphs of reciprocal curves

If a number could be divided by zero the answer would be infinite.

So $8 \div 0$ is said to be $\infty$.

Think about $\frac{8}{x}$.

If $x = 1$, $\frac{8}{x} = 8$

$x = 0.01$, $\frac{8}{x} = 800$

$x = 0.001$, $\frac{8}{x} = 8000$

as $x$ approaches zero $\frac{8}{x}$ will approach $\infty$.

If $x = {}^{-}1, {}^{-}0.01, {}^{-}0.001, \ldots$ then $\frac{8}{x}$ approaches $^{-}\infty$.

An asymptote to a curve is a straight line to which a curve continuously draws nearer but without ever touching it.

Displaying this graph on a graphical calculator will help you to investigate how the graph of a curve approaches an asymptote.

Reciprocal graphs can be identified from their equation.

There will be a term in the form $\frac{n}{x}$

**Example** Draw the graph of $y = \frac{8}{x}$ for $x$: $^{-}4 \leqslant x \leqslant 4$

When $x = {}^{-}4$, $y = {}^{-}2$
$x = {}^{-}3$, $y = {}^{-}2.7$ (1dp)
$x = {}^{-}2$, $y = {}^{-}4$
$x = {}^{-}1$, $y = {}^{-}8$
$x = 0$, $y$ is $\infty$
$x = 1$, $y = 8$
$x = 2$, $y = 4$
$x = 3$, $y = 2.7$ (1dp)
$x = 4$, $y = 2$

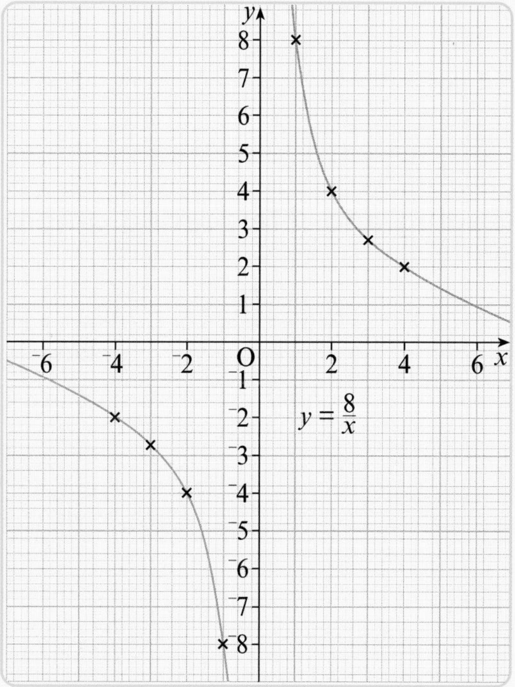


From these values, the curve is broken at $x = 0$.

Also, as $x$ becomes very large, $\frac{8}{x}$ becomes very small, but $\frac{8}{x}$ will never equal zero.

So the graph will get very close to each of the axes, but it will never touch either of them.

Here the $x$-axis ($y = 0$) and the $y$-axis ($x = 0$) are called asymptotes to the curve.

**Exercise 16.4**
**Reciprocal graphs**

**1** **a** **i** For $x$: $^{-}5 \leqslant x \leqslant 5$, draw the graph of $y = \frac{5}{x}$.

**ii** Give the equation of any asymptotes to the graph of $y = \frac{5}{x}$.

**b** On the same axes draw and label the graph of $y = \frac{5}{2x}$.

**2** **a** For $x$: $^{-}5 \leqslant x \leqslant 5$ draw the graph of $y = \frac{4}{x} + 3$.

**b** Give the equation of any asymptotes to the graph of $y = \frac{4}{x} + 3$.

3 Give the equation of any asymptotes to the curve $y = \frac{6}{x} - 8$.

4 **a** Sketch the curve of $y = \frac{6}{x} + 2$.

One asymptote to the curve is the line $x = 0$.

**b** On your sketch draw and label the other asymptote with its equation.

5 **a** For $x$: $1 \leqslant x \leqslant 12$, draw the graph of $y = \frac{12}{x}$.

**b** Use the graph to find approximate solutions to these equations.

**i** $\frac{12}{x} = x + 1$ **ii** $\frac{12}{x} = x - 3$ **iii** $x^2 = 12$

6 Draw graphs of $y = \frac{20}{x}$, and $y = x$ on the same axes.

From your graphs show that $\sqrt{20} \approx \pm 4.5$.

7 **a** For $x$: $1 \leqslant x \leqslant 10$, draw the graph of $y = \frac{10}{x} - x$.

**b** As the value of $x$ increases, explain what value $y$ will approach.

**c** Give the equation of two asymptotes to the graph of $y = \frac{10}{x} - x$.

8 Explain why $y = {}^{-}2x$ is an asymptote to the graph of $y = \frac{25}{x} - 2x$.

Include a labelled sketch graph in your explanation.

9 For $x$: ${}^{-}5 \leqslant x \leqslant 4$, draw the graph of $y = \frac{6}{x - 4}$.

## Reading and interpreting graphs

When a graph is read or interpreted, it is important to remember that any values taken from the graph can only be approximate, say to 1 dp.

**Example** For $x$: $1 \leqslant x \leqslant 10$ draw the graph of $y = \frac{15}{x} + x - 5$.

Use the graph to find:

**a** the minimum value for $y$

**b** the values of $x$ when $y = 5.2$.

For the graph, calculate values for $y$ from values for $x$

i.e. when $x = 1$, $y = 11$

$x = 2$, $y = 4.5$

$x = 3$, $y = 3$

…, …

From the graph:

**a** the minimum value for $y$ is approximately 2.7

**b** when $y = 5.2$, the values of $x$ are: $x \approx 1.8$, and $x \approx 8.4$

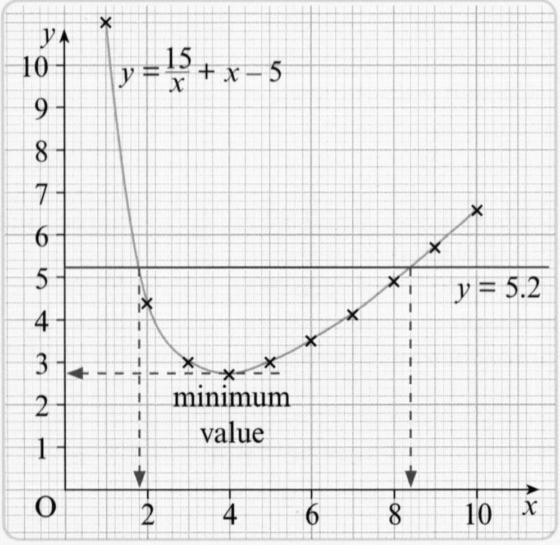


**Exercise 16.5**
Reading and interpreting graphs

1 For $x$: $1 \leqslant x \leqslant 10$, draw the graph of $y = \frac{12}{x} + x - 10$. From the graph find:

**i** the minimum value of $y$ **ii** values of $x$ when $y = {}^{-}2.5$

**2** Draw the graph of $y = 4 - 3x - x^2$.

- **a** Use the graph to find an approximate maximum value for $y$.
- **b** Give two values of $x$ for which $y = 2.5$.

**3** For $x$: ${}^-6 \leqslant x \leqslant 2$, draw the graph of $y = 9 - 5x - x^2$.

- **a** From your graph give the approximate maximum value of $y$.
- **b** When $y = 11.5$, give approximate values for $x$.
- **c** Solve the equation $6 - 2x = 9 - 5x - x^2$.
- **d** **i** Explain why, by drawing the graph of the line $y = x + 15$, you should be able to find approximate solutions to $5x + 8 = 2 - x - x^2$.
  - **ii** Find approximate solutions to $5x + 8 = 2 - x - x^2$.
- **e** On the same axes draw the graph of $y = x^2 + x + 5$.
  - **i** What is the approximate minimum value of $y$?
  - **ii** Use your graphs to solve $x^2 + x + 5 = 9 - 5x - x^2$.
  - **iii** Give an equation for the line joining the points where the two curves intersect.

**4** Draw the graph of $y = \frac{1}{2}(x^2 - 3x - 5)$ for $x$: ${}^-3 \leqslant x \leqslant 4$.

- **a** From the graph:
  - **i** give an approximate minimum value of $y$
  - **ii** find approximate values for $x$ when $y = {}^-1$.
- **b** On the same axes draw the graph of $y = \dfrac{4}{x} - 6$.
- **c** Explain why neither of the curves intersect with the line $y = {}^-6$.

**5** On the same axes draw the graphs of $y = \dfrac{2x}{3 - x}$, and $y = \frac{1}{2}x^2 + 2$.
Use $x$: ${}^-4 \leqslant x \leqslant 4$.

- **a** Use the graphs to solve the equation $2x = (0.5x^2 + 2)(3 - x)$.
- **b** Only one curve will intersect with the line $x = 3$.
  Which curve will not intersect? Explain why.

## Graphs of higher-order polynomial equations

An equation of higher order than a quadratic has a highest power of the variable greater than 2, as in the following polynomial equations.

$$3a^3 + 4a \qquad a^3 + a^2 - 1 \qquad y^5 - 4y = 3 \qquad k^6 = 1$$

**Example**

Draw the graph of $y = x^3 + 2$ for $x$: ${}^-3 \leqslant x \leqslant 3$.

The graph, when drawn, will be a smooth curve.

When
$x = {}^-3$, $y = {}^-25$
$x = {}^-2$, $y = {}^-6$
$x = {}^-1$, $y = 1$
$x = 0$, $y = 2$
$x = 1$, $y = 3$
$x = 2$, $y = 10$
$x = 3$, $y = 29$

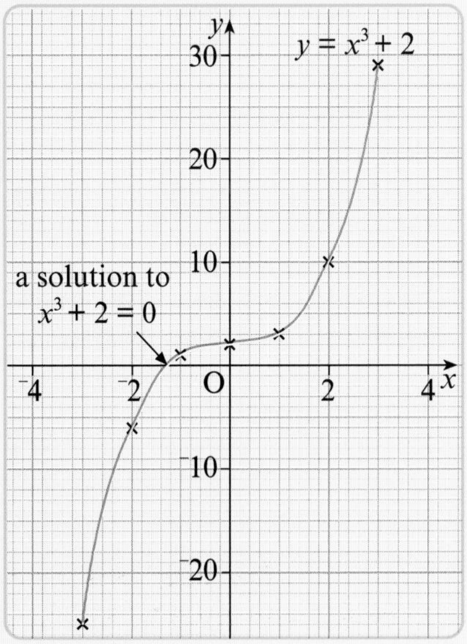


The graph can be used to solve $x^3 + 2 = 0$
From the graph if $x^3 + 2 = 0$ then $x \approx {}^-1.3$

$x^3 + 2 = 0$ is known as a **cubic** equation.

Remember values read from the graph will only be approximate.

**Exercise 16.6**
Graphs of polynomial equations

1 On the same axes draw graphs of $y = x^3 + 4$ and $y = 4 - x^3$.

a Describe any common features of the graphs.
b Describe any differences between the graphs.
c For each graph:
  i describe what happens as $x$ becomes increasingly large
  ii describe what happens as $x$ becomes increasingly negative.

2 For $x$: $^-4 \leqslant x \leqslant 3$ draw the graph of $y = x^3 + 2x^2$.

a From your graph solve the equation $x^3 + 2x^2 = 0$.
b By drawing an additional graph, find an approximate solution to the equation $12x + 5 = 2x^2 + x^3$.
c i On the same axes draw the graph of $y = 4x^2 - x^3$ for $x$: $^-2 \leqslant x \leqslant 4$.
  ii Use your graph to solve $4x^2 - x^3 = 0$.
  iii Find approximate values for $x$ when $y = 7$.
  iv Find an approximate value for $y$ when $x = {}^-1.5$.

3 Draw the graph of $y = x^3 - 3x^2$ for values of $x$ from $^-3$ to 4.

a Use your graph to find an approximate value for $x$ when $y = 8$.
b Find values for $x$ when $x^3 - 3x^2 = {}^-4$.

For $x$: $^-3 \leqslant x \leqslant 4$, the equation $x^3 - 3x^2 = n$ is satisfied by only one value of $n$.

4 Draw the graph of $y = \frac{1}{2}(x^3 - 3x - 8)$ for values of $x$ from $^-4$ to 4.

Use the graph to find:

a an approximate value for $y$, when $x = 3.4$
b an approximate value for $x$ when $y = {}^-9$
c an approximate solution to the equation $\frac{1}{2}(x^3 - 3x - 8) = 0$
d three approximate values of $x$ that satisfy $\frac{1}{2}(x^3 - 3x - 8) = 5x - 2$.

5 Draw the graph of $y = x^3 - x^2 - 4x + 4$ for $x$: $^-3 \leqslant x \leqslant 4$.

a Use the graph to solve $x^3 - x^2 - 4x + 4 = 0$.
b When $x = {}^-1.5$ give an approximate value for $y$.
c Find three approximate values for $x$ that give $x^3 - x^2 - 4x + 4 = 2$.

6 The width of a rectangle is given as $w$ (cm), and the length as $w + 8$ (cm). An expression for the area $A$ ($cm^2$) of the rectangle is $A = w(w + 8)$.

a Draw a graph of $A = w(w + 8)$ for values of $w$ from zero to 8. (Plot $w$ on the horizontal axis.)
b From the graph find approximate dimensions of the rectangle when the area is:
  i 55 $cm^2$  ii 18 $cm^2$  iii 100 $cm^2$
c What is the approximate area of the rectangle when $w = 3.8$ cm?
d If the rectangle has an area of 75 $cm^2$, find the approximate value of $w$.

7 The dimensions of a rectangle of area $A$ ($cm^2$) are given in this way.

- let the width of the rectangle be $w$ (cm)
- the length is 4 cm less than twice the width of the rectangle.

a Write an expression for the area of the rectangle.
b Draw a graph of $A$ against $w$, for values of $w$: $0 \leqslant w \leqslant 8$.
c From the graph, give the approximate dimensions of the rectangle when the area is:
  i 35 $cm^2$  ii 62 $cm^2$  iii 80 $cm^2$
d What is the approximate area of the rectangle when the width is 5.6 cm?
e Give the dimensions of the rectangle (to 1 dp) when the area is 44 $cm^2$.

## Sketching graphs by completing the square

A full explanation of completing the square can be found on page 105 of this book.

If you can write a quadratic expression as a complete square, it can help to sketch the graph.

**Example**

Sketch the graph of $y = x^2 + 4x + 1$.

Completing the square gives:

$y = (x + 2)^2 - 4 + 1$

i.e. $y = (x + 2)^2 - 3$

For the graph of $y = (x + 2)^2 - 3$
$(x + 2)^2$ cannot be negative,
so the minimum value of $y$
is when $(x + 2) = 0$, i.e. $x = {}^-2$

When $x = {}^-2$, $y = {}^-3$, so,
the minimum point is $({}^-2, {}^-3)$.

The intersect with the $y$-axis
is where $x = 0$ i.e.
when $x = 0$, $y = 1$
so, the $y$-intercept is at $(0, 1)$.

The graph can now be sketched.

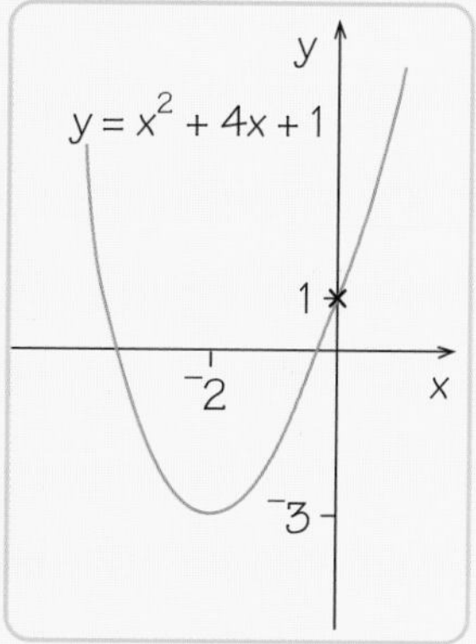


**Exercise 16.7**
**Sketching graphs by completing the square**

1 For the graph of $y = x^2 + 6x - 2$:
  - **a** Rearrange the equation by completing the square.
  - **b** What is the minimum value of $y$? Explain your answer.
  - **c** Give the coordinates of the intercept with the $y$-axis.
  - **d** Sketch the graph of $y = x^2 + 6x - 2$.

2 For the graph of $y = x^2 + 10x + 4$:
  - **a** Show that the minimum value of $y$ is ${}^-21$.
  - **b** Sketch the graph of $y = x^2 + 10x + 4$.

3 Sketch the graph of $y = x^2 + 14x - 5$.

4 Sketch the graph of $y = x^2 - 8x - 6$.

5 **a** On the same axes sketch the graphs of $y = x^2 + 8x + 4$, and $y = x - 9$.
  **b** Explain, with the help of your sketch graphs, why you cannot find a value for $x$ that satisfies $x^2 + 8x + 4 = x - 9$.

6 The diagram shows a sketch of a quadratic graph in the form:

$y = ax^2 + bx + c$

Find an equation for the graph that is sketched and list values for $a$, $b$, and $c$.

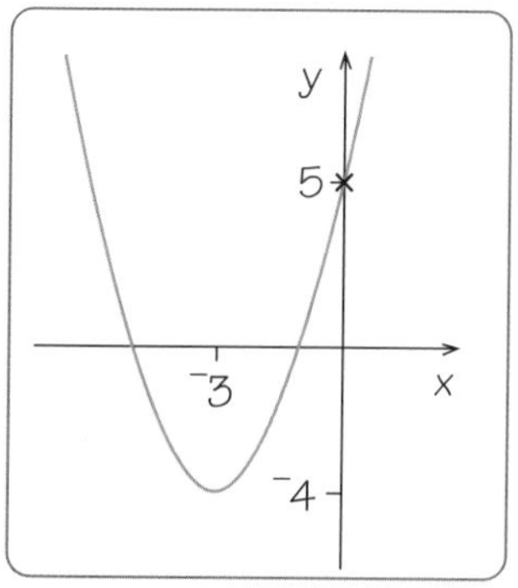

# End points

You should be able to ... ... so try these questions

**A** Solve simultaneous linear equations graphically

**A1** By drawing suitable graphs solve these equations for $x$ and $y$.

**a** $5x - 3y = 13$
$2x + 5y = {}^{-}1$

**b** $2x - 3y = 3$
$7x - 2y = 19$

**A2** Two groups of people bought tickets on the Cliff Railway. The first group of five adults and two children paid a total of £32. The second group of two adults and three children paid a total of £20.50.

By drawing suitable linear graphs, find the charges for adult and child tickets for the Cliff Railway.

**B** Work with, and solve equations from quadratic graphs

**B1** Give the equation of the line of symmetry of the graph $y = x^2 - 4x$.

**B2** Solve these quadratic equations by drawing suitable graphs.

**a** $x^2 + 2x - 8 = 0$
**b** $x^2 - 3x - 7 = 3$
**c** $x^2 - x + 12 = 16$
**d** $2x^2 - 2x - 12 = 0$

**B3** **a** Draw the graph of $y = 2x^2 + x - 5$.

Use the graph of $y = 2x^2 + x - 5$ together with any additional graphs you choose to draw to solve these equations.

**b** $2x^2 + 4x - 9 = 5x - 5$
**c** $5x + 7 = 2x^2 + 5x + 1$
**d** $2x - 5 = 2x^2 + 5x - 7$

**C** Draw and interpret reciprocal graphs

**C1** For $x$: ${}^{-}5 < x < 5$ draw the graph of $y = \frac{6}{x} - 2$.

**C2** For $x$: ${}^{-}6 \leqslant x \leqslant 4$ draw the graph of $y = \frac{2}{x} + 5$.

Give the equation of any asymptotes to the graph of $y = \frac{2}{x} + 5$.

**C3** Sketch the curve of $y = \frac{5}{x} + 4$.

On your sketch draw and label any asymptotes.

**D** Read and interpret graphs

**D1** For $x$: $1 \leqslant x \leqslant 10$, draw the graph of $y = \frac{12}{x} + x - 8$.

**a** From the graph find an approximate minimum value for $y$.
**b** Give approximate values for $x$ when $y = {}^{-}0.6$.

**D2** For $x$: ${}^{-}6 \leqslant x \leqslant 2$ draw the graph of $y = 8 - 5x - x^2$.

**a** From the graph give an approximate maximum value for $y$.
**b** When $y = 3.8$ give two approximate values for $x$.
**c** Use the graph and any suitable additional graph find values of $x$ that satisfy the equation $2x + 4 = 9 - x - x^2$.

**D3** The width of a rectangle is given as $w$ (cm) and the length as $w + 6$ (cm). An expression for the area $A$ (cm$^2$) of the rectangle is given as $A = w(w + 6)$.

**a** Draw a graph of $A = w(w + 6)$ for values of $w$ from zero to 6. (Plot $w$ on the horizontal axis.)
**b** From the graph find approximate dimensions of the rectangle when the area is:
**i** 58 cm$^2$ **ii** 32 cm$^2$.
**c** Calculate the approximate area of the rectangle when $w = 5.4$ cm.
**d** If $A = 35$ cm$^2$, calculate a value for $w$ correct to 1 dp.

**E** Sketch graphs and use completing the square

**E1** On the same axes sketch and label these graphs.

**a** $y = x^2 + 3x + 1$
**b** $y = 2x + 3$

**E2** Sketch and label these graphs on the same axes.

**a** $y = x^2 - 4$
**b** $y = 4 - x^2$
**c** $y = 4 - 3x - x^2$

**E3** On the same axes sketch and label these graphs.

**a** $y = \frac{9}{x} + 1$
**b** $y = x^3 + 4$
**c** $y = 4 - x^3$

**E4** For the graph of $y = x^2 + 8x - 7$.

**a** Show that the graph has a minimum value of $^-23$.
**b** Sketch the graph of $y = x^2 + 8x - 7$.

**E5** The diagram shows a sketch of a quadratic graph in the form

$$y = ax^2 + bx + c$$

Find an equation for the graph and list values for $a$, $b$, and $c$.

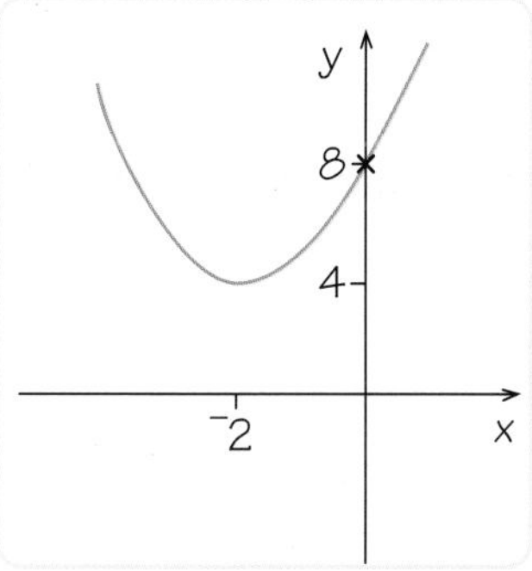


## Some points to remember

- The equation of any graph gives you data about the graph before you plot any points or sketch a line on a pair of axes. This data can indicate shape, intersects with axes and any maximum or minimum values.
- The graphs of curves should be drawn as a smooth curve, with axes clearly graduated, points accurately plotted and the final curve labelled with its equation.
- Data that you read from a graph is only an approximation. For more exact values use trial and improvement starting with any approximate value read from a graph.

## Starting points

You need to know about ...

### A Trigonometric ratios

- These are definitions of three trigonometric ratios (trig ratios).

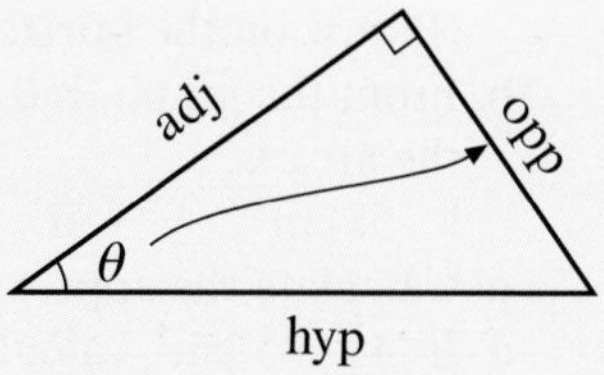


$\sin\theta = \frac{\text{opp}}{\text{hyp}}$ $\cos\theta = \frac{\text{adj}}{\text{hyp}}$ $\tan\theta = \frac{\text{opp}}{\text{adj}}$

opp = hyp × $\sin\theta$ adj = hyp × $\cos\theta$ opp = adj × $\tan\theta$

- These trig ratios can be used to calculate sides or angles in right-angled triangles.

**Example** In ΔABC AB = AC = 5 cm and BC = 7 cm. Calculate the angle CÂB.

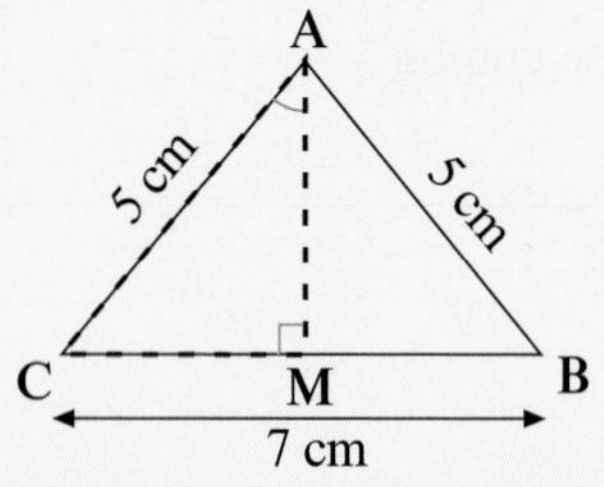


ΔABC is isosceles.
CM = MB = 3.5 cm
CÂM = BÂM

In ΔAMC:
$\sin \text{C}\hat{\text{A}}\text{M} = \frac{\text{CM}}{\text{AC}} = \frac{3.5}{5} = 0.7$

CÂM = 44.42 ...°

CÂB = 88.9° to 1 dp.

### B Graphs of $y = \sin x$ and $y = \cos x$

- This diagram shows the graphs of $y = \sin x$ and $y = \cos x$ for values of $x$ from 0° to 90°.

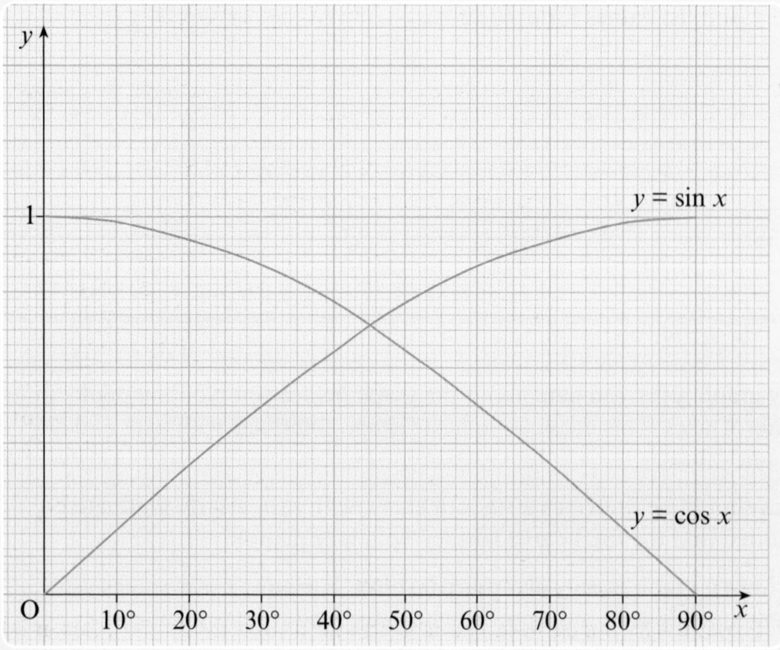


... so try these questions

**A1** The points B and C lie on the circumference of a circle with centre O.

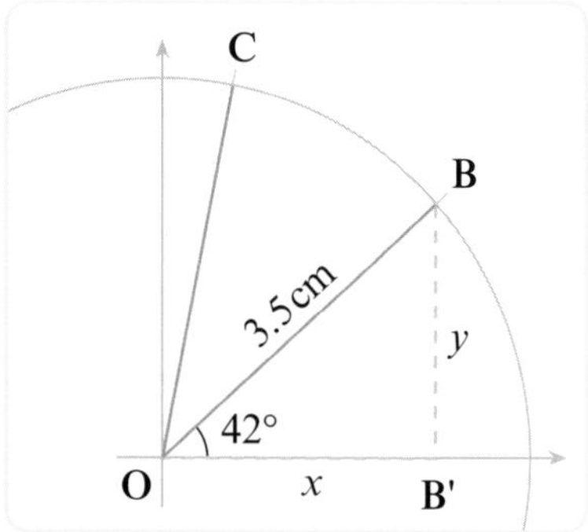


**a** Calculate, correct to 2 dp:
**i** the coordinates of B
**ii** the area of ΔOBB′

**b** C is the point (0.88, 3.39). Calculate the angle between OC and the horizontal axis, correct to 1 dp.

**A2** Calculate the area of a regular pentagon of side 8 cm.

**A3** If $\tan\theta = \frac{3}{4}$ find the value of:
**a** $\cos\theta$ **b** $\sin\theta$.
Do not use a calculator.

Use the graphs of $y = \sin x$ and $y = \cos x$ to answer Questions **B1** – **B3**.

**B1** Estimate the value of:
**a** cos 40°
**b** sin 40°

**B2** Estimate the angle $x$ for each of these:
**a** $\sin x = 0.4$
**b** $\cos x = 0.4$

**B3** Find the angle $x$ for each of these:
**a** $\sin 20° = \cos x$
**b** $\cos 48° = \sin x$
**c** $\cos x = \sin x$

## Notation for trigonometric ratios

- $y = \cos 4x$ can also be written as $y = \cos(4x)$ to show that you need to find the value of $4x$ before finding the cosine.

  When $x = 10$ $\quad y = \cos 4x$
  $= \cos(4 \times 10)°$
  $= \cos 40°$
  $= 0.766$ (3 sf)

- $y = 4\cos x$ can also be written as $y = 4 \times \cos x$

  When $x = 10$ $\quad y = 4\cos x$
  $= 4 \times \cos 10°$
  $= 3.94$ (3 sf)

- In the same way
  $y = \sin 3x$ can be written as $y = \sin(3x)$ and
  $y = 3\sin x$ can be written as $y = 3 \times \sin x$

**Exercise 17.1**
Evaluating trigonometric functions

**1** Evaluate these expressions, correct to 2 dp, when $x = 20$.

**a** $2\sin x$ **b** $\sin 3x$ **c** $\cos 2x$
**d** $3\cos 2x$ **e** $2\sin 4x$ **f** $5\cos 3x$

**2** If $y = 4\cos 3x$:

**a** find the value of $y$ when
**i** $x = 5$ **ii** $x = 7.5$ **iii** $x = 12.35$
**b** for what value of $x$ does
**i** $y = 0$ **ii** $y = 2$?

## Polar and cartesian coordinates

$\theta$ is a Greek letter pronounced 'theta'. It is used commonly to indicate angles.

- Polar coordinates give the position of a point in a plane by stating:
  - its distance, $r$ from a fixed point, the pole
  - the angle, $\theta$ measured anticlockwise, between a line drawn from the pole to the point and a fixed line, the polar axis.

  These are written $(r, \theta)$ the distance followed by the angle, for example (4, 50°).
- Cartesian coordinates give the position of a point in a plane by stating
  - its distance from two perpendicular axes, for example (6, 5).

**Example** This diagram shows the position of a boat at S on a radar screen with rings 10 km apart.

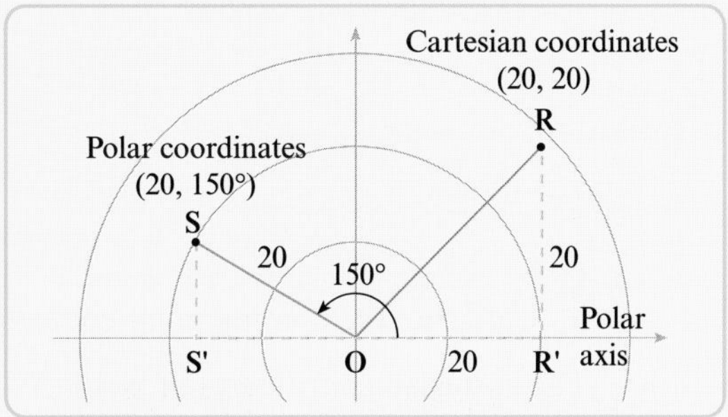


The cartesian coordinates for R are (20, 20)
In $\Delta ORR'$
$R\hat{O}R' = 45$
$OR = \sqrt{800}$ (by Pythagoras)
$= 28.28$ (2dp)
So the polar coordinates for R are (28.28, 45°).

**Exercise 17.2**
Using polar and cartesian coordinates

**1** The polar coordinates of S are (20, 150°).

**a** Explain why the cartesian coordinates for S are ($^-20\cos 30°$, $20\sin 30°$).
**b** Calculate the cartesian coordinates of S.

2 Each of the points A to D on the radar screen shows the position of an aeroplane. The rings are 10 km apart.
AOD and BOC are straight lines.

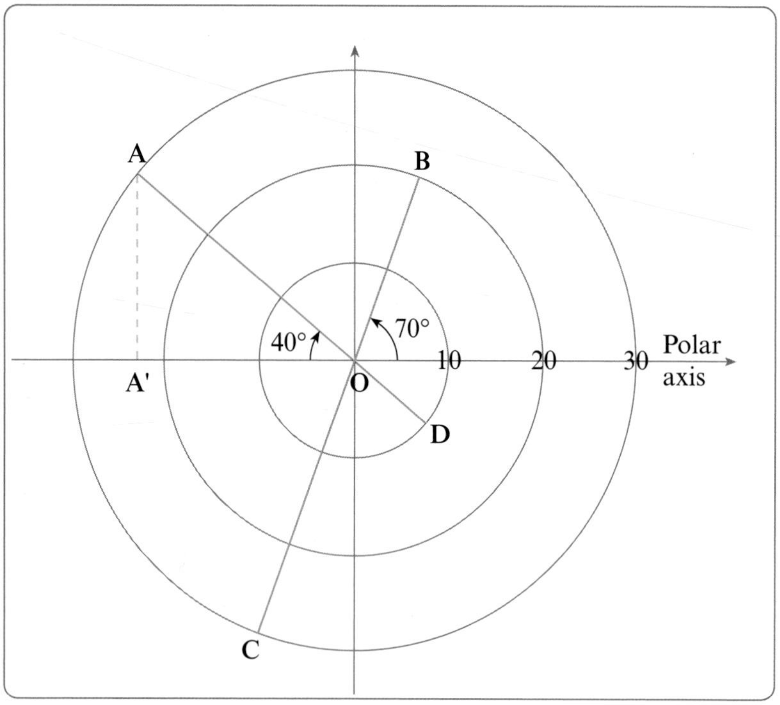


**a** Explain why the polar coordinates for the point A are (30, 140°).
**b** Calculate these distances, correct to 2 dp.
  **i** OA′  **ii** AA′.
**c** Give the cartesian coordinates of A.
**d** For each of the points B, C and D:
  **i** give the polar coordinates
  **ii** calculate the cartesian coordinates correct to 2 dp.

For the cartesian coordinates each distance is given correct to 2 dp.

It may help to mark each position on a sketch.

3 For three other aeroplanes the polar coordinates and cartesian coordinates are given below. For two more aeroplanes only the cartesian coordinates are given.

**a** Match each of the polar coordinates to its cartesian coordinates.

| Polar coordinates | Cartesian coordinates |
|---|---|
| (13, 225°) | (⁻16.27, 36.54) |
| (40, 114°) | (⁻9.19, ⁻9.19) |
| (25, 291°) | (21.53, ⁻22.30) |
| | (⁻16.19, 3.24) |
| | (8.96, ⁻23.34) |

**b** Calculate the missing polar coordinates, to the nearest whole number.

4 The cartesian coordinates of point A are $(x, y)$, and the polar coordinates are $(r, \theta)$.

a Write an expression for $r$ in terms of $x$ and $y$.
b Write the cartesian coordinates in terms of $r$ and $\theta$.
c If $r = 1$, what are the cartesian coordinates in terms of $\theta$?

# Trigonometric functions for any angle

A unit circle is a circle of radius 1 unit.

- For any angle the values of $\sin\theta$ and $\cos\theta$ are defined as the cartesian coordinates of the point on a unit circle with polar coordinates $(1, \theta)$.

**Example** Find the value of cos 128° and sin 128°.

The point A lies on a unit circle so the coordinates of A are (cos 128°, sin 128°).

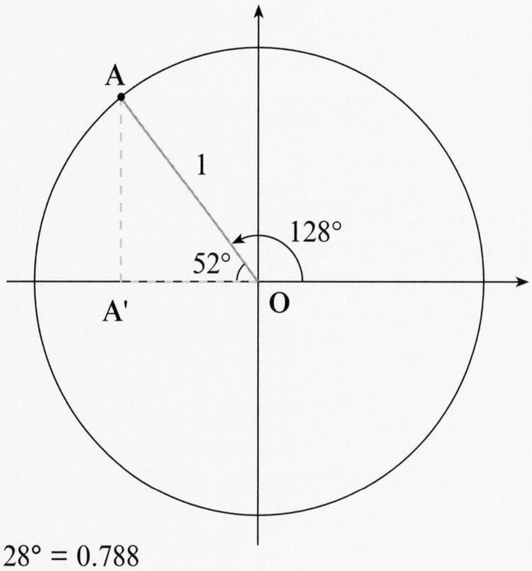


So $\cos 128° = {}^{-}OA'$
$\sin 128° = AA'$

In $\Delta$ OAA′:

$OA' = 1 \times \cos 52°$
$= \cos 52°$
$= 0.616$

$AA' = 1 \sin 52°$
$= \sin 52°$
$= 0.788$

So the coordinates of A are $({}^{-}0.616, 0.788)$

So, $\cos 128° = {}^{-}0.616$ and $\sin 128° = 0.788$

**Exercise 17.3**
Evaluating trigonometric functions for any angle

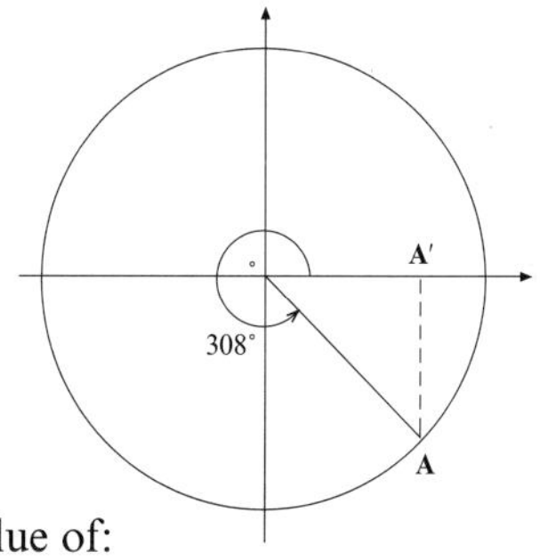


1 Find another angle whose cosine = $^{-}0.616$.

2 Use the sketch to help you find the value of:
**a** cos 308° **b** sin 308°.

3 Use diagrams to explain why:
**a** sin 21° = sin 159° **b** sin 201° = $^{-}$sin 21°
**c** cos 5° = cos 355° **d** cos 185° = $^{-}$cos 5°.

It may help to sketch a unit circle for each question.

For example for **3b**:

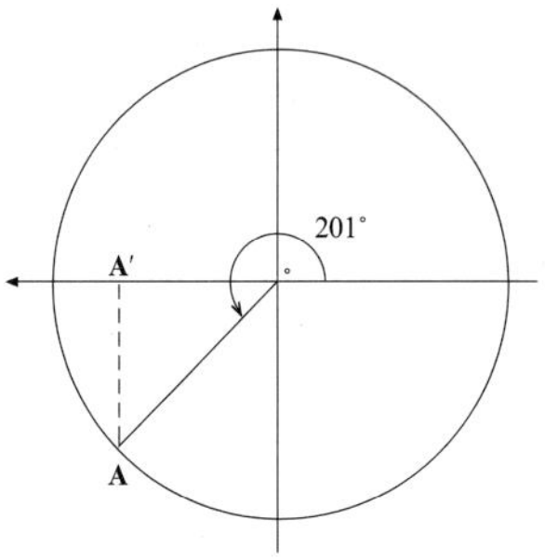


4 The value of cos 36° is 0.809 (3sf). What is the value of:
**a** cos 144° **b** cos 324° **c** cos 216°?

5 **a** Sort these into pairs that have the same value.

| | | |
|---|---|---|
| sin 52° | sin 232° | sin 128° |
| cos 233° | cos 128° | sin 308° |

**b** Find the value for each pair correct to 3 sf.

6 For what values of $\theta$, from 0° to 360°, is:
**a** $\cos\theta < 0$ **b** $\sin\theta < 0$

7 A negative angle is measured clockwise from the polar axis.
a Draw a diagram to find the value of $\cos\theta$ and $\sin\theta$ for:
i $\theta = {}^{-}90°$ ii $\theta = {}^{-}52°$
b Find a negative value for $\theta$ which satisfies:
i $\sin\theta = \sin 54°$ ii $\sin\theta = \sin 100°$ iii $\cos\theta = \cos 200°$

8 a Explain why cos 450° = 0.
b Give three values of $\theta$ for which $\sin\theta = 1$.

# Graphs of $y = \sin x$ and $y = \cos x$

- The diagram shows the graphs of $y = \sin x$ and $y = \cos x$ for $0° \leqslant x \leqslant 360°$
- If you use your calculator to find $\sin x$ for values of $x$ from 0° to 90°, you can use the symmetry of the graph to find other solutions.

**Example** **Solve $\sin x = \sin 20°$ and $\sin x = {}^-\sin 20°$ for $0° < x < 360°$**

From a calculator $\sin 20° = 0.34$ (2 dp), so $^-\sin 20° = {}^-0.34$ (2dp).

The graph shows that at the points A and B the value of $\sin x$ is 0.34. This corresponds to the angles 20° and 160°.

In the same way at P and Q $\sin x = {}^-0.34$ for angles of 200° and 340° respectively.

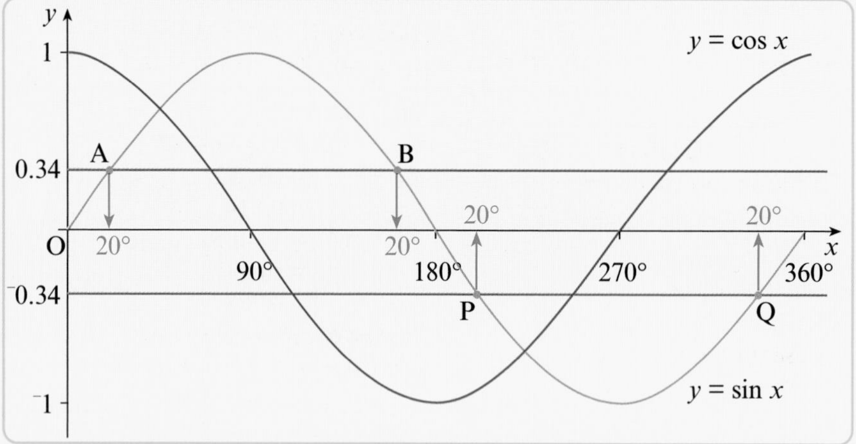


**Solve $\sin x = \sin 20°$**

From a calculator:
$\sin 20° = 0.34$ (2 dp)
$\sin x = 0.34$

From the graph:
$\mathbf{x = 20°}$

or $x = 180° - 20°$
$\mathbf{x = 160°}$

**Solve $\sin x = {}^-\sin 20°$**

From a calculator:
$\sin 20° = 0.34$ (2 dp)
so $^-\sin 20° = {}^-0.34$ (2dp)

From the graph:
$x = 180° + 20°$
$\mathbf{x = 200°}$

or $x = 360° - 20°$
$\mathbf{x = 340°}$

**Exercise 17.4**
Using the graphs of $y = \sin x$ and $y = \cos x$

To avoid marking the page put tracing paper over the graph each time you use it.

1 What is the maximum and minimum value for:

**a** $\sin x$ **b** $\cos x$?

2 Use the graph of $\sin x$ to estimate for which values of $x$, between 0° and 360°, $\sin x = 0.5$. [Check your answers using a calculator]

3 Solve each of these for values of $x$ between 0° and 360°.

**a** **i** $\cos x = \cos 40°$ **ii** $\cos x = {}^-\cos 35°$ **iii** $\sin x = {}^-\sin 72°$
**b** **i** $\sin x = 0.9$ **ii** $\cos x = {}^-0.9$
**c** **i** $\cos x < 0.5$ **ii** $\sin x < {}^-0.5$ **iii** $\cos x = \sin x$
**iv** $\cos x < \sin x$

4 Two solutions for $\cos x = 0.5$ are $x = 60°$ and $x = 300°$.

**a** Find two other solutions with $x > 300°$.
**b** Give another solution with $x \leqslant 0°$.

5 One solution for $\sin x = 1$ is $x = 90°$.

Find another solution with:

**a** $x > 90°$ **b** $x \leqslant 0°$.

An example which shows that a statement is false is called a counter-example.

Finding a value for which a statement is true does not prove that it is true for all values of $x$.

6 Three of these statements are false for some values of $x$.

a Give a counter-example for the three statements that are false.

A $\sin(180° - x) = \sin(180° + x)$
B $\cos x = \sin(90° + x)$
C $\sin(x + x) = \sin x + \sin x$
D $\cos x = \cos(90° + x)$
E $\cos x = \cos(360° - x)$

b Explain how you know that the other two statements are true for all values of $x$.

## Transformations of the graphs of $y = \sin x$ and $y = \cos x$

**Exercise 17.5**
Drawing trigonometric graphs

We say the movement is modelled by this equation because it does not describe the movement exactly but it is reasonably accurate.

1 The movement of the tip of the needle in this sewing machine can be modelled by the equation

$$h = 9 \cos t$$

where:

$t$ is the time in milliseconds (thousandths of a second) that the machine has been running.
$h$ is the height above the footplate in millimetres.

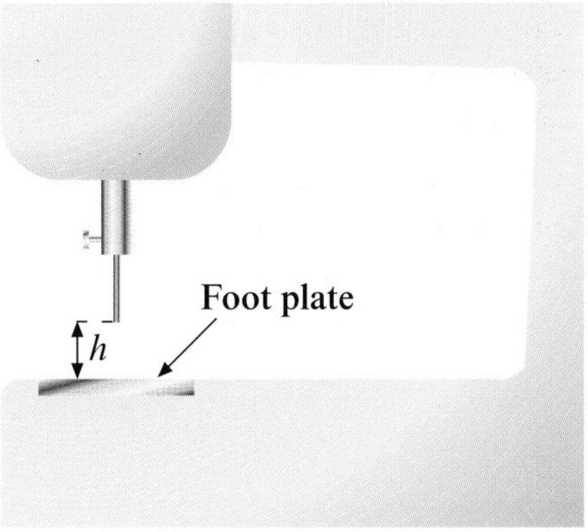


These are some values for $h = 9 \cos \theta$.

$t = 0 \quad h = 9 \times \cos 0° = 9$

$t = 60 \quad h = 9 \times \cos 60° = 4.5$

a Copy and complete this table for values of $h = 9 \cos t$ over the first 600 milliseconds.

| $t$ (ms) | 0 | 60 | 120 | 180 | 240 | 300 | 360 | 420 | 480 | 540 | 600 |
|---|---|---|---|---|---|---|---|---|---|---|---|
| $h$ (mm) | 9 | 4.5 | | | | | | | | | |

b i Explain why the graph should be a smooth curve.
ii Draw the graph of $h = 9 \cos t$.

c At what height above the footplate did the tip of the needle start?

d Each time the tip of the needle returns to its maximum height above the footplate it completes one stitch.
i How long does the machine take to complete one stitch?
ii For how long is the needle below the footplate on each stitch?

e How many stitches will the machine complete in one second?

Use the same pair of axes as you used for Question **1b**.

2 The movement of the tip of the needle on another sewing machine can be modelled by the equation

$$h = 12 \cos t$$

a Sketch the graph of $h = 12 \cos t$
b Does this machine complete stitches at the same rate?
c What is the difference between the two machines?

- Each of the graphs below is a transformation of the graph of $y = \sin x$.

**A** **A graph of $y = \sin x + 2$**

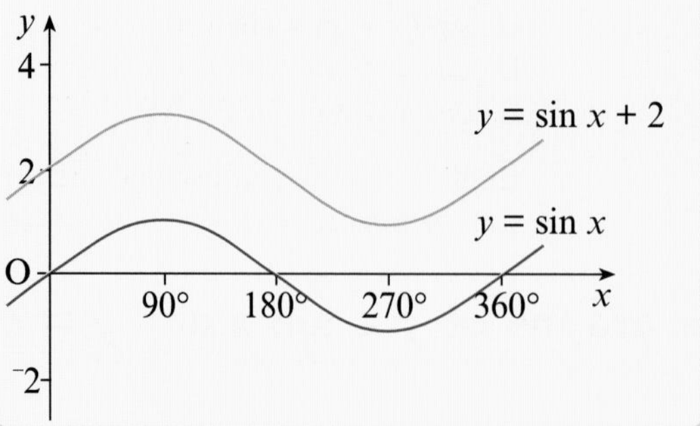


**B** **A graph of $y = 2 \sin x$**

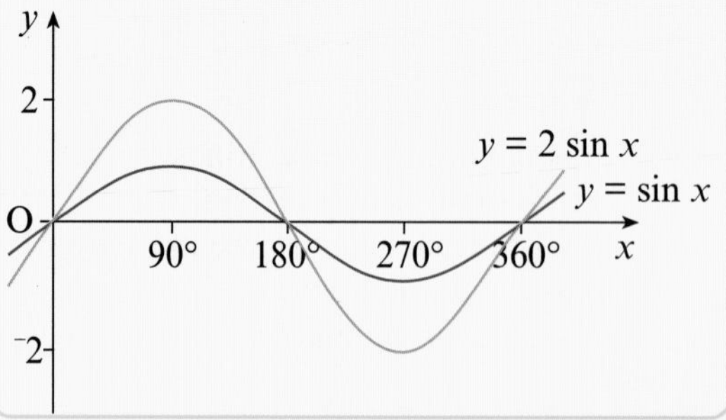


**C** **A graph of $y = \sin 2x$**

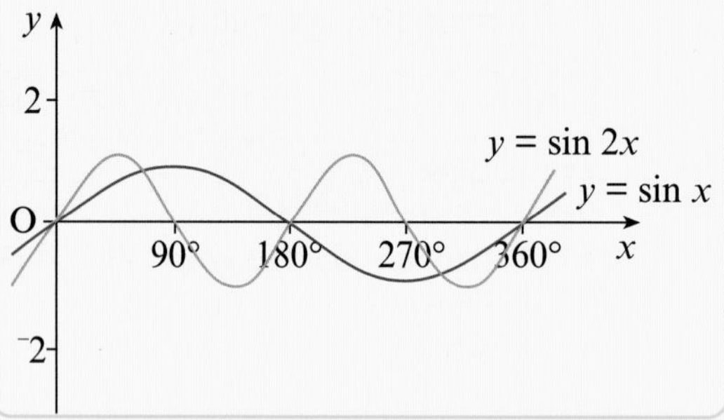


**Exercise 17.6**
**Sketching and using trigonometric graphs**

1 Using the diagrams above:
  a state the maximum and minimum values of $y$ on each graph
  b describe the link between the graphs of $y = \sin x$ and:
    i $y = \sin x + 2$  ii $y = 2 \sin x$  iii $y = \sin 2x$.

It may help to sketch the graph of $y = \cos x$ first.

2 a For the values $0 \leqslant x \leqslant 360°$ sketch the graphs of:
    i $y = \cos x - 2$  ii $y = 3 \cos x$  iii $y = \cos 3x$.
  b For each graph state the maximum and minimum values of $y$.

3 What are the maximum and minimum values of $y$ for:
  a $y = \sin 4x$  b $y = 5 \cos x$  c $y = \sin x - 4$
  d $y = \cos 2x + 1$  e $y = 2 \sin x - 2$  f $y = 3 \sin 2x + 2$.

4 On one pair of axes sketch the graph of:
  a $y = \sin 3x$  b $y = 2 \sin 3x$  c $y = 2 \sin 3x - 4$.

**5** Each of these graphs is a transformation of $y = \sin x$ or $y = \cos x$.

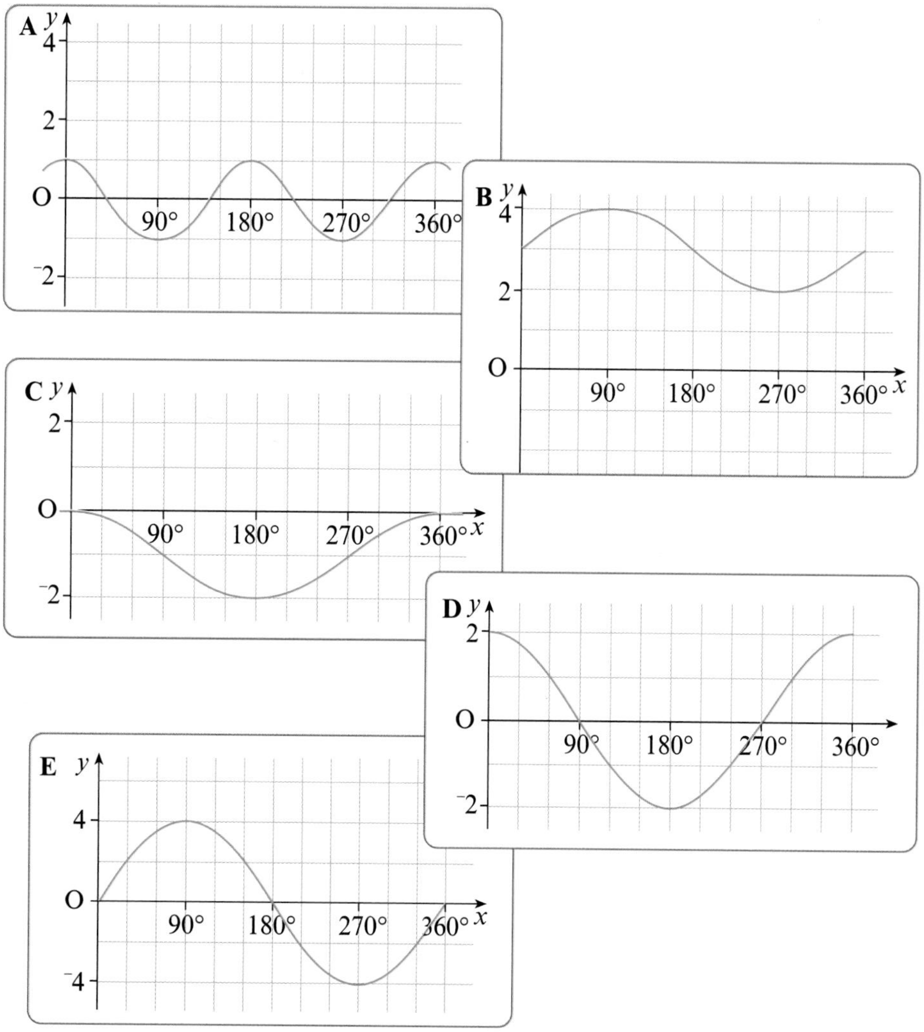


In each equation $a$ is a constant.

These are the five equations for the graphs above.

$y = a \sin x$ $y = \cos x + a$ $y = a \cos x$

$y = \sin x + a$ $y = \cos ax$

For each graph:

**a** write down the matching equation
**b** find value of $a$ in its equation.

These are some values for $y = \cos (x + 45°)$:

| $x$ | $y = \cos (x + 45°)$ |
|---|---|
| 0 | $\cos (0 + 45°) \approx 0.71$ |
| 45 | $\cos (45° + 45°) = 0.00$ |

**6** For the graph of $y = a \sin bx + c$, where $a$, $b$ and $c$ are constants, write an expression for the maximum value of $y$.

**7** Sketch the graphs of:

a $y = \cos (x + 45°)$ b $y = \cos (x - 45°)$.

**8** Explain why the graph of $y = \sin (x + 90°)$ and $y = \cos x$ are the same.

**9** Which of these gives the same graph as $y = \cos x$?

A $y = \sin (x + 90°)$ B $y = \sin x - 90°$ C $y = \cos (x + 360°)$
D $y = \sin x + 90°$ E $y = \sin (x - 90°)$ F $y = \cos x + 1$

## The graph of $y = \tan x$

- You can write $\tan x$ in terms of $\sin x$ and $\cos x$.
  For values of $x$ in the first quadrant

$$\frac{\sin x}{\cos x} = \frac{\text{opp}}{\text{hyp}} \div \frac{\text{adj}}{\text{hyp}}$$
$$= \frac{\text{opp}}{\text{hyp}} \times \frac{\text{hyp}}{\text{adj}}$$
$$= \frac{\text{opp}}{\text{adj}}$$
$$= \tan x$$

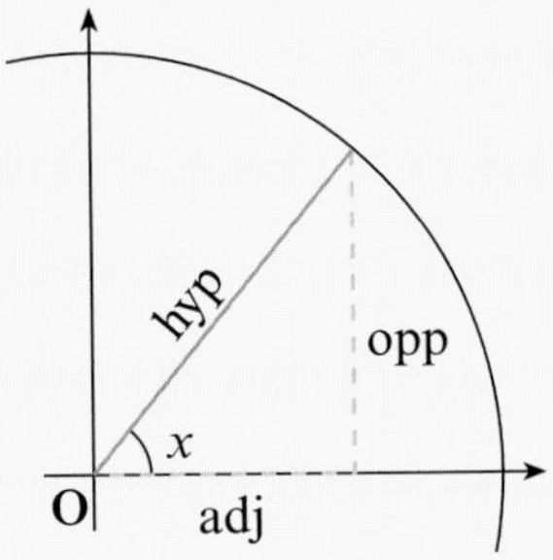


So $\frac{\sin x}{\cos x} = \tan x$ for $0 \leqslant x < 90$

This can be extended to values of $x$ in the other quadrants.

- This is a graph of $y = \tan x$ for $^-90^\circ < x < 450^\circ$

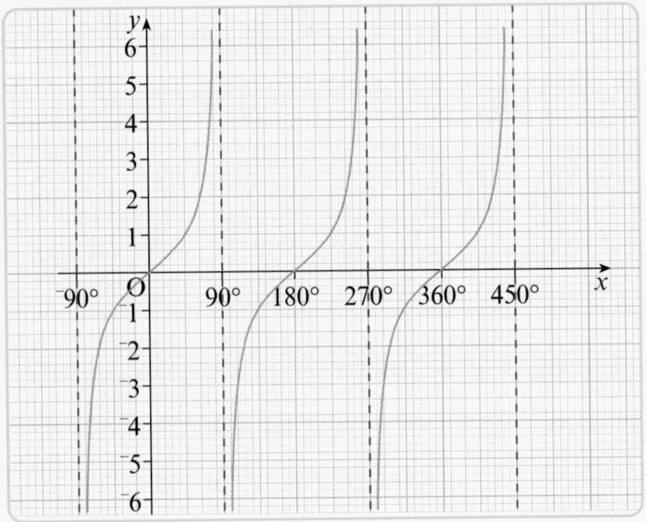


**Exercise 17.7**
**Using the graph of $y = \tan x$**

1 Explain why the value of $\tan 100^\circ$ is negative.

2 Solve each of these for values of $x$ between $0^\circ$ and $360^\circ$.

**a** **i** For what acute angle $x$ does $\tan x = 1$?
**ii** Give another value of $x$ that satisfies this equation.

**b** Give two values of $x$ that satisfy the equation $\tan x = {}^-1$.

3 Explain why:

**a** $\tan 56^\circ = \tan 236^\circ$ **b** $\tan 304^\circ = \tan ({}^-56)^\circ$.

4 **a** Use your calculator to find a value for $x$ such that $\tan x \approx 1.1918$.
**b** Use the graph to find other values of $x$ which satisfy this equation.

5 Give three values of $x$ between $^-90^\circ$ and $450^\circ$ for which:

**a** $\tan x = \tan 60^\circ$ **b** $\tan x = {}^-\tan 25^\circ$ **c** $\tan x = {}^-\tan 32^\circ$

6 **a** Use your calculator to find the value of:
i $\tan 89^\circ$ ii $\tan 89.5^\circ$ iii $\tan 89.9^\circ$ iv $\tan 89.99^\circ$

**b** Use your calculator to find the value of:
i $\tan 91^\circ$ ii $\tan 90.5^\circ$ iii $\tan 90.1^\circ$ iv $\tan 90.01^\circ$

**c** Why do you think your calculator does not give a value for $\tan 90^\circ$?

**d** Lines drawn at $^-90^\circ$, $90^\circ$, $270^\circ$ and $450^\circ$ are called asymptotes.
At what value of $x$ would the next asymptote occur?

# End points

You should be able to ...   ... so try these questions

**A** Solve trigonometric equations

**A1** Solve each of these for $0° \leqslant x \leqslant 360°$.
- **a** $\cos x = \cos 70°$
- **b** $\cos x = {}^{-}\cos 70°$
- **c** $\tan x = {}^{-}\tan 40°$
- **d** $\sin x = {}^{-}\sin 28°$.

**A2** Sort these into pairs that have the same value.

| | |
|---|---|
| cos 41° | cos 139° |
| cos 229° | cos 319° |
| cos 131° | cos 221° |

**B** Transform trigonometric graphs

**B1** Sketch the graph of:
- **a** $y = 3\cos x$
- **b** $y = \sin 3x$
- **c** $y = \sin x + 4$
- **d** $y = 2\cos x - 3$

**B2** Which of these equations matches this graph?
- A $y = \sin 2x$
- B $y = \cos x + 2$
- C $y = \cos (x + 2°)$
- D $y = \sin (x + 90°) + 2$
- E $y = 2\cos x + 2$

Give reasons for your answers.

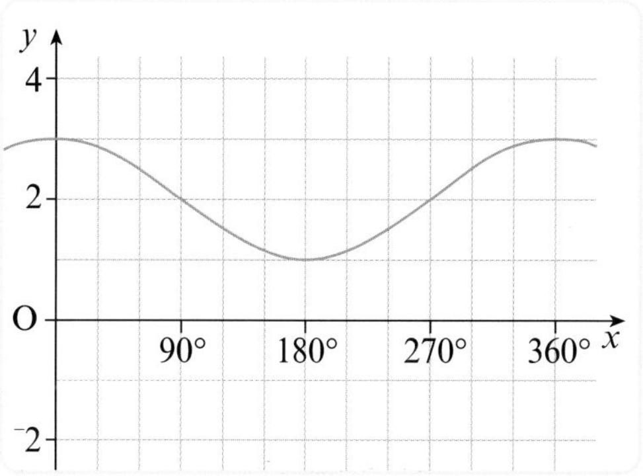


## Some points to remember

- The graphs of $y = \sin x$ and $y = \cos x$

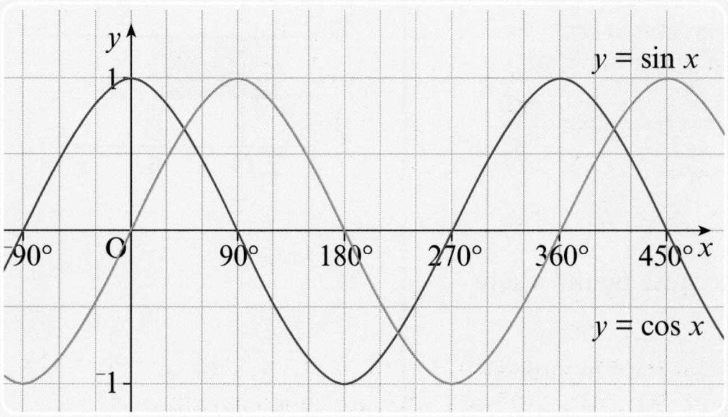


- The graph of $y = \tan x$

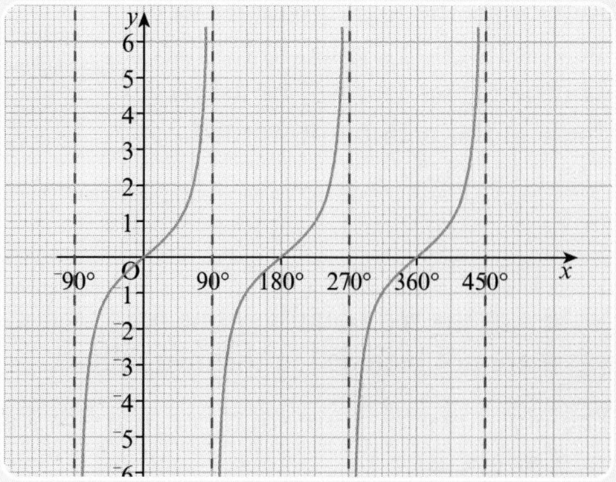

## Starting points

You need to know about ... / ... so try these questions

### A Inequalities that describe a set of integers or range of values

- The inequality signs:
  - $>$ **greater than**
  - $<$ **less than**
  - $\geqslant$ **greater than or equal to**
  - $\leqslant$ **less than or equal to**

  can be used to define a range of values for a variable.
  For instance $^{-}3 < c \leqslant 4$ means the variable $c$ can have any value greater than $^{-}3$ but less than or equal to 4.
  Values for $c$ such as $^{-}2$, 3.2, $^{-}1\frac{1}{2}$ are all acceptable.
  This is sometimes called the **solution set** for $c$.

- The solution set $^{-}3 < c \leqslant 4$ can be shown on a number line like this.

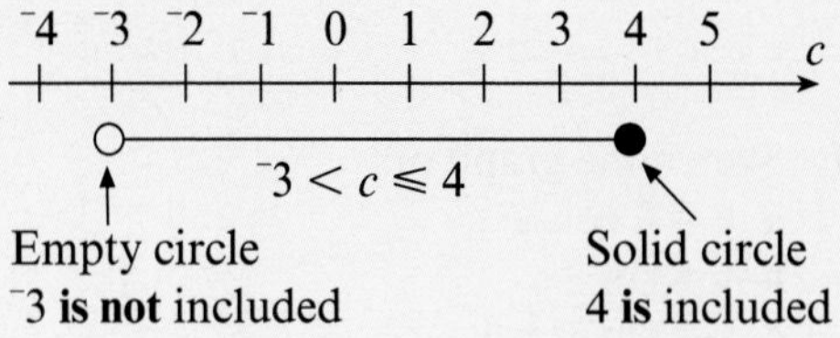


- Sometimes only integer values are considered. Two different inequalities can then describe the same set of integers.

  For instance, both these solution sets have the integers 0, 1, 2, 3, 4, 5

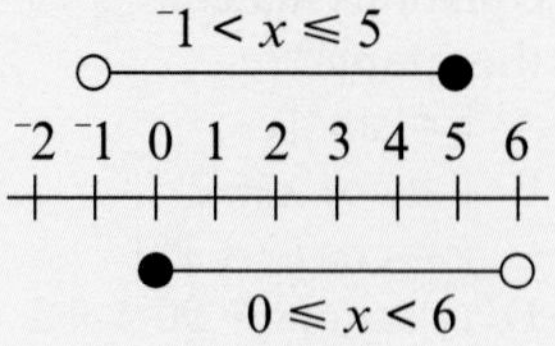


### B Identifying and showing simple regions on a graph

- The conditions satisfied by two or more inequalities can be shown by a region on a graph.

  **Example**
  Sketch the region where $x > 4$ **and** $^{-}5 \leqslant y \leqslant 2$.

- The region where $x > 4$ is shown by the red shaded area and where $^{-}5 \leqslant y \leqslant 2$ by the blue one. Where both regions overlap both conditions are satisfied.

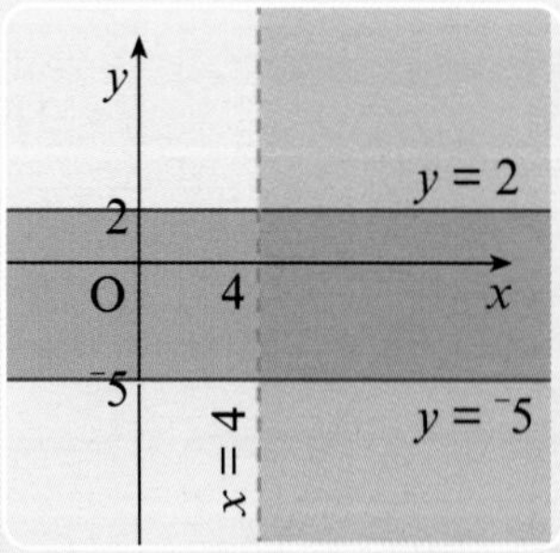


- Sometimes the region which meets the condition is shaded in (as here). At other times you will be asked to shade in the region not wanted. Check carefully.

- Note that the red boundary line for $x$ is dotted because points where $x = 4$ are not included in the inequality $x > 4$.

  The blue lines for $y$ are solid because $y = 2$ and $y = {^-5}$ are included in $^{-}5 \leqslant y \leqslant 2$.

**A1** **a** Show the inequality $^{-}2 \leqslant x < 5$ on a number line.
**b** Which of these values for $x$ satisfy the inequality: 2, $^{-}5$, 5, $^{-}2$, 1.634, $^{-}4$, $4\frac{3}{4}$ ?

**A2** For each of these, what integer values of $n$ satisfy the inequality?
**a** $4 < n \leqslant 9$
**b** $^{-}2 \leqslant n \leqslant 3$
**c** $71 > n \geqslant 68$
**d** $^{-}4 \geqslant n > {^-12}$

**A3** Which one of these inequalities describe a different set of integer values from the others?
$^{-}4 \leqslant h < 6$, $\quad ^{-}5 < h < 6$,
$^{-}4 \leqslant h \leqslant 5$, $\quad ^{-}4 < h < 6$,
$^{-}5 < h \leqslant 5$

**A4** Write two different inequalities involving $g$ which are satisfied by these integers: $^{-}3$, $^{-}2$, $^{-}1$, 0, 1.

**A5** Write a single inequality to show those integers common to all three of these inequalities.
$^{-}5 < f \leqslant 8 \quad ^{-}24 \leqslant f < 7$
$^{-}3 \leqslant f \leqslant 7$

**B1** Sketch graphs to show the regions satisfied by each pair of inequalities.
**a** $2 \leqslant x < 5 \quad y \geqslant 3$
**b** $x < {^-4} \quad 3 \leqslant y < 7$
**c** $^{-}3 < x \leqslant 3 \quad ^{-}8 \leqslant y \leqslant {^-1}$

**B2** What two inequalities are satisfied by the shaded region on this sketch graph?

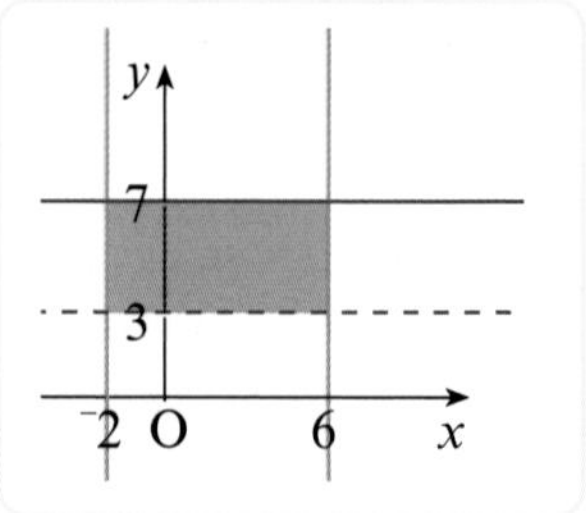

## C The graphical link between a linear equation and an inequality

**Example**

Show the region on a graph where both the inequalities $y > 10 - 2x$ and $y \leqslant 2x - 1$ are satisfied.

- Lightly draw the linear graphs of $y = 10 - 2x$ and $y = 2x - 1$ to show the **boundaries** of the regions.

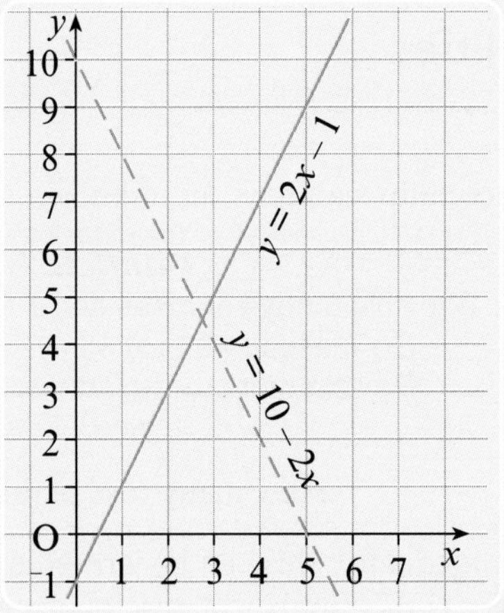


- Decide if the lines should be dotted or solid.

  $y = 2x - 1$ is solid since points on the line are included in $y \leqslant 2x - 1$.

  $y = 10 - 2x$ is dotted since points on the line are not included.

- For each boundary line in turn, choose a point P on one side of it, e.g. (6, 7). Substitute P's coordinates into the inequality and check if the inequality is true or not for that point.

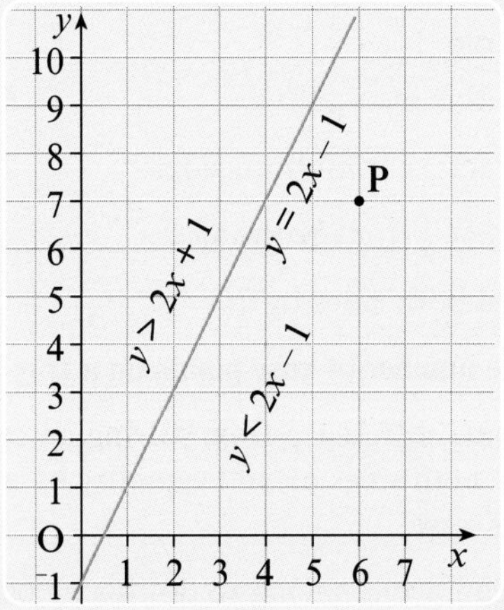


  Substituting into $y \leqslant 2x + 1$ gives $7 \leqslant (2 \times 6) + 1$
  $7 \leqslant 13$ true:
  so for this side of the line $y \leqslant 2x + 1$ is true.

  Then choose a point to check which side of its boundary line the other inequality applies to.

- Shade out the regions where the inequalities are **not true**.

  The part of the graph **not shaded** meets both conditions.

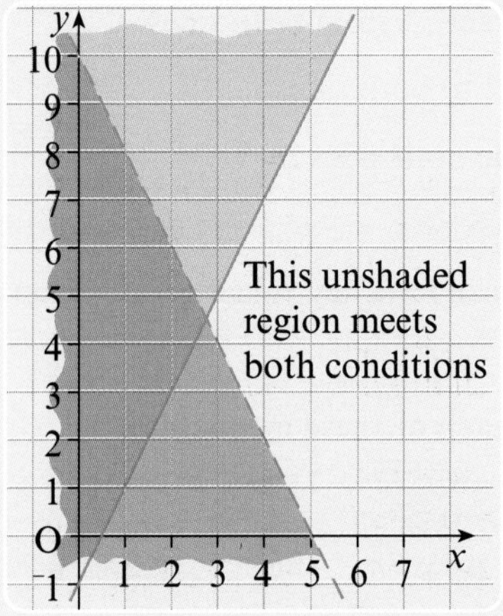


**C1** Use inequalities to describe the shaded region on each graph.

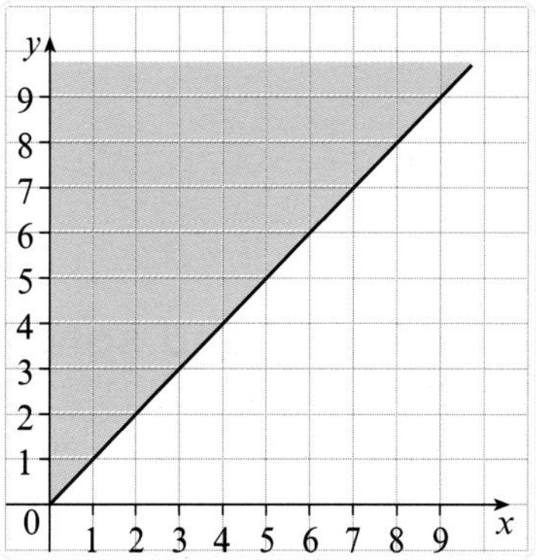

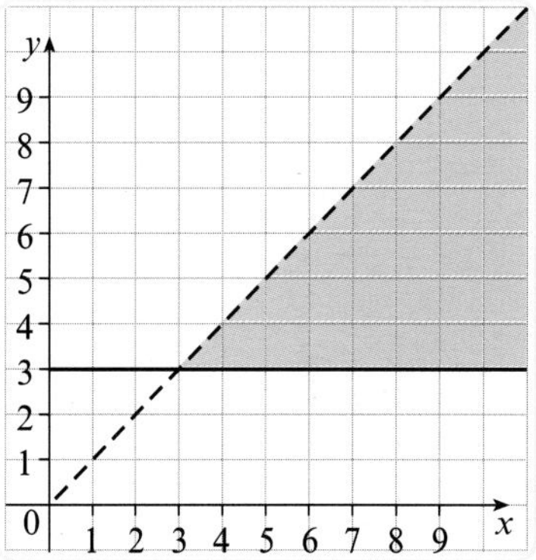

**C2** Show the region on a graph where both the inequalities $y \geqslant x + 1$ and $y < 8 - \frac{1}{2}x$ are satisfied. Shade out the regions **not required**.

**C3** Use a graph to show the region where all of these inequalities are satisfied:

$y < x + 2$
$y \geqslant x$
$^{-}1 \leqslant x < 4$

**C4** What two inequalities describe the unshaded region?

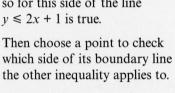

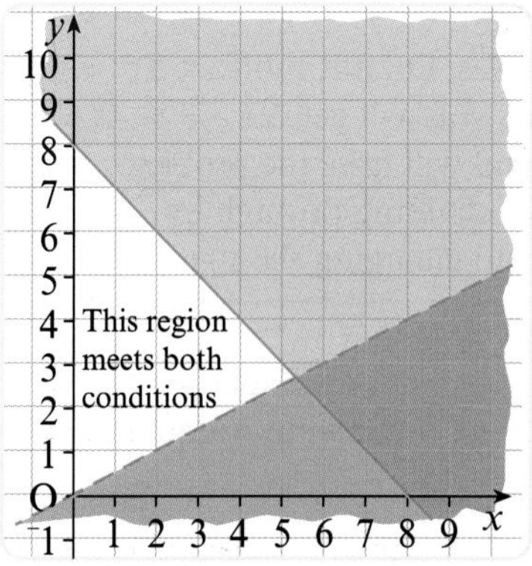

**Thinking ahead to ...**
**solving inequalities**

**A** Start with the inequality **8 > 4** which is true.
Do each of the following operations to the inequality and state if the result is still true.

**a** Add 7 to both sides.
**b** Subtract 5 from both sides.
**c** Multiply both sides by 8.
**d** Divide both sides by 2.
**e** Add $^-8$ to both sides.
**f** Multiply both sides by $^-3$.
**g** Divide both sides by $^-2$.

## Rearranging and solving inequalities

- An inequality can have an unknown on both sides – for instance:

> ***Danfield Nurseries***
> We sold 5 trays of roses and 3 single rose plants on Monday.
> On Tuesday we sold 3 trays and 13 single rose plants.
> We sold more rose plants on Monday than on Tuesday.

> How many rose plants are on a tray?

As an inequality this can be written as $\mathbf{5x + 3 > 3x + 13}$
where $x$ is the number of rose plants on a tray.

- To solve an inequality you can treat it like an equation.

**Example 1**

| | |
|---|---|
| Solve this inequality: | $\mathbf{5x + 3 > 3x + 13}$ |
| Subtract $3x$ from both sides: | $\mathbf{2x + 3 > 13}$ |
| Subtract 3 from both sides: | $\mathbf{2x > 10}$ |
| Divide both sides by 2: | $\mathbf{x > 5}$ |

**So the number of rose plants on a tray must be greater than 5, i.e. at least 6.**

- **The only difference from solving an equation is that if you multiply or divide both sides of an inequality by a negative value the inequality signs will reverse.**

You can avoid having to do this by keeping the coefficient of $x$ positive.

**Example 2**

| | |
|---|---|
| Solve this inequality | $\mathbf{3x - 8 \leqslant 5x - 2}$ |
| Subtract $3x$ from both sides | $\mathbf{^-8 \leqslant 2x - 2}$ |
| Add 2 to both sides | $\mathbf{^-6 \leqslant 2x}$ |
| Divide both sides by 2 | $\mathbf{^-3 \leqslant x}$ |
| So | $\mathbf{x \geqslant {}^-3}$ |

**So $x$ can have any values greater than or equal to $^-3$.**

> When you rearrange an inequality try to keep the number in front of the variable positive. (i.e. try to keep the coefficient of $x$ positive).
>
> **Example**
>
> Solve
> $3x - 8 \leqslant 5x - 2$
>
> Subtract $3x$ from both sides. Do not subtract $5x$ or you will get $^-2x - 8 \leqslant {}^-2$ and will have the problem of dividing through by $^-2$ and changing the sign.

**Exercise 18.1**
**Solving inequalities**

**1** Solve each of these inequalities.

**a** $3y + 6 < 27$
**b** $13 \geqslant s + 5$
**c** $5p - 3 \geqslant 27$
**d** $10 < 2b + 3$
**e** $2 \geqslant 6a + 50$
**f** $5x - 2 > 40$
**g** $8 - 3k \geqslant {}^-6$
**h** $22 \leqslant 4a - 6$
**i** $13 - 4d > 33$
**j** $^-32 \geqslant 3a + 7$

Where there are brackets, multiply them out first.

For inequalities like $^-9 < 2x + 1 < 5$ treat them as two separate inequalities, e.g. $^-9 < 2x + 1$ and $2x + 1 < 5$

So $^-5 < x < 2$

2 Solve each of these inequalities.

**a** $2a - 5 \geqslant 6 - a$ **b** $2k - 6 > 7k + 8$
**c** $3x + 7 > x - 11$ **d** $2n - 2 \leqslant 4n - 9$
**e** $3q + 5 < 1 - 2q$ **f** $6 < 2(2k - 1)$
**g** $7h + 3 > 13h + 5$ **h** $2(3c - 2) < 11$
**i** $7h - 4 < 5(4 - 3h)$ **j** $2 - 4(2 - 3p) \geqslant 5(p + 2)$

3 Rearrange these inequalities into the form $A < x < B$, where $A$ and $B$ are integers.

**a** $^-3 < x + 1 < 2$ **b** $^-7 < 2x - 3 < 7$
**c** $^-14 < 1 - x < 2$ **d** $6 < 2 - 4x < 14$

## Inequalities with squared terms

- Where an inequality involves an index power, the range of values for the variable may be continuous or have several separate parts.

**Example 1** For which values of $x$ is $x^2 < 9$ true?

Since $x^2 = 9$ has two solutions $x = 3$ and $x = {}^-3$ the inequality $x^2 < 9$ has two values between which $x$ must lie $\mathbf{^-3 < x < 3}$.
This is a continuous range between $^-3$ and 3.

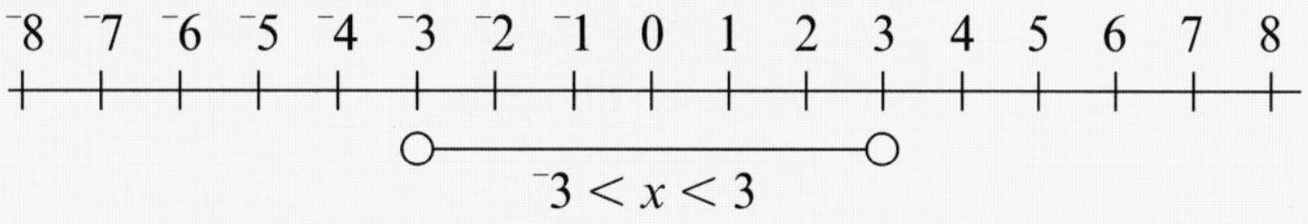


**Example 2** For which values of $x$ is $x^2 > 9$ true?

Here the inequality $x^2 > 9$ has two values outside of which $x$ must lie so $\mathbf{x < {}^-3}$ or $\mathbf{x > 3}$
This is not a continuous range but two separate parts.

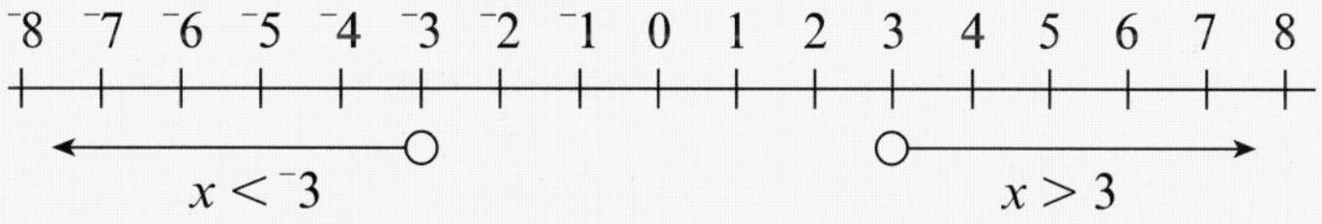


**Exercise 18.2 Inequalities with squared terms**

1 Solve each of these inequalities to give the range of values for the variable.

a $d^2 \geqslant 36$ b $17 > g^2$
c $a^2 + 7 \leqslant 56$ d $4 + 2h^2 > 36$
e $64 - 3t^2 < 16$ f $14 - 4c^2 \geqslant 6 - 2c^2$
g $3c^2 - 8 \geqslant 1 - 6c^2$ h $2(f^2 + 2) < 4(2f^2 - 5)$

2 What set of integers satisfies each of these inequalities?

a $k^2 \leqslant 36$ b $5 + b^2 < 30$
c $1 - 2d^2 > d^2 - 47$ d $5 \geqslant 4(x^2 - 5)$
e $7s^2 - 45 < 5(s^2 - 6.5)$ f $3(x^2 - 3) \geqslant 2(3x^2 - 7)$

3 Why is there no value of $h$ which satisfies $2h^2 + 40 < 0$?

4 Which of these inequalities has no solution?

a $34 + 2d^2 > 6$ b $34 - 2d^2 > 6$
c $34 + 2d^2 < 6$ d $34 - 2d^2 < 6$

5 What set of integers satisfies the inequality $14p - 4p^2 > 0$?

# Linear programming (An extension activity)

The optimum solution is the best possible solution which meets all the conditions imposed.
It is often the solution which gives the maximum profit.

Some mathematical problems can be described by a set of inequalities and the optimum solution found by examining regions on a graph.
This is known as linear programming.

## Example

A poultry keeper wants to keep ducks and turkeys in the same run.
The keeper wants fewer than 20 birds and no more than 4 turkeys.
She wants to spend £60 or more for a discount from her supplier.
At discount prices, a duck costs £3.60 and a turkey costs £4.80.
Let $d$ the number of ducks and $t$ be the number of turkeys.

**a** Write three different inequalities from the conditions above.
**b** Draw a graphs of $t$ against $d$ to show the inequalities.
**c** The profit from a duck is £3 and from a turkey is £7.
How many of each bird should she buy for maximum profit?

**a** For the number of turkeys: $t \leqslant 4$ ... (1)

For the total number of birds: $d + t < 20$ ... (2)

For the cost: $360d + 480t \geqslant 6000$
divide both sides by 120 $3d + 4t \geqslant 50$ ... (3)

**b** The regions for these inequalities can be drawn by looking at where each boundary line crosses the $t$ and $d$ axes.
For (2), when $d = 0$, $t = 20$ and when $t = 0$, $d = 20$
For (3), when $d = 0$, $t = 12.5$ and when $t = 0$, $d = 16.7$

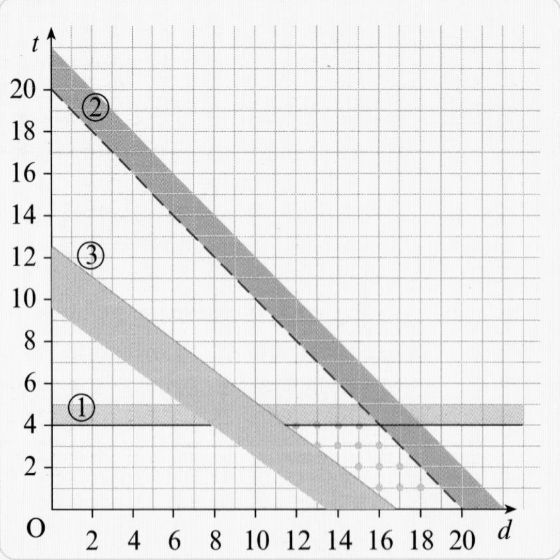


**c** At the point (19, 0) the profit is (19 × £3) + (0 × £7) = £57
At the point (17, 0) the profit is (17 × £3) + (0 × £7) = £51
At the point (15, 4) the profit is (15 × £3) + (4 × £7) = £73
At the point (12, 4) the profit is (12 × £3) + (4 × £7) = £64

**So 15 ducks and 4 turkeys gives the greatest profit.**

The largest or smallest values of the profit will lie at one vertex (corner) of the region which meets the other conditions. This means it is only necessary to test points nearest the vertices.

**Exercise 18.3**
Linear programming

1 For the turkey and duck problem:

**a** Why is the point (19, 1) not in the solution set, i.e. not shown as a gold dot?
**b** Why is the point (16.5, 2.5) not in the solution set?
**c** The point (13, 3) appears to be in the solution set.
How can you be certain that it is in the solution set?
**d** What profit is represented by the point (17, 2)?

2 Seascape Ferries have two types of ferry, Viking and Neptune.

They have 7 Vikings and 8 Neptunes in their fleet.
A Viking carries 80 cars and 3 lorries and a Neptune 50 cars and 6 lorries.
On one day Seascape must transport at least 600 cars and at least 45 lorries.

a If $v$ Vikings and $n$ Neptunes are used, explain why:

**i** for lorries: $3v + 6n \geqslant 45$ or $v + 2n \geqslant 15$ **ii** for cars: $8v + 5n \geqslant 60$

b Give two other inequalities which apply.
c Draw a graph of $v$ against $n$ and show the region which meets all the conditions.
Mark with dots all the possible combinations in the solution set.
d What is the smallest number of ferries that can be used?
How many of each type could be used?

3 Canfield Exhibition Hall hire out stands to exhibitors.
There are prominent stands by the entrance, each with an area of $8\,m^2$, and stands in the centre with an area of $10\,m^2$.
The charge is £280 a day for an entrance stand and £120 for a centre one.
The total area for all stands is $120\,m^2$ and the rental per day must be at least £1680.
The manager has calculated the ratio of centre stands to entrance stands.
Let $E$ be the number of entrance stands and $C$ the number of centre stands.

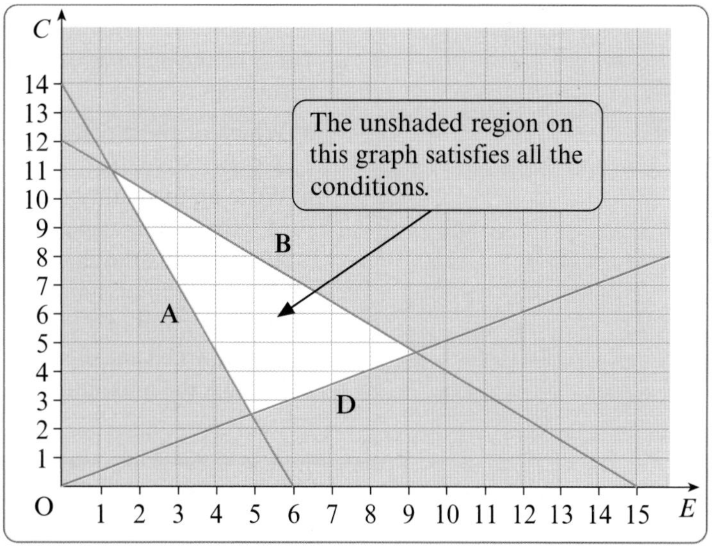


a From the information given, write an inequality which applies to:
**i** the stand areas **ii** the rental.
b What do you think are the equations of lines A and B on the graph?
c Line D shows the boundary of the ratio of large stands to entrance stands.
Write an inequality which describes this ratio.
d The number of visitors to the hall is in direct proportion to the number of stands. How many centre stands and entrance stands should there be for the maximum number of visitors?

## End points

| You should be able to ... | ... so try these questions |
|---|---|
| **A** Find integer values which satisfy an inequality | **A1** For each of these what integer values of $g$ satisfy the inequality? <br> **a** $^-2 < g < 0$ **b** $^-5 \leqslant g < 4$ <br> **c** $7 \geqslant g \geqslant 4$ **d** $^-56 \geqslant g > {}^-60$ <br> **e** $d^2 < 16$ **f** $4c^2 - 6 \geqslant 6c^2 - 32$ |
| **B** Solve an inequality to find the range of values for $x$ | **B1** Solve these inequalities. <br> **a** $5x \leqslant 30$ **b** $12 < 4f$ <br> **c** $4a + 5 \geqslant 53$ **d** $5k + 28 > 3$ <br> **e** $3(4s - 2) \leqslant 42$ **f** $7d - 10 > 5 - 3d$ <br> **g** $x^2 > 100$ **h** $2x^2 - 6 \leqslant 2$ |
| **C** Draw a region to illustrate where several inequalities are satisfied | **C1** On a graph with values of $x$ and $y$ between 0 and 12 show the region where the inequalities $y \geqslant x + 3$ and $y \leqslant 12 - 3x$ are both satisfied. |
| **D** Use linear programming to find an optimum solution to a problem | **D1** Cinema 2 can seat up to 200 people and the owners need takings per performance of at least £720 to make a profit. <br> The majority of seats are standard grade at £4 each and the remainder are luxury grade at £5 each. For different performances there are different numbers of standard and luxury seats, but there are never more than three times as many standard seats as luxury seats. <br> Let $S$ be the number of standard seats and $L$ the number of luxury seats. <br> **a** Which of these conditions fits the data given? <br> $S < L$  $S \leqslant L$  $S = L$  $S > L$  $S \geqslant L$ <br> **b** Write an inequality for: <br> **i** the seating capacity of the cinema <br> **ii** the seats that have to be sold to make a profit <br> **iii** the ratio of standard to luxury seats. <br> **c** Draw a graph of $L$ against $S$ with axes from 0 to 200 to show the region which satisfies all the conditions. <br> **d** What is the maximum possible takings on one performance? |

## Some points to remember

- Always check carefully to see if $\geqslant$ rather than $>$, or $\leqslant$ rather than $<$, is used.
- When you solve an inequality you can treat it like an equation except that:
  - if you multiply or divide both sides by a negative value you must reverse the signs.
- Often in linear programming you have to find the optimum solution to a further condition. In this case test the points nearest to the vertices of the region which matches the other conditions.

# SECTION 19 WORKING WITH PERCENTAGES

## Starting points

You need to know about ... ... so try these questions

### A Writing fractions as decimals and percentages

- To write a fraction as a decimal:
  - divide the numerator by the denominator.

$\frac{3}{8} = 3 \div 8 = 0.375 = 37.5\%$ $\frac{5}{6} = 5 \div 6 = 0.8\dot{3} = 83.\dot{3}\%$

**A1** Write each of these as a decimal and as a percentage:

**a** $\frac{3}{4}$ **b** $\frac{5}{8}$ **c** $\frac{8}{5}$ **d** $\frac{5}{4}$

**e** $\frac{6}{5}$ **f** $\frac{13}{20}$ **g** $\frac{8}{32}$ **h** $\frac{19}{6}$

### B Calculating a percentage of a given amount

To calculate a percentage of a given amount:

- Multiply the amount by the appropriate decimal.

  For instance, to calculate 65% of 420 kg

  $420 \times 0.65 = 273$

  **So 65% of 420 kg is 273 kg**

**B1** Calculate:

- **a** 55% of 650 miles
- **b** 72.5% of £400
- **c** 46% of 3450
- **d** 28% of £415
- **e** 4.5% of 360 metres
- **f** 38% of £4.50
- **g** 0.5% of £25
- **h** 8% of 3 miles.

### C Writing one number as a percentage of another

Compare the numbers 18 and 36:

- 18 is half of 36 or $18 \div 36 = 0.5$
- 18 is 50% of 36 or $0.5 \times 100 = 50\%$

As a single calculation: $(18 \div 36) \times 100 = 50\%$

This method can be used with any two numbers $p$ and $t$

To **calculate $p$ as a percentage of $t$: $(p \div t) \times 100$**

**Example**

Jenny and Bruce walked from John O'Groats to Lands End, a distance of 868 miles. By the end of Day 4 they had walked 113 miles. What percentage of the total distance is this?

To calculate 113 as a percentage of 868

$(113 \div 868) \times 100 = 13.018 \ldots$

They had travelled 13.0% (1 dp) of the total distance.

**C1** Calculate, giving your answer correct to 2 dp:

- **a** 52 as a percentage of 80
- **b** 85 as a percentage of 184
- **c** 1.5 as a percentage of 12
- **d** 15 as a percentage of 12
- **e** 0.75 as a percentage of 25.

**C2** Of a total bill for £65.70, only £9.20 of this was for parts.

Calculate the charge for parts as a percentage of the total bill (correct to 2 sf).

### D Interpreting a calculator display in calculations with money

Interpreting a calculator display correctly, is important.
When you calculate 24% of £15 the display shows:

The calculation is in pounds so the answer is **£3.60**

When you calculate 24% if £1.50 the display shows:

The calculation is in pounds so the answer is **£0.36**
It is more likely that the answer will be given as **36 pence**.

**D1** In each of these, interpret your calculator display appropriately.

- **a** Calculate 16% of £85 (answer in pounds)
- **b** Calculate 14% of 4 cm (answer in millimetres).
- **c** Find 44% of 3.5 kg (answer in grams).
- **d** Find 9% of £2.27 (answer in pence).
- **e** What is 78% of 65 mm:
  - **i** in millimetres
  - **ii** in centimetres?

**Thinking ahead to ...
percentage changes**

Often percentages are used to describe a change in an amount.
For example, in supermarkets special offers can be shown as:

**Extra 15% FREE !**

For the customer:

**A** How much extra do you get free?
Is it more or less than a quarter of the amount for the normal price?

**B** What fraction would you use to describe the extra free amount?

**C** How many ml of the product are in the special offer pack?

## Percentage changes

For an increase by a certain percentage, a final value can be found in this way.

**Example**
Toothpaste is sold in tubes containing 150 ml.
In a special offer, 12% extra toothpaste is in the tube for the same price.
How much toothpaste is in the special offer tube?

> Think of the toothpaste in this way: 100% of the contents is 150 ml.
>
> The special offer tube has 12% extra, so it must contain: 112% of 150 ml.
>
> 112 % as a decimal is 1.12
>
> To calculate 112% of 150: $150 \times 1.12 = 168$
>
> The special offer tube contains **168 ml** of toothpaste.

Similarly you can decrease by a given percentage.

**Example**
A fast food store decided to decrease by 18% the weight of packaging for their regular meals, which weighed 40 grams.
Calculate the weight of the new packaging.

> Think of the packaging in this way: 100% of the contents weighs 40 g.
>
> The new packaging weighs 18% less, so it must weigh 82% of the 40 grams.
>
> 82% as a decimal is 0.82
>
> To calculate 82% of 40: $40 \times 0.82 = 32.8$
>
> The new regular meal packaging weighs **32.8 grams.**

**Exercise 19.1
Percentage changes**

**1** In a special offer, the 440 gram of coffee in a jar is to be increased by 15%.
- **a** What percentage of the 440 grams of coffee are in the special offer jar?
- **b** Calculate the amount of coffee in the special offer jar.

**2**
- **a** Increase 350 kg by 18%.
- **b** Increase 447 km by 16%.
- **c** Increase 4.50 metres by 22%.
- **d** Increase £35 000 by 8%.
- **e** Increase £27.50 by 32%.
- **f** Increase 25800 tonnes by 17%.
- **g** Increase 3 500 000 litres by 9%.
- **h** Increase 0.6 cm by 15%.

Depreciate means to lose value.

3 A manufacturer buys a machine for £56 800. The machine is expected to depreciate by 12% in the first year and by 8.5% each future year.
What will be the value of the machine after ten years to the nearest £1?

4 Baz paid £2685 for a new keyboard. The salesman told him to expect that, at the end of every year, the value of the keyboard would fall by 15% of its value at the start of that year.

What do you expect to be the value of the keyboard at the end of five years?

5 Last year a company spent £3.4 million on advertising. Next year they have set a target of saving 3.5% on advertising spending.

Next year, how much do they expect to spend on advertising?

6 In 1997 Wyvern Water lost water from their underground pipes at the rate of 600 litres per minute. Between 1998 and the end of 2004 they have a programme of repairs that will save, on average, 4.8% per year, on water loss.

In total, how much water do Wyvern Water expect to lose in 2005?

7 Each year for five years, a company plans to use 20% less paper.
Will they use no paper in five years time? Explain your answer.

## Percentages and VAT

To calculate the total price of goods or services including VAT is the same as: increasing the cost price by the percentage VAT.

**Example** A garden shed is advertised for: **£114.99** + VAT at 17.5%
Calculate the total charge for the shed.

This example is with VAT at 17.5%.

In all questions that involve VAT you must use the standard rate of VAT at the time.

If you are unsure, ask for the rate of VAT.

Think of the total charge (cost price + VAT) for the shed in this way:

The total charge for the shed is: **117.5%** of its cost price

To calculate 117.5% of £114.99:

$$114.99 \times 1.175 = 135.113$$

The total charge for the shed (including VAT) is:

**£135.11** (to the nearest penny)

**Exercise 19.2**
**Calculating with VAT**

1 The cost of these items is given without VAT (ex VAT).
Calculate the charge, including VAT, for each item.

| | | |
|---|---|---|
| **a** camera £16 | **b** trainers £44.25 | **c** bike £185 |
| **d** toaster £19.40 | **e** pen 72 pence | **f** TV £368.42 |
| **g** fridge £262 | **h** mower £24.55 | **i** CD £9.35 |
| **j** video £188.70 | **k** battery 34 pence | **l** pencil 11 pence |

Traders have to charge VAT. The VAT is then paid by the traders to the Customs and Excise.

These payments are made every three months.

2 To hire a coach for a hockey tour cost £675 + VAT.
The cost of the coach was shared equally by the 23 people on the tour.

Calculate to the nearest fifty pence how much each person on the tour had to pay for the hire of the coach.

3 Fraser needs four new tyres for his car. He has a maximum of £150 in total to spend on the tyres. All tyre prices are given ex VAT.

To the nearest 10 pence, what is the most Fraser can pay for a tyre ex VAT?

## Reverse percentages

When you divide by 117.5, you are calculating 1%.

You then multiply by 100% to calculate the 100%.

This is for VAT at 17.5%

Check on the current rate of VAT to find the number to divide by to calculate 1%.

Calculating the original value of something, before an increase or decrease took place, is called 'calculating a reverse percentage'.

**Example** The total price of a bike (inc. VAT at 17.5%) is £146.85
Calculate the cost price of the bike ex VAT.

The total price of the bike = 117.5% of the ex VAT price
$146.85 = 117.5 \times$ ex VAT price
$146.85 \div 1.175 =$ the ex VAT price
$124.978... =$ the ex VAT price

**The price of the bike ex VAT is £124.98** (to the nearest penny)

**Exercise 19.3**
**Calculating reverse percentages**

**1** These prices include VAT, calculate each price ex VAT.
Give your answers correct to the nearest penny.

**a** CD player £135.50 **b** camera £34.99 **c** ring £74.99
**d** trainers £65.80 **e** TV £186.75 **f** phone £14.49
**g** calculator £49.99 **h** PC £799.98 **i** tent £98.99

**2** Ella bought a pair of climbing boots for £45.60 in a sale that gave '20% off!'.

**a** What was the non-sale price of the boots?
**b** How much did Ella save buying the boots in the sale?

**3** When Mike sold his bike for £35, he said he made a profit of about 45%.
To the nearest pound, how much did Mike pay for the bike?

**4** A 675 gram box of cereal is said to hold 35% more than the regular box.
How many grams of cereal are in the regular box?

**5** The Bay Point Hotel charges £65.45 per person per night for bed and breakfast. This charge includes VAT.
How much does the hotel take per person per night, at this charge?

**6** In 1997 Reynards Fast Foods made a profit of £146 455.
The profits for 1997 were 6.5% above the profits made in 1996.
Calculate the profit made in 1996, correct to the nearest £5.

**7** A light-bulb manufacturer makes clear and pearl light-bulbs.
Each week they make 45 360 clear bulbs, which is 64% of the total production of the bulbs made in a week.

**a** What is the total number of bulbs produced in a week?
**b** How many pearl bulbs are produced in a week?
**c** For each week give the ratio in its lowest terms, of clear bulbs made : pearl bulbs made.

**8** Asif and Nina make and sell vegetarian snack foods. They sell at prices to give them a profit of 315%. Last year their total sales were £26 855.

**a** To the nearest pound, how much profit did they make last year?
**b** Write a formula that they might use to calculate their profit ($p$) from their total sales ($s$).

**9** In 1996, a total of 126 788 people visited the Butterfly Sanctuary.
This was estimated to be 14% fewer visitors than 1995.
Give an estimate of the total number of visitors in 1995.
Explain the degree of accuracy you worked to.

**10** Write a formula to calculate the cost of an item without VAT ($c$) from the price of the item with VAT ($p$).

# Percentages and interest

**Interest** is explained in the *Oxford Mathematics Study Dictionary* in this way:

Interest The interest is the amount of extra money paid in return for having the use of someone else's money.

When you borrow, or save money with a Bank, Building Society, the Post Office, a Credit Union, or a finance company, interest is *charged* or *paid.*

Interest is: *charged* on money you borrow and *paid* on money you save.

Interest is: *charged* or *paid* pa, at a fixed rate, e.g. 6% pa.

pa means *per annum*, or each year.

**Simple interest** is a type of interest not often used these days.

- Simple interest is fixed to the sum of money you originally borrow or save.

Simple *interest* can be calculated using this *simple interest formula*:

In words, the formula is:

$I = P \times R \times T$ ***I***nterest = ***P***rincipal × ***R***ate × ***T***ime

***P***rincipal is the amount of money you borrow or save.
***R***ate is the interest rate pa. **as a decimal**.
***T***ime is for how long, usually in years.

**Example**
Calculate the interest charged on a loan of £750, for 4 years at 7% pa.
Using the formula gives:

$I = P \times R \times T$, with $P = 750$, $R = 0.07$ (7% = 0.07) and $T = 4$
$I = 750 \times 0.07 \times 4$
$I = 210$

**The interest charged on this loan is £210.**

(If the £750 had been saved for 4 years at 7% pa,
**the interest paid on the savings would be £210.**)

## Exercise 19.4 Calculating with simple interest

1 Calculate the simple interest charged, or paid for each of these:

**a** £450 borrowed for 6 years with interest at 12% pa
**b** £280 saved for 8 years with a rate of interest of 3% pa
**c** £1500 saved for 15 years at 6% pa interest
**d** £6000 borrowed for 10 years at an interest rate of 17% pa

2 To buy new kit a band borrows £3500 over 10 years at 17% interest pa.

**a** Calculate the amount of interest paid on the loan.
**b** At the end of the loan, how much in total will have been paid for the kit?

3 A new bridge will cost an estimated £44 million, and take six years to build. If, at the start, all £44 million is borrowed at a simple interest rate of 8.5%, estimate the total cost of the bridge.

4 A sum of £350 was saved for six years. Over this time the simple interest earned was a total of £96.60. Calculate the rate of interest paid.

# Compound interest

Compound interest is used by banks, building societies, credit card companies and high street shops.

Compound interest is either *charged* or *paid*, but not just on the original sum.

**Example**

Calculate the interest on £100 saved for 2 years at 4%.

- At the end of year 1, the total is £104 (£100 + £4 interest)
- At the end of year 2, the total is £108.16 (£104 + £4.16 interest)

In short, compound interest includes interest on interest already *paid* or *charged*.

Working year-by-year, you can calculate compound interest and see that it builds to a greater total than simple interest over the same period of time.

**Example**

Calculate the interest on £480 saved for 3 years at 7%.

- Interest paid at the end of year 1 is: **£33.60** (£480 × 0.07)
  Interest for year 2 will be calculated on £513.60 (£480 + £33.60)
- Interest paid is: **£35.95** (£513.60 × 0.07)
  Interest for year 3 will be calculated on £549.55 (£513.60 + £35.95)
- Interest paid is: **£38.47**
  **The total interest paid is:** £33.60 + £35.95 + £38.47 = **£108.02**

**Exercise 19.5 Calculating with compound rates**

1 Calculate the total interest paid on these savings at compound interest:

**a** £350 for 3 years at 9% pa
**b** £1400 for 4 years at 3% pa
**c** £3600 for 5 years at 6% pa
**d** £12 250 for 3 years at 4% pa
**e** £4050 for 2 years at 12% pa
**f** £35 250 for 2 years at 9.5% pa
**g** £565 for 3 years at 10.5% pa
**h** £185 for 4 years at 0.3% pa
**i** £45 000 for 4 years at 7.2% pa
**j** £1224 for 2 years at 2.3% pa.

2 Shelley won £5000, and put it in a savings scheme for five years. The savings scheme pays interest at 6.8% pa compound.

**a** Calculate the total interest paid on her savings.
**b** At the end of five years what was the total in Shelley's saving scheme?

3 Roughly how long must a sum of money be saved at 8% pa compound, for the value of the savings to double? Explain your answer.

4 On average, the price of new cars has increased by 3.8% pa over the last five years. At this rate, what would you expect a model that cost £18 955 five years ago to be priced at today to the nearest £1?

5 The world population of a species of bird is estimated to be decreasing at the rate of 18% pa. There were estimated to be 4200 birds in 1995.

**a** If the rate of decrease remains the same, estimate the year in which the species would become extinct. Explain your answer.

In 1995 a programme was put in place to halt the decrease, and increase the world population of the species by 4% pa.

**b** If this rate remains the same, when would you expect the population of this species to reach 6000? Explain your answer.

Check that your formula works by using it with Questions **2**, **3**, and **4** above.

6 The sum of £$A$ is saved in a scheme that pays interest at $r$% pa compound. Write, and simplify, an expression for the total vaule of the savings at the end of five years.

Here the rate of interest is:

given as a percentage but used as a **decimal** for calculations.

There is a compound interest formula.

- For a principal of £150, invested at a rate of 6% for a period of 4 years

At the end of year 1, the value of the principal is:

$$150 \times 1.06$$

At the end of year 2, the value of the principal is:

$$150 \times 1.06 \times 1.06 \quad \text{or} \quad 150 \times 1.06^2$$

At the end of year 3:

$$150 \times 1.06^2 \times 1.06 \quad \text{or} \quad 150 \times 1.06^3$$

At the end of year 4:

$$150 \times 1.06^3 \times 1.06 \quad \text{or} \quad 150 \times 1.06^4$$

The total for the interest £$I$ is therefore: $I = (150 \times 1.06^4) - 150$

For any principal, a general formula can be used. This is based on:

Interest £$I$, on a principal sum £$P$, invested at $r$%, for $n$ years.

The compound interest can be found by using this formula:

$$I = [P \times \left(1 + \frac{r}{100}\right)^n] - P$$

## Exercise 19.6 Using the compound interest formula

1 Calculate the total interest paid on a principal of £8500 invested for twelve years at a rate of 4% pa compound.

2 W&J claim that the value of their Unit Trust is likely to grow by 8.5% pa compound for the next five years.

If you invest £15 250 with the W&J Unit Trust, what would you expect your investment to be worth at the end of five years?

3 Two sisters, Rhian and Celine, were each left £2500 in a will.

Rhian decided to spend £5 per week on the National Lottery, and at the end of three years she had won a total of £1216.50.

Celine decided to invest her share at an interest rate of 7.5% pa compound.

Who do you think made the best decision? Explain your answer.

4 A building society offered the following terms for a minimum saving of £25 000 over a 10-year period.

For the first three years interest is paid at 5.8% pa compound
For the next four years interest is paid at 6.6% pa
For the remaining time interest is paid at 7.2% pa.

At the end of ten years, what will be the value of a £25 000 investment?

5 A village is given a grant of £185 000 to build a new village hall. Once the grant is given it cannot be invested for more than five years.

The village hall committee decide to invest the grant in total for five years, at the end of which they would need **at least** £232 000 in the account.

**a** What is the lowest rate of interest (1 dp) they need for this investment? Explain your answer.

**b** How will your figures change for a maximum period of four years?

Problems involving compound growth or reduction such as Question 6 can be calculated in a similar way to compound interest.

6 There are 436 rabbits in a colony and their population increases by 8% per year. What will be the size of the colony after 16 years?

## Buying on credit

Here the deposit is an amount you must pay in cash.

The deposit paid can be an amount of money (£30), or a percentage of the full price, e.g. 20% deposit.

There are rules for buying on credit. These can be changed by governments, and it may be possible to buy on credit with no deposit.

If you buy on credit, you pay a deposit on purchase and the remainder you pay off with a loan.
The total for the loan is paid off in a number of equal *instalments*.

### Example

This advert is for a colour television.

**only £179.99**

*CREDIT PLAN:* **You pay £20 deposit the balance at 23% compound interest in 24 equal instalments (2 years)**

Calculate the monthly instalment for this credit plan.

The amount of the loan is £159.99 (i.e. £179.99 – £20 deposit)
With interest the total to be repaid on the loan is £242.05 (£159.99 $\times$ $1.23^2$)
With 24 instalments, each instalment = £242.05 ÷ 24
= £10.085 ...
**The monthly instalment will be £10.09** (to the nearest penny)

**Exercise 19.7**
**Buying on credit**

1 With the same Credit Plan as for the television above, calculate the monthly instalment for a washing machine costing £399.99 over three years.

2 Eric bought a camcorder advertised for £469.99.
He paid no deposit, was charged interest at 26% pa compound, and had the loan over 4 years (48 equal instalments).
**a** Calculate the total Eric will pay for the camcorder.
**b** Calculate the monthly instalment.

3 Dinah decides to buy a second-hand car priced at £6995.
She is allowed £1150 for her old car and takes out a loan for the difference.
The interest rate is 13.5% pa compound, and the loan is over 30 months.
**a** Calculate her monthly instalment.
**b** How much, in total, will this *new* car have cost Dinah?

4 An extension to an hotel has to be carpeted.
The total floor area is $586\text{m}^2$, and the carpet costs £34.99 ex VAT per $\text{m}^2$.
The hotel pays a deposit of £8 500, and borrows the remainder of the money over five years at an interest rate of 14.2% pa compound.
**a** How much in total will the hotel pay for the carpet?
**b** What will the hotel have paid for each square metre of carpet?

5 Copy and complete this repayment table for up to 60 months.

**Repayment table Interest rate 9.6% pa (compound)**

| Amount borrowed | Repayments (months) 12 | 24 | 36 | 48 | 60 |
|---|---|---|---|---|---|
| £100 | | | | | |
| £500 | | | | | |
| £1000 | | | | | |
| £2000 | | | | | |
| £5000 | | | | | |
| £10000 | | | | | |

## End points

| You should be able to ... | ... so try these questions |
|---|---|
| **A** Calculate percentage changes | **A1** Increase 480 km by 16%.<br>**A2** A CJ regular size cola is 380 ml. CJ decide to increase the size of their regular cola by 12%. To the nearest millilitre, give the size of the larger regular cola.<br>**A3** Decrease £55.80 by 6%. Give your answer to the nearest penny.<br>**A4** In 1992, there were a total of 42 154 breakdowns on a motorway section. In 1993 the total number of breakdowns fell by an estimated 7%. Estimate the number of breakdowns in 1993, on this motorway section. |
| **B** Work with VAT | **B1** A fridge/freezer is advertised for £849.99 ex VAT. Calculate the price of the fridge/freezer with VAT.<br>**B2** To the nearest penny, the VAT (at 17.5%) added to a bill was £84.88. Calculate the total for the bill including VAT. |
| **C** Calculate reverse percentages | **C1** In 1996 Buslinks *UK* made a profit of £2.42 million pounds. This was a increase of 6.5% on the profits for 1995. Calculate the profits for 1995 correct to 4 sf.<br>**C2** Jetstream have been providing charter flights to Spain for three years. In their third year they flew a total of 425 620 passengers to Spain. This was an increase of 8% on their second year, which was an increase of 3.5% on their first year. Estimate the total number of passengers they flew to Spain in their first year. |
| **D** Calculate simple interest | **D1** Calculate the simple interest on £235 invested at 4.5% for 7 years.<br>**D2** Asif had savings of £650 for a period of 5 years. He was paid simple interest, and at the end of the 5 years he had £796.25 in the account. Calculate the rate of interest he was paid. |
| **E** Calculate compound interest | **E1** Calculate the interest paid on a principal of £45 250 invested for 6 years at 4.25% pa compound. |
| **F** Calculate with buying on credit | **F1** Lauren buys an electric guitar priced at £499.95. She pays £100 and pays off the rest at 18% p.a. compound interest, with monthly instalments over 2 years.<br>**a** How much are the monthly instalments?<br>**b** How much will Lauren have paid in total for her guitar? |

## Some points to remember

- Working with a calculator it is easier if you think of, and use, percentages as decimals.
- Check that your answer to a calculation is of the right size, and to a sensible degree of accuracy.
- When you are calculating interest, paid or charged, make sure you know whether it is at a rate of simple interest, or compound interest.

## Starting points

You need to know about ...

### A The terminology for parts of a circle

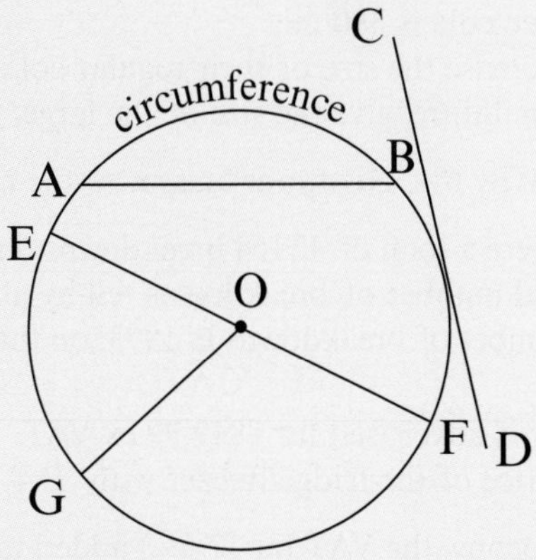


In this circle with centre O:

| | |
|---|---|
| AB is a **chord** | A chord is a straight line that joins two points on the circumference of a circle. |
| CD is a **tangent** | A tangent to a circle is a straight line which touches the circle at one point only. |
| EF is a **diameter** | A diameter is a chord that passes through the centre. |
| **Circumference** | The circumference is the distance measured around the curved edge of a circle. |
| OG is a **radius** | A radius is any straight line from the centre to the circumference. |

The curved part of the circle between A and B is known as an **arc**.
There are two **arcs** AB:
the larger is the **major arc**
the smaller is the **minor arc**.

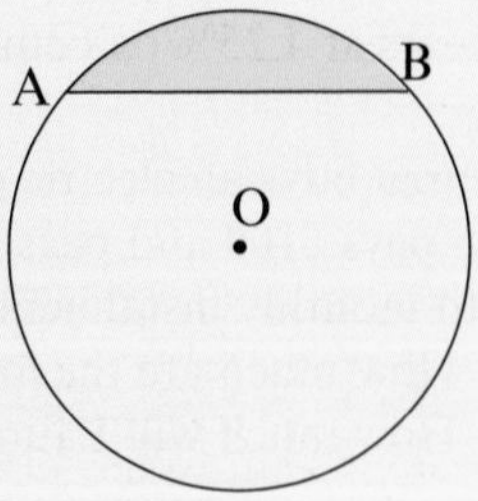


The chord AB divides the circle into two **segments**.
There are two segments created by the chord AB: the **major segment** and the **minor segment**.

AB, PR, and GH are all chords.
The **perpendicular bisector** of each chord has been drawn to show that:
**The perpendicular bisector of any chord passes through the centre of its circle.**

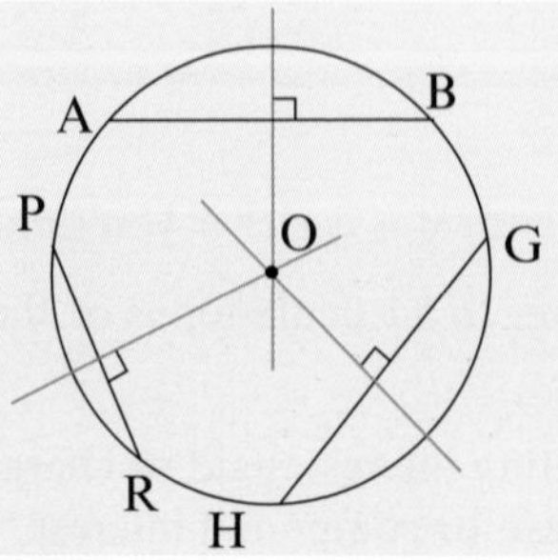


... so try these questions

**A1**
- Draw a circle, centre O with a radius of 4 cm.
- Mark any point P on your page 8 cm from O.
- From P draw two tangents to the circle.
- Label the point where one tangent touches the circle A, and where the other tangent touches B.

**a** Measure the distance PA and PB.
**b** What do you notice?
**c** Is this the case for other points P?
- Join PO.

**d** Measure angles APO and BPO.
**e** What do you notice?
**f** Is this always the case for other points P?
- Draw in the lines OA and OB.

**g** Measure angles OAP and OBP.
**h** What do you notice?
**i** Is this always the case?
- Draw in the chord AB.

**j** Is OP the perpendicular bisector of AB?
Give reasons for your answer.

**A2** Draw any circle centre O.
The chord AB is such that the ratio
minor arc AB : major arc AB
is 1 : 2.
**a** Mark a position for A and B on your circle.
**b** Explain how you were able to fix points for A and B.

**A3 a** Draw around a circular object. By drawing and bisecting chords find the centre of the circle you have drawn.
**b** Two straight lines meet at right angles at B so that AB = 4 cm and BC = 6 cm. AB and BC are chords of the same circle, centre O. Find the distances OA, OB and OC.

## The angle subtended at the centre

For a chord AB and any point C on the circumference of a circle, joining A to C, and B to C gives angle ACB.

Angle ACB is said to be subtended by the chord AB.

In this circle, centre O, AB is a chord and C is a point on the circumference.
Angle ACB is:

- subtended by the chord AB
- at the circumference of the circle.

Angle AOB is:

- subtended by the chord AB
- at the centre of the circle.

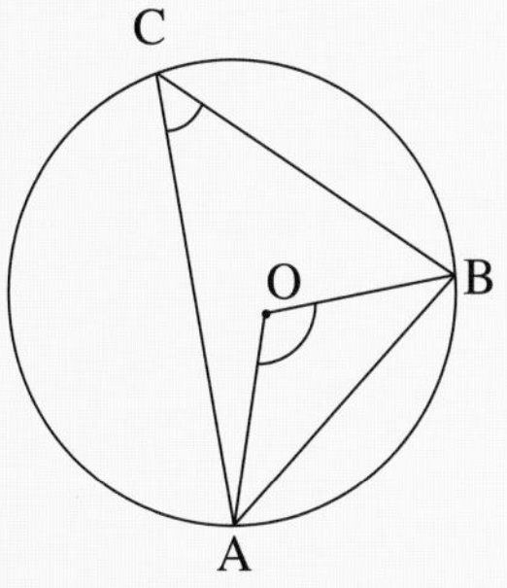


A geometrical property of the circle is that:
the angle AOB is twice the angle ACB ($A\hat{O}B = 2 \times A\hat{C}B$).

The general statement is:
**For the same chord, the angle subtended at the centre is twice the angle subtended at the circumference in the major segment.**

When the angle subtended at the centre is in the minor segment:

The general statement is:
**For the same chord, the angle subtended at the circumference is half the reflex angle subtended at the centre.**

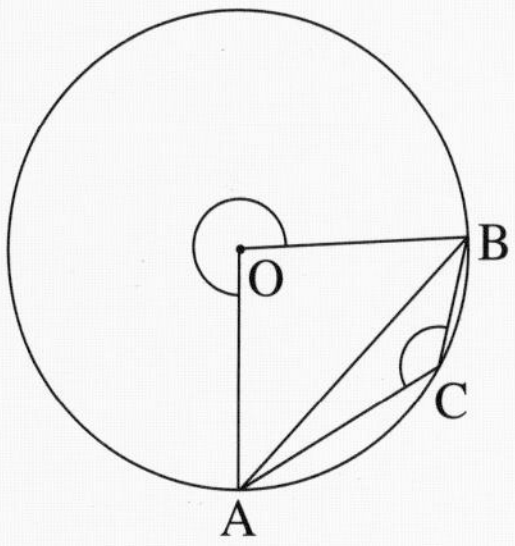


**Proof:**

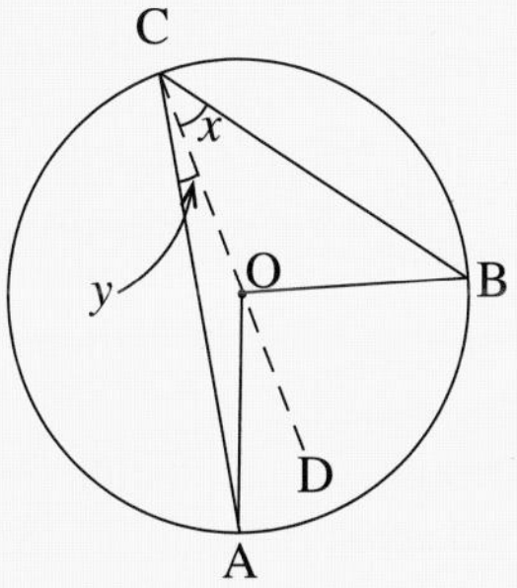


When you show a proof or answer questions on circle theorems you need to give reasons at each stage – see the statements in brackets here.

let $B\hat{C}O = x$ and $A\hat{C}O = y$

| | |
|---|---|
| In $\triangle$ OCB, $O\hat{C}B = x$ | **(isoscoles $\triangle$)** |
| So $B\hat{O}C = (180° - 2x)$ | **(angle of sum $\triangle$)** |
| In $\triangle$ OCA, $O\hat{A}C = y$ | **(isoscoles $\triangle$)** |
| So $A\hat{O}C = (180° - 2y)$ | **(angle sum of $\triangle$)** |

Angle $B\hat{O}D = 180° - B\hat{O}C$ **(angle sum of straight line)**
$= 180° - (180° - 2x)$
$= 2x$

Angle $A\hat{O}D = 180° - A\hat{O}C$ **(angle sum of straight line)**
$= 180° - (180° - 2y)$
$= 2y$

So $A\hat{O}B = 2x + 2y$
Also $A\hat{C}B = x + y$
So $\mathbf{A\hat{O}B = 2 \times A\hat{C}B}$

## Angles in the same segment

In this circle, AB is a chord. The points C, D, and E are other points on the circle.

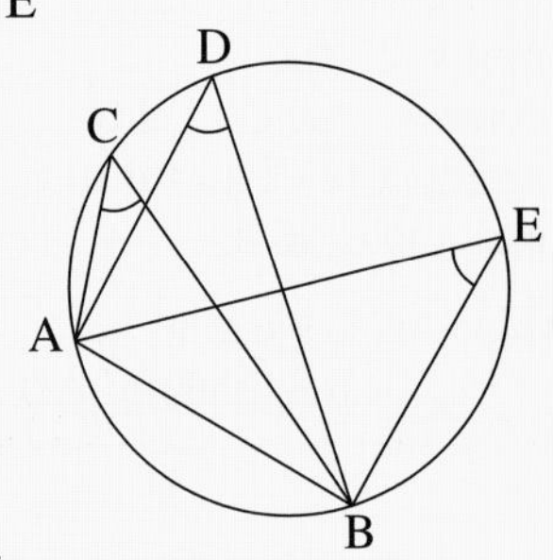


The angles ACB, ADB, and AEB are all:

- subtended by the chord AB
- at the circumference of the circle
- in the same segment (in this case the major segment).

A geometrical property of the circle is that:
the angles ACB, ADB, and AEB are all equal in size.
The general statement is:
**Angles subtended at the circumference by the same chord in the same segment are equal.**

**Proof:**

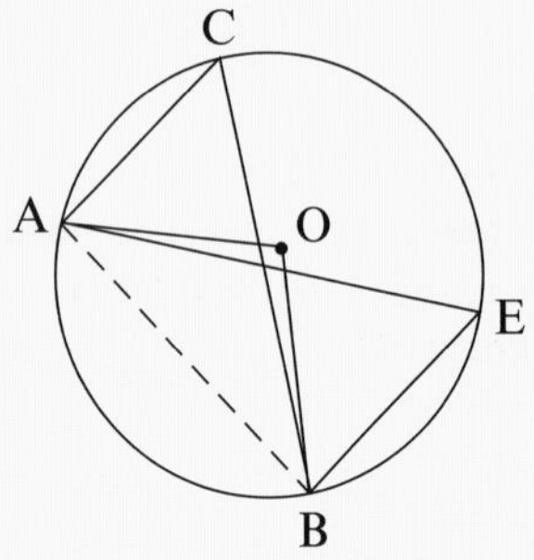


O marks the centre of the circle.

$A\hat{O}B = 2 \times A\hat{C}B$ **(angle at the centre is 2 × angle at circum.)**

Also $A\hat{O}B = 2 \times A\hat{E}B$
So $2 \times A\hat{C}B = 2 \times A\hat{E}B$
So $\mathbf{A\hat{C}B = A\hat{E}B}$
Similarly $A\hat{C}B = A\hat{D}B$

## Angles in a semicircle

When a chord becomes a diameter, then each segment is a semicircle. The angle subtended at the circumference in a semicircle is a right angle.

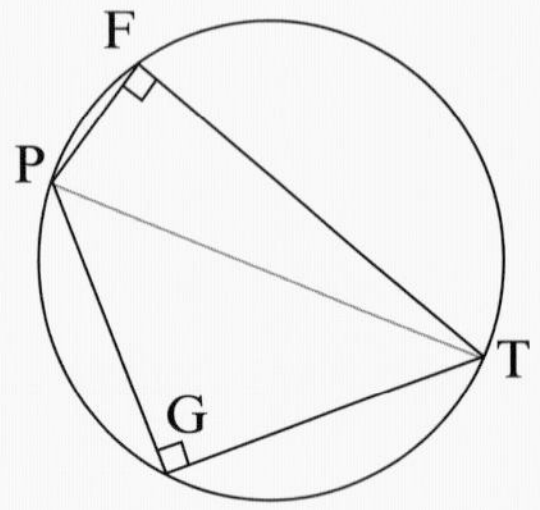


In this case $P\hat{G}T = P\hat{F}T = 90°$

The general statement is:
**Any angle subtended at the circumference in a semicircle is a right angle.**

**Proof:**

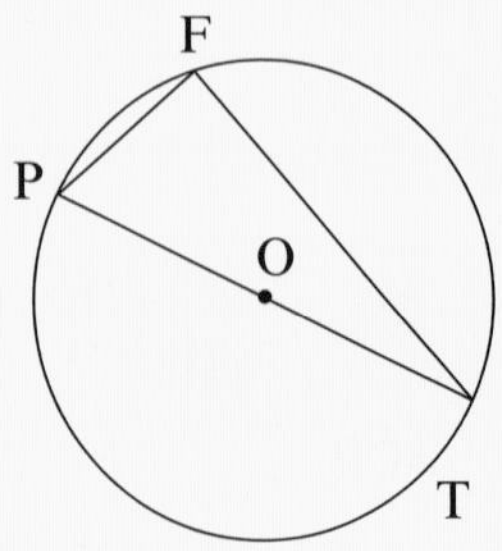


PT is a diameter and passes through the centre O.

$P\hat{O}T = 180°$ **(straight line)**

$P\hat{F}T = \frac{1}{2}P\hat{O}T$ **(angle at centre = 2 × angle at circum.)**

So $\mathbf{P\hat{F}T = 90°}$

## Secant

Any straight line that cuts across a circle at two points is called a **secant**.

A tangent is a special case of a secant as the two cuts become a single point of contact.

The two cuts are said to become **coincident**, giving a single point.

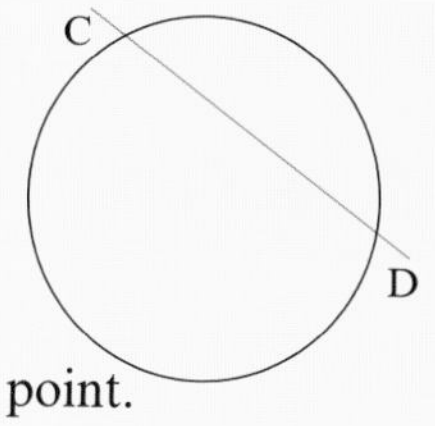


## Tangents to a circle

**From any point P, outside the circle, two tangents to the circle can be drawn; the distance from P to the point of contact being the same for each tangent,**
i.e. PS = PT.

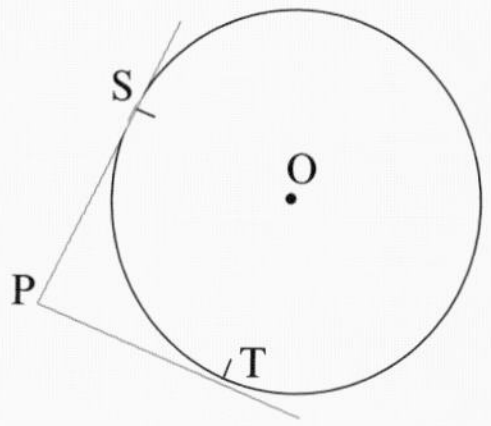


**The angle between a radius, drawn to the point of contact of a tangent, and the tangent is a right angle,**
i.e. $P\hat{S}O = P\hat{T}O = 90°$.

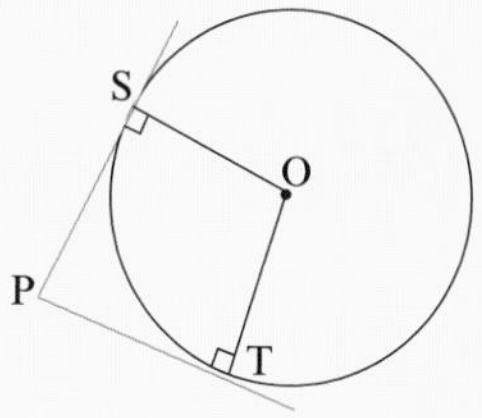


Any chord (ST) drawn in a circle creates two segments. One of these two segments can be described as being **alternate** to the other.

When a tangent (PT) is drawn at one end of the chord, then an angle is created between the tangent and the chord ($P\hat{T}S$).

This angle, measured on one side of the chord, is equal to the angle in the alternate segment ($S\hat{C}T$).

In this case $P\hat{T}S = S\hat{C}T$

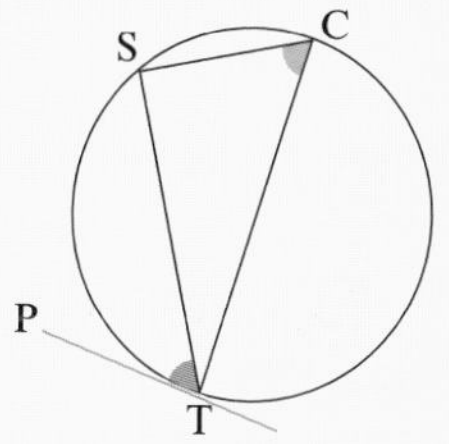


The general statement is:

**The angle between a chord and the tangent, at its point of contact is equal to the angle subtended at the circumference in the alternate segment.**

**Proof:**

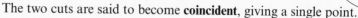

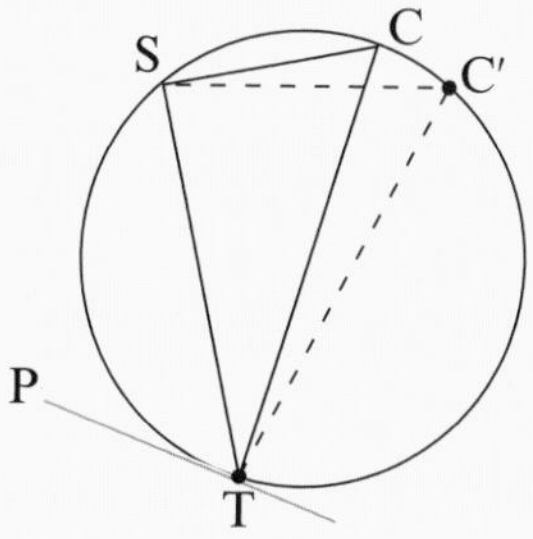


Let CT be the diameter.

$T\hat{S}C = 90°$ **(angles in a semicircle)**

So $S\hat{C}T + S\hat{T}C = 90°$ **(angle of sum △)**

Also $P\hat{T}C = 90°$ **(tangents meet diameter at 90°)**

So $P\hat{T}S + S\hat{T}C = 90°$

So $\mathbf{S\hat{C}T = P\hat{T}S}$

For the general case when CT is not the diameter, imagine point C moves to the new position C′.

$S\hat{C'}T = S\hat{C}T$ since angles in the same segment are equal.

**So $P\hat{T}S$ always equals $S\hat{C}T$ for all positions of C in that segment.**

## Using circle theorems

**Example 1**

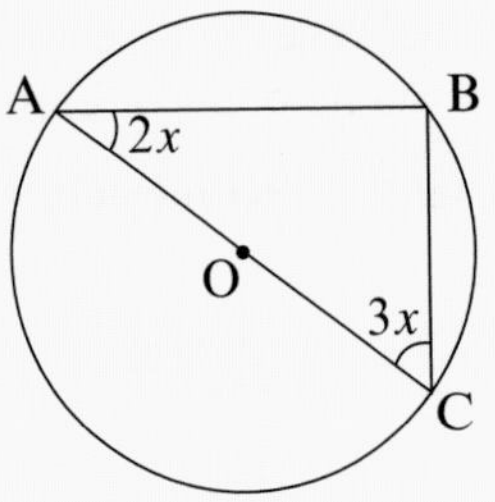


AC is a diameter and O is the centre.
Find the value of $x$.

$\hat{ABC} = 90°$ **(angle in a semicircle)**
So $2x + 3x = 90°$ **(angle sum of △)**
$5x = 90°$
$x = 18°$

**Example 2**

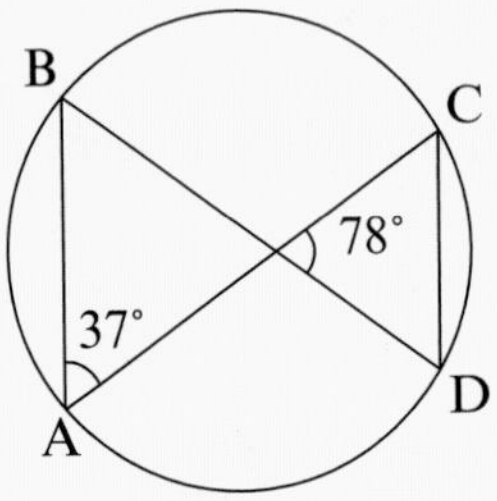


Calculate $C\hat{D}B$ and $A\hat{C}D$.

$C\hat{D}B = 37°$ **($C\hat{D}B$ and $C\hat{A}B$ are in the same segment on chord BC)**

$A\hat{C}D = 180° - 37° - 78°$ **(angle sum of △)**
$= 65°$

**Example 3**

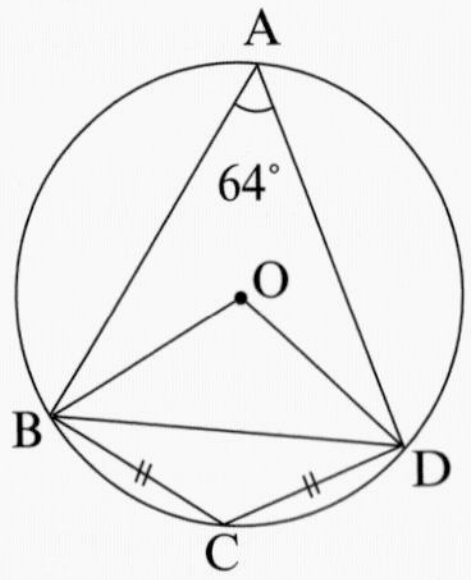


Calculate $D\hat{B}C$.

$B\hat{O}D = 2 \times 64°$ **(angle at centre)**
$= 128°$
So reflex $B\hat{O}D = 360° - 128°$
$= 232°$

$B\hat{C}D = \frac{1}{2}$ of reflex $B\hat{O}D$ **(angle at centre)**
$= \frac{1}{2} \times 232 = 116°$

Triangle BCD is isoscele so $D\hat{B}C = B\hat{D}C$.
So $D\hat{B}C = \frac{1}{2}(180 - 116)$ **(angle sum of △)**
$D\hat{B}C = 32°$

**Exercise 20.1**
**Using circle theorems**

**1** Find the angles shown by letters in each diagram giving your reasons. (In each case O is the centre.)

**a**

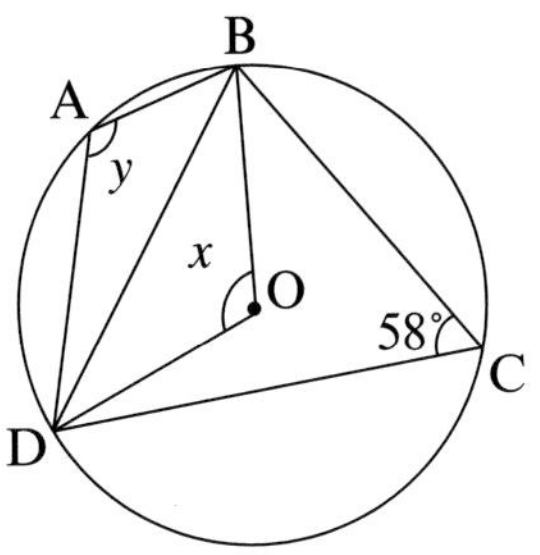


**b**

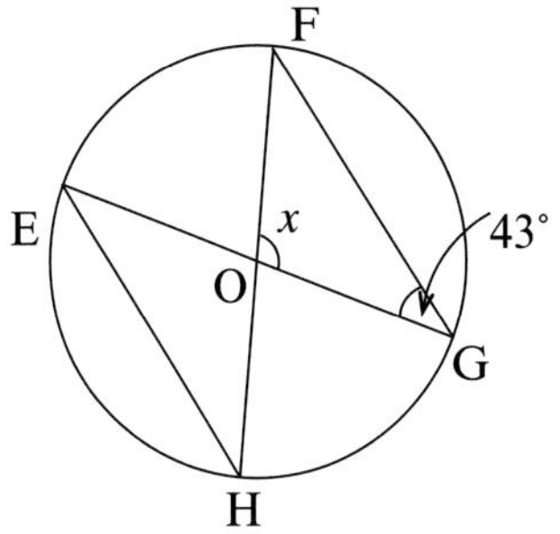


**c**

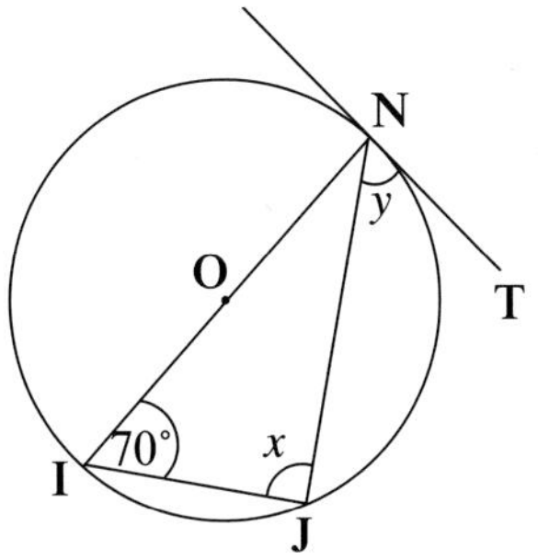


TN is a tangent at N.

**d**

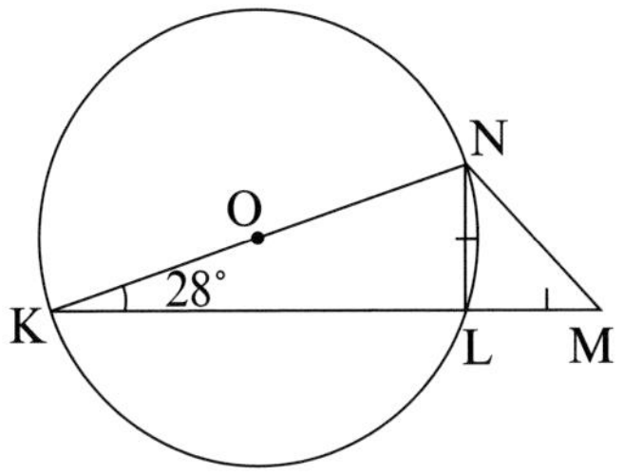


Find $K\hat{N}M$.

**2** **a** Name two angles subtended at the circumference by the chord AC.
**b** Name two angles subtended at the circumference by the chord CD.
**c** What is the size of $C\hat{A}D$?
**d** What is the size of $C\hat{E}A$?
**e** BC is a tangent to the circle.
What is the size of $B\hat{C}A$?
Explain the reason for your answer.

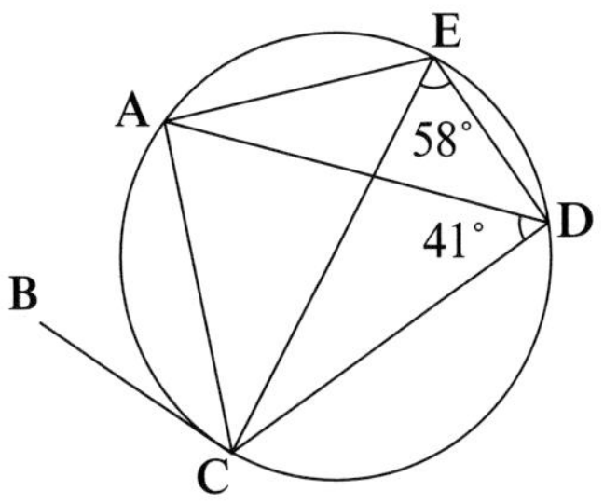


**3** In this diagram, the circle has centre O.
BE is a chord, AB a tangent,and $B\hat{D}E = 37°$.

**a** What is the size of $E\hat{O}B$?
Explain your answer.
**b** What is the size of $B\hat{C}E$?
**c** What is the size of $A\hat{B}O$?
Explain the reason for your answer.
**d** What is the size of $O\hat{B}E$?
Explain the reason for your answer.
**e** What is the size of $E\hat{B}A$?

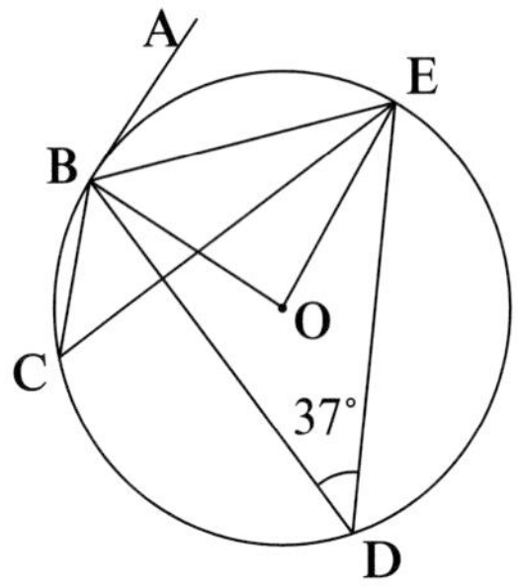


**4** In the diagram, the circle has centre O.
ED is a chord.

**a** Name an angle at the circumference in the major segment subtended by ED.
**b** Calculate the size of $E\hat{O}D$ (reflex).
**c** Calculate the size of $D\hat{F}E$.
**d** Calculate the size of $O\hat{E}D$.
Give the reason for your answer.

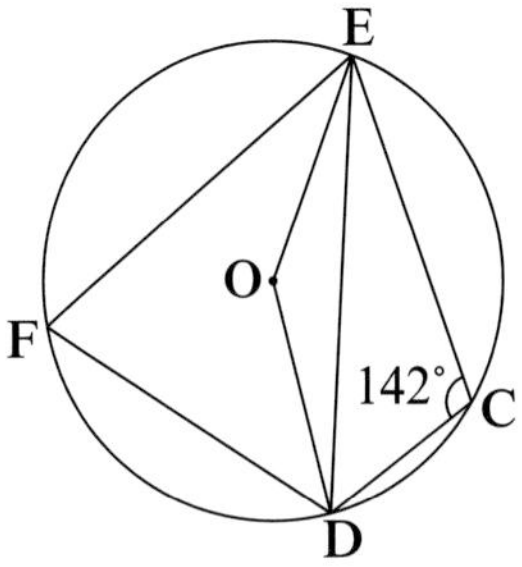

**5** In the diagram, the circle has centre O.
DE is a diameter, FE is a chord, and FC a tangent.

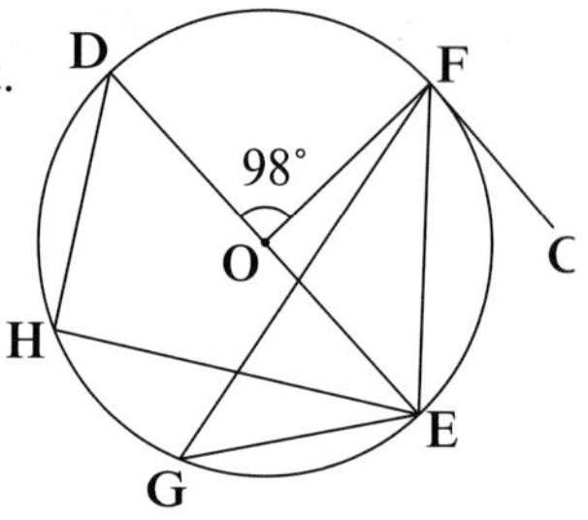


**a** Calculate the size of $F\hat{G}E$.
Explain the reasons for your answer.

**b** Calculate the size of $O\hat{E}F$.
Explain the reasons for your answer.

**c** Explain why $E\hat{F}C$ is 41°.

**d** DEH is 29°. Calculate the size of $H\hat{D}E$.
Explain your answer.

**6** In the diagram, the circle has centre O.
GC is a diameter and DF a tangent.

DF = 7.2 cm and CF = 2.8 cm.

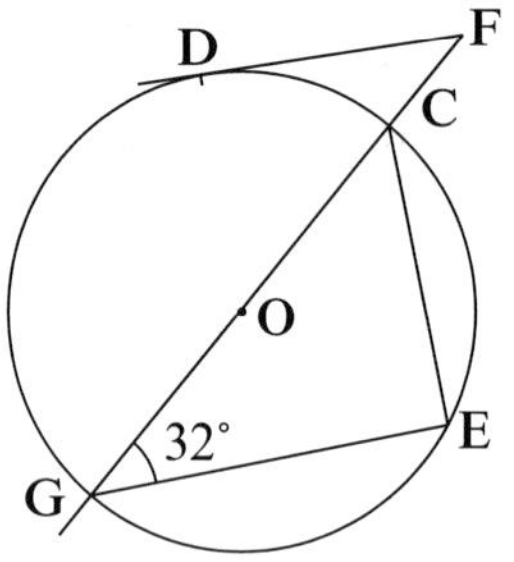


**a** Calculate the length GF.

**b** Calculate the length CE.

**c** Calculate the area of triangle GCE.

7 In the diagram, the circle has centre O and BC is a tangent.
The radius of the circle is $b$ cm.

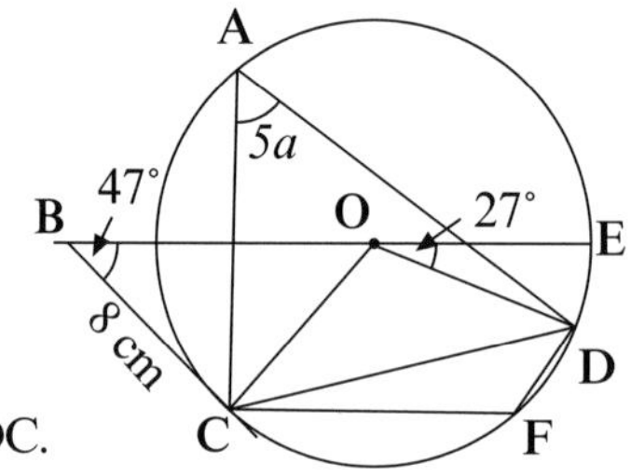


a Write an expression in $a$ for:
i $D\hat{O}C$ ii $O\hat{C}D$

b By considering the angles at O:
i Write an equation in $a$ for the sum of the angles at the centre.
ii Solve your equation, and calculate the size of $C\hat{A}D$, $D\hat{O}C$ and $O\hat{D}C$.

c Write an expression in $b$ for the length OB.

d Calculate the size of $C\hat{F}D$. Explain your calculation.

8 In the diagram the circle has centre O and FB this is a tangent.
DE is a diameter.

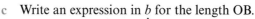

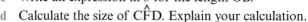

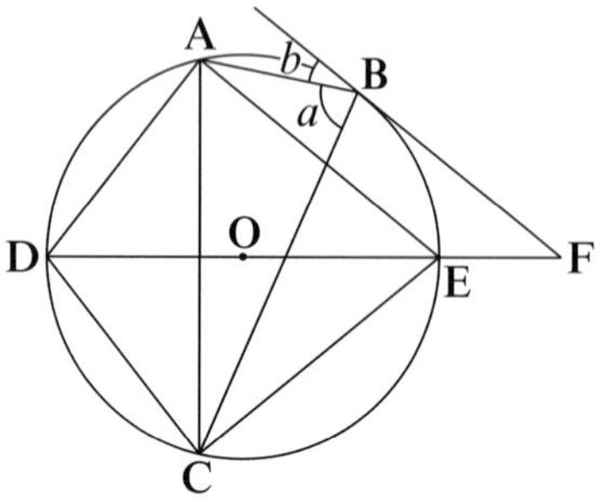


a Give an expression for each of these angles:
i $A\hat{E}C$ ii $A\hat{C}B$ iii $A\hat{O}C$

b Write an expression for $C\hat{A}B$.

c Write an expression for $A\hat{D}C$.
Explain your answer.

d AE = EC. Write an expression for $B\hat{C}E$.

e AD = DC. Write an expression $D\hat{C}A$.

# The cyclic quadrilateral

RSTU is a cyclic quadrilateral.
A quadrilateral is said to be cyclic when:

- the four vertices of the quadrilateral lie on the circumference of the same circle.

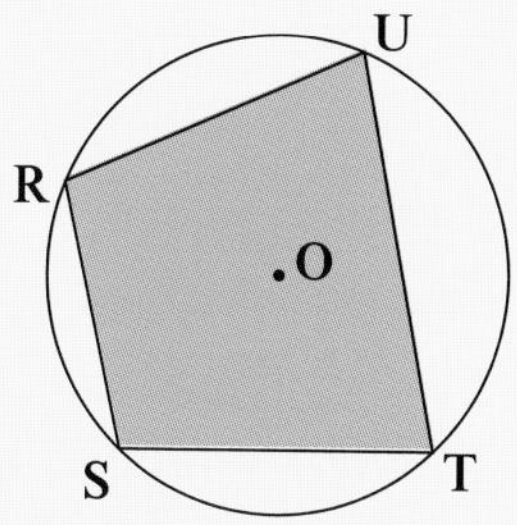


The other property all cyclic quadrilaterals have is that:

- the angles at opposite vertices are supplementary.

Supplementary angles are a pair of angles that, when added together, give a total of 180°.

As RSTU is cyclic then: $R\hat{S}T + T\hat{U}R = 180°$
and $S\hat{R}U + U\hat{T}S = 180°$

**Proof:**

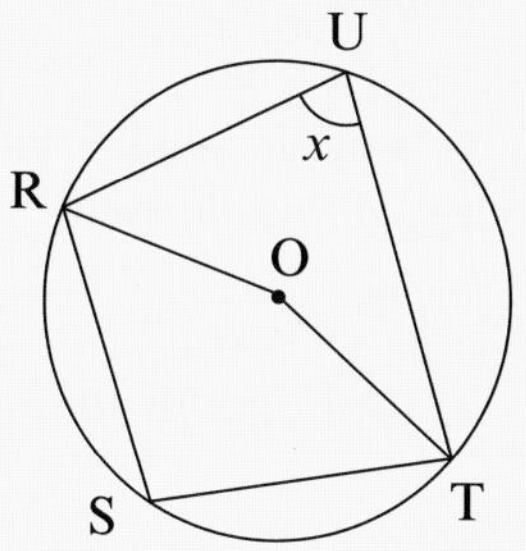


Let $R\hat{U}T = x$
So obtuse angle $R\hat{O}T = 2x$ **(angle at centre)**
So reflex angle $R\hat{O}T = 360° - 2x$
But angle $R\hat{S}T = \frac{1}{2}$ of reflex $R\hat{O}T$
**(angle at centre)**

So $R\hat{S}T = \frac{1}{2}(360° - 2x)$
$= 180° - x$
So $R\hat{S}T = 180° - R\hat{U}T$
or $\mathbf{R\hat{S}T + R\hat{U}T = 180°}$

**Exercise 20.2**
Cyclic quadrilaterals

1 Explain why all rectangles are cyclic quadrilaterals.

2 Is it possible to draw a parallelogram that is cyclic?
Explain your answer.

3 Apart from the rectangle, what other quadrilateral is always cyclic?
Explain your answer with diagrams.

4 CDEF is a cyclic quadrilateral with angles of $v$, $w$, $x$ and $y$ degrees as shown by the diagram.
If $p = 15°$, explain why:
$v + x = 12p$

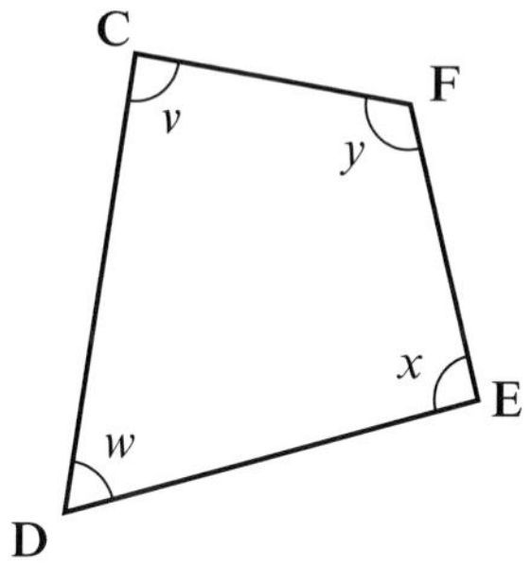


5 In the diagram the circle has centre O.
The size of $N\hat{K}L$ is given as $a$.
Use circle properties to explain why $N\hat{K}L$ and $L\hat{M}N$ are supplementary.

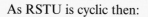

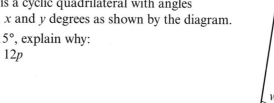

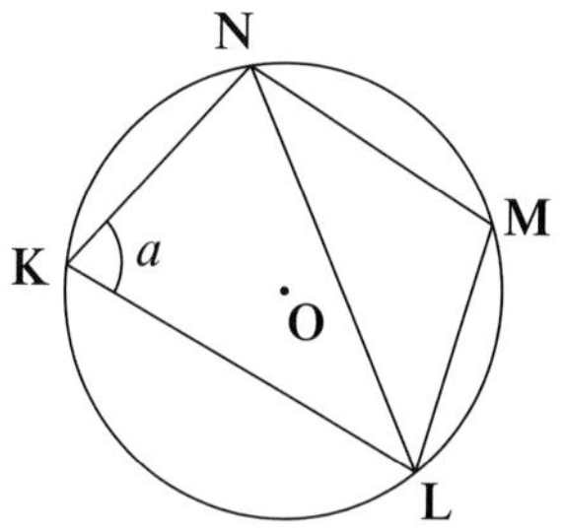

**6** In the diagram the circle has centre O.
EOG is the diameter.
Calculate $F\hat{G}H$.

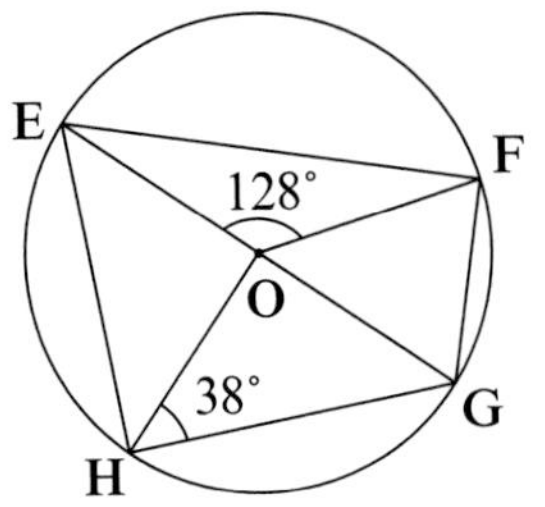


**7** In the diagram the circle has centre O.
A tangent is drawn to the circle at G.
FH = GF.

**a** Write and solve an equation in $x$ to find the size of angles:
**i** $H\hat{E}G$ **ii** $H\hat{F}G$.
Explain your answer.

**b** Calculate the size of $G\hat{O}H$.

**c** Could HOGF be a cyclic quadrilateral? Give two reasons for your answer.

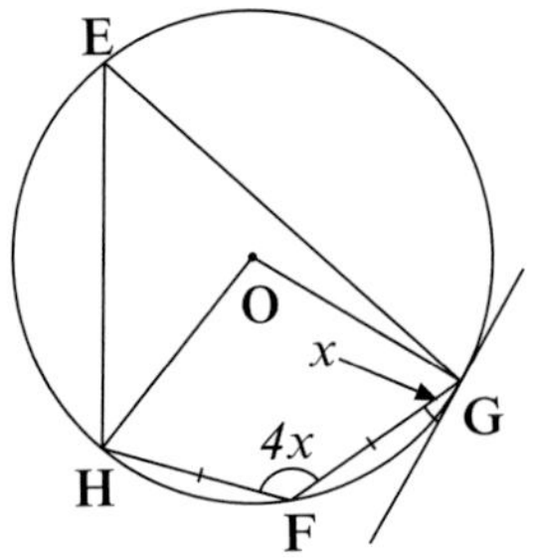


**8** In the diagram, O is the centre of the circle and FB is a chord.
The ratio $B\hat{C}E : B\hat{F}E$ is 3 : 5.

**a** Calculate the size of:
**i** $B\hat{C}E$ **ii** $B\hat{F}E$ **iii** $F\hat{B}C$

**b** Calculate the size of $F\hat{O}C$.
Explain your answer.

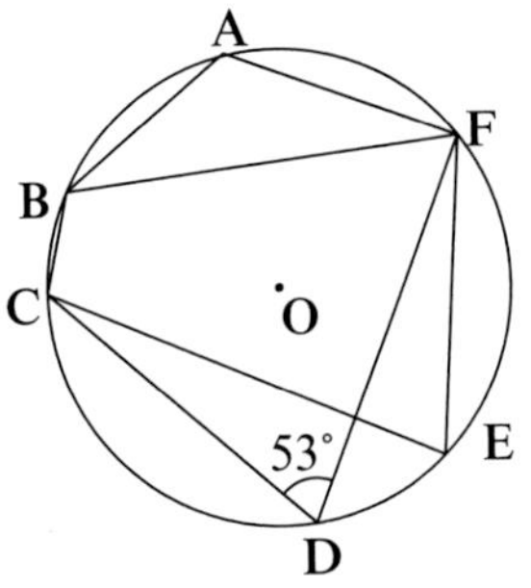


**9** In the diagram O is the centre of the circle.
AB is a chord and DC = DA.
Angle $A\hat{O}B$ is $v$ degrees.

**a** Write an expression in $v$ for each of these angles:
**i** $B\hat{C}A$ **ii** $O\hat{B}A$ **iii** $D\hat{C}B$

**b** Explain why an expression for $D\hat{A}B$ is:
$$135° - \tfrac{1}{2}v$$

**c** Show that an expression for $D\hat{E}B$ is:
$$\tfrac{1}{2}(90° + v)$$

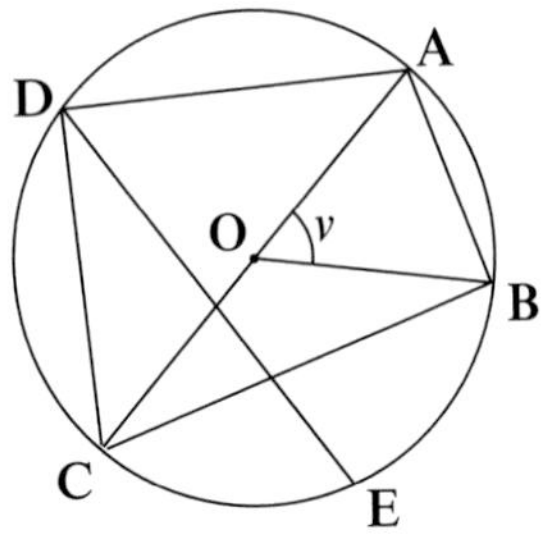


**10** In the diagram O is the centre of the circle.
BD is a tangent to the circle.

**a** Give the size of $C\hat{E}A$.
Explain your answer.

**b** Calculate $C\hat{D}E$.
Give reasons to support your calculations.

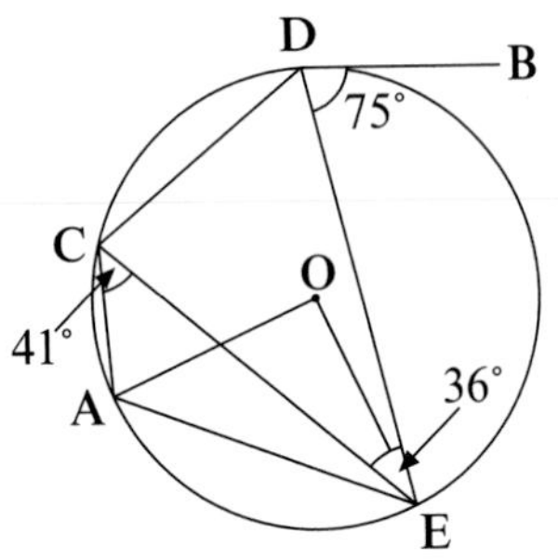

# End points

You should be able to ... ... so try these questions

**A** Identify angles in the same segment

**A1** **a** What is the size of $D\hat{A}B$?
**b** What is the size of $E\hat{A}C$?

**A2** Give an angle that is the same size as:
**a** $E\hat{A}D$ **b** $C\hat{E}B$

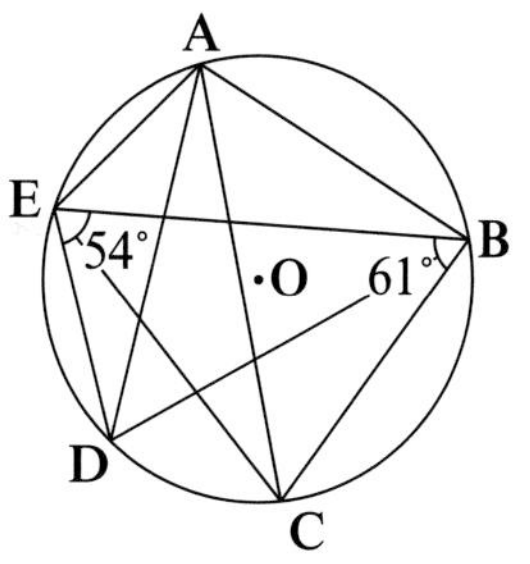


**B** Link angles at the centre and at the circumference on the same chord

**B1** Calculate the size of:
**a** $A\hat{B}C$ **b** $A\hat{D}C$

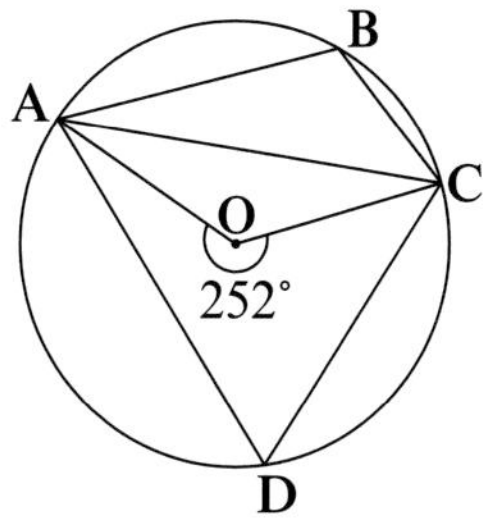


**C** Work with angles in a semicircle

**C1** In the circle the centre is O and $C\hat{A}B = 28°$
Calculate the size of $A\hat{C}B$.
Explain your calculation.

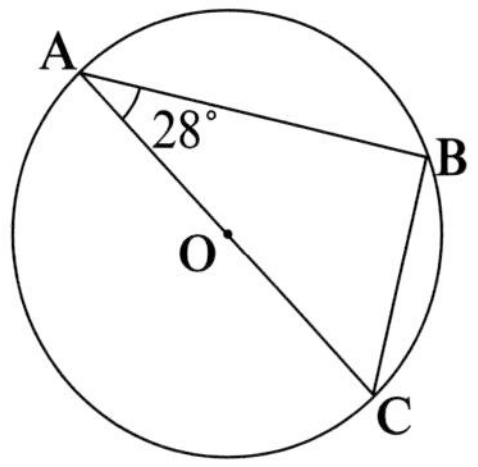


**D** Solve problems when tangents are involved

**D1** Calculate the size of $D\hat{A}E$.
Explain your working.

**D2** Calculate the size of $D\hat{E}A$.
Give reasons for your answer.

**D3** Calculate the size of $D\hat{A}O$.
Explain your answer.

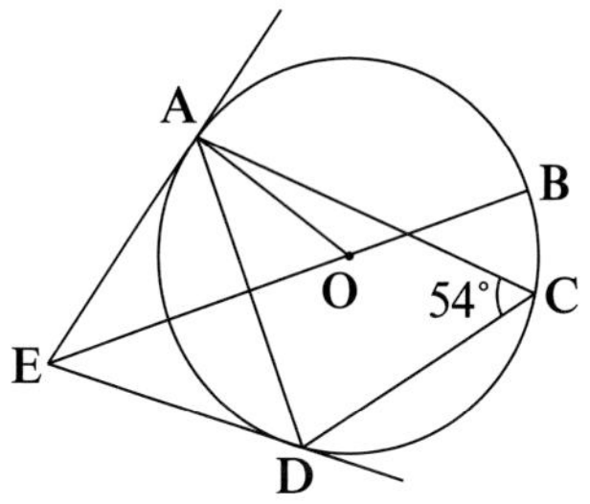


**E** Use the properties of cyclic quadrilaterals

**E1** List the properties of all cyclic quadrilaterals.

**E2** The angles of a cyclic quadrilateral are given as $3k$, $4k$, $5k$, and $6k$.
**a** On a diagram show possible positions of these angles.
**b** Write and solve an equation in $k$ to find the size of each angle.

Southampton Evening Chronicle *Monday 15 April 1912*

# TRAGEDY AT SEA

It is with great regret that we bring you the news that last night at 10:40 pm the 'unsinkable liner' the Titanic hit an iceberg on her way to New York. The Titanic later sank at 2:20 am with the loss of many lives. It was the liner's maiden voyage and on board were 331 first class passengers, 273 second class, 712 third class and a full crew – only 32.2 % of those on board survived. Each first-class passenger had paid £870 for the privilege of making the voyage in this luxury floating palace. To reassure the passengers the orchestra was still playing as the liner was going down and many passengers were so sure the ship could not sink that they refused to board the lifeboats.

The captain had been given repeated warnings of icebergs ahead but chose to steam on at 22.5 knots. It was calm weather with good visibility but the lookouts had not been issued with binoculars. The iceberg is thought to have had a height of about 100 feet showing above the water and a weight of about 500 000 tons. The sea water temperature was only 28° Fahrenheit and this took its toll on those jumping overboard. It is thought that the capacity of the lifeboats was insufficient for the number of people on board.

Survivors were picked up by the liner Carpathia which had heard the SOS when it was 58 miles away. The Carpathia steamed at a staggering 17.5 knots to reach the sinking Titanic. The ship's engineer said this was 25% faster than her usual speed.

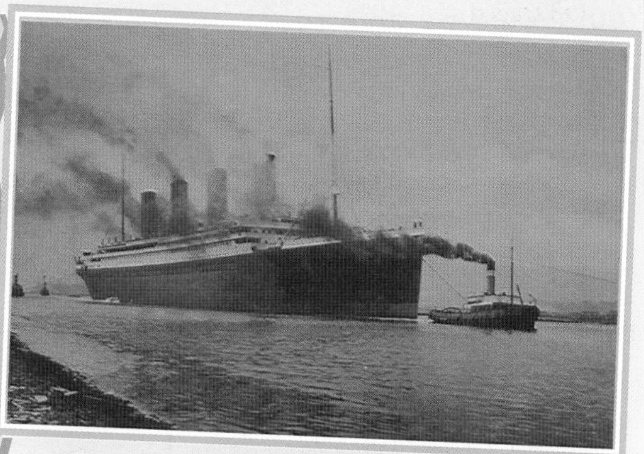

## HOW FAIR WAS THE RESCUE?

Reports coming in give the final casualty figures from the Titanic. There was not enough lifeboat space for all on board because the Titanic was considered to be the first unsinkable ship. The owners White Star admit that lifeboats could only hold 33% of the full capacity of the liner and 53% of those on board on that fateful night. Breaking the survival figure down by class we find 203 first-class, 118 second-class and 178 third-class passengers were rescued. Nearly a quarter of all crew were saved. Analysis of these figures is taking place to see if all people on board had an equal chance of being rescued.

## *Strange* BUT *true*

Fourteen years before the disaster, and before the Titanic had been built, a story was published which described the sinking of an enormous ship called the Titan after it had hit an iceberg on its maiden voyage.
The comparisons between ships is even more amazing.

| | *Titan* (Fiction 1898) | *Titanic* (True 1912) |
|---|---|---|
| Flag | British | British |
| Month of sailing | April | April |
| Displacement (tons) | 70 000 | 66 000 |
| Propellers | 3 | 3 |
| Max. speed | 24 knots | 24 knots |
| Length | 800 feet | 882 feet |
| Watertight bulkheads | 19 | 15 |
| No. of lifeboats | 24 | 20 |
| No. on board (inc crew) | 2000 | 2208 |
| What happened? | Starboard hull split by iceberg | Starboard hull split by iceberg |
| Full capacity | 3000 | |

Distances at sea are measured in nautical miles and ships' speeds are given in knots.
1 nautical mile is 1852 metres.
1 knot is a speed of 1 nautical mile/hour.

**1** Calculate approximately how many people died in the tragedy.

**2** If all those who survived were in lifeboats, what was the mean number of people per boat?

**3** In 1912 the price of a small house was about £200. In 1996 the same house would cost about £68 000. If fares on a cruise liner increased in the same ratio what would have been the first class ticket price in 1996?

**4** About $\frac{1}{9}$th of the volume of an iceberg is above water level.
The part of the iceberg showing above the water line can be approximated to a cone with base diameter of 85 metres.

**a** Estimate the volume of the iceberg showing in cubic metres. Take 1 foot as 0.305 metres.

**b** The density of an iceberg is about 950 kg per metre$^3$.

Estimate the total mass (in tonnes) of the iceberg that the Titanic hit.

**5** **a** What was the Carpathia's usual speed?

**b** At 17.5 knots, how long would it take the Carpathia to steam the 58 nautical miles to the Titanic?

**c** The Carpathia received the SOS message at 12:30 am. Approximately how long after the Titanic sank did she arrive at the scene?

**6** You can convert a temperature from °Celsius ($C$) to °Fahrenheit ($F$) with this formula:

$$F = \frac{9}{5}C + 32$$

**a** Make $C$ the subject of the formula.

**b** Calculate a water temperature of 28 °F in degrees Celsius.

**c** An approximation to convert °F to °C is "Subtract 30 from the temperature in °F then halve the result."
For a temperature of 28°F find the percentage error that this approximation gives.

**7** Show, with working, that the statement

Nearly a quarter of all crew were saved

is correct.

**8** **a** How many people could the lifeboats have held in total?

**b** Use your answer to part **a** to calculate an approximate value for the full capacity of the Titanic.

**9** Calculate the relative frequency of survival for:

**a** first-class passengers

**b** second-class passengers

**c** third-class passengers.

**10** Calculate the total percentage of the passengers aboard who were rescued.

**11** What is the displacement of the Titanic to 1 sf?

**12** A nautical mile is defined as "the length of arc along the equator subtended by an angle of $\frac{1}{60}$ of a degree at the centre of the Earth". The Earth has a diameter of about 12 756 kilometres.

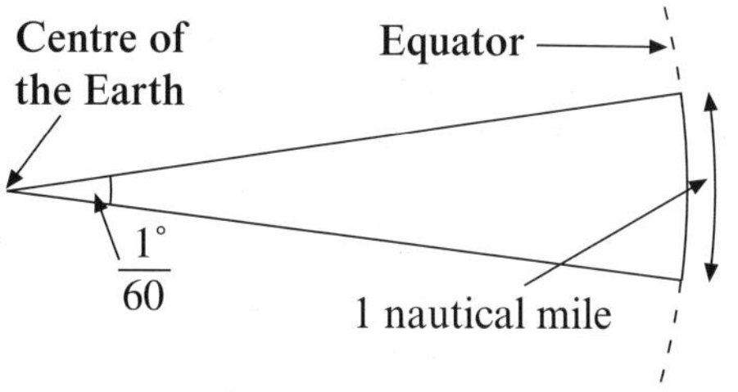


Show that a nautical mile is about 1850 metres.

**13** The Titanic had cranes, known as "derricks", for lifting the cargo on to the ship. This diagram shows a derrick in one position. The tower and part of the cable are vertical.

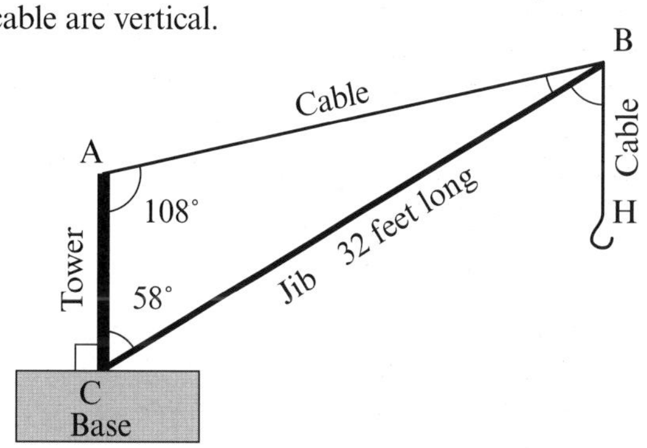


**a** What is the angle ABH? Give your reasons.

**b** Calculate the length AB on the derrick.

**c** Calculate how high point B is above point A?

At the enquiry after the sinking, White Star, the owners, said that in the previous ten years they had carried 21 79 594 passengers with the loss of only 2 lives.

**14** Give the number of passengers carried to:

**a** 1 sf **b** 4 sf

**15** In the ten years between 1981 and 1990 about $1.2 \times 10^7$ people from the UK crossed the Atlantic by plane. With the same death rate as White Star gave, roughly how many people would have died in that time?

## Starting points

You need to know about ...

### A Bearings

- All bearings are:
  - measured clockwise from North
  - written using three figures.

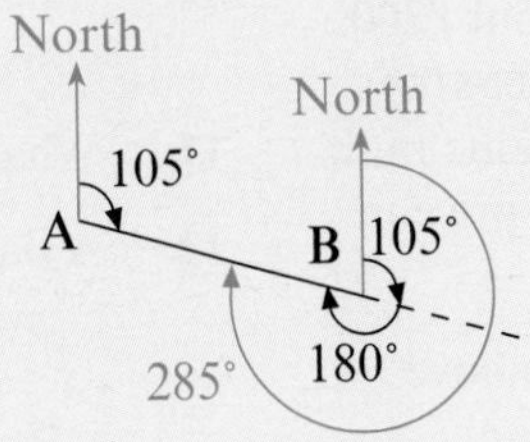


**Example**

The bearing of B from A is 105°.
The bearing of A from B is 285°.

- You can fix a position by:
  - giving a bearing and distance from one point
  - giving a bearing from two different points.

**Example**

The point B is on a bearing of 105° from A and is 10 km from A.

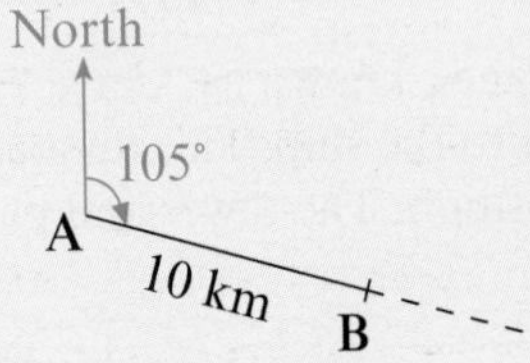


**Example**

The point C is on a bearing of 078° from A and on a bearing of 320° from B.

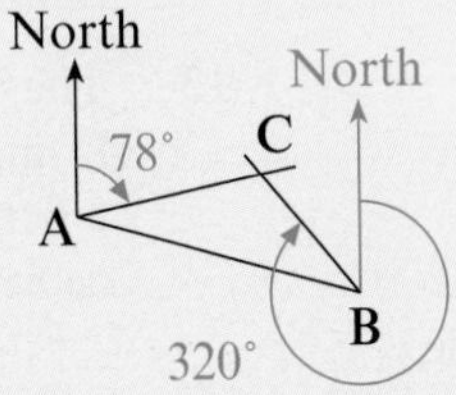


### B Pythagoras' rule

- In any right-angled triangle, the area of the square on the hypotenuse is equal to the sum of the area of the squares on the other two sides.

  AB is the hypotenuse,
  so $AB^2 = AC^2 + BC^2$
  i.e. $AB = \sqrt{(AC^2 + BC^2)}$

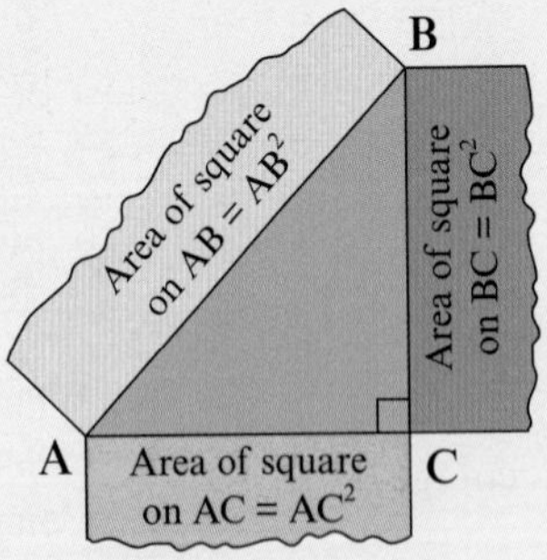


### C Trigonometric ratios

- The trigonometric ratios can be used to calculate sides or angles in right-angled triangles.
- Each trigonometric ratio can be written in different ways.

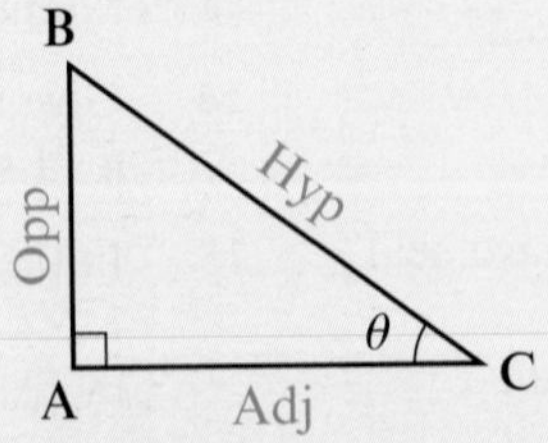


| $\sin\theta = \dfrac{\text{Opp}}{\text{Hyp}}$ | $\cos\theta = \dfrac{\text{Adj}}{\text{Hyp}}$ | $\tan\theta = \dfrac{\text{Opp}}{\text{Adj}}$ |
|---|---|---|
| $\text{Opp} = \text{Hyp} \times \sin\theta$ | $\text{Adj} = \text{Hyp} \times \cos\theta$ | $\text{Opp} = \text{Adj} \times \tan\theta$ |

... so try these questions

**A1** This diagram shows some towns on a radar screen positioned at H.

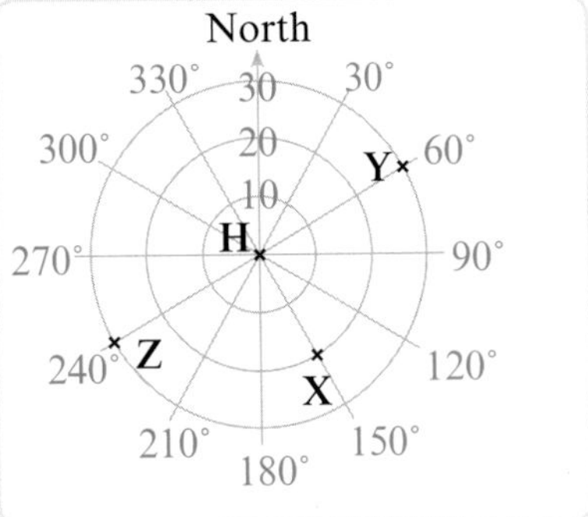


a What is the bearing of:
  i Y from H
  ii H from Y
  iii Z from Y?

b Estimate the bearing of X from Y.

c The rings on the screen are 10 km apart. Point P is due North of X and on a bearing of 030° from H. How far from H is P?

**B1** a From the radar screen above sketch and label ΔXHY.

b Calculate the distance XY to the nearest 0.1 km.

**C1** Using triangle XHY above calculate the angle $H\hat{X}Y$ to the nearest degree.

**C2** Z is south and west of H.

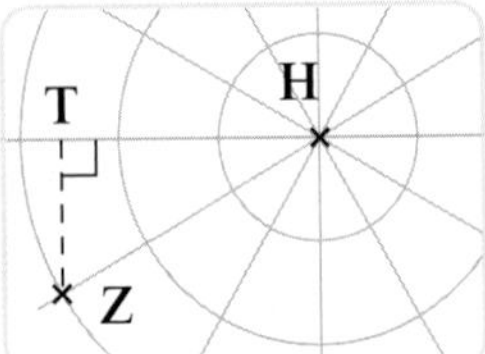


Calculate, to the nearest km:
a how far Z is west of H
b how far Z is south of H.

## D Sine and cosine of angles from 0° to 180°

- Graph y = sin $x$ and $y = \cos x$ from 0° to 180°

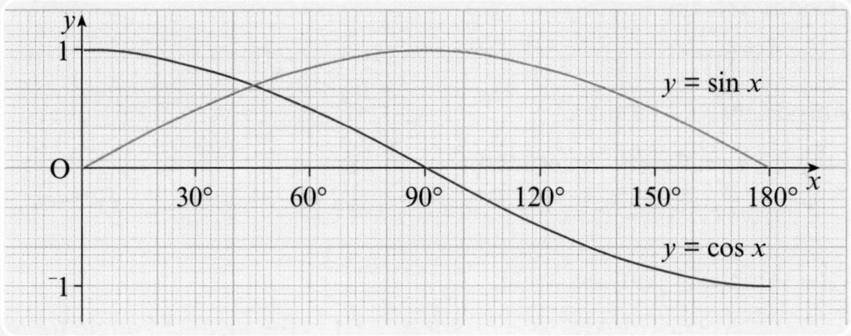


**D1** Solve each of these for:
$0 \leqslant \theta \leqslant 180°$.

Give the answer to the nearest degree.

**a** **i** $\sin \theta = 0.5$

**ii** Explain why there are two possible values.

**b** $\cos \theta = {}^{-}0.2$

**c** $\cos \theta = 0.2$

**d** $\cos \theta = \sin \theta$

## E Constructing triangles

- To construct a triangle with sides of $a$ cm, $b$ cm and $c$ cm.
  - draw one line as the base e.g. $a$
  - with one end of the base as centre:
  - draw an arc of radius $b$
  - with the other end of the base as centre draw an arc of radius $c$
  - join the ends of base to the intersection of the arcs to give the triangle.

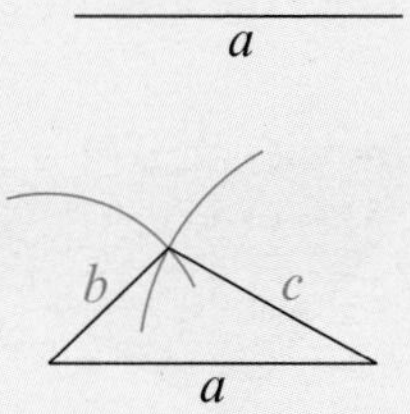


- To construct a triangle when you know two sides $v$ and $t$, and one non-included angle 54°.
  - draw one side as the base e.g. $v$
  - at one end of the base measure and draw an angle of 54°
  - with the other end of the base as centre draw an arc of radius $t$
  - where the arc and the line at 54° to the base intersect is the third vertex of the triangle.

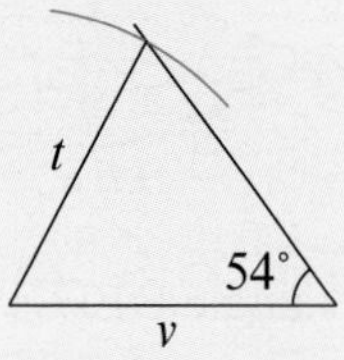


(Note:- the triangle depicted is only one of two possible triangles).

**E1** **a** Construct a triangle with sides of length 6 cm, 4 cm and 8 cm.

**b** Measure the largest angle.

**E2** **a** Construct two different triangles ABC with $A = 42°$ AB = 8 cm and BC = 6 cm.

**b** For each of your triangles measure the angle $B$.

## F Triangles

- When sides of a triangle are labelled, it is normal to use a small letter that matches the label of the opposite vertex.

  For example:
  side ST is labelled $w$.

- The largest side is always opposite the largest angle.

  For example:
  largest angle SWT, longest side $w$, or ST.

- Angle SWT is called:
  the **included angle** between sides SW and WT.

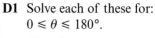

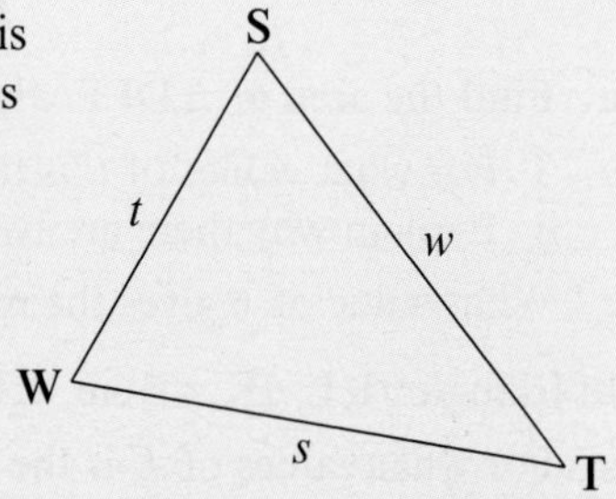


**F1** In triangle PQR $p = 8$ cm, $q = 4$ cm and $R = 90°$. Calculate the size of the smallest angle to the nearest degree.

**F2** In triangle DEF $f = 5$ cm, $d = 6$ cm and $E = 58°$. Calculate, to the nearest millimetre, the shortest distance from F to the line DE.

# The area of a triangle

The included angle for two sides of a triangle is the angle between those two sides.
Example

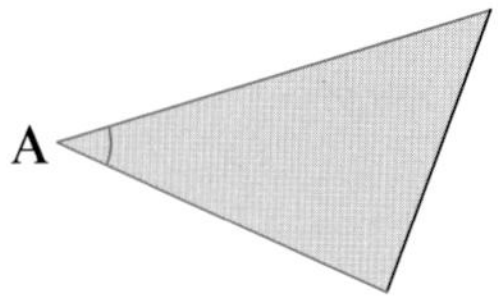


$\hat{A}$ is the included angle for sides AB and AC.

- The area of any triangle can be written in terms of two sides and the included angle.

Area of $\Delta ABC = \frac{1}{2}ah$

Using $\Delta CAO$ $h = b \sin C$

So the area of $\Delta ABC = \frac{1}{2}ab \sin C$

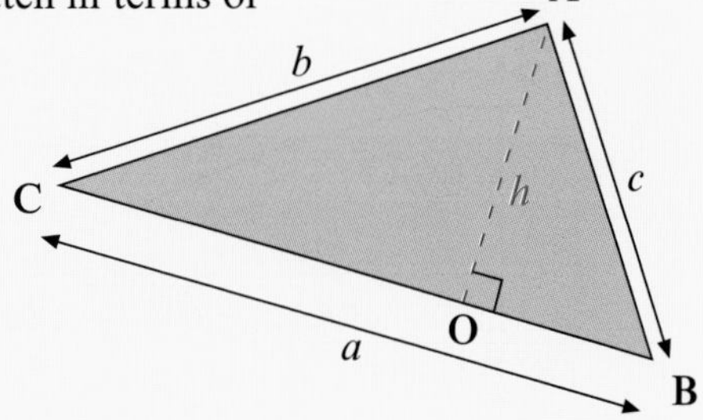


In the same way you can write:

Area of $\Delta ABC = \frac{1}{2}bc \sin A$ ($A$ is the included angle for sides BA and CA)

Area of $\Delta ABC = \frac{1}{2}ca \sin B$ ($B$ is the included angle for sides CB and AB)

## Exercise 21.1
The area of a triangle

**Accuracy**
For this exercise:
- give each length and area correct to 2 sf
- give each angle to the nearest degree.

1 Calculate the area of each of these triangles.

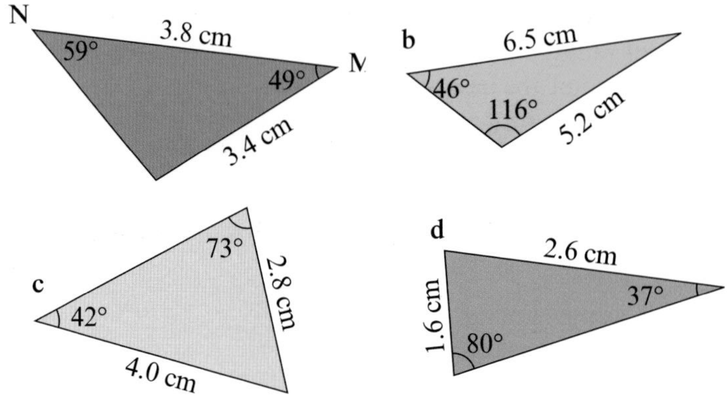


2 Triangle DEF is isosceles with DE = EF = 4 cm and $D\hat{E}F = \theta$.

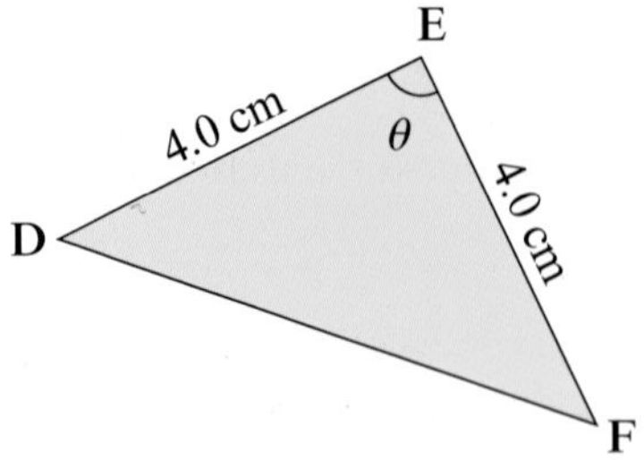


**a** Find the area of $\Delta DEF$ when $\theta = 48.5°$.

**b** **i** For what values of $\theta$ is the area $4\,cm^2$?
**ii** Explain why there are two possible values for $\theta$.

**c** What value of $\theta$ gives the maximum area of $\Delta DEF$?

3 In triangle JKL, JK = 5 cm, KL = 6 cm.

**a** For what values of $K$ is the area of $\Delta JKL = 10.8\,cm^2$?
**b** Calculate the maximum area of $\Delta JKL$?

4 Jamie drew $\Delta PQR$ with PQ = 3.4 cm, PR = 2.8 cm and QPR = 52°.
He drew lengths to the nearest millimetre and angles to the nearest degree.

**a** Find the maximum and minimum length of PQ.
**b** Calculate the maximum and minimum area possible for $\Delta PQR$.

# Sine rule

◆ The **sine rule** links the sides and angles of a triangle.
This is one way to prove the **sine rule**.

In $\Delta$ABD $h = c \sin A$
In $\Delta$BCD $h = a \sin C$
Therefore $a \sin C = c \sin A$

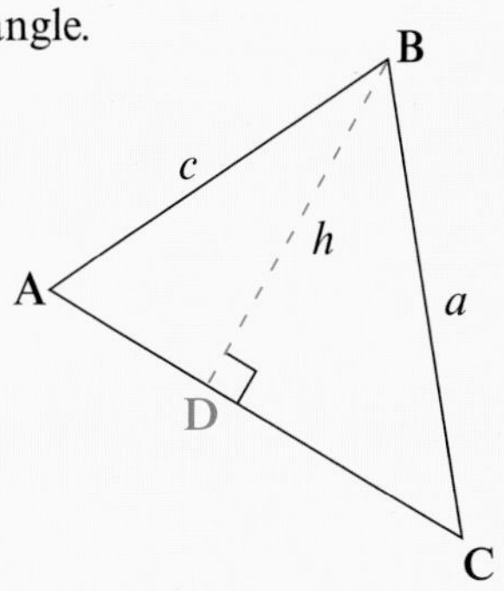


This can be rearranged to give the **sine rule**.

$$\frac{a}{\sin A} = \frac{c}{\sin C} \text{ and } \frac{\sin A}{a} = \frac{\sin C}{c}$$

In the same way you can show that:

$$\frac{b}{\sin B} = \frac{c}{\sin C} \text{ and } \frac{\sin B}{b} = \frac{\sin C}{c}$$

and $$\frac{a}{\sin A} = \frac{b}{\sin B} \text{ and } \frac{\sin A}{a} = \frac{\sin B}{b}$$

**Exercise 21.2**
**The sine rule**

**1** **a** Which of these expressions is true for $\Delta$PQR?

(A) $\frac{p}{\sin Q} = \frac{q}{\sin P}$ (B) $r = \frac{\sin P \times q}{\sin Q}$

(C) $p \sin P = q \sin Q$ (D) $p \sin R = r \sin P$

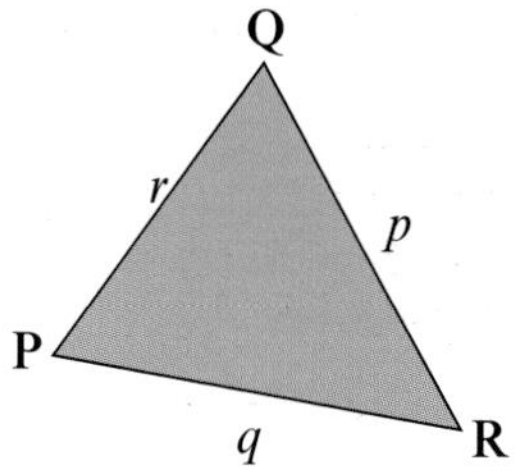


**2** Copy and complete these statements for $\Delta$LMN.

**a** $m = \frac{3.4 \times \sin \square}{\sin \square}$

**b** $\sin N = \frac{\square \times \sin \square}{\square}$

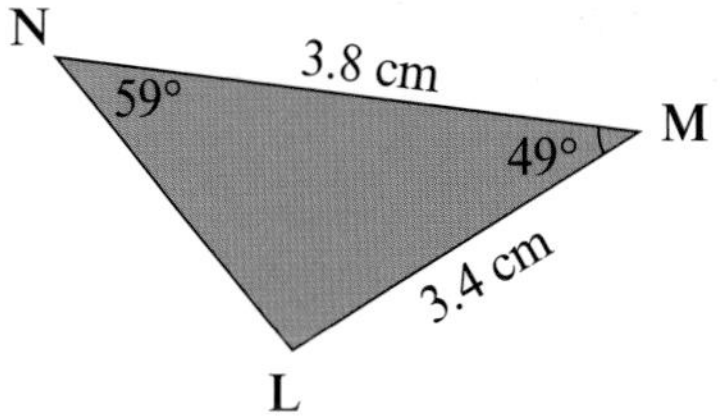


It may help to label the sides from the vertices first.

In some triangles you will need to calculate the missing angle before you can use the sine rule.

◆ The sine rule can be used to calculate the length of a side in a triangle, given any two angles and the length of one side.

**Example** In triangle RST calculate the length of RS.

❖ Use the angle sum of a triangle for the missing angle.
$S = 180° - (53° + 48°) = 79°$
❖ Use the sine rule to calculate a side.

$$\frac{t}{\sin T} = \frac{s}{\sin S}$$

$$t = \frac{s \times \sin T}{\sin S} = \frac{6.5 \times \sin 48°}{\sin 79°} = 4.9 \text{ to 1 dp}$$

So the length of RS is 4.9 cm (1 dp).

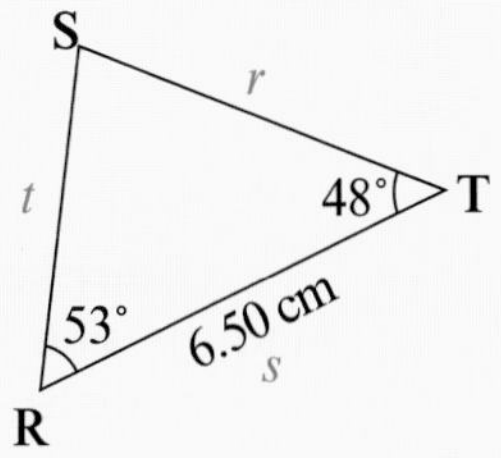


**Exercise 21.3**
**Calculating lengths using the sine rule**

**1** **a** Find the sequence of key presses on your calculator to calculate the length of RS.
**b** Calculate the length of ST correct to 1 dp.

> **Accuracy**
> For this exercise give each answer correct to 1 dp.

**2** In these triangles calculate the length of the side marked with a letter.

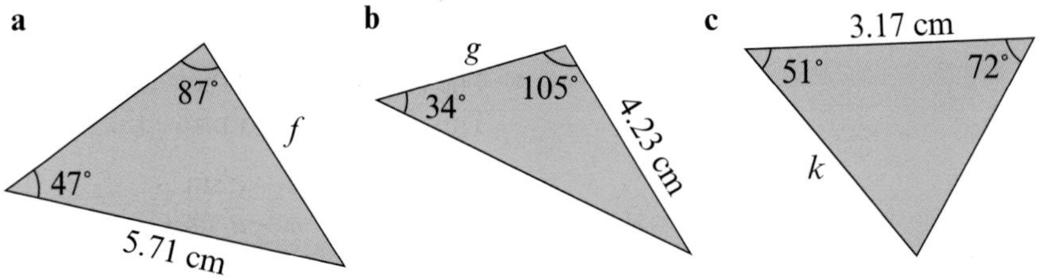


**3** Triangle PQR is isosceles with PQ = 8 cm and QPR = 56°.

**a** Draw two different triangles, and show that QR has two possible lengths.
**b** Calculate the two possible lengths for QR.

> Use the full calculator value for each calculation

**4** In triangle TUV the angle $T = 104°$, $U = 37°$ and TU = 5 cm.
Calculate the area of $\Delta$TUV.

**5** To construct $\Delta$XYZ Eamonn drew XZ, to the nearest millimetre, and angles $X$ and $Z$ to the nearest degree.

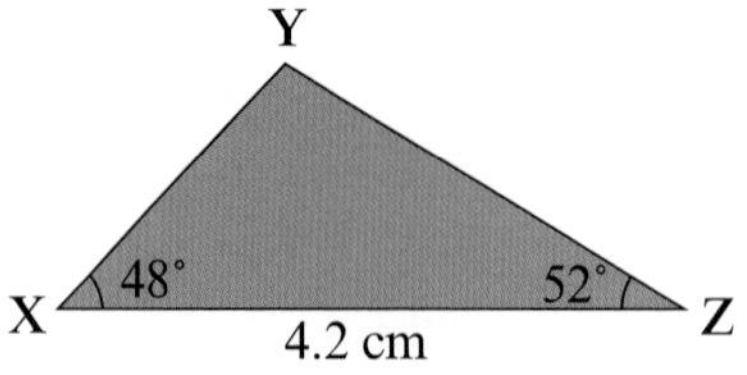


**a** What is the largest possible value of $X$?
**b** Explain why the smallest possible value of $Y$ is 79°.
**c** Calculate the maximum possible length of YZ.

> On some formula sheets the sine rule is only given as:
>
> $$\frac{a}{\sin A} = \frac{b}{\sin B} = \frac{C}{\sin C}$$
>
> For each problem use one pair of expressions.
> For instance
>
> $$\frac{a}{\sin A} = \frac{c}{\sin C}$$
>
> can be rearranged to give:
>
> $$\frac{\sin A}{a} = \frac{\sin C}{c}$$

- With the sine rule, given two sides and one opposite angle, a second angle in the triangle can be calculated.

**Example** Calculate angle $B$ in $\Delta$ABC

❖ Use the sine rule with the two sides given.

In $\Delta$ABC $\dfrac{\sin C}{3.7} = \dfrac{\sin 24°}{4.8}$

$$\sin C = \frac{3.7 \times \sin 24}{4.8}$$

$$\sin C = 0.313\,52° \ldots$$

$$C = 18.2718° \ldots$$

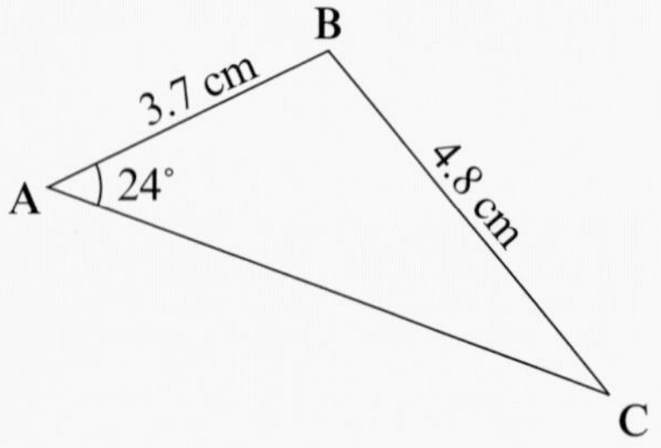


❖ Calculate the required angle:
$B = 180° - (24 + 18.2718 \ldots)°$
$B = 137.73°$ (2 dp)

## Exercise 21.4
## Using the sine rule to calculate angles

> **Accuracy**
> For this exercise give each answer correct to 1 dp.

**1** Calculate the angle $\theta$ in each of these triangles.

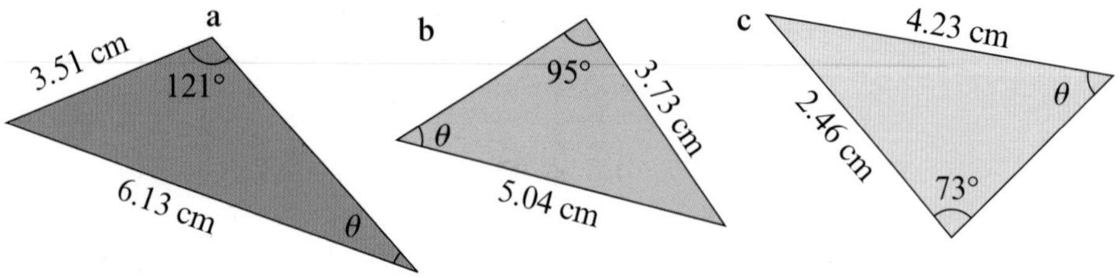

**2** This pentagon is right-angled at A and E.

**a** The angle $B\hat{D}C = 37°$.
Calculate the angle $A\hat{B}C$.

**b** Calculate the area of ABCDE.

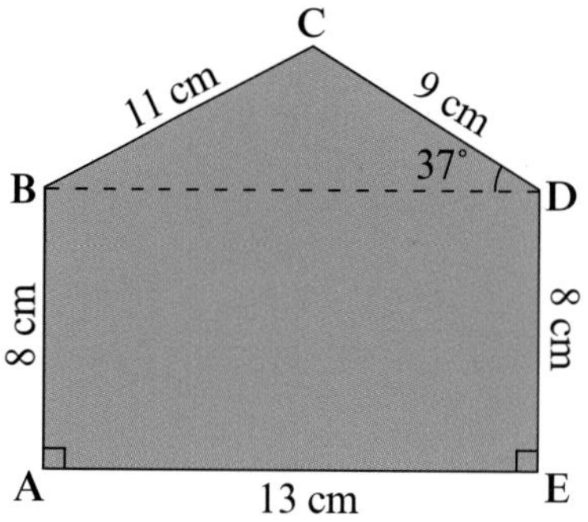


Use the full calculator value in each calculation.

**3** In triangle ABC, AC = 6.2 cm, AB = 5.1 cm and **C** = 28°.
This diagram shows that in Δ ABC, there are two possible positions for B, shown as $B_1$ and $B_2$.

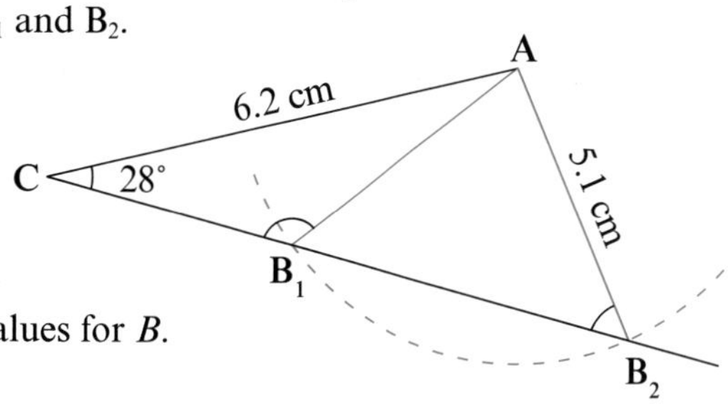


**a** Use the sine rule to find the value of sin *B*.

**b** Calculate two possible values for *B*.

**4** **a** Sketch two possible triangles for each of these.

**i** In triangle XYZ XZ = 8.3 cm, XY = 7.3 cm and Z = 58°

**ii** In triangle XYZ XZ = 9.2 cm, YZ = 8.0 cm and X = 43°

**b** Give two possible values for *Y* in each of these triangles.

## Cosine rule

◆ The cosine rule links an angle and three sides of a triangle.
This is one way to prove the **cosine rule**.

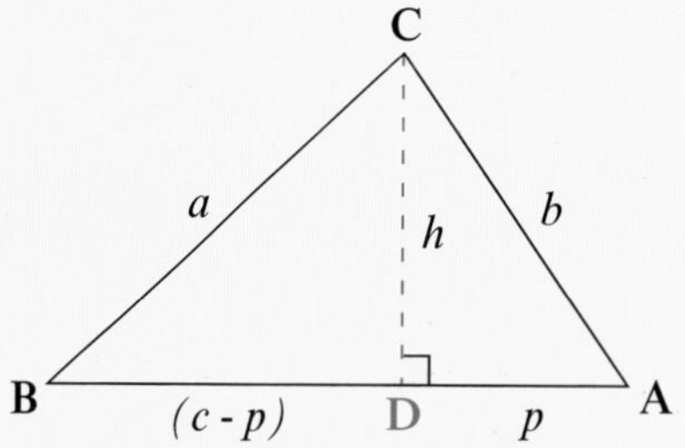


In ΔCAD $h^2 = b^2 - p^2$

In ΔBCD $a^2 = h^2 + (c - p)^2$

$$a^2 = b^2 - p^2 + (c - p)^2$$
$$a^2 = b^2 - p^2 + c^2 - 2cp + p^2$$
$$a^2 = b^2 + c^2 - 2cp$$
$$a^2 = b^2 + c^2 - 2c \times b \cos A \quad \text{(in } \Delta\text{CAD, } p = b\cos A\text{)}$$

This gives the **cosine rule**:

$$\boldsymbol{a^2 = b^2 + c^2 - 2bc \cos A}$$

The cosine rule gives an expression for one side in terms of **the opposite angle** and the other two sides of the triangle.

◆ The cosine rule can be rearranged to give an expression for the cosine of one angle in terms of the lengths of the sides.

$$a^2 = b^2 + c^2 - 2bc \cos A$$
$$2bc \cos A = b^2 + c^2 - a^2$$
$$\cos A = \frac{b^2 + c^2 - a^2}{2bc}$$

**Exercise 21.5**
**The cosine rule**

1 a Copy and complete these for $\Delta$ DEF.

i $d^2 = e^2 + f^2 - 2ef \cos \square$

ii $f^2 = e^2 + \square - 2e \square \cos \square$

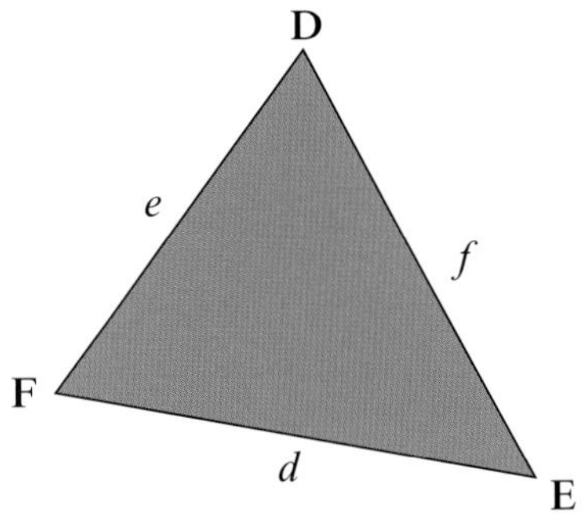


b Which of these is true for $\Delta$ DEF?

i $e^2 = d^2 + f^2 - 2de \cos E$

ii $e^2 + d^2 - f^2 = 2de \cos F$

iii $\cos D = \dfrac{e^2 + f^2 - d^2}{2ef}$

iv $\cos E = \dfrac{e^2 + f^2 - d^2}{2df}$

2 In this diagram AB is the diameter of the circle.

Use the cosine rule to explain why:

a $AB^2 > AC^2 + CB^2$

b $AB^2 = AD^2 + DB^2$

c $AB^2 < AE^2 + EB^2$

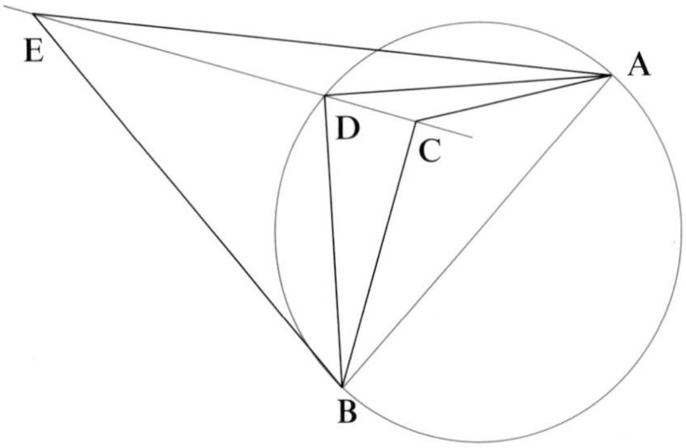


- The cosine rule can be used to calculate the length of a side in a triangle given the length of two sides and the included angle.

**Example**

Calculate the length of PQ in triangle PQR.

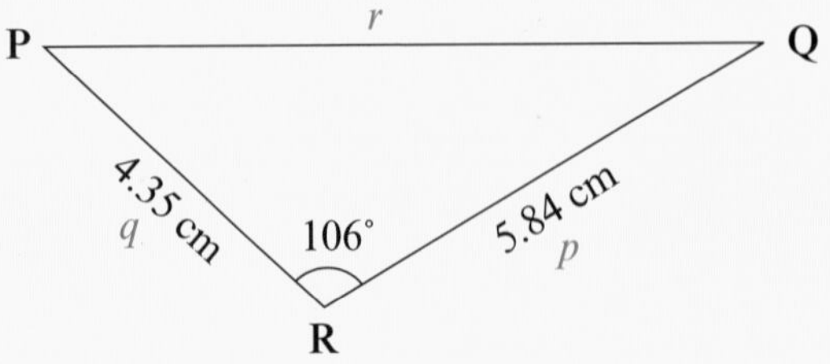


From the vertices label the sides $p$, $q$ and $r$.

Use the cosine rule to calculate the length of the side.

$$r^2 = p^2 + q^2 - 2pq \cos R$$
$$r^2 = 4.35^2 + 5.84^2 - 2 \times 4.35 \times 5.84 \times \cos 106°$$
$$r^2 = 67.032683 \ldots$$
$$r = 8.2 \text{ (1dp)}$$

- Given the lengths of all three sides, any angle of a triangle can be calculated using the cosine rule.

**Example** Calculate the angle $F$ in triangle EFG.

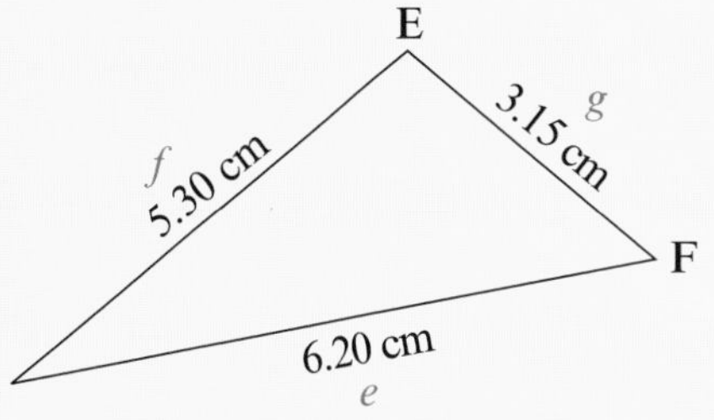


From the vertices label the sides $e$, $f$ and $g$.

Use the cosine rule to calculate the angle.

$$\cos F = \frac{e^2 + g^2 - f^2}{2eg}$$

$$\cos F = \frac{6.20^2 + 3.15^2 - 5.30^2}{2 \times 6.20 \times 3.15}$$

$\cos F = 0.519\,0092\ \ldots$

Therefore $F = 58.7°$ (1 dp)

**Exercise 21.6**
**Calculating angles using the cosine rule**

**Accuracy**
For this exercise give each length and area correct to 3 sf.

1 For triangle EFG above:
  **a** list the correct sequence of key presses to evaluate $F$.
  **b** explain why there is only one possible value for $F$.

2 In each of these triangles find the side or angle marked in blue.

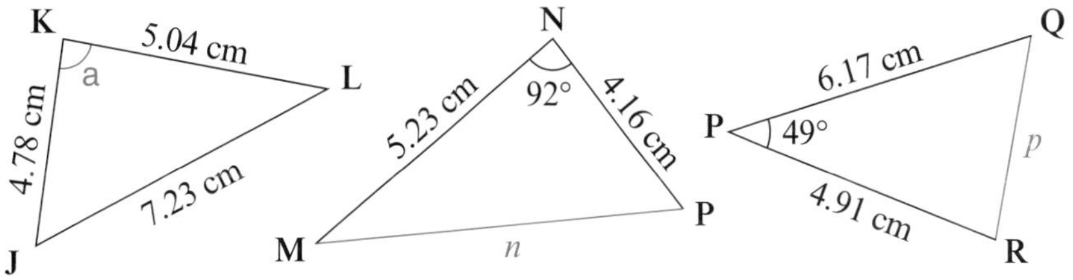


3 Find the largest angle in a triangle with sides of 8 cm, 12 cm, and 15 cm.

4 Calculate the area of a triangle with sides of length 6 cm, 9 cm and 10 cm.

5 Triangle ACE is one triangle that can be drawn by joining the vertices of ABCDE, a regular pentagon of side 4 cm.
  **a** What is the size of:
    **i** $A\hat{B}C$
    **ii** $C\hat{A}B$
    **iii** $C\hat{A}E$?
  **b** Calculate the length of AC.
  **c** Calculate the perimeter of Δ ACE.
  **d** Calculate the area of the Δ ACE.
  **e** How many different triangles can you draw by joining the vertices of ABCDE?

Use the full calculator value in each calculation.

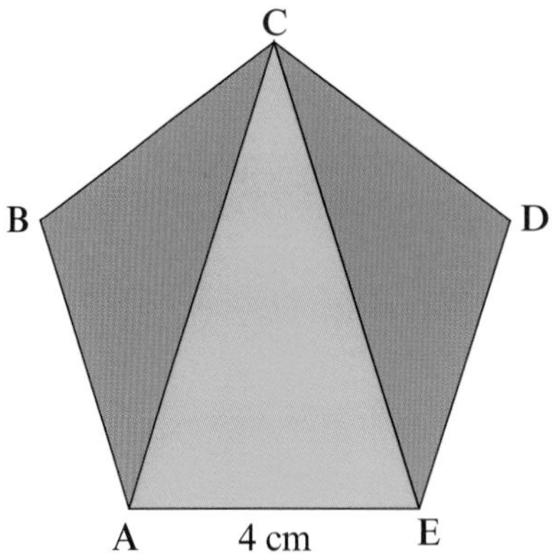

**6** FGHIJKLMN is a regular nonagon. Each side is 6 cm.

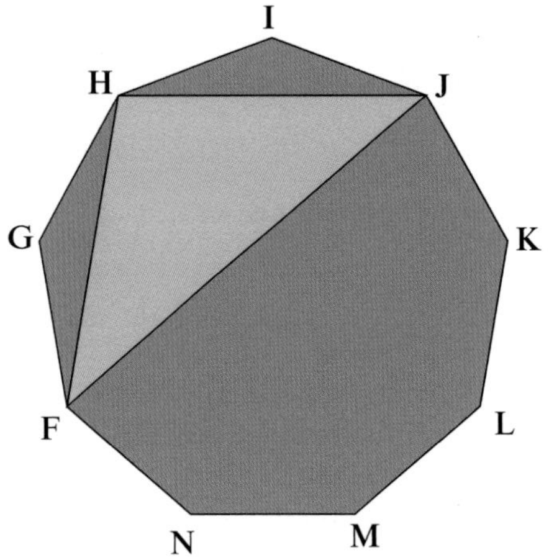


**a** Calculate the length HF.

**b** FHJ is one triangle that can be drawn by joining the vertices of the nonagon. Calculate its area and perimeter.

**c** Calculate the perimeter of triangle FMJ.

**7** Any polygon is cyclic if all its vertices lie on the circumference of a circle.
This circle is called its circumcircle.

This is a regular decagon.

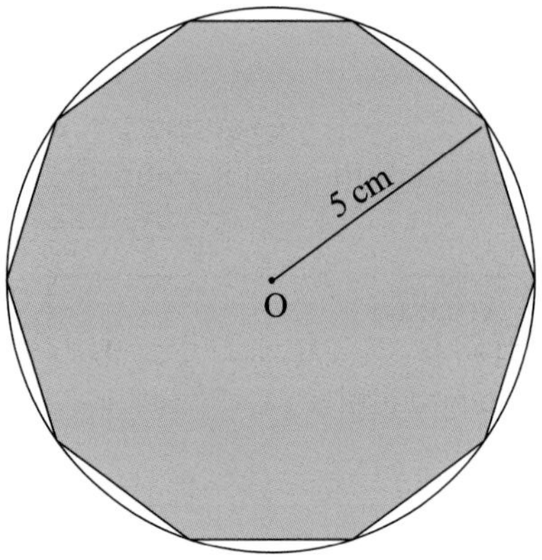


The radius of its circumcircle is 5 cm and its centre is at O.

**a** Calculate the area of the decagon.

**b** Calculate the perimeter of the decagon.

The circumcircle of a shape is a circle that passes through all its vertices.
**Example**
This is the circumcircle of triangle ABC.

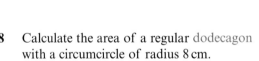

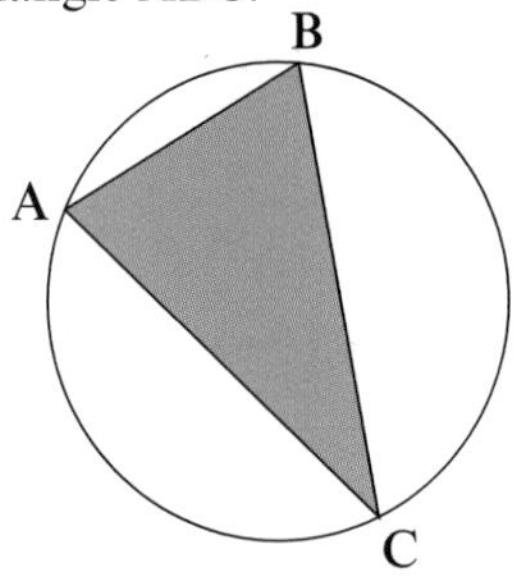


**8** Calculate the area of a regular dodecagon with a circumcircle of radius 8 cm.

A dodecagon is the name given to a 12-sided polygon.

**9** **a** A regular polygon (with $n$ sides) has a circumcircle of radius $r$. Write an expression for the area of the polygon in terms of $r$ and $n$.

**b** If the radius of the circumcircle is 1 cm, calculate the area of the regular polygon when:

**i** $n = 20$ **ii** $n = 40$ **iii** $n = 100$ **iv** $n = 1000$

What do you notice about the area?

**10** Find the radius of the circumcircle of a regular hexagon of perimeter 54 cm.

# Bearings

**Exercise 21.7**
**Calculating bearings and distances**

For questions like this split the route into triangles

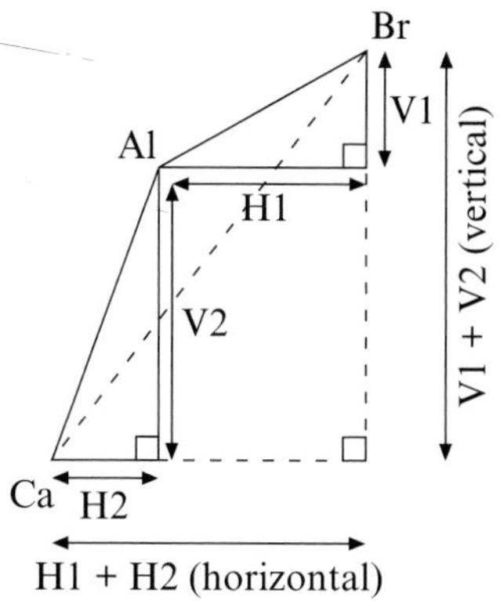


**Accuracy**
For this exercise:
- give each distance correct to 4 sf
- round each angle to the nearest degree.

**1** This diagram shows the position of three lighthouses.
Braydon is 56 km from Alington on a bearing of 071°.
Caster is 78 km from Alington on a bearing of 216°.

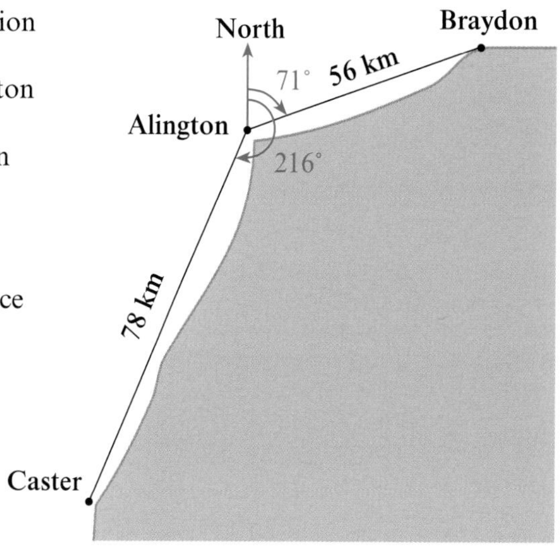


**a** What is the bearing of Alington from Caster?
**b** Calculate the direct distance from Braydon to Caster.
**c** Calculate the bearing of Braydon from Caster.

**2** This diagram shows the route taken by a boat which starts at P.
The boat sails on a bearing of 143° for 24 km to a buoy at Q, then sails due south for 18 km to a buoy at R.

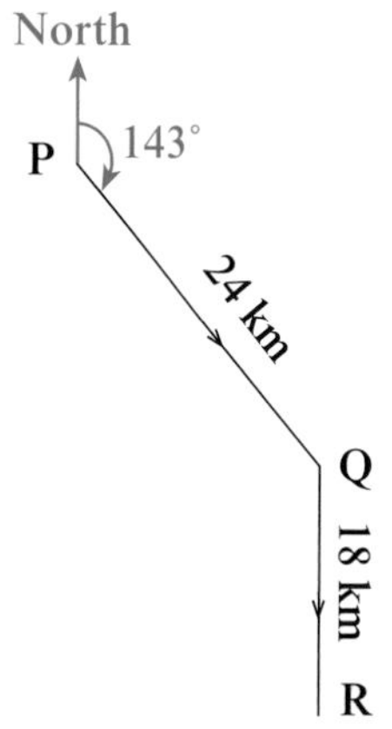


**a** What is the bearing of P from Q?
**b** Calculate:
  **i** the distance between P and R
  **ii** the bearing of R from P.
**c** **i** How far is Q east of P?
  **ii** How far is Q south of P?
**d** How far is Q east of PR?

**3** A boat starts at A and sails on a bearing of 048° for 12.6 km to B, where it changes course to a bearing of 154° to sail 18.5 km to C.
Calculate how far C is due east of the starting point A.

**4** From his boat Salamander Harry fixes the position of two other boats.
Enya is at a distance of 2300 metres on a bearing of 146°.
Cosmos is at a distance of 4200 metres on a bearing of 062°.
**a** Calculate the distance (in a straight line) between Enya and Cosmos.
**b** What is the bearing of Cosmos from Enya.

**5** The boat Wija sails due north between two headlands A and B. The bearing of B from A is 303°. From W, the bearing of A is 055°, and B 329°.
Wija is 3.4 Km from A and 7.5 km from B, calculate the distance AB.

**6** Two boats leave harbour at the same time.
One sails on a bearing of 308° at a speed of 18 knots, the other on a bearing of 204° at a speed of 22 knots.
What is the distance between the boats after 30 minutes?

## Angles of elevation and depression

- From a point A the angle of elevation of a point B is the angle between the horizontal and the line of sight from A to B.

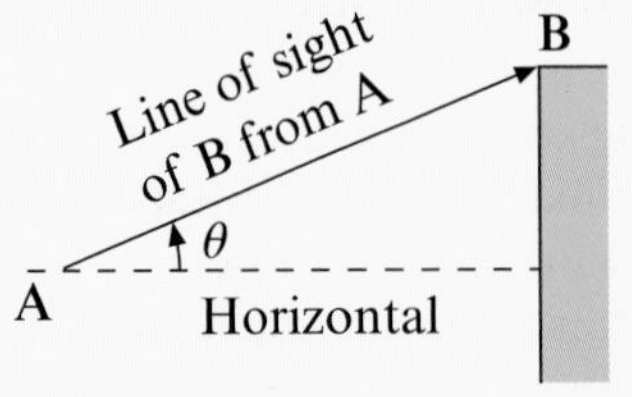


$\theta$ is the angle of elevation of B from A.

- From a point A the angle of depression of a point C is the angle between the horizontal and the line of sight from A to C.

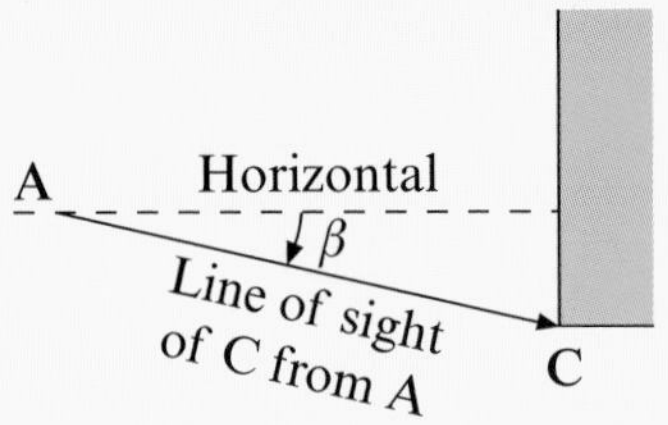


$\beta$ is the angle of depression of C from A.

### Example

A man of height 1.6 m is standing on a tower.
From this position he can see a ball which is 30 m from the base of the tower. He measures the angle of depression to the ball as 62°. Calculate the height of the tower.

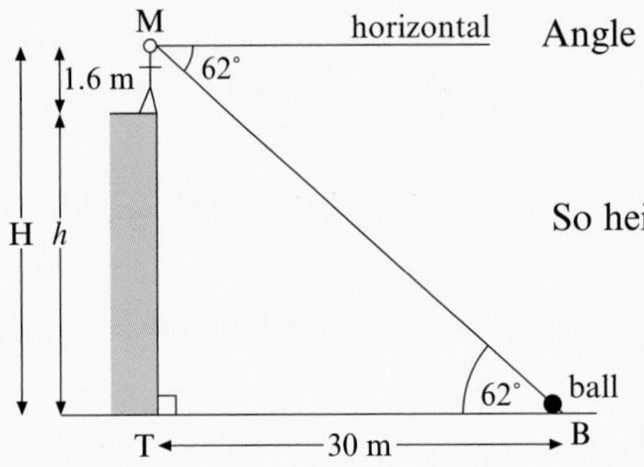


Angle $T\hat{B}M = 62°$ **(alternate angles)**
So $H = 30 \times \tan 62°$
$H = 56.4$ m (to 1 d.p.)

So height of tower $h = 56.4 - 1.6$
$= 54.8$ m to 1 dp

**Exercise 21.8**
**Angles of elevation and depression**

> **Accuracy**
> For this exercise round your answers to 2 dp.

1 Jan uses a theodolite at T to measure the distance and the angle of elevation or depression to points A and B.

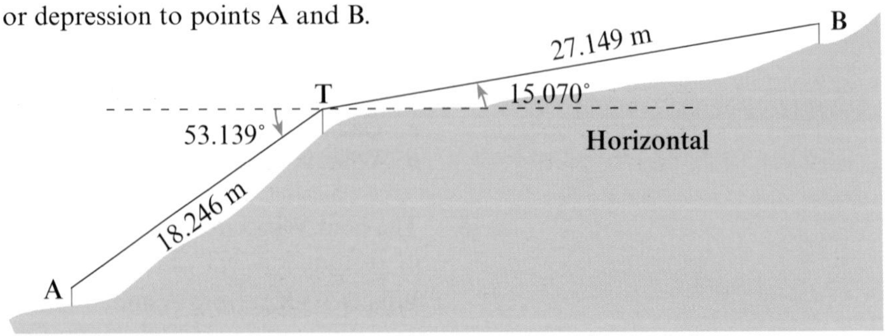


The angle of depression of A from T is 53.139° and TA is 18.246 metres.
The angle of elevation from T to B is 15.070° and TB is 27.149 metres.

**a** What is the obtuse angle $A\hat{T}B$?
**b** Calculate:
  **i** the distance AB
  **ii** the angle of depression from B to A
  **iii** the angle of elevation from A to B.

2 This diagram shows a loading ramp used for ferries. It is supported by cables that are fed out from a tower and fixed half way along the ramp at A.
The ramp must be fixed so that it is horizontal at high tide.
The maximum angle of depression allowed, at low tide, on the ramp is 18°.

a At which of these ports could this ramp be used?

| Harbour | Tidal range |
|---|---|
| Arden | 4.28 m |
| Jameston | 2.86 m |
| Daleen | 3.85 m |
| Palter | 3.54 m |

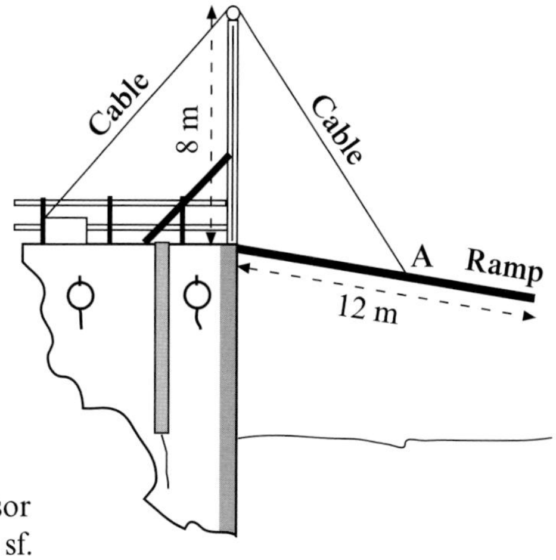


b A sensor is to be fitted to the cable sounding an alarm when the slope reaches 18°.
How far from A should the sensor be fixed? Give your answer to 3 sf.

## Three-dimensional trigonometry

Perpendicular indicates that two lines are at right angles to each other.

When you work in 3-D, it is essential to identify angles correctly.

The diagram shows a pyramid.
The vertical height is shown by WY.
(The line WY is perpendicular to the base.)

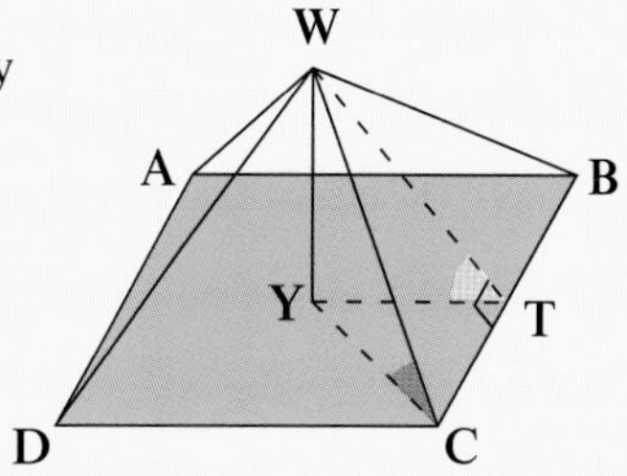


Identifying angles:

The angle between a side (WCB) and the base is angle WTY where WT is perpendicular to BC.

The angle between the edge CW and the base is angle WCY.

Note that angle WTY and angle WCY are **not** the same size.

**Exercise 21.9 Three-dimensional trigonometry**

1 The diagram shows a phone mast and points A and B due west and due south respectively.
The angle of elevation from A to the point T is 35°.
The distance AB is 740 m and the mast is 48 m high

Calculate correct to 2 sf:

**a** the distance AM.

**b** the angle of elevation of the top of the mast from point B.

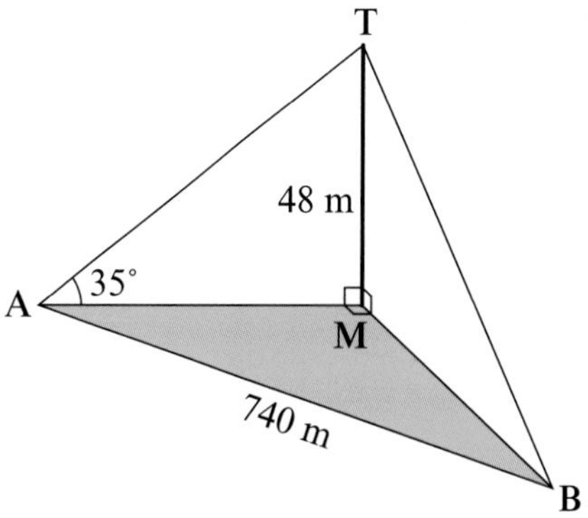

**2**

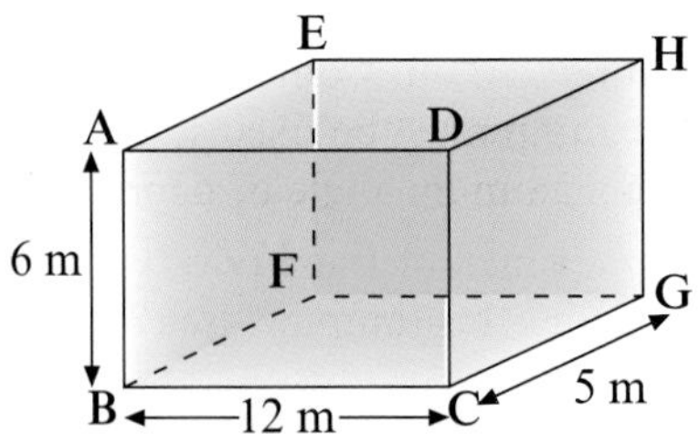


The diagram shows a cuboid with dimensions of 6 m, 12 m and 5 m.

Calculate, correct to 2 sf:

**a** the distance BG.
**b** the diagonal distance GA.
**c** the angle $C\hat{E}D$.

**3** The diagram shows an 8 cm cube.
Cutting through the midpoints, B, C and D of three edges forms a pyramid ABCD which has a triangular base.

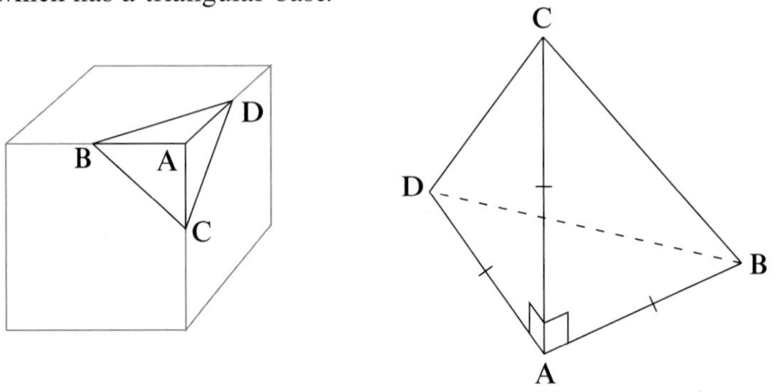


**a** Calculate the volume of the pyramid cut from the cube
**b** What is the angle between the edge BC and the base ABD?

The pyramid is placed to stand on the cut face, Δ BCD.

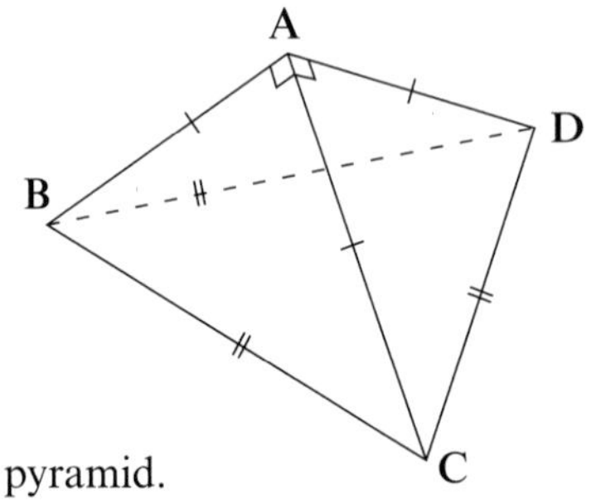


**c** Calculate the length of each side of the base BCD.
**d** What is the angle BCD?
**e** Calculate the area of Δ BCD.
**f** Calculate the perpendicular height of the pyramid.
**g** Calculate the angle between the line AB and the base BCD.

Give your answers correct to 2 s.f.

## End points

| You should be able to ... | ... so try these questions |
|---|---|
| **A** Use the sine rule | **A1** In triangle RST calculate, correct to 2sf: **a** the length RS **b** the length of RT. |
| | **A2** In triangle CDE calculate, to 2sf: **a** angle DEC **b** the length of ED. |
| **B** Calculate the area of a triangle | **B1** In Δ EFG calculate, to 3 sf: **a** the size of angle GEF **b** the length EG **c** the area of triangle EFG. |
| **C** Use the cosine rule | **C1** Using the cosine rule in Δ RST, calculate, to 3 sf: **a** the size of angle TRS **b** the size of angle RST. |
| | **C2** Using the cosine rule in Δ KLM calculate, to 3 sf: **a** the length of KL **b** the size of angle MKL. |

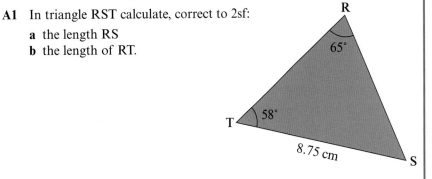

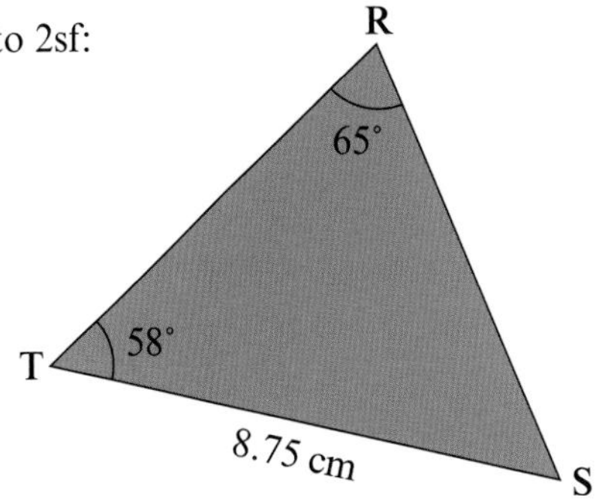


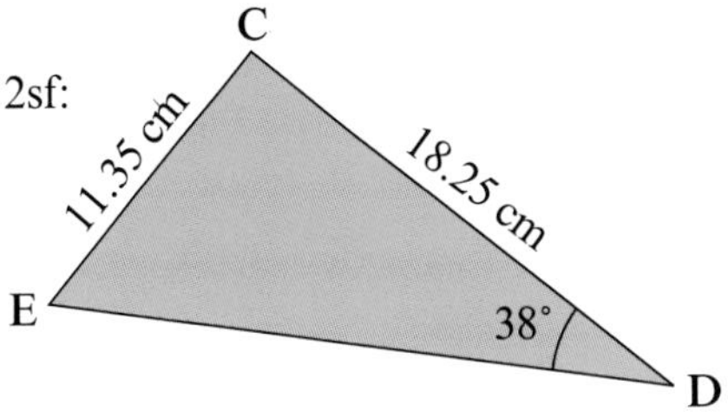


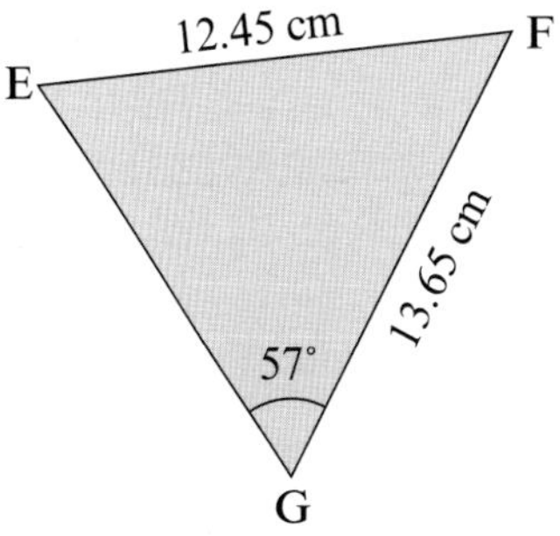


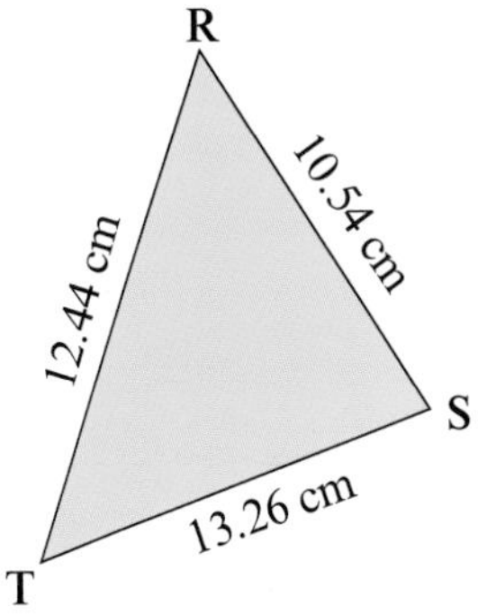


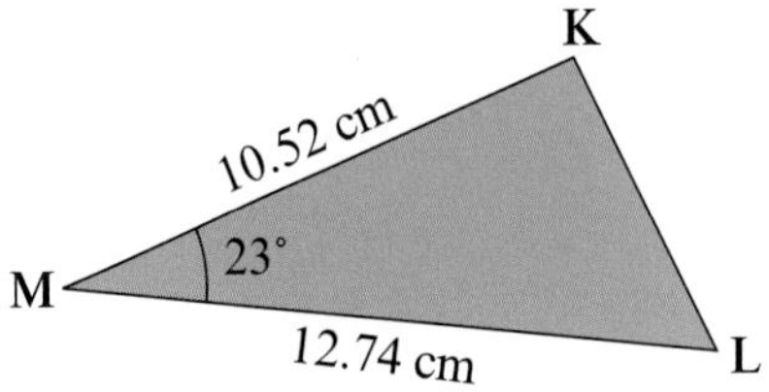

**D** Use bearings

**D1** A boat sails from point A for 4.5 kilometres on a bearing of 072° to point B. From B the boat sails on a bearing of 134° to point C. C is 6 kilometres due east of A.

Calculate, correct to 3 sf., how far the boat sailed from point B to point C.

**D2** Two boats leave harbour at the same time:

Seeker Me sails at a speed of 15 knots on a bearing of 192°.
Bold Over sails at a speed of 18 knots on a bearing of 316°.

Calculate the distance between the two boats after 150 minutes.

Give your answer correct to 3 sf.

**E** Use an angle of elevation or depression

**E1** From the top of a church tower, an observer notes that the church gate and a signpost are in line.

The gate is 65 metres from the base of the tower and at an angle of depression of 60° from the top of the tower.

The angle of depression of the signpost from the top of the tower is 18°.

Calculate, to the nearest metre, the distance between the gate and the signpost.

**E2** A radio mast is 75 metres tall is erected on level ground.

A camera is to be installed, at ground level, 20 metres from the base of the mast.

Calculate, to the nearest degree, the angle of elevation of the camera if it is to monitor the top of the radio mast.

**F** Solve three-dimensional problems

**F1** The diagram is of a growing unit sold by a garden centre.

The unit is in the shape of a square base with four triangular frames on top, in the shape of a pyramid.

When in position the triangular frames are at an angle of 55° to the base.

Calculate, correct to 2 sf, the full height of the growing unit.

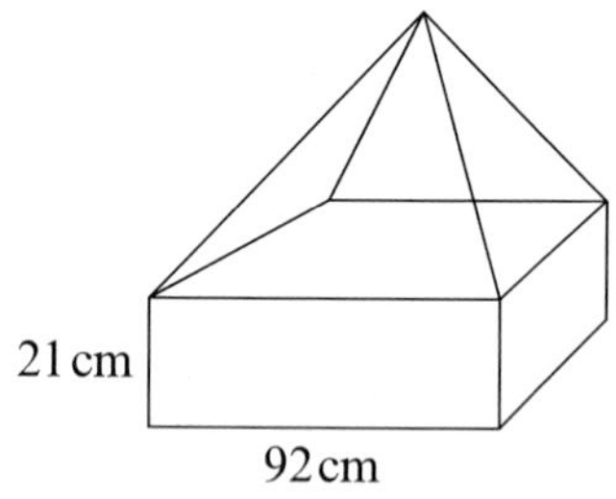


## Some points to remember

For any triangle ABC

- Sine rule

$$\frac{a}{\sin A} = \frac{b}{\sin B} = \frac{c}{\sin C}$$

$$\frac{\sin A}{a} = \frac{\sin B}{b} = \frac{\sin C}{c}$$

- Cosine rule

$$a^2 = b^2 + c^2 - 2bc \cos A$$

$$\cos A = \frac{b^2 + c^2 - a^2}{2bc}$$

- Area of $\Delta$ABC

$$\tfrac{1}{2}ab \sin C$$

## Starting points

You need to know about ... | ... so try these questions

### A Investigations using data

- The reason for carrying out an investigation using data is to either test a **hypothesis** or answer a **question**.
  The start of an investigation into sleep could be:
  Hypothesis – most people sleep at least 7 hours a night
  Question – how long do people sleep at night?
- You can carry out an investigation in four stages:

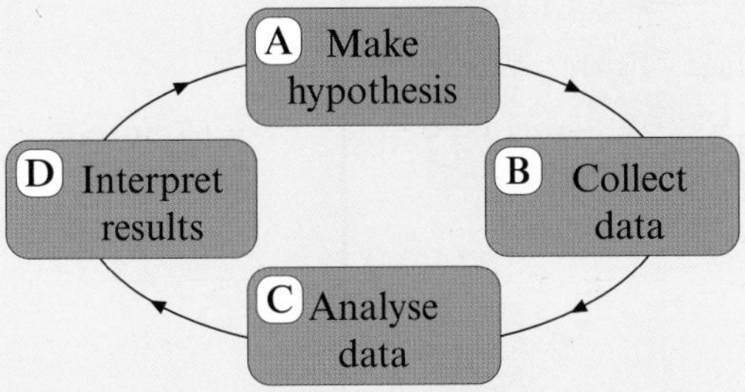


A, make a hypothesis (or ask a question)
B, collect the data you need
C, analyse the data you have collected
D, use the results of your analysis to decide whether the hypothesis is true or false.

- Stage D may give you an idea for a follow-up investigation.
  For example, the start of a follow-up into sleep could be:
  Hypothesis – people need more sleep in the winter
  Question – is there a link between sleep and time of year?

**A1** **a** Make your own hypothesis about sleep.
**b** Decide what data you need to collect to test your hypothesis.

### B Collecting data

- A **data collection sheet** is any form or table used to collect data.

To the BDA,
10 Queen Anne Street, London W1M 0BD
Tel: 0171-323 1531

*A charity helping people with diabetes and supporting diabetes research.*

I enclose a cheque/postal order* payable to the BDA £ ______
Debit my Access/Visa* card by the amount of £ ______
Card number □□□□□□□□□□□□□□□□
Expiry data □□□□
Please send me more information and membership details
Name ______
Address ______
Signature ______
*Delete which is inapplicable
Reg. Charity no. 215199

**Body Matters**

| Name | Length of thumb (cm) | Length of foot (cm) | Height (cm) |
|---|---|---|---|
| Sam | | | |
| Wasim | | | |
| Liz | | | |
| Shane | | | |
| Des | | | |
| Linda | | | |
| Dean | | | |
| Nisha | | | |

- A **questionnaire** collects data by asking questions.

***Sleep Questionnaire***

1 What is your name? ______
2 How old are you? ______ years
3 What time do you usually go to bed? ______

- A data collection sheet or questionnaire is also called a **survey**.

**B1** Design a data collection sheet to use for a traffic survey.

## C Types of question

- You can use different types of questions on a questionnaire:
  - multi-choice questions

> **7** What do you sleep on?
> *Please tick one box only* ☐ Back ☐ Front ☐ Side

  - multi-choice questions with a scale

> **8** How heavy a sleeper are you? HEAVY ⟶ LIGHT
> *Please tick one box only* Very ☐ Fairly ☐ Average ☐ Fairly ☐ Very ☐

  - branching questions

> **9** Do you suffer from regular sleepless nights?
> *Please tick one box only* ☐ Yes ☐ No
> *If your answer is* NO *then go to Question 12*

  - questions with more than one answer.

> **12** What helps you get a good night's sleep?
> *Put a 1 in the box for the most helpful, a 2 for the next most helpful, and so on*
> ☐ A hot drink just before bed
> ☐ Eating just before bed
> ☐ Exercise during the evening
> ☐ Relaxation breathing
> ☐ Other *(please state)* ______

## D Using a scatter diagram

- You can use a scatter diagram to investigate if there is a link between two sets of data.

Question – does the time taken to get to sleep depend on the amount of light in the room?

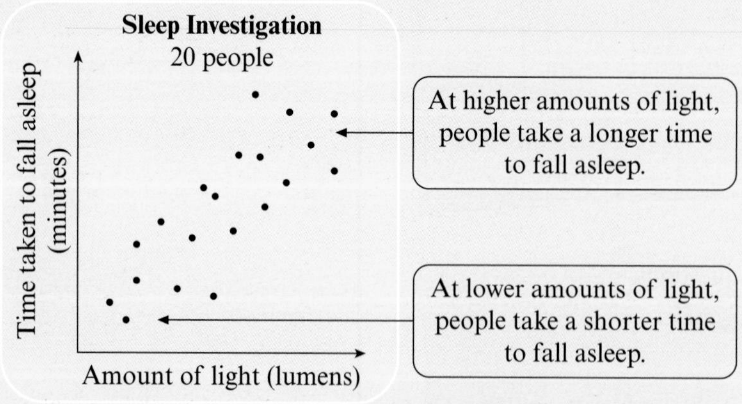
Sleep Investigation
20 people
Time taken to fall asleep (minutes)
Amount of light (lumens)
At higher amounts of light, people take a longer time to fall asleep.
At lower amounts of light, people take a shorter time to fall asleep.

The time taken to fall asleep does depend on the amount of light: as the amount of light increases, the time taken to fall asleep also increases.

**C1** Design a questionnaire about sleep which includes different types of question.

**D1**

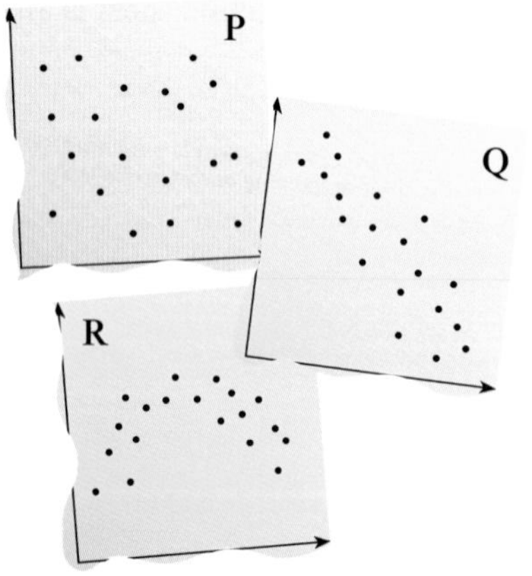
P
Q
R

Match each scatter diagram to each pair of sets of data.

| | | | |
|---|---|---|---|
| **a** | Time taken to get to sleep | against | Time since drank coffee |
| **b** | Amount of sleep | against | Room temperature |
| **c** | Amount of sleep | against | Height of bed off floor |

## Designing and criticising questions

- When you create a questionnaire, your questions must be carefully designed to:
  - make them easy to answer
  - make sure the answers give you the data you need.

| Poor question | | Improved question |
|---|---|---|
| How much sleep did you get last night? ☐ Less than average ☐ About average ☐ More than average | This question is **not clear**: the words used need to be more exact. (Different people are likely to have different ideas of what is meant by 'average'.) | How much sleep did you get last night? ☐ Less than 8 hours ☐ About 8 hours ☐ More than 8 hours |
| Do you agree that we need at least 8 hours sleep each night? | This is a **leading** question: it leads people into giving a certain answer. (The question seems to expect the answer 'Yes'.) | Do you think we need at least 8 hours sleep each night? ☐ Yes ☐ No ☐ Not sure |
| What do you sleep on? | This question is **ambiguous**: it could have more than one meaning. (The question is meant to be about sleeping position, but could be answered 'a bed'!) | What do you sleep on? ☐ Back ☐ Front ☐ Side |

**Exercise 22.1**
Designing and criticising questions

**1** **a** Explain why this question is not clear.
**b** Write an improved question.

When do you usually go to bed?
☐ Early ☐ Late

**2** **a** Explain why this is a leading question.
**b** Write an improved question.

You get a worse night's sleep on a soft bed, don't you?

**3** **a** Explain why this question is ambiguous.
**b** Write an improved question.

Where do you sleep best?

**4**

***Leisure Centre Survey***

1 Do you agree that the town needs a new leisure centre? ☐ Yes ☐ No
2 Would you be a frequent user of the centre? ☐ Yes ☐ No
3 Would you use the courts? ☐ Yes ☐ No
4 How much would you be prepared to pay to use the pool? ☐ Less than £1.50 ☐ More than £2.50

This questionnaire has been written to survey local people about a new leisure centre.

**a** Criticise each of the questions.
**b** Write an improved question for each one.

# Experiments

- A survey asks people to give an opinion about something, or asks about facts which are easy to remember.

Example

***TV Survey***

**1** What is your favourite TV channel? ☐ BBC1 ☐ BBC2 ☐ ITV ☐ Channel 4 ☐ Channel 5

**2** Did you watch TV last night? ☐ Yes ☐ No

- The data needed for some investigations can only be collected:
  - over a period of time

    How much time do you spend in a week watching each TV channel?

  - by designing an experiment.

    People take longer to get to sleep the more light there is in the room.

The data collection sheet used for these types of investigation can be called an **observation sheet**.

## Exercise 22.2 Experiments

To design your experiment:
- decide what data you need
- decide how to collect it
- design an observation sheet.

**1** Design an observation sheet to collect data on how much time people spend in a week watching each TV channel.

**2** Design an experiment to answer this question.

***Body Matters***
How many times do people blink in a day?

**3** **a** Carry out your experiment.
**b** Analyse the data you collect.
**c** Interpret your results to answer the question.

**4** Design an experiment to test this hypothesis.

***Body Matters***
Taller people do not have as good a sense of balance as shorter people.

**5** **a** Carry out your experiment.
**b** Analyse the data you collect.

**6** Do you think the hypothesis is true or false? Explain why.

## Thinking ahead to ... sampling

**A** Rachel is investigating how much sleep students in her school get each night.

This is the distribution of students in the school.

| Yr 7 | Yr 8 | Yr 9 | Yr 10 | Yr 11 | Total |
|---|---|---|---|---|---|
| 150 | 155 | 161 | 140 | 146 | 752 |

Rachel decides to take a random sample of 150, i.e. about 20%.

Which of these possible selections do you think will give the best sample? Explain your answer.

Ⓐ

| Yr 7 |
|---|
| 150 |

Ⓑ

| Yr 7 | Yr 8 | Yr 9 | Yr 10 | Yr 11 |
|---|---|---|---|---|
| 30 | 30 | 30 | 30 | 30 |

Ⓒ

| Yr 7 | Yr 8 | Yr 9 | Yr 10 | Yr 11 |
|---|---|---|---|---|
| 30 | 31 | 32 | 28 | 29 |

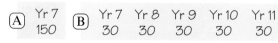

## Stratified random sampling

- When you choose a random sample from a population, the sample should model the characteristics of the population as closely as possible.
- If you think there are certain groups, or strata, within the population that are likely to give very different responses, then you can model this by taking a random sample from each group: a **stratified random sample**.
- The size of the sample you take from each group, or stratum, should be proportional to the size of the group in the population.

**Example** Show how a sample of 150 is chosen from this population, taking into account the students' year group and gender.

| | Yr 7 | Yr 8 | Yr 9 | Yr 10 | Yr 11 | Total |
|---|---|---|---|---|---|---|
| Boys | 69 | 74 | 85 | 74 | 76 | 378 |
| Girls | 81 | 81 | 76 | 66 | 70 | 374 |
| Total | 150 | 155 | 161 | 140 | 146 | 752 |

Number of Yr 7 girls in sample = $\frac{81}{752} \times 150 =$ **16** to nearest whole number

Number of Yr 11 boys in sample = $\frac{76}{752} \times 150 =$ **15** to nearest whole number

**Exercise 22.3**
**Sampling**

1 Copy and complete the following table to show how the stratified random sample is chosen for the example above.

| | Yr 7 | Yr 8 | Yr 9 | Yr 10 | Yr 11 | Total |
|---|---|---|---|---|---|---|
| Boys | | | | | 15 | |
| Girls | 16 | | | | | |
| Total | | | | | | 150 |

2 The Great Britain Team for the 1996 Olympic Games had 500 members: 312 competitors and 188 support staff.
Show how a stratified random sample of 120 members is chosen.

3 The gender breakdown of the Great Britain Olympic Team was:

| | Competitors | Support Staff | Total |
|---|---|---|---|
| Male | 189 | 132 | 321 |
| Female | 123 | 56 | 179 |
| Total | 312 | 188 | 500 |

**a** Show how a stratified random sample of 120 members is chosen.
**b** What is the quickest way to calculate how many females would be in a stratified random sample of 75 members?
Explain your answer.

4 **a** What was the total population of the UK in 1995?
**b** Calculate how many people from each country would be in a stratified random sample of 500 000.

**1995 UK Population (millions)**

| England | Northern Ireland | Scotland | Wales |
|---|---|---|---|
| 48.9 | 1.7 | 5.1 | 2.9 |

Make sure that the number of people in your sample adds up to 500 000.

# Correlation

◆ You can describe the link between two sets of data using the term **correlation**.

**Sleep Experiment 1**
Does the length of time you take to fall asleep depend on how light the room is?

These results show **positive** correlation: an *increase* in one set of data tends to be matched by an *increase* in the other set.

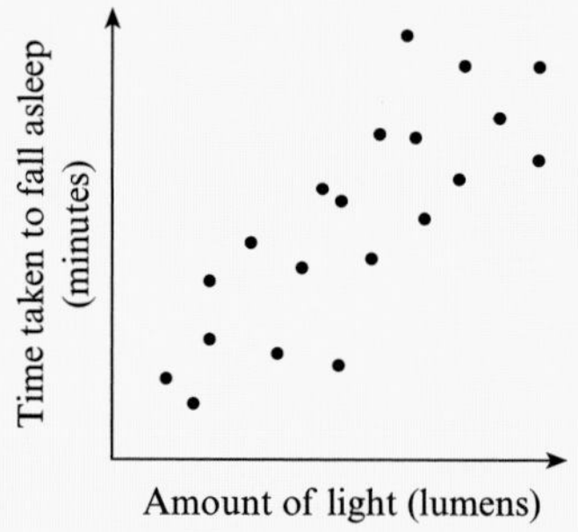


**Sleep Experiment 2**
Does the length of time you sleep depend on the length of time since you last slept?

These results show **negative** correlation: an *increase* in one set of data tends to be matched by an *decrease* in the other set.

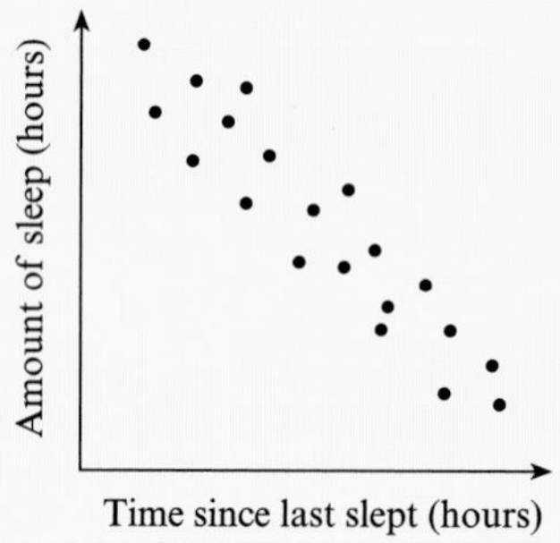


The result of experiment 2 may not be what you expect. It happens because sleep is part of your daily rhythm of sleeping and waking.

Going to bed late means that you will soon reach your time for waking, and vice versa.

A daily rhythm, like this sleep/wake example, is called a **circadian rhythm**.

## Exercise 22.4
Correlation

**1**

**Mean semi-detached house prices in towns near London – 2nd Quarter 1996**

| | | | | | | | | | | | | |
|---|---|---|---|---|---|---|---|---|---|---|---|---|
| Distance from London (miles) | 44 | 66 | 41 | 75 | 86 | 47 | 68 | 36 | 62 | 77 | 57 | 53 |
| Mean house price (£000's) | 93 | 78 | 98 | 72 | 63 | 97 | 71 | 104 | 86 | 64 | 78 | 88 |

**a** Draw these axes: horizontal 0 to 90, vertical 50 to 130.
**b** Plot the house price data on your diagram.
**c** Is the correlation positive or negative?

**2**

**Eye Tests for 10 people**

| | | | | | | | | | | |
|---|---|---|---|---|---|---|---|---|---|---|
| Pressure in eye (mmHg) | 12.1 | 11.7 | 15.2 | 19.1 | 11.2 | 18.9 | 15.9 | 17.3 | 13.0 | 16.6 |
| Refractive power of lens (dioptres) | 3.6 | ¯3.9 | 5.1 | 10.4 | ¯6.4 | 3.0 | ¯6.9 | 6.5 | ¯8.4 | 0.8 |

A negative number of dioptres shows short-sightedness; a positive number of dioptres shows long-sightedness.

**a** Draw these axes: horizontal 10 to 20, vertical ¯12 to 12
**b** Plot the eye test data on your diagram.
**c** Is the correlation positive or negative?

**3** For each of your scatter diagrams, draw a line through the middle of the plots, like this:

This is called drawing a line **by eye** or **by inspection**.

**4** Which scatter diagram did you find it easier to draw the line on? Explain why.

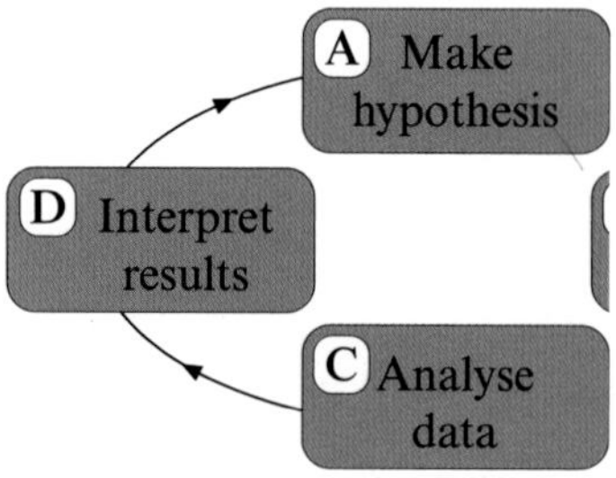

## Using a line of best fit

- A line drawn through the middle of the plots on a scatter diagram is called a **line of best fit**. The stronger the correlation, the easier it is to draw this line.

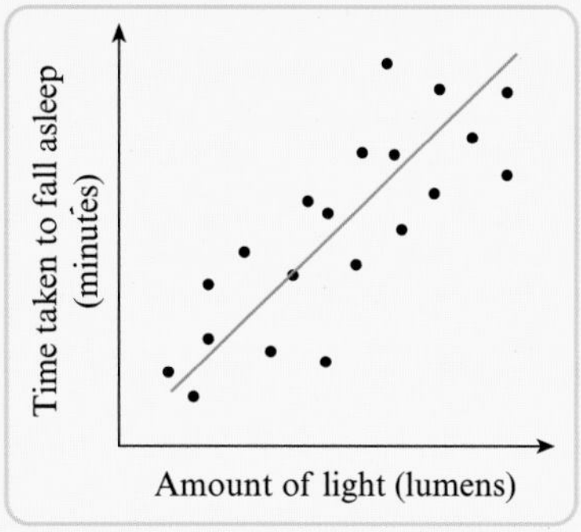


This is **moderate** positive correlation because the plots are well scattered around the line of best fit.

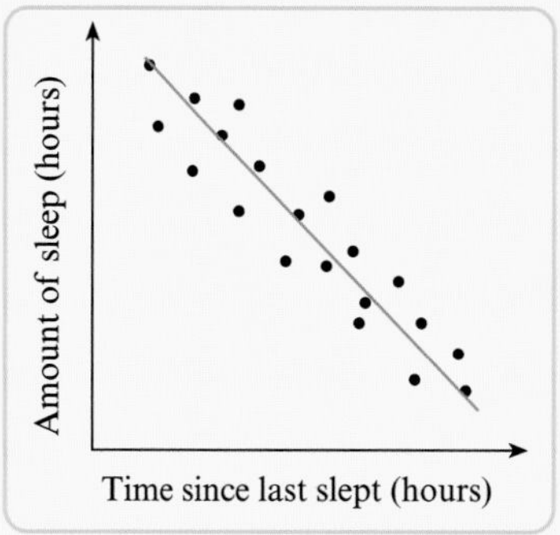


This is **strong** negative correlation because the plots are quite close to the line of best fit.

- When it is not possible to draw a line of best fit, there is no link between the two sets of data: there is **no correlation**.

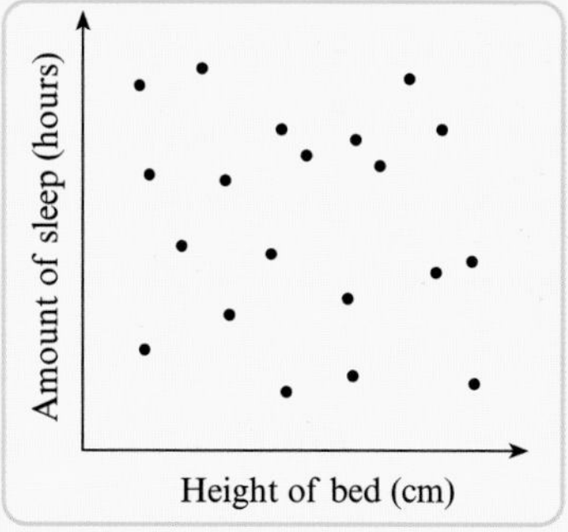


**Exercise 22.5**
Describing correlation

1 Use this scatter diagram to describe the correlation between income and percentage of income given to charity.

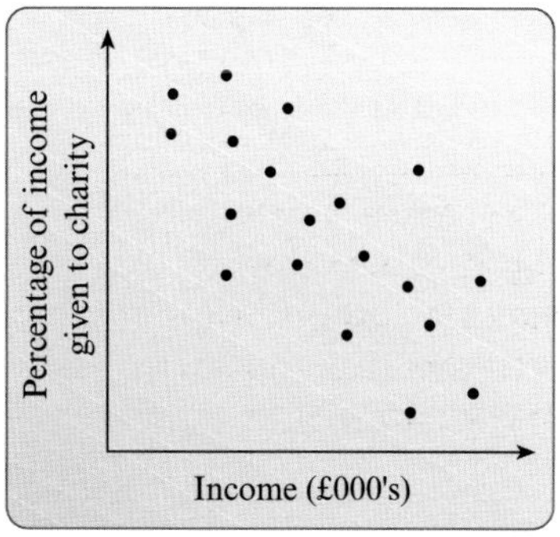


2 Use your scatter diagram from Exercise 22.4 Question **1** to describe the correlation between distance of town from London and mean house price.

3 Use your scatter diagram from Exercise 22.4 Question **2** to describe the correlation between pressure in eye and refractive power of lens.

4 Design an experiment to answer this question:
'Is there any correlation between your fathom and your height?'

5 **a** Carry out your experiment.
**b** Plot the data you collect on a scatter diagram.
**c** Draw a line of best fit.
**d** Use your scatter diagram to describe any correlation.

Your fathom is the distance between the ends of your fingers when your arms are stretched as wide as possible. This distance is roughly six feet for an adult.

# Estimating values from a line of best fit

- It is possible to estimate values from a line of best fit.

Example

**a** Estimate the height of a person with head circumference 56 cm.
**b** Estimate the head circumference of a person 195 cm tall.

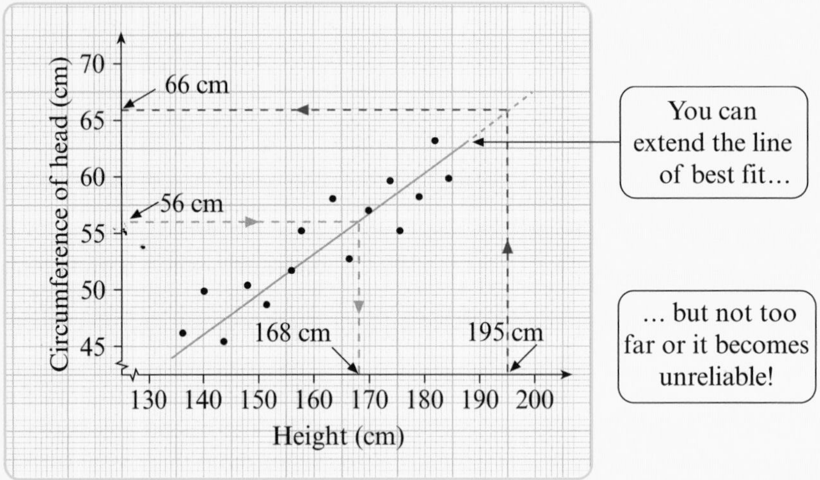


**a** Estimated height of person with head circumference 56 cm = **168 cm**
**b** Estimated head circumference of person 195 cm tall = **66 cm**

**Exercise 22.6**
Estimating values

Use your scatter diagram from Exercise 22.4 Question **1** for Questions **1** to **3**.

**1** Estimate the distance from London of a town with a mean house price of:
**a** £90 000 **b** £65 000

**2** Estimate the mean house price for a town:
**a** 70 miles from London **b** 55 miles from London

**3** Extend your line of best fit to estimate:
**a** the distance from London of a town with a mean house price of £120 000
**b** the mean house price for a town 25 miles from London.

**4**

**Natural Births – Length of Pregnancy & Weight of Baby**

| | | | | | | | | | | |
|---|---|---|---|---|---|---|---|---|---|---|
| Length of pregnancy (days) | 271 | 287 | 283 | 274 | 271 | 279 | 263 | 276 | 283 | 270 |
| Weight of baby (kg) | 2.5 | 4.2 | 3.8 | 3.3 | 4.5 | 3.4 | 2.9 | 4.1 | 4.3 | 3.5 |

This data has been collected to test the hypothesis:
'A longer pregnancy leads to a heavier baby.'
Plot this data on a scatter diagram.

**5** **a** Draw a line of best fit on your scatter diagram.
**b** Describe the correlation between length of pregnancy and weight of baby.
**c** Use your line of best fit to estimate:
**i** the length of pregnancy for a baby that weighs 3.5 kg
**ii** the weight of a baby with a length of pregnancy of 280 days.

**6** Do you think it would make sense to extend this line of best fit?
Give reasons for your answer.

## End points

**You should be able to …**   **… so try these questions**

**A** Design and criticise questions for a questionnaire

**A1**

*Bypass Survey*

1 Are you a local? ☐ Yes ☐ No
2 What do you think of the traffic in the village? ____________
3 Do you agree that the village needs a bypass? ☐ Yes ☐ No

**a** Criticise each of these questions.
**b** Write an improved question for each one.

**B** Design experiments

**B1** Design an experiment to answer this question.

*Body Matters*
How long can people hold their breath for?

**C** Choose a stratified random sample

**C1**

**1991 Population of capital Cities (000's)**

| Belfast | Cardiff | Edinburgh | London |
|---|---|---|---|
| 279 | 272 | 402 | 2343 |

Calculate how many people from each capital city would be in a stratified random sample of 10 000.

**Petrol Cars – Size of Engine & Petrol Consumption**

| | | | | | | | | | | | | |
|---|---|---|---|---|---|---|---|---|---|---|---|---|
| Size of engine (litres) | 1.6 | 2.6 | 1.2 | 4.0 | 2.5 | 1.1 | 3.2 | 4.0 | 3.2 | 1.8 | 2.4 | 3.5 |
| Petrol consumption (mpg) | 45 | 37 | 40 | 22 | 43 | 48 | 29 | 29 | 33 | 39 | 32 | 25 |

**D** Use a scatter diagram and line of best fit to describe correlation

**D1** **a** Draw axes: horizontal 0 to 6.0, vertical 0 to 50
**b** Plot the petrol car data on your diagram.
**c** Draw a line of best fit.
**d** Describe any correlation between size of engine and petrol consumption.

**E** Estimate values from a line of best fit

**E1** Use your scatter diagram to estimate:
**a** the petrol consumption of a car with a 3.0 litre engine
**b** the engine size of a car with petrol consumption of 40 mpg.

**E2** Extend your line of best fit to estimate:
**a** the petrol consumption of a car with a 5.0 litre engine.
**b** the engine size of a car with petrol consumption of 20 mpg.

## Some points to remember

- When you choose a stratified random sample, check that the sum of the sample sizes from each stratum equals the total sample size.
- When you draw a line of best fit on a scatter diagram, make sure the line goes through the middle of the points.
- In some cases, it does not make sense to extend the line of best fit on a scatter diagram.

## Starting points

You need to know about ... | ... so try these questions

### A Calculations with time, distance and speed

**Speed = Distance ÷ Time**

**Time = Distance ÷ Speed**

**Distance = Speed × Time**

**Example 1** How long does it take a car travelling at an average speed of 54 mph to cover a distance of 350 miles?

Time = Distance ÷ Speed
= 350 ÷ 54
= 6.48 hours (to 3 sf)
= **6 hours 29 minutes** (to the nearest minute)
(since 0.48 hours is 0.48 × 60 = 28.8 minutes)

**Example 2** An arrow from a long bow flew a distance of 196 metres in 2 seconds. What was its average speed in mph?

Speed = Distance ÷ Time
= 196 ÷ 2 = 98 metres per second
= 98 × 60 × 60 = 352 800 metres per hour
= 352 800 ÷ 1000 = 352.8 km per hour
= 352.8 ÷ 1.609 = **219.3 mph** (1 mile = 1.609 km)

**A1** What is 0.68 hours to the nearest minute?

**A2** A car made the journey from Doncaster to Oxford (135 miles) in 2 hours 52 minutes. What was its average speed? Give your answer to 2 sf.

**A3** How long does it take a hot air balloon to fly 100 metres at a speed of 41 m $\text{min}^{-1}$? Give your answer in minutes and seconds.

**A4** Convert the following:
**a** 14 metres per second into miles per hour
**b** 38 miles per hour into metres per second.

### B The area of a trapezium

**Area = $\frac{1}{2}(a + b)h$**

where $h$ is the perpendicular height.

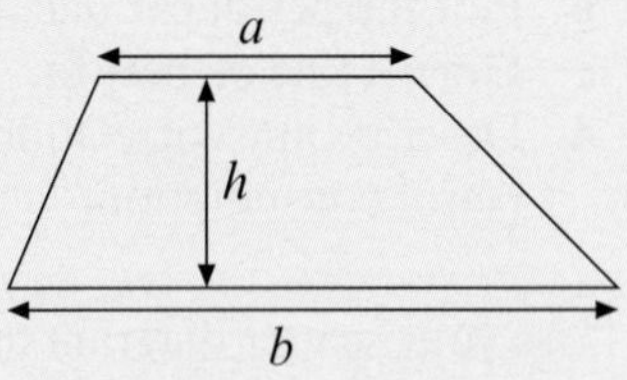


**Example** Find the area of this trapezium.

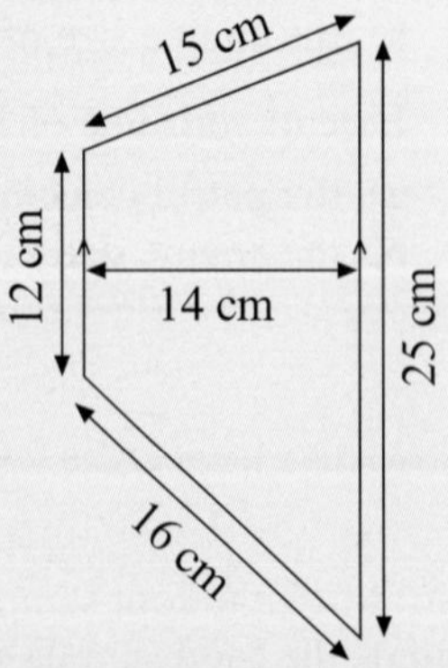


First identify the dimensions $a$, $b$ and $h$:

$a$ = 12 cm
$b$ = 25 cm
$h$ = 14 cm

Note that the lengths 15 cm and 16 cm are not used.

Area = $\frac{1}{2}(a + b)h$
= $\frac{1}{2}$ × (12 + 25) × 14
= 0.5 × 37 × 14
= 259

**The area of the trapezium is 259 cm².**

**B1** What is the area of this trapezium?

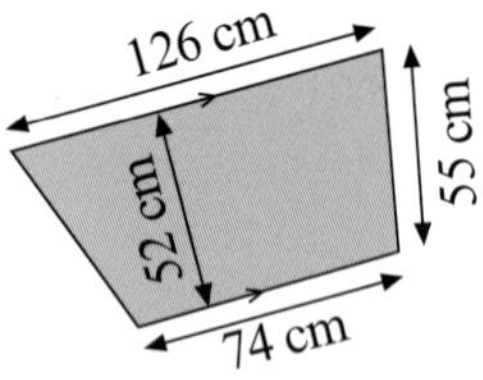


**B2** Calculate the area of this trapezium in square units.

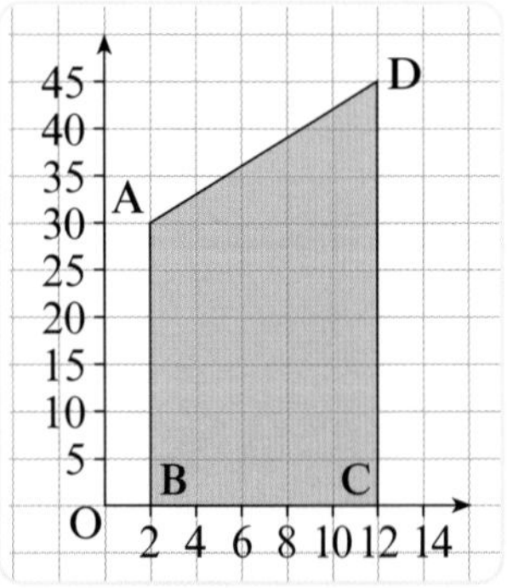

## C Using distance–time graphs

This graphs shows journeys made by four people on one day. They all travelled on the same road.

Joe left home in his Citröen and got back home some time after 4 pm.

Sally took the Jeep and stopped at a friends for lunch then continued her journey.

Ravi drove without a break all day on his Honda motorbike.

Charlie pedalled non-stop on his Claude Butler bike.

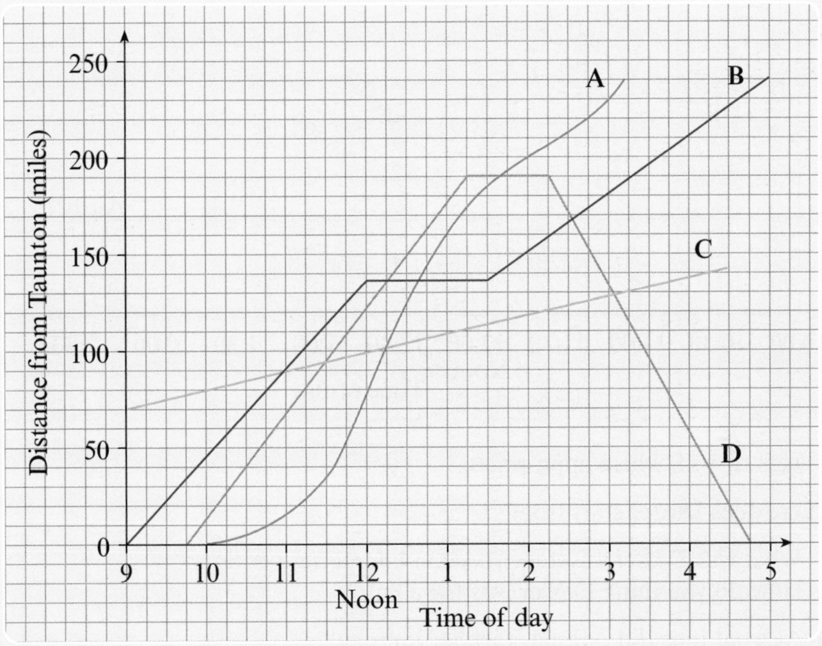


$$\text{Average speed} = \frac{\textbf{Total distance}}{\textbf{Total time}}$$

The total time includes any breaks.

**Example 1** Calculate the average speed of A over the journey.

$$\text{Average speed} = \frac{240}{(10\text{ am to }3{:}15\text{ pm})}$$

$$= \frac{240}{5.25} \qquad (5\text{ h }15\text{ min is }5.25\text{ h})$$

$$= \mathbf{45.7\ mph}$$

**Example 2** Calculate the average speed of D over the day.

$$\text{Average speed} = \frac{(190 + 190)}{(9{:}45\text{ to }4{:}45)}$$

$$= \frac{380}{7}$$

$$= \mathbf{54.3\ mph}$$

**C1** Identify the line on the graph for each person's journey. Give reasons for your answers.

**C2** Which two journeys finished at the same place? Explain your answer.

**C3** Which two journeys started at the same time?

**C4** At what time did D's journey start?

**C5** At noon how far away from Taunton was:
**a** A **b** B
**c** C **d** D?

**C6** How long did Sally spend at her friend's house? How can you tell?

**C7** Who did not start at Taunton? How far was this person from Taunton at 9 am?

**C8** How far from Taunton was A when overtaking C?

**C9** From the graph: D passed C twice.
- **a** At what times did they pass ?
- **b** What was the difference about the directions on the two occasions?

**C10** Calculate C's average speed over the whole journey.

**C11** What was B's average speed over the whole journey ?

**C12** How does B's speed, up to noon, compare with B's speed after 1:30 pm? Give your reasons.

**C13**
- **a** Who appears to reach the highest top speed at any time on their journey?
- **b** Roughly when was this?
- **c** Why is it impossible to tell for certain from the graph?

# Finding speeds from distance–time graphs

This distance–time graph shows journeys taken by four people who arrange to meet in Haddington. Julie and Simon are having problems with their cars.

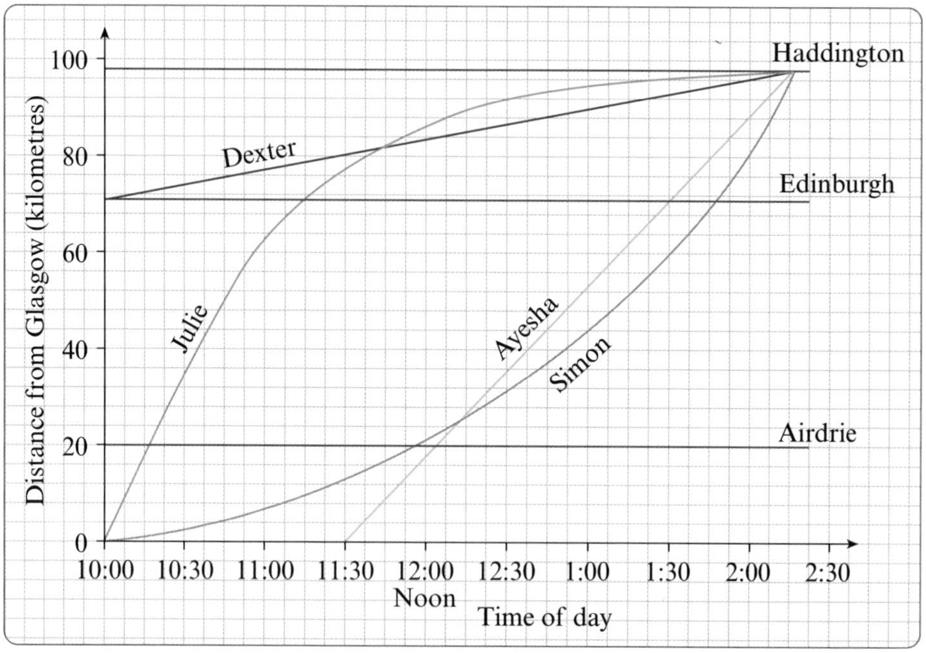


**Exercise 23.1**
Speeds from graphs

You may see
**kilometres per hour**
in any of these forms

**kph**
**km h$^{-1}$**
**km/h**

Give answers to 3 s.f. where appropriate.

**1** Where does Dexter start from?

**2** **a** What was Dexter's average speed in kilometres per hour?
**b** How do you think Dexter was travelling?

**3** **a** At what time did Ayesha start her journey?
**b** What was her average speed in km h$^{-1}$?

**4** Describe how the speeds during Julie's and Simon's journeys differed from each other.

**5** What was the average speed over the journey for:
**a** Simon
**b** Julie?

**6** How did Simon's, Ayesha's and Julie's speeds compare when they passed through Airdrie?

For question 7 you need to check where the slope of Ayesha's curve is roughly the same as the slope of Dexter's graph.

**7** At what time was Simon travelling at roughly the same speed as
**a** Ayesha
**b** Dexter?
Explain how you decided.

**8** How far from Glasgow was Julie when she was travelling at roughly the same speed as:
**a** Ayesha
**b** Dexter?

**9** At roughly what time were Simon and Julie travelling at the same speed?

**10** Julie and Simon passed the same point on the road at the same speed. How far from Glasgow is this point?

**Exercise 23.2**
**Drawing and interpreting graphs**

> **Accuracy**
> In this exercise, give all average speeds to 1 dp.

1 Ken and his sister Amy go on a cycling holiday.
The graph shows part of Ken's journey for the first day.

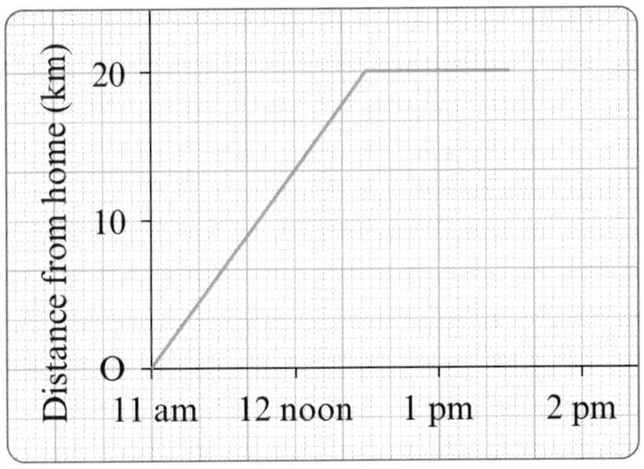


**a** After 1.30 pm he cycles at 15 km $h^{-1}$ to reach a camp-site that is 50 km from home.
  **i** Copy and complete the graph for Ken's first day.
  **ii** Find Ken's average speed for his whole journey.

**b** Amy leaves their home at 1pm and cycles along the same route at 22 km $h^{-1}$.
  **i** Draw a line on your graph to show Amy's journey.
  **ii** When did she reach the camp-site?
  **iii** At about what time did Amy pass Ken on her way to the camp-site?

2 Alice lives in Cupar and Emily lives 10 miles away in St Andrews.
Alice cycles to St Andrews. This graph shows part of her journey.

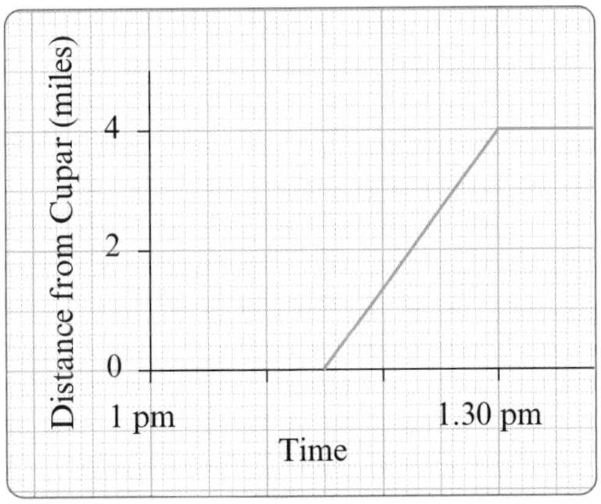


**a** At 1.30 pm, she stops for 10 minutes to buy cakes and then cycles at a constant speed to reach Emily's house at 2.20 pm.
  **i** Copy and complete the graph for Alice's journey.
  **ii** What is her speed in mph after her stop to buy cakes?
  **iii** Calculate Alice's average speed in mph for her whole journey.

**b** On the same day, Emily decides to visit Alice. She leaves at 1 pm and cycles at a speed of 12 mph for 4 miles. She stops for five minutes and then carries on at a constant speed to reach Alice's house at 1.47 pm.
  **i** How far is Emily from Cupar at the start of her journey?
  **ii** Show her journey on your graph.
  **iii** Find Emily's average speed in mph for her journey to Cupar.
  **iv** Explain why you think Emily did not see Alice on her journey.

**c** After ringing Alice's door bell for 3 minutes, Emily decides to cycle home. She reaches St Andrews at the same time as Alice does.
  **i** Show Emily's journey home to St Andrews on your graph.
  **ii** How fast does she cycle home?

## The link between distance–time and speed–time graphs

A distance–time graph can give an indication of changing speed when the changing gradient of the curve is considered. It is also possible to draw a speed–time graph from it. Here are two graphs of the same short bicycle journey.

On this distance–time graph some tangents have been drawn to show the speed at these points. It is possible, though, to draw a tangent at any point on the curve. There are an infinite number of tangents possible.

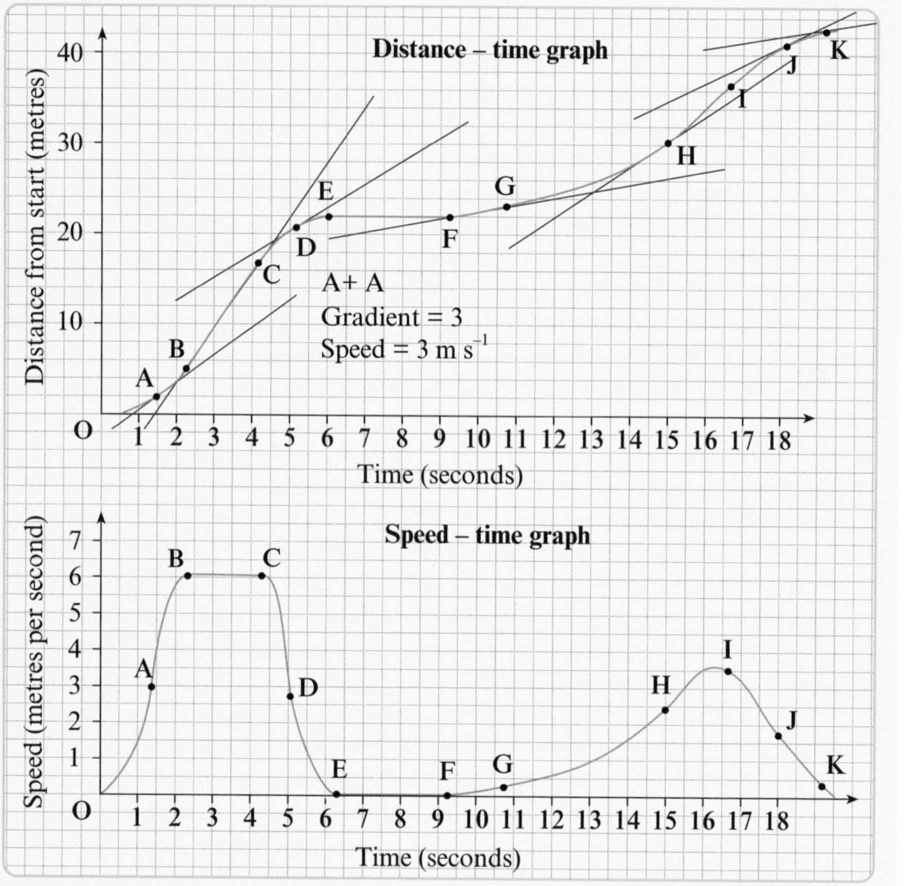


You may see **metres per second** in any of these forms:

**m per s**
**m/s**
**m s$^{-1}$**

At any point on the distance–time graph a gradient can be drawn and the speed calculated. For example, at point A (after about 1.3 seconds) the speed of the bike is 3 metres per second. Point A can therefore be plotted on the speed–time graph as a speed of 3 m s$^{-1}$ after 1.3 seconds. The more points at which the gradient is found, the more accurate will be the speed–time graph.

### Exercise 23.3 Reading a speed–time graph

1 How can you tell from the distance–time graph that the bike is travelling faster at point B than at point A?

2 What does a horizontal line indicate on:

**a** a distance–time graph **b** a speed–time graph?

3 The following questions are about the speed–time graph, but you may need to look at the corresponding points on the distance–time graph.

**a** Why does the graph show a speed of zero between points E and F?
**b** Why are the speeds equal at points B and C?
**c** Why is the second peak on the graph lower than the first peak?
**d** Give two labelled points where the bike is:
**i** accelerating (speeding up) **ii** decelerating (slowing down).

## End points

You should be able to ... | ... so try these questions

**A** Interpret distance–time graphs

**A1** This graph shows the journey of a car and a cycle.
The cyclist travelled from Leeds to York.
The car driver took the same route to York but returned to Leeds.

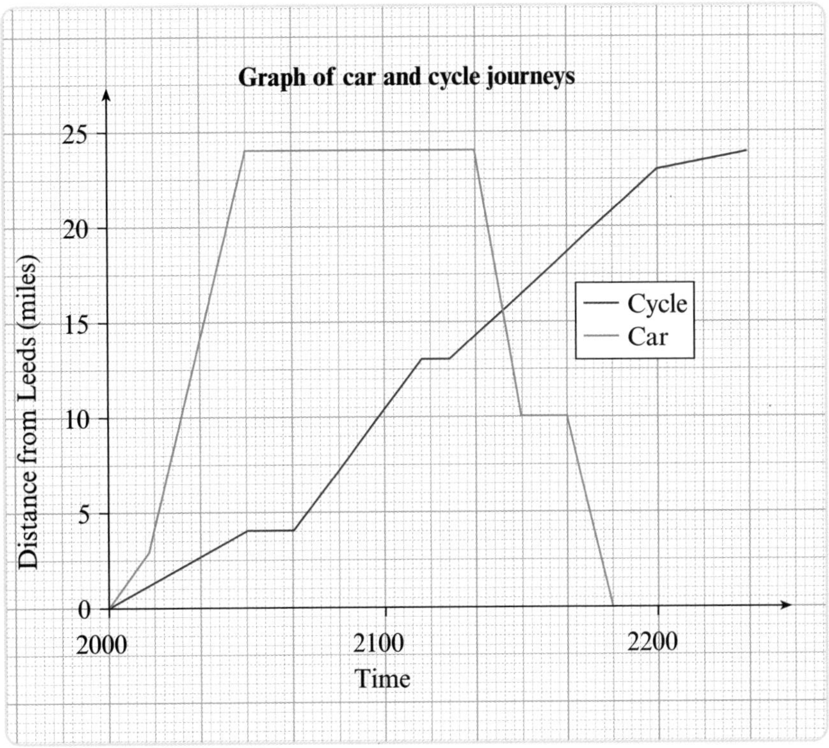


**a** How far did the cyclist travel in the first 20 minutes?

**b** How long did it take the car to travel the first 20 miles?

**c** How many times did the cyclist stop on her journey to York?

**d** What was the cyclist's average speed for this journey in mph correct to 1 dp?

**B** Interpret a speed–time graph

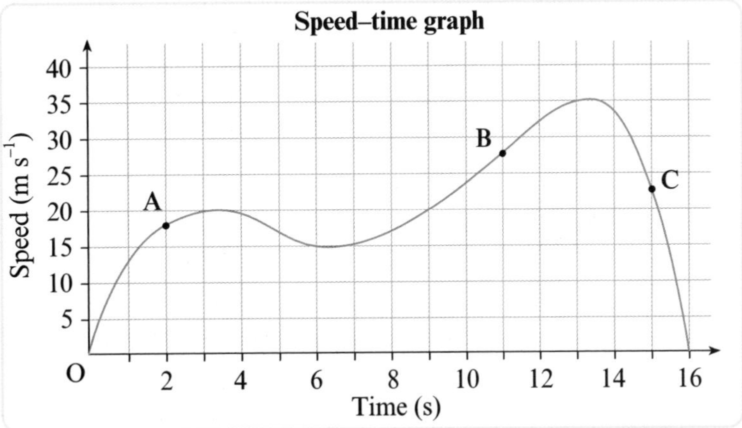


**B1** **a** At what times was this vehicle travelling at 18 m $s^{-1}$

**b** Give two times when the vehicle was starting to accelerate.

**c** At which times was the vehicle stationary?

**d** Describe what the vehicle was doing 4 seconds from the start.

## Starting points

You need to know about ... ... so try these questions

### A Using vectors to describe a translation

- A vector can be used to describe the movement in a translation.

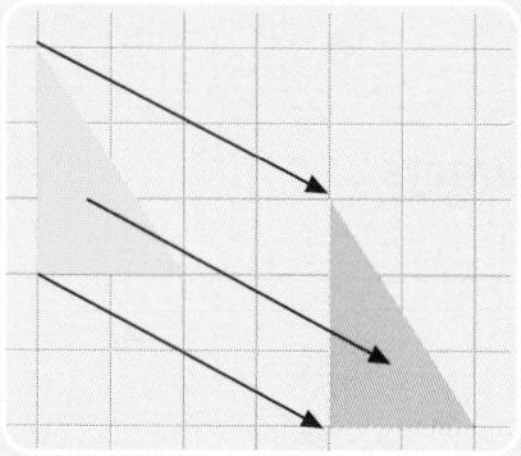

Every point on the triangle has moved 4 squares right and 2 squares down.

The vector which describes this translation is: $\begin{pmatrix} 4 \\ ^{-}2 \end{pmatrix}$

**A1** Copy triangle ABC and translate it using the vector:

**a** $\begin{pmatrix} 3 \\ ^{-}1 \end{pmatrix}$ **b** $\begin{pmatrix} 5 \\ 2 \end{pmatrix}$

**c** $\begin{pmatrix} ^{-}2 \\ ^{-}3 \end{pmatrix}$ **d** $\begin{pmatrix} ^{-}4 \\ 2 \end{pmatrix}$

### B Resultant vectors

- The total effect of two or more vectors can be shown as a single vector called the **resultant**.
  So if you translate,
  first by $\begin{pmatrix} 3 \\ 1 \end{pmatrix}$ then by $\begin{pmatrix} ^{-}5 \\ 3 \end{pmatrix}$,
  the resultant vector is given by:
  $\begin{pmatrix} 3 + ^{-}5 \\ 1 + 3 \end{pmatrix} = \begin{pmatrix} ^{-}2 \\ 4 \end{pmatrix}$

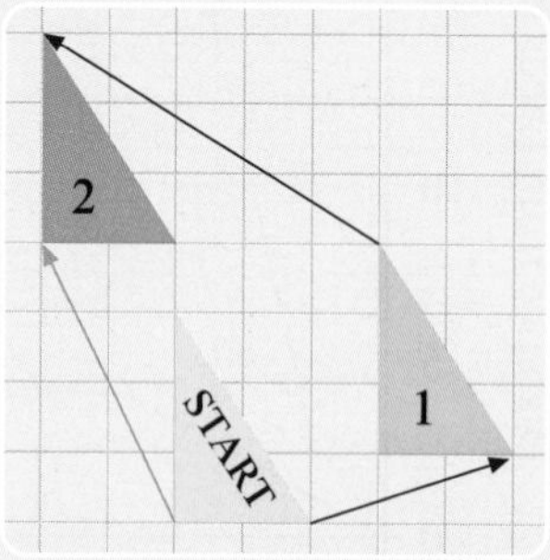


**B1** Give the resultant vector for each of these combined translations.

**a** $\begin{pmatrix} 3 \\ 1 \end{pmatrix} \begin{pmatrix} 2 \\ ^{-}2 \end{pmatrix}$

**b** $\begin{pmatrix} ^{-}2 \\ ^{-}4 \end{pmatrix} \begin{pmatrix} 3 \\ 2 \end{pmatrix}$

**c** $\begin{pmatrix} 2 \\ 5 \end{pmatrix} \begin{pmatrix} 0 \\ ^{-}2 \end{pmatrix} \begin{pmatrix} ^{-}3 \\ 1 \end{pmatrix}$

### C Using algebra

- Expressions that include negative coefficients can be factorised in two ways.
  To factorise $6y - 2x$:

| | | |
|---|---|---|
| $6y - 2x$ <br> $\mathbf{= 2(3y - x)}$ | or | $6y - 2x$ <br> $= {}^{-}2({}^{-}3y + x)$ <br> $\mathbf{= {}^{-}2(x - 3y)}$ |

- Expressions that include fractional coefficients can be factorised.

To factorise: $\frac{1}{2}x + y$ $\frac{1}{2}(x - y) + x$

$\frac{1}{2}x + y$
$\mathbf{= \frac{1}{2}(x + 2y)}$

$\frac{1}{2}(x - y) + x$
$= \frac{1}{2}x - \frac{1}{2}y + x$
$= 1\frac{1}{2}x - \frac{1}{2}y$
$\mathbf{= \frac{1}{2}(3x - y)}$

**C1** Copy and complete:

**a** $4x - 2y = 2(\square)$
**b** $4x - 2y = {}^{-}2(\square)$
**c** $9y - 3x = {}^{-}3(\square)$
**d** $9y - 3x = 3(\square)$

**C2** Factorise:

**a** $x + \frac{1}{2}y$
**b** $\frac{1}{3}x - y$
**c** $x + \frac{1}{2}(x + y)$
**d** $\frac{1}{3}(x - y) + x$
**e** $2x + \frac{2}{3}(y - 2x)$
**f** $2y - \frac{2}{3}(x + y)$

### D Ratios and fractions of straight lines

- A straight line can be divided into two parts, like this:

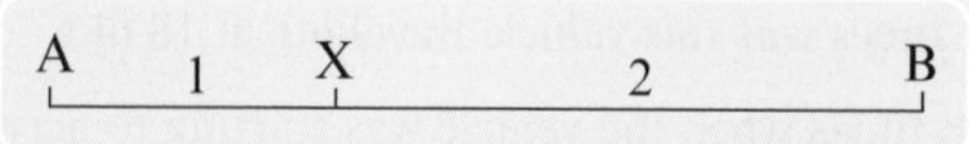


Point X divides line AB in the ratio 1 : 2, and AX is $\frac{1}{3}$ of the line.

AX : AB = 1 : 2 AX = $\frac{1}{3}$AB

**D1** Give these ratios.
**a** PY : YQ **b** QY : YP

**D2** What fraction of PQ is:
**a** YQ **b** PY?

# Defining vectors

- A **vector** can be defined by two quantities:
  - its **size**
  - its **direction**.

- A vector can be placed anywhere in a plane, and is usually represented by a straight line:
  - the size of the vector is the length of the line,
  - the direction of the vector is how the line is pointing, in the direction of the arrow.

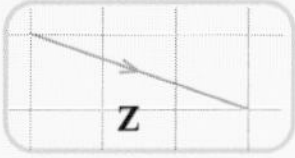


Other ways of writing the vector **z** include:

$\underline{z}$ $\underset{\sim}{z}$ $\bar{z}$ $\tilde{z}$

- The **negative** of a given vector is a vector that is:
  - equal in size but
  - opposite in direction.

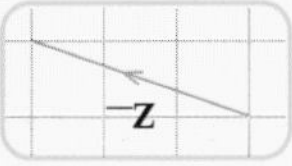


- A **scalar multiple** of a given vector is found by:
  - multiplying its size by a single number
  - leaving its direction unchanged.

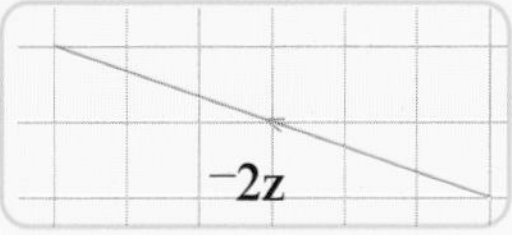


- When a vector is placed on a coordinate grid, it can also be represented by a **column vector**.

$$\mathbf{z} = \begin{pmatrix} 3 \\ ^-1 \end{pmatrix} \qquad ^-\mathbf{z} = \begin{pmatrix} ^-3 \\ 1 \end{pmatrix} \qquad ^-2\mathbf{z} = \begin{pmatrix} ^-6 \\ 2 \end{pmatrix}$$

**Exercise 24.1**
Vectors on a grid

Use these vectors for this exercise.

$$\mathbf{p} = \begin{pmatrix} 1 \\ 3 \end{pmatrix} \qquad \mathbf{q} = \begin{pmatrix} 4 \\ 1 \end{pmatrix} \qquad \mathbf{r} = \begin{pmatrix} ^-1 \\ 2 \end{pmatrix}$$

1 Draw vectors **p**, **q**, and **r** on a coordinate grid.

2 Write each of these as a column vector, and draw it on a coordinate grid.
**a** ⁻**p** **b** 2**q** **c** 3**r** **d** ⁻2**p** **e** ⁻3**r**

3 This diagram shows the effect of the combined translation, **q** then **p**.
Give the resultant vector.

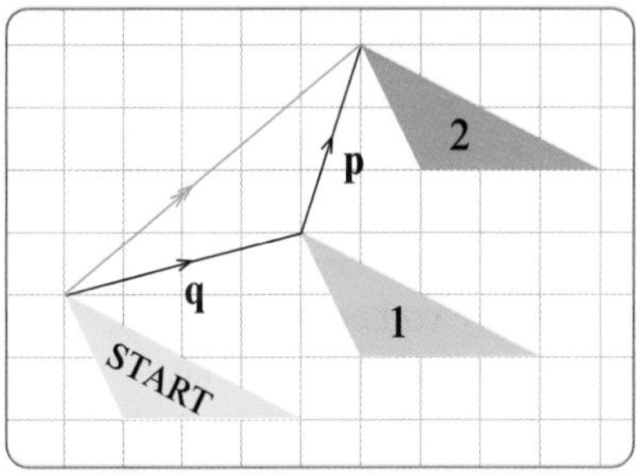


4 **a** Draw a diagram to show the combined translation, **q** then **p**.
**b** Give the resultant vector.

5 Write down what you notice about the effect of combined translations.

6 This diagram shows the effect of the combined translation, **q** then ⁻**r**.
**a** Give the resultant vector.
**b** Explain how you can find the resultant from column vectors **q** and **r**.

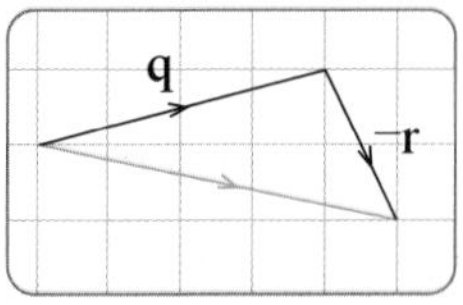

# Adding vectors

- You can add vectors by combining them end to end, so that their direction 'follow on' from each other.

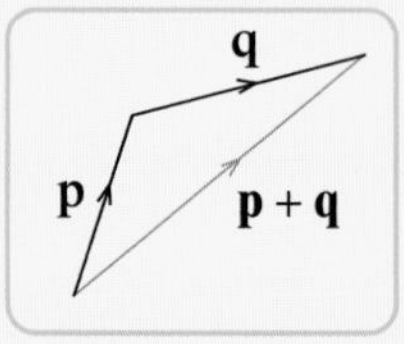


**Example**

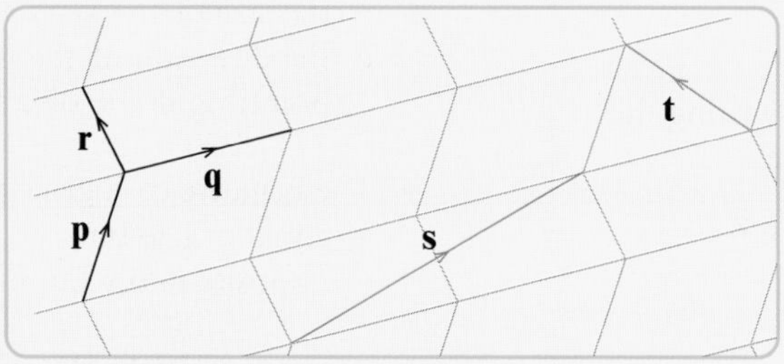


Give **s** and **t** in terms of **p**, **q**, and **r**.

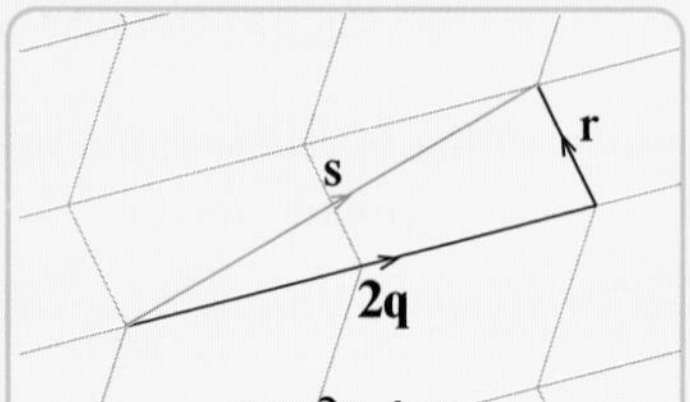


$\mathbf{s} = 2\mathbf{q} + \mathbf{r}$

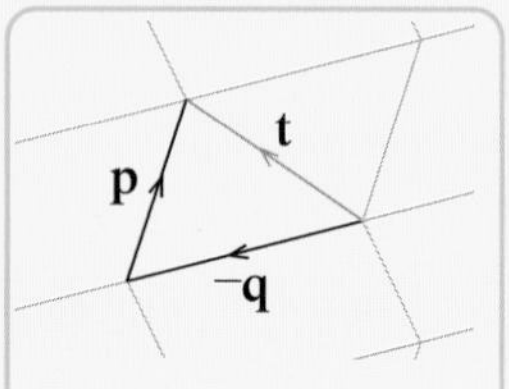


$\mathbf{t} = {}^-\mathbf{q} + \mathbf{p}$

**Exercise 24.2**
**Adding vectors**

1 a This sketch to show $\mathbf{x} = \mathbf{q} + \mathbf{r}$ is incorrect. Explain why.
  b Make a sketch to show the correct **x**.

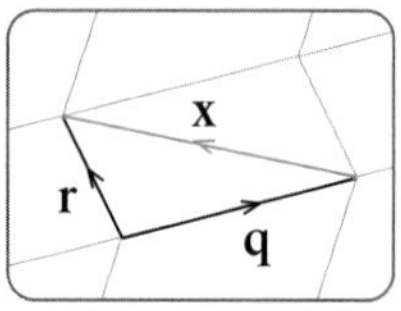


2 Give each of these vectors in terms of **p**, **q**, and **r**.

a

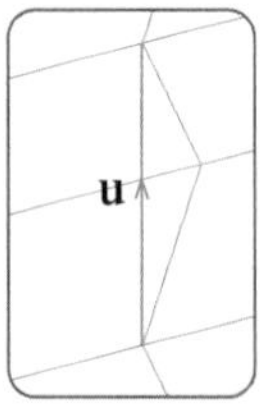


b

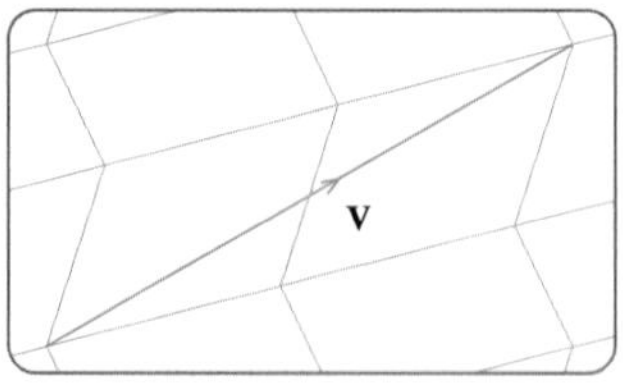


c

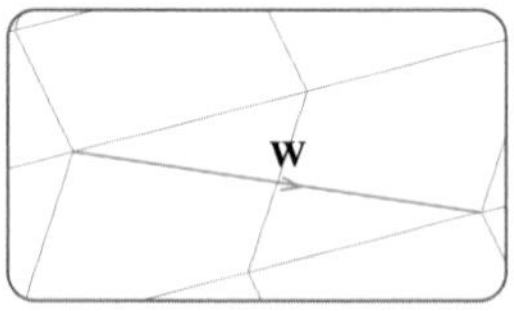


Adding vectors is commutative because, however you combine them, the resultant is the same.

3 a Explain why this diagram shows that adding vectors is commutative:
$$\mathbf{p} + \mathbf{q} = \mathbf{q} + \mathbf{p}$$
  b Make a sketch to show that:
$${}^-\mathbf{q} + \mathbf{p} = \mathbf{p} + {}^-\mathbf{q}$$

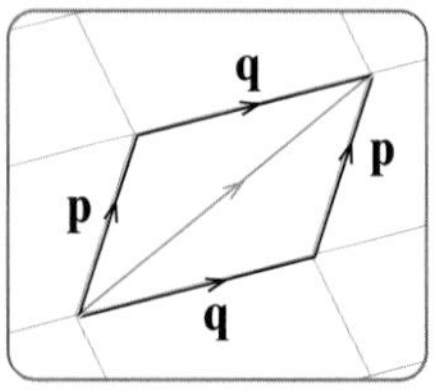


Adding vectors is associative because, however pairs of vectors are grouped, the resultant is the same.

4 a Sketch this diagram.
  b Label your sketch to show that adding vectors is associative:
$$\mathbf{p} + (\mathbf{q} + \mathbf{r}) = (\mathbf{p} + \mathbf{q}) + \mathbf{r}$$

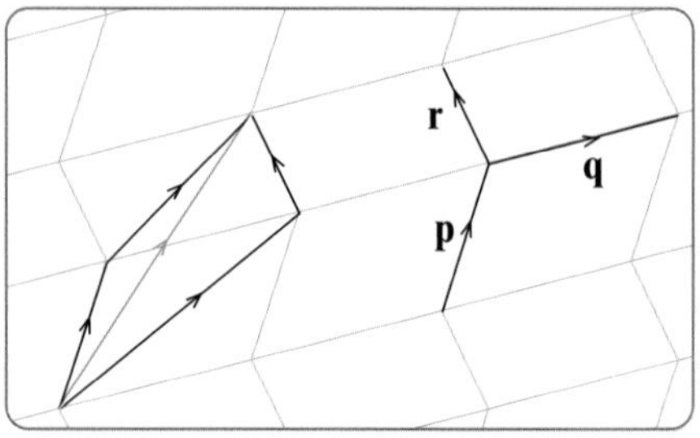

# Subtracting vectors

- Adding a negative vector can be written as a subtraction.

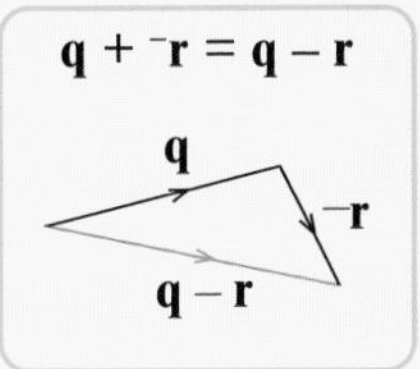


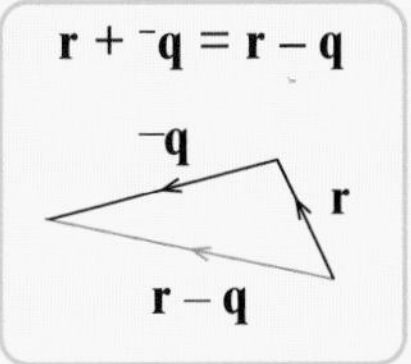


- In general, vectors $\mathbf{a} - \mathbf{b}$ and $\mathbf{b} - \mathbf{a}$ are:
  - the same in size,
  - opposite in direction.

Therefore, $\mathbf{b} - \mathbf{a}$ is the negative vector of $\mathbf{a} - \mathbf{b}$.

$$\begin{aligned} &{}^-(\mathbf{a} - \mathbf{b}) \\ &= {}^-\mathbf{a} + \mathbf{b} \\ &= \mathbf{b} + {}^-\mathbf{a} \\ &= \mathbf{b} - \mathbf{a} \end{aligned}$$

## Exercise 24.3
Subtracting vectors

1 Show that $\mathbf{p} - \mathbf{q}$ is the negative vector of $\mathbf{q} - \mathbf{p}$:

**a** using a sketch **b** using algebra

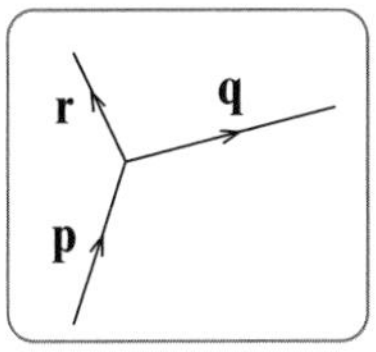


2 Make a sketch to show the vector $(\mathbf{p} - \mathbf{q}) - \mathbf{r}$.

3 Make a sketch to show:

**a** $\mathbf{q} - \mathbf{r}$ **b** $\mathbf{p} - (\mathbf{q} - \mathbf{r})$

4 Use algebra to explain why subtracting vectors is not associative:

$\mathbf{p} - (\mathbf{q} - \mathbf{r})$ is not equal to $(\mathbf{p} - \mathbf{q}) - \mathbf{r}$.

5

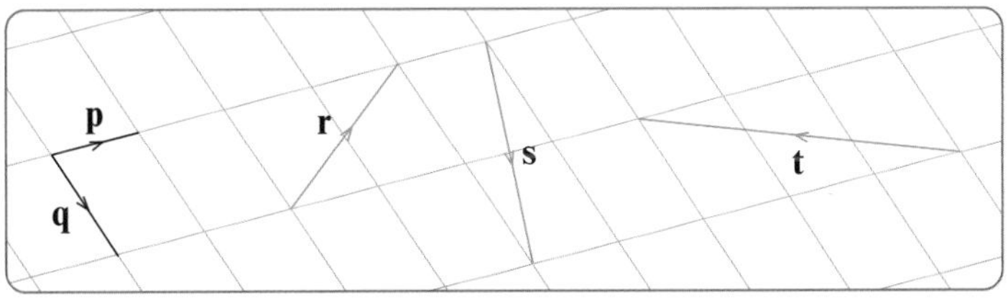


Give vectors **r**, **s**, and **t** in terms of **p** and **q**.

6 **a** Sketch a large grid like the one in Question **5**.

**b** Use your grid to draw:

**i** $3\mathbf{p} - 2\mathbf{q}$ **ii** $2\mathbf{q} - 3\mathbf{p}$ **iii** $3\mathbf{p} - \mathbf{q}$ **iv** $6\mathbf{p} + 2\mathbf{q}$

**c** Use algebra to explain why $3\mathbf{p} - \mathbf{q}$ and $2\mathbf{q} - 6\mathbf{p}$ are parallel vectors.

7 These vectors include three pairs of parallel vectors.

$^-2\mathbf{p} - \mathbf{q}$ | $2\mathbf{p} - \mathbf{q}$ | $2\mathbf{q} - \mathbf{p}$ | $2\mathbf{p} + 4\mathbf{q}$ | $2\mathbf{q} - 4\mathbf{p}$ | $2\mathbf{p} + \mathbf{q}$ | $\mathbf{p} + 2\mathbf{q}$

List each pair.

8 **a** Give each side of this parallelogram of vectors in terms of **p** and **q**.

**b** Add your vectors.

**c** Explain your answer.

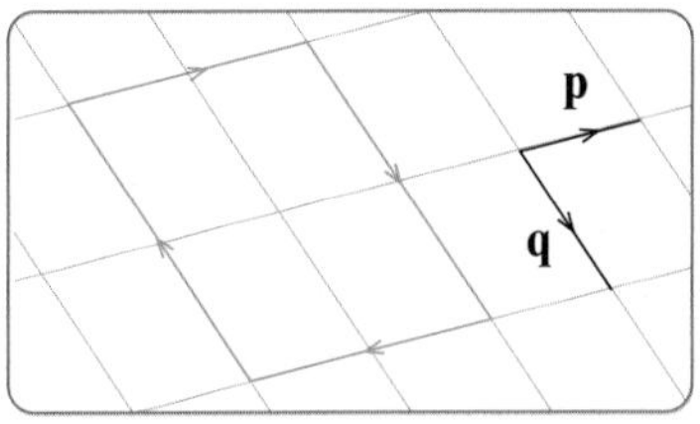

# Vectors in space

- You can refer to a vector by labelling each end.

  **Example** $\overrightarrow{KL}$ is a vector with:

  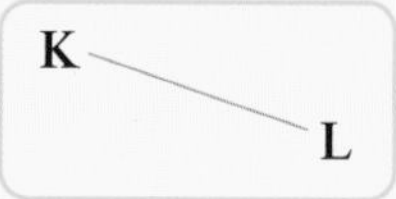


  - size equal to the distance from K to L
  - direction from K to L.

  $\overrightarrow{LK}$ is the same size as $\overrightarrow{KL}$, and opposite in direction.

- This notation is useful when giving a vector as a **vector sum**.

  **Example**

  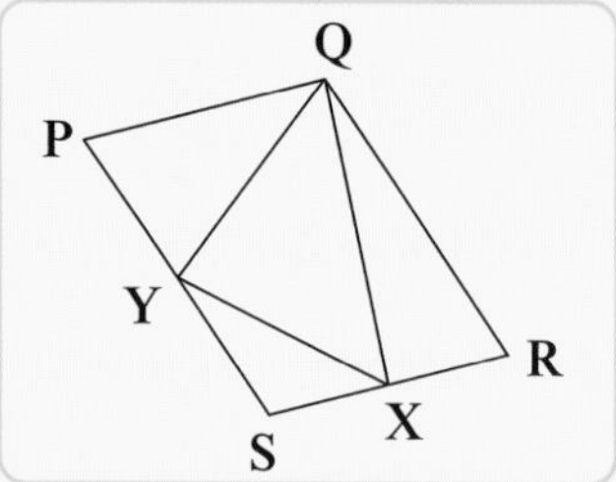


  $\overrightarrow{PY} = \overrightarrow{PQ} + \overrightarrow{QY}$
  $\overrightarrow{PS} = \overrightarrow{PQ} + \overrightarrow{QR} + \overrightarrow{RS}$
  $\overrightarrow{YX} = \overrightarrow{YQ} + \overrightarrow{QX}$
  etc.

- You can use vector sums to give other vectors.

  **Example**

  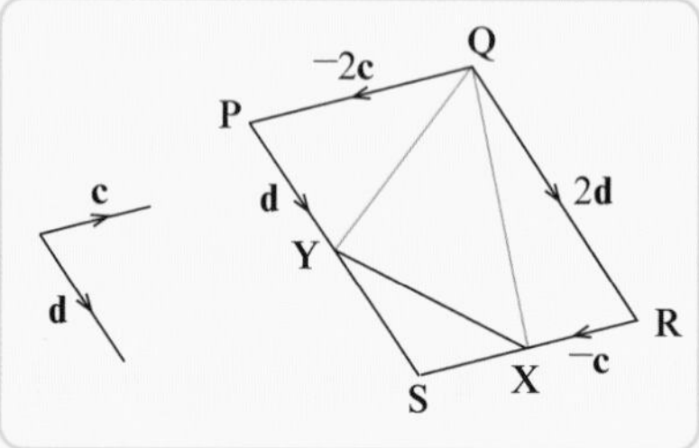


  Y is the midpoint of PS;
  X is the midpoint of SR.

  Give $\overrightarrow{QY}$ and $\overrightarrow{QX}$ in terms of **c** and **d**.

  $\overrightarrow{QY} = \overrightarrow{QP} + \overrightarrow{PY}$
  $= {}^{-}2\mathbf{c} + \mathbf{d}$
  $= \mathbf{d} - 2\mathbf{c}$

  $\overrightarrow{QX} = \overrightarrow{QR} + \overrightarrow{RX}$
  $= 2\mathbf{d} + {}^{-}\mathbf{c}$
  $= 2\mathbf{d} - \mathbf{c}$

**Exercise 24.4**
Vectors in space

1 **a** Give $\overrightarrow{YQ}$ in terms of **c** and **d**.

**b** Use $\overrightarrow{YX} = \overrightarrow{YQ} + \overrightarrow{QX}$ to show that $\overrightarrow{YX} = \mathbf{c} + \mathbf{d}$.

2 **a** Give the resultant vector for:

**i** $\overrightarrow{EH} + \overrightarrow{HD}$

**ii** $\overrightarrow{EF} + \overrightarrow{FG}$

**iii** $\overrightarrow{HG} + \overrightarrow{GD} + \overrightarrow{DE}$

**iv** $\overrightarrow{EF} + \overrightarrow{FD} + \overrightarrow{DG}$

**b** List three vector sums that each give the vector $\overrightarrow{FH}$.

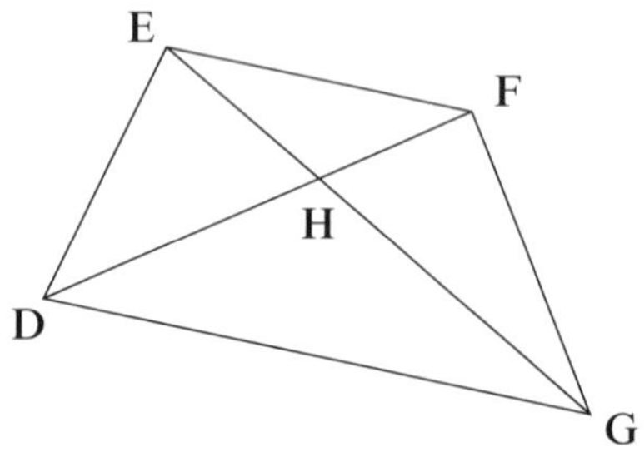


3 **a** Use $\overrightarrow{FG} = \overrightarrow{FE} + \overrightarrow{ED} + \overrightarrow{DG}$ to show that $\overrightarrow{FG}$ is equal to $\mathbf{q} - \mathbf{p}$.

**b** Z is the midpoint of FG, so:

$\overrightarrow{FZ} = \frac{1}{2}(\mathbf{q} - \mathbf{p})$

**i** Show that:

$\overrightarrow{EZ} = \frac{1}{2}(3\mathbf{q} - \mathbf{p})$

**ii** Give $DZ$ in terms of **p** and **q**.

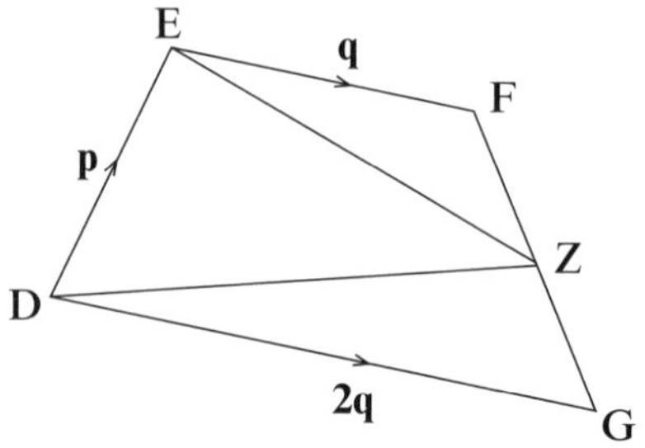

# Using vectors

Plane geometry is the study of the properties and relationships of points and lines in two-dimensional space.

- You can use vectors to prove facts in plane geometry.

**Example**

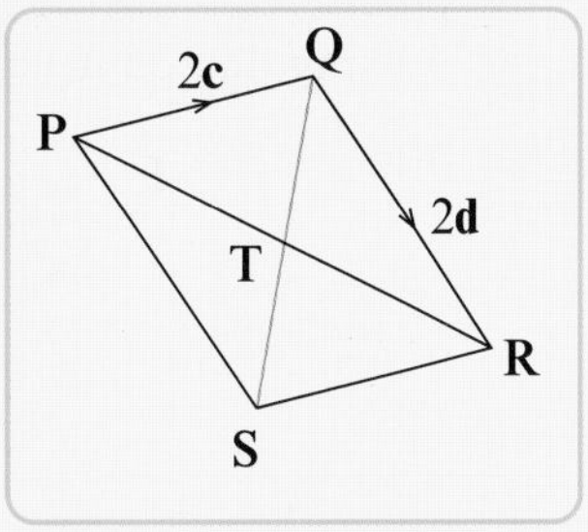


PQRS is a parallelogram.
T is the midpoint of PR:
$PT = \frac{1}{2}PR$

Prove that T is also the midpoint of QS.

$\overrightarrow{PR} = \overrightarrow{PQ} + \overrightarrow{QR}$
$= 2\mathbf{c} + 2\mathbf{d}$
$= 2(\mathbf{c} + \mathbf{d})$

So $\frac{1}{2}\overrightarrow{PR} = \mathbf{c} + \mathbf{d}$

Writing $\overrightarrow{QS}$ and $\overrightarrow{QT}$ in terms of **c** and **d**:

$\overrightarrow{QS} = \overrightarrow{QR} + \overrightarrow{RS}$
$= 2\mathbf{d} + {}^{-}2\mathbf{c}$
$= 2\mathbf{d} - 2\mathbf{c}$
$= 2(\mathbf{d} - \mathbf{c})$

$\overrightarrow{QT} = \overrightarrow{QP} + \overrightarrow{PT}$
$= \overrightarrow{QP} + \frac{1}{2}\overrightarrow{PR}$
$= {}^{-}2\mathbf{c} + \mathbf{c} + \mathbf{d}$
$= \mathbf{d} - \mathbf{c}$

The distance QS is twice the distance QT so, T must be the midpoint of QS.

**Exercise 24.5**
Using vectors

1 D is the point so that $PD = 3\mathbf{d} - \mathbf{c}$.

Use $\overrightarrow{QD} = \overrightarrow{QP} + \overrightarrow{PD}$ to prove that D lies on an extension of QS.

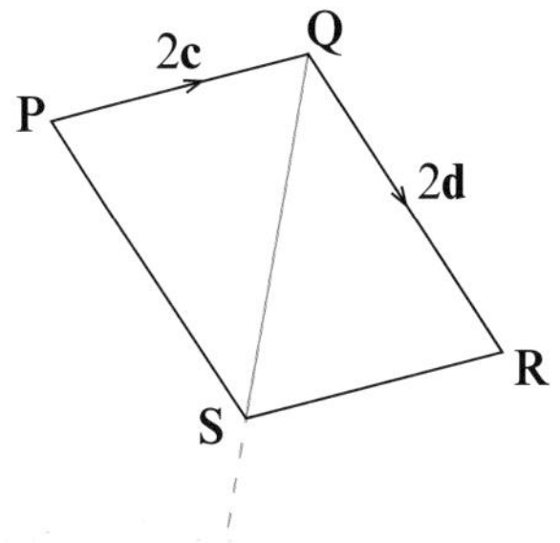


2 In this diagram $DH = \frac{2}{3}DF$.

**a** Give these in terms of **p** and **q**.

**i** $\overrightarrow{DH}$ **ii** $\overrightarrow{EG}$

**b** Prove that $EH = \frac{1}{3}EG$.

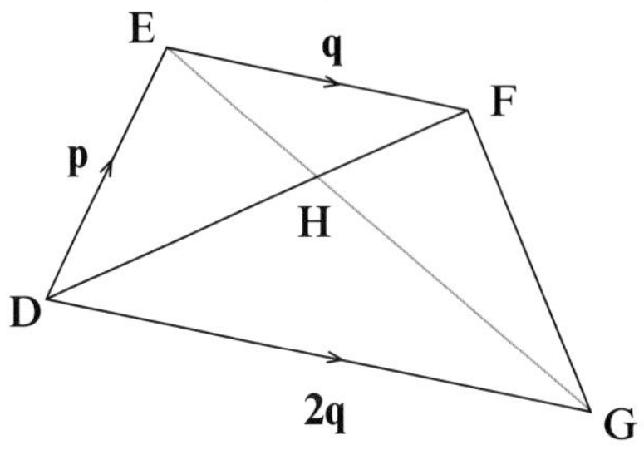


3 OJMN is a parallelogram.
K is the midpoint of JM.
The ratio KL : LN is 1 : 2.

**a** Give these in terms of **x** and **y**.

**i** $\overrightarrow{KN}$ **ii** $\overrightarrow{KL}$

**b** Prove that the ratio OL : LM is 2 : 1.

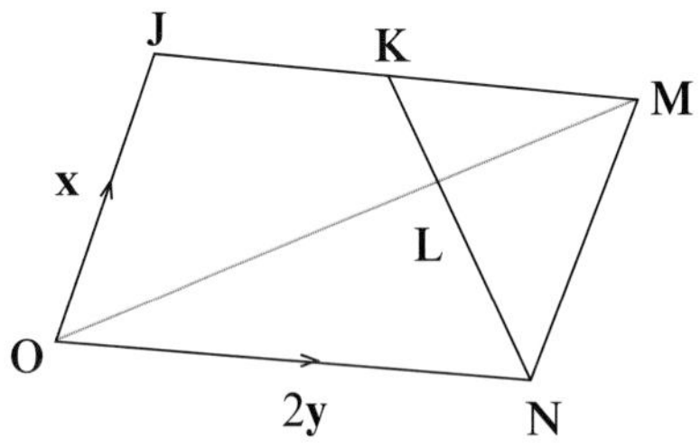

## Using parallel vectors

Points are collinear when they all lie in the same straight line.

- Other uses of vectors in plane geometry include:
  - proving that lines are parallel,
  - proving that points are collinear.

**Example**

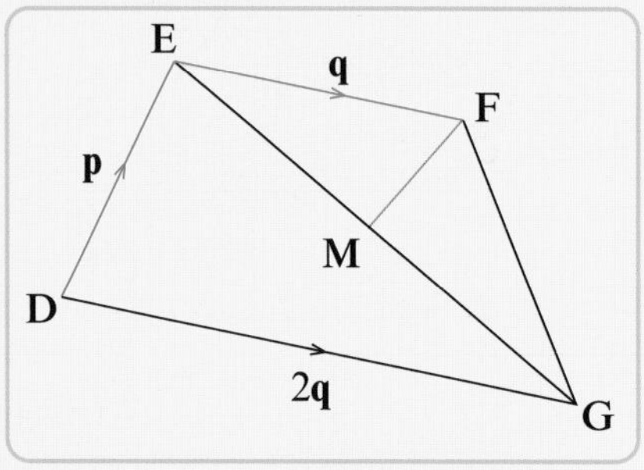


M is the midpoint of EG.

**a** Prove that MF is parallel to DE.

**b** If N is the midpoint of DG, prove that N, M, and F are collinear.

**a** $\overrightarrow{MF} = \overrightarrow{ME} + \overrightarrow{EF}$

$= \frac{1}{2}\overrightarrow{GE} + \overrightarrow{EF}$

$= \frac{1}{2}(\mathbf{p} - 2\mathbf{q}) + \mathbf{q}$

$= \frac{1}{2}\mathbf{p} - \mathbf{q} + \mathbf{q}$

$= \frac{1}{2}\mathbf{p}$

$\overrightarrow{MF}$ is in the same direction as $\overrightarrow{DE}$, so MF must be parallel to DE.

$\overrightarrow{GE} = \overrightarrow{GD} + \overrightarrow{DE}$

$= {}^{-}2\mathbf{q} + \mathbf{p}$

$= \mathbf{p} - 2\mathbf{q}$

So $\frac{1}{2}\overrightarrow{GE}$

$= \frac{1}{2}(\mathbf{p} - 2\mathbf{q})$

**b** $\overrightarrow{NF} = \overrightarrow{ND} + \overrightarrow{DE} + \overrightarrow{EF}$

$= {}^{-}\mathbf{q} + \mathbf{p} + \mathbf{q}$

$= \mathbf{p}$

$\overrightarrow{NF}$ is in the same direction as $\overrightarrow{MF}$. As point F lies on NF and MF, the points N, M, and F must be collinear.

**Exercise 24.6**
**Using parallel vectors**

**1** In this diagram, OV = 2ON and OW = 2OP. Prove that VW is parallel to NP, and twice its size.

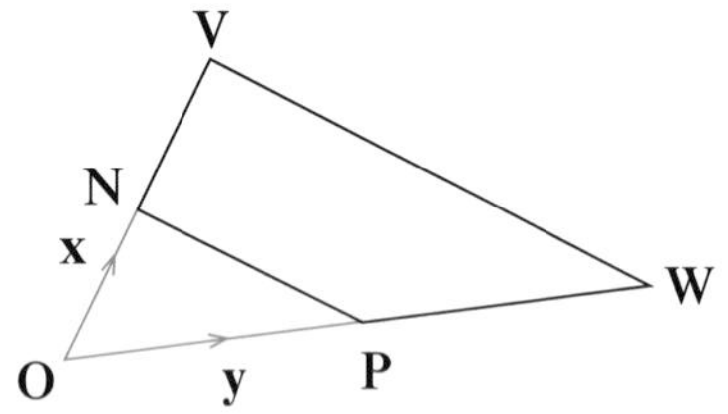


**2** HIJK is a parallelogram, and HL = $\frac{1}{3}$HK. The ratio JK:KM is 1 : 2.

Prove that points I, L, and M are collinear.

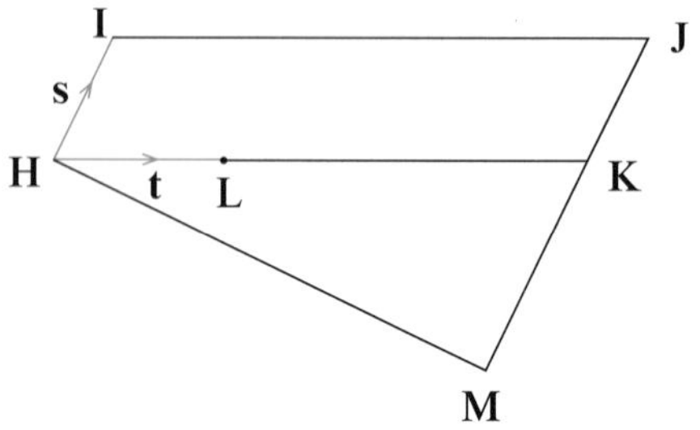


**3** In this diagram,
OU = $\frac{1}{4}$OT,
OX = $\frac{1}{2}$OS,
UW = $\frac{1}{2}$US,
V is the midpoint of TS.

Prove that:

**a** XW is parallel to OT,
**b** V, W, and X are collinear.

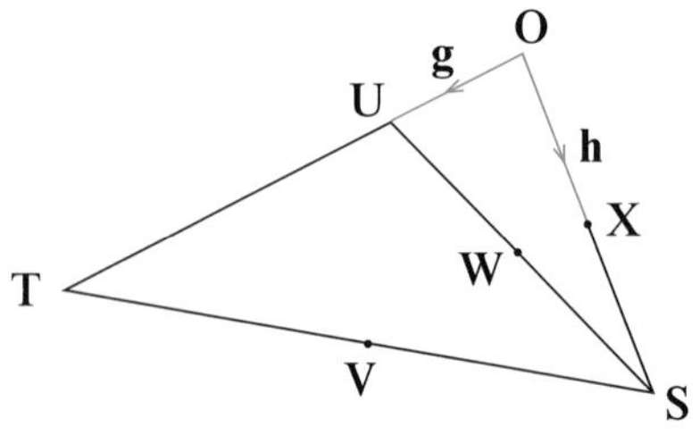

## End points

| You should be able to ... | ... so try these questions |
|---|---|
| **A** Add and subtract column vectors | $\mathbf{r} = \begin{pmatrix} 2 \\ ^{-}1 \end{pmatrix}$ $\mathbf{s} = \begin{pmatrix} 1 \\ 4 \end{pmatrix}$ $\mathbf{t} = \begin{pmatrix} ^{-}2 \\ 2 \end{pmatrix}$ <br> **A1** Write each of these as a column vector, and draw it on squared paper. <br> **a** 2**s** **b** $^{-}$3**r** **c** **r** + **s** **d** **s** – **t** **e** 2**t** – **r** **f** 2**s** + **t** |
| **B** Add and subtract vectors in space | **B1** **a** Write CD as a vector sum. <br> **b** Use your vector sum to give CD in terms of **p** and **q**. <br> **B2** M is the midpoint of CD. <br> **a** Show that BM = $\frac{1}{2}$(4**p** + **q**) <br> **b** Give AM in terms of **p** and **q**. <br> 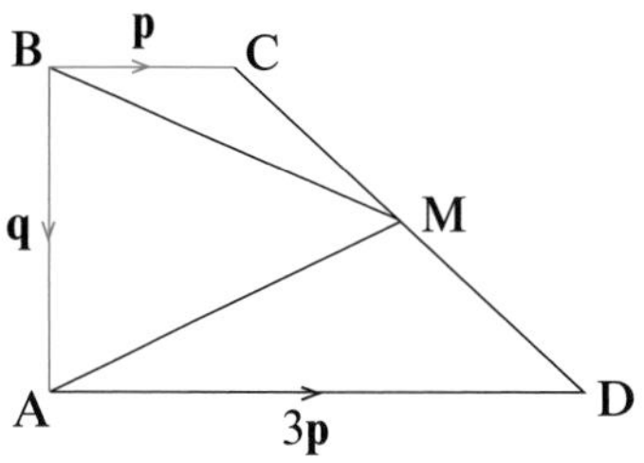 |
| **C** Recognise when vectors are parallel | **C1** These vectors include two pairs of parallel vectors. <br> 3**r** + **s** 3**s** – **r** 2**r** + 6**s** 6**r** + 2**s** 2**r** – 6**s** <br> List each pair. <br> **C2** Use algebra to explain why 2**s** – **t** and 3**t** – 6**s** are parallel vectors. |
| **D** Use vectors to prove facts in plane geometry | **D1** OPQR is a quadrilateral. EFGH is another quadrilateral, formed by joining the midpoints of each side. <br> **a** Give these in terms of **a**, **b**, and **c**. <br> **i** $\overrightarrow{RQ}$ **ii** $\overrightarrow{PG}$ <br> **b** Prove that $\overrightarrow{EH} = \overrightarrow{FG}$. <br> **c** Prove that EFGH is a parallelogram. <br> 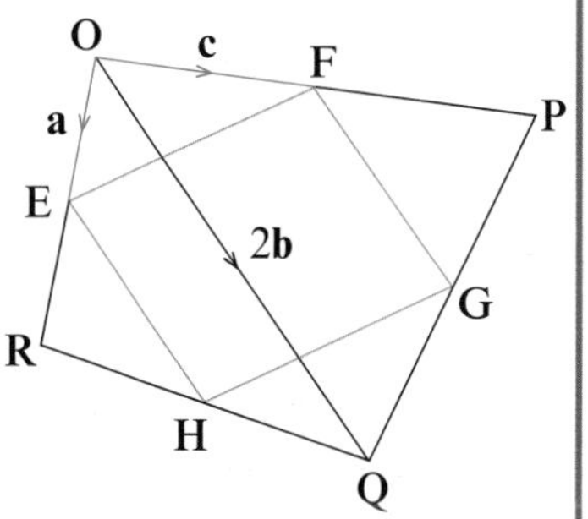 <br> **D2** XYZ is a triangle. L, M, and N are the midpoints of each of the sides. The ratio YW:WM is 2:1 <br> **a** Give these in terms of **c** and **d**. <br> **i** $\overrightarrow{YM}$ **ii** $\overrightarrow{YW}$ **iii** $\overrightarrow{YZ}$ <br> **b** Prove that X, W, and N are collinear. <br> 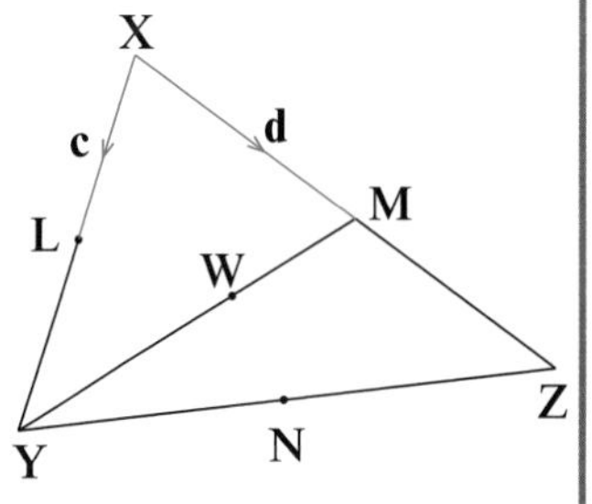 |







### Some points to remember

- A vector can be represented in different ways, for example: $\overrightarrow{AB}$ **a** $\begin{pmatrix} x \\ y \end{pmatrix}$
- Adding vectors is:
  - commutative **a** + **b** = **b** + **a**
  - associative **a** + (**b** + **c**) = (**a** + **b**) + **c**
- Subtracting a vector has the same effect as adding its negative vector.

## Starting points

You need to know about ... | ... so try these questions

### A Finding a linear rule from a set of data

The speed of a moving object is measured at different times.

| Time $t$ (seconds) | 0 | 1 | 2 | 3 | 4 | 5 |
|---|---|---|---|---|---|---|
| Speed $v$ (m $s^{-1}$) | 2.4 | 3.0 | 3.6 | 4.0 | 4.6 | 5.2 |

- To test for a linear rule, draw a graph of $v$ against $t$.

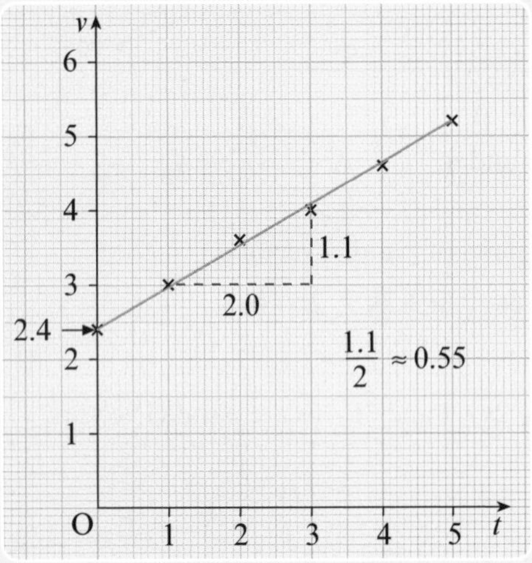


- The points lie approximately on a straight line.
- To 2 significant figures, the gradient is 0.55 and the $y$-intercept is 2.4.
- So a rule that fits the data approximately is: $v = 0.55t + 2.4$

**A1**

A

| $x$ | 0 | 1 | 2 | 3 | 4 |
|---|---|---|---|---|---|
| $y$ | 4.1 | 4.3 | 4.9 | 5.5 | 5.6 |

B

| $x$ | 2 | 4 | 6 | 8 | 10 |
|---|---|---|---|---|---|
| $y$ | 2.1 | 1.9 | 19 | 50 | 97 |

C

| $x$ | 1 | 2 | 4 | 5 | 9 |
|---|---|---|---|---|---|
| $y$ | 12 | 18 | 30 | 36 | 60 |

For each table of values:

**a** Draw a graph of $y$ against $x$.

**b** **i** Decide if $y$ and $x$ could be linked by a linear rule.

**ii** Find an approximate equation for any linear rule.

### B Sketching quadratic graphs

- It is easier to find the minimum or maximum point of a quadratic graph when its equation is given in completed square form.

**Example** Sketch the graph of $y = (x + 3)^2 - 4$.

- When $x = {}^-3$, $(x + 3)^2 = 0$, so the minimum value of $y$ occurs when $x = {}^-3$.
- When $x = {}^-3$, $y = ({}^-3 + 3)^2 - 4 = {}^-4$, so the minimum point is $({}^-3, {}^-4)$.
- The graph of $y = x^2$ is mapped to the graph of $y = (x + 3)^2 - 4$ by a translation of $\begin{pmatrix} {}^-3 \\ {}^-4 \end{pmatrix}$

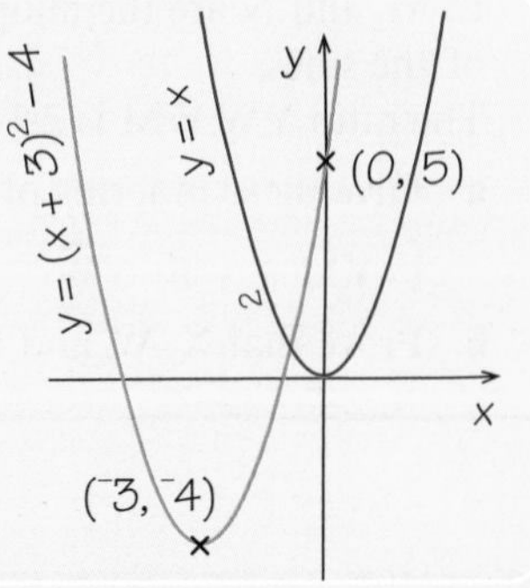


**B1** For each equation, sketch a graph and show clearly the coordinates of the minimum or maximum point and the $y$-intercept.

**a** $y = (x + 2)^2 + 3$

**b** $y = (x - 1)^2 + 5$

**c** $y = (x + 3)^2$

**d** $y = (x + 5)^2 - 1$

**e** $y = (x - 8)^2 - 4$

**B2** What translation maps the graph of $y = x^2$ to the graph of $y = (x - 6)^2 + 5$?

### C Function notation

- A rule that maps one number to another is called a function.

**Example** $f(x) = x^2 + 6$ is a function.
Calculate the value of $f(3)$.
$f(3) = 3^2 + 6 = 9 + 6 = 15$
We say '3 maps to 15' and '15 is the image of 3'.

**C1** A function $f$ is defined so that $f(x) = 3x^2 - 5$.

**a** Calculate the value of:

**i** $f(4)$ **ii** $f({}^-1)$

**iii** $f(2.5)$ **iv** $f(0)$

**b** Find a value for $x$ so that $f(x) = {}^-3.31$.

## Matching graphs

A symbol that is used for a constant represents a number whose value is fixed.

- The equation of this curve is of the form $y = kx^2 + c$, where $k$ and $c$ are constant.
- The point (0, 0.5) is on the curve so, for these values of $x$ and $y$:

$$y = kx^2 + c$$
$$0.5 = k \times 0^2 + c$$
$$0.5 = c$$

So the equation is of the form:

$$y = kx^2 + 0.5$$

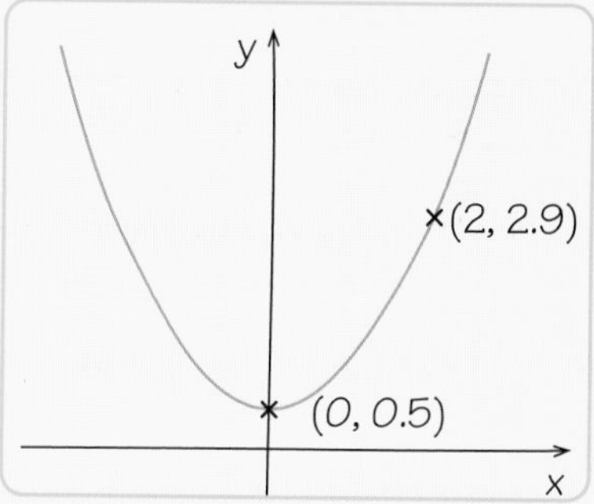


- The point (2, 2.9) is also on the curve so, for these values of $x$ and $y$:

$$y = kx^2 + 0.5$$
$$2.9 = k \times 2^2 + 0.5$$
$$2.4 = 4k$$
$$0.6 = k$$

So the equation of the curve is: $y = 0.6x^2 + 0.5$

**Exercise 25.1**
**Matching graphs**

**1**

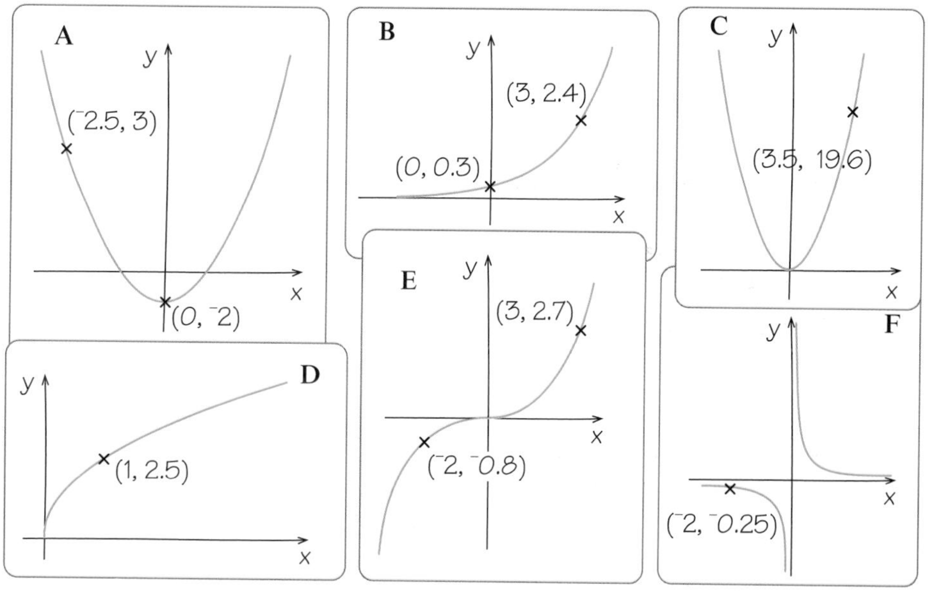


In all equations in this exercise, $p$, $q$ and $k$ are constants.

$y = kx^2$ $\quad$ $y = kx^2 + p$ $\quad$ $y = \frac{k}{x}$ $\quad$ $y = qx^k$ $\quad$ $y = k\sqrt{x}$ $\quad$ $y = pq^x$

Each equation above matches one of the graphs.

**a** For each graph:
  **i** Write down its matching equation.
  **ii** Find the value of each constant in its equation.

**b** Explain why the equation of curve D could not be of the form $y = qx^4$.

**2** The equation of a curve is of the form $y = px^2 + q$.
The curve passes through the points (1.5, 15.3) and ($^-0.5$, 3.3).
Find the values of $p$ and $q$.

**3** A curve has an equation of the form $y = kq^x$.
It passes through the points (1, 2) and (2, 8).
Find the values of $k$ and $q$.

## Looking for relationships

Many different equations give curved graphs; for example:
$C = ar^3 + b$, $C = ar^2 + b$.

For circular tables, the cost could be linked to the area ($\pi r^2$) of the table so $C = ar^2 + b$ is a likely rule.

- This shows the cost of some circular tables.

| Radius $r$ (m) | 0.4 | 0.6 | 0.8 | 1.0 | 1.2 | 1.4 |
|---|---|---|---|---|---|---|
| Cost $C$ (£) | 175 | 210 | 260 | 320 | 395 | 480 |

- Graph $C$ against $r$ to test for a linear rule.

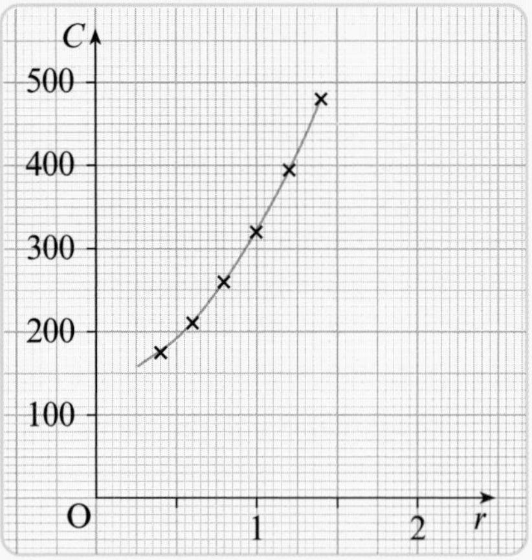


- Points on the graph of $C$ against $r$ lie on a curve, so any rule linking $C$ and $r$ is not linear.
- $C$ and $r$ could be linked by an equation of the form $C = ar^2 + b$

- Graph $C$ against $r^2$ to test for a rule of the form $C = ar^2 + b$

| $r^2$ | 0.16 | 0.36 | 0.64 | 1.0 | 1.44 | 1.96 |
|---|---|---|---|---|---|---|
| $C$ | 175 | 210 | 260 | 320 | 395 | 480 |

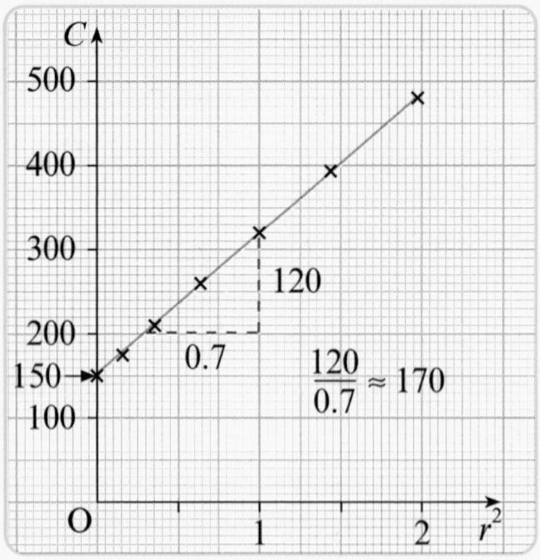


- The points lie approximately on a straight line.
- To 2 significant figures the gradient is 170 and the $y$-intercept is 150.
- So a rule that fits the data approximately is: $C = 170r^2 + 150$

If possible, use the table to check your answers are sensible.

**Exercise 25.2**
**Finding non-linear rules**

1 A stone is dropped and the distance it has fallen is measured at different times.

| Time $t$ (seconds) | 1 | 2 | 3 | 4 | 5 |
|---|---|---|---|---|---|
| Distance fallen $d$ (metres) | 5 | 20 | 44 | 78 | 123 |

**a** Draw a graph of $d$ against $t^2$.
**b** Explain why your graph shows that a rule of the form $d = kt^2 + p$ approximately fits the data.
**c** Find a rule of the form $d = kt^2 + p$ that approximately fits the data.
**d** Use your rule to estimate how far the stone had fallen after 2.5 seconds.
**e** About how long did it take the stone to fall 100 metres?

2 This table shows the cost of some circular rugs.

| Radius $r$ (m) | 0.5 | 1.0 | 1.5 | 2.0 | 2.5 | 3.0 |
|---|---|---|---|---|---|---|
| Cost $C$ (£) | 85 | 145 | 245 | 385 | 565 | 785 |

**a** Draw a graph of $C$ against $r^2$.
**b** Use your graph to find a rule that links $C$ and $r$.
**c** Estimate the cost of a rug with a radius of 4.5 m.

3 For pendulums of different lengths,
these are the times taken to complete 100 swings.

| Length of pendulum $l$ (cm) | Time for 100 swings $t$ (seconds) |
|---|---|
| 5 | 45 |
| 10 | 63 |
| 15 | 77 |
| 20 | 89 |
| 25 | 100 |
| 30 | 110 |
| 35 | 118 |
| 40 | 126 |
| 45 | 134 |
| 50 | 141 |

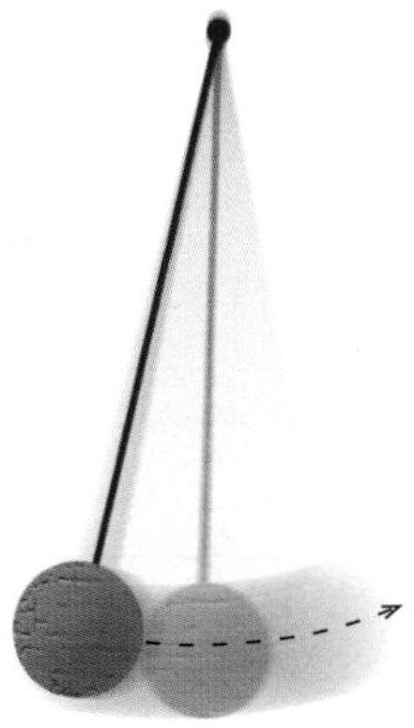

**a** Draw a graph of $t$ against $\sqrt{l}$.
**b** Explain why your graph shows that a rule of the form $t = k\sqrt{l}$ approximately fits the data.
**c** Find the value of $k$ correct to 2 significant figures.
**d** Estimate how long a pendulum of length 60 cm would take to complete 100 swings.

4 The table below gives information about the planets in our Solar System.

| Planet | Average distance from the Sun, $R$ (millions of miles) | Speed at which it travels through space, $V$ (mph) | Time taken to go once round the Sun, $T$ (days) |
|---|---|---|---|
| Earth | 93 | 66 641 | 365 |
| Jupiter | 484 | 29 216 | 4329 |
| Mars | 142 | 53 980 | 687 |
| Mercury | 36 | 107 132 | 88 |
| Neptune | 2794 | 12 147 | 60 150 |
| Pluto | 3674 | 10 604 | 90 670 |
| Saturn | 887 | 21 565 | 10 753 |
| Uranus | 1784 | 15 234 | 30 660 |
| Venus | 67 | 78 364 | 225 |

Hypotheses are statements that are thought to be true but have not yet been proved.

These are some hypotheses about relationships between $R$, $V$ and $T$, where $p$, $q$ and $n$ are constants.

(A) $V \approx \dfrac{p}{\sqrt{R}}$ (B) $V \approx n\sqrt{T}$ (C) $T \approx q\sqrt{R^3}$

**a** Draw suitable straight-line graphs to show which hypotheses are correct.
**b** Find the value of the constants in the correct hypotheses.

# Transforming graphs

**Exercise 25.3**
Transforming graphs

Each straight line continues for ever.

**1**

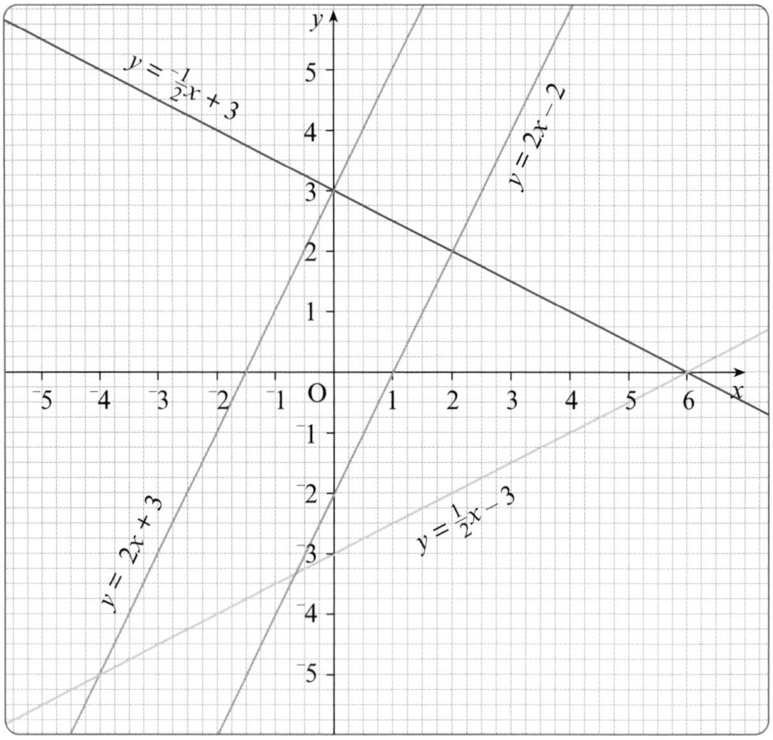


**a** Which line is the image of $y = 2x + 3$ after a translation of $\begin{pmatrix} 1 \\ ^{-}3 \end{pmatrix}$?

**b** Give three different translations that map $y = 2x - 2$ to $y = 2x + 3$.

**c** Which line is the image of $y = 2x + 3$ after a rotation of $^{-}90°$ about (0, 3)?

**d** Describe a reflection that maps $y = \frac{1}{2}x - 3$ to $y = {}^{-}\frac{1}{2}x + 3$.

**2** **a** On a set of axes, draw graphs of:

**i** $y = 2x + 1$ **ii** $y = {}^{-}3x - 4$ **iii** $y = \frac{1}{2}x - 1$.

**b** For each line, find the equation of the image after reflection in the $y$-axis.

**c** Without drawing, give the equation of the image of $y = 4x - 3$ after reflection in the $y$-axis. Explain your method.

**3** Give the equation of the image of $y - x = 8$ after reflection in the $y$-axis.

**4** **a** On a set of axes, draw graphs of:

**i** $y = x + 1$ **ii** $y = {}^{-}2x + 3$ **iii** $y = \frac{1}{4}x - 1$.

**b** For each line, find the equation of the image after reflection in the $x$-axis.

**c** Without drawing, give the equation of the image of $y = {}^{-}5x + 1$ after reflection in the $x$-axis. Explain your method.

**5** **a** Find three different translations that map $y = x - 1$ to $y = x + 5$.

**b** Describe the link between $p$ and $q$ when $y = x - 1$ is mapped.

to $y = x + 5$ after a translation of $\begin{pmatrix} p \\ q \end{pmatrix}$.

6 What is the equation of the image of $y = 3x + 1$ after a translation of $\begin{pmatrix} v \\ w \end{pmatrix}$?

7 Find the image of $y = kx + l$ after a translation of $\begin{pmatrix} a \\ b \end{pmatrix}$.

- If a function $f$ is defined so that $f(x) = x^2$, then $y = x^2$ can be written as $y = f(x)$.
- Other equations can be expressed in terms of this function $f$, and their graphs can be drawn.

**Example 1**

Draw a graph for the equation $y = f(x) + 2$.

| $x$ | $y = f(x) + 2$ | $y$ |
|---|---|---|
| $^-3$ | $f(^-3) + 2 = (^-3)^2 + 2$ | $= 11$ |
| $^-2$ | $f(^-2) + 2 = (^-2)^2 + 2$ | $= 6$ |
| $^-1$ | $f(^-1) + 2 = (^-1)^2 + 2$ | $= 3$ |
| 0 | $f(0) + 2 = (0)^2 + 2$ | $= 2$ |
| 1 | $f(1) + 2 = (1)^2 + 2$ | $= 3$ |
| 2 | $f(2) + 2 = (2)^2 + 2$ | $= 6$ |
| 3 | $f(3) + 2 = (3)^2 + 2$ | $= 11$ |

Graph of $y = f(x) + 2$ and $y = f(x)$

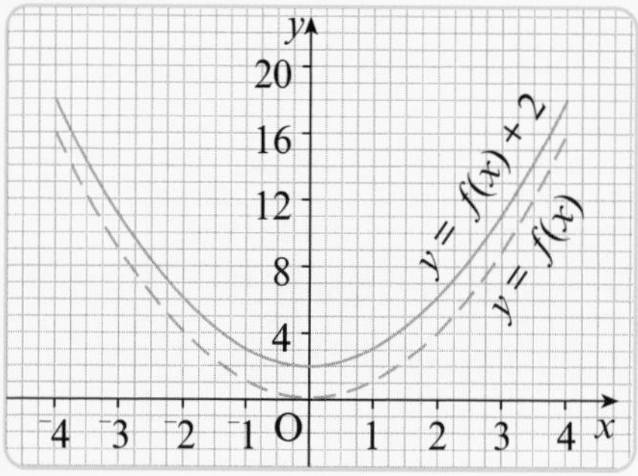


**Example 2**

Draw a graph for the equation $y = f(x + 2)$.

| $x$ | $y = f(x + 2)$ | $y$ |
|---|---|---|
| $^-3$ | $f(^-3 + 2) = f(^-1) = (^-1)^2$ | $= 1$ |
| $^-2$ | $f(^-2 + 2) = f(0) = 0^2$ | $= 0$ |
| $^-1$ | $f(^-1 + 2) = f(1) = 1^2$ | $= 1$ |
| 0 | $f(0 + 2) = f(2) = 2^2$ | $= 4$ |
| 1 | $f(1 + 2) = f(3) = 3^2$ | $= 9$ |
| 2 | $f(2 + 2) = f(4) = 4^2$ | $= 16$ |
| 3 | $f(3 + 2) = f(5) = 5^2$ | $= 25$ |

Graph of $y = f(x + 2)$ and $y = f(x)$

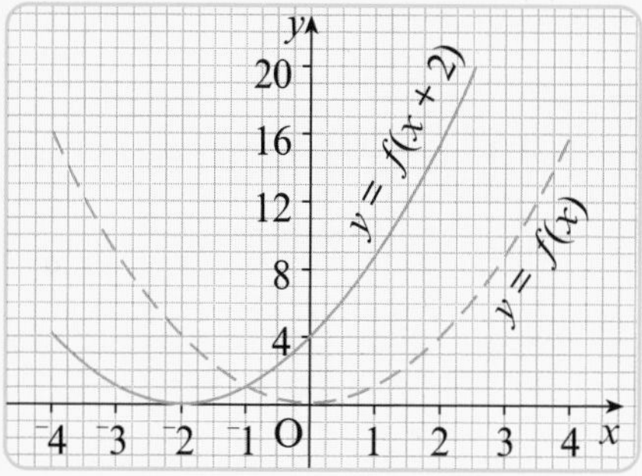


**Exercise 25.4**
**Using function notation**

1 A function $g$ is defined so that $g(x) = 2x - 1$.
On one set of axes, draw graphs of:

**a** $y = g(x)$ **b** $y = g(x) + 5$ **c** $y = g(x + 5)$

2 A function $h$ is defined so that $h(x) = x^2 + 3$.

**a** On one set of axes, draw graphs of:
**i** $y = h(x)$ **ii** $y = h(x - 2)$ **iii** $y = h(x) - 2$

**b** Describe a transformation that maps $y = h(x)$ to $y = h(x - 2)$.

3 This is the graph of a function $y = g(x)$.

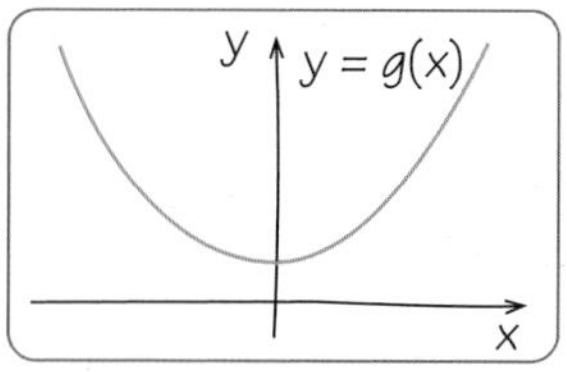


Which of these could be the graph of $y = g(x) - 4$?

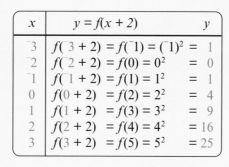

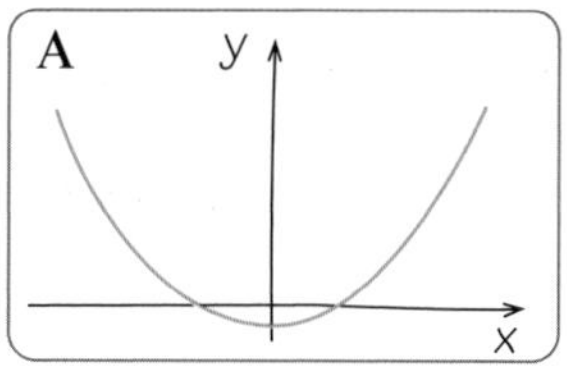


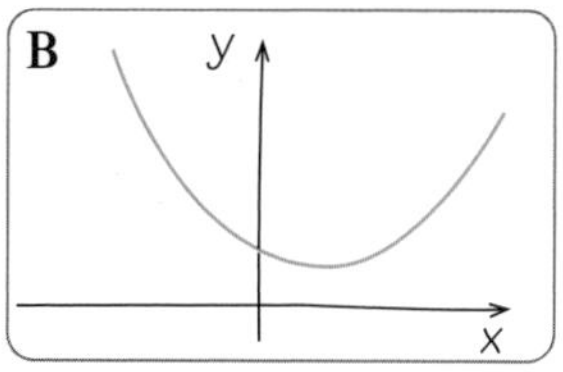


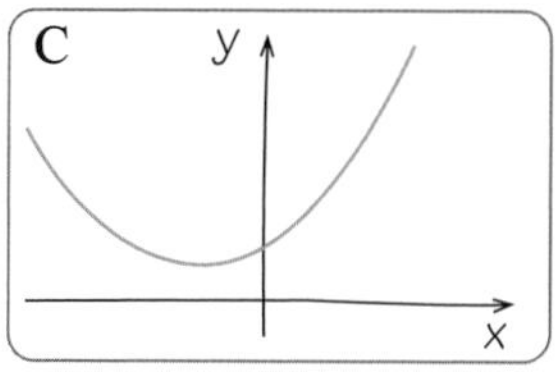

**4** A function $f$ is defined so that $f(x) = x^2$.

**a** On one set of axes, draw graphs of:

**i** $y = f(x)$ **ii** $y = f(x + 3)$ **iii** $y = f(x + 3) + 2$

**b** Describe a transformation that maps $y = f(x)$ to $y = f(x + 3) + 2$.

**5** A student makes the following statement:

> For any function f, a translation of $\begin{pmatrix} ^-4 \\ 5 \end{pmatrix}$ will map the graph of $y = f(x)$ to $y = f(x - 4) + 5$. ✗

Explain what is wrong with this statement.

Your sketches should show:
- the shape of the graph
- the coordinates of any $y$-intercepts.

**6** Two functions are defined so that $g(x) = x^2 + 1$ and $h(x) = x - 4$

**a** On one set of axes, sketch graphs of:

**i** $y = g(x)$ **ii** $y = h(x)$

**iii** $y = {}^-g(x)$ **iv** $y = {}^-h(x)$

**b** **i** Describe a transformation that maps $y = g(x)$ to $y = {}^-g(x)$ and also maps $y = h(x)$ to $y = {}^-h(x)$.

**ii** Do you think this transformation will map $y = f(x)$ to $y = {}^-f(x)$ for any function $f$? Explain your answer fully.

**7** A function $k$ is defined so that $k(x) = 4x - 5$.

**a** On one set of axes, draw graphs of $y = k(x)$ and $y = k({}^-x)$.

**b** Describe a transformation that maps $y = k(x)$ to $y = k({}^-x)$.

**c** Do you think this transformation will map $y = k(x)$ to $y = k({}^-x)$ for any function $k$? Explain your answer fully.

**8** For $g(x) = x^2$, explain why the graphs of $y = g(x)$ and $y = g({}^-x)$ are the same.

**9** This is a sketch of the graph of a function $y = p(x)$.

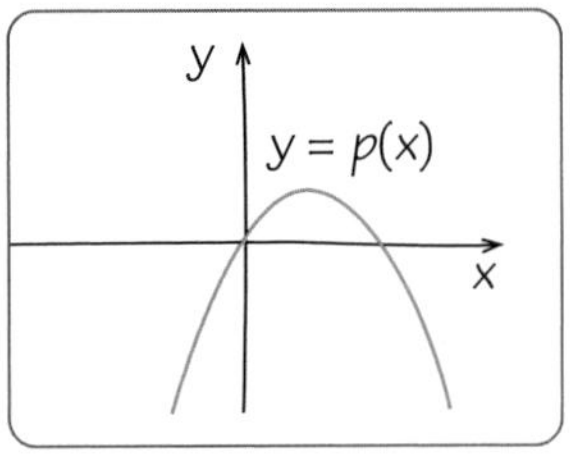


Which of the sketches below could be the graph of:

**a** $y = {}^-p(x)$ **b** $y = p({}^-x)$?

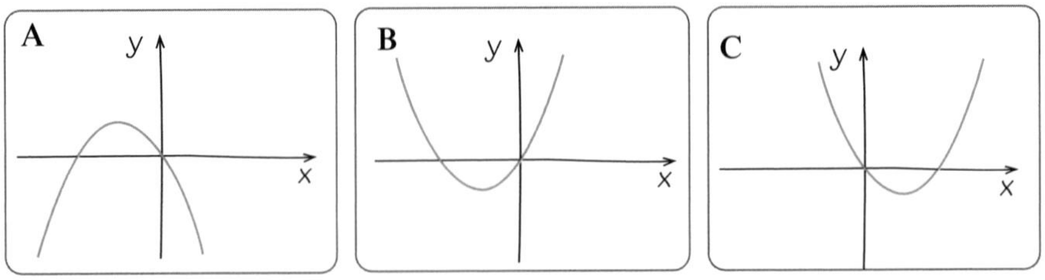


**10** A function $g$ is defined so that $g(x) = x^2 - 3$.

**a** Draw sketch graphs of:

**i** $y = g(x)$ **ii** $y = g(2x)$ **iii** $y = 2g(x)$.

**iv** $y = \left(\frac{1}{4}\right)g(x)$ **v** $y = g\left(\frac{1}{4}x\right)$ **vi** $y = {}^-g(2x)$

**b** Comment on your results.

Remember for question 11a
$y = {}^-f(x) + 5$ means
$y = {}^-x^2 + 5$ when
$f(x) = x^2$

**11** Sketch $y = f(x)$ where $f(x) = x^2$.
On separate diagrams, sketch graphs of:

**a** $y = {}^-f(x) + 5$ **b** $y = f({}^-3x)$ **c** $y = f\left(\frac{1}{2}x\right) - 4$

**12** This is a sketch graph of the function $y = f(x)$ where $f(x) = x^2 - 4x + 5$.

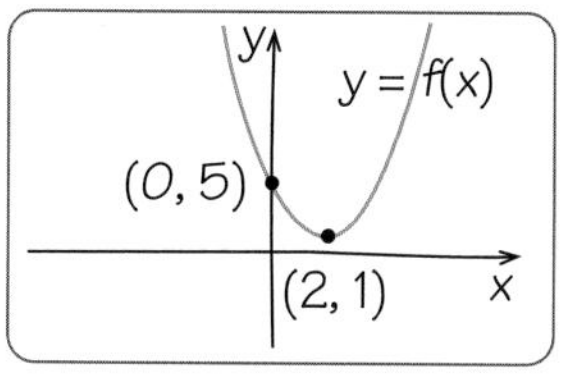


For the following sets of equations and graphs, match each equation to a graph.

**a** $y = f(^{-}x)$ **b** $y = f(x + 5)$ **c** $y = f(2x)$
**d** $y = f(\frac{1}{2}x)$ **e** $y = {}^{-}f(x) + 10$ **f** $y = f(x - 3) - 2$

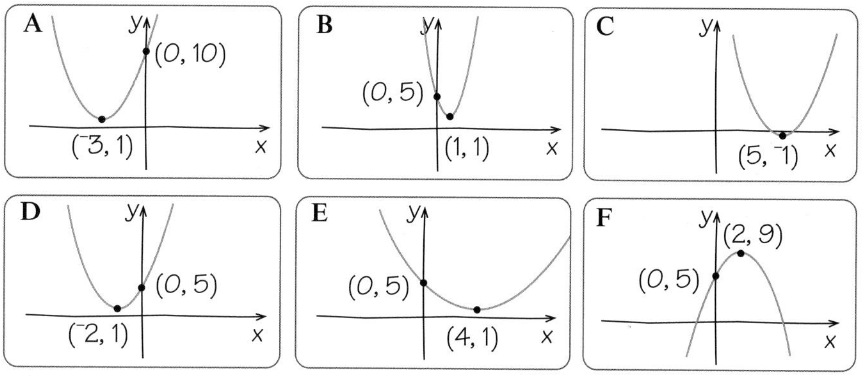


## Using graphs only

This is part of the graph of a function $y = f(x)$.
From this you can draw other related graphs.

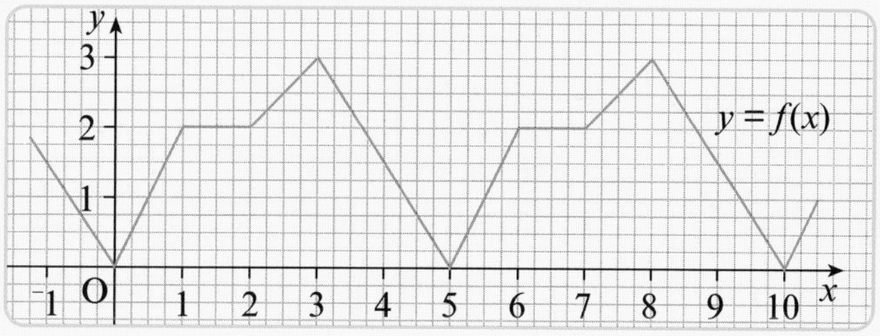


**Example** Draw the graph of $y = f(2x)$ for $0 \leqslant x \leqslant 5$.

> In this example, you can only find values for $y = f(2x)$ from the graph of $y = f(x)$.

One way to draw the graph is to find values for $y = f(2x)$.

- Choose some values for $x$ and find the corresponding $y$ values.

| $x$ | $y = f(2x)$ | $y$ |
|---|---|---|
| 0 | $f(2 \times 0) = f(0)$ | $= 0$ |
| 1 | $f(2 \times 1) = f(2)$ | $= 2$ |
| 2 | $f(2 \times 2) = f(4)$ | $= 1.5$ |
| 3 | $f(2 \times 3) = f(6)$ | $= 2$ |
| 4 | $f(2 \times 4) = f(8)$ | $= 3$ |
| 5 | $f(2 \times 5) = f(10)$ | $= 0$ |

- Plot the points.
- Decide how the graph of $y = f(2x)$ is linked to the graph of $y = f(x)$.
  (Find more values for $y$ if needed:
  e.g. for $x = 0.5$, $y = f(1) = 2$
  for $x = 1.5$, $y = f(3) = 3$.)
- Draw the graph.

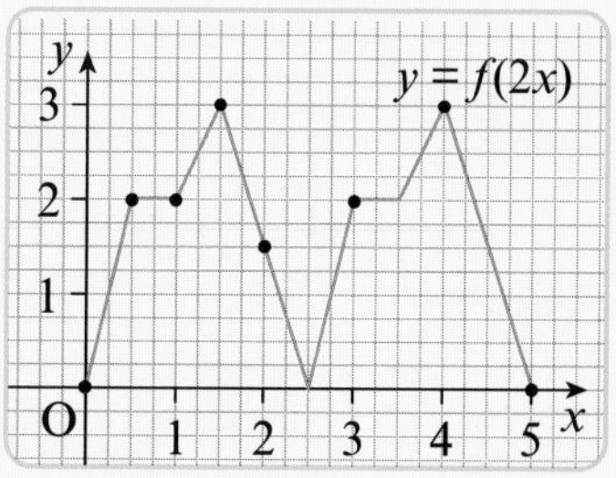


> When transforming graphs, draw the transformed graph on axes that use the same scales as the original graph.

**Exercise 25.5**
**Transforming graphs**

**1** Using the graph of $y = f(x)$ above, draw the graph of:

**a** $y = f(x + 2)$ for $0 \leqslant x \leqslant 5$ **b** $y = 2f(x)$ for $0 \leqslant x \leqslant 5$.

**2** The graph of $y = p(x)$ is drawn for $^-2 \leqslant x \leqslant 6$.

**a** For $0 \leqslant x \leqslant 3$ draw graphs of:

**i** $y = \frac{1}{2}p(x)$
**ii** $y = p(x - 2)$
**iii** $y = -p(x)$
**iv** $y = p(2x)$

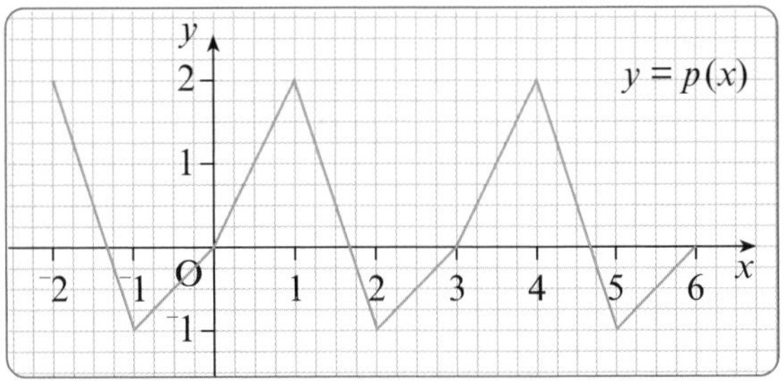


**b** Describe the transformation that maps the graph of $y = p(x)$ to $y = {}^-p(x)$.
**c** Find two values of $x$ for which $p(x + 2) = 0$.

**3** The graph of $y = g(x)$ is drawn for $^-5 \leqslant x \leqslant 5$.

It may help to work out where the asymptotes will be first.

**a** For $^-4 \leqslant x \leqslant 4$ draw graphs of:

**i** $y = g(x) + 3$
**ii** $y = g(x + 1) - 2$
**iii** $y = g(-x)$
**iv** $y = g\left(\frac{x}{2}\right)$.

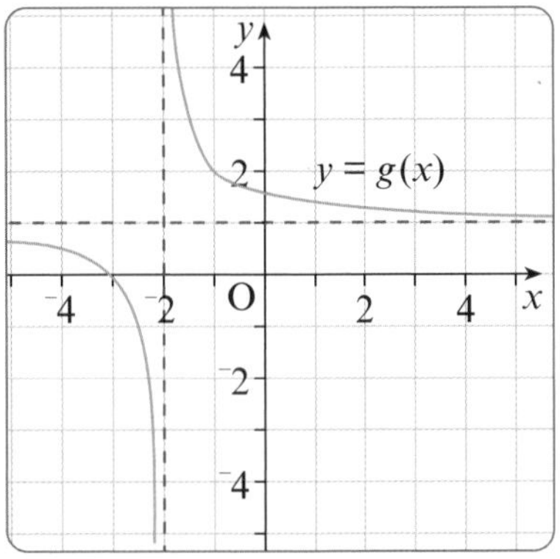


**b** Which transformation maps the graph of $y = g(x)$ to $y = g(x + 1) - 2$?
**c** Find a value of $x$ for which $g(x) = {}^-g(x)$.

**4** This is the graph of $y = h(x)$.

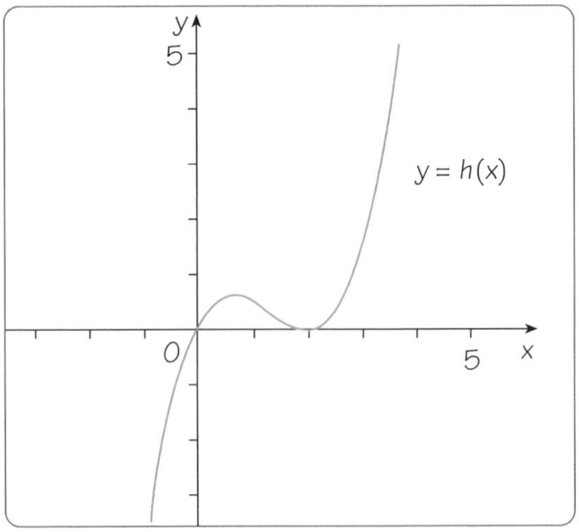


**a** Estimate values of $x$ for which:

**i** $h(x) = 2$
**ii** $h(^-x) = -2$
**iii** $h(x + 3) = 3$.

**b** Sketch graphs of:

**i** $y = h\left(\frac{x}{3}\right)$
**ii** $y = {}^-h(x) - 1$
**iii** $y = h(x - 1) + 1$.

**5** The graph of $y = k(x)$ is drawn for $0 \leqslant x \leqslant 4$.

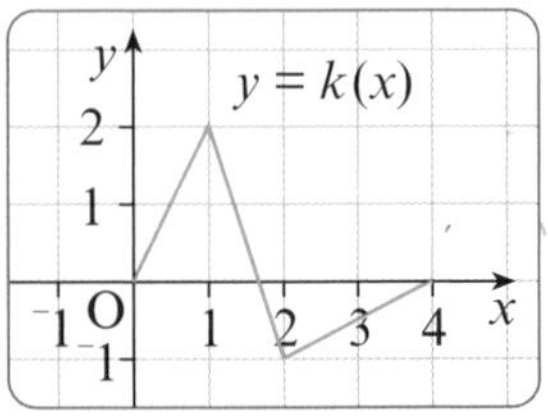


Write down, in terms of $k$, the equation of each graph below.

**A**

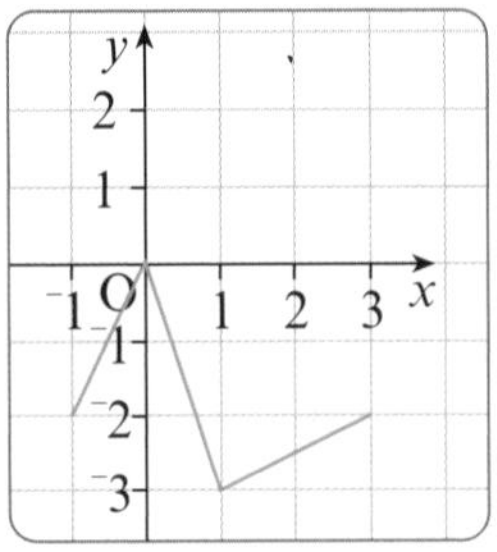

**B**

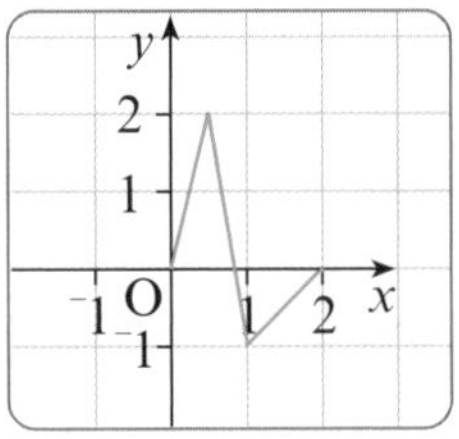

## End points

You should be able to ... | ... so try these questions

**A** Calculate constants in equations from coordinates on graphs

**A1** The equation of this curve is of the form $y = pq^x$, where $p$ and $q$ are constants.

Find the value of $p$ and $q$.

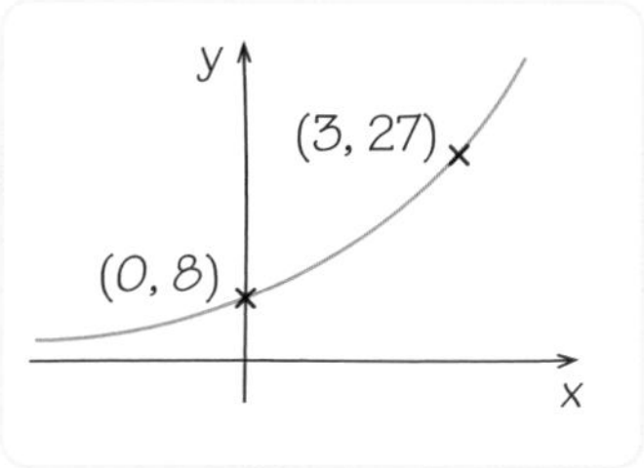


**B** Find rules by drawing linear graphs

**B1** This table shows the cost of some square tablecloths.

| Width ($x$ m) | 0.6 | 0.8 | 1.0 | 1.2 | 1.4 | 1.6 |
|---|---|---|---|---|---|---|
| Cost (£ $C$) | 7.50 | 8.50 | 9.80 | 10.40 | 13.25 | 15.40 |

**a** Draw a graph of $C$ against $x^2$.
**b** It is claimed that $C$ and $x$ are linked by a formula of the type:

$$C = ax^2 + b$$

**i** Explain why your graph supports this claim.
**ii** Estimate the values of $a$ and $b$.
**c** Estimate the cost of a tablecloth that is 1.5 m wide.

**C** Transform graphs of functions

**C1** This is the graph of a function $y = f(x)$ for $^-4 \leqslant x \leqslant 6$.

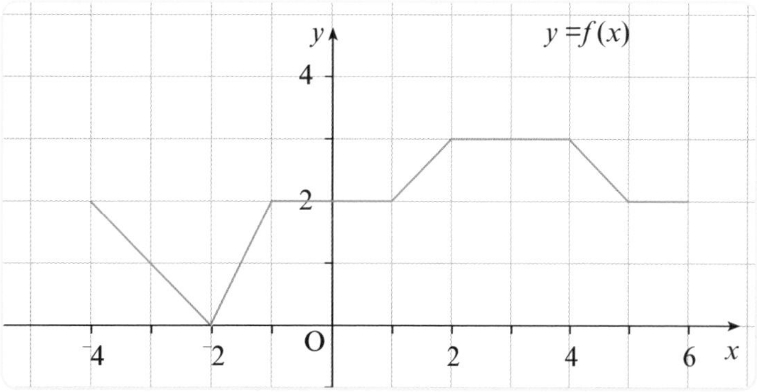


**a** For $0 \leqslant x \leqslant 4$, draw the graph of:

**i** $y = f\left(\frac{x}{2}\right)$ **ii** $y = f(x + 1)$ **iii** $f(^-x)$.

**b** Write down, in terms of $f$, the equation of this graph.

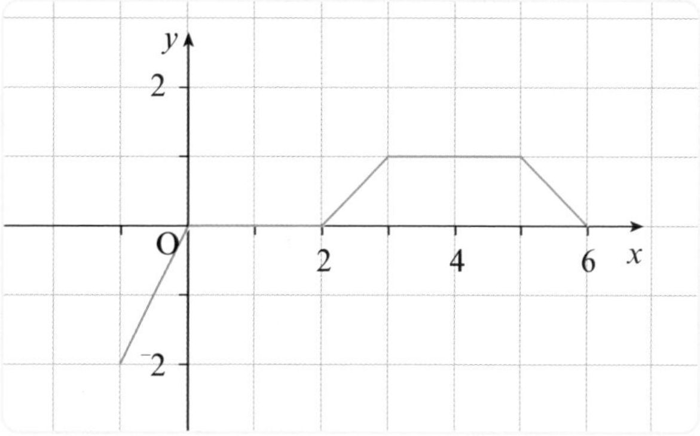

# In Focus 1

## Fractions

**1** For each of the following pairs of numbers, find:

**i** the highest common factor
**ii** the lowest common multiple.

**a** 2 and 7 **b** 25 and 30
**c** 28 and 63 **d** 10 and 100
**e** 36 and 126 **f** 660 and 1260

**2** Give each answer as a fraction in its lowest terms. Write fractions greater than 1 as mixed numbers.

**a** $\frac{1}{4} + \frac{1}{9}$ **b** $2\frac{1}{2} - 1\frac{1}{3}$ **c** $\frac{3}{5} + \frac{1}{10}$
**d** $3\frac{1}{5} - 2\frac{3}{10}$ **e** $\frac{2}{3} - \frac{1}{24}$ **f** $1\frac{1}{2} + \frac{4}{9}$
**g** $1\frac{5}{12} + 2\frac{7}{8}$ **h** $\frac{3}{4} - \frac{1}{22}$ **i** $5\frac{1}{4} - 3\frac{5}{8}$

**3** Give each answer as a fraction in its lowest terms. Write fractions greater than 1 as improper fractions.

**a** $\frac{2}{3} \times 2\frac{1}{2}$ **b** $\frac{5}{6} \div \frac{2}{3}$ **c** $\frac{3}{4} \div \frac{1}{2}$
**d** $1\frac{1}{3} \times 2\frac{1}{4}$ **e** $3\frac{4}{5} \div \frac{3}{4}$ **f** $\frac{8}{9} \times \frac{9}{10}$
**g** $4\frac{1}{6} \div \frac{3}{5}$ **h** $\frac{1}{2} \times \frac{2}{5} \times \frac{3}{10}$

**4** Find the value of these, in fractional form, when $x = \frac{1}{3}$, $y = \frac{2}{5}$ and $z = \frac{3}{10}$.

**a** $5(x + y)$ **b** $\frac{x}{z}$ **c** $z^2x$
**d** $\frac{y - z}{x}$ **e** $\frac{1}{x} + \frac{1}{y}$ **f** $\frac{z}{y - x}$

**5** Find the value of these, in fractional form, when $p = 2$, $q = \frac{1}{4}$ and $r = \frac{2}{3}$.

**a** $pqr$ **b** $(r - q)^2$ **c** $q(p - r)$
**d** $\frac{q}{p}$ **e** $\frac{1}{p} + \frac{1}{q}$ **f** $\frac{1}{p - q}$

**6** If $\frac{1}{f} = \frac{2}{g} + \frac{3}{k}$

what is the value of $f$, in fractional form, when:

**a** $g = 4$ and $k = 8$ **b** $g = 3$ and $k = 5$
**c** $g = 6$ and $k = 7$ **d** $g = 5$ and $k = 2$

**7** Find two different unit fractions that add to give:

**a** $\frac{2}{3}$ **b** $\frac{1}{4}$ **c** $\frac{1}{9}$

**8** In a magic square, the sums of the numbers in each row, column and diagonal are equal.

**Square A**

| | | |
|---|---|---|
| $\frac{3}{2}$ | $\frac{1}{3}$ | |
| $\frac{2}{3}$ | | |
| $\frac{5}{6}$ | | |

**Square B**

| | | |
|---|---|---|
| $2\frac{1}{4}$ | $\frac{1}{2}$ | $2\frac{2}{3}$ |
| $2$ | $1\frac{11}{12}$ | $1\frac{5}{6}$ |
| $1\frac{1}{6}$ | $3\frac{1}{3}$ | $1\frac{1}{4}$ |

**a** Copy and complete Square A so that it is a magic square.
**b** Square B is not a magic square.
How could you change one of the fractions to make a magic square?

**9** Find four pairs of equivalent expressions.

(A) $\frac{x}{x(x + 5)}$ (B) $\frac{x + 5}{x(x + 5)}$ (C) $\frac{2}{x + 5}$
(D) $\frac{1}{x + 5}$ (E) $\frac{1}{x(x + 5)}$ (F) $\frac{3}{x}$
(G) $\frac{4}{2(x + 5)}$ (H) $\frac{1}{x}$ (I) $\frac{3x}{x^2}$

**10** Write each of these as a single fraction. Give each answer in its simplest from.

**a** $\frac{x}{3} + \frac{x}{9}$ **b** $\frac{2}{x} \times \frac{x}{5}$ **c** $\frac{7}{x} - \frac{4}{x}$
**d** $\frac{7x}{3} \div \frac{x}{9}$ **e** $\frac{2x}{5} - \frac{x}{10}$ **f** $\frac{3}{x} + \frac{1}{3x}$
**g** $\frac{x}{2} + \frac{y}{7}$ **h** $\frac{3}{x} - \frac{1}{x + 4}$ **i** $\frac{x}{4} - \frac{y}{2}$
**j** $\frac{5}{x} \times \frac{x}{x - 1}$ **k** $\frac{x}{x + 2} \div \frac{x}{x - 1}$ **l** $\frac{1}{x} + \frac{2}{y}$
**m** $\frac{4}{x} + \frac{1}{x - 1}$ **n** $\frac{1}{x} - \frac{x}{y}$ **o** $\frac{1}{x + 4} \div \frac{1}{x - 1}$
**p** $\frac{2}{x + 1} + \frac{3}{x + 2}$ **q** $\frac{2}{x + 1} + \frac{3}{x - 2}$
**r** $\frac{5}{x + 3} + \frac{1}{x + 3}$ **s** $\frac{2}{x} - \frac{1}{x + 2}$
**t** $\frac{1}{x - 1} + \frac{5}{x + 2}$ **u** $\frac{5}{x + 1} + \frac{1}{y + 2}$

**11** Copy and complete:

**a** $\frac{\square}{x} - \frac{3}{x - 3} = \frac{3(x - 6)}{x(x - 3)}$

**b** $\frac{1}{x + 1} + \frac{\square}{x - 1} = \frac{2(5x + 4)}{(x + 1)(x - 1)}$

## Properties of shapes

**1**

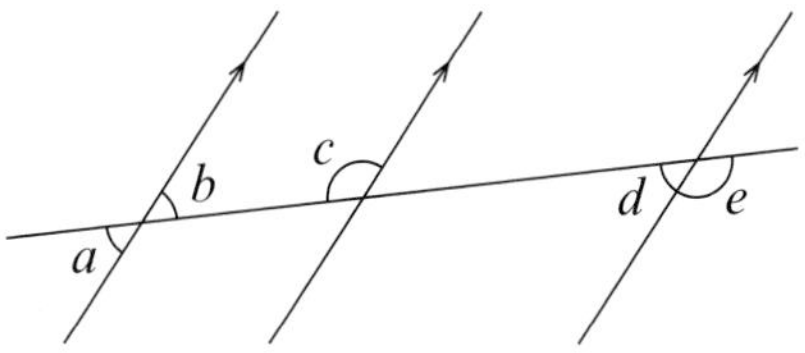


**a** If $a = 47°$, calculate angles $b$ to $e$.
**b** If $a = 51.5°$, calculate the angles $b$ to $e$.
**c** If $e = 169°$, calculate angles $a$ to $d$.

**2** These polygons are drawn on a grid of parallel lines. The diagonals of PQRS are marked in red.

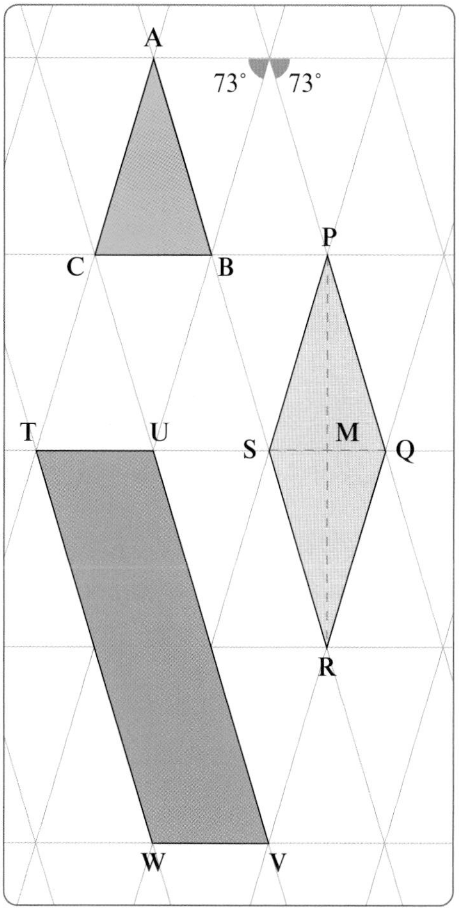


**a** Work out each interior angle of ΔABC.
**b** What type of triangle is ABC?
**c** Work out each interior angle of:
  **i** PQRS  **ii** TUVW
**d** Explain why PQRS is a rhombus.
**e** Explain why TUVW cannot be a rhombus.
**f** The diagonals of PQRS intersect at M.
Work out each of these angles:
  **i** $P\hat{M}Q$  **ii** $P\hat{Q}M$  **iii** $M\hat{Q}R$
  **iv** $Q\hat{R}M$  **v** $S\hat{R}M$

**3** These quadrilaterals are on an isometric grid.

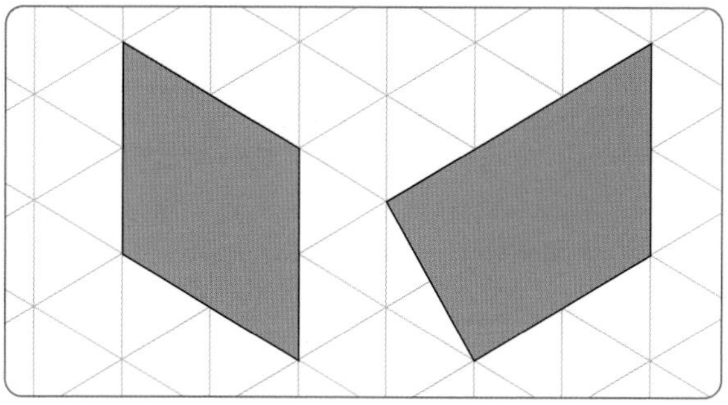

**a** Show how each of these quadrilaterals will tessellate on its own.
**b** Show how they will tessellate together.

**4** Calculate the angles $a$ to $m$ in the diagram below.

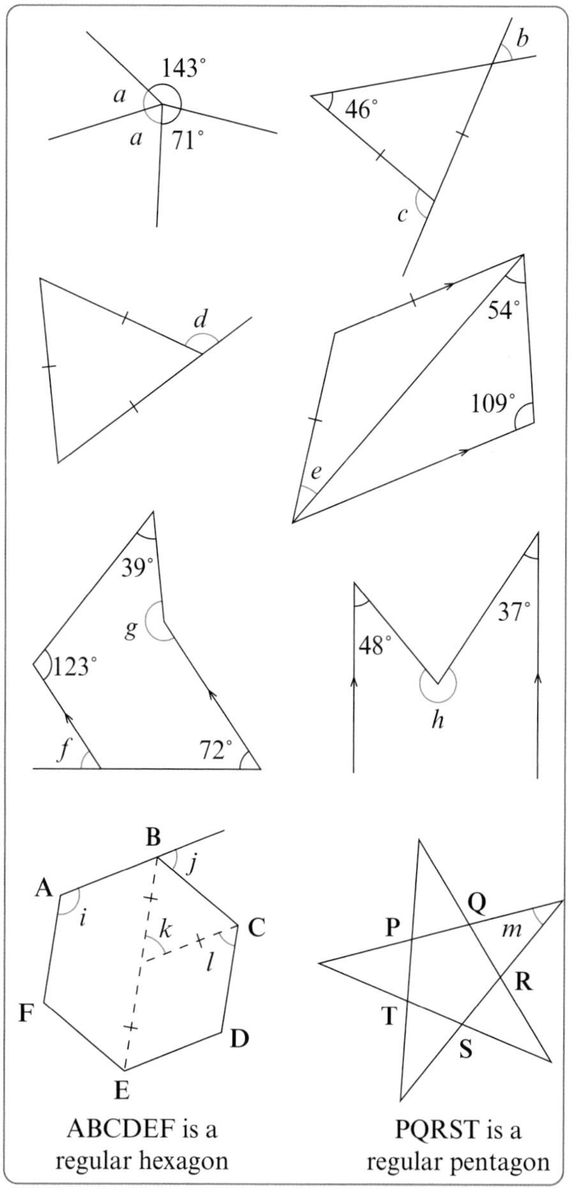

## Manipulation

**1** Multiply these terms.

**a** $mn \times mn$ **b** $pq \times 4p$ **c** $9b \times 2a^2$
**d** $a^3 \times g^3$ **e** $4p \times 5p^2$ **f** $7a^3 \times 4b$

**2** Multiply out these brackets.

**a** $6(4a - 3b)$ **b** $p(n - p)$ **c** $s(s + t)$
**d** $6m(n - m)$ **e** $5x(x + y)$ **f** $u(4u + 3)$
**g** $p(5q + 3r)$ **h** $3n(4m + 7n)$ **i** $9a(3b + a)$

**3** Simplify these.

**a** $8n + 4m - 6n + 3m$ **b** $12a^2 - a + 9a^2$
**c** $b^3 + 2b - b^3$ **d** $8x - 4xy - 2y + 6xy$
**e** $p + 8q - 5q + 3p$ **f** $8p^2 + 3pq + 5q^2 - pq$
**g** $4m^2 + mn - 6n + nm$ **h** $x + 2xy + x - 4xy$

**4** Which of these expressions is equivalent to $4a(2ab + 3a) + a(5b - a) + 4b(2a^2 + 5a)$?

A $35a^2b + 11a^2$

B $12a^2b + 14ab + 7a^2$

C $41ab + 10a$

D $16a^2b + 11a^2 + 25ab$

**5** Multiply these out and simplify

**a** $x(6x - 2) + 2x(5x + 3)$
**b** $8x(2y + 3x) + x(6x - 3y)$
**c** $6(p + 2q) - 3(4p + 2q)$
**d** $4(3a + b) - (3a - 2b)$
**e** $p(5p - 4) - 2(2p^2 - 3)$

**6** Write each of these as a single fraction.

**a** $\frac{3}{p+1} + \frac{5}{p-1}$ **b** $\frac{5}{p+4} - \frac{3}{p+2}$

**c** $\frac{1}{2p+1} + \frac{3}{3p-2}$ **d** $\frac{p+4}{3} - \frac{p-3}{6}$

**7** Write an expression for the width of rectangles A and B.

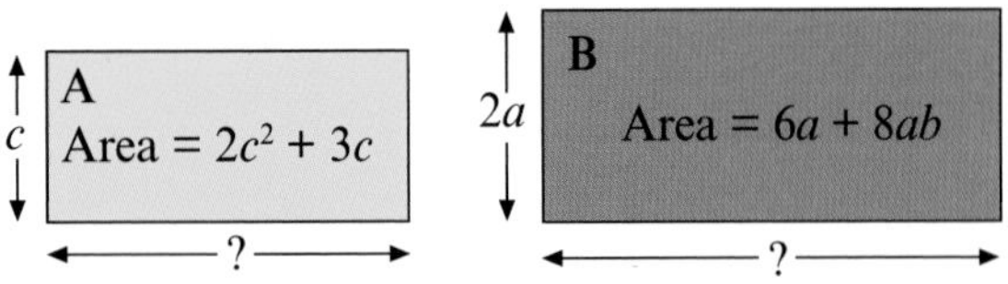


**8** Factorise these fully.

**a** $7p + 3pq$ **b** $m + 3mn$ **c** $8pq + 4q$
**d** $3y - 12xy$ **e** $2b^2 + 10b$ **f** $18ab + 24bc$
**g** $x^2y - 2xy$ **h** $9a^2b + 6ab^2$ **i** $12m^2n - 9mn$

**9** Multiply out these brackets and simplify.

**a** $(w + 2)(w + 6)$ **b** $(y + 9)(y + 5)$
**c** $(x - 3)^2$ **d** $(v - 2)(v - 3)$
**e** $(b - 3)(b + 5)$ **f** $(b - 3)(b - 5)$
**g** $(h + 5)(h - 5)$ **h** $(t - 3)(t + 3)$

**10** Multiply out these brackets and simplify.

**a** $(2w - 3)(2w + 3)$ **b** $(2p + 1)^2$
**c** $(5x + 6)(3x - 2)$ **d** $(3g - 7)(4g - 5)$

**11**

| | | | | | |
|---|---|---|---|---|---|
| $x + 1$ | $x + 2$ | $x + 3$ | $x + 4$ | $x + 6$ | $x + 12$ |

Which pair of expressions multiply to give:

**a** $x^2 + 7x + 12$ **b** $x^2 + 8x + 12$ **c** $x^2 + 7x + 6$
**d** $x^2 + 5x + 6$ **e** $x^2 + 13x + 12$ **f** $x^2 + 6x + 8$?

**12** Factorise the following expressions.

**a** $x^2 + 4x + 4$ **b** $x^2 + 9x + 20$ **c** $x^2 + 3x - 4$
**d** $x^2 + 2x - 15$ **e** $x^2 - x - 6$ **f** $x^2 - 11x + 10$
**g** $x^2 - 2x - 8$ **h** $x^2 - 3x + 2$ **i** $x^2 - 4x + 4$

**13** Simplify each of these

**a** $\frac{3x - 6}{x^2 + x - 6}$ **b** $\frac{6(x - 3) - 2(x - 3)}{x^2 + 3x - 18}$

**14**

| | | | | | |
|---|---|---|---|---|---|
| $x - 6$ | $x - 3$ | $x - 2$ | $x + 1$ | $x + 2$ | $x + 3$ |

Which pair of expressions multiply to give:

**a** $x^2 + x - 6$ **b** $x^2 - 2x - 3$ **c** $x^2 - 5x + 6$
**d** $x^2 - 4$ **e** $x^2 + 5x + 6$ **f** $x^2 - 9$?

**15** Factorise these fully.

**a** $3x^2 + 13x + 12$ **b** $3x^2 - 15x + 12$
**c** $2x^2 - 9x + 4$ **d** $6x^2 - 3x - 3$
**e** $3x^2 - 6x$ **f** $9x^2 - 25y^2$

**16** Find the values of $p$ and $q$ such that for all values of $x$:

**a** $x^2 - px + 16 = (x - q)^2$ **b** $x^2 - 6x + p = (x - q)^2$

**17** Write each of these as a single fraction in its simplest form.

**a** $\frac{4}{p-3} + \frac{3}{p+1}$ **b** $\frac{2}{(a-b)^2} - \frac{3}{2(a-b)}$

**c** $\frac{15p^2}{8a} \div \frac{10p}{12a^3}$ **d** $\frac{2}{p+1} \times \frac{3}{p-1}$

**18** **a** Factorise $p^2 - q^2$.
**b** In the equation $p^2 - q^2 = 149$, $p$ and $q$ are positive integers.
**i** Explain why there is only one solution.
**ii** Explain why $(p - q)$ must equal 1.
**iii** Find the value of $p$ and $q$.

**19** Use the difference of two squares to calculate the exact value of:
$132\,413\,241\,324^2 - 132\,413\,241\,323^2$

**20** Copy and complete:

$$\frac{\square\square\square}{x+1} + \frac{3}{x} = \frac{x^2 + 5x + 3}{x(x+1)}$$

## Comparing data

**1** These are the round scores from a competition.

Dina: 60 68 52 68 64 72 64

Cian: 57 61 59 68 61 75

Joe: 58 59 63 65 57 70 68 74

For each set of scores, calculate:

**a** the mean
**b** the standard deviation.

**2** If Joe had scored 15 less in each round, what would the mean and standard deviation of his scores have been?

**3** Dina improved each of her round scores by 25% in the next competition.
Give the new mean and standard deviation.

**4**

Sally

| Ring score | 2 | 3 | 4 | 5 | 6 | 7 | 8 | 9 |
|---|---|---|---|---|---|---|---|---|
| Frequency | 2 | 1 | 3 | 2 | 3 | 6 | 4 | 3 |

Gavin

| Ring score | 2 | 3 | 4 | 5 | 6 | 7 | 8 | 9 |
|---|---|---|---|---|---|---|---|---|
| Frequency | 1 | 1 | 3 | 7 | 14 | 13 | 6 | 7 |

**a** For each distribution, calculate:
**i** the mean
**ii** the standard deviation
**b** Compare the two distributions.

**5**

Peggy

| Ring score | 2 | 3 | 4 | 5 | 6 | 7 | 8 |
|---|---|---|---|---|---|---|---|
| Frequency | 1 | 0 | 2 | 3 | 5 | 4 | 1 |

Toby

| Ring score | 3 | 4 | 5 | 6 | 7 | 8 | 9 |
|---|---|---|---|---|---|---|---|
| Frequency | 2 | 4 | 5 | 4 | 6 | 0 | 1 |

Sue

| Ring score | 2 | 3 | 4 | 5 | 6 | 7 | 8 | 9 |
|---|---|---|---|---|---|---|---|---|
| Frequency | 2 | 1 | 3 | 2 | 3 | 6 | 4 | 3 |

**a** For each distribution:
**i** find the median
**ii** calculate the interquartile range.
**b** Compare the three distributions.

| | Number of chips | | | | | | | | |
|---|---|---|---|---|---|---|---|---|---|
| Restaurant | 34 | 35 | 36 | 37 | 38 | 39 | 40 | 41 | Total |
| P | 8 | 10 | 14 | 9 | 6 | 5 | 4 | 4 | 60 |
| Q | 8 | 7 | 5 | 9 | 12 | 11 | 8 | 0 | 60 |
| R | 0 | 3 | 5 | 10 | 12 | 14 | 10 | 6 | 60 |

**6** For each restaurant:

**a** find the median number of chips
**b** calculate the interquartile range
**c** draw a box-and-whisker plot.

**7** **a** Draw frequency polygons for restaurant P and restaurant R on the same diagram.
**b** Compare the two distributions.

**8**

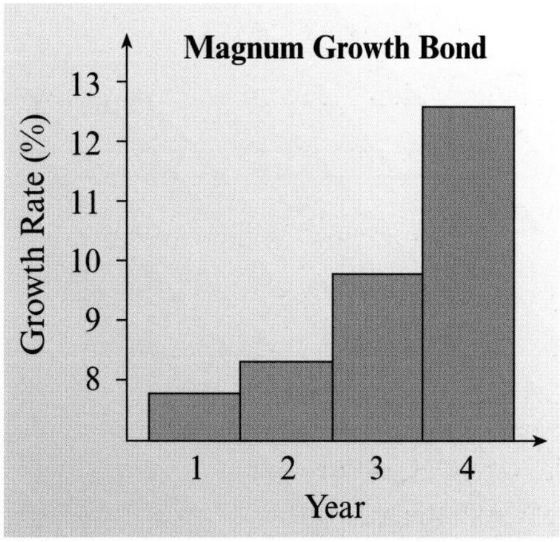


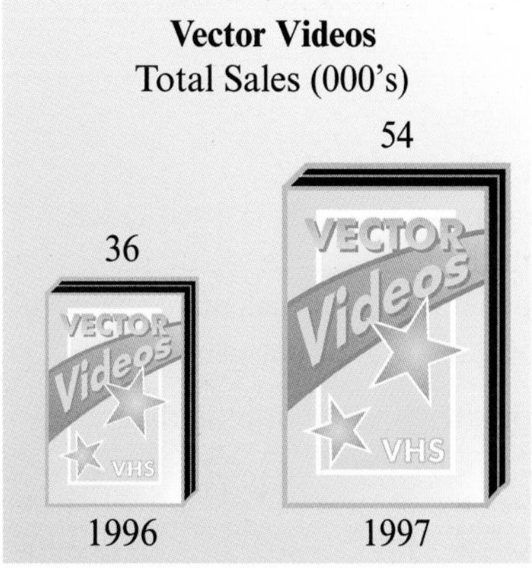


Explain why each of these diagrams is misleading.

**9**

*Wilton Dale Theme Parks*

| | 1994 | 1996 |
|---|---|---|
| A Number of injuries | 21 | 14 |
| B Number of rides | 6 | 15 |

Draw a misleading diagram to show each of these sets of data.

## Sequences

1 Each diagram shows the first three patterns in a sequence.

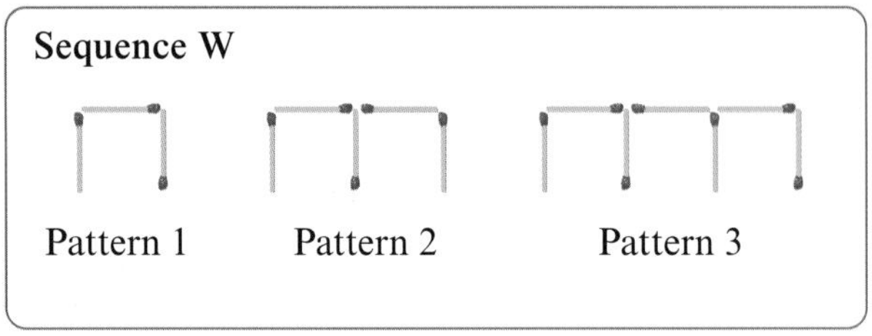


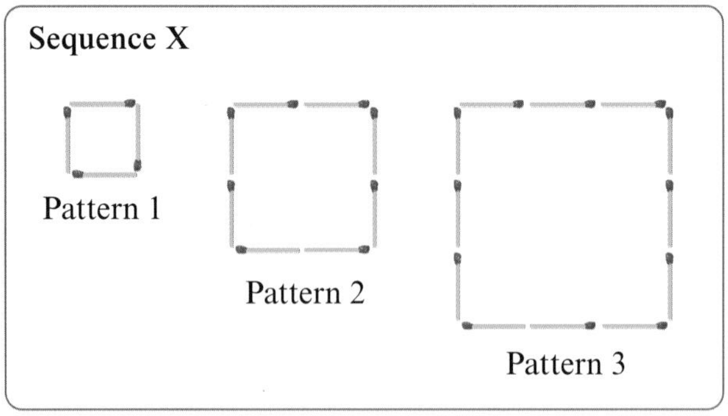


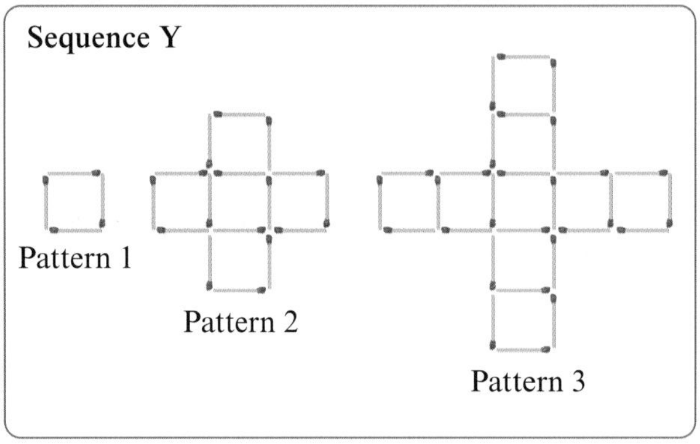


For each sequence:

**a** Draw the 4th pattern.
**b** Find the general term for the sequence. Write it in the form $a_n = \dots$ .
**c** Use the general term to calculate the number of matches in the 10th pattern.

2 Copy and complete each mapping diagram.

**a**

1 ⟶ 9
2 ⟶ 12
3 ⟶ 15
4 ⟶ 18
⋮
20 ⟶ ☐
⋮
$n$ ⟶ ☐

**b**

1 ⟶ 3
2 ⟶ $^{-}1$
3 ⟶ $^{-}5$
4 ⟶ $^{-}9$
⋮
40 ⟶ ☐
⋮
$t$ ⟶ ☐

3 These are the first three patterns in a sequence.

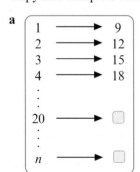

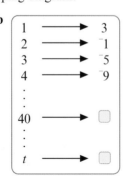

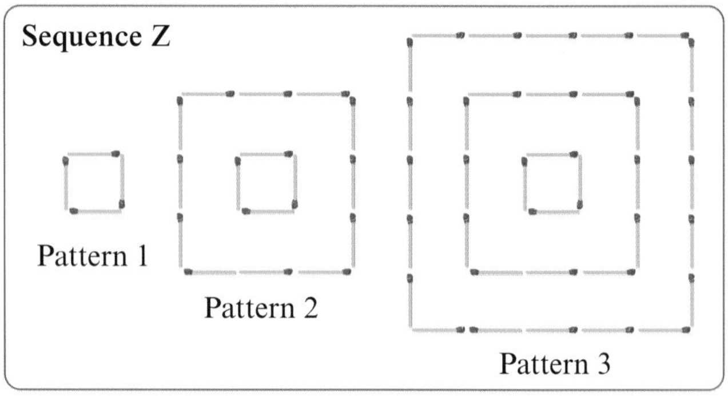


**a** Find a rule for the number of matches in the $n$th pattern in the form $n \longrightarrow \dots$ .
**b** Show how you found your rule.

4 **A** 1, 4, 7, 10, 13, …
**B** 7, 13, 19, 25, 31, …
**C** 2, 6, 10, 14, 18, …
**D** 7, 12, 17, 22, 27, …
**E** 11, 8, 5, 2, $^{-}1$, …
**F** 14, 5, $^{-}4$, $^{-}13$, $^{-}22$, …

For each of the sequences A to F:

**a** Find the general term, $g_n$.
**b** Calculate $g_{30}$.

5 Give the general term for each of these sequences.

**a** 2, 5, 10, 17, 26, …
**b** 11, 8, 3, $^{-}4$, $^{-}13$, …
**c** $^{-}1$, 2, 7, 14, 23, …
**d** 1, 4, 9, 16, 25, …
**e** 6, 12, 22, 36, 54, …
**f** $^{-}898$, $^{-}895$, $^{-}890$, $^{-}883$, $^{-}874$, …

6 For each of the following sequences where $u_1 = 1$:

**a** Calculate $u_4$.
**b** State if the sequence diverges or converges and any limit it approaches.

**A** $u_{n+1} = u_n(u_n + 4)$

**B** $u_{n+1} = \dfrac{u_n + 1}{2}$

**C** $u_{n+1} = \dfrac{u_n + 2}{u_n}$

**D** $u_{n+1} = \dfrac{9u_n + 5}{2u_n}$

# Constructions and loci

**1** Construct these triangles accurately. Show all your construction lines.

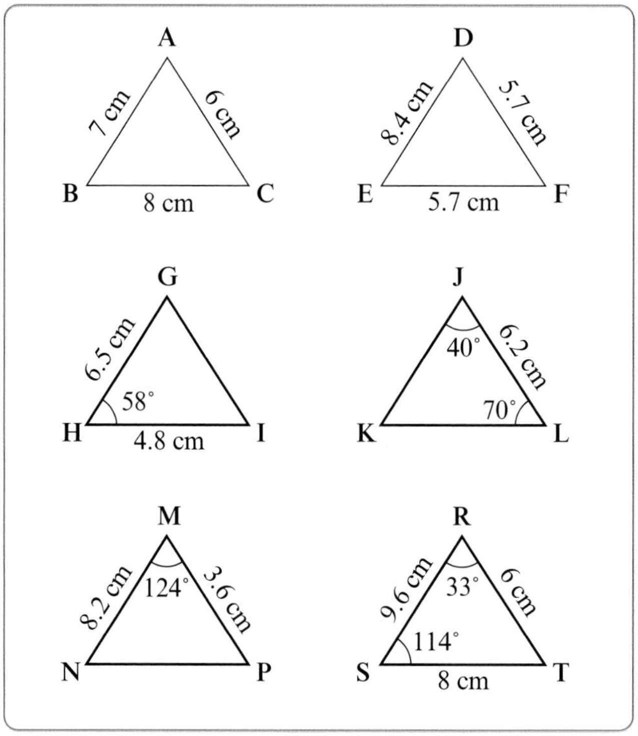


**2** Copy each line full size and contruct a perpendicular bisector of it.

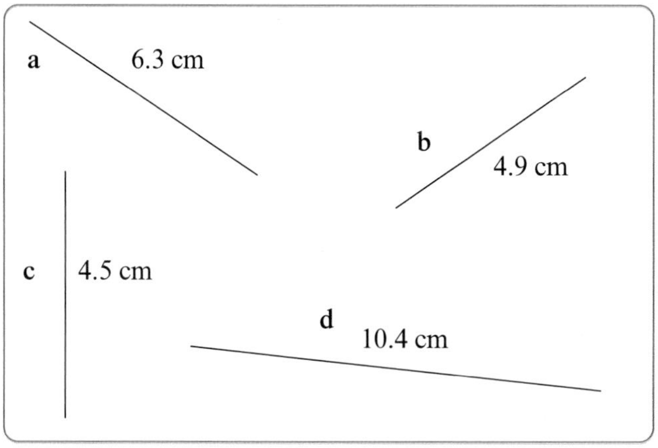


**3** Draw each angle accurately and construct its bisector.

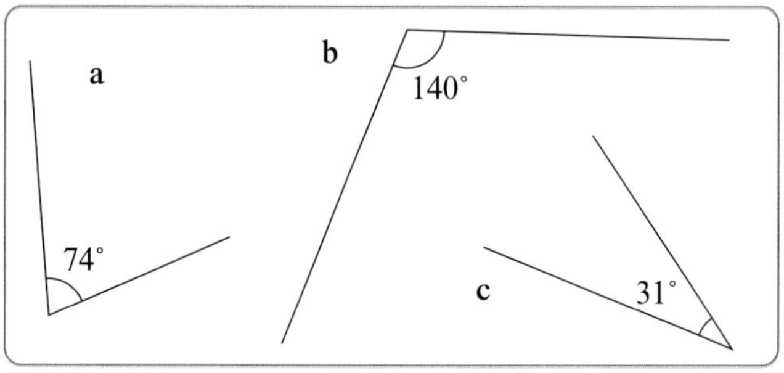


**4** For each pair of triangles, state if they are congruent or not necessarily so and, if congruent, state the case of congruence.

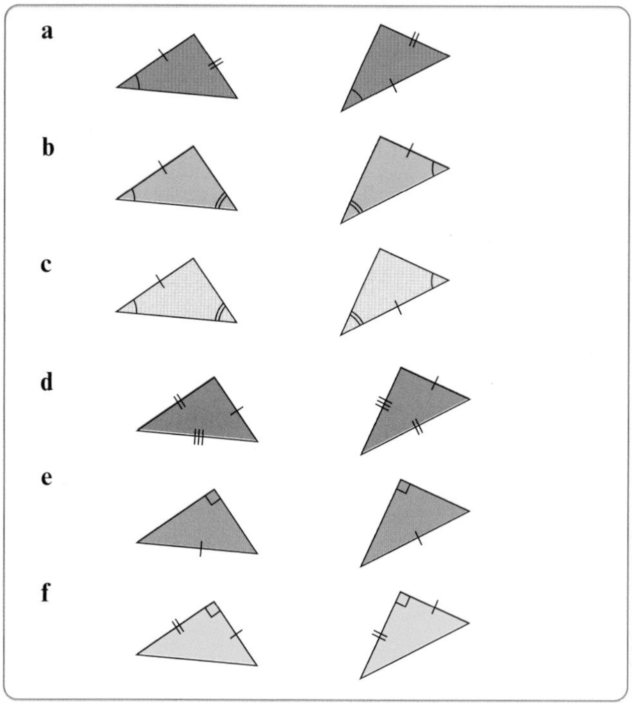


**5** A field for the village fete is shaped as a triangle with these dimensions.

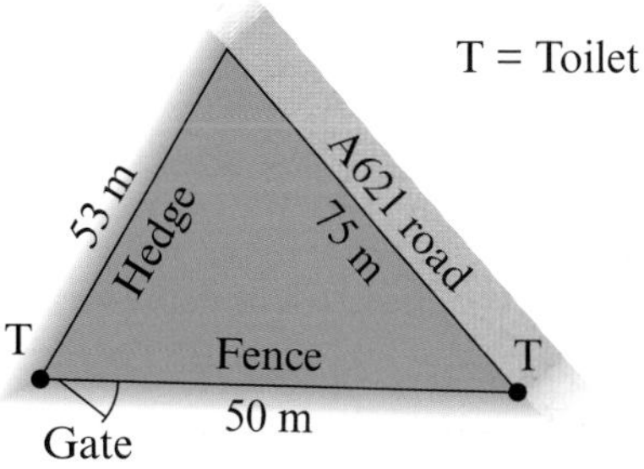


Colonel Briggs Shilton has set out these conditions for the position of the drinks tent.

| | |
|---|---|
| **Condition 1** | The tent must be equidistant from the hedge and the fence. |
| **Condition 2** | The distance from each toilet to the tent must be equal. |

**a** Construct a scale drawing of the field to a scale of 1 to 1000.
**b** Find by construction the position of the tent.
**c** Why is this position not a sensible one?
**d** Give some different conditions which you think puts the tent in a better position.

## Powers and Roots

**1** Which of these are not integers?

**a** $\sqrt[4]{40}$ **b** $\sqrt[5]{7776}$ **c** $\sqrt[3]{3}$
**d** $\sqrt[7]{(^-1)}$ **e** $\sqrt[5]{10}$ **f** $\sqrt[9]{1}$

**2** Evaluate:

**a** $\sqrt[3]{64}$ **b** $\sqrt[6]{0.046\,656}$
**c** $\sqrt[5]{(^-243)}$ **d** $\sqrt[5]{1.610\,51}$

**3** Evaluate correct to 1 decimal place:

**a** $\sqrt[3]{15}$ **b** $\sqrt[6]{20}$ **c** $\sqrt[4]{24}$

**4** Evaluate as a fraction:

**a** $4^{^-3}$ **b** $\sqrt[4]{\frac{81}{625}}$ **c** $\sqrt[3]{\frac{1}{343}}$

**5** Find the value of $x$ when:

**a** $\sqrt[5]{x} = 1.3$ **b** $\sqrt[x]{10.4976} = 1.8$
**c** $\sqrt[x]{512} = 2$ **d** $\sqrt[7]{x} = {}^-0.000\,0128$

**6** Give the answer to each of these using index notation.

**a** $4^5 \times 4^3$ **b** $3^2 \times 3^{^-3}$ **c** $2^4 \times 2^{^-4}$
**d** $(2^3)^4$ **e** $(6^7)^{^-2}$ **f** $6^5 \div 6^3$
**g** $6^3 \div 6^5$ **h** $(6^{^-2})^7$

**7** To what power must $2^3$ be raised to give:

**a** $2^9$ **b** $2^{^-6}$ **c** $\frac{1}{2^3}$?

**8** Simplify as far as you can:

**a** $y^2 \times y^3$ **b** $(y^6)^5$ **c** $(3y^2)^4$
**d** $3y^5 \times 4y^2$ **e** $\frac{y^5}{y^4}$ **f** $\frac{4y^3}{2y^7}$
**g** $\frac{2y^3 \times 6y^5}{4y^2}$ **h** $\frac{4y \times 3y^4}{6y^7}$ **i** $\frac{11y^5}{2y^2 \times 5y^3}$

**9** Find two sets of equivalent expressions.

(A) $3^{\frac{2}{5}}$ (B) $\frac{1}{\sqrt{3^5}}$ (C) $3^{\frac{5}{2}}$ (D) $\sqrt[5]{3^2}$

(E) $\sqrt{3^5}$ (F) $\frac{1}{\sqrt[3]{3^2}}$ (G) $(\sqrt{3})^{-5}$ (H) $3^{-\frac{2}{5}}$

**10** Evaluate these as integers or in fractional form.

**a** $49^{\frac{1}{2}}$ **b** $81^{\frac{1}{2}}$ **c** $27^{\frac{1}{3}}$
**d** $343^{\frac{1}{3}}$ **e** $1^{\frac{1}{4}}$ **f** $625^{\frac{1}{4}}$
**g** $243^{-\frac{1}{5}}$ **h** $(^-1024)^{\frac{1}{5}}$ **i** $32^{\frac{1}{5}}$
**j** $16^{-\frac{1}{2}}$ **k** $216^{-\frac{1}{3}}$ **l** $729^{-\frac{1}{6}}$
**m** $343^{\frac{2}{3}}$ **n** $256^{\frac{3}{4}}$ **o** $32^{\frac{2}{5}}$
**p** $64^{\frac{5}{6}}$ **q** $512^{\frac{2}{3}}$ **r** $625^{\frac{3}{4}}$
**s** $32^{\frac{3}{5}}$ **t** $243^{\frac{4}{5}}$ **u** $729^{\frac{5}{6}}$
**v** $27^{\frac{4}{3}}$ **w** $64^{-\frac{2}{3}}$ **x** $256^{-\frac{3}{4}}$
**y** $243^{-\frac{3}{5}}$ **z** $64^{-\frac{5}{2}}$

**11** Solve these equations.

**a** $121^{\frac{1}{x}} = 11$ **b** $3^x = \frac{1}{27}$
**c** $27^x = 3$ **d** $729^{\frac{1}{x}} = 3$
**e** $2^{3x+1} = 16$ **f** $3^{2x-1} = \frac{1}{81}$
g $125^x = 25$ h $81^x = 729$
i $x^{\frac{4}{3}} = 16$ j $25^x = \frac{1}{5}$
k $343^x = \frac{1}{49}$ l $x^{-\frac{2}{5}} = \frac{1}{16}$

**12** Simplify as far as you can:

**a** $(\sqrt{6})^2$ **b** $(\frac{1}{5}\sqrt{5})^2$
**c** $\sqrt{2} \times \sqrt{7}$ **d** $\sqrt{5} \times \sqrt{20}$
**e** $(\sqrt{3} + \sqrt{5})^2$ **f** $(\sqrt{7} - \sqrt{2})^2$
**g** $\sqrt{14} \div \sqrt{7}$ **h** $(2\sqrt{5} + 1)^2$
**i** $(\sqrt{11} - 3)^2$

**13** Write each of these recurring decimals as a fraction in its lowest terms.

**a** 0.333333 … **b** 0.363636 …
**c** 0.010101 … **d** 0.126126 …
**e** $0.\dot{5}$ **f** $0.\dot{0}\dot{9}$
**g** $0.\dot{8}5714\dot{2}$ **h** $0.\dot{2}\dot{7}$

**14**

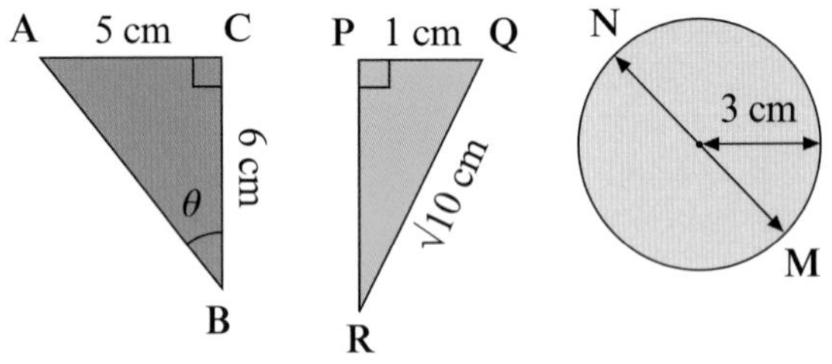


Which of these lengths are rational?

**a** AB
**b** PR
**c** MN
**d** the perimeter of triangle ABC
**e** the circumference of the circle

# IN FOCUS 8

## Approximation and Errors

**1** Round each number to the degree of accuracy stated:

**a** 5.674 (2 dp)
**b** 12.652 (1 dp)
**c** 12.652 (1 sf)
**d** 2143 (nearest ten)
**e** 15.77777 (4 sf)
**f** 93747656 (5 sf)
**g** 15.986 (2 sf)
**h** 456.21345 (1 sf)

**2** By approximating each number to 1 sf calculate approximate answers to each of these.

**a** $84.3 \times 452.53$
**b** $4.876 \times 37.71$
**c** $5683.2 \times 0.0372$
**d** $458.12 \times 518$
**e** $734.6 \div 2.316$
**f** $0.005\,682\,43 \div 7.8931$
**g** 419 52 + 77442
**h** $\dfrac{34.6296 + 87.3}{0.003\,21}$
**i** $\dfrac{6834 \times 1939.453}{54.26}$
**j** $\dfrac{45.95 \times (2.943\,56)^2}{(0.067)^2}$

**3** Work out approximate answers by rounding each part first to 1 sf then calculate the percentage error produced by doing this.

**a** $56.6 \times 21.5$
**b** $4924 \times 3.7$
**c** $246.3 \div 9.67$
**d** $54.27 \times (21.6)^3$
**e** $\dfrac{17.63 \times 1839}{(54.64)^2}$

**4** State both the upper and lower bound for each number when it is given correct to the degree of accuracy stated.

**a** 5.6 given to 1 dp
**b** 54 given to nearest whole number
**c** 690 given to nearest ten
**d** 56.372 given to 3 dp
**e** 65.00 given to 2 dp
**f** 100 given to 1 sf
**g** 100 given to 2 sf
**h** 160 000 given to 4 sf
**i** 12.9 given to 3 sf
**j** 16 million given to nearest ten thousand
**k** 1000.00 given to 2 dp
**l** 7.0001 given to 5 sf

**5** The dimensions of a large room are shown on this plan and are given correct to the nearest ten centimetres.

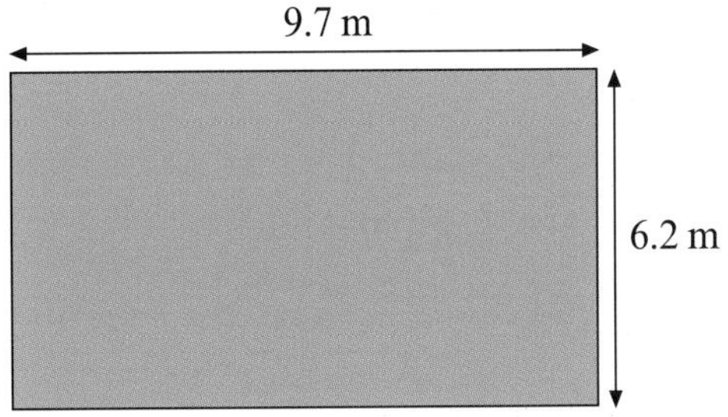


**a** Calculate the range of values that the perimeter could have.
**b** Calculate the upper and lower bounds for the floor area.
**c** The floor is to be covered with carpet at £11 per square metre (to the nearest pound).
Calculate the range of values for the cost of the carpet used.

**6** A scooter covers a distance of 1320 metres (to the nearest ten metres) in 60 seconds (to the nearest 10 seconds).
Calculate the range of values for its speed in metres per second.

**7** The volume $V$ of a sphere is given by $V = \frac{4}{3}\pi r^3$ where $r$ is its radius.
A spherical gas tank has a radius of 3.2 metres, to the nearest ten centimetres.

Calculate the upper and lower bounds for the volume of the tank.

**8** Convert these numbers into standard form.

**a** 435 000 **b** 12 million
**c** 0.000 321 **d** 54.267 45
**e** $654 \times 10^6$ **f** $654 \times 10^{-4}$

**9** Convert these into ordinary numbers.

**a** $5.23 \times 10^3$ **b** $1.5 \times 10^{-6}$
**c** $1.094 \times 10^{-1}$ **d** $3 \times 10^{-2}$

**10** Calculate the value of each of these giving your answers in standard form.

**a** $(3.42 \times 10^3)^2 \times (1.35 \times 10^{-2})$
**b** $\dfrac{(4 \times 10^3) + (5 \times 10^2)}{2 \times 10^{-5}}$
**c** $\dfrac{(4.2 \times 10^{-4}) - (4.2 \times 10^{-5})}{(4.2 \times 10^3) - (4.2 \times 10^2)}$
**d** $(9 \times 10^{-2})^3 - (9 \times 10^3)^{-2}$
**e** $(9 \times 10^{-2})^3 - (9 \times 10^{-3})^2$

## Solving equations algebraically

**1** Solve these equations.

**a** $3(2x-5)=12$ **b** $4(5x+1)=3(4x+12)$

**c** $5(6x+9)=12x$ **d** $3(x+1)=9(5x-9)$

**e** $\frac{2}{3}(x-1)=8$ **f** $\frac{4}{5}(2x+3)=14$

**g** $\frac{3}{4}(2x-4)=2(x+5)$

**h** $\frac{2}{3}(x-1)=\frac{3}{5}(x+4)$

**i** $\dfrac{5}{2x-1}=10$

**j** $\dfrac{3}{x-1}=\dfrac{5}{2x-3}$

**k** $\dfrac{3(2x-1)}{4}=\dfrac{2(6x+3)}{5}$

**2** The perimeter of a rectangle is 86 cm.
The long side is 9 cm longer than the short side.

Write and solve an equation to find the dimensions of the rectangle.

**3** Remove the brackets from these expressions.

**a** $(x+3)(x-5)$ **b** $(2x-1)(x-3)$
**c** $(4-5x)(3x+2)$ **d** $(x-3)(3-4x)$
**e** $(x+4)(x-4)$ **f** $(2x-3)(2x+3)$
**g** $x^2(x-4)$ **h** $2x(x^2+3x-8)$
**i** $3x(x^3+5x^2+4x-1)$
**j** $(x+3)(x^2+6x-7)$
**k** $(x-5)(3+5x-x^2)$
**l** $(x^2+8x-15)(4+x)$
**m** $(x+1)(x+2)(x-3)$

**4** Solve each pair of simultaneous equations.

**a** $2x+3y=9$, $5x-4y=11$ **b** $4x-3y=5$, $5x+7y=17$

**c** $6x+3y=3$, $5x+2y=3$ **d** $8x-3y=30$, $5x-2y=19$

**e** $7x-2y=^{-}11$, $3x+5y=7$ **f** $3x+9y=15$, $4x+5y=6$

**g** $2x+7y=29$, $3x-4y=^{-}29$ **h** $3x-5y=10$, $5x-7y=6$

**i** $4x+y=^{-}11$, $3x+5y=^{-}21$ **j** $13x-7y=66$, $9x+15y=6$

**5** Two numbers are such that when you add three times the first number to three times the second number the answer is 3. The difference between the first and second numbers is 9.

What are the two numbers?

**6** Gareth pays for a holiday with a mixture of £5 and £10 notes. the holiday costs £365 and this is paid with a total of 55 notes.

How many of each type of note were used to pay for the holiday?

**7** A shop sells two types of calculator:
Scientific and Graphical.
Customer A orders 35 Scientific and 72 Graphical calculators at a total cost of £1228.15
Customer B orders 124 Scientific and 15 Graphical calculators at a total cost of £1242.05

Calculate the price of each type of calculator.

**8** A bag contains a mixture of 5 pence and 20 pence coins. There are 2250 coins in the bag. If the monetary value of the coins in the bag is £183.30, how many of each type of coin is in the bag?

**9** A bag contains large and small marbles, and there are 378 marbles in the bag.
Each small marble weighs 2.5 g and each large marble weighs 4.8 g. The total weight of the marbles in the bag is 1.4349 kg.

Calculate how many of each type of marble is in the bag.

**10** Solve these quadratic equations by factorising.

**a** $x^2+5x-84=0$ **b** $x^2-8x+7=0$
**c** $x^2+3x-54=0$ **d** $x^2-15x+36=0$
**e** $x^2+8x+12=0$ **f** $x^2-4x-320=0$
**g** $x^2-11x-60=0$ **h** $x^2-14x+48=0$
**i** $x^2+6x=135$ **j** $x^2-8x=65$
**k** $x^2+20=9x$ **l** $x^2+28=16x$
**m** $x^2=3x$ **n** $x^2-64=0$
**o** $x^2-196=0$ **p** $x^2-10\,000=0$
**q** $2x^2-5x-3=0$ **r** $2x^2+6x-20=0$
**s** $3x^2+x-2=0$ **t** $3x^2+x-4=0$
**u** $6x^2-x-2=0$ **v** $6x^2+35x-6=0$
**w** $8x^2+14x-4=0$ **x** $18x^2+3x-6=0$

**11** Solve these equations to 2 dp.

**a** $x^2-5x+3=0$ **b** $x^2+6x-2=0$
**c** $x^2+8x+1=0$ **d** $x^2-3x-1=0$
**e** $x^2+x-5=0$ **f** $x^2+2x-7=0$
**g** $x^2-3x-2=0$ **h** $x^2+6x-8=0$
**i** $x^2-4x-3=0$ **j** $x^2-x-3=0$
**k** $x^2+6x+3=0$ **l** $x^2+x-1=0$
**m** $2x^2+3x-6=0$ **n** $3x^2+x-3=0$
**o** $4x^2+x-1=0$ **p** $7x^2+3x=5$

**12** Solve these equations using iteration.

**a** $x^2+6x-15=0$ **b** $x^2-6x+4=0$
**c** $x^2-4x-30=0$ **d** $x^2+2x-16=0$

## Transformations

**1** Draw the pentagon A on axes with: $^{-}12 \leqslant x \leqslant 6$ and $^{-}6 \leqslant y \leqslant 6$.

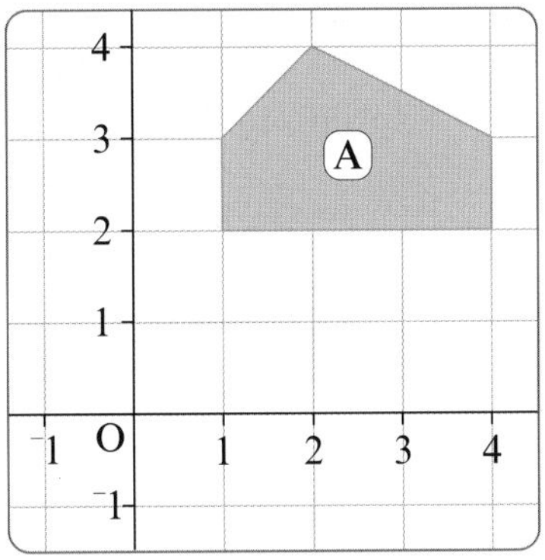


**2** Draw the image of A after:

**a** an enlargement SF $^{-}2$ with centre $(^{-}1, 1)$
**b** an enlargement SF $\frac{1}{2}$ with centre $(7, ^{-}4)$.

**3** These transformations map A on to B, C, D, E and F.

| Object | Transformation | Image |
|---|---|---|
| A | Rotate $^{+}90°$ about (0, 0) | B |
| A | Rotate $^{-}90°$ about (1, 1) | C |
| A | Reflect in $x = 5$ | D |
| A | Rotate $^{+}90°$ about (6, 2) | E |
| A | Reflect in $y = {^{-}x}$ | F |

**a** On a new diagram, on axes with $^{-}6 \leqslant x \leqslant 10$ and $^{-}6 \leqslant y \leqslant 6$, draw and label the images B, C, D, E and F.
**b** Describe the transformation that maps:
**i** C on to A **ii** D on to A.
**c** Describe the inverse of the transformation that maps:
**i** A on to E **ii** A on to F.
**d** For each of these mappings which pentagon is the image of B?

| | Object | Transformation | Image |
|---|---|---|---|
| **i** | B | Enlarge SF 1 with centre (0, 1) | |
| **ii** | B | Reflect in $y = 0$ | |
| **iii** | B | Translate $\begin{pmatrix} 8 \\ ^{-}4 \end{pmatrix}$ | |

**e** Describe a transformation that maps C on to E.

**4**

| Object | Transformations: First | Second | Image |
|---|---|---|---|
| A | Rotate $^{+}90°$ about (0, 1) | Reflect in $y = 0$ | G |
| A | Reflect in $y = 0$ | Rotate $^{+}90°$ about (0, 1) | H |

**a** On a new diagram draw the images G and H.
**b** What single transformation maps G on to H?
**c** In the table below, each pair of transformations maps G on to H.
Describe each of the second transformations.

| | Object | Image | First transformation | Second transforma |
|---|---|---|---|---|
| **i** | G | H | Translate $\begin{pmatrix} 6 \\ 0 \end{pmatrix}$ | |
| **ii** | G | H | Rotate $^{-}90°$ about $(1, ^{-}2)$ | |
| **iii** | G | H | Rotate 180° about $(0, ^{-}1)$ | |

**5** These transformations map A on to J, K L and M.

| Object | Transformations: First | Second | Image |
|---|---|---|---|
| A | Enlarge SF 2 with centre (0, 1) | Reflect in $y = 1$ | J |
| A | Reflect in $x = 1$ | Enlarge SF 2 with centre (5, 0) | K |
| A | Enlarge SF 2 with centre (5, 0) | Reflect in $x = 4$ | L |
| A | Enlarge SF $^{-}2$ with centre (5, 0) | Reflect in $x = 2$ | M |

**a** **i** Draw A on a new diagram with axes: $^{-}10 \leqslant x \leqslant 12$ and $^{-}10 \leqslant y \leqslant 8$.
**ii** Draw and label images J, K, L and M.
**b** What single transformation maps:
**i** J on to M **ii** L on to J **iii** J on to K?

**6** Give three different types of transformation that are their own inverse.

## Using formulas

**1** Rearrange each formula to make $x$ the subject:

**a** $3x + 5a = 4$ **b** $2(3 - x) = a + 1$
**c** $3(a - 2x) = b$ **d** $3x - 1 = 2(a - 2x)$
**e** $3a(x + a) = 2a$ **f** $5(1 - 2x) = 3(x - a)$
**g** $3a^2x - 1 = n + 2$ **h** $y(2 + ax) = 1 - 3y$
**i** $ay + ax = 2 - y^2$ **j** $c^2(3a - ac + bx) = 1$
**k** $wx - wy = 2bc$ **l** $n(2a + ax - c) = ab$
**m** $b^2(a - 3x) = a^2b$ **n** $\pi(3a - xy) = 2a(\pi - 1)$
**o** $w(3x - 4) = w - 1$ **p** $3a + 1 = 2a(a - x)$
**q** $a(1 - ax) = 5a$ **r** $2(3 - 2ax) = a(1 + 3x)$
**s** $h = ab + 0.5cx$ **t** $ac(w^2 - 3x) = acx$

**2** The formula or the cost (£$C$) of hiring a bicycle for $n$ days is: $C = 5.75 + 2.4n$

Zina paid £32.15 to hire a bike while on holiday. For how many days hire was this?

**3** The cost (£$C$) of hiring a car for $n$ days travelling $t$ miles is: $C = 21n + 0.15(t - 120n)$

**a** Calculate the cost of hiring a car to travel a total of 1654 miles over four days.

**b** **i** Make $t$ the subject of the formula.
**ii** £145.50 was paid for 6 days hire. Calculate the number of miles travelled.

**c** The hire charge is a fixed price per day, plus so much a mile travelled. A number of free miles are allowed each day. From the formula:
**i** What is the fixed price per day?
**ii** What is the charge per mile?
**iii** How many free miles are allowed a day? Explain your answer.

**4** Make $n$ the subject of each formula:

**a** $\frac{1}{3}n = a + 2$ **b** $\frac{3n}{5} = 2(1 - x)$
**c** $\frac{1}{n} = \frac{2}{a}$ **d** $\frac{3x - 2}{2n} = 2a$
**e** $\frac{3n}{a^2 - b} = \frac{1}{2}$ **f** $2x + 3 = \frac{3}{5}(n - 1)$
**g** $\frac{2}{n - 3} = ax$ **h** $\frac{a}{2n} = \frac{3}{5}$
**i** $x^2 = \frac{3}{2n - 1}$ **j** $w = \frac{x^2}{3 - n}$
**k** $\frac{a}{n \cos 30°} = b$ **l** $\frac{a}{x^2} = \frac{3}{2n}$
**m** $\frac{2}{3n - 4} = a^2$ **n** $\frac{1}{n} = \frac{2}{a + 1}$
**o** $\frac{a}{b} = \frac{c}{n}$ **p** $\frac{bn}{3n - 2a} = 5$
**q** $\cos 30° = \frac{3}{n}$ **r** $a = \frac{x}{n - b}$
**s** $x^2 = \frac{2n - 3}{5}$ **t** $\frac{a}{n} + \frac{1}{n} = w$

**5** Given the formula: $\frac{a}{2n} = b + c$

Explain why $n \neq \frac{1}{2}a - \frac{1}{2}(b + c)$.

**6** Rewrite each of these as a formula in $p$.

**a** $px - py = 3$ **b** $pt^2 = 1 + 3p$
**c** $3p - 2 = ap + c$ **d** $2(3 - 2p) = a(p - 1)$
**e** $2(3p - 5) = a^2p$ **f** $\frac{2p + 3}{p + 1} = ab$
**g** $3 = \frac{p - 2x}{p}$ **h** $ap + ax = b^2p + y$
**i** $2a(3p - y) = p$ **j** $p \sin x + p \sin y = w$
**k** $3pw + 1 = p - 3$ **l** $a^2bp + 1 = b^2p$
**m** $1 + 2cp = ap + b$ **n** $3n(p + b) = a(p - 2b)$
**o** $3a + \frac{1}{p} = bc$ **p** $\frac{2}{3}(p + 1) = 2(1 - ap)$
**q** $\frac{1}{4}(2p + 1) = an$ **r** $b(3 + ap) = \frac{1}{3}c^2n$
**s** $\frac{2}{5}p + a = py$ **t** $a^2p + 2 = \frac{1}{2}(p + 2)$

**7** The surface area $S$ of a cylinder is: $S = 2\pi rh + 2\pi r^2$

Rearrange the formula to make $\pi$ the subject.

**8** A formula for the cost (£$C$) of supplying and laying $n$ paving slabs is:

$C = 1.85n + y(n - 50)$ with a materials fee of $y$

**a** Make $n$ the subject of the formula.

**b** What do you think is the cost of one slab? Explain your answer.

**9** Rewrite each formula with $v$ as the subject:

**a** $\sqrt{v} = (1 + x)$ **b** $3a = \sqrt{(1 - v)}$
**c** $2x = v\sqrt{3}$ **d** $\sqrt{(2v - 3)} = a - b$
**e** $a^2 = 3\sqrt{v} - 1$ **f** $2x - 1 = \sqrt{5v}$
**g** $4 = \sqrt{(v - 3a^2)}$ **h** $x + 3 = \frac{1}{\sqrt{v}}$
**i** $y^2 = v^2 + 1$ **j** $2v^2 + 1 = ax$
**k** $3v^2 + av^2 = 4$ **l** $3x - v^2 = v^2y$
**m** $ab + av^2 = bv^2$ **n** $3(2 - v^2) = a(v^2 - 2)$
**o** $a^2 + 1 = \frac{1}{v^2}$ **p** $kv^2 + 1 = a^2(v^2 - k)$
**q** $\frac{1}{3v^2} = 2a - 1$ **r** $\frac{1}{v} = \frac{v}{x + 2}$

**10** A formula for the area $A$ of a disc of inner radius $r$ and outer radius $R$ is: $A = \pi(R^2 - r^2)$

**a** Make $r$ the subject of the formula.

**b** A disc has an outer radius of 7.5 cm, and an area of 91.8 cm². To 1 sf, what is the inner radius of the disc?

**11** A formula linking $t$, $a$, and $g$ is given as: $t = 2\pi\sqrt{\frac{a}{g}}$

Make $g$ the subject of the formula.

## Ratio and Proportion

**1** This recipe makes 25 biscuits.

**Chocolate Biscuits**
*50 g icing sugar • 225 g margarine*
*100 g plain chocolate • 225 g flour*

Calculate how much of each ingredient is needed to make:

**a** 45 biscuits **b** 10 biscuits

**2** Tropical Fruit Juice is a mix of pineapple and grapefruit in the ratio 5 : 2. Calculate:

**a** the amount of pineapple juice to mix with 180 ml of grapefruit juice
**b** how much Tropical Fruit Juice is produced.

**3** On a map, 3.6 cm stands for a distance of 108 m.

**a** What length on the map stands for 87 m?
**b** Give the map scale as a unitary ratio.
**c** Explain why the total number of parts has no meaning in this ratio.

**4**

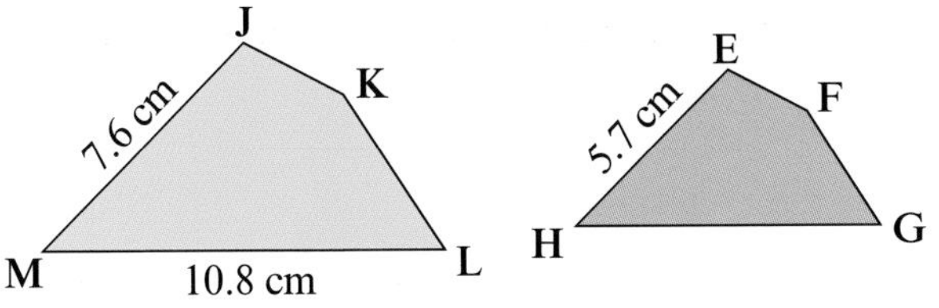


EFGH is an enlargement of JKLM.

**a** Calculate the scale factor of the enlargement.
**b** Find the length HG.
**c** Calculate the ratio: **i** $\frac{MJ}{ML}$ **ii** $\frac{HE}{HG}$
**d** Explain why the ratios are equal.

**5**

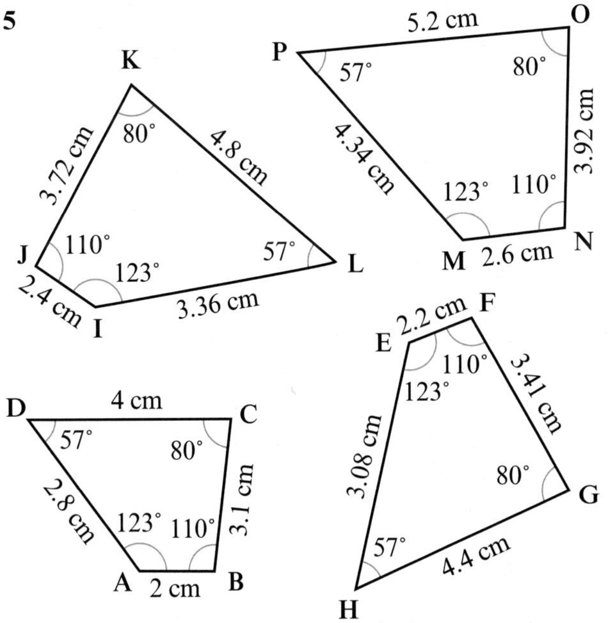


Which of these trapeziums are similar to ABCD?

**6**

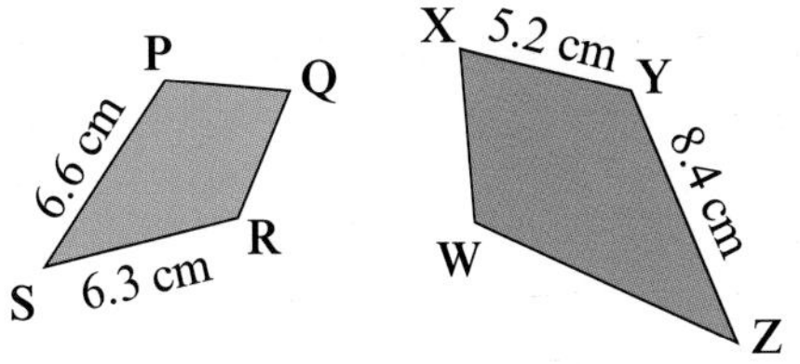


PQRS and WXYZ are similar.
Find the length: **i** QR **ii** WZ

**7**

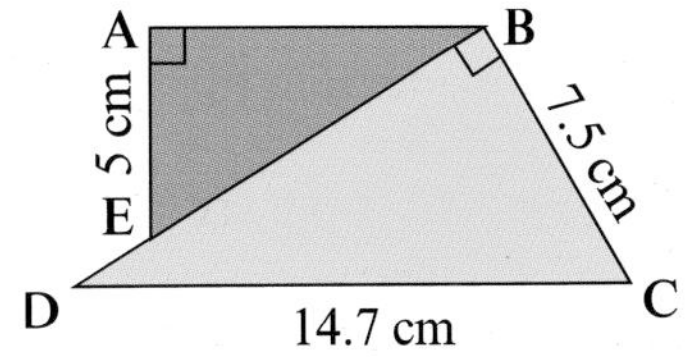


AB and DC are parallel.

**a** Explain why ABE and BDC are similar triangles.
**b** Give the corresponding side to: **i** AB **ii** AE
**c** Find the length BE.

**8**

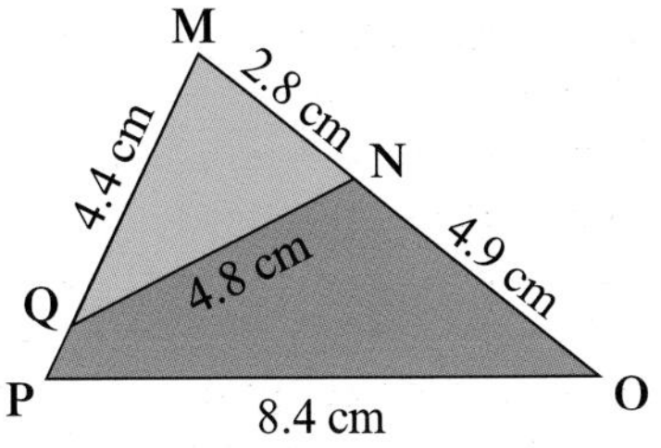


**a** Calculate the ratio: **i** $\frac{PO}{QN}$ **ii** $\frac{MO}{QM}$
**b** Explain why MOP and MQN are similar triangles.
**c** Find the length PQ.

**9** $z$ is in direct proportion to the square of $p$, and $z = 21.6$ when $p = 6$.

**a** Write down the relationship between $z$ and $p$.
**b** Find the constant.
**c** Give an equation connecting $z$ and $p$.
**d** Calculate $z$ when $p = 5$.
**e** Rearrange your equation to give $p$ in terms of $z$.
**f** Calculate $p$ when $z = 86.4$.

**10** Chunkie is a chocolate bar that comes in four different sizes of square prism. Each bar has the same mass of chocolate.
The length of a bar, $l$ cm, is inversely proportional to the square of its width, $w$ cm[2].
The 1.5 cm bar is 16 cm long.

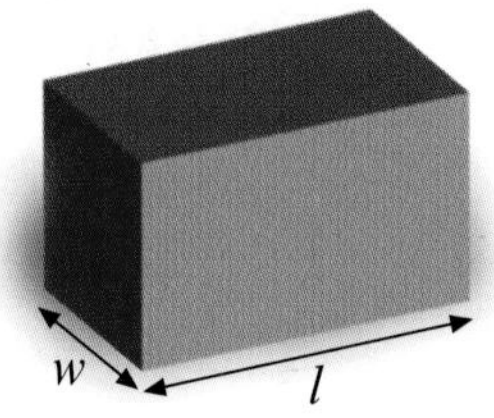


**a** Give an equation connecting $l$ and $w$.
**b** Calculate the length of a 2.5 cm Chunkie.
**c** Calculate the width of a Chunkie 25 cm long.

## Working in 3D

1 Each of these solids is made from three cubes.

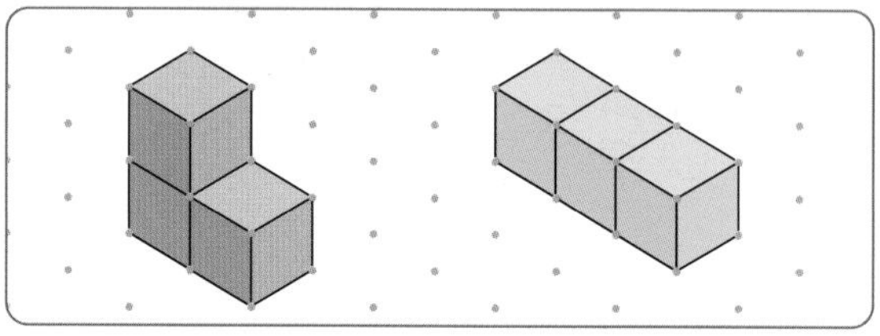

How many planes of symmetry has each solid?

2 **a** Draw a solid made from five cubes with 1 plane of symmetry.
**b** Draw a solid made from six cubes with 3 planes of symmetry.

3 How many planes of symmetry has:
**a** a cuboid with no square faces
**b** a cube?

4 Each prism P and Q has a capacity of one litre.

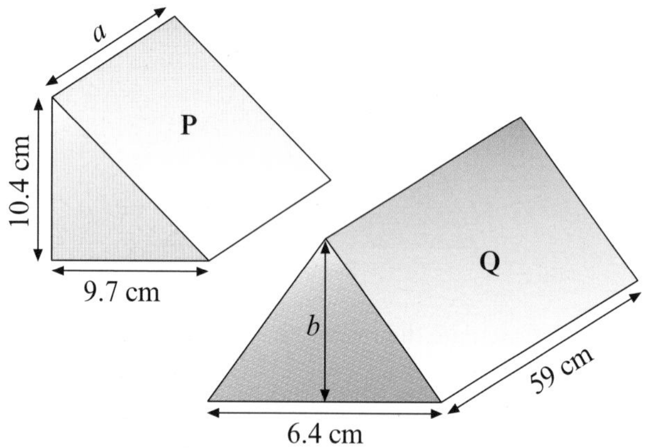


Find the values of $a$ and $b$ correct to 1 dp.

5 Calculate the area of the shaded ring.

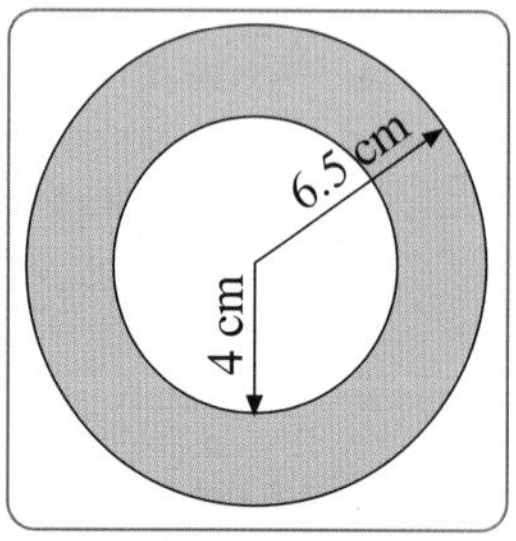


6 This is a company logo. Calculate:
**a** its total area
**b** its perimeter.

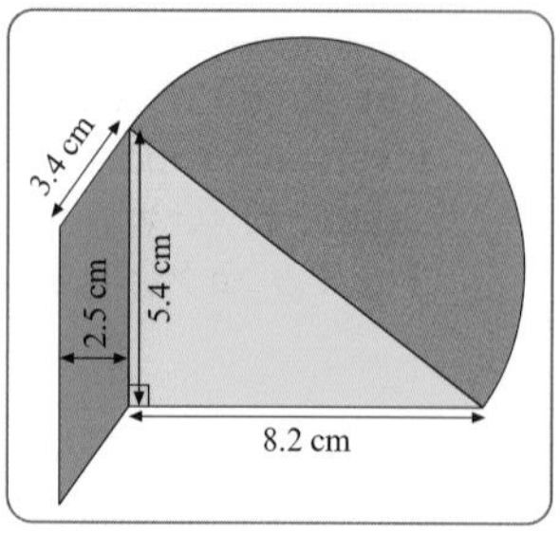


7 A cylindrical cheese has a sector of angle 135° cut out vertically from one end.
**a** Calculate the volume of the remaining piece of cheese.
**b** Calculate the surface area of the piece removed.

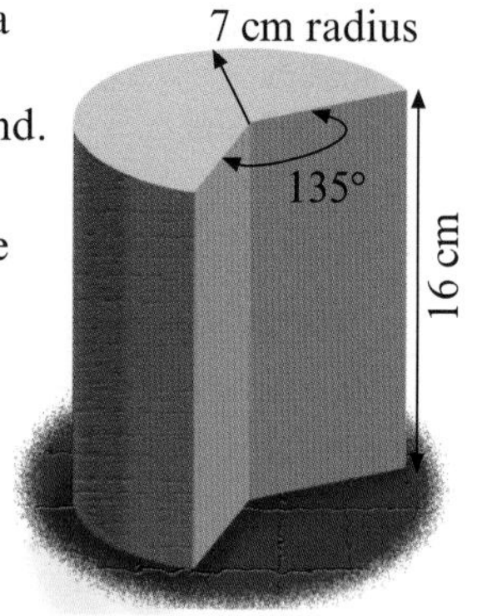


8 A bubblegum display cabinet is shaped like a frustum of a cone with a hemisphere on top.

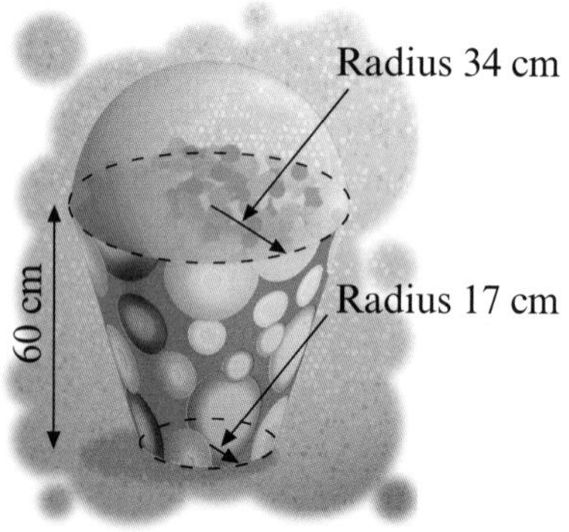


**a** Calculate the volume of the cabinet.
**b** Calculate its surface area.

9 The capacity of a hollow Russian doll is 1024 cm³. A smaller similar doll from the set has a capacity of 640 cm³. The height of the larger doll is 16 cm.
Calculate the height of the smaller doll.

10 In the following expressions, $r$, $h$ and $l$ each represent a length.
Decide if each expression represents: a length, area, volume or none of these.

**a** $4h$
**b** $h^2$
**c** $rh$
**d** $rhl$
**e** $2(h + l)$
**f** $\pi r^2$
**g** $hl + r$
**h** $\frac{1}{3}rh^2$
**i** $\sqrt{h^2 + l^2}$
**j** $\pi rh - 3l^2$
**k** $r^3 + h^2 + rl$
**l** $h^3 + rl^2$
**m** $5(rh + hl + rl)$
**n** $\pi rh(h + l)$
**o** $2h(r - l)$
**p** $\frac{4}{3}\pi(r + h)$

## Probability

1 A 1 to 6 dice is rolled.
What is the probability of getting:

a a multiple of 4
b a prime number which is a factor of 6
c a square number or a triangle number?

2 a If you have six cards with the numbers 1 to 6 on them what is the probability of getting a Fibonacci number by picking a card
b If you have each of the first six Fibonacci numbers on six cards what is the probability of picking a card with 1 on it?

3 You roll two dice at the same time.
What is the probability of getting:

a a total of 3
b a total score of 8
c a 3 on only one dice
d a 5 on both dice
e a total which is not 6?

4 In an experiment an object could land in three ways, on a face, on an edge or on a side. These are the results from a number of trials: face 68, edge 25, side 18. Calculate an estimate of the probability that the object will land:

a on a face
b not on a side
c on a side or a face.

5 Dom's top drawer has 2 red T-shirts, 3 blue ones and 1 yellow one. The draw below has 3 red socks 4 blue socks and 2 yellow ones. Dom is unconcerned with fashion and always picks his clothes from a drawer at random.

a He picks a T-shirt and one sock.
What is the probability that:
i they are both blue
ii one is yellow and one red
iii they are the same colour?

b He picks two socks.
What is the probability that:
i they are both red
ii they are different colours
iii neither of them is blue?

c He picks three socks.
What is the probability that he has a pair the same colour amongst them?

d He picks a T-shirt and two socks.
What is the probability that:
i they are all blue
ii they are all the same colour?

6 Parveen is a disc jockey. She has 18 reggae and 14 jungle CDs to play. She chooses which CD to play at random and only plays any CD once.

a Draw a tree diagram to show the probabilities when she plays the first three CDs in her programme.

b What is the probability that the first three CDs Parveen plays:
i are all reggae CDs
ii include exactly two jungle CDs
iii include only one reggae CD
iv have no two of the same type played consecutively.

c Parveen plays five CDs in the first twenty minutes of her programme. What is the probability that they are:
i all jungle
ii all the same type?

7 The probability that snow will fall on the roof of the Weather Centre in London on a day in December is 0.07. If snow falls on one day, the probability that it will snow on the next day increases to 0.15.

a Draw a tree diagram to show the probabilities for two consecutive days in December.
What is the probability that:
i there will be snow on both days
ii there will be snow on only one day
iii there will be no snow on either day?

b For three consecutive days in December calculate the probability that there will be snow:
i on only one day
ii on all three days
iii on exactly two days.

8 Amongst a group of 30 British students, 18 speak some French and 14 speak some German but 6 speak neither language.

a Draw a Venn diagram to show the information.
b What is the probability that a person chosen at random:
i speaks only German
ii speaks both foreign languages
iii speaks only one foreign language
iv does not speak French?

## Grouped data

**1996 Olympic Games – Men's Marathon**
**Top 100 finishers**

| Time (min) | 132 – | 137 – | 139 – | 142 – | 146 – | 150 – 157 |
|---|---|---|---|---|---|---|
| Frequency | 15 | 18 | 18 | 20 | 11 | 18 |

**1** **a** Draw a histogram to show these times.
**b** (Do **not** use the statistical function on your calculator to answer this question.) Calculate an estimate of the mean time.
**c** Draw a cumulative frequency curve for the men's marathon.
**d** Estimate:
**i** the median time
**ii** the interquartile range of the times.
**e** Estimate how many times were:
**i** less than 135 minutes
**ii** greater than 155 minutes
**iii** between 140 and 150 minutes.

**1996 Olympic Games**
**Women's Marathon**
**Top 60 finishers**

| Time | Frequency |
|---|---|
| 146 – | 5 |
| 151 – | 9 |
| 154 – | 8 |
| 156 – | 12 |
| 158 – | 9 |
| 162 – | 10 |
| 167 – 176 | 7 |

**2** **a** Draw a histogram to show these times.
**b** Use your calculator to estimate the mean time.
**c** Draw a cumulative frequency curve for the women's marathon.
**d** Estimate:
**i** the median time
**ii** the interquartile range of the times.
**e** Estimate how many times were:
**i** less than 150 minutes
**ii** greater than 165 minutes
**iii** between 150 and 160 minutes.

**3** Use your answers to Questions **1** and **2** to compare the marathon times for men and women.

**1996 Olympic Games**
**GB Athletics Team**

| Age | Frequency | |
|---|---|---|
| | Track | Field |
| 18 – | 7 | 2 |
| 23 – | 27 | 10 |
| 28 – | 20 | 7 |
| 33 – 42 | 5 | 4 |
| Totals | 59 | 23 |

**4** **a** Draw a histogram to show the age distribution of:
**i** track athletes **ii** field athletes.
**b** Explain why drawing frequency polygons on the same diagram would not give a fair comparison of the ages of track athletes and field athletes.
**c** Use your calculator to estimate the mean for:
**i** track athletes **ii** field athletes.
**d** Compare the two distributions.

**1996 Olympic Games – Great Britain Team**
**Ages of Competitors at 20/7/96**

| | | | | | | | | | | | | | | | |
|---|---|---|---|---|---|---|---|---|---|---|---|---|---|---|---|
| 22 | 21 | 29 | 18 | 26 | 16 | 23 | 23 | 28 | 31 | 26 | 29 | 27 | 20 | 27 | 20 |
| 20 | 24 | 19 | 23 | 21 | 24 | 25 | 18 | 21 | 20 | 21 | 23 | 26 | 20 | 21 | 19 |
| 21 | 31 | 30 | 24 | 27 | 31 | 27 | 23 | 30 | 35 | 27 | 25 | 23 | 22 | 27 | 36 |
| 27 | 24 | 20 | 23 | 27 | 27 | 26 | 30 | 35 | 23 | 30 | 24 | 29 | 24 | 24 | 31 |
| 31 | 26 | 23 | 19 | 29 | 29 | 27 | 24 | 32 | 32 | 25 | 23 | 24 | 21 | 38 | 39 |
| 28 | 25 | 32 | 31 | 27 | 31 | 24 | 21 | 30 | 32 | 24 | 29 | 38 | 29 | 34 | 22 |
| 29 | 23 | 24 | 29 | 32 | 24 | 40 | 30 | 24 | 21 | 23 | 27 | 31 | 22 | 23 | 29 |
| 35 | 31 | 24 | 24 | 22 | 32 | 23 | 29 | 27 | 30 | 27 | 26 | 21 | 27 | 29 | 26 |
| 23 | 21 | 22 | 25 | 27 | 32 | 26 | 26 | 28 | 29 | 22 | 25 | 37 | 25 | 18 | 22 |
| 29 | 35 | 34 | 32 | 30 | 25 | 29 | 28 | 27 | 35 | 25 | 28 | 34 | 29 | 29 | 28 |
| 23 | 27 | 22 | 27 | 34 | 35 | 30 | 30 | 41 | 37 | 40 | 31 | 27 | 26 | 35 | 31 |
| 33 | 38 | 42 | 35 | 36 | 40 | 36 | 29 | 20 | 16 | 22 | 16 | 26 | 23 | 21 | 29 |
| 29 | 27 | 27 | 23 | 28 | 25 | 34 | 30 | 27 | 28 | 30 | 23 | 33 | 20 | 33 | 26 |
| 26 | 29 | 30 | 29 | 29 | 25 | 28 | 27 | 25 | 24 | 28 | 31 | 32 | 27 | 25 | 26 |
| 26 | 31 | 23 | 23 | 21 | 30 | 20 | 32 | 28 | 35 | 31 | 28 | 24 | 29 | 24 | 27 |
| 24 | 30 | 27 | 28 | 27 | 26 | 28 | 29 | 34 | 22 | 27 | 32 | 26 | 24 | 27 | 20 |
| 27 | 22 | 29 | 28 | 31 | 26 | 24 | 25 | 28 | 30 | 34 | 27 | 24 | 30 | 21 | 30 |
| 21 | 29 | 21 | 27 | 27 | 28 | 48 | 19 | 34 | 47 | 31 | 36 | 25 | 29 | 28 | 29 |
| 21 | 27 | 22 | 28 | 31 | 26 | 23 | 25 | 19 | 29 | 31 | 30 | 25 | 32 | 24 | 29 |
| 28 | 28 | 25 | 29 | 30 | 26 | 29 | 34 | | | | | | | | |

**5** **a** Group this age data using a class interval of 2.
**b** Combine some of your classes at each end of the distribution.
**c** Use your new groupings to draw a histogram.
**d** The age data is listed at random.
Choose a systematic sample of:
**i** 20 **ii** 40 **iii** 80
**e** For each of your random samples:
**i** group the data **ii** draw a histogram
**f** Compare the shape of your distributions.

## Solving equations graphically

**1** Solve each pair of equations graphically.

**a** $x + 4y = 6$, $2x + 5y = 9$
**b** $5x - 3y = 8$, $3x + y = 2$
**c** $5x + 7y = 29$, $x + 3y = 9$
**d** $3x + 2y = 15$, $4x - 2y = 13$
**e** $5x - 4y = 14$, $3x + 2y = 4$
**f** $3y + 2x = 2$, $2y - 6x = {}^-17$
**g** $4y + 2x = 23$, $2y - 6x = 1$
**h** $4x - 5y = 3$, $2x + 3y = 18$
**i** $2x + 3y = 16$, $4x - y = 11$
**j** $2x - 3y = {}^-2$, $2x + 2y = 13$

**2** Give the line of symmetry for each of these graphs.

**a** $y = x^2 + 8$
**b** $y = x^2 + 4x$
**c** $y = x^2 - 8x$
**d** $y = x^2 + 6x - 4$
**e** $y = 12x - x^2$
**f** $y = 5 - 8x - x^2$

**3** For $x$: ${}^-4 \leqslant x \leqslant 4$ draw graphs of these curves on the same axes.

**a** $y = x^2 - 2x - 5$
**b** $y = 4x - x^2$

**4** For $x$: ${}^-5 \leqslant x \leqslant 5$ draw graphs of these curves on the same axes.

**a** $y = 2x^2 + 3x$
**b** $y = 3x^2 - x$

**5** For $x$: ${}^-8 \leqslant x \leqslant 6$ draw the graph of $y = x^2 + 3x - 5$.

**a** Use the graph to solve these equations.
  **i** $x^2 + 3x - 5 = 4$
  **ii** $x^2 + 3x - 5 = {}^-3$
  **iii** $x^2 + 3x - 5 = 12.5$
**b** What can you say about any values you read from the graph?

**6** For values of $x$: ${}^-3 \leqslant x \leqslant 5$ draw the graph of $y = x^2 - 2.5$.

**a** On the same axes draw the graph of $y = 1.5x$.
**b** Use the graphs to solve the equation

$$x^2 - 2.5 = 1.5x$$

**7** Draw a suitable graph and solve the equation

$$(x - 4)^2 - 12 = 0$$

**8** For values of x from ${}^-2$ to 6 draw the graph of $y = 2x(x - 4) - 10$.

Use the graph to solve $2x(x - 4) = 10$.

**9** For $x$: ${}^-5 \leqslant x \leqslant 5$ draw the graph of $y = x^2 + 6x - 2$.

**a** By drawing an additional graph use the graph to solve $x^2 + 4x + 1 = x + 4$.
**b** Explain how you can use your graph to solve the equation $x^2 + x + 2 = 7 - 7x$.
**c** Find an approximate value for $x$ that will satisfy $x^2 + x + 2 = 7 - 7x$.

**10** On a pair of axes draw a graph of $y = \frac{6}{x} - 4$. Use $x$: ${}^-5 \leqslant x \leqslant 5$.

**11** For $x$: ${}^-6 \leqslant x \leqslant 6$ draw the graph of $y = \frac{12}{x} + 8$.

Give the equation of any asymptotes to the graph.

**12** For $x$ : $1 \leqslant x \leqslant 10$ draw the graph of $y = x + \frac{10}{x} - 12$.

Use the graph to find approximate values for:

**a** the minimum value of $y$
**b** $x$ when $y = {}^-3.4$.

**13** For values of $x$ from ${}^-4$ to 5 draw the graph of $y = x^3 + 4$.

**14** On the same axes draw the graphs of:

**a** $y = x^3 + 5$
**b** $y = 5 - x^3$

**15** Draw the graph of $y = x^3 - 2x^2$ for values of $x$ from ${}^-4$ to 4.

**a** Use the graph to solve $x^3 - 2x^2 = 0$.
**b** From the graph find an approximate value for $x$ when $y = 10.4$.

**16** Sketch and label these graphs.

**a** $y = 5x + 2$
**b** $y = x^2 - 6$
**c** $y = x^3 + 1$

**17** Show how completing the square helps to sketch the graph of $y = x^2 + 6x - 1$.

**18** For the graph of $y = x^2 + 4x - 6$ give:

**a** the minimum value of $y$
**b** the point at which the minimum value of $y$ occurs
**c** the intercept with the $y$-axis.
**d** Sketch the graph of $y = x^2 + 4x - 6$.

## Trigonometric graphs

**1** The radial lines on the diagram below are drawn at intervals of 30°.

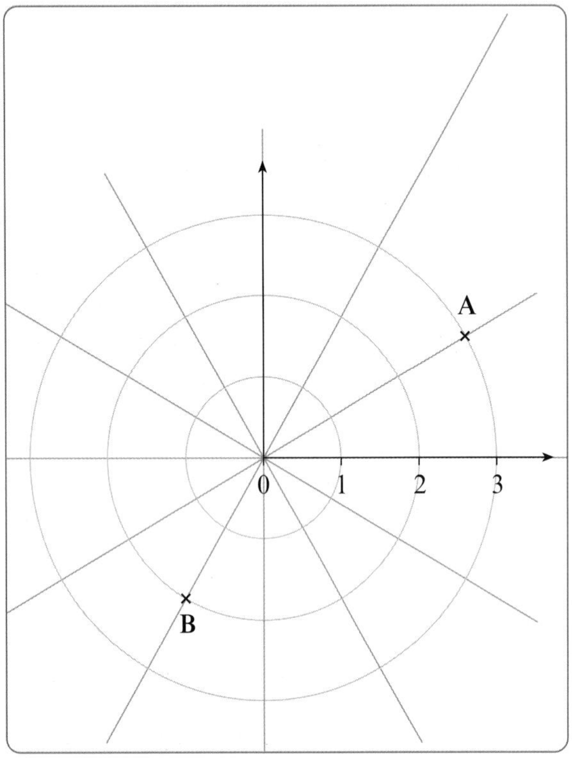


Calculate the Cartesian coordinates of points A and B.

**2a** Make a table of values for $y = 4 \cos 20t$ for $t = 0, 1, 2, \ldots, 20$.

**b** Draw the graph of $y = 4 \cos 20t$ on axes with $0 \leqslant t \leqslant 20$ and $^-8 \leqslant y \leqslant 8$.

**c** For what values of $t$ is:
  **i** $y \leqslant 0$ **ii** $y \leqslant 2$?

**d** What is the maximum value of $y$?

**e** For what values of $t$ is:
  **i** $y = 0$ **ii** $y = {}^-2$?

**f** **i** On the same axes sketch the graph of $y = 4 \cos 20t - 2$.
  **ii** Give the maximum and minimum value of $4 \cos 20t - 2$.

**3** Solve each of these for $0° \leqslant x \leqslant 360°$.
Find two values of $x$ such that:

| | | | |
|---|---|---|---|
| **a** | $\cos x = 0.5$ | **b** | $\sin x = {}^-0.5$ |
| **c** | $\tan x = {}^-1$ | **d** | $\sin x = \cos x$ |
| **e** | $\cos x = \cos 205°$ | **f** | $\sin x = \sin 205°$ |
| **g** | $\tan x = \tan 205°$ | **h** | $\sin x = \sin 105°$ |
| **i** | $\cos x = \cos 105°$ | **j** | $\tan x = \tan 105°$ |

**4** Solve each of these in the range $^-90 \leqslant p \leqslant 450$.

**a** List all the values of $p$ for which:

| | | | |
|---|---|---|---|
| **i** | $\sin p = \sin 54°$ | **ii** | $\cos p = \cos 54°$ |
| **iii** | $\tan p = \tan 54°$ | **iv** | $\sin p = \sin 154°$ |
| **v** | $\cos p = \cos 154°$ | **vi** | $\tan p = \tan 154°$ |

**b** Find all the solutions of these equations.

| | | | |
|---|---|---|---|
| **i** | $\cos p = {}^-\cos 62°$ | **ii** | $\sin p = {}^-\sin 62°$ |
| **iii** | $\sin p = 1$ | **iv** | $\cos p = 0$ |

**5** Sketch each of these graphs for $0° \leqslant x \leqslant 360°$.

| | | | |
|---|---|---|---|
| **a** | $y = \frac{1}{2} \cos x$ | **b** | $y = \frac{1}{2} \sin x$ |
| **c** | $y = 5 \sin x$ | **d** | $y = \cos 3x$ |
| **e** | $y = \sin x - 5$ | **f** | $y = 2 \cos x + 1$ |

**6** Sketch each of these graphs for $^-90° \leqslant x \leqslant 540°$.

**a** $y = \sin (x + 45°)$
**b** $y = \cos (x - 90°)$

**7** **a** On one pair of axes, with $0° \leqslant x \leqslant 360°$ draw the graphs of:
  **i** $y = \cos x$
  **ii** $y = \frac{1}{2} \sin x$.

**b** Use your graphs to solve the equation:
$\cos x = \frac{1}{2} \sin x$

**8** Give the maximum and minimum value of $y$ for each of these.

| | | | |
|---|---|---|---|
| **a** | $y = 6 \cos x$ | **b** | $y = \sin x + 6$ |
| **c** | $y = 8 \cos x$ | **d** | $y = \sin 2x$ |
| **e** | $y = 3 \sin x + 6$ | **f** | $y = 2 \cos 3x - 4$ |

**9** This is the graph of $y = a \sin bx + c$, where $a$, $b$ and $c$ are constants.

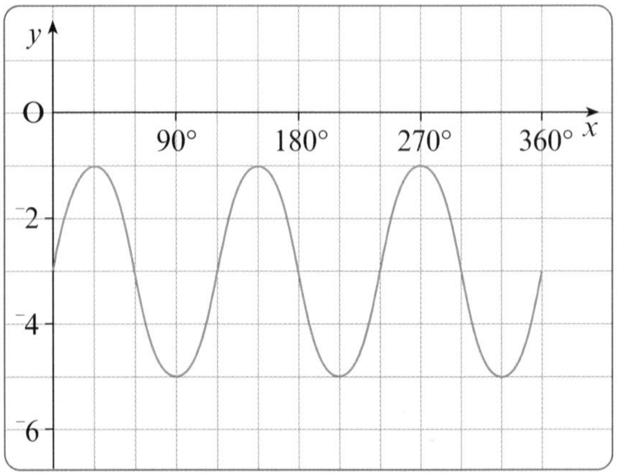


What are the values of $a$, $b$ and $c$?

**10** Solve each of these equations in the range $^-90° \leqslant p \leqslant 450°$.

| | | | |
|---|---|---|---|
| **a** | $\sin p = \cos p$ | **b** | $\sin 2p = \sin 80°$ |
| **c** | $\cos 3p = 0.5$ | **d** | $\tan 3p = 1$ |

## Inequalities

1 Which of these numbers is not a possible value for $t$ where $t \leqslant {}^{-}7$?
${}^{-}8$, 4, 71, ${}^{-}3.2$, ${}^{-}7.003$, ${}^{-}28.6$, ${}^{-}7$, ${}^{-}0.33$

2 What are the integer values for $k$, where

- **a** $9 > k \geqslant 4$
- **b** ${}^{-}3 \leqslant k \leqslant {}^{-}1$
- **c** ${}^{-}5 \leqslant k < 1$
- **d** ${}^{-}4 \geqslant k > {}^{-}7$
- **e** ${}^{-}57 < k < {}^{-}56$
- **f** $24 \leqslant k \leqslant 24$
- **g** ${}^{-}14.2 \geqslant k > {}^{-}16$
- **h** $6 > k > 0$
- **i** $k^2 \leqslant 16$
- **j** $9 < k^2 + 8$
- **k** $12 - k^2 \geqslant 3$
- **l** $4(k^2 + 3) \geqslant 3(2k^2 - 2)$?

3 Write two other inequalities in $h$ which describe the same integer values as ${}^{-}2 < h < 2$.

4 Explain why these two inequalities are different types of inequality from each other.

> This chair is suitable for people with weights given by $5 \leqslant w \leqslant 15$, where $w$ is their weight in stones.

> The waiters in a restaurant are given by $5 \leqslant w \leqslant 15$ where $w$ is the number of waiters.

5 Draw graphs and label the regions described by these inequalities:

- **a** $7 \leqslant x \leqslant 10$
- **b** ${}^{-}4 \leqslant x < 1$
- **c** ${}^{-}5 < x < {}^{-}2$
- **d** $0 < x \leqslant 4$
- **e** ${}^{-}1 \leqslant y < 3$ and $x \geqslant 4$
- **f** $0 \leqslant x \leqslant 6$ and ${}^{-}6 < y < {}^{-}2$
- **g** $y \geqslant x$ and $9 \geqslant x \geqslant 6$
- **h** $2x + y < 10$ and $y \geqslant \frac{1}{2}x + 1$

6 What two inequalities describe the shaded region on this graph?

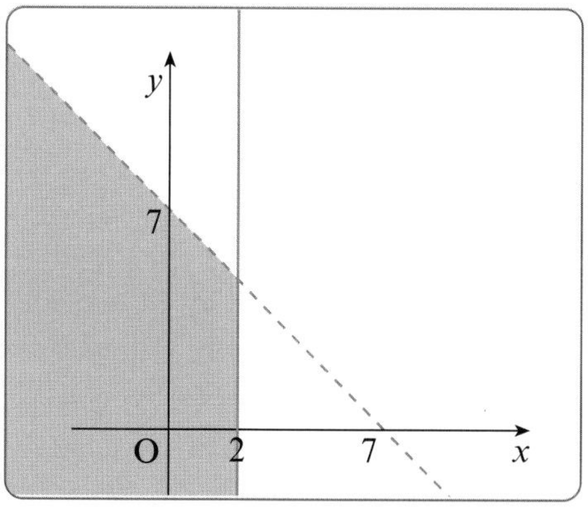


7 The conditions on $x$ are given by the inequality ${}^{-}4 \leqslant x < 3$.
What is:

- **a** the smallest possible value of $x^2$
- **b** the largest possible value of $x^2$?

8 Solve the following inequalities.

- **a** $3x \geqslant 27$
- **b** $6t < {}^{-}42$
- **c** $52 \geqslant 4p$
- **d** $k^2 \leqslant 121$
- **e** $1 + 3w < 7$
- **f** $5q + 7 > 53.5$
- **g** $19 \leqslant 3t - 5$
- **h** $7 - 2h \geqslant 3$
- **i** $4(3f - 2) < 28$
- **j** $c^2 - 5 < 76$
- **k** $36 \leqslant 3(2g + 3)$
- **l** $3x + 5 > x - 1$
- **m** $3j - 2 < 2j + 17$
- **n** $5d + 20 \geqslant 6 - 2d$
- **o** $24 - 8v > 6 - 4v$
- **p** $2(x + 3) \leqslant x - 7$
- **q** $5 - 3s > 5 + 3s$
- **r** $9u + 6 \leqslant 7u$
- **s** $5(4 - 3j) < 7j - 4$
- **t** $5 + 2r^2 > 23$
- **u** $2(3x^2 + 2) > 3(4x^2 - 3) + 1$

9 If the limits for $x$ and $y$ are given by $4 < x < 7$ and $7 < y < 9$, what can you say about the limits for:

**a** $xy$ **b** $x + y$ **c** $\frac{y}{x}$?

10 Sam is planting an orchard with a mixture of apple bushes and pear trees. An apple bush needs an area of 16 m$^2$ and a pear tree needs 24 m$^2$. An apple bush costs £8 and a pear tree £10. She has £200 to spend and her orchard can have a maximum area of 432 m$^2$.
Sam wants at least half of her plants to be apple bushes but at least 6 must be pear trees.
Let $a$ represent the number of apple bushes and $p$ the number of pear trees.
This graph shows only some of the boundaries.

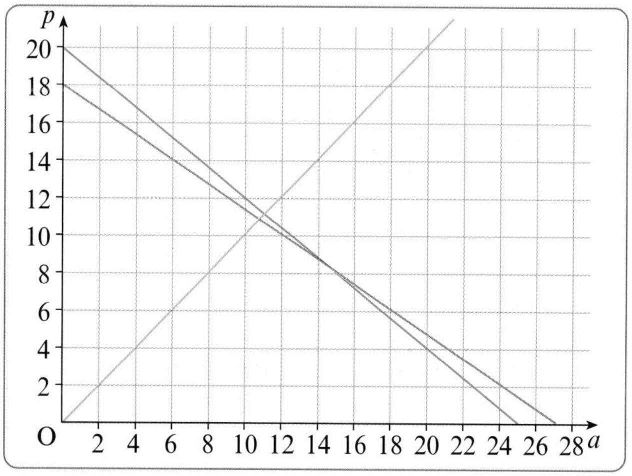


- **a** Write inequalities for:
  - **i** the land area
  - **ii** the cost of plants
  - **iii** the number of pear trees.
- **b** Explain why $p < a$.
- **c** Draw the complete graph, label each line, and shade out all the regions not required.
- **d** How many of each type gives the maximum total number of plants.
  What is the total cost of these plants?

## Working with percentages

1 Calculate, giving answers correct to 2 dp:

- a 38 as a percentage of 60
- b £14 as a percentage of £55
- c 68 as a percentage of 24
- d 1550 as a percentage of 2500.

2 Nina bought a secondhand bicycle for £45.
She later sold the bike for £58.
Calculate her percentage profit to 2 sf.

3
- a Increase 25 kg by 18%.
- b Increase 1400 km by 62%.
- c Increase 3560 miles by 4%.
- d Increase 1350 yards by 36%.

4 Last year one company sold 19.6 million CDs.
Next year they plan to increase sales by 18.5%.
How many CDs do they plan to sell next year?

5
- a Decrease 485 ml by 12%.
- b Decrease £45.75 by 24%.
- c Decrease 65 mm by 65%.
- d Decrease 0.8 cm by 8%.

6 This year an arable farm used 12.4 tonnes of herbicides. They plan to decrease the herbicides used by 8.5% per year.
Calculate, to 2 sf, the herbicide they plan to be using three years from now.

7 The price of each item is given ex.VAT.
Calculate the price including VAT at $17\frac{1}{2}\%$,
to the nearest penny.

- a crash helmet £185.85
- b cycle tyre £11.69
- c fishing rod £44.86
- d steam iron £21.75
- e microwave oven £268.55
- f VCR £159.99
- g personal CD player £135.38
- h CD £9.24
- i phone £49.99
- j multimedia PC £1499

8 A printer is advertised for £132 + VAT.
Calculate the total price of the printer.

9 The VAT (at 17.5%) paid on a bill was £15.
Calculate the total for the bill including VAT.

10 Jamal has a maximum of £10 000 to spend on building a garage. Calculate the maximum he can spend ex.VAT.

11 Callum sees the same model TV advertised by two shops in this way:

| | |
|---|---|
| TV World | £199.99 inc. VAT |
| Price busters | £169.99 ex.VAT |

From which shop would you advise Callum to buy the TV?
Give reasons for your answer.

12 Calculate the simple interest charged, or paid, on each of these:

- a £675 borrowed for 5 years at 12% pa
- b £12 400 borrowed for 2 years at 17% pa
- c £170 saved for 4 years at 3% pa

13 Rearrange the simple interest formula,
$I = P \times R \times T$, to make $R$ the subject.

14 £1400 was saved at 3.6% simple interest.
The total interest paid was £378, for how long was the saving?

15 Calculate the compound interest charged, or paid on each of these:

- a £500 saved for 8 years at 6% pa
- b £1400 borrowed for 3 years at 19% pa
- c £250 saved for 12years at 7.5% pa
- d £105 saved for 7 years at 3%
- e £12500 borrowed for 12 years at 16%
- f £32 saved for 13 years at 4%
- g £750 borrowed for 3 years at 21%
- h £3675 borrowed for 5 years at 17%
- i £125 saved for 9 years at 8%
- j £500 saved for 30 years at 4.5%.

16 These prices include VAT. Calculate each price ex.VAT.

- a freezer £299.99
- b CD £12.99
- c camera £44.99
- d phone £9.99
- e climbing boots £75
- f tent £89.95
- g kettle £26.99
- h PC £129.99
- i calculator £18.99
- j TV £139.99

17 Jo bought a ski jacket in a '15% off sale' for £85.45.
Calculate the pre-sale price of the jacket?

18 In 1995 Ferrykink made a profit of £3 600 000.
This was 14% more than the profit for 1994.

Calculate the profit made in 1994.

## Circle properties

**1** Draw any circle, and a chord AB.

**a** Label:
 **i** the minor segment
 **ii** the major segment.

A chord CD is such that the circle is divided into two segments that are the same size.

**b** What can you say about the chord CD?

**2**

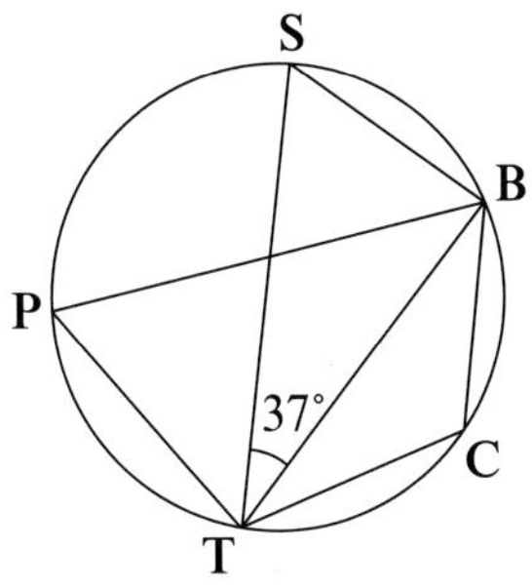


**a** Name an angle in the minor segment.

**b** Calculate the size of $T\hat{B}S$ when $T\hat{P}B = 75°$. Explain your answer.

**3** In the diagram PT is a chord and O the centre of the circle.

Calculate the size of $T\hat{P}O$.
Give reasons for your answer.

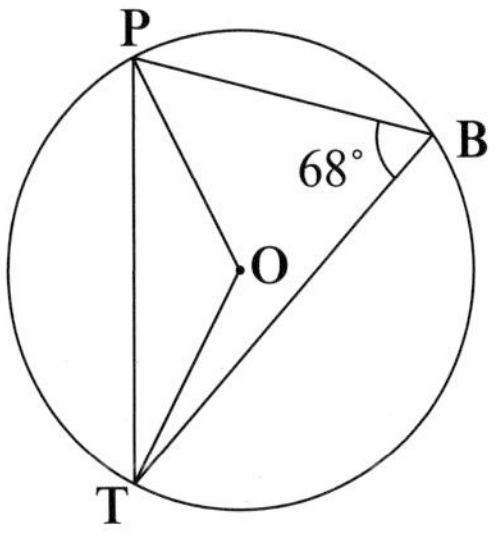


**4** The circle has a centre O.

**a** Calculate the size of $P\hat{T}B$. Explain your answer.

**b** What is the size of $A\hat{C}B$? Explain your answer.

**c** Explain why $A\hat{B}P$ is 76°.

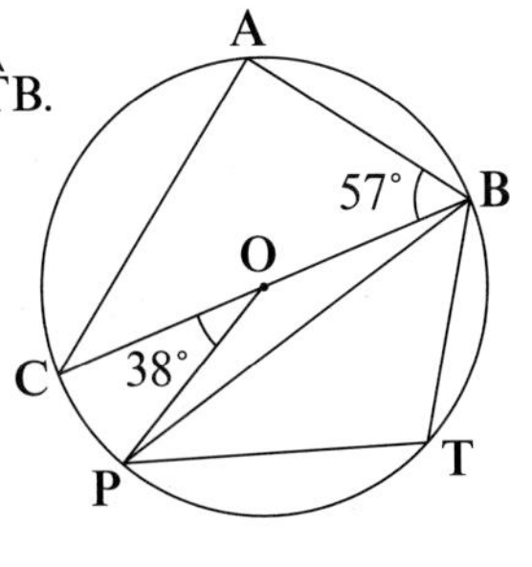


**5** The circle has a centre O.

**a** Explain why $P\hat{A}T$ is 48°.

**b** Calculate $P\hat{C}T$. Explain your answer.

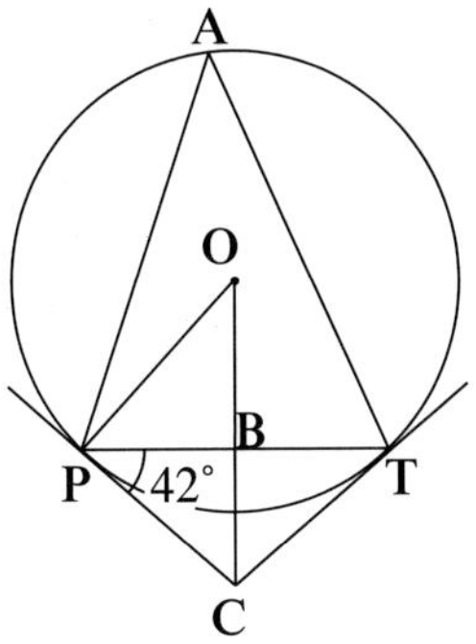


**6** **a** Explain why $T\hat{B}P$ is 35°.

**b** Calculate $B\hat{P}T$. Give reasons for your answer.

**c** Explain why $A\hat{C}B$ is 73°.

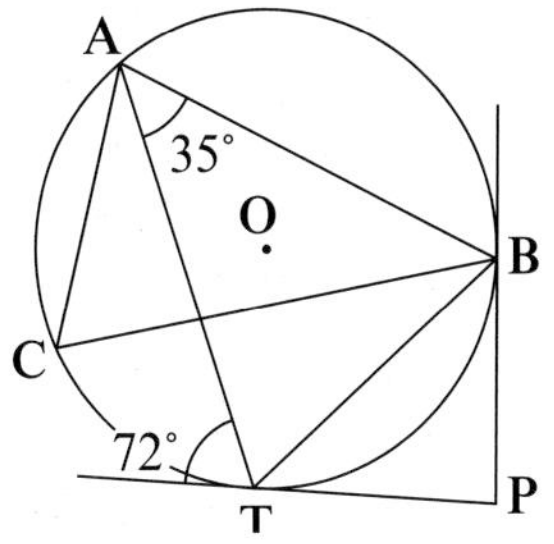


**7** AC is a tangent to the circle with centre O. AC is 12.5 cm long and B is the midpoint.

**a** If AE = 5 cm, calculate the length of EG to 1 dp.

**b** If DF = 11.2 cm, calculate to 1 dp the length of CD.

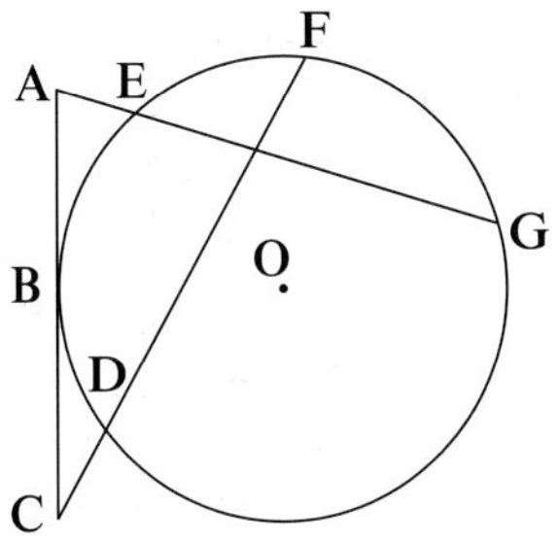


**8** AC is a tangent to the circle with centre O. AC is 16.2 cm long and B is the midpoint.

**a** If AD is 4 cm, and EF is 7.5 cm calculate the size of $F\hat{D}E$ to 2 sf.

**b** Calculate the area of triangle FDE to 3 sf.

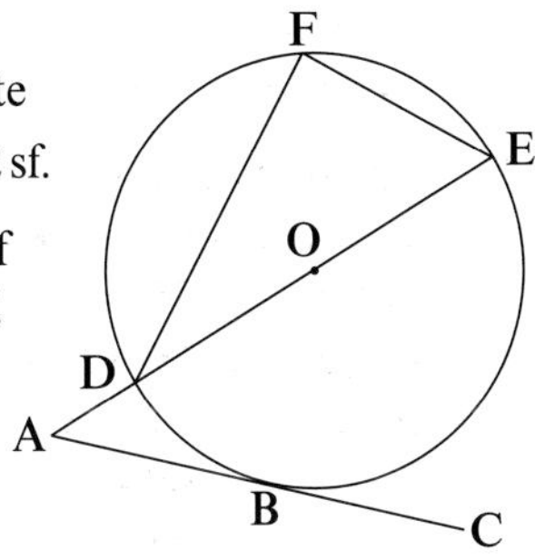


**9** **a** Explain why $E\hat{C}D = 31°$.

**b** Calculate the size of $A\hat{B}C$. Explain your answer.

**c** $E\hat{A}B = 52°$
Give two different explanations to show this.

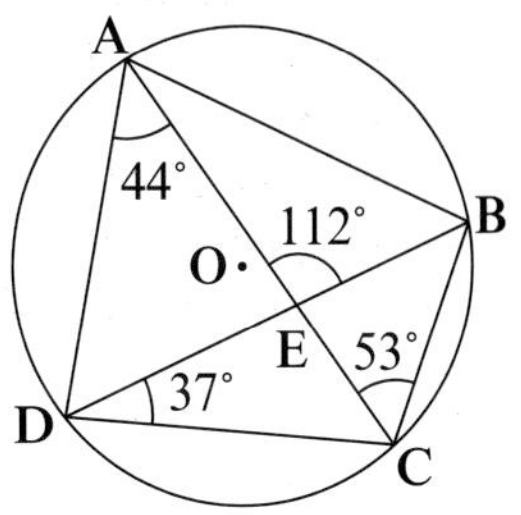


**10** Show why an expression for $D\hat{A}B$ is given as:

$180 - a$

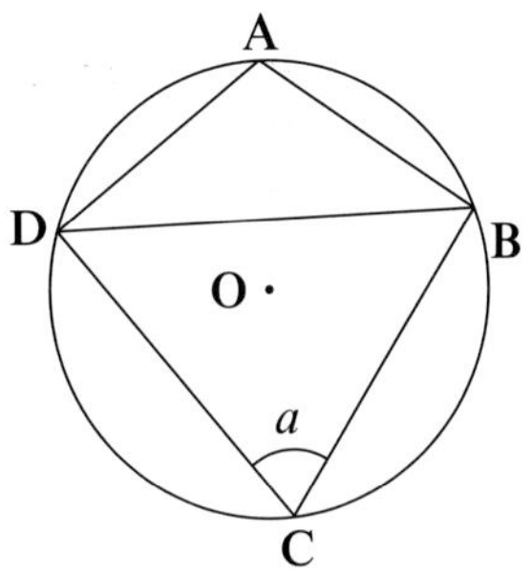

## Working with angles

**1** Calculate the area of each of these triangles:

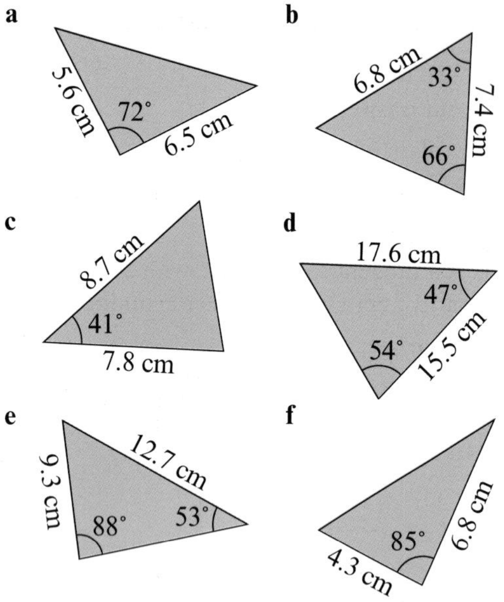


**2** Triangle ABC has an area of 30.4 cm$^2$.

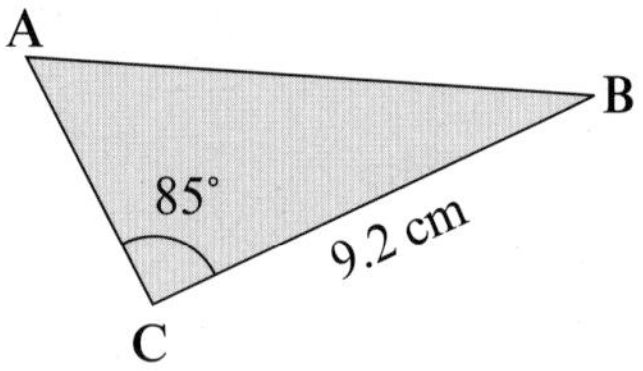


Calculate the length of AC.

**3** Use the sine rule to calculate ? in each of these triangles.

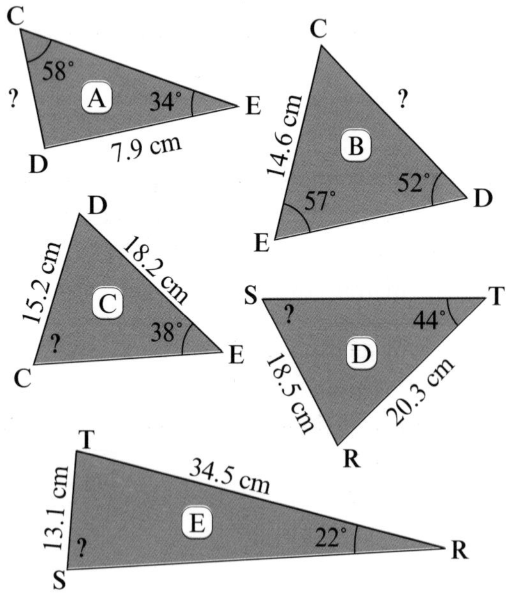


**4** Calculate the length of CD in each of these triangles:

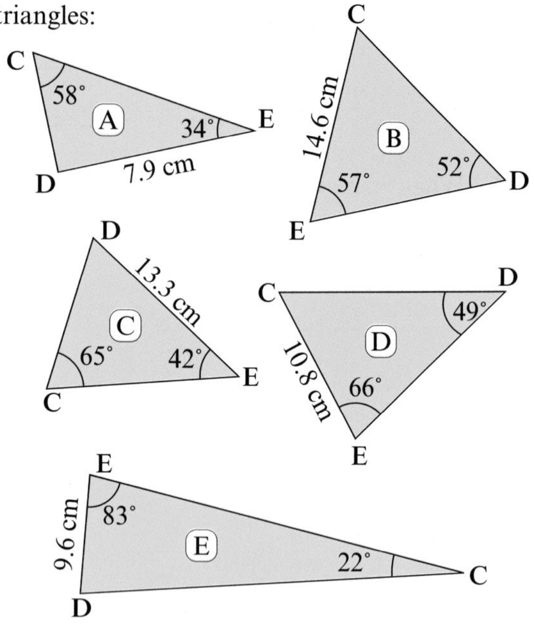


**5** Calculate $B\hat{A}C$ in each of these triangles:

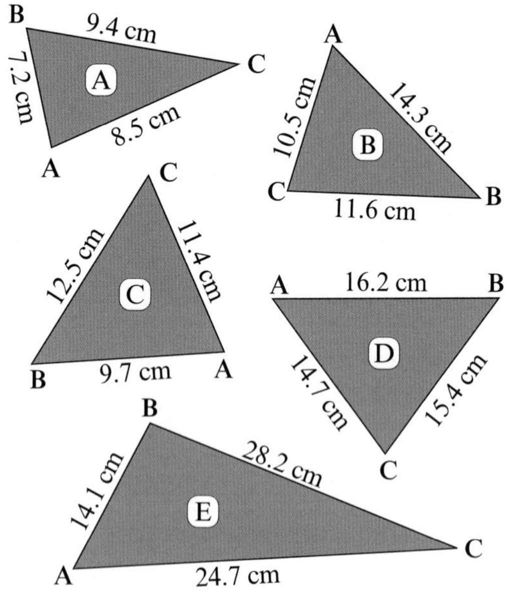


**6** For triangle PQR,
copy and complete the table.

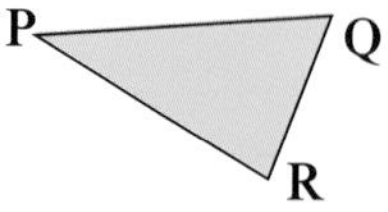


| Angle | $p$(cm) | $q$(cm) | $r$(cm) |
|---|---|---|---|
| $P = 54°$ | **a** | 7.3 | 8.5 |
| $Q = 63°$ | 8.4 | **b** | 9.1 |
| $R = 49°$ | 11.5 | 9.8 | **c** |
| $P = 74°$ | **d** | 3.5 | 4.5 |
| $Q = 57°$ | 37.5 | **e** | 31.8 |

## Processing data

**1**

*Town Centre Survey*

1 Do you come into the town centre often? ☐ Yes ☐ No

2 Do you agree that the town centre should be pedestrianised? ☐ Yes ☐ No

3 What do you think about buses? ______________

**a** Criticise each of these questions.
**b** Write an improved question for each one.

**2**

***Body Matters***
**Right-handed people are more likely to have a stronger left eye than right eye.**

Design an experiment to test this hypothesis.

**3** **a** Carry out your experiment.
**b** Analyse the data you collect.

**4** **a** Do you think the hypothesis is true or false?
**b** Explain why.

**5**

***Body Matters***
**Do right-handed people fold their arms differently to left-handed people?**

Design an experiment to answer this question.

**6** **a** Carry out your experiment.
**b** Analyse the data you collect.
**c** Interpret your results to answer the question.

**7**

**Motorbikes – Size of Engine & Price**

| | | | | | | | | |
|---|---|---|---|---|---|---|---|---|
| Engine size (cc) | 250 | 900 | 600 | 125 | 650 | 900 | 500 | 750 |
| Price (£) | 4500 | 8100 | 6700 | 2400 | 5100 | 6700 | 3400 | 9200 |

**a** Draw axes: horizontal 0 to 1300
vertical 0 to 12 000
**b** Plot the motorbike data on your diagram.
**c** Draw a line of best fit.
**d** Describe any correlation between size of engine and price.

**8** Estimate the price of a motorbike with engine size:
**a** 800 cc **b** 450 cc

**9** Estimate the engine size of a motorbike costing:
**a** £4000 **b** £6500

**10** Extend the line of best fit on your scatter diagram to estimate:
**a** the price of a motorbike with a 1000 cc engine
**b** the engine size of a motorbike costing £10 000.

**11** The GB Team for the 1996 Olympic Games had 123 female competitors, and 189 male competitors. Show how a stratified random sample of 60 competitors is chosen.

**12** The ages of female and male competitors, listed at random, are:

**1996 Olympic Games – Great Britain Team**
**Ages of 123 Female Competitors at 20/7/96**

| | | | | | | | | | | | | | | | |
|---|---|---|---|---|---|---|---|---|---|---|---|---|---|---|---|
| 21 | 26 | 16 | 23 | 31 | 27 | 20 | 20 | 24 | 24 | 18 | 23 | 26 | 21 | 31 | 24 |
| 27 | 20 | 23 | 27 | 23 | 24 | 29 | 24 | 24 | 31 | 26 | 23 | 29 | 32 | 23 | 24 |
| 38 | 28 | 32 | 31 | 31 | 24 | 21 | 30 | 38 | 29 | 22 | 24 | 32 | 40 | 30 | 31 |
| 23 | 32 | 29 | 27 | 21 | 27 | 23 | 27 | 25 | 25 | 22 | 35 | 29 | 28 | 25 | 28 |
| 34 | 35 | 30 | 37 | 31 | 26 | 35 | 35 | 36 | 16 | 16 | 33 | 20 | 33 | 26 | 26 |
| 29 | 30 | 29 | 29 | 25 | 28 | 27 | 25 | 24 | 28 | 31 | 32 | 26 | 31 | 23 | 30 |
| 20 | 28 | 31 | 28 | 24 | 29 | 30 | 27 | 27 | 29 | 22 | 29 | 28 | 30 | 27 | 28 |
| 47 | 25 | 29 | 27 | 28 | 31 | 26 | 31 | 29 | 28 | 34 | | | | | |

**1996 Olympic Games – Great Britain Team**
**Ages of 183 Male Competitors at 20/7/96**

| | | | | | | | | | | | | | | | |
|---|---|---|---|---|---|---|---|---|---|---|---|---|---|---|---|
| 22 | 29 | 18 | 23 | 28 | 26 | 29 | 27 | 20 | 19 | 23 | 21 | 25 | 21 | 20 | 21 |
| 20 | 21 | 19 | 30 | 31 | 27 | 23 | 30 | 35 | 27 | 25 | 23 | 22 | 27 | 36 | 27 |
| 24 | 27 | 26 | 30 | 35 | 30 | 31 | 19 | 29 | 27 | 24 | 32 | 25 | 21 | 39 | 25 |
| 27 | 32 | 24 | 29 | 34 | 29 | 23 | 29 | 24 | 24 | 21 | 23 | 27 | 22 | 29 | 35 |
| 31 | 24 | 24 | 22 | 23 | 30 | 27 | 26 | 29 | 26 | 21 | 22 | 25 | 32 | 26 | 26 |
| 28 | 29 | 22 | 37 | 18 | 29 | 34 | 32 | 30 | 25 | 27 | 35 | 29 | 29 | 28 | 23 |
| 27 | 22 | 27 | 34 | 30 | 41 | 40 | 27 | 31 | 33 | 38 | 42 | 36 | 40 | 29 | 20 |
| 22 | 26 | 23 | 21 | 29 | 29 | 27 | 27 | 23 | 28 | 25 | 34 | 30 | 27 | 28 | 30 |
| 23 | 27 | 25 | 26 | 23 | 21 | 32 | 35 | 24 | 27 | 24 | 28 | 26 | 28 | 34 | 22 |
| 27 | 32 | 26 | 24 | 27 | 20 | 27 | 28 | 31 | 26 | 24 | 25 | 30 | 34 | 27 | 24 |
| 30 | 21 | 21 | 29 | 21 | 27 | 48 | 19 | 34 | 31 | 36 | 28 | 29 | 21 | 22 | 23 |
| 25 | 18 | 29 | 30 | 25 | 32 | 24 | 28 | 25 | 29 | 30 | 26 | 29 | | | |

**a** Use your answer to Question **11** and choose a systematic sample from each group.
**b** For the 60 competitors in your sample:
**i** group the data **ii** draw a histogram.

**13** For your grouped data, calculate an estimate of the mean age.

**14** **a** Choose another stratified random sample of 60 competitors by allocating a number to each competitor in the group, and using a random number table.
**b** **i** Group the data for your second sample using the same groups as in Question **12**
**ii** Draw a histogram.

**15** For your second sample, calculate an estimate of the mean age.

**16** Compare your two samples.

## Distance, speed and time

1 Write these speeds in miles per hour.

a 0.6 miles per minute
b 1.2 miles per minute
c 0.45 miles per minute
d 0.1 miles per second

2 How far does a plane travel in 3 h 20 min at a constant speed of 960 kilometres per hour?

3 How far can a car travel at a constant speed of 51 mph in:

a 1 h 30 min b 45 min
c 20 min d 55 min
e 2 h 25 min f 3 h 45 min?

4 Calculate the time taken in hours and minutes, to the nearest minute, to travel 60 km at a speed of:

a 120 km/h b 30 km/h
c 75 km/h d 9 km/h
e 55 km/h f 49 km/h
g 50 km/h h 150 km/h.

5 Convert the following:

a 4.38 hours to hours and minutes
b 12.76 hours to hours and minutes
c 5 hours 16 minutes to hours
d 8 hours 51 minutes to hours
e 24 $m\,s^{-1}$ to $km\,h^{-1}$
f 16 $m\,s^{-1}$ to mph
g 58 mph to $m\,s^{-1}$
h 152 $km\,h^{-1}$ to $m\,s^{-1}$.

6 A bike travels 96 miles in $8\frac{1}{2}$ hours. Calculate its average speed in mph.

7 How many miles does a car travelling at an average speed of 65 mph cover in 3 hours 45 minutes?

8 How long does it take Alison, walking at an average speed of $3\frac{1}{2}$ mph, to travel $8\frac{1}{2}$ miles?

9 Between traffic lights a car covered a distance of 132 metres in 9 seconds. What was its average speed in mph?

10 A balloon drifting at 6 $m\,s^{-1}$ travelled a total distance of 16.8 miles. How long did this flight last?

11 How many metres would an arrow fly in 15 seconds at an average speed of 96 $km\,h^{-1}$?

12 This is a graph of the distance of a dog from its owner who sits still on a bench.

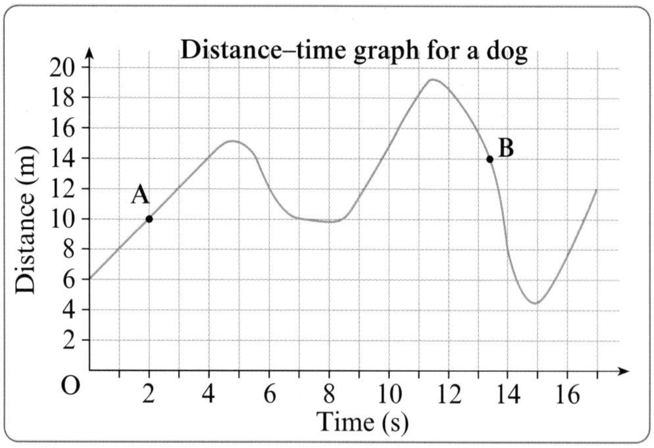


a Describe what the dog did over the 16 seconds.
b At what times was the dog 12 metres from its owner?
c How far from the owner did the dog start?
d When did the dog stop to sniff?
e At what time was the dog moving fastest?
f Estimate how fast the dog was running at:
i point A ii point B.

13 Calculate the average speed (in mph or km/h to 2 dp) of a car which travels:

a 40 miles in 50 min
b 50 km in 35 min
c 70 miles in 1 h 20 min
d 100 km in 1 h 12 min
e 400 miles in 7 h 30 min
f 500 metres in 1 min.

14 Mandy left Glasgow at 10:00 am and cycled 28 miles at a speed of 10 mph. Amin left Glasgow at 10:20 am and travelled along the same route at a speed of 12 mph.

a Show both their journeys on one graph.
b About what time did Amin pass Mandy?

15 Andy left Taunton at 3:00 pm and took 45 minutes to drive 50 miles to Bristol. He stayed there for 2 h 30 min. He then returned along the same route and arrived home at 7:05 pm.

a Draw a graph to show Andy's complete journey.
b At what time did he begin his return journey to Taunton?
c Calculate his average speed for the journey to Bristol.
d What was his average speed on the return journey?

# IN FOCUS 24

## Vectors

**1** $\mathbf{u} = \begin{pmatrix} ^-2 \\ 5 \end{pmatrix}$ $\quad \mathbf{v} = \begin{pmatrix} 1 \\ 4 \end{pmatrix}$ $\quad \mathbf{w} = \begin{pmatrix} 5 \\ 6 \end{pmatrix}$

Write each of these vectors as a column vector, and draw it on a coordinate grid.

**a** $2\mathbf{u}$ **b** $3\mathbf{v}$ **c** $^-2\mathbf{w}$ **d** $\mathbf{v} + \mathbf{w}$
**e** $\mathbf{u} + \mathbf{v}$ **f** $2\mathbf{w} - 3\mathbf{v}$ **g** $2\mathbf{v} - \mathbf{u}$ **h** $3\mathbf{w} - 5\mathbf{v} + 2\mathbf{u}$

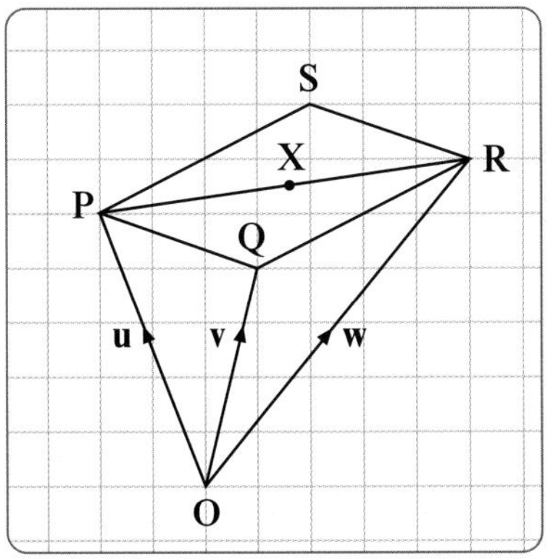


**2** PQRS is a parallelogram.
Write each of these in terms of **u**, **v**, and **w**.

**a** $\overrightarrow{PQ}$ **b** $\overrightarrow{QR}$ **c** $\overrightarrow{SP}$ **d** $\overrightarrow{RS}$ **e** $\overrightarrow{OS}$

**3** $\mathbf{u} = \begin{pmatrix} ^-2 \\ 5 \end{pmatrix}$ $\quad \mathbf{v} = \begin{pmatrix} 1 \\ 4 \end{pmatrix}$ $\quad \mathbf{w} = \begin{pmatrix} 5 \\ 6 \end{pmatrix}$

**a** Write each of your answers to Question **2** as a column vector.
**b** Use the diagram to check your answers.

**4** Give the resultant vector for:

**a** $\overrightarrow{PQ} + \overrightarrow{QR}$ **b** $\overrightarrow{PS} + \overrightarrow{SR}$ **c** $\overrightarrow{PQ} + \overrightarrow{QR} + \overrightarrow{RS}$

**5** What is special about the vector $\overrightarrow{RS} + \overrightarrow{SP} + \overrightarrow{PR}$?

**6** Use your answers to Question **2** to give the resultant vectors in Question **4** in terms of **u**, **v**, and **w**.

**7** Show that $\mathbf{v} - \mathbf{u}$ is the negative vector of $\mathbf{u} - \mathbf{v}$:

**a** using a sketch
**b** using column vectors
**c** using algebra.

**8** X is the midpoint of PR.
Prove that X is also the midpoint of QS.

**9** These vectors include three pairs of parallel vectors.

| | | | |
|---|---|---|---|
| $2\mathbf{v} - 3\mathbf{u}$ | $6\mathbf{u} + 4\mathbf{v}$ | $2\mathbf{u} + 3\mathbf{v}$ | $6\mathbf{u} - 9\mathbf{v}$ |
| $3\mathbf{v} - 2\mathbf{u}$ | $4\mathbf{v} - 6\mathbf{u}$ | $^-3\mathbf{u} - 2\mathbf{v}$ | |

List each pair.

**10**

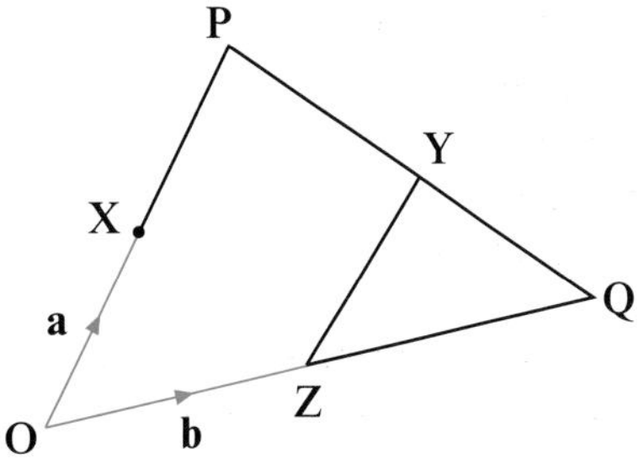


OPQ is a triangle.
X, Y, and Z are the midpoints of each of the sides.

**a** Give $\overrightarrow{PQ}$ in terms of **a** and **b**.

**b** Prove that ZY is parallel to OP and half its size.

**11**

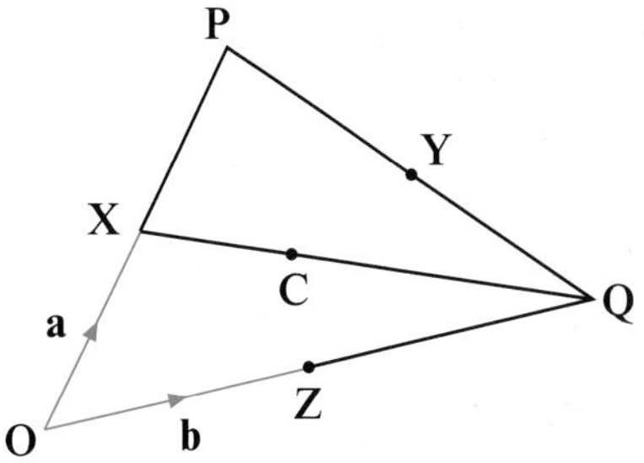


OPQ is a triangle.
X, Y, and Z are the midpoints of each of the sides.
The ratio XC : CQ is 1 : 2.

**a** Give these in terms of **a** and **b**.
**i** $\overrightarrow{XQ}$ **ii** $\overrightarrow{XC}$
**b** Prove that P, C, and Z are collinear.
**c** Give the ratio PC : CZ
**d** Prove that O, C, and Y are collinear.
**e** What do your answers tell you about point C?

**12**

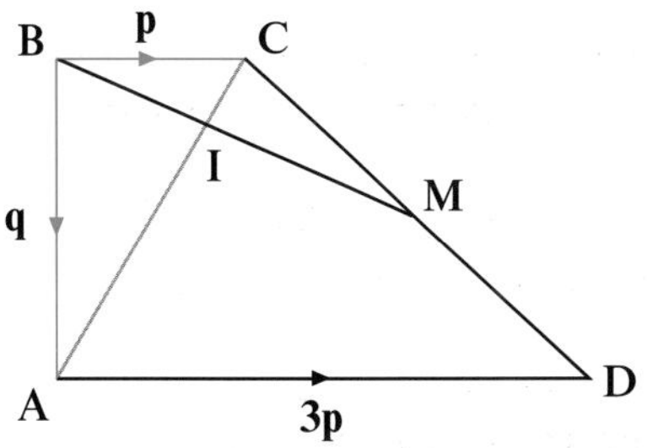


In this diagram, M is the midpoint of CD.
The ratio BI : IM is 2 : 3.

**a** Give these in terms of **p** and **q**.
**i** $\overrightarrow{CD}$ **ii** $\overrightarrow{CM}$ **iii** $\overrightarrow{BI}$
**b** Prove that the ratio AI : IC is 4 : 1.

## Transforming graphs

**1** Functions $f$, $g$ and $h$ are defined so that:

$f(x) = \dfrac{1}{x-5}$

$g(x) = (x+3)^2 - 5$

$h(x) = \dfrac{3}{x} + 4$

**a** Calculate:
**i** $f(6)$ **ii** $g(1)$ **iii** $h(2)$
**iv** $f(^-3)$ **v** $g(^-3)$ **vi** $h(^-3)$

**b** Find values of $x$ for which:
**i** $f(x) = 0.1$ **ii** $g(x) = 4$ **iii** $h(x) = 7.75$

**2** An experiment is carried out to investigate the bounciness of a ball. The ball is dropped from different heights and the maximum height of the rebound is measured.

The results are shown in the table below.

| Drop height $D$ (cm) | 50 | 100 | 150 | 200 |
|---|---|---|---|---|
| Rebound height $R$ (cm) | 46 | 92 | 139 | 180 |

**a** Draw a graph to show that $R$ and $D$ are linked by a formula of the form $R = aD + b$, where $a$ and $b$ are constants.
**b** Estimate the values of $a$ and $b$.
**c** **(i)** What do you think the rebound height will be when the drop height is 350 cm?
**(ii)** Why might your answer be unreliable?

**3** A small weight is dropped from the top of a building. Its position is recorded at intervals of 0.5 seconds using an electronic camera.

The results are shown in the table below, giving distance from ground $d$ (m) and time taken $t$ (seconds).

| $t$ | 0 | 0.5 | 1 | 1.5 | 2 | 2.5 | 3 |
|---|---|---|---|---|---|---|---|
| $d$ | 50.0 | 48.8 | 45.1 | 40.0 | 30.4 | 19.4 | 5.9 |

**a** Draw the graph of $d$ against $t^2$.
**b** **i** Does your graph support the claim that $d$ and $t$ are connected by an equation of the form $d = pt^2 + q$?
**ii** If it does, estimate the values of $p$ and $q$.

**4**

| $x$ | 0 | 2 | 4 | 6 | 8 | 10 | 12 |
|---|---|---|---|---|---|---|---|
| $y$ | $^-2.4$ | $^-0.8$ | 4.2 | 11.7 | 23.1 | 37.5 | 55.0 |

**a** Draw a graph to show that $x$ and $y$ are linked by a formula of the form $y = mx^2 + n$, where $m$ and $n$ are constants.
**b** Estimate the values of $m$ and $n$.
**c** Find the value of $y$ when $x = 5$.

**5** This is a sketch of the graph of $y = f(x)$ where $f(x) = x^2 - 6x + 10$.

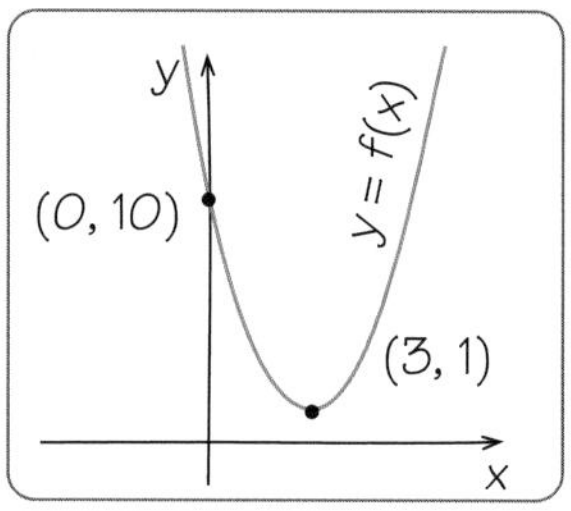


**a** Sketch graphs of:
**i** $y = -f(x)$ **ii** $y = f(-x)$ **iii** $y = f(x) + 3$
**iv** $y = f(x+3)$ **v** $y = f(x-2) + 1$
**b** Describe a transformation that maps $y = f(x)$ to each graph.

**6** This is the graph of the function $y = g(x)$.

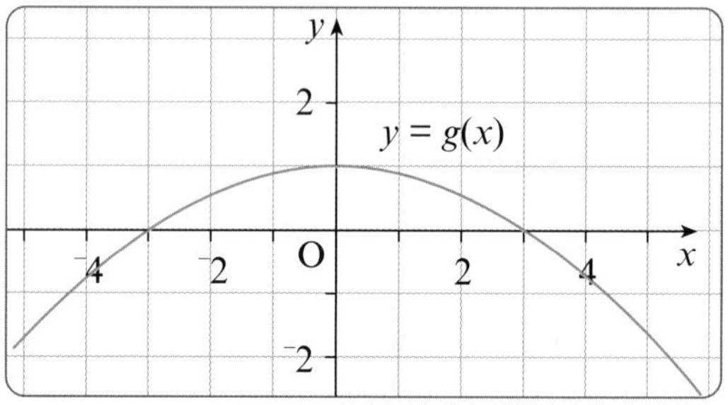


Sketch graphs of:
**a** $y = 2g(x)$ **b** $y = g(2x)$ **c** $y = g(\frac{1}{2}x)$

**7** This is the graph of the function $y = k(x)$.

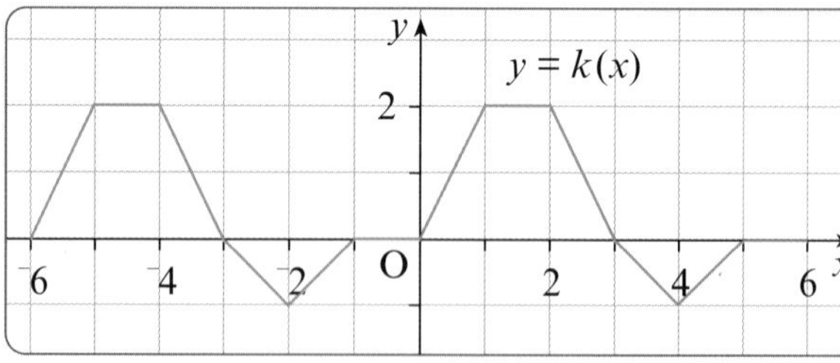


**a** For $^-3 \leqslant x \leqslant 3$, sketch graphs of:
**i** $y = k(^-x)$ **ii** $y = k(2x)$ **iii** $y = k(\frac{1}{2}x)$
**iv** $y = k(\frac{1}{3}x)$ **v** $y = \frac{1}{2}k(x)$ **vi** $y = 2k(x)$
**vii** $y = k(x) + 3$ **viii** $y = k(x+3)$
**ix** $y = k(x-2)$ **x** $y = k(x-3) - 4$

**b** In terms of $k$, write the equation of the graph below.

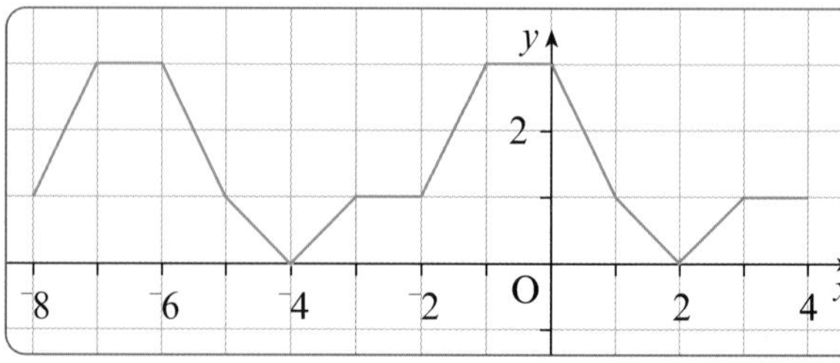

For your GCSE Maths you will need to produce two pieces of coursework:

1. Investigative
2. Statistical

Each piece is worth 10% of your exam mark.
This unit gives you guidance on approaching your coursework and will tell you how to get good marks for each piece.

## Investigative coursework

In your investigative coursework you will need to:

- Say how you are going to carry out the task and provide a plan of action
- Collect results for the task and consider an appropriate way to represent your result
- Write down observations you make from your diagrams or calculations
- Look at your results and write down any observations, rules or patterns – try to make a general statement
- Check out your general statements by testing further data – say if your test works or not
- Develop the task by posing your own questions and provide a conclusion to the questions posed
- Extend the task by using techniques and calculations from the content of the higher tier
- Make your conclusions clear and link them to the original task giving comments on your methods.

### Assessing the task

Before you start, it is helpful for you to know how your work will be marked. This way you can make sure you are familiar with the sort of things which examiners are looking for.

Investigative work is marked under three headings:

1. **Making and monitoring decisions to solve problems**
2. **Communicating mathematically**
3. **Developing skills of mathematical reasoning**

Each strand assesses a different aspect of your coursework. The criteria are:

**1. Making and monitoring decisions to solve problems**

This strand is about deciding what needs to be done then doing it. The strand requires you to select an appropriate approach, obtain information and introduce your own questions which develop the task further.

For the higher marks you need to analyse alternative mathematical approaches and, possibly, develop the task using work from the higher tier GCSE syllabus content.

For this strand you need to:

- solve the task by collecting information
- break down the task by solving it in a systematic way
- extend the task by introducing your own *relevant* questions
- extend the task by following through alternative approaches

- develop the task by including a number of mathematical features
- explore the task extensively using higher level mathematics

**2. Communicating mathematically**

This strand is about communicating what you are doing using words, tables, diagrams and symbols. You should make sure your chosen presentation is accurate and relevant.

For the higher marks you will need to use mathematical symbols accurately, concisely and efficiently in presenting a reasoned argument.

For this strand you need to:

- illustrate your information using diagrams and calculations
- interpret and explain your diagrams and calculations
- begin to use mathematical symbols to explain your work
- use mathematical symbols consistently to explain your work
- use mathematical symbols accurately to argue your case
- use mathematical symbols concisely and efficiently to argue your case

**3. Developing skills of mathematical reasoning**

This strand is about testing, explaining and justifying what you have done. It requires you to search for patterns and provide generalisations. You should test, justify and explain any generalisations.

For the higher marks you will need to provide a sophisticated and rigorous justification, argument or proof which demonstrates a mathematical insight into the problem.

For this strand you need to:

- make a general statement from your results
- confirm your general statement by further testing
- make a justification for your general statement
- develop your justification further
- provide a sophisticated justification
- provide a rigorous justification, argument or proof

This unit uses a series of investigative tasks to demonstrate how each of these strands can be achieved.

## Task 1

**TRIANGLES**

Patterns of triangles are made as shown in the following diagrams:

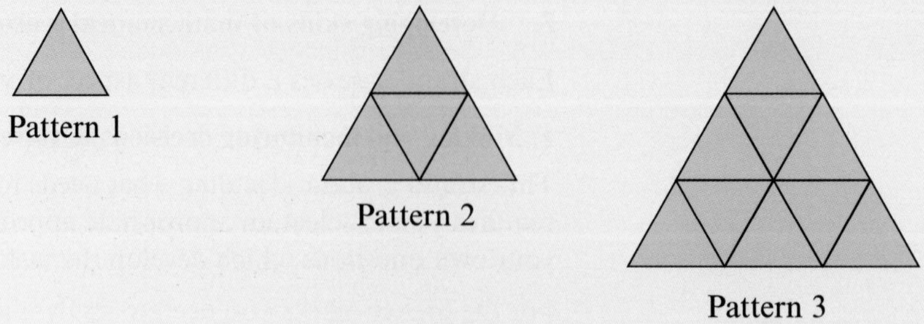


What do you notice about the pattern number and the number of triangles?

Investigate further.

Note: You are reminded that any coursework submitted for your GCSE examination must be your own. If you copy from someone else then you may be disqualified from the examination.

# Planning your work

A straightforward approach to this task is to continue the pattern further. You can record your results in a table and then see if you can make any generalisations.

## Collecting information

A good starting point for any investigation is to collect information about the task.

You can see that:

- Pattern 1 shows 1 triangle
- Pattern 2 shows 4 triangles
- Pattern 3 shows 9 triangles

You can then extend this by considering further patterns.

Pattern 4

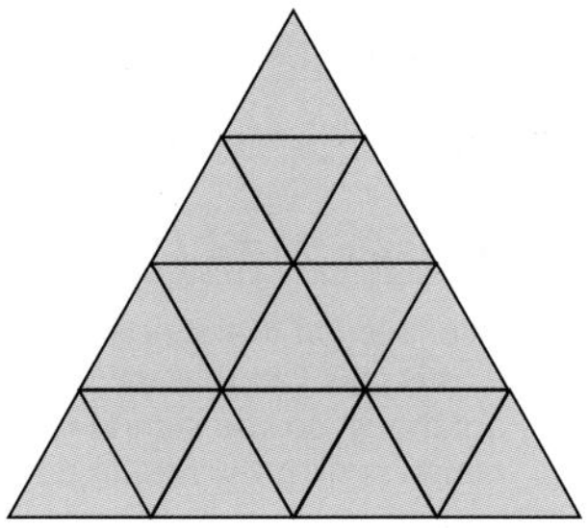

16 triangles

Pattern 5

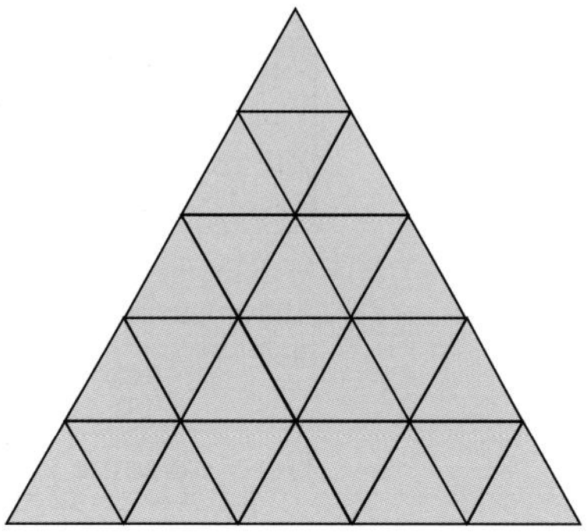

25 triangles

**Moderator comment**

It is a good idea not to collect too much data as this is time consuming ... aim to collect 4 or 5 items of data to start with.

## Drawing up tables

The information is not easy to follow so it is **always** a good idea to illustrate it in a table.

A table to show the relationship between the pattern number and the number of triangles

| Pattern number | 1 | 2 | 3 | 4 | 5 |
|---|---|---|---|---|---|
| No. of triangles | 1 | 4 | 9 | 16 | 25 |

It is important to give your table a title and to make it quite clear what the table is showing.

**Now you try ...**

Using the table:

- Can you see anything special about the number of triangles?
- Ask yourself: are they odd numbers, even numbers, square numbers, triangle numbers, are they multiples, do they get bigger, do they get smaller ...
- What other relationships might you look for?

You should now try and explain what your table tells you:

From my table I can . . .

## Finding a relationship – making a generalisation

To make a generalisation, you need to find a relationship from your table of results. A useful method is to look at the differences between terms in the table.

Here is the table of results from task 1:

A table to show the relationship between the pattern number and the number of triangles

| Pattern number | 1 | 2 | 3 | 4 | 5 |
|---|---|---|---|---|---|
| No of triangles | 1 | 4 | 9 | 16 | 25 |
| | +3 | +5 | +7 | +9 | +11 |
| | +2 | +2 | +2 | +2 | |

In this table the first "differences" are going up in two's.

The "second differences" are all the same ... this tells you that the relationship is quadratic, so you should compare the numbers with the square numbers

Here are some generalisations you could make:

From my table I notice that the number of triangles are all square numbers.

From my table I notice that the number of triangles are the pattern numbers squared.

**Moderator comment**

Remember that the idea is to identify some relationship from your table or, graph so that there is some point in you using it in your work ... a table or graph without any comment on the findings is of little use.

A graph may help see the relationship:

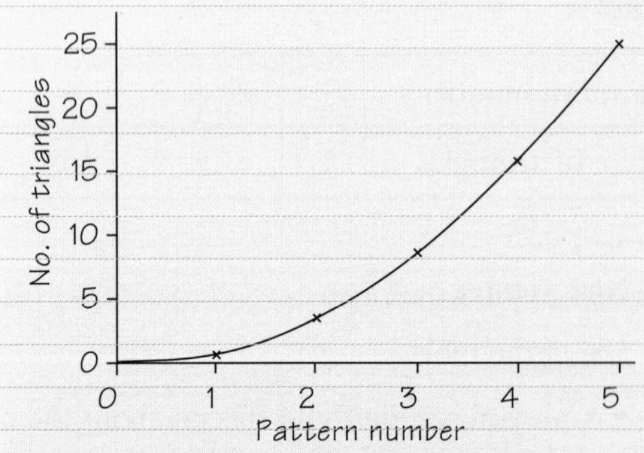


From my graph I notice that there is a quadratic relationship between the number of triangles and the pattern number.

# Using algebra

You will gain marks if you can express your generalisation or rule using algebra.

The rule:

I notice that the number of triangles are the pattern numbers squared.

can be written in algebra:

Remember that you must explain what $t$ and $p$ stand for.

$t = p^2$ where $t$ is the number of triangles and $p$ is the pattern number

## Testing generalisations

You need to confirm your generalisation by further testing.

From my table I notice that the number of triangles are the pattern numbers squared so that the sixth pattern will have $6^2 = 36$ triangles.

You can confirm this with a diagram:

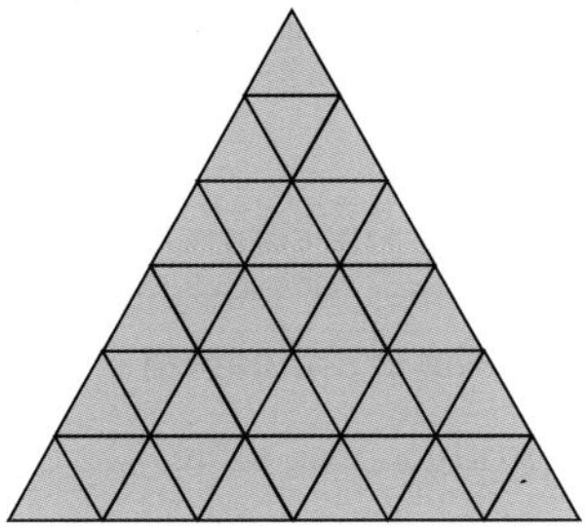

Always confirm your testing by providing some comment, even if your test has not worked!

My diagram confirms the generalisation for the sixth pattern.

Now try to use all of these ideas by following task 3.

**Now you try ...**

For task 3: 'Perfect tiles' given below:

- Collect the information for different arrangements
- Make sure you are systematic
- Draw up a table of your results.

### Task 3

**PERFECT TILES**

Floor spacers are used to give a perfect finish when laying tiles on the kitchen floor.

Three different spacers are used including

**L** spacer
**T** spacer
**+** spacer

Here is a 3 × 3 arrangement of tiles

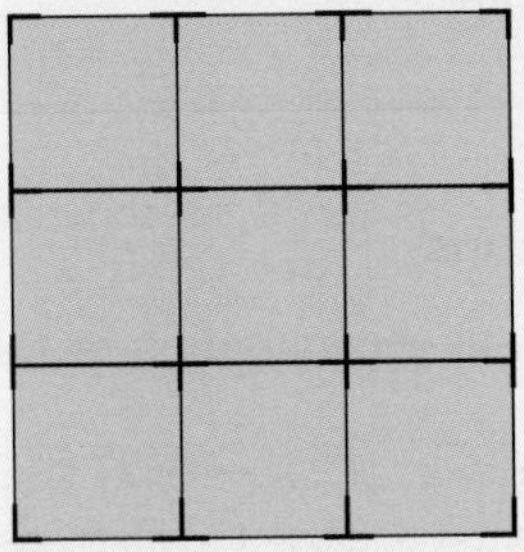

Draw picture of a 3 × 3 arrangement

This uses 4 **L** spacers
8 **T** spacers and
4 **+** spacers

Investigate different arrangements of tiles.

Note: You are reminded that any coursework submitted for your GCSE examination must be your own. If you copy from someone else then you may be disqualified from the examination.

You should be able to produce this table:

Number of spacers for different arrangements of tiles

| Size of square | 1 × 1 | 2 × 2 | 3 × 3 | 4 × 4 |
|---|---|---|---|---|
| Number of L spacers | 4 | 4 | 4 | 4 |
| Number of T spacers | 0 | 4 | 8 | 12 |
| Number of + spacers | 0 | 1 | 4 | 9 |

**Now you try ...**

What patterns do you notice from the table?
What general statements can you make?
Now test your general statements.

Generalisations for the 'Perfect tiles' task might include:

$L = 4$ where $L$ is the number of $L$ spacers

$T = 4(n - 1)$ where $T$ is the number of $T$ spacers and $n$ is the size of the arrangement $(n \times n)$

The formula $T = 4(n - 1)$ can also be written as $T = 4n - 4$.

**Test**:

For a 5 × 5 arrangement (ie $n = 5$)

$T = 4(n - 1)$
$T = 4(5 - 1)$
$T = 4 \times 4$
$T = 16$

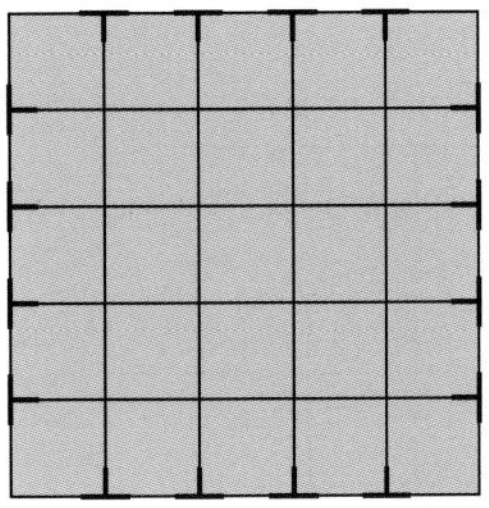

The number of $T$ spacers is 16 so the formula works for a 5 × 5 arrangement.

$+ = (n - 1)^2$ where + is the number of + spacers and $n$ is the size of the arrangement ($n \times n$)

**Test**:
For a 6 × 6 arrangement (ie $n = 6$)

$+ = (n - 1)^2$
$+ = (6 - 1)^2$
$+ = (5)^2$
$+ = 25$

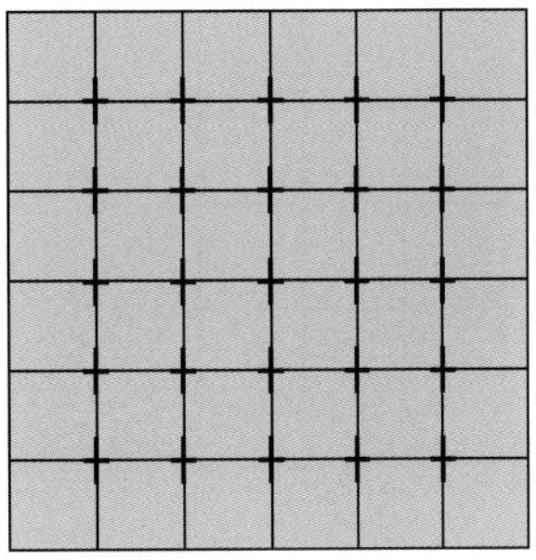

The number of + spacers is 25 so the formula works for a 6 × 6 arrangement.

## Making justifications

Once you have found the general statement and tested it then you need to justify it. You justify the statement by explaining WHY it works.

For example, here are possible justifications for the rules in task 3:

WHY is the number of L spacers always 4?

The number of L spacers is always 4 because there are 4 corners to each arrangement.

WHY are all of the T spacers multiples of 4?

The number of T spacers are multiples of 4 because each time you add another tile to the side then you add another T spacer.

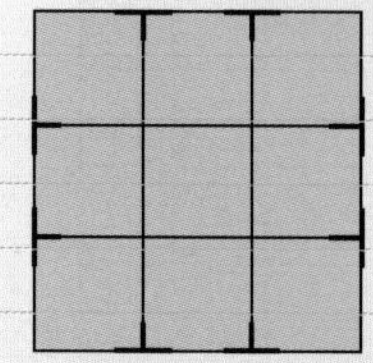

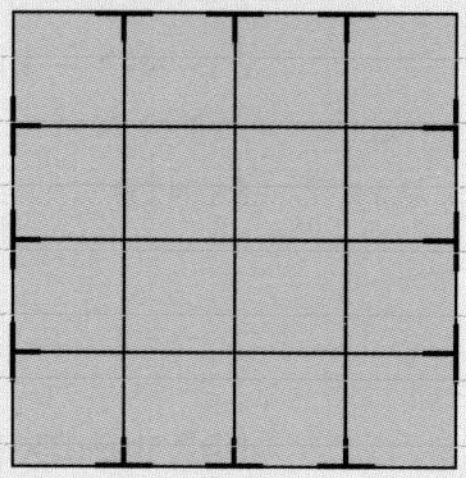

As there are 4 sides to the arrangement then the number of T spacers will go up by 4.

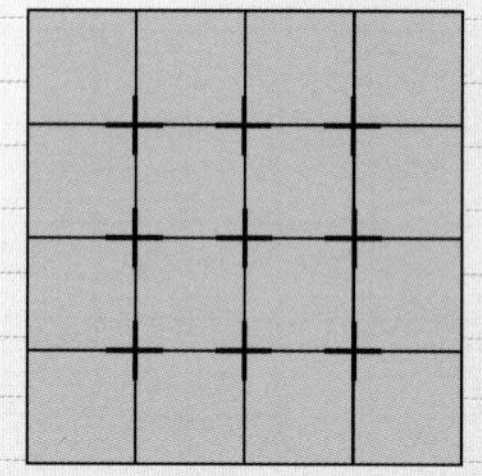

WHY are all of the + spacers square numbers?

**Now you try ...**

Explain why the number of + spacers are **always** square numbers.

## Extending the problem – investigating further

Once you have understood and explained the basic task, you should extend your work to get better marks.

To extend a task you need to pose your own questions.
This means that you must think of different ways to extend the original task.

### Extending task 1: The triangle problem

You could ask and investigate:

What about different patterns?

For example, triangles

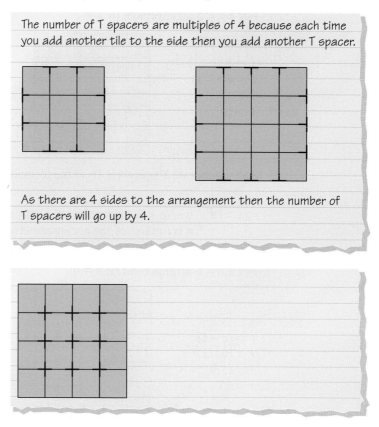

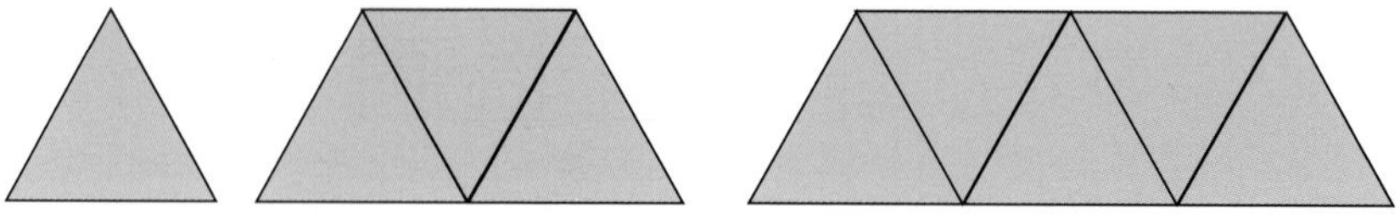

What about patterns of different shapes?

For example, squares

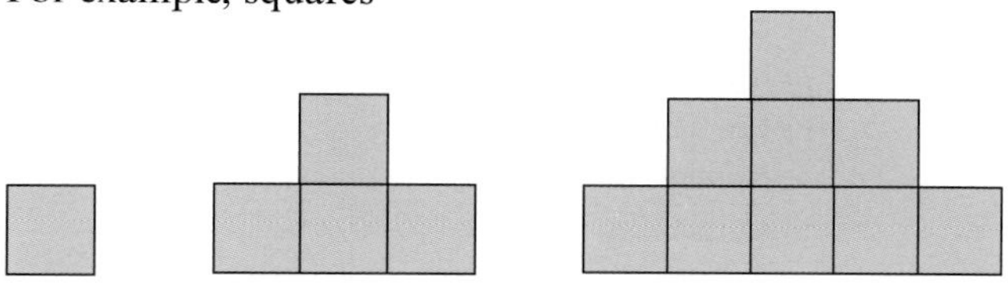

What about 3 dimensional patterns?

For example cubes

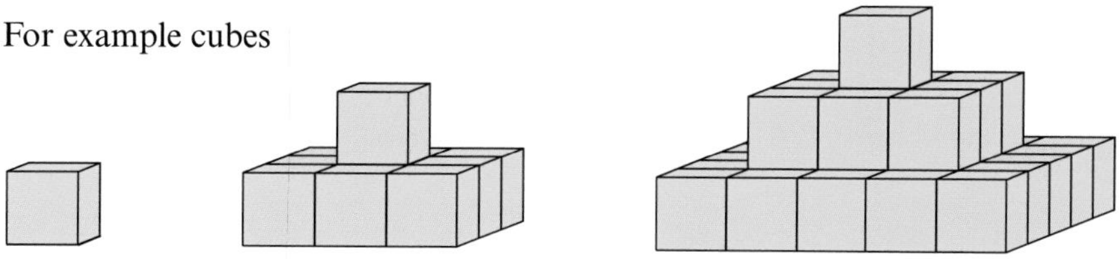

### Extending task 2: Arrange a letter

What about different numbers of letters? For example, ABCDE

What happens when:
- a letter is repeated? For example, AABCD
- a letter is repeated more than once? For example, AAABC
- more than one letter is repeated? For example, AAABB

Of course, you will have to do more than just pose a question – you will have to carry out the investigation and come to a conclusion.

**Now you try ...**

Think of some different ways to extend task 3: Perfect Tiles.
What sort of questions might you ask?
What other areas might you explore?

### Extending task 3: Perfect Tiles

You could ask:

What about rectangular arrangements of tiles?

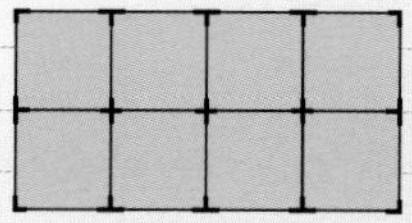

What about different shaped tiles?
For example, triangles

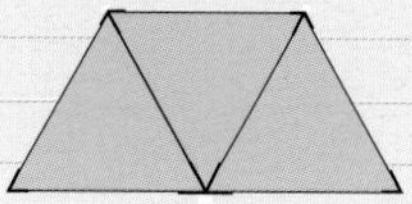

Remember to work systematically.

Consider rectangular arrangements of tiles:

Number of spacers for different arrangements of tiles

| Size of arrangement | 2 × 1 | 2 × 2 | 2 × 3 | 2 × 4 | 2 × 5 |
|---|---|---|---|---|---|
| Number of L spacers | 4 | 4 | 4 | 4 | 4 |
| Number of T spacers | 2 | 4 | 6 | 8 | 10 |
| Number of + spacers | 0 | 1 | 2 | 3 | 4 |
| Number of spacers | 6 | 9 | 12 | 15 | 18 |

Number of spacers for different arrangements of tiles

| Size of arrangement | 3 × 1 | 3 × 2 | 3 × 3 | 3 × 4 | 3 × 5 |
|---|---|---|---|---|---|
| Number of L spacers | 4 | 4 | 4 | 4 | 4 |
| Number of T spacers | 4 | 6 | 8 | 10 | 12 |
| Number of + spacers | 0 | 2 | 4 | 6 | 8 |
| Number of spacers | 8 | 12 | 16 | 20 | 24 |

Rules for the 'Perfect tiles' task extension might include:

$L = 4$ where $L$ is the number of $L$ spacers

$T = 2(x-1) + 2(y-1)$ where T is the number of T spacers and x is the length of the arrangement and y is the width of the arrangement ($x \times y$)

Note that the formula $T = 2(x - 1) + 2(y - 1)$ can also be written
$T = 2x - 2 + 2y - 2$
$T = 2x + 2y - 4$

$P = (x-1)(y-1)$ where $P$ is the number of + spacers and x is the length of the arrangement and y is the width of the arrangement ($x \times y$)

It is sensible to replace + by $P$ so that $P$ stands for the number of + spacers.

$N = (x+1)(y+1)$ where N is the total number of spacers and x is the length of the arrangement and y is the width of the arrangement ($x \times y$)

**Moderator comment**

The use of higher order algebra can result in high marks awarded on the middle strand as well as an opportunity to provide a sophisticated justification

The total number of spacers should equal the number of L spacers plus the number of T spacers plus the number of + spacers

Proof:
$$\begin{aligned} N &= L + T + P \\ &= 4 + [2(x-1) + 2(y-1)] + [(x-1)(y-1)] \\ &= 4 + 2x - 2 + 2y - 2 + xy - y - x + 1 \\ &= xy + x + y + 1 \\ &= (x+1)(y+1) \end{aligned}$$

Since $N = (x+1)(y+1)$ is true, then my theory is proved

## Task 4

### SQUARE SEA SHELLS

Square sea shells are formed from squares whose pattern of growth is shown as follows:

Day 1

Day 2

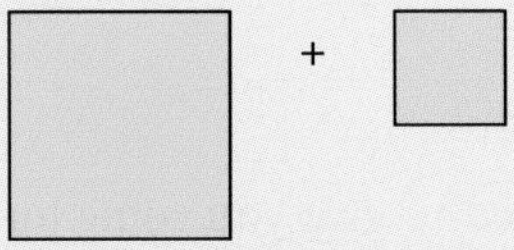

The length of the new square is half that of the previous day

Day 3

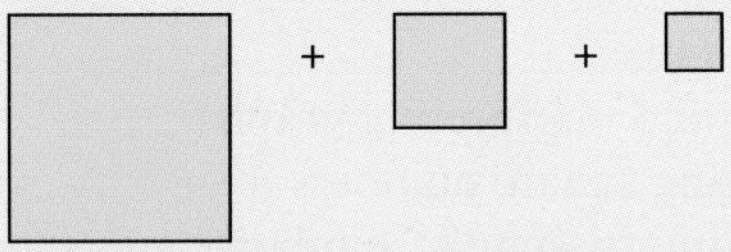

The length of the new square is half that of the previous day

Day 4

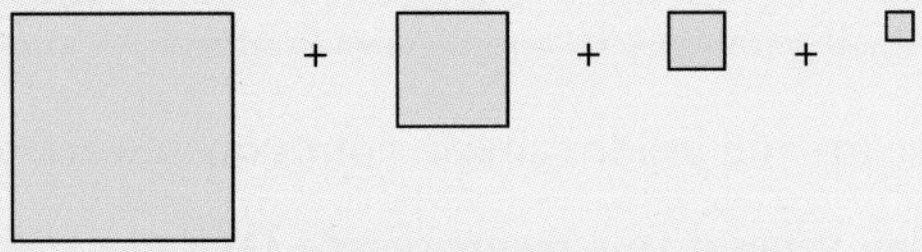

The length of the new square is half that of the previous day

This pattern of growth continues

Investigate the area covered by square sea shells on different days.

Investigate further.

For a square of side $a$

Day 1 $\quad$ Area $= a^2$

Day 2 $\quad$ Area $= a^2 + \frac{a^2}{4}$

Day 3 $\quad$ Area $= a^2 + \frac{a^2}{4} + \frac{a^2}{16}$

Day 4 $\quad$ Area $= a^2 + \frac{a^2}{4} + \frac{a^2}{16} + \frac{a^2}{64}$

$$\text{Day} \ldots \quad \text{Area} = a^2 + \frac{a^2}{4} + \frac{a^2}{16} + \frac{a^2}{64} + \ldots$$
$$= a^2\left(1 + \frac{1}{4} + \frac{1}{16} + \frac{1}{64} + \ldots\right)$$
$$= \ldots$$
$$= \ldots$$
$$= \frac{4}{3}a^2$$

This work involves summing an infinite series which is beyond the mathematical content of the GCSE examination and so will get you higher marks.

**Moderator comment**

Remember that for the higher levels, you will need to explore the task extensively using higher level mathematics, use mathematical symbols concisely and efficiently to argue your case and provide a rigorous justification, argument or proof.

**Extending task 4: square sea shells**

For an equilateral triangle of side $a$

Day 1 $$\text{Area} = \frac{1}{2} a^2 \sin 60^\circ = \frac{1}{2} a^2 \sqrt{\frac{3}{2}} = \sqrt{\frac{3}{4}} a^2$$

or

For a regular $n$ sided polygon of length $a$

Day 1 $$\text{Area} = \frac{na^2}{4 \tan\left(\frac{180}{n}\right)^\circ}$$

In this unit we have tried to give you some hints on approaching investigative coursework to gain your best possible mark.

This mathematics is often useful in investigative tasks:

- Creating tables
- Drawing and interpreting graphs
- Recognising square numbers
- Recognising triangular numbers
- Finding the $n$th term of a linear sequence

## Summary

These are the grade criteria your coursework will be marked by:

**Testing general statements (grade D)**

***To achieve this level you must:***
- break down the task by solving it in an orderly manner
- interpret and explain your diagrams and calculations
- confirm your general statement by further testing

**Posing questions and justifying (grade C)**

***To achieve this level you must:***
- extend the task by introducing your own relevant questions
- begin to use mathematical symbols to explain your work
- make a justification for your general statement
- provide a sophisticated justification

**Making further progress (grade B)**

***To achieve this level you must:***
- extend the task by following through alternative approaches
- use mathematical symbols consistently to explain your work
- develop you justification further

**Justifying a number of mathematical features (grade A)**

***To achieve this level you must:***
- develop the task by including a number of mathematical features
- use mathematical symbols accurately to argue your case
- provide a sophisticated justification

**Exploring extensively (grade A*)**

***To achieve this level you must:***
- explore the task extensively using higher level mathematics
- use mathematical symbols concisely and efficiently to argue your case and provide a rigorous justification, argument or proof

# Statistical coursework

In your statistical coursework you will need to:

- Provide a well considered hypothesis and provide a plan of action to carry out the task
- Decide what data is needed and collect results for the task using an appropriate sample size and sampling method
- Consider the most appropriate way to represent your results and write down any observations you make
- Consider the most appropriate statistical calculations to use and interpret your findings in terms of the original hypothesis
- Develop the task by posing your own questions – you may need to collect further data to move the task on
- Extend the task by using techniques and calculations from the content of the higher tier
- Make your conclusions clear – always link them to the original hypothesis recognising limitations and suggesting improvements.

## Assessing the task

It may be helpful for you to know how the work is marked. This way you can make sure you are familiar with the sort of thing that examiners are looking for.

Statistical work is marked under three headings:

**1. Specifying the problem and planning**
**2. Collecting, processing and representing the data**
**3. Interpreting and discussing the results**

Each strand assesses a different aspect of your coursework as follows:

### 1. Specifying the problem and planning

This strand is about choosing a problem and deciding what needs to be done, then doing it. The strand requires you to provide clear aims, consider the collection of data, identify practical problems and explain how you might overcome them.

For the higher marks you need to decide upon a suitable sampling method, explain what steps were taken to avoid possible bias and provide a well structured report.

### 2. Collecting, processing and representing the data

This strand is about collecting data and using appropriate statistical techniques and calculations to process and represent the data. Diagrams should be appropriate and calculations mostly correct.

For the higher marks you will need to accurately use higher level statistical techniques and calculations from the higher tier GCSE syllabus content.

### 3. Interpreting and discussing the results

This strand is about commenting, summarising and interpreting your data. Your discussion should link back to the original problem and provide an evaluation of the work undertaken.

For the higher marks you will need to provide sophisticated and rigorous interpretations of your data and provide an analysis of how significant your findings are.

This unit uses a series of statistical tasks to demonstrate how each of these strands can be achieved.

## Planning your work

Statistical coursework requires careful planning if you are to gain good marks. Before undertaking any statistical investigation, it is important that you plan your work and decide exactly what you are going to investigate – do not be too ambitious!

Before you start you should:

- decide what your investigation is about and why you have chosen it
- decide how you are going to collect the information
- explain how you intend to ensure that your data is representative
- detail any presumptions which you are making

## Getting started – setting up your hypothesis

**Moderator comment**

Your statistical task must always start with a 'hypothesis' where you state exactly what you are investigating and what you expect to find.

A good starting point for any statistical task is to consider the best way to collect the data and then to write a clear hypothesis you can test.

**Task 1**

**WHAT THE PAPERS SAY**

Choose a passage from two different newspapers and investigate the similarities and differences between them.

Write down a hypothesis to test.

Design and carry out a statistical experiment to test the hypothesis.

Investigate further.

Note: You are reminded that any coursework submitted for your GCSE examination must be your own. If you copy from someone else then you may be disqualified from the examination.

First consider different ways to compare the newspapers:

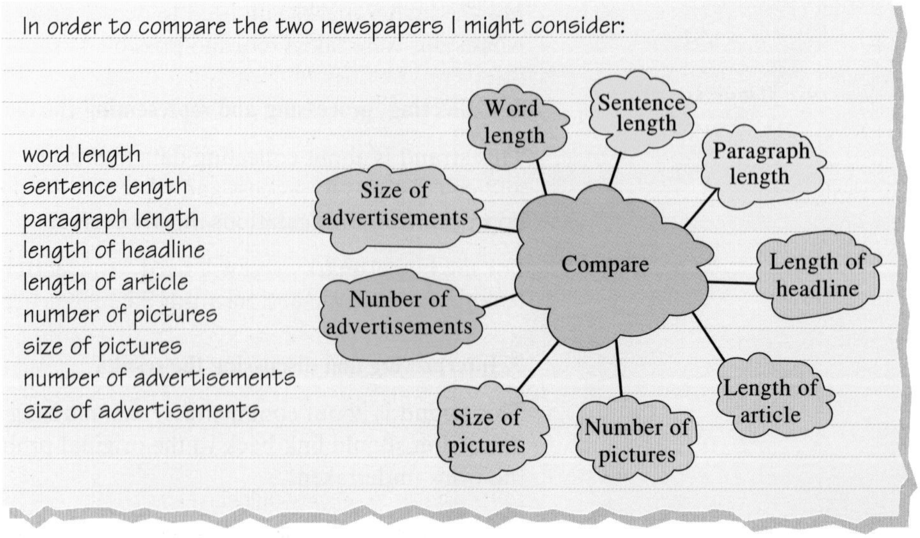


Now formulate your hypothesis.

Here are some possible hypotheses:

For my statistics coursework, I am going to investigate the hypothesis that 'tabloid' papers use shorter words than 'broadsheet' newspapers.

My hypothesis is that word lengths in the tabloid newspaper will be shorter than word lengths in the broadsheet newspaper.

My hypothesis is that sentence lengths in the tabloid newspaper will be shorter than sentence lengths in the broadsheet newspaper.

My hypothesis is that the number of advertisements in the tabloid newspaper will be greater than the number of advertisements in the broadsheet newspaper.

**Remember:**
it does not matter whether your hypotheses are true or false and you will still gain marks if your hypothesis turns out to be false.

**Now you try ...**

See if you can add some hypotheses of your own.

Explain how you would proceed with the task.

## Choosing the right sample

Moderator comment

It is important to choose an appropriate sample, give reasons for your choice and explain what steps you will take to avoid bias.

Once you have decided your aims and set up a hypothesis then it is important to consider how you will test your hypothesis.

Sampling techniques include:

- **Random sampling** is where each member of the population has an equally likely chance of being selected. An easy way to do this would be to give each person a number and then choose the numbers randomly.
- **Systematic sampling** is the same as random sampling except that there is some system involved such as numbering each person and then choosing every 20th number to form the sample.
- **Stratified sampling** is where each person is placed into some particular group or category (stratum) and the sample size is proportional to the size of the group or category in the population as a whole.
- **Convenience sampling** or opportunity sampling is one which involves simply choosing the first person to come along ... although this method is not particularly random!

Moderator comment

You should always say why you choose your sampling method.

Here are some ways you could test the hypotheses for task 1:

| Sampling method | Reason |
| --- | --- |
| To investigate my hypothesis I am going to choose a similar article from each type of newspaper and count the lengths of the first 100 words. | I decided to choose similar articles because the words will be describing similar information and so will be better to make comparisons. |
| To investigate my hypothesis I will choose every tenth page from each type of newspaper and calculate the percentage area covered by pictures. | I decided to choose every tenth page from each type of newspaper because the types of articles vary throughout the newspaper (for example headlines at the front and sports at the back of the paper). |

# Collecting primary data

**Moderator comment**

It is important to carefully consider the collection of reliable data. Appropriate methods of collecting primary data might include observation, interviewing, questionnaires or experiments.

If you are collecting primary data then remember:

- **Observation** involves collecting information by observation and might involve participant observation (where the observer takes part in the activity), or systematic observation (where the observation happens without anyone knowing).
- **Interviewing** involves a conversation between two or more people. Interviewing can be formal (where the questions follow a strict format) or informal (where they follow a general format but can be changed around to suit the questioning).
- **Questionnaires** are the most popular way of undertaking surveys. Good questionnaires are
  - simple, short, clear and precise,
  - attractively laid out and quick to complete

  and the questions are
  - not biased or offensive
  - written in a language which is easy to understand
  - relevant to the hypothesis being investigated
  - accompanied by clear instructions on how to answer the questions.

**Moderator comment**

It is always a good idea to undertake a small scale 'dry run' to check for problems. This 'dry run' is called a pilot survey and can be used to improve your questionnaire or survey before it is undertaken.

## Avoiding bias

You must be very careful to avoid any possibility of bias in your work. For example, in making comparisons it is important to ensure that you are comparing like with like.

Jean undertook task 1: 'What the papers say'.
She collected this data from two newspapers by measuring (observation).

| | Tabloid | Broadsheet |
|---|---|---|
| Number of advertisements | 35 | 30 |
| Number of pages | 50 | 30 |
| Area of each page | 1000cm$^2$ | 2000cm$^2$ |

Her hypothesis is:

*The tabloid newspaper has more advertisements than the broadsheet newspaper.*

A quick glance at the table may make her claim look true: the broadsheet has 30 adverts but the tabloid has 35, which is more.

However, the sizes of the newspapers are different so Jean is not really comparing like with like.
To ensure Jean compares like with like she should take account of:

- the area of the pages
- the number of pages

and so on.

Percentage coverage per page:
Tabloid $= 35 \div 50 \times 100\% = 70\%$
Broadsheet $= 30 \div 30 \times 100\% = 100\%$

This shows that:

The broadsheet newspaper has more advertisements per page than the tabloid newspaper.

The total area of the pages is:
Tabloid = $50 \times 1000\text{cm}^2 = 50\,000\text{cm}^2$
Broadsheet = $30 \times 2000\text{cm}^2 = 60\,000\text{cm}^2$

Note: this still doesn't take account of the size of the adverts!

So the percentage coverage is:
Tabloid = $35 \div 50\,000 \times 100\% = 0.07\%$
Broadsheet = $30 \div 60\,000 \times 100\% = 0.05\%$

This shows:

The tabloid newspaper has more advertisements per area of coverage than the broadsheet newspaper.

## Methods and calculations

You can represent the data using statistical calculations such as the mean, median, mode, range and standard deviation.

Once you have collected your data, you need to use appropriate statistical methods and calculations to process and represent your data.

**Task 2**

**GUESSING GAME**

Dinesh asked a sample of people to estimate the length of a line and the weight of a package.

Write down a hypothesis about estimating lengths and weights and carry out your own experiment to test your hypothesis.

Investigate further

Note: You are reminded that any coursework submitted for your GCSE examination must be your own. If you copy from someone else then you may be disqualified from the examination.

Maurice and Angela decide to explore the hypothesis that:

Students are better at estimating the length of a line than the weight of a package.

To test the hypothesis they collect data from 50 children, detailing their estimations of the length of a line and the weight of a package.

Here are their findings:

A table to show the estimations for the length of a line and the weight of a package

| | Length (cm) | Weight (g) |
|---|---|---|
| Mean | 15.9 | 105.2 |
| Median | 15.5 | 100 |
| Mode | 14 | 100 |
| Range | 8.6 | 28 |
| SD | 1.2 | 2.1 |

**Moderator comment**

It is important to consider whether information on all of these statistical calculations is essential.

Note: the actual length of the line is 15cm and the weight of the package is 100g.

**Now you try ...**

Using the table:

- What do you notice about the average of the length and weight?
- What do you notice about the spread of the length and weight?
- Does the information support the hypothesis?

## Graphical representation

Other statistical representations might include pie charts, bar charts, scatter graphs and histograms.

**Moderator comment**

You should only use appropriate diagrams and graphs.

Remember to consider the possibility of bias in your data:

The percentage error for the lengths is $\frac{0.9}{15} \times 100\% = 6\%$

The percentage error for the weights is $\frac{5.2}{100} \times 100\% = 5.2\%$

The data for task 2 was sorted into different categories and represented as a table.

It may be useful to show your results using graphs and diagrams as sometimes it is easier to see trends.

Graphical representations might include stem and leaf diagrams and cumulative frequency graphs and box-and-whisker diagrams.

Once you have drawn a graph you should say what you notice from the graph:

From my representation, I can see that the estimations for the line are generally more accurate than the estimations for the weight because:

- the mean is closer to the actual value for the lengths and
- the standard deviation is smaller for the lengths

However, on closer inspection:

- the percentage error on the mean is smaller for the weights
- so the median and mode value are better averages to use for the weights

## Using secondary data

**Moderator comment**

If you use secondary data there must be enough 'to allow sampling to take place' – about 50 pieces of data.

You may use secondary data in your coursework.

Secondary data is data that is already collected for you.

**Task 3**

**GENDER DIFFERENCES IN EXAMINATIONS**

Jade is conducting a survey in the GCSE examination results for Year 11 students at her school.

She has collected data on last year's results, and wants to compare the performance of boys and girls at the school.

Jade explores the hypothesis that:

> Year 11 girls do better in their GCSE examinations than boys

To test the hypothesis she decided to concentrate on the core subjects and her findings are shown in the table:

A table to show the performance of Year 11 girls and boys in their GCSE examinations

| | | %A*-C | %A*-G | APS |
|---|---|---|---|---|
| English | Girls | 62 | 93 | 4.9 |
| | Boys | 46 | 89 | 4.3 |
| | All | 54 | 91 | 4.6 |
| Mathematics | Girls | 47 | 91 | 4.2 |
| | Boys | 45 | 89 | 4.2 |
| | All | 46 | 90 | 4.2 |
| Science | Girls | 48 | 91 | 4.4 |
| | Boys | 45 | 88 | 4.3 |
| | All | 47 | 90 | 4.4 |

**Now you try ...**

Using the table:

- What do you notice about the percentages of A*-C grades?
- What do you notice about the percentages of A*-G grades?
- What do you notice about the average point scores?
- Does the information support the hypothesis?

The data has been sorted into different categories and represented as a table.

Jade could use comparative bar charts to represent the data as it will allow her to make comparisons more easily.

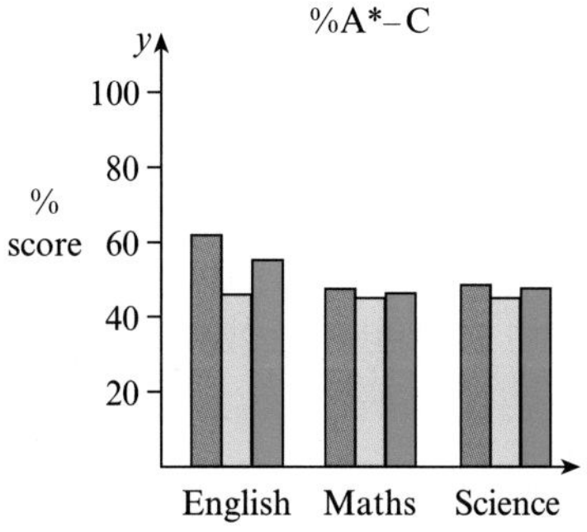


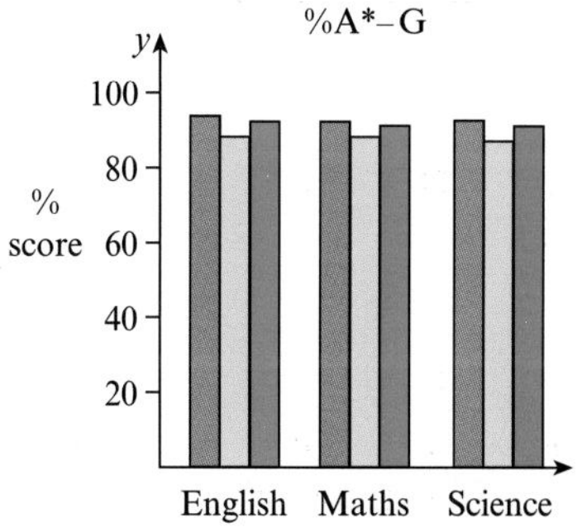


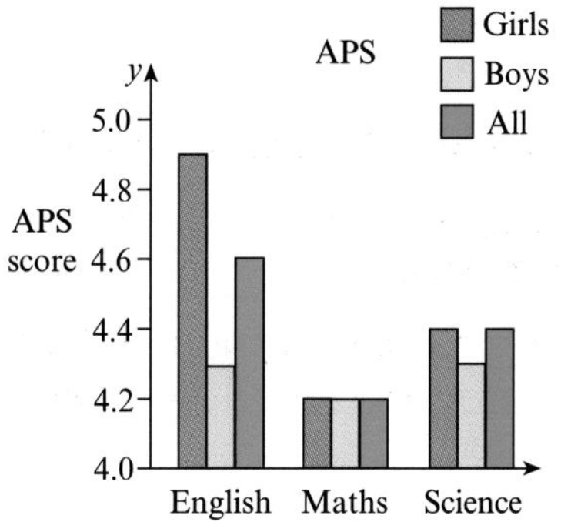

## Summarising and interpreting data

Moderator comment

You should refer to your original hypothesis when you summarise your results

**Hint:**
You need to appreciate that the data is more secure if the sample size is 500 rather than 50.

Jade summarises her findings like this:

*From my graph, I can see that Year 11 girls do better in their GCSE examinations than boys in terms of A*-C grades, A*-G grades and average point scores.*

*The performance of Year 11 girls is significantly better than boys in English although less so in mathematics.*

Moderator comment

In your conclusion you should also suggest limitations to your investigation and explain how these might be overcome.

You may wish to discuss:

- sample size
- sampling methods
- biased data
- other difficulties

## Extending the task

To gain better marks in your coursework you should extend the task in light of your findings.

In your extension you should:

- Give a clear hypothesis
- Collect further data if necessary
- Present your findings using charts and diagrams as appropriate
- Summarise your findings referring to your hypothesis

### Extending task 3: Gender differences in examinations

Jade extends the task by looking at the performance of individual students in combinations of different subjects.

*I am now going to extend my task by looking at the performance of individual students in English and mathematics. My hypothesis is that there will be no correlation between the two subjects.*

She presents the data on a scattergraph.

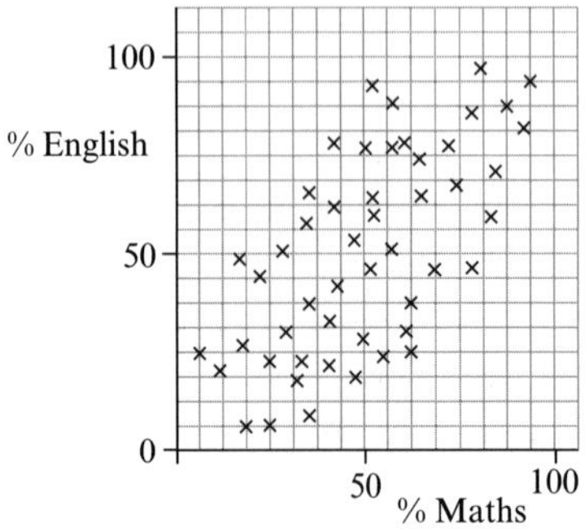


**Note:**
You should only draw a line of best fit on the diagram to show the correlation if you make some proper use of it (for example to calculate a students' likely English mark given their mathematics mark.)

She summarises her findings:

From my scattergraph, I can see that there is some correlation between the performance of individual students in English and mathematics.

She extends the task further:

I shall now look at the performance of individual students in mathematics and science. My hypothesis is that there will be a correlation between the two subjects.

She presents the data on a scattergraph.

**Note:**
The strength of the correlation could be measured using higher level statistical techniques such as Spearman's Rank Correlation.

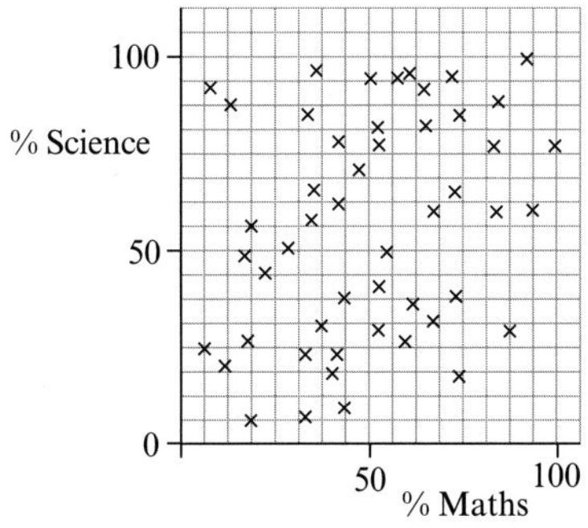


From my scattergraph, I can see that there is no correlation between the performance of individual students in mathematics and science.

## Extending task 2: Guessing game

Maurice extends the 'Guessing game' like this:

I am now going to extend my task by looking to see whether people who are good at estimating lengths are also good at estimating weights. My hypothesis is that there will be a strong correlation between peoples' ability at estimating lengths and estimating weights.

He draws a scattergraph:

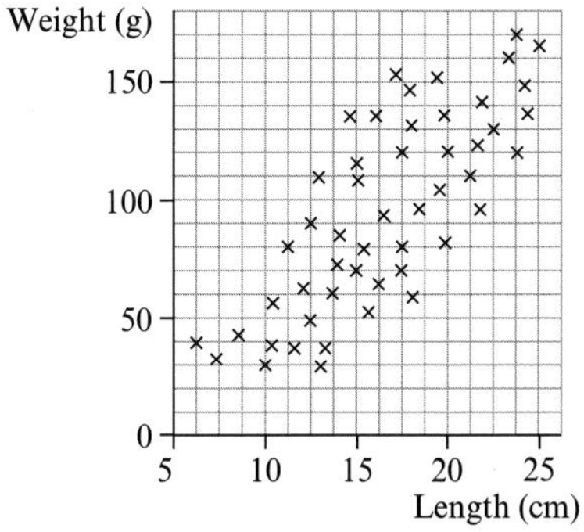


He summarises his findings:

From my scattergraph I can see that there is a strong correlation between peoples' ability at estimating lengths and estimating weights.

**Moderator comment**

The development of the task to include higher level statistical analyses must be 'appropriate and accurate'.

50 people were in the survey so $n = 50$

The strength of the correlation can be measured using higher level statistical techniques such as Spearman's Rank Correlation or Product Moment Correlation.

I am now going to extend my investigation by using Spearman's Rank Correlation to calculate the rank correlation coefficient.

| Person | Length | Weight | Rank Length | Rank Weight | Difference D | $D^2$ |
|---|---|---|---|---|---|---|
| Jane | 15 cm | 102g | 22 | 18 | 4 | 16 |
| Suresh | 16 cm | 108g | 11 | 12 | -1 | 1 |
| | | | | | Total | 2250 |

$$r = 1 - \frac{6(\sum D^2)}{n(n^2 - 1)}$$

$$r = 1 - \frac{6(2250)}{50(50^2 - 1)}$$

$$r = 1 - \frac{13500}{50(2499)}$$

$$r = 1 - \frac{13500}{124950}$$

$r = 1 - 0.108043.....$

$r = 0.891956.....$

$r = 0.89$ (2dp)

The value for the rank correlation coefficient is quite close to 1 so that there is a strong positive correlation between my results.

This tells me that there is a strong correlation between peoples' ability at estimating lengths and estimating weights and confirms my original hypothesis.

## Using a computer

It is quite acceptable that calculations and representations are generated by a computer, as long as any such work is accompanied by some analysis and interpretation.

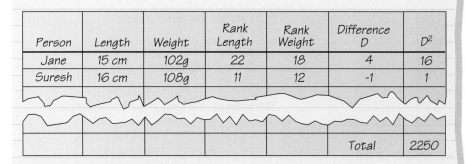

**Remember:**
make sure that your computer generated scattergraph has labelled axes and a title to make it quite clear what it is showing

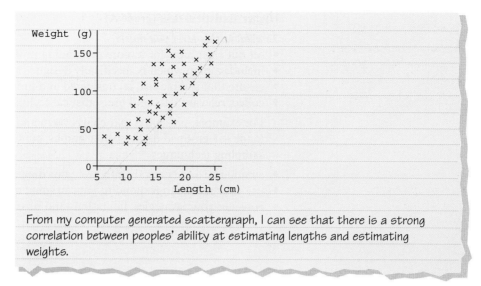

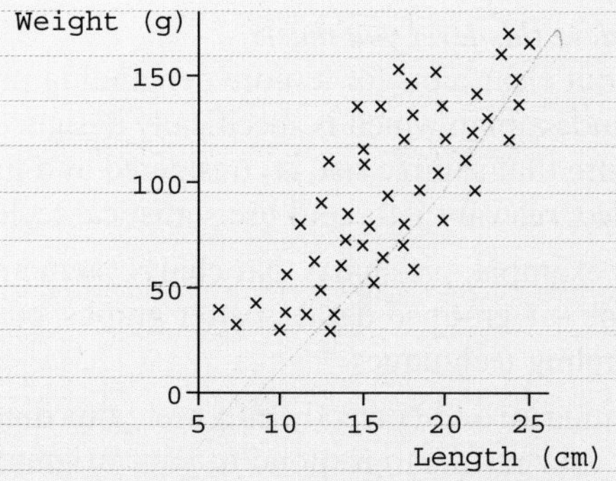


From my computer generated scattergraph, I can see that there is a strong correlation between peoples' ability at estimating lengths and estimating weights.

In this unit we have tried to give you some hints on approaching statistical coursework to gain your best possible mark.

This statistics is often useful in investigative tasks:

- Calculating averages (mean, median and mode)
- Finding the range
- Pie charts, bar charts, stem and leaf diagrams
- Constructing a cumulative frequency graph
- Finding the interquartile range
- Histograms
- Calculating the standard deviation
- Drawing a scatter graph and line of best fit
- Sampling techniques
- Discussing bias

## Summary

These are the grade criteria your coursework will be marked by:

**Foundation statistical task (grade E/F)**

***To achieve this level you must:***
- set out reasonably clear aims and include a plan
- ensure that the sample size is of an appropriate size (about 25)
- collect data and make use of statistical techniques and calculations

  For example: pie charts, bar charts, stem and leaf diagrams, mean, median, mode and scattergraphs
- summarise and interpret some of your diagrams and calculations

**Intermediate statistical task (grade C)**

***To achieve this level you must:***
- set out clear aims and include a plan designed to meet those aims
- ensure that the sample size is of an appropriate size (about 50)
- give reasons for your choice of sample
- collect data and make use of statistical techniques and calculations

  For example: pie charts, bar charts, stem and leaf diagrams, mean, median, mode (of grouped data), scatter graphs and cumulative frequency
- summarise and correctly interpret your diagrams and calculations
- consider your strategies and how successful they were

### Higher statistical task (grade A)

***To achieve this level you must:***

- set out clear aims for a more demanding problem
- include a plan which is specifically designed to meet those aims
- ensure that sample size is considered and limitations discussed
- collect relevant data and use statistical techniques and calculations

  For example: pie charts, bar charts, stem and leaf diagrams, mean, median, mode (of grouped data), scatter graphs, cumulative frequency, histograms and sampling techniques

- summarise and correctly interpret your diagrams and calculations
- use your results to respond to your original question
- consider your strategies, limitations and possible improvements

## Formula sheet

In the GCSE examination you will be given a formula sheet like this one.
You should use it as an aid to memory, and it will be useful to become familiar with the information on the sheet.
The formula sheet is the same for all Examining Groups.

**Volume of a prism** = area of cross section × length

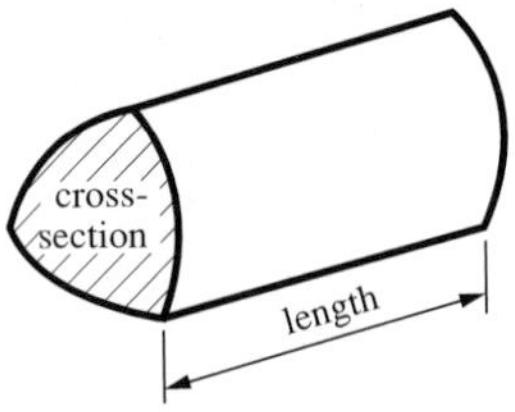


**Volume of sphere** = $\frac{4}{3}\pi r^3$

**Surface area of sphere** = $4\pi r^2$

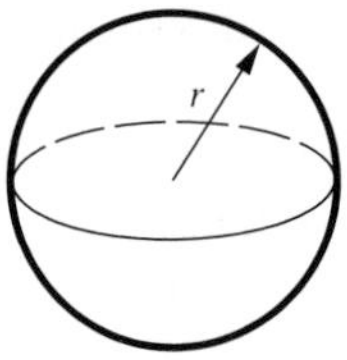


**Volume of cone** = $\frac{1}{3}\pi r^2 h$

**Curved surface area of cone** = $\pi r l$

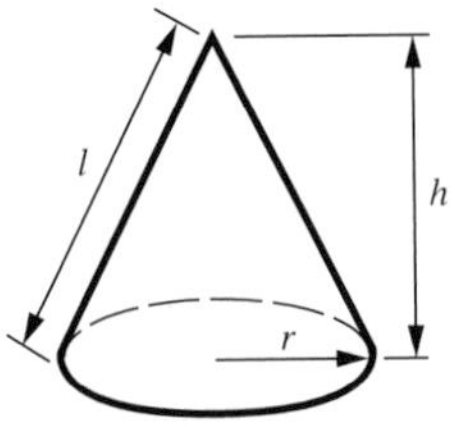


**In any triangle** $ABC$

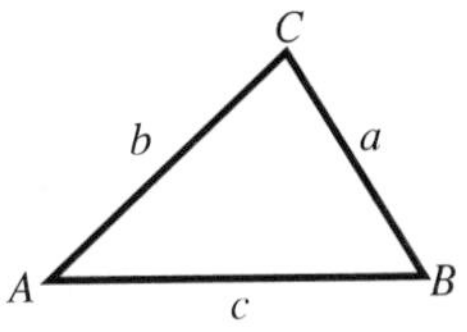


**Sine Rule** $\dfrac{a}{\sin A} = \dfrac{b}{\sin B} = \dfrac{c}{\sin C}$

**Cosine Rule** $a^2 = b^2 + c^2 - 2bc \cos A$

**Area of triangle** = $\frac{1}{2}ab \sin C$

**The Quadratic Equation**

The solutions of $ax^2 + bx + c = 0$ where $a \neq 0$, are given by

$$x = \frac{-b \pm \sqrt{(b^2 - 4ac)}}{2a}$$

# Number

## N1
Units, conversions and compound measures

**N1.1** Kay is going to make a dress.
She has 3 metres of material.
The instructions say that she needs 3.5 yards of material to make the dress.

Has Kay got enough material?
Explain your answer.

| You may wish to use 1 yard = 3 feet or 1 yard = 36 inches |
|---|

(3 marks) (NEAB, 1998)

**N1.2** Jane cycles from *A* to *B* and then from *B* to *C*.

Details of each stage of her journey are given below.

| | |
|---|---|
| *A* to *B* | Distance 55 km.<br>Average speed 22 km per hour. |
| *B* to *C* | Time taken 1 hour 30 minutes.<br>Average speed 30 km per hour. |

Calculate Jane's average speed over the whole of her journey from *A* to *C*.

(4 marks) (SEG, 1999)

**N1.3** A tiger runs at a speed of 50 kilometres per hour for 9 seconds.

How many metres does the tiger run?

(4 marks) (SEG, 2000)

**N1.4** Paul rides his bike to school, a distance of 3 km.

On Monday he left home at 0820 and cycled to school at a steady speed of 15 km/hour.
At what time did he arrive at school?

(3 marks) (SEG, 2000)

**N1.5** **a** Angus drives from Southampton to Durham, a distance of 310 miles.
His average speed for the journey is 64 mph.

How long does his journey take?

Give your answer in hours and minutes.

**b** On a mountain walk it takes approximately:

| 1 hour for every 5 kilometres travelled horizontally,<br>*plus*<br>1 hour for every 600 metres climbed vertically. |
|---|

Ben goes on a mountain walk.
He estimates that he will travel 14.5 km horizontally and climb 750 m vertically.
He plans to stop once for 30 minutes.

Estimate how long he will take in hours.
Give your answer to an appropriate degree of accuracy.

(6 marks) (SEG, 2000)

## N2 Estimation, accuracy and bounds

**N2.1** Work out the following.

**a** $20\ 000 \times 600$

**b** $\dfrac{40\,000}{800}$

**c** $\dfrac{20-90}{60^2}$

(4 marks) (NEAB, 2000)

**N2.2** A newspaper reported 58 000 people attended an International Football match.

This figure is correct to the nearest 1000.

**a** What are the largest and smallest possible attendances at the match?

Another newspaper gave the attendance as 57 500.
This figure is correct to the nearest 500.

**b** Give a possible attendance figure that agrees with **both** newspapers.

(3 marks) (SEG, 1998)

**N2.3** Use a calculator to work out the following.
In each case write down the full calculator display.

**a** $\dfrac{10}{\tan 50°}$

**b** $\sqrt{5-\pi}$.

(2 marks) (NEAB, 1998)

**N2.4** Jonathan uses his calculator to work out the value of $42.2 \times 0.027$

The answer he gets is 11.394

Use approximation to show that his answer is wrong.

(2 marks) (NEAB, 2000)

**N2.5** A survey counted the number of visitors to a website on the Internet.
Altogether it was visited 30 million times.
Each day it was visited 600 000 times.

Based on this information, for how many days did the survey last?

(2 marks) (NEAB, 2000)

**N2.6** In the United Kingdom there are 28.7 million males of which 4.13 million are over 65.

Calculate the percentage of males in the United Kingdom who are over 65.

Give your answer to an appropriate degree of accuracy.

(3 marks) (AQA)

**N2.7** Do **not** use a calculator in this question.

Use approximations to estimate the value of $\dfrac{82.3}{0.042 \times 4.8}$.

You **must** show all your working.

(3 marks) (SEG, 1999)

**N3**
**Types of number**

**N3.1** Find the values of

**a** $\sqrt{(25)} + \sqrt{(144)}$

**b** $\sqrt{(25 \times 144)}$

(3 marks) (NEAB, 2000)

**N3.2** **a** Work out the highest common factor of 48 and 72.

**b** Work out the least common multiple of 48 and 72.

(4 marks) (AQA)

**N3.3** **a** Find the value of $x$ if $2x^3 = 54$.

**b** Write 216 as a product of its prime factors.

(3 marks) (AQA)

**N3.4** Write the number $0.\dot{3}2\dot{5}$ as a fraction.

(2 marks) (SEG, 2000)

**N3.5** **a** Write $\sqrt{6} \times \sqrt{3}$ in the form $a\sqrt{b}$ where $a$ and $b$ are prime numbers.

**b** Simplify fully $5\sqrt{2} + \sqrt{8}$.

(4 marks) (SEG, 2000)

**N3.6** **a** Show clearly that $(\sqrt{3} + \sqrt{12})^2 = 27$.

**b** Given that $x = \sqrt{6}$, $y = \sqrt{18}$ and $z = \sqrt{27}$, evaluate the following.

**i** $y \div x$

**ii** $xyz$

(7 marks) (NEAB, 2000)

**N3.7** **a** **i** Expand and simplify $(2 - \sqrt{3})^2$.

**ii** Hence, or otherwise, show that $x = (2 - \sqrt{3})$ is a solution to the equation

$$x^2 - 4x + 1 = 0$$

**b** There are two solutions to the equation $x^2 - 4x + 1 = 0$.
The sum of the two solutions is 4.
What is the second solution to the equation?

(5 marks) (NEAB, 2000)

**N3.8** What is the area of a square of side $(1 + \sqrt{2})$ units?
Give your answer in the form $a + b\sqrt{2}$.

(2 marks) (AQA)

**N3.9** **a** Simplify $(3 + \sqrt{2})^2$.

**b** A curve has the equation $y = x^2 - 6x + 7$.

**i** Find the value of $y$ when $x = (3 + \sqrt{2})$.

**ii** Explain what your answer tells you about the graph of $y = x^2 - 6x + 7$.

(5 marks) (NEAB, 1998)

## N4
## Fractions, decimals and percentages

**N4.1** **a** What is the reciprocal of $\frac{3}{2}$?

**b** How many $1\frac{1}{2}$ m lengths of rope can be cut from a length of rope measuring 25 m?

(3 marks) (NEAB, 2000)

**N4.2** Evaluate

**a** $3 - 1\frac{1}{5}$

**b** $4 \times 1\frac{1}{3}$

**c** $7\frac{1}{2} \div 1\frac{1}{2}$

(5 marks) (NEAB, 2000)

**N4.3** Find the exact value of

$\frac{1}{a} + \frac{1}{b}$ when $a = \frac{1}{2}$ and $b = \frac{1}{3}$

(3 marks) (NEAB, 1999)

**N4.4** **a** A car cost £14 000 when it was new.
Now it is worth £9100.
Express its value now as a fraction of its value when it was new.
Give your answer in its simplest form.

**b** The value of another car has dropped by 20%.
Its value is now £8200.
What was its original value?

(4 marks) (NEAB, 1998)

**N4.5** A garden centre buys plants and resells them at a profit of 28%.
How much was the original price of a rose bush which is sold for £4.80?

(3 marks) (NEAB, 2000)

**N4.6** In a sale a dress costs £32.40.
The original price has been reduced by 10%.
What was the original price?

(3 marks) (NEAB, 1998)

**N4.7** A computer is advertised at £1116.25 including $17\frac{1}{2}$% VAT.
How much is the computer before VAT is added?

(3 marks) (NEAB, 2000)

**N4.8** Sanjay gets a 4.5% wage increase.
His mew wage is £8.36 per hour.
How much did he earn per hour before his wage increase?

(3 marks) (SEG, 1999)

**N4.9** On average, 5 litres of paint covers 22.5 m$^2$.
A decorator buys 96 litres of paint to paint an area of 405 m$^2$.
What percentage of the paint is used?

(4 marks) (SEG, 1999)

**N4.10** In the United Kingdom there are 26 million vehicles.

Approximately 5.8 million vehicles are 10 years old or more.

What percentage of vehicles are less than 10 years old?

Give your answer to an appropriate degree of accuracy.

(3 marks) (SEG, 2000)

**N4.11** **a** Martin weighs 58 kg.
A year ago Martin weighed 51 kg.

Calculate the percentage increase in his weight.
Give your answer to an appropriate degree of accuracy.

**b** The length of Martin's hand is 17 cm correct to the nearest centimetre.

What is the minimum length his hand could be?

(4 marks) (AQA)

**N4.12** In 1982 it was estimated that there were only 20 000 Minke whales left in the world.
The hunting of Minke whales was banned in 1982.
After 1982 the population increased by 45% each year.

**a** How many Minke whales were there in 1983 (1 year after the ban)?

**b** How many Minke whales were there in 1985 (3 years after the ban)?
Give your answer to a suitable degree of accuracy.

**c** It was agreed that when the Minke whale population reached 250 000 some hunting of Minke whales would be allowed again.
In what year did this happen?

(8 marks) (NEAB, 1998)

**N4.13** Phil has 80 birds; some are blue, the rest are yellow.

Phil sells 30% of his birds.
The new ratio of blue birds to yellow birds is 4 : 3.

How many blue birds has he got left?

(4 marks) (AQA)

**N4.14** Jack invests £2000 at 7% per annum compound interest.

Calculate the value of his investment at the end of 2 years.

(3 marks) (SEG, 2000)

**N4.15** The length of a rectangle is increased by 10%.

The width of the rectangle is decreased by 30%.

Calculate the percentage change in the area of the rectangle.

(3 marks) (SEG, 1998)

**N4.16** When a ball is dropped onto the floor, it bounces and then rises.

This is shown in the diagram.

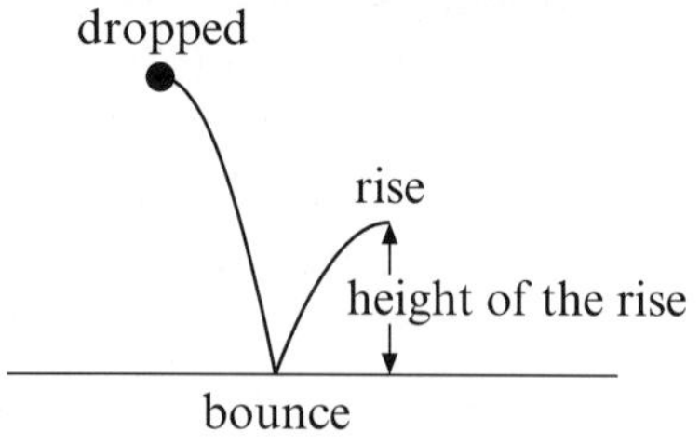


The ball rises to 80% of the height from which it was dropped.

It was dropped from a height of 3 metres.

**a** Calculate the height of the rise after the first bounce.

The ball bounces a second time.

It rises 80% of the height of the first rise.

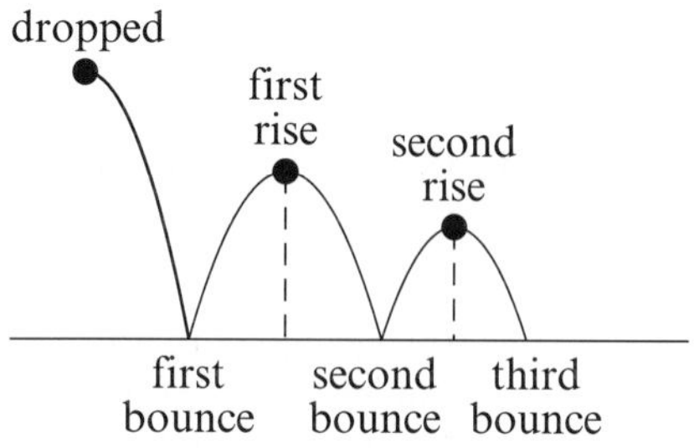


**b** Calculate the height of the second rise.

**c** The ball carries on bouncing in this way.

Each time it rises to 80% of the last rise.

For how many bounces does it rise to a height greater than 1 metre?

(4 marks) (NEAB, 1998)

**N4.17** Express $0.\dot{4}\dot{8}$ as a fraction in its simplest terms.

(3 marks) (NEAB, 2000)

**N4.18** Write the number $0.\dot{2}\dot{1}$ as a fraction in its simplest form.

(3 marks) (SEG, 1999)

## N5 Indices and standard form

**N5.1** In the box are six numbers written in standard form.

| | | |
|---|---|---|
| $8.3 \times 10^4$ | $3.9 \times 10^5$ | $6.7 \times 10^{-3}$ |
| $9.245 \times 10^{-1}$ | $8.36 \times 10^3$ | $4.15 \times 10^{-2}$ |

**a** **i** Write down the largest number.
**ii** Write your answer as an ordinary number.
**b** **i** Write down the smallest number.
**ii** Write your answer as an ordinary number.
**c** Evaluate $(8.3 \times 10^4) \div (4.15 \times 10^{-2})$.

(5 marks) (NEAB, 2000)

**N5.2** **a** Work out

$$4 \times 10^7 \times 6 \times 10^{-5}$$

giving your answer in standard form.

**b** Work out

$$\frac{3.2 \times 10^8}{4 \times 10^4}$$

giving your answer in standard form.

**c** Work out

$$(4 \times 10^5)^2$$

giving your answer in standard form.

(6 marks) (NEAB, 2000)

**N5.3** **a** Evaluate
**i** $2^{-2}$
**ii** $8^0$
**iii** $10^1$
**b** Simplify

$$\frac{5a^2b^2 \times 4a^2b^3c}{2a^2b}$$

(7 marks) (NEAB, 2000)

**N5.4** Approximate figures for the amount of carbon dioxide entering the atmosphere from artificial sources are shown below.

| | |
|---|---|
| Total amount (world wide) | $7.4 \times 10^9$ tonnes |
| Amount from the United Kingdom | $1.59 \times 10^8$ tonnes |

**a** What percentage of the total amount of carbon dioxide entering the atmosphere comes from the United Kingdom?

**b** Approximately 19% of the amount of carbon dioxide from the United Kingdom comes from road transport.
How many million tonnes of carbon dioxide is this?

(6 marks) (SEG, 1999)

**N5.5** Last year the population of the United Kingdom was approximately $5.3 \times 10^7$.

**a** An average of £680 per person was spent on food last year in the United Kingdom.

What was the total amount spent on food last year in the United Kingdom?

Give your answer in standard form.

**b** Last year there were $1.4 \times 10^7$ car drivers in the United Kingdom.

They spent a total of £$1.5 \times 10^{10}$ on their cars.

What was the average amount spent by each car driver?
Give your answer to a suitable degree of accuracy.

(6 marks) (NEAB, 1998)

**N5.6** Light travels at $1.862\,84 \times 10^5$ miles per second.

The planet Jupiter is 483.6 million miles from the Sun.

Using suitable approximations, estimate the number of **minutes** light takes to travel from the Sun to Jupiter.

(5 marks) (AQA)

**N5.7** **a** Calculate $\sqrt{\frac{3.9}{(0.6)^3}}$.

**b** The number $p$ written in standard form is $8 \times 10^5$.

The number $q$ written in standard form is $5 \times 10^{-2}$.

**i** Calculate $p \times q$.

Give your answer in standard form.

**ii** Calculate $p \div q$.
Give your answer in standard form.

(6 marks) (SEG, 2000)

**N5.8** **a** Calculate $81^{-\frac{1}{2}}$, giving your answer as a fraction.

**b** Calculate $\frac{2^3}{2^{-2}}$ leaving your answer in the form $2^p$.

**c** Write $\frac{\sqrt{3}}{\sqrt{2}}$ in the form $\frac{\sqrt{a}}{b}$ where $a$ and $b$ are whole numbers.

(5 marks) (AQA)

**N5.9** For each of the following equations, write down the value of $n$.

**a** $2^n = 32$

**b** $n^3 = 125$

**c** $8^n = 8$

(3 marks) (NEAB, 1998)

**N5.10** Find the value of

**a** $4^{-2}$

**b** $36^{\frac{1}{2}}$

**c** $27^{\frac{2}{3}}$

(3 marks) (NEAB, 2000)

**N5.11** Calculate the value of $2^a \times a^2$, when $a = -2$.

(2 marks) (SEG, 1999)

**N5.12** **a** Write $2^{-2} \times 4$ as a single power of 2.

**b** Write $2^3 \div \frac{1}{8}$ as a single power of 2.

(4 marks) (SEG, 1998)

**N5.13** **a** Use your calculator to find the decimal values of the expressions in the table.

The first one has been done for you.

Copy the table and write down the whole calculator display for each expression.

| Expression | Decimal value |
|---|---|
| $1 + \frac{1}{16}$ | 1.0625 |
| $1 + \frac{1}{16} + \frac{1}{16^2}$ | |
| $1 + \frac{1}{16} + \frac{1}{16^2} + \frac{1}{16^3}$ | |
| $1 + \frac{1}{16} + \frac{1}{16^2} + \frac{1}{16^3} + \frac{1}{16^4}$ | |

**b** What do you notice about your results?

**c** Evaluate $\dfrac{1}{(1 - \frac{1}{16})}$.

Express your answer as a decimal.

(4 marks) (NEAB, 1999)

**N5.14** **a** Simplify $\dfrac{(x^{-\frac{1}{2}})^3}{x^{-2}}$ leaving your answer as a power of $x$.

**b** Find the value of $n$ such that $25^n = 5^{\frac{1}{4}}$.

(5 marks) (NEAB, 1999)

**N5.15** In the formula $y = ab^x$, $a$ and $b$ are positive rational numbers.

When $x = 0$, $y = 10$ and when $x = -2$, $y = 40$.

Calculate the value of $x$ when $y = 160$.

(5 marks) (SEG, 1999)

## N6 Ratio and proportion

**N6.1** Sam and Anna share £94.50 in the ratio 11 : 4.

How much does each of them receive?

(3 marks) (SEG, 1998)

**N6.2** **a** Phil has 50 birds; some are blue, the rest are yellow.
The ratio of blue birds to yellow birds is 3 : 7.

How many yellow birds are there?

**b** Phil sells some of his birds and buys some others.
The new ratio of blue birds to yellow birds is 5 : 2.
There are 16 yellow birds.

How many blue birds are there?

(5 marks) (NEAB, 1998)

**N6.3** Sam and Tom share 100 counters in the ratio 2 : 3.
They use the counters to keep score in a game.
At the end of the game Sam and Tom have counters in the ratio 7 : 13.

Calculate the change in the number of Sam's counters.

(3 marks) (SEG, 1999)

**N6.4** Box $A$ contains black and white counters in the ratio 3 : 5.

**a** What percentage of the counters are black?

Box $B$ contains black counters and white counters in the ratio 7 : 5.
Box $A$ and Box $B$ each contain the same number of counters.

**b** What is the smallest possible number of counters in each box?

(4 marks) (SEG, 2000)

**N6.5** $y$ is inversely proportional to the square root of $x$.
When $y = 6$ then $x = 4$.

**a** What is the value of $y$ when $x = 9$?

**b** What is the value of $x$ when $y = 10$?

(5 marks) (NEAB, 1999)

**N6.6** $y$ is directly proportional to the square root of $x$.
Copy and complete the table.

| | | | |
|---|---|---|---|
| $x$ | 100 | 25 | |
| $y$ | 3 | | 6 |

(4 marks) (NEAB, 1999)

**N6.7** $C$ is inversely proportional to $t^2$.
When $C = 16$, $t = 3$.
Find $t$ when $C = 9$.

(3 marks) (NEAB, 2000)

# Algebra

## A1 Formulae

**A1.1** This formula can be used to convert temperatures from degrees Celsius, °C, to degrees Fahrenheit, °F.

$$F = \frac{9C + 160}{5}$$

**a** Use the formula to convert −7 °C to degrees Fahrenheit.

**b** Use the formula to find at what temperature $F = C$.

**c** Rearrange the formula to give $C$ in terms of $F$.

(8 marks) (SEG, 1998)

**A1.2** The diagram shows the dimensions of a rectangle.

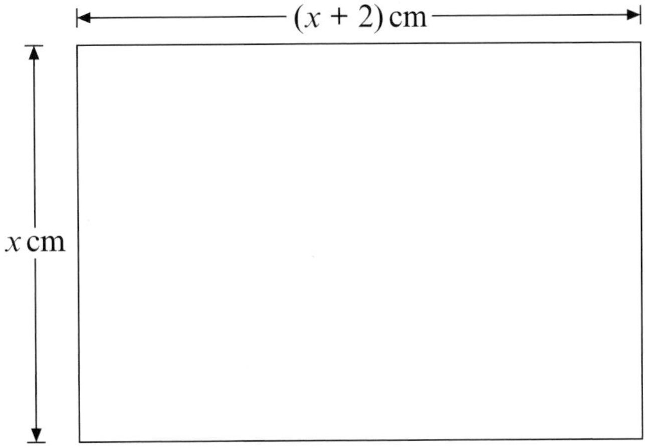


The perimeter of the rectangle is 18 cm.

Find the value of $x$.

(3 marks) (NEAB, 1998)

**A1.3** $s$ is given by the formula $s = ut + \frac{1}{2}at^2$.

Find the value of $s$ when $u = 2.8$, $t = 2$ and $a = -1.7$

(2 marks) (NEAB, 2000)

**A1.4** Sam earns £4 per hour and James earns £5 per hour.

Sam works for $x$ hours per week.

James works for 3 hours per week more than Sam.

**a** Write down an expression for the total amount earned by Sam and James.

Mary earns £6 per hour and works for 5 hours per week more than Sam.

Mary earns as much as James and Sam do in total.

**b** Write down and solve an equation to find $x$.

(6 marks) (SEG, 2000)

## A2 Solving linear and simultaneous linear equations

**A2.1** Solve the equations.

**a** $7x - 13 = 5(x - 3)$

**b** $\frac{10 - x}{4} = x - 1$

(? marks) (NEAB, 2000)

**A2.2** **a** Solve the equation $7(x - 1) = 3x + 2$.

**b** Multiply out and simplify $(x + 3)(2x - 1)$.

(5 marks) (SEG, 1999)

**A2.3** **a** Solve the equation $7x - 3 = 3x - 2$.

**b** Multiply out and simplify $(3x + 1)(x - 2)$.

**c** Solve the equation $x^2 - 8x = 0$.

(6 marks) (SEG, 1998)

**A2.4** Solve the simultaneous equations

$$2x + y = 2$$
$$4x - 3y = 9$$

(? marks) (NEAB, 2000)

**A2.5** This graph shows the line $y - 3x = -2$.

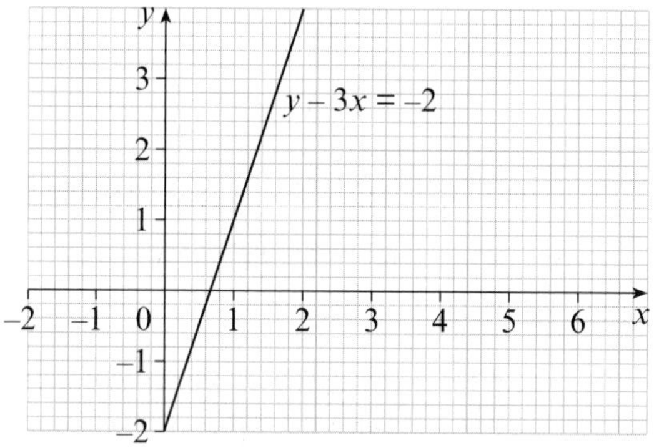


By drawing another line, use the graph to solve the simultaneous equations:

$$y - 3x = -2$$
$$\text{and} \quad 3y + 2x = 9.$$

(4 marks) (SEG, 1998)

**A2.6** Two adults and three children pay £10.75 to go to the cinema.
One adult and two children pay £6.00 to go to the same cinema.
Let an adult pay £$x$ and a child pay £$y$.

**a** Write down two equations connecting $x$ and $y$.

**b** Solve these simultaneous equations to find $x$ and $y$.

(5 marks) (SEG, 1998)

**A2.7** The diagram shows a sketch of the line $2y = x - 2$.

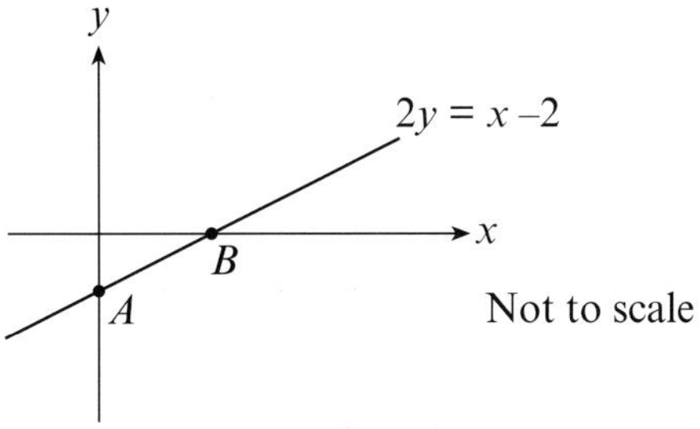


**a** Find the coordinates of the points $A$ and $B$.

**b** Find the gradient of the line $2y = x - 2$.

**c** Explain why the simultaneous equations $2y = x - 2$ and $2y = x - 3$ have no solution.

(4 marks) (SEG, 1998)

**A2.8** **a** Factorise $2x^2 + 4x$.

**b** Multiply out and simplify $(x + 1)(x - 7)$.

**c** Solve the quadratic equation $x^2 + 3x - 10 = 0$.

**d** Solve the simultaneous equations

$$4x - y = 9$$
$$x + 2y = 0$$

(10 marks) (SEG, 2000)

**A2.9** Solve the following equations.

**a** $\dfrac{x-3}{2} - \dfrac{4x+1}{3} = 0$

**b** $3x^2 - 13x - 10 = 0$

(7 marks) (SEG, 1999)

**A2.10** **a** Solve the equation $6x + 3 = 2(x + 1)$.

**b** Factorise fully $10y - 15y^2$.

**c** The diagram shows a sketch of the line $y = ax + b$.

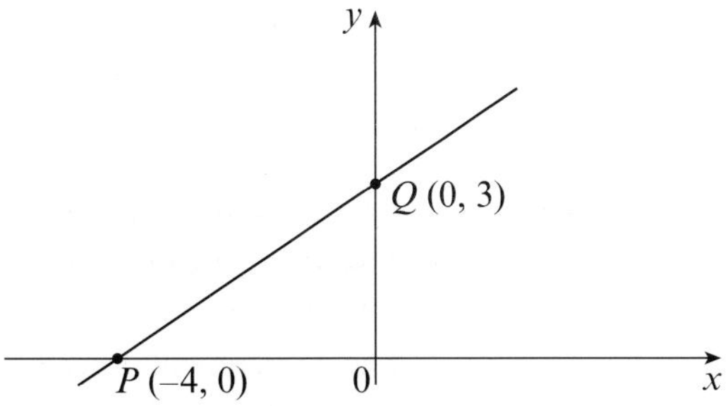


The line $y = ax + b$ goes through the points $P(-4, 0)$ and $Q(0, 3)$. Work out the values of $a$ and $b$.

(8 marks) (SEG, 2000)

## A3
## Quadratic and polynomial equations

**A3.1** A solution of the equation $x^3 + x^2 = 4$ lies between $x = 1$ and 2.
Use the method of trial and improvement to find this solution.
Give your answer to one decimal place.
You **must** show all your trials.
(3 marks) (SEG, 1999)

**A3.2** Katy is using trial and improvement to find an answer to the equation

$$x^3 - x = 35$$

This table shows her first two tries.

| $x$ | $x^3 - x$ | Comment |
|---|---|---|
| 3 | 24 | Too low |
| 4 | 60 | Too high |

Copy and continue the table to find a solution to the equation.
Give your answer correct to 1 decimal place.
(4 marks) (NEAB, 1998)

**A3.3** **a** Use trial and improvement to solve the equation $x^3 + x^2 = 500$.
One trial has been completed for you.
Show all your trials on a copy of the table.
Give your answer correct to 1 decimal place.

| $x$ | $x^3 + x^2$ | |
|---|---|---|
| 7 | $7^3 + 7^2 = 392$ | Too small |

Copy and complete this sentence:
The answer is $x =$ .......... (correct to 1 decimal place).

**b** A company makes cartons for fruit juice.

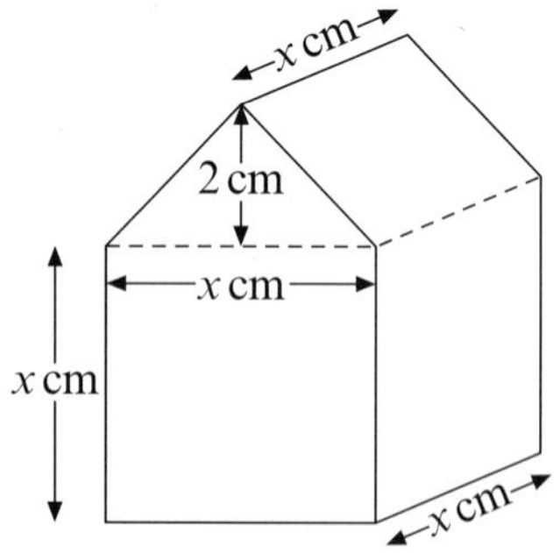


Each carton is a pentagonal prism as shown.
The cross-section of the prism is formed from a square and an isosceles triangle.
The perpendicular height of the triangle is 2 cm and the base of the triangle is $x$ cm.
The prism is $x$ cm long.

**i** Show that the volume of the carton, $V$ cm$^3$, is given by the formula

$$V = x^3 + x^2$$

**ii** The volume of the carton is 500 cm$^3$.
Use your answer to part **a** to find the height of the carton.

(8 marks) (NEAB, 1998)

**A3.4** **a** Factorise $6x^2 - x - 12$.

**b** Solve the equation $x^2 - 7x - 6 = 0$ giving your answers to 2 decimal places.

(5 marks) (NEAB, 1998)

**A3.5** Solve the equation $x^2 + 2x - 35 = 0$.

(3 marks) (NEAB, 1998)

**A3.6** Solve the equation $2x^2 + 7x - 1 = 0$.

(3 marks) (SEG, 1999)

**A3.7** Factorise

**a** $2x^2 - 7x - 15$

**b** $x^2 - 25y^2$

(4 marks) (NEAB, 2000)

**A3.8** The diagram below represents a flag which measures 80 cm by 60 cm.

The flag has two axes of symmetry.

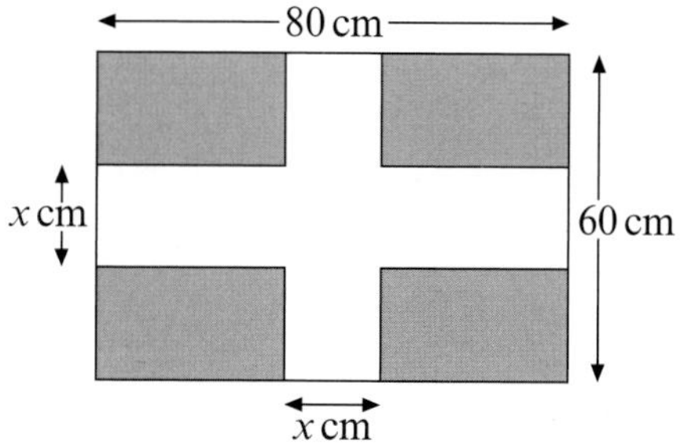


Not drawn accurately

The flag has two colours, grey and white.

The bands of white are of equal width, $x$ cm.

The white area forms half the area of the flag.

**a** Show that $x$ satisfies the equation $x^2 - 140x + 2400 = 0$.

**b** Solve the equation $x^2 - 140x + 2400 = 0$ to find the width of the white bands.

(6 marks) (NEAB, 1998)

**A3.9** Find, correct to 2 decimal places, the solutions of the equation $x^2 - 2x - 1 = 0$.

Working must be shown. Do not use a trial and improvement method.

(3 marks) (NEAB, 1998)

**A3.10** If $\dfrac{1}{x-1} - \dfrac{1}{x+5} = 2$ show that $x^2 + 4x - 8 = 0$.

(4 marks) (SEG, 1999)

**A3.11** Solve the equation $\dfrac{4}{2x-1} - \dfrac{1}{x+1} = 1$.

(6 marks) (SEG, 1999)

**A3.12** **a** The expression $x^2 - 10x + a$ can be written in the form $(x + b)^2$. find the values of $a$ and $b$.

**b** Solve the equation $x^2 - 10x + 20 = 0$.
Give your answers to two decimal places.

**c** State the minimum value of $y$ if $y = x^2 - 10x + 20$.

(7 marks) (AQA)

**A3.13** **a** Write the expression $x^2 - 10x + 7$ in the form $(x - a)^2 + b$, where $a$ and $b$ are integers.

**b** Solve the equation $x^2 - 10x + 7 = 0$.

(5 marks) (SEG, 1998)

**A3.14** The expression $4x^2 - 12x - 2$ can be written in the form $(ax + b)^2 - 11$.
Find the values of $a$ and $b$.

(2 marks) (SEG, 2000)

**A3.15** **a** Simplify the expression

$$\frac{2x^2 - 5x + 2}{x^2 - 4}$$

**b** The expression $x^2 - 8x + 17$ can be written in the form $(x - p)^2 + q$.
Calculate the values of $p$ and $q$ such that

$$x^2 - 8x + 17 = (x - p)^2 + q.$$

(5 marks) (SEG, 1998)

**A3.16** Solve the equation $\frac{1}{x} + \frac{3x}{x - 1} = 3$.

(5 marks) (AQA)

**A3.17** The diagram shows the circle $x^2 + y^2 = 25$ and the line $y = x + 7$.
The line and the circle intersect at the points $A$ and $B$.

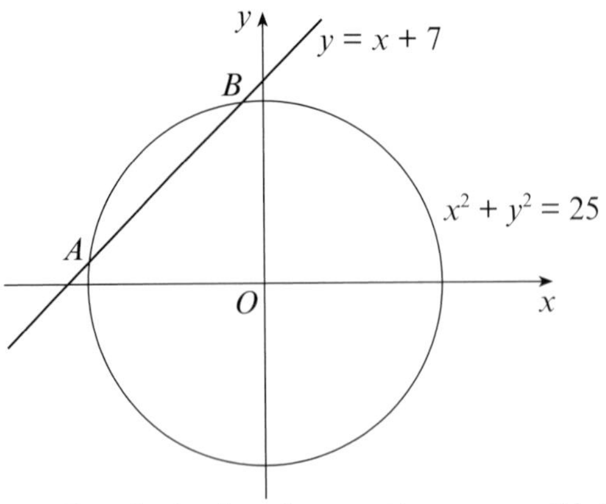


Not drawn accurately

**a** By substituting the equation $y = x + 7$ into equation $x^2 + y^2 = 25$, show that

$$x^2 + 7x + 12 = 0$$

**b** Hence find the coordinates of $A$ and $B$.

(6 marks) (AQA)

## A4
## Linear, quadratic and polynomial graphs

**A4.1** The graph shows the line $AB$.

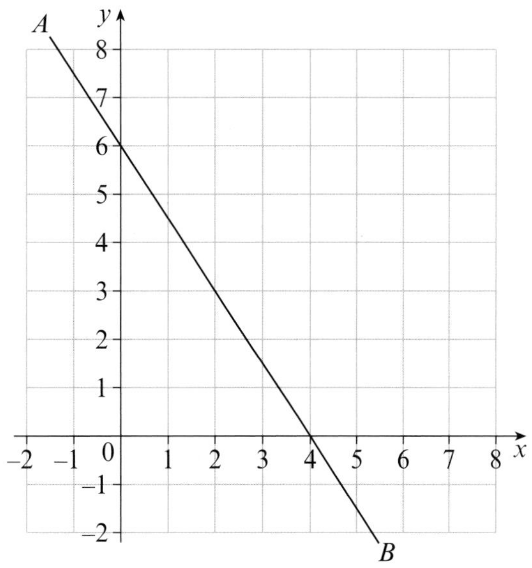


Work out the equation of the line $AB$.
(4 marks) (NEAB, 2000)

**A4.2** **a** Copy and complete the table for the equation $y = x^2 - 2x + 2$.

| $x$ | $-1$ | 0 | 1 | 2 | 3 |
|---|---|---|---|---|---|
| $y$ | | | | | |

**b** Draw the graph of $y = x^2 - 2x + 2$ for $-1 \leqslant x \leqslant 3$.

**c** Use your graph to find the values of $x$ when $y = 3$.
(4 marks) (NEAB, 2000)

**A4.3** The graph of $y = x^2 - 7$ is drawn on the grid.

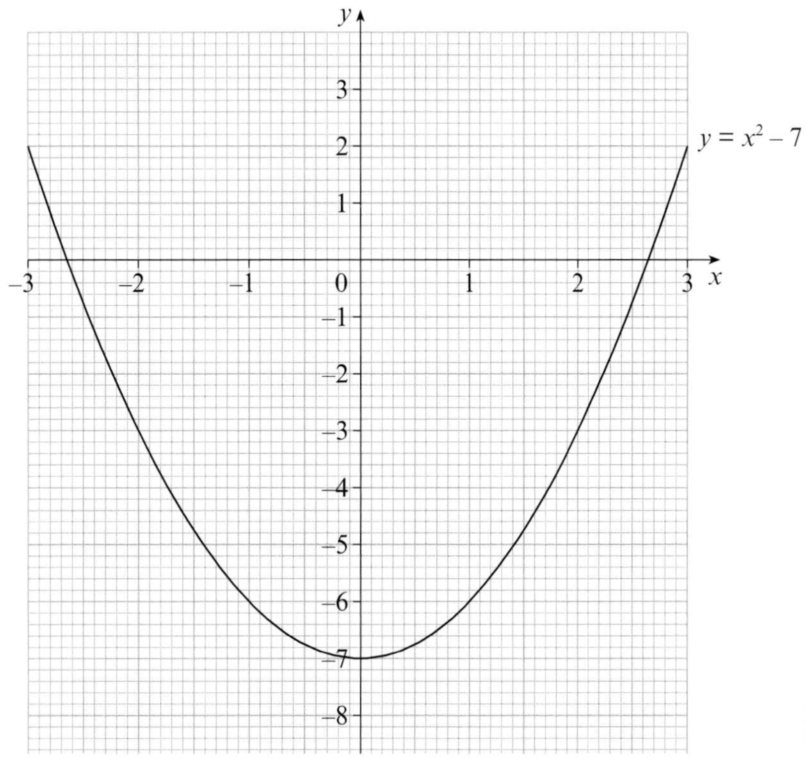

**a** Use the graph to solve the equation $x^2 - 7 = 0$.

**b** State the minimum value of $y$.

(3 marks) (NEAB, 2000)

**A4.4** The graph of $y = x^2 + 2x - 7$ is drawn below for values of $x$ between −5 and +5.

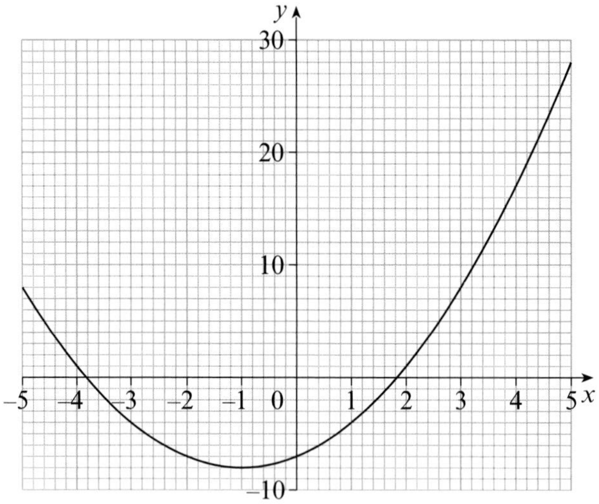


Copy the graph. By drawing an appropriate linear graph, write down the solutions of $x^2 - x - 11 = 0$.

(3 marks) (AQA)

**A4.5** Use a graphical method to solve the simultaneous equations

$$y = x + 7 \qquad \text{and} \qquad y + 3x = 5$$

(4 marks) (NEAB, 2000)

**A4.6** The graph shows the line $4x + 5y = 20$.

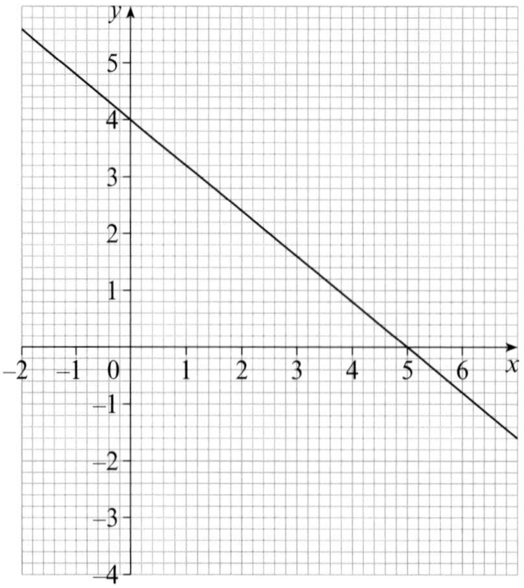


Copy the graph. By drawing another line, use the graph to solve the simultaneous equations

$$4x + 5y = 20$$
$$\text{and } 3x - 2y = 6$$

(3 marks) (AQA)

**A4.7** The total monthly bill for my mobile phone is made up of two parts: the fixed network charge and the cost of calls.
The graph shows the monthly bill, £$B$, for calls up to a total of 60 minutes.

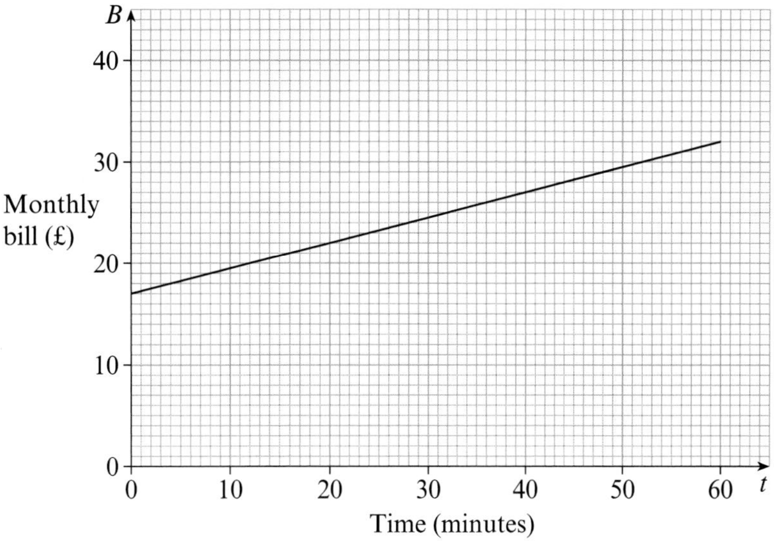


**a** What is the fixed network charge?

**b** Calculate the gradient of the line.

**c** Write down the equation of the line.

(5 marks) (SEG, 1999)

**A4.8** $P$ is the point (1, 8) and $Q$ is the point (11, 3).
$M$ is the mid-point of the line $PQ$.

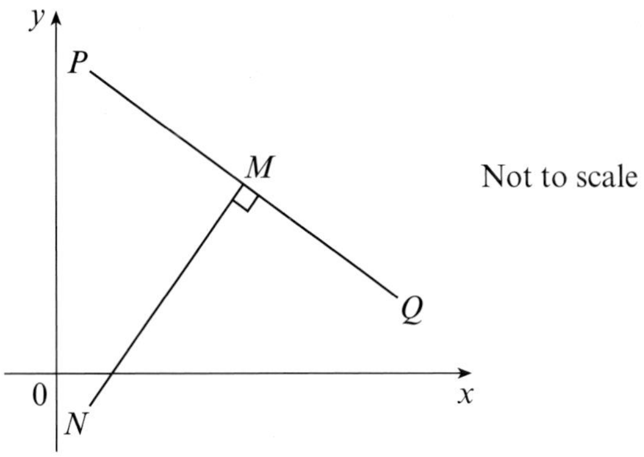


**a** Find the coordinates of the point $M$.

The gradient of the line $PQ$ is $-\frac{1}{2}$.

The line $MN$ is perpendicular to the line $PQ$.

**b** Write down the gradient of the line $MN$.

(3 marks) (AQA)

**A4.9** Gary has produced this set of experimental data in his science lesson.

| $x$ | 1.2 | 2.5 | 3.6 | 4.0 | 5.2 |
|---|---|---|---|---|---|
| $y$ | −3 | 7 | 20 | 26 | 48 |

Gary knows that the two variables are connected by a formula $y = ax^2 + b$ where $a$ and $b$ are constants.

**a** Calculate the values of $x^2$ and plot the graph of $y$ against $x^2$.

**b** Use your graph to estimate the values of $a$ and $b$.

(7 marks) (SEG, 1999)

**A4.10** There are three values of $x$ which are solutions to the equation

$$x^3 - 22x + 24 = 0$$

**a** Show that $x = 4$ is a solution to the equation.

**b** A second solution lies between 1 and 2.
Use trial and improvement to find this second positive solution of $x$.
Copy and complete the table. Give your answer correct to 1 decimal place.

| $x$ | $x^3 - 22x + 24$ |
|---|---|
| 1 | $1^3 - 22 \times 1 + 24 = 3$ |
| 2 | $2^3 - 22 \times 2 + 24 = -12$ |

**c** Part of the graph of $y = x^3 - 22x + 24$ is shown below.

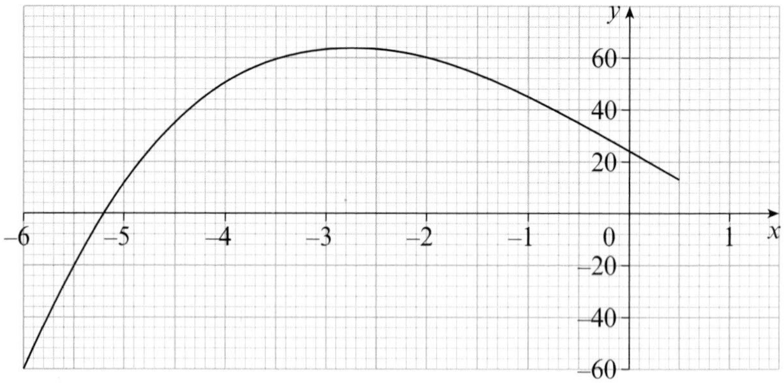


Use the graph to write down the third solution of $x^3 - 22x + 24 = 0$.

(6 marks) (NEAB, 1999)

## A5 Inequalities

**A5.1** Solve the inequality $x + 20 < 12 - 3x$.

(2 marks) (NEAB, 2000)

**A5.2** **a** Solve the inequality $3x - 7 > x - 3$.

**b** Solve the inequality $x^2 > 16$.

(4 marks) (NEAB, 2000)

**A5.3** **a** Write down all the solutions of the inequality $3 < x \leqslant 8$ where $x$ is an integer.

**b** Solve the inequality $4y - 3 \leqslant 2 + y$.

(5 marks) (SEG, 1999)

**A5.4** **a** Solve the inequality $7x + 3 > 17 + 5x$.

**b** Simplify the following.

**i** $2x^3 \times 6x^2$

**ii** $(3y^3)^2$

**c** Multiply out and simplify $(2x - 1)(x - 3)$.

(6 marks) (SEG, 1998)

**A5.5** **a** Factorise $5a - 10$.

**b** Solve the inequality $3x - 5 \leqslant 16$.

**c** On a copy of the grid below, shade the region where $x + y \leqslant 4$.

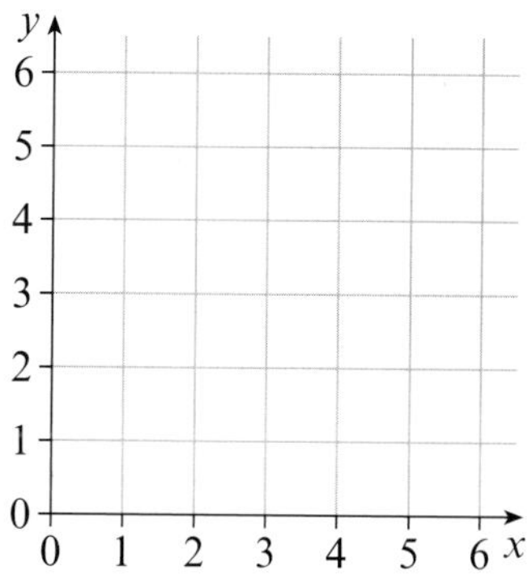


(5 marks) (AQA)

**A5.6** On copies of the grid below shade the region where

**a** $x \geqslant 2$

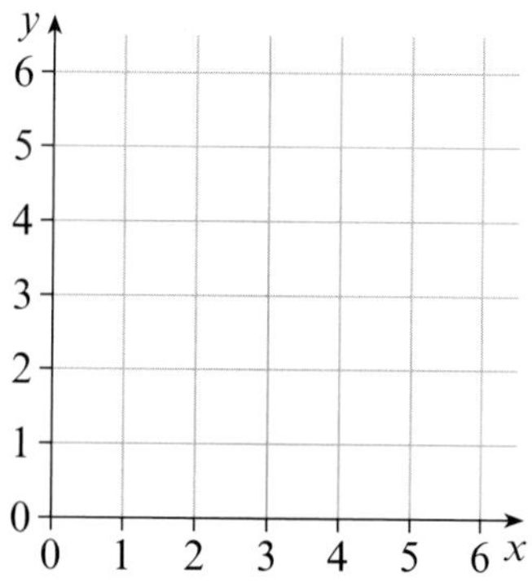


**b** $y \leqslant 4$

**c** $x + y \leqslant 4$

(3 marks) (NEAB, 1999)

**A5.7** The coordinates of the points in the region ***R*** satisfy the inequalities:

$y > (x+1)(x-3) \qquad 2y - x < 2 \qquad 2x + y < 4 \qquad x > 0$

Represent these inequalities on a graph.
Label the region ***R***.

(4 marks) (SEG, 2000)

**A5.8** Solve the inequality $(x+3)^2 < x^2 + 2x + 7$.
Do not use a trial and improvement method.

(3 marks) (NEAB, 1998)

**A5.9** **a** Solve the inequality $3x + 4 \leqslant 7$.

**b** **ii** Solve the inequality $3x + 11 \geqslant 4 - 2x$.

**ii** If $x$ is an integer what is the smallest possible value of $x$?

**c**

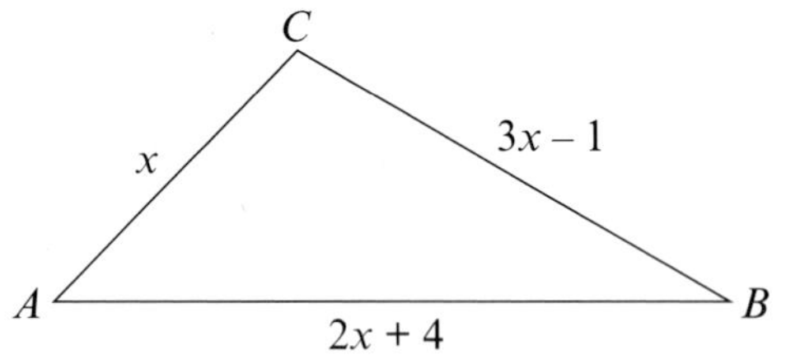


A triangle $ABC$ has sides of $x$, $3x - 1$ and $2x + 4$.

A mathematics books says:

*'The sum of the lengths of any two sides of a triangle must be greater than the length of the third side.'*

Use this fact to find the lower bound of the side $AB$.

(10 marks) (NEAB, 1999)

**A5.10** Josie manages the Segville Minimarket which has 8 checkout tills.
She can staff the tills with part-timers or full-timers.
The number of part-timers is $x$.
The number of full-timers is $y$.

**a** Write down the equation of the line drawn on the grid.

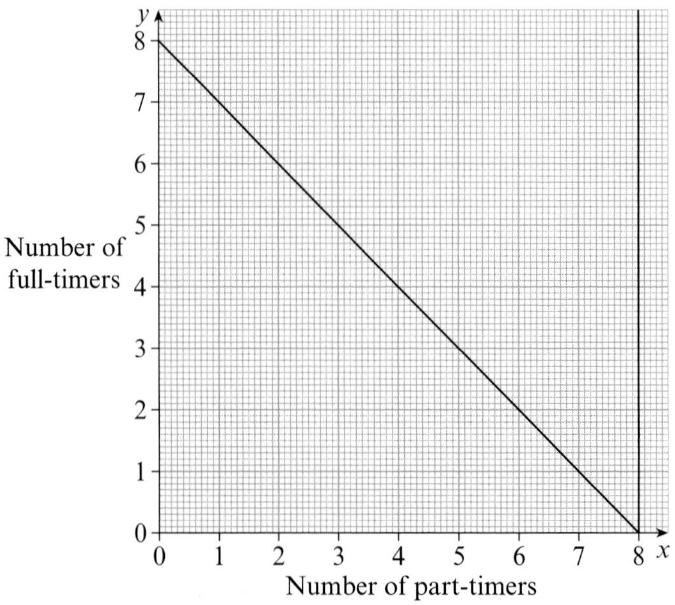

**b** Explain why the points representing possible values of $x$ and $y$ must lie below the line or on the line.

Part-timers check through 30 customers per hour and full-timers check through 24 customers per hour.
Josie has to make sure she can get at least 180 customers through the tills each hour.

**c** Use this information to write down an inequality.

**d** Draw a line on a copy of the grid to show the region which satisfies this inequality.

Josie makes sure there are always more part-timers than full-timers on the tills.

**e** Draw a line on the grid to show this inequality.

**f** Show clearly the region within which the values of $x$ and $y$ must lie. Label this region with the letter $R$.

(9 marks) (SEG, 1999)

**A5.11** Zeenat is buying some computer discs.
She can buy two types of disc.
She buys $x$ SuperDrive discs and $y$ BasicDrive discs.

**a** A SuperDrive disc costs 80p.
A BasicDrive disc costs 48p.

Zeenat does not want to spend more than £12 on computer discs.
Explain why $5x + 3y \leqslant 75$.

**b** A SuperDrive disc holds 1.5 Mb of information.
A BasicDrive disc holds 0.5 Mb of information.

Zeenat needs enough discs to hold at least 15 Mb of information.
Explain why $3x + y \geqslant 30$.

**c** Zeenat must have at least 20 separate discs.

**i** Show clearly why 4 SuperDrive discs and 16 BasicDrive discs do not satisfy all of Zeenat's requirements.

**ii** Find **one** combination of SuperDrive discs and BasicDrive discs that does satisfy all of the requirements.

(5 marks) (NEAB, 1998)

## A6 Manipulation

**A6.1** Expand and simplify

**a** $5(x + 2) + 3(2x - 1)$

**b** $(3x - 2)^2$

(4 marks) (NEAB, 2000)

**A6.2** **a** Expand and simplify

**i** $3(x + 2) + 5(2x - 1)$

**ii** $(x + 4)(x - 2)$

**b** **i** Factorise the expression $x^2 + 8x + 15$.

**ii** Hence solve the equation $x^2 + 8x + 15 = 0$.

(7 marks) (NEAB, 1998)

**A6.3** Simplify

**a** $2a \times 3b \times 4c$

**b** $x^3 \div x^3$

(2 marks) (NEAB, 2000)

**A6.4** **a** Expand $x(3x^2 - 5)$.

**b** Expand and simplify $(2x + 1)(3x - 2)$.

(5 marks) (NEAB, 1998)

**A6.5** Expand and simplify $(2x + 3y)^2$.

(3 marks) (NEAB, 2000)

**A6.6** **a** Factorise completely $3x^2 - 6x$.

**b** Expand and simplify $(3x + 2)(x - 4)$.

**c** Make $t$ the subject of the formula.

$$W = \frac{5t - 3}{4}$$

(8 marks) (NEAB, 1998)

**A6.7** Simplify

**a** $t^3 \times t^5$

**b** $p^6 \div p^2$

**c** $\frac{a^3 \times a^2}{a}$

(3 marks) (NEAB, 1998)

**A6.8** **a** Simplify these expressions.

**i** $\frac{a^2}{a^5}$

**ii** $(2pq^2)^5$

**b** Multiply out and simplify $(y - 7)^2$.

**c** Calculate the values of $p$ and $q$ in the identity

$$x^2 - 6x + 14 = (x - p)^2 + q.$$

(7 marks) (AQA)

**A6.9** **a** Multiply out and simplify

**i** $3a^3 \times 2a^2$

**ii** $2(3x - 1) - 3(x - 5)$

**b** You are given that $v = u + at$.

**i** Make $t$ the subject of this formula.

**ii** Find the value of $t$ when $a = -10$, $u = 12$ and $v = -18$.

(8 marks) (SEG, 1999)

**A6.10** Rearrange this formula to make $x$ the subject.

$$y = \frac{x}{5 + x}$$

(4 marks) (AQA)

**A6.11** **a** A formula is given as $t = 7p - 50$.

Rearrange the formula to make $p$ the subject.

**b** Make $x$ the subject of $y = kx + 3x$.

(4 marks) (NEAB, 1998)

**A6.12** Simplify $\dfrac{x^2 - 3x}{x^2 - 9}$.

(3 marks) (AQA)

**A6.13** Make $h$ the subject of the formula $S = 2\pi r^2 h + 2\pi rh$.

(3 marks) (NEAB, 2000)

**A6.14** Simplify $(2y^2x^4)^3$.

(2 marks) (NEAB, 2000)

**A6.15** The diagram shows a cuboid.

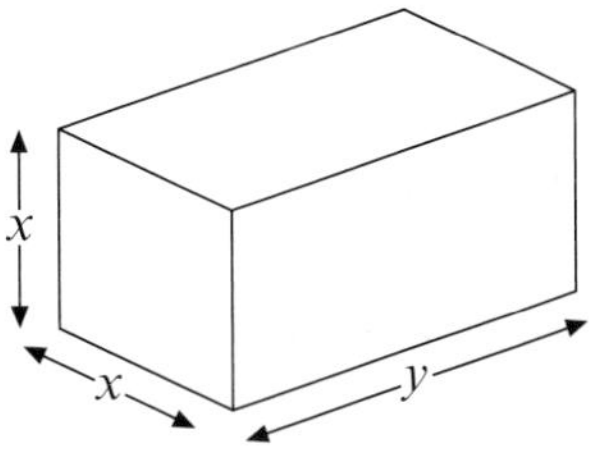


The end of the cuboid is a square of side $x$ centimetres.

The length of the cuboid is $y$ centimetres.

The surface area of the cuboid is 80 cm$^2$.

**a** Show $y = \dfrac{40 - x^2}{2x}$.

**b** Calculate $x$ when $y = 8$.

Hence, calculate the volume of the cuboid.

Give your answer correct to one decimal place.

(8 marks) (AQA)

**A6.16** You are given the formula $P = \dfrac{V^2}{R}$.

**a** Work out the value of $P$ when $V = 3.85$ and $R = \frac{8}{5}$.

**b** Rearrange the formula to give $V$ in terms of $P$ and $R$.

(4 marks) (SEG, 1999)

**A6.17** **a** Expand and simplify $(x + a)(x - a)$.

**b** Simplify $\dfrac{x^2 + 2x - xy - 2y}{x^2 - 4}$.

(5 marks) (NEAB, 2000)

**A6.18** This formula shows the relationship between $x$ and $A$.

$$A = \frac{8}{2x - 1} - \frac{4}{x}$$

**a** Show clearly that this formula can be simplified to

$$A = \frac{4}{2x^2 - x}$$

**b** Find the values of $x$ when $A = 2$.

(7 marks) (SEG, 1998)

**A6.19** Simplify fully the following expression.

$$\frac{6x - 18}{x^2 - 5x + 6}$$

(3 marks) (SEG, 1998)

**A6.20** Rearrange this formula to make $c$ the subject.

$$a = \frac{c}{2 - c}$$

(4 marks) (NEAB, 1998)

**A6.21** **a** Simplify to a single fraction in its lower terms $\dfrac{4a}{5} + \dfrac{a}{15} - \dfrac{3a}{10}$.

**b** Simplify this expression $\dfrac{2x - 5}{6x - 15}$.

**c** Solve the simultaneous equations

$$x^2 + y^2 = 100$$
$$x - y = 2$$

(10 marks) (AQA)

## A7 Number patterns and sequences

**A7.1** Here are the first four lines of a number pattern.

*Line 1* $1^2 + 2 = 1 \times 3$
*Line 2* $2^2 + 4 = 2 \times 4$
*Line 3* $3^2 + 6 = 3 \times 5$
*Line 4* $4^2 + 8 = 4 \times 6$

Write down the $n$th line of this pattern.

(2 marks) (NEAB, 2000)

**A7.2** Look at this pattern:

$$15^2 - 14^2 = 29$$
$$14^2 - 13^2 = 27$$
$$13^2 - 12^2 = 25$$
$$12^2 - 11^2 = 23$$

Copy and complete this line to give the general rule for this pattern.

$$r^2 - \ldots\ldots = \ldots\ldots$$

(2 marks) (NEAB, 2000)

**A7.3** John and Sarah are each asked to continue a sequence which begins

2, 5, ...

**a** John writes 2, 5, 8, 11, ...
Write down the $n$th term of John's sequence.

**b** Sarah writes 2, 5, 10, 17, ...
Write down the $n$th term of Sarah's sequence.

(4 marks) (SEG, 1998)

**A7.4** **a** Write down the $n$th term for each of the following sequences.

**i** 1, 4, 9, 16, ...

**ii** 4, 16, 36, 64 ...

**b** The $n$th term of another sequence is $(n + 1)(n + 2)$.
Explain why every term of the sequence is an even number.

(5 marks) (NEAB, 1998)

**A7.5** Write down the $n$th term of the following number sequences.

**a** 1, 4, 9, 16, 25, ...

**b** 6, 9, 14, 21, 30, ...

(2 marks) (NEAB, 1999)

**A7.6** Give the $n$th term of the following sequences.

**a** 1, 4, 9, 16, 25, 36, ...

**b** 2, 5, 10, 17, 26, 37, ...

**c** 3, 7, 13, 21, 31, 43, ...

(3 marks) (NEAB, 2000)

## A8 Functions and graphs

**A8.1** A graph of the equation $y = ax + b$ is shown.

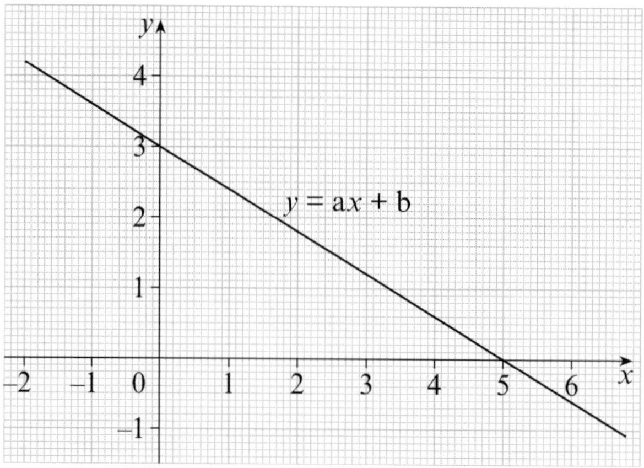


Find the values of $a$ and $b$.

(3 marks) (SEG, 1998)

**A8.2** **a** Below are three graphs.

Match the graph with one of the following equations.

Equation A: $y = 3x - p$

Equation B: $y = x^2 + p$

Equation C: $3x + 4y = p$

Equation D: $y = px^3$

In each case $p$ is a positive number.

**i**

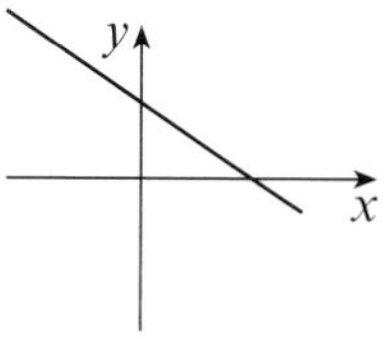


**ii**

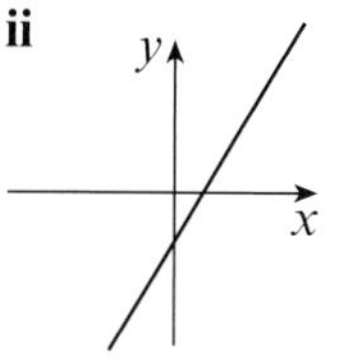


**iii**

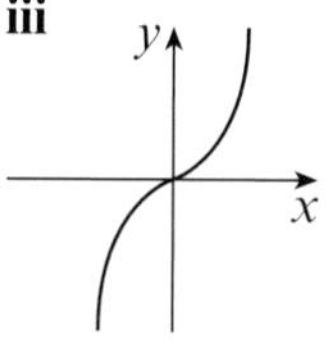


**b** Sketch a graph of the equation you have not yet chosen.

(5 marks) (NEAB, 1998)

**A8.3** The following data are obtained from an experiment.

| $x$ | 1 | 2 | 3 | 4 | 5 | 6 |
|---|---|---|---|---|---|---|
| $y$ | 0.7 | 1.2 | 2.1 | 3.3 | 4.7 | 6.7 |

It is thought that they may obey the rule $y = ax^2 + b$.

**a** By plotting values of $y$ against $x^2$, find approximate values of $a$ and $b$.

**b** Find the value of $y$ when $x = 4.5$

(5 marks) (NEAB, 2000)

**A8.4** The graph shows $y = x^2 + 5x - 6$.

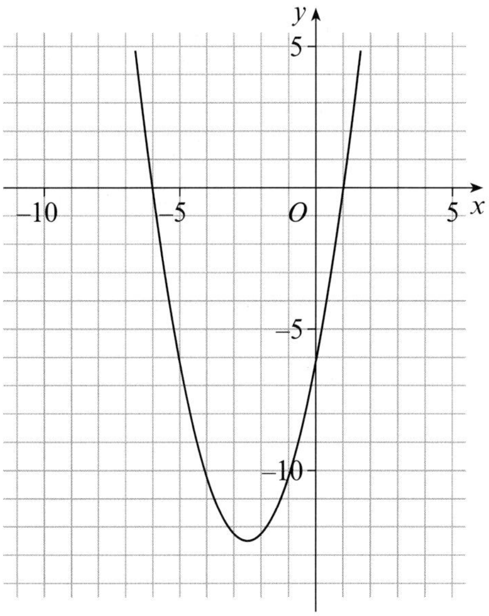


**a** What is the equation of the **straight** line you would need to draw to find the solution of this equation?

$$x^2 + 5x - 7 = 0$$

**b** Copy the graph, draw this line and use your graph to write down solutions of $x^2 + 5x - 7 = 0$ to one decimal place.

(3 marks) (AQA)

**A8.5** The graph of $y = x^2 - 2x - 4$ is drawn on the grid.

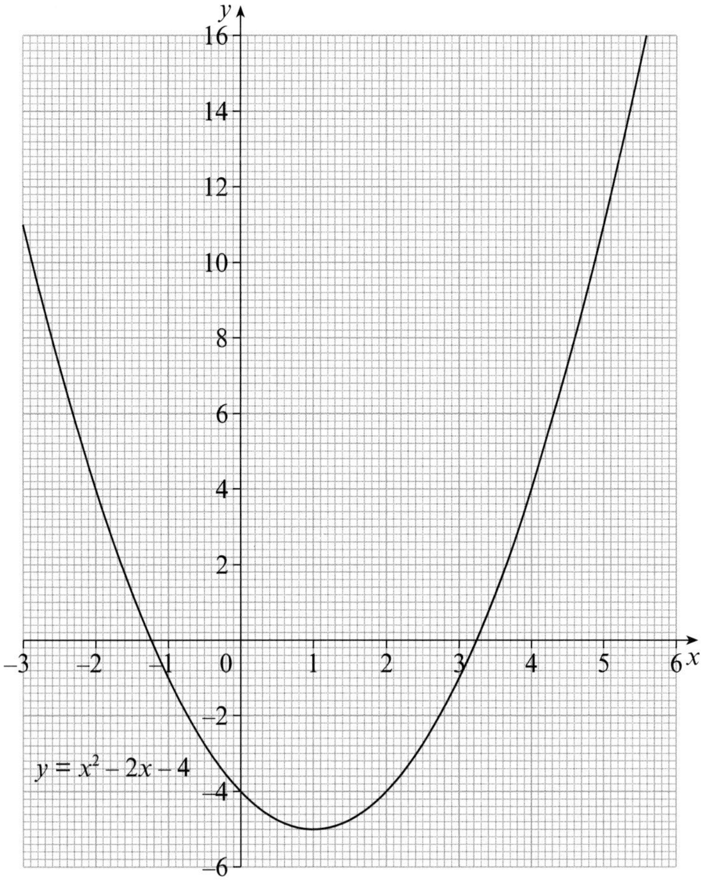

**a** Copy the graph. On the same axes, draw the line with equation $y = 2x + 1$.

**b** Hence solve the equation $x^2 - 2x - 4 = 2x + 1$.

**c** Show that $x^2 - 2x - 4 = 2x + 1$ can be simplified to $x^2 - 4x - 5 = 0$.

**d** You are going to use the graph of $y = x^2 - 2x - 4$ to solve the equation $x^2 - x - 8 = 0$.

What is the equation of the line you would need to draw?

(5 marks) (NEAB, 1998)

**A8.6** The curve drawn on the graph has equation $y = x^2 - x - 6$.

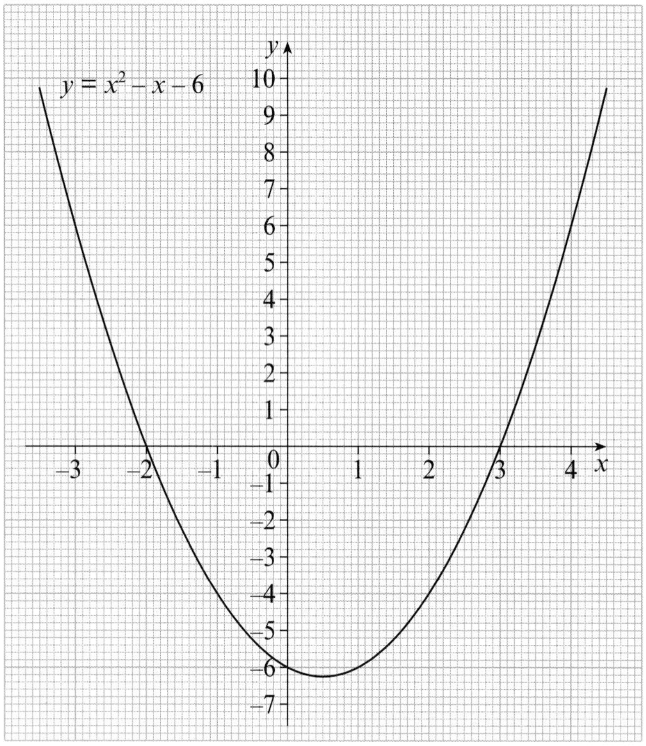


**a** Copy the graph. On the same axes draw the line with equation $y = 2x - 1$.

**b** **i** Show that $x^2 - x - 6 = 2x - 1$ is the same equation as

$$x^2 - 3x - 5 = 0.$$

**ii** Hence, using the graph, solve the equation $x^2 - 3x - 5 = 0$.

(4 marks) (NEAB, 2000)

**A8.7** Part of the graph of $y = \frac{5}{x}$ is shown on the grid.

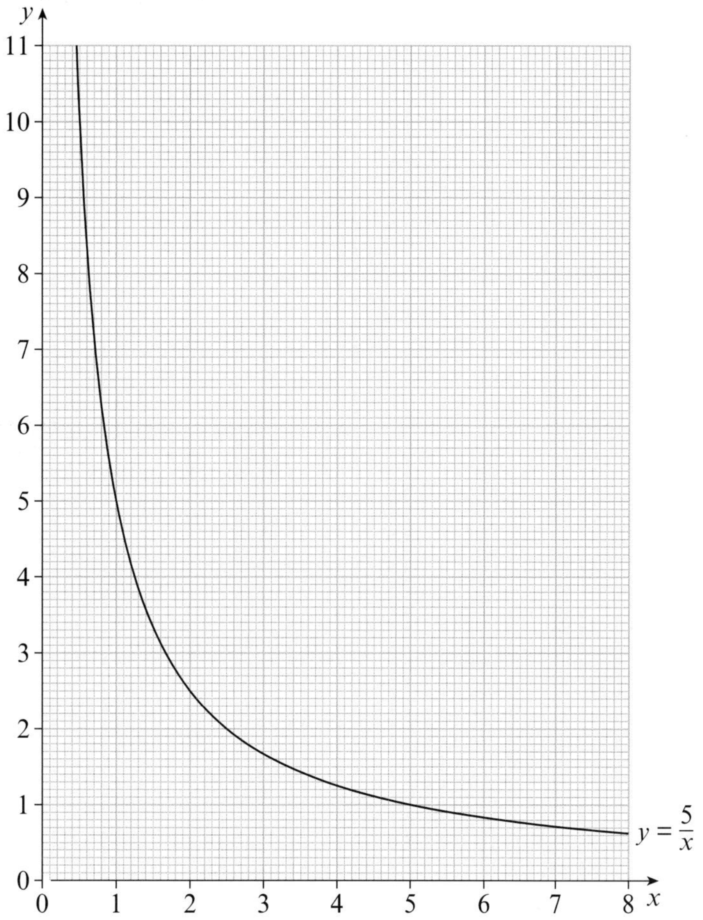


**a** **i** Copy the graph. Draw the graph of $y = 11 - 2x$ on the same grid.

**ii** Use the graphs to solve the equation $\frac{5}{x} = 11 - 2x$.

**b** **i** Show that $\frac{5}{x} = 11 - 2x$ can be written as $2x^2 - 11x + 5 = 0$.

**ii** Explain how your answer to part **a** can help you sketch the graph of $y = 2x^2 - 11x + 5$.

**iii** Sketch the graph of $y = 2x^2 - 11x + 5$.
Write down the coordinates of any points where the graph cuts the axes.

(8 marks) (NEAB, 2000)

**A8.8** You are given the equation $y = 5x^n$.

**a** Find the value of $y$ when $x = -4$ and $n = 0$.

**b** Find the value of $n$ when $y = 10$ and $x = 4$.

**c** Rearrange the formula to express $x$ in terms of $y$ and $n$.

(5 marks) (SEG, 1998)

**A8.9** Kate carried out an experiment to find the time for one swing of a pendulum. She made the pendulum by tying a weight to the end of a piece of string. She repeated the experiment using different lengths of string. Her results are given in the table.

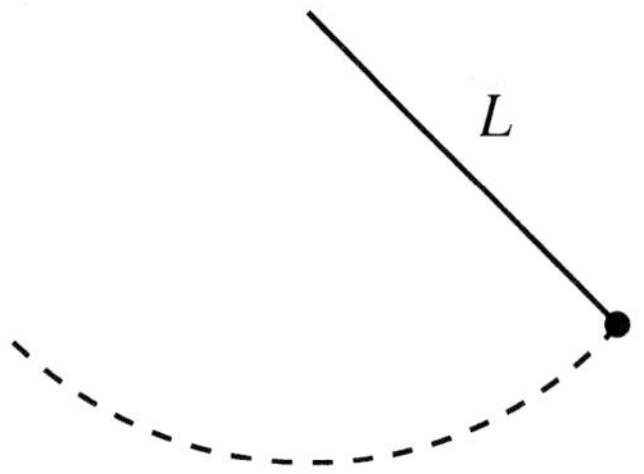


| Length of string, $L$, metres | 0.4 | 0.6 | 0.8 | 1.0 | 1.2 | 1.4 | 1.6 |
|---|---|---|---|---|---|---|---|
| $\sqrt{L}$ | 0.63 | | | | | | |
| Time for one swing, $T$, seconds | 1.27 | 1.55 | 1.79 | 2.01 | 2.20 | 2.37 | 2.54 |

**a** **i** Copy and complete the table to show the values of $\sqrt{L}$, correct to 2 decimal places.

**ii** On the grid, draw the graph of $T$ against $\sqrt{L}$.

**iii** Explain why the graph suggests that the relationship between $T$ and $L$ is of the form $T = k\sqrt{L}$ where $k$ is a constant.

**iv** Use your graph to estimate the value of $k$.

**b** It can be shown that $k = \dfrac{2\pi}{\sqrt{g}}$.

Use your answer to part **a iv** to find the value of $g$ correct to 1 decimal place.

(9 marks) (NEAB, 1999)

**A8.10** The values of $x$ and $y$ given in the table were obtained from an experiment.

| $x$ | 0.6 | 1.0 | 1.4 | 1.6 | 1.8 | 2.0 |
|---|---|---|---|---|---|---|
| $y$ | 3.7 | 4.0 | 4.7 | 5.2 | 5.9 | 6.8 |

It is known that $y$ is approximately equal to $px^3 + q$ where $p$ and $q$ are constants.

The graph shows $y$ plotted against $x^3$.

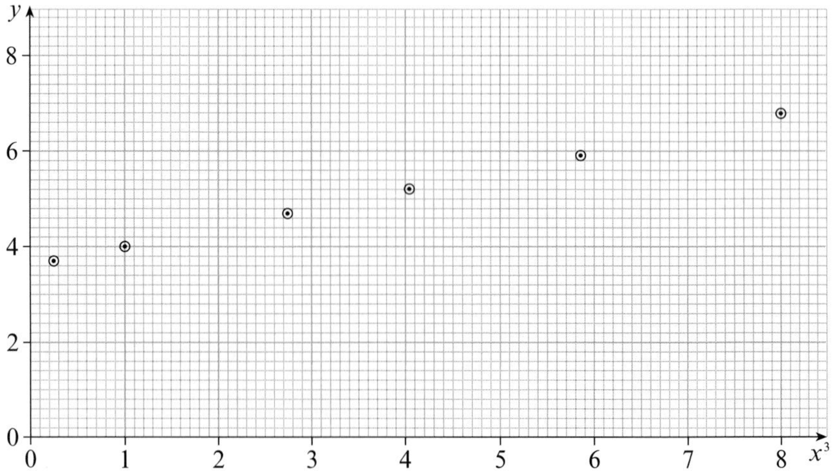


Use the graph to estimate the values of $p$ and $q$.

(3 marks) (SEG, 1998)

**A9**
**Transformations of graphs**

**A9.1** The graph of $y = f(x)$ is shown below.

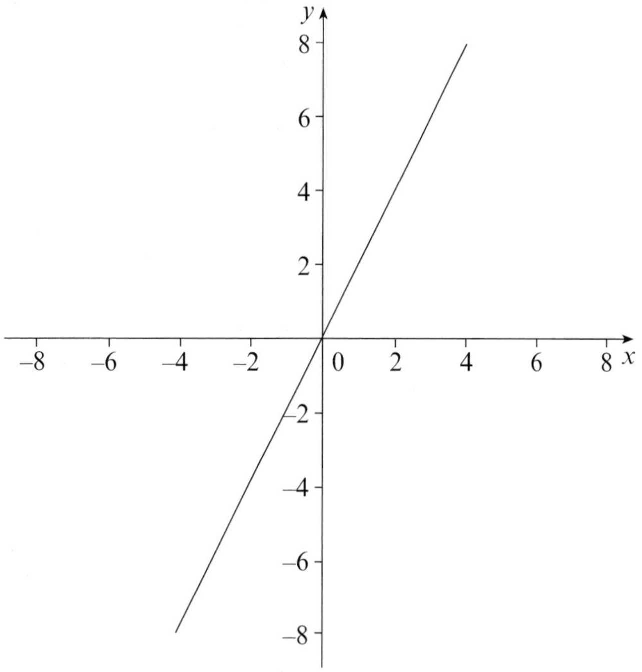


Copy the graph and sketch the graph of $y = f(x - 4)$ on the same axes.
(2 marks) (SEG, 2000)

**A9.2** The graph of $y = \sin x$ is shown below

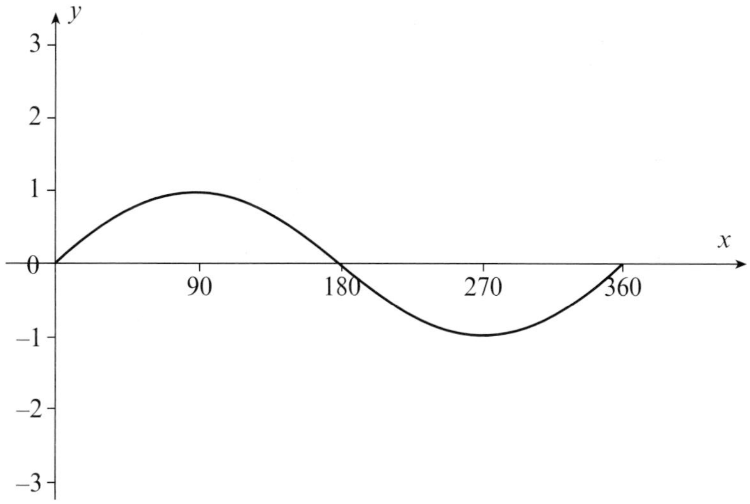


On a similar grid sketch the graph of

**a** $y = \sin 2x$.

**b** $y = \sin (x + 90)$

**c** $y = 2 + \sin x$

(6 marks) (NEAB, 2000)

**A9.3** The graph of the function $y = f(x)$ is shown on the grid.
The point $P(-1, 5)$ lies on the curve.

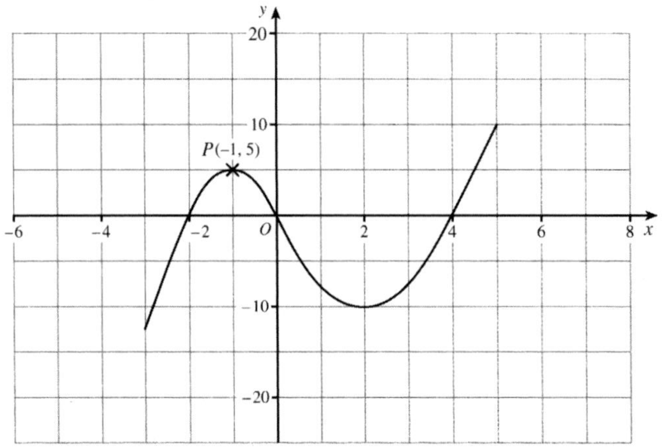


On similar grids draw the graph of the transformed function.
In each case write down the coordinates of the transformed point $P$.

**a** $y = = f(x + 3)$

**b** $y = 2f(x)$

**c** $y = -f(x)$

(6 marks) (NEAB, 2000)

**A9.4** **a** The graph of $y = x^3 - 4x$ is shown on the following grid.

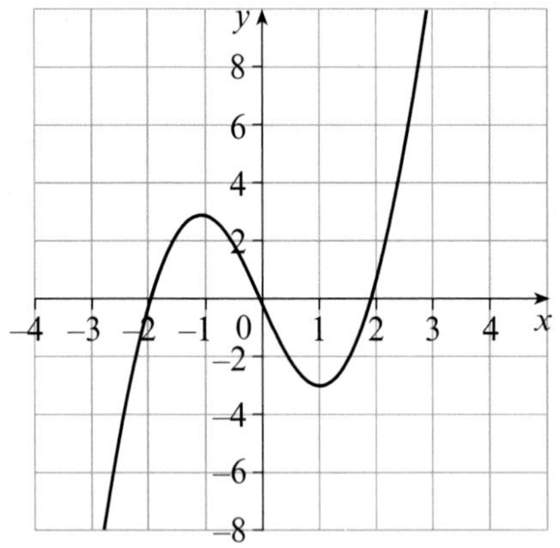


Copy the graph and on the same grid, sketch the graph of $y = x^3 - 4x - 2$.

**b** The graph of $y = x^2$ is shown below.

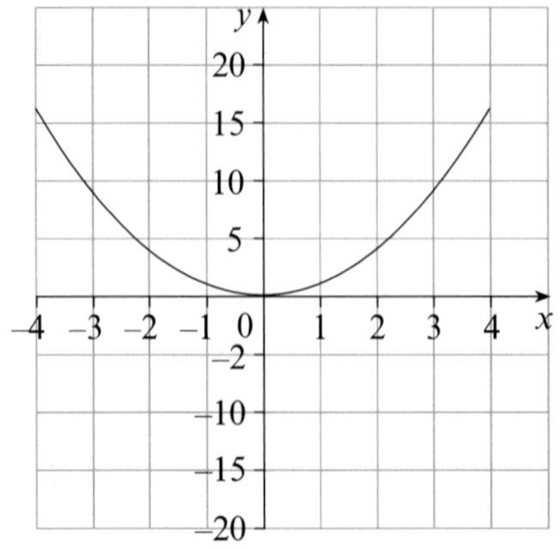


On a copy of this grid, sketch the graph of $y = (x - 1)^2$.

(4 marks) (SEG, 1999)

**A9.5** The sketch below is of $y = x^2$.

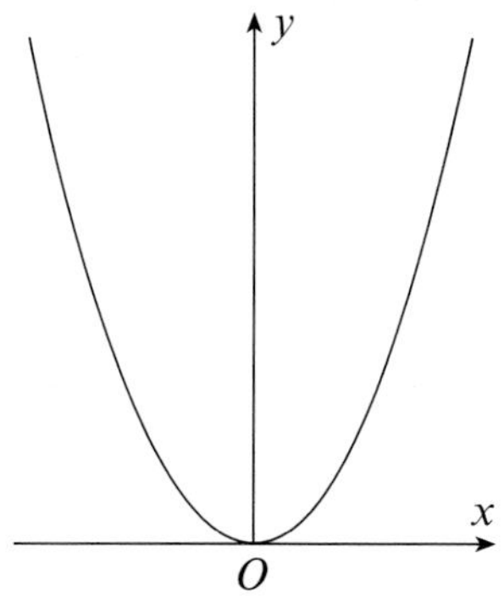


Sketch the following graphs.
In each case $p$ is a positive integer greater than 1.

**a** Sketch $y = px^2$

**b** **i** Sketch $y = x^2 + p$

**ii** Write down the coordinates of the point where the graph crosses the $y$-axis.

**c** **i** Sketch $y = (x + p)^2$

**ii** Write down the coordinates of the point where the graph touches the $x$-axis.

(6 marks) (NEAB, 1999)

**A9.6** **a** The diagrams, which are **not drawn to scale**, show three transformations of the graph $y = x^2$.

The new positions of five points $O(0, 0)$, $A(-3, 9)$, $B(-1, 1)$, $C(1, 1)$ and $D(3, 9)$ are shown on the diagrams.

Use this information to write down the equations of the transformed graphs.

**i**

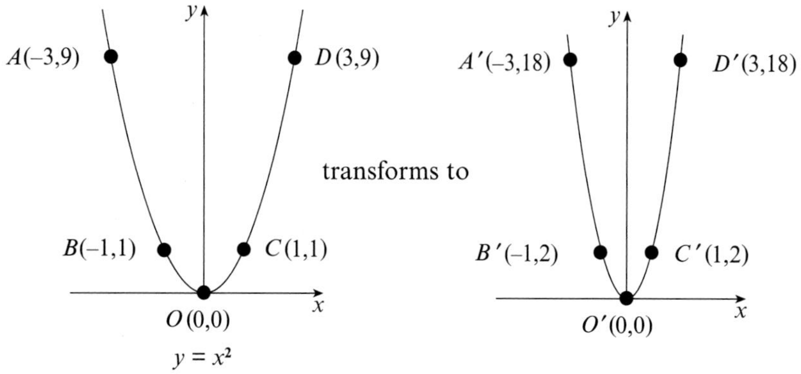


**ii**

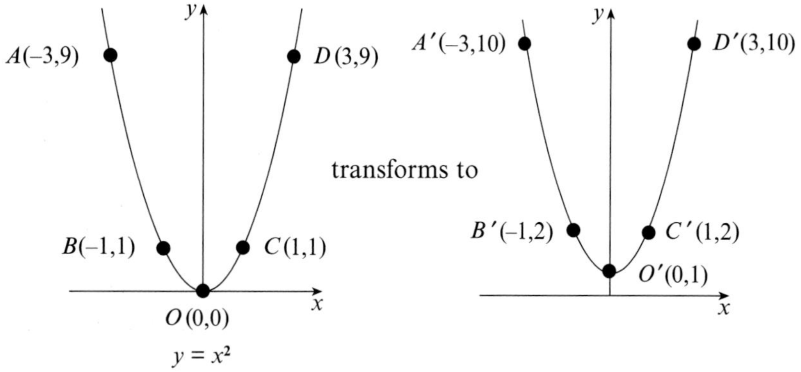

**iii**

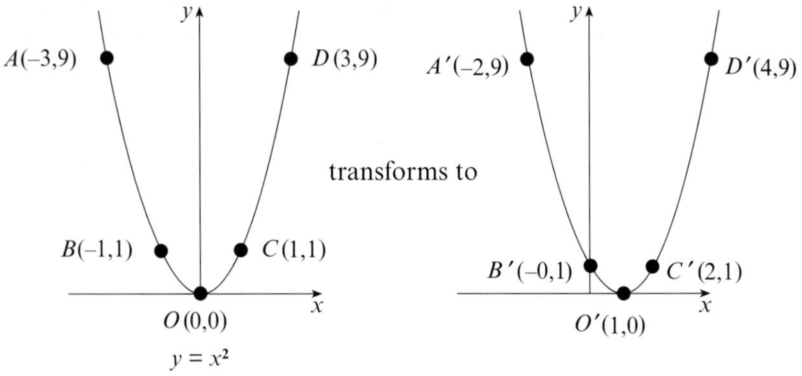


**b** This is the graph of $y = \cos x$.
Below are shown three transformations of the graph $y = \cos x$.
Give the equation of each graph.

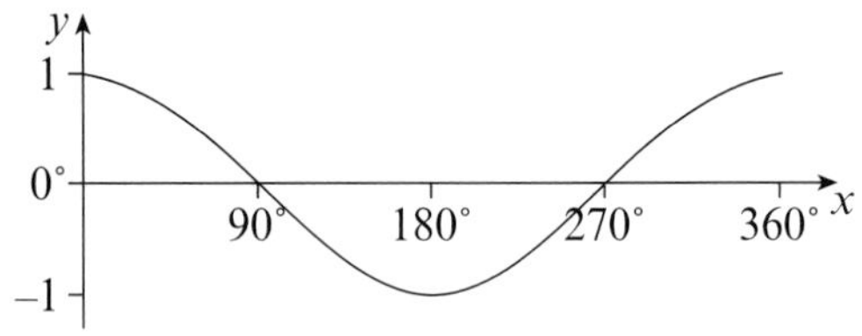


**i**

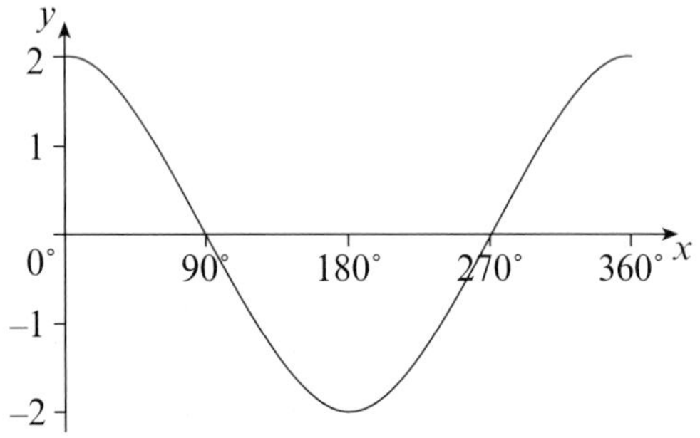


**ii**

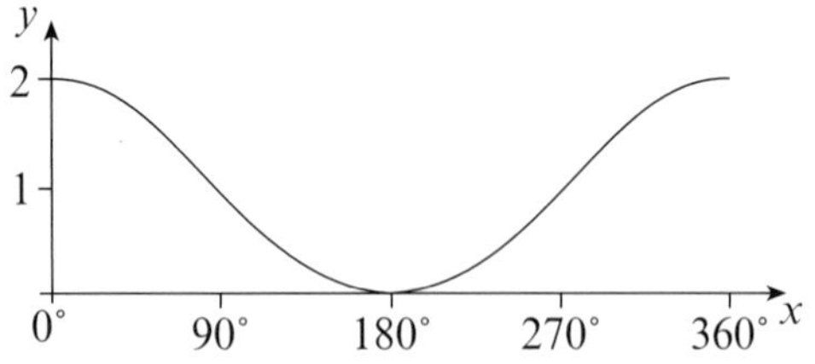


**iii**

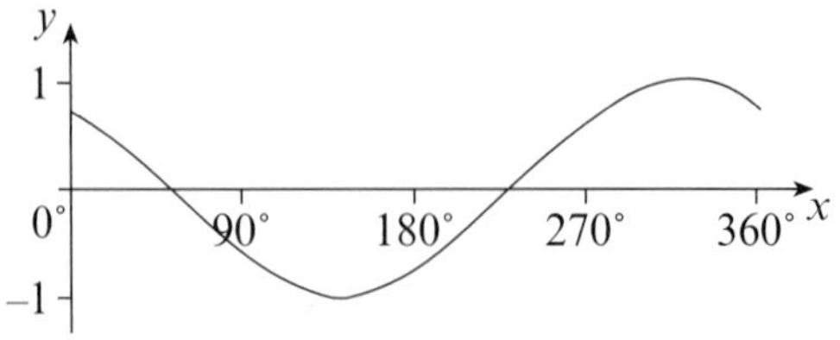


(6 marks) (NEAB, 1999)

## A10 Miscellaneous

**A10.1** **a** $x$ is a number which is greater than 1.
List the following four terms in order of size, smallest first.

| | | | |
|---|---|---|---|
| $x^{-2}$ | $x$ | $x^{\frac{1}{2}}$ | $\frac{1}{x}$ |

**b** If $0 < x < 1$, how should your list in part **a** be rearranged, if at all?

(3 marks) (NEAB, 1998)

**A10.2** The graph shows the curve $y = x^2 + x$.

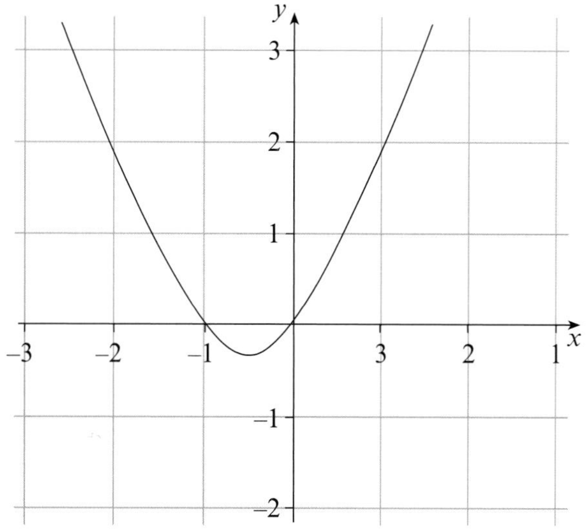


Use the graph, or otherwise, to find the value of $k$ such that the solutions of $x^2 + x + k = 0$ differ by 3.

(3 marks) (AQA)

**A10.3** The curve $y = x^2 = 3$ passes through the points $A(1, 4)$ and $B(1.2, 4.44)$.

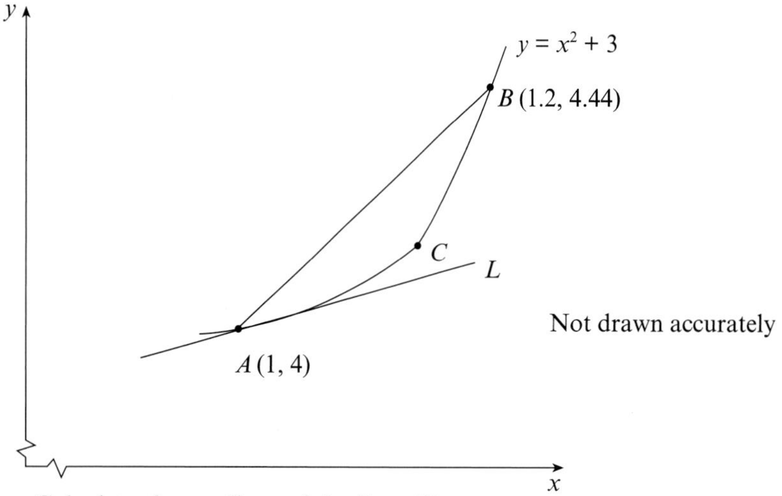


**a** Calculate the gradient of the line $AB$.

**b** The point $C$ also lies on the curve $y = x^2 + 3$.
The $x$-coordinate of $C$ is 1.1.

Calculate the gradient of the line $AC$.

**c** By choosing another point, calculate a better estimate for the gradient of the line $L$ which is the tangent to the curve at $x = 1$.

(6 marks) (AQA)

## Shape, space and measures

**S1**
**Angles and polygons**

**S1.1** Sketch a tetrahedron.
Show any hidden edges as dotted lines.
(2 marks) (NEAB, 2000)

**S1.2** **a** The diagram shows two regular polygons.

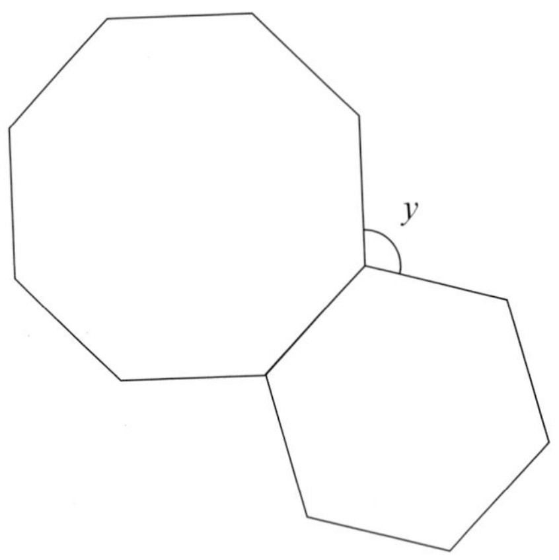


Work out the size of angle $y$.

**b** This diagram shows a kite $ABCD$.

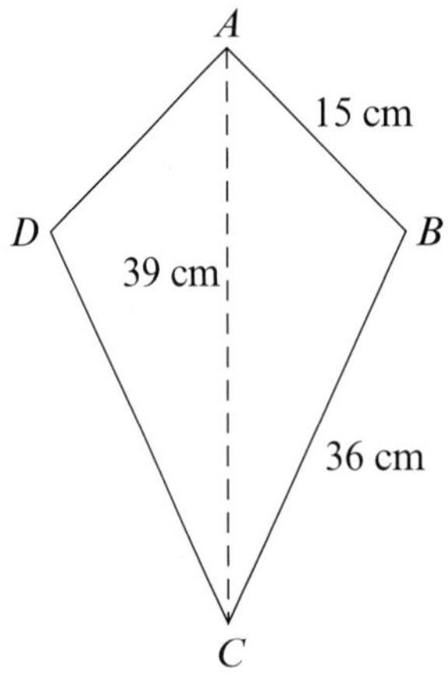


Not drawn accurately

$AB = 15$ cm, $BC = 36$ cm and $AC = 39$ cm.
Explain why angle $B = 90°$.
(6 marks) (AQA)

**S1.3** These regular polygons are similar.

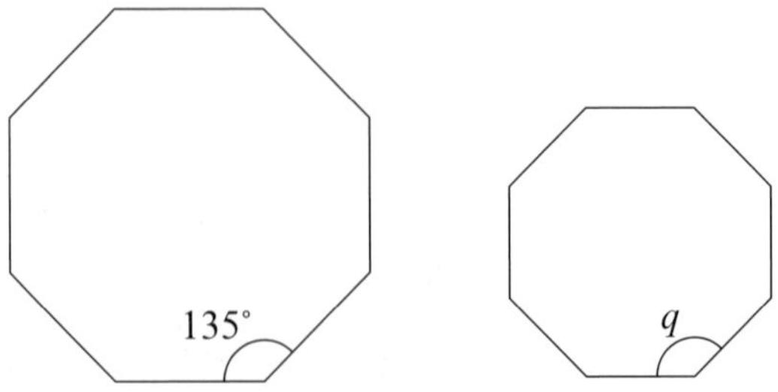


Not to scale

The lengths of the sides are in the ratio 4 : 3.
The larger polygon has a perimeter of 48 cm.
Find the perimeter of the smaller polygon and the size of angle $q$.
(3 marks) (SEG, 1999)

**S1.4** A regular pentagon is drawn below.

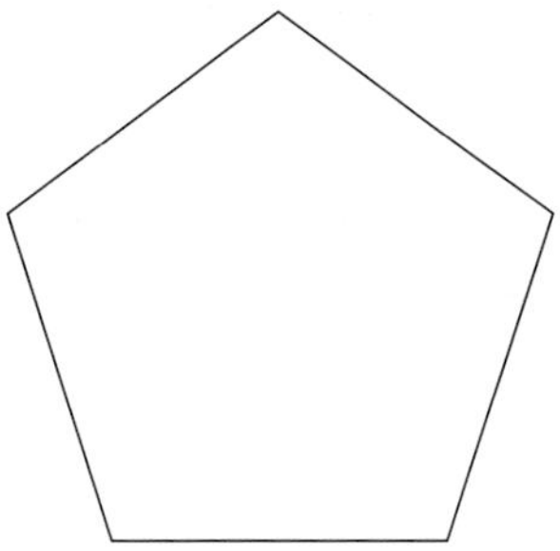

Use angles to explain why you cannot make a tessellation from regular pentagons.

(3 marks) (SEG, 1999)

**S1.5** **a** Calculate the size of an interior angle of a regular octagon.

**b** Explain why a regular polygon cannot have an interior angle equal to 110°.

(4 marks) (SEG, 2000)

**S1.6** The diagram shows a regular pentagon *ABCDE*.
*BD* is parallel to *XY*. Angle *CBD* = 36°.

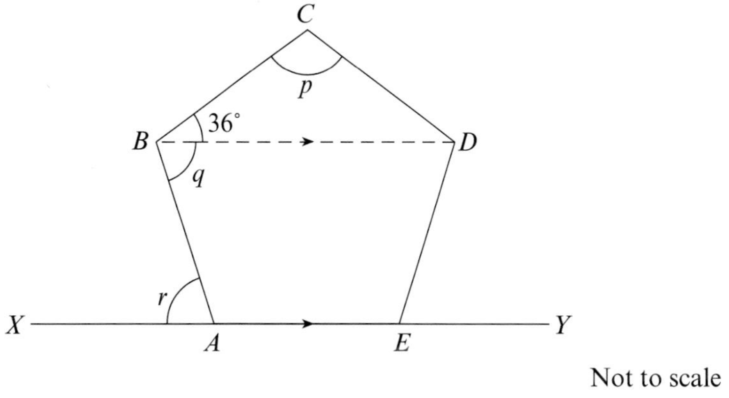


Work out the sizes of the angles *p*, *q* and *r*.

(4 marks) (SEG, 2000)

**S1.7**

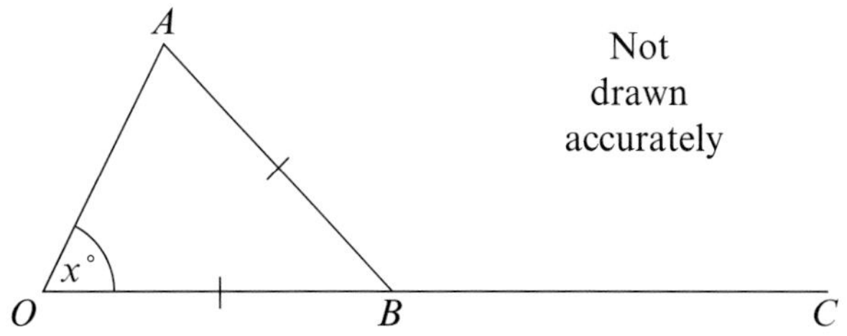


*OBC* is a straight line.
*AOB* is an isosceles triangle with *OB* = *AB*.
Angle *AOB* = $x$°.

**a** Write down, in terms of $x$

**i** angle *OAB*

**ii** angle *ABC*

**b** Angle *OBA* is $(x - 12)$ degrees.
Find the value of $x$.

(4 marks) (NEAB, 2000)

**S1.8** The diagram shows a kite *PQRS*.
Angle $QRS = 93°$ and angle $PQR = 118°$.

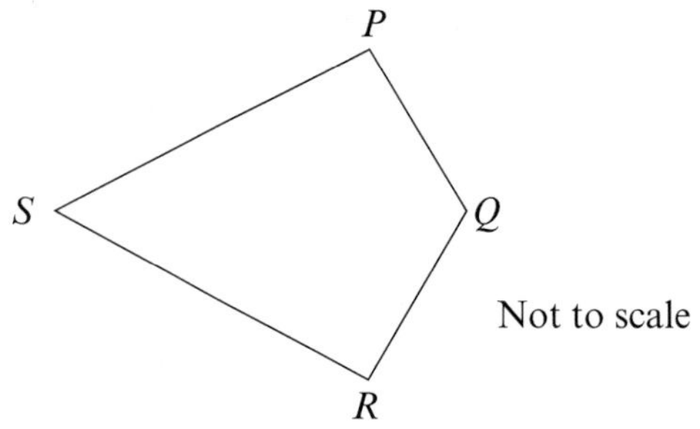


**a** Work out the size of angle *PSR*.

**b** *WXYZ* is similar to *PQRS*.
$PQ = 1.8$ cm, $SR = 3.3$ cm and $WX = 2.4$ cm.

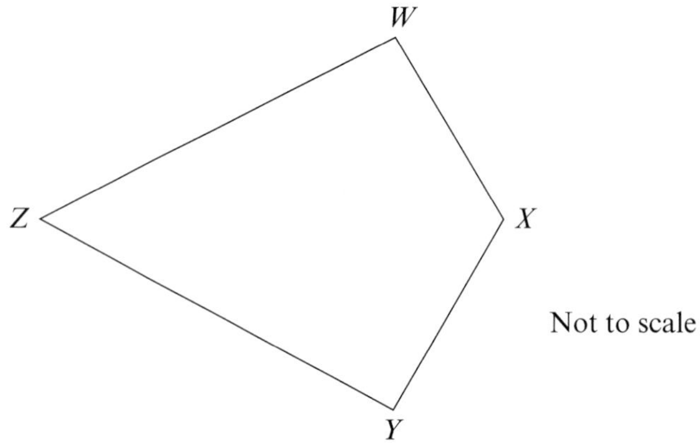


**i** Find the size of angle *XWZ*.

**ii** Calculate the length of *ZY*.

(5 marks) (SEG, 1998)

**S1.9** *ABCDEFGHIJ* is a regular decagon.

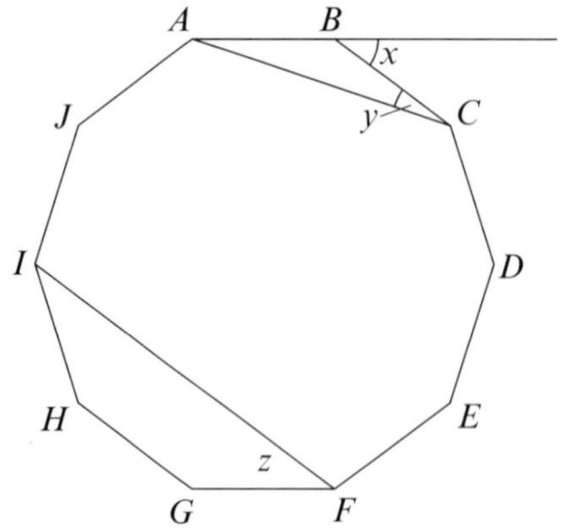


**a** **i** Calculate the size of the angle marked $x$.

**ii** Calculate the size of the angle marked $y$.

**b** **i** What can you say about the lines *GH* and *FI*?

**ii** Calculate the size of the angle marked $z$.

(7 marks) (NEAB, 2000)

## S2 Angles in circles

**S2.1** In the diagram, which is not drawn to scale, *L*, *M* and *N* are three points on the circumference of a circle.

*MN* is a diameter of the circle.

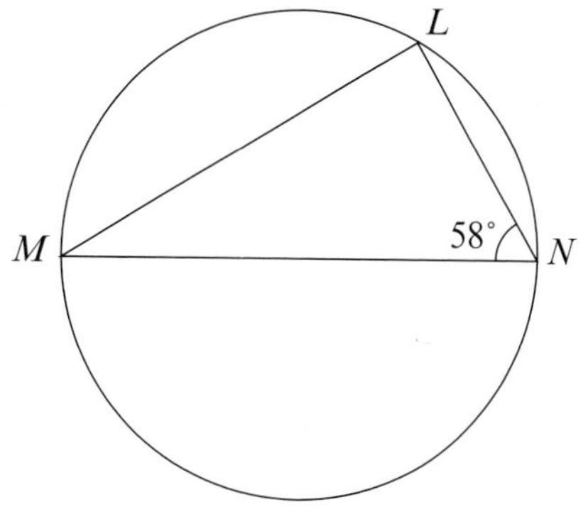


Calculate the size of angle *LMN*.
You must write down all the steps of your working, giving a reason for each step.

(3 marks) (SEG, 1999)

**S2.2** In the diagram, *O* is the centre of a circle and *P*, *Q*, *R*, *S* are points on the circumference.
Angle *POR* = 112°.

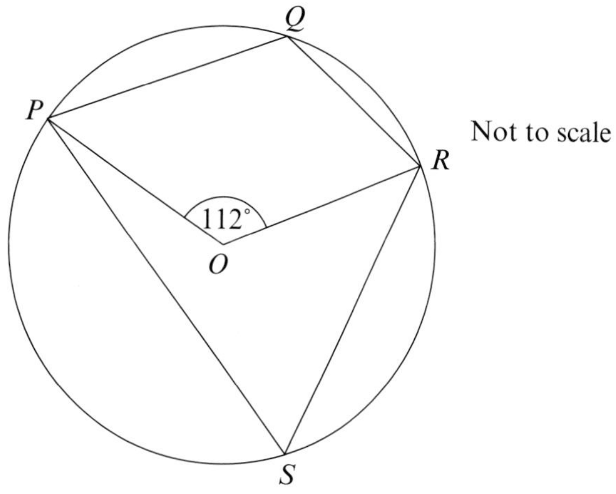


**a** Calculate the size of angle *PSR*, giving a reason for your answer.

**b** Calculate the size of angle *PQR*, giving a reason for your answer.

(4 marks) (SEG, 2000)

**S2.3** The line $PQR$ is a tangent to a circle with centre $O$.
$QS$ is a diameter of the circle.
$T$ is a point on the circumference of the circle such that $POT$ is a straight line.
The angle $OPQ$ is 34°.

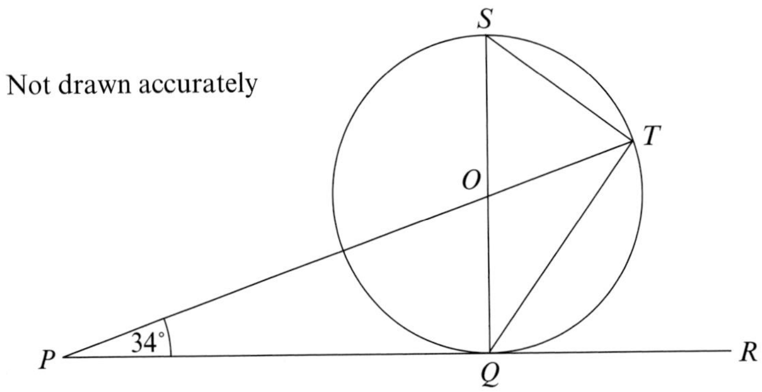


Calculate the size of angle $TQR$.

(5 marks) (NEAB, 1998)

**S2.4** In the diagram, $O$ is the centre of the circle.
Angle $COA = 100°$

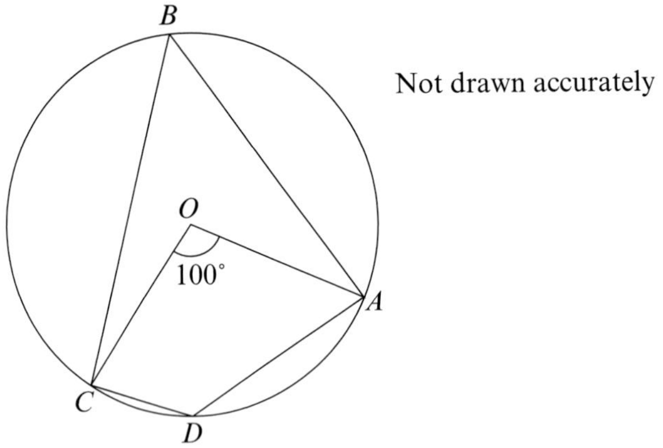


Calculate

**a** angle $CBA$

**b** angle $CDA$

(2 marks) (NEAB, 1999)

**S2.5** $A$, $B$, $C$ and $D$ are points on the circumference of a circle.
$BD$ is a diameter of the circle and $PA$ is a tangent to the circle at $A$.
Angle $ADB = 25°$ and angle $CDB = 18°$.

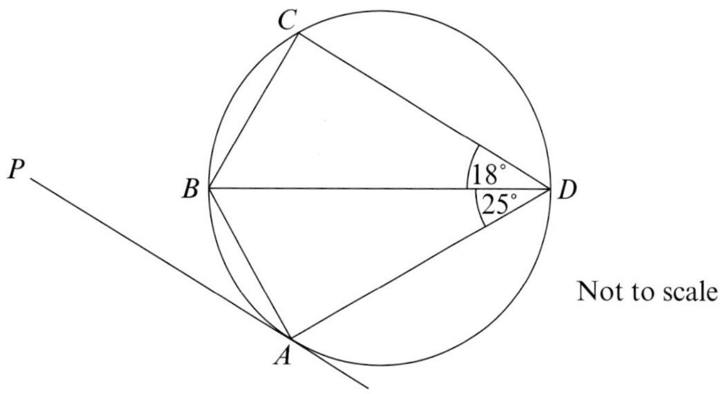


**a** Write down the value of angle

**i** $BCD$

**ii** $PAB$

**b** Calculate angle $ABC$.

(3 marks) (SEG, 2000)

**S2.6** In the diagram below, the chords $AB$ and $CD$ are perpendicular.

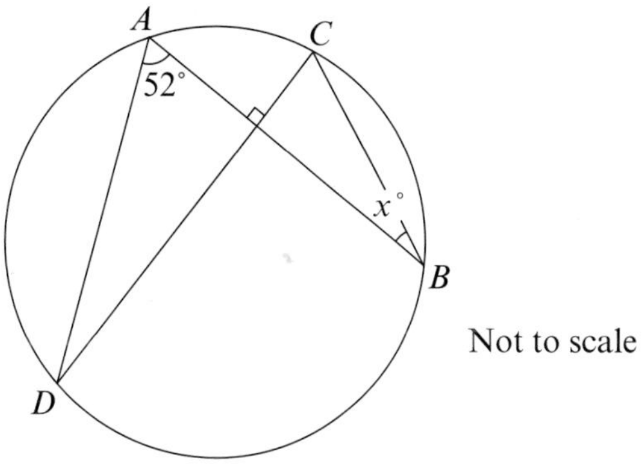


Angle $DAB = 52°$

**a** Calculate the value of $x$.

You must give a reason for your answer.

**b** $P$ is a point on the minor arc $DB$.

Explain, with reference to circle theorems, why angle $DPB$ is 128°.

(4 marks) (AQA)

**S2.7** Angle $BAT = 55°$ and angle $DBT = 30°$.
$AT$ and $BT$ are tangents to the circle.
$BC$ is a diameter of the circle.

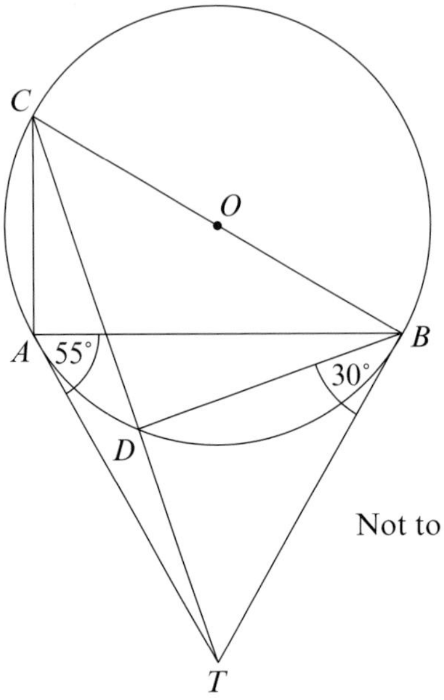


Not to scale

**a** Calculate the size of angle $BCT$.

**b** Calculate the size of angle $ABC$.

**c** Calculate the size of angle $ATC$.

(4 marks) (SEG, 1998)

**S2.8** $ABCD$ is a cyclic quadrilateral and $PA$ is a tangent to the circle at $A$.
Angle $BAP = 20°$ and angle $ADC = 62°$.

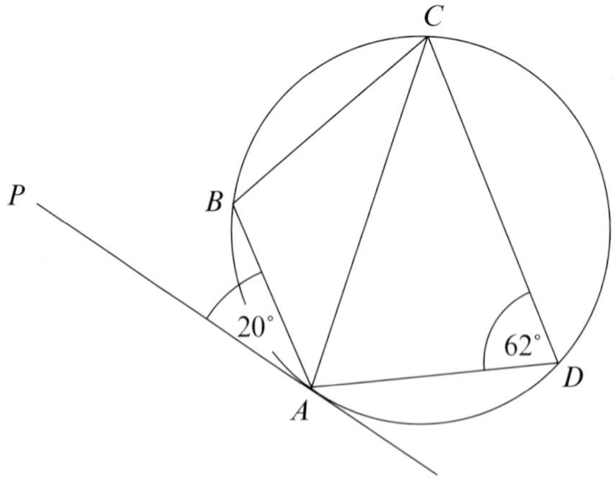


Not to scale

Find angles $ABC$ and $BAC$.

(3 marks) (SEG, 2000)

**S2.9** In the diagram, $O$ is the centre of the circle.
$SAT$ is a tangent to the circle at $A$.
Angle $BAC = 80°$ and angle $SAB$ = angle $TAC$.

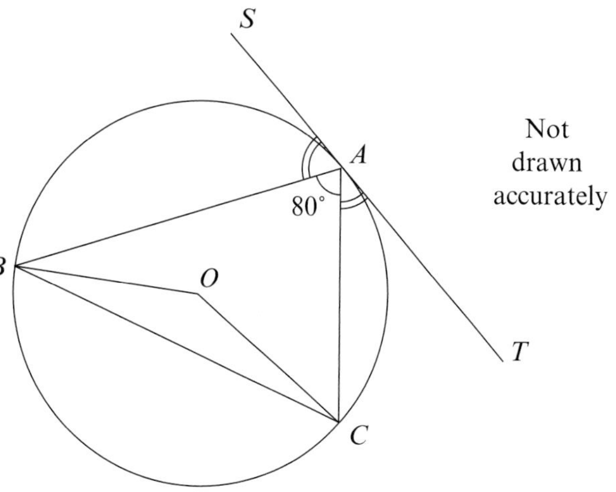


**a** Calculate
- **i** angle $BOC$
- **ii** angle $OBC$
- **iii** angle $ABO$
- **iv** angle $ACO$

**b** Explain why triangles $OAB$ and $OAC$ are congruent.

(7 marks) (NEAB, 1998)

**S2.10** $BDEF$ is a cyclic quadrilateral.
$EFA$ is a straight line
$CBA$ is a tangent to the circle
Angle $EDB = 63°$ and angle $FBA = 38°$

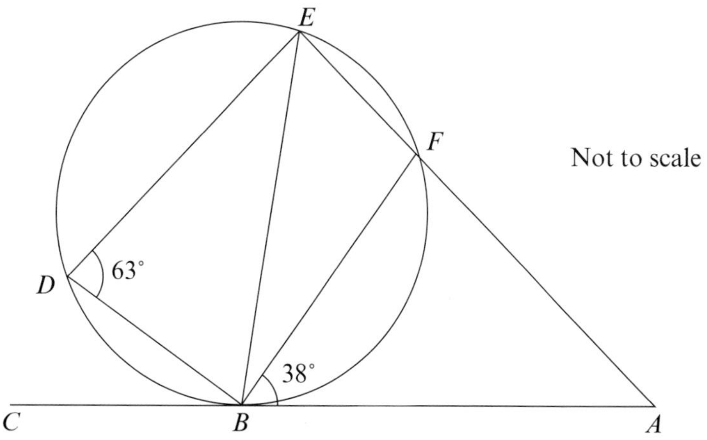


**a** Calculate the size of angle $BAF$

**b** Calculate the size of angle $EBF$.

(4 marks) (SEG, 2000)

**S2.11** The perimeter of triangle $PQR$ is 40 cm.
The circle touches $PQ$ at $A$, $QR$ at $B$ and $PR$ at $C$.
$PA = 9.6$ cm and $AQ = 4.2$ cm.

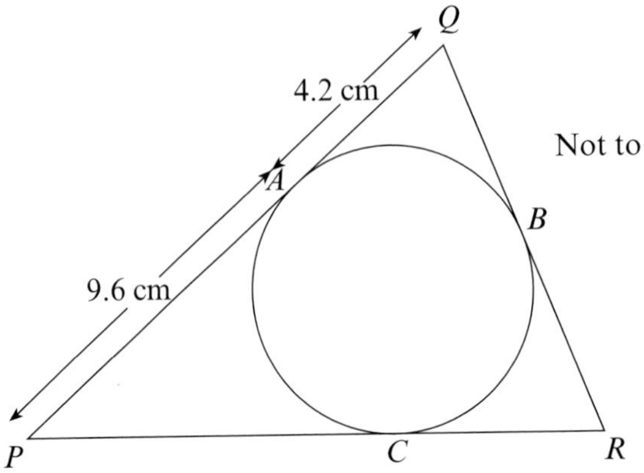


Find the length of $PR$.

(3 marks) (SEG, 2000)

**S2.12** In the diagram, $LM = LN$ and angle $LNM = 73°$.

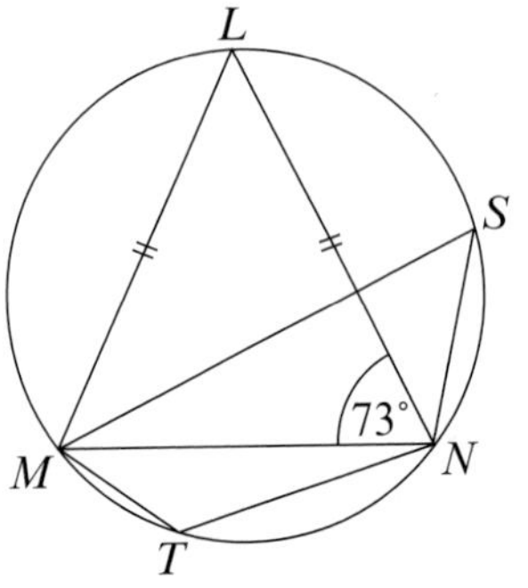


**a** Work out the size of angle $MSN$.
Give a reason for each step of your working.

**b** Work out the size of angle $MTN$.

(5 marks) (SEG, 1999)

**S2.13**

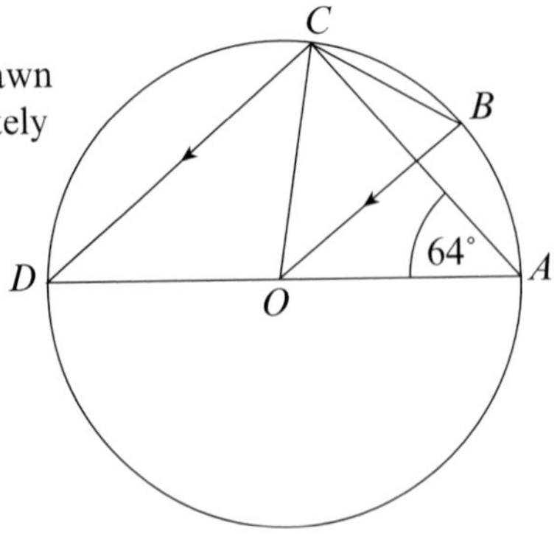


$AD$ is a diameter of a circle centre $O$.
$B$ and $C$ are points on the circumference such that $DC$ is parallel to $OB$.
The angle $OAC = 64°$.

**a** Calculate the size of angle $ODC$.

**b** Calculate the size of angle $AOB$.

**c** Calculate the size of angle $ACB$.

**d** Calculate the size of angle $OBC$.

(5 marks) (NEAB, 2000)

## S3 Dimensional analysis

**S3.1** One of these formulae gives the volume of an egg of height $H$ cm and width $w$ cm.

Formula A: $\frac{1}{6}\pi H^2 w^2$

Formula B: $\frac{1}{6}\pi H w^2$

Formula C: $\frac{1}{6}\pi H w$

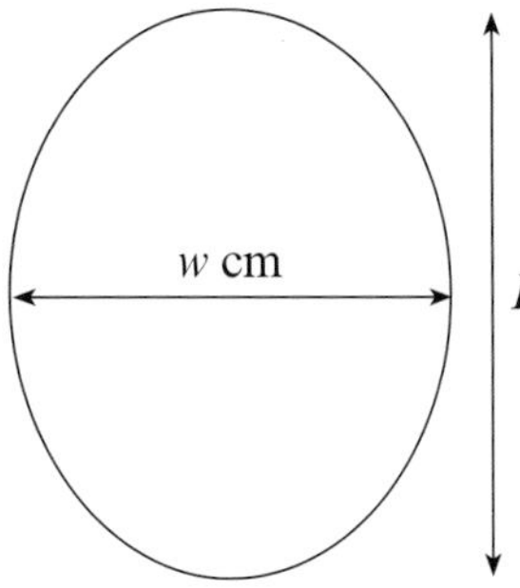


**a** Which is the correct formula?

**b** Explain how you can tell that this is the correct formula.

(2 marks) (NEAB, 1998)

**S3.2** David is writing down some formulae for some shapes.
The letters $r$, $h$ and $l$ represent lengths.

$\frac{1}{3}\pi r^2 h \quad \pi r(l+r) \quad \sqrt{r^2+h^2} \quad 2\pi rh \quad \pi r(2h+r^2) \quad \pi r\sqrt{r^2+h^2}$

Which of these formulae could represent area?

(2 marks)

**S3.3** In the following expressions $r$, $a$ and $b$ represent lengths.

For each expression state whether it represents

a **length**,
an **area**,
a **volume**,
or **none** of these.

**a** $\pi ab$

**b** $\pi r^2 a + 2\pi r$

**c** $\dfrac{\pi r a^3}{b}$

(3 marks) (NEAB, 2000)

**S3.4** The diagram shows a prism.

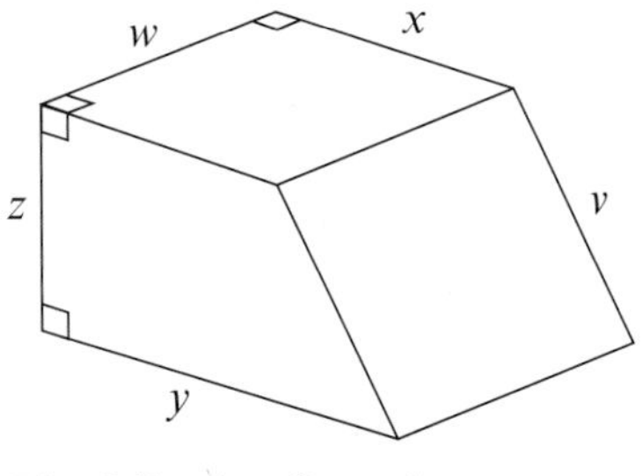


Not to scale

The following formulae represent certain quantities connected with the prism.

$wx + wy \qquad \frac{1}{2}z(x+w)w \qquad \dfrac{z(x+y)}{2} \qquad 2(v+2w+x+y+z)$

**a** Which of these formulae represents length?

**b** Which of these formulae represents volume?

(2 marks) (SEG, 2000)

**S4**
Length, area and volume

**S4.1** Calculate the area of a circle with a diameter of 15 cm.
Give your answer to an appropriate degree of accuracy.
(3 marks) (AQA)

**S4.2** The diagram shows the view of the entrance to a tunnel.

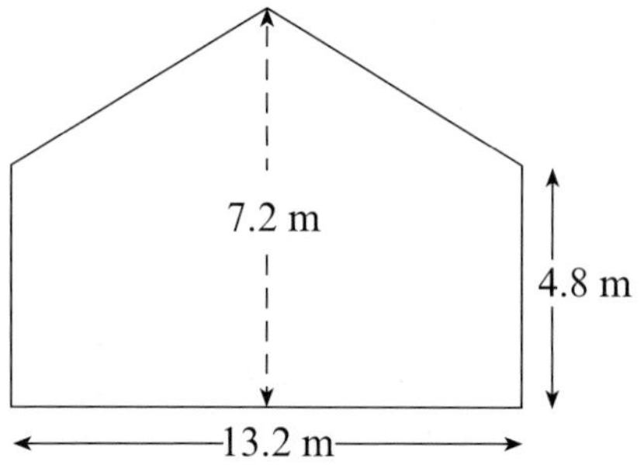


Calculate the area of the entrance to the tunnel.
Remember to give the units of your answer.
(4 marks) (NEAB, 2000)

**S4.3** Pamela has a fish-tank which is 30 cm long and has a capacity of 4.8 litres.

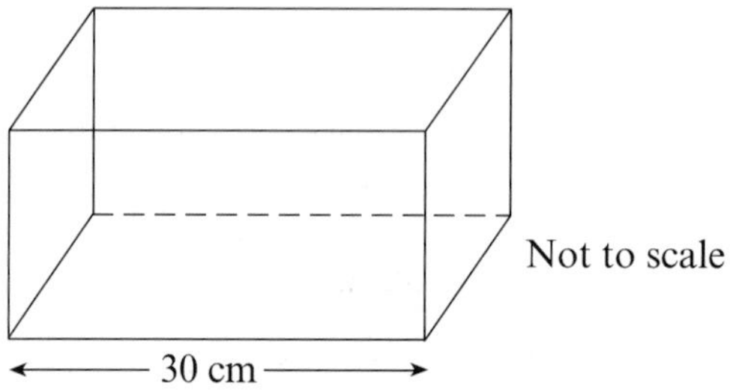


Rani has a similar fish-tank which is 45 cm long.
What is the capacity of Rani's fish-tank?
Give your answer in litres.
(2 marks) (SEG, 2000)

**S4.4** A solid plastic cuboid has dimensions 3 cm by 5 cm by 9 cm.
The density of the plastic is 0.95 grams per $cm^3$.

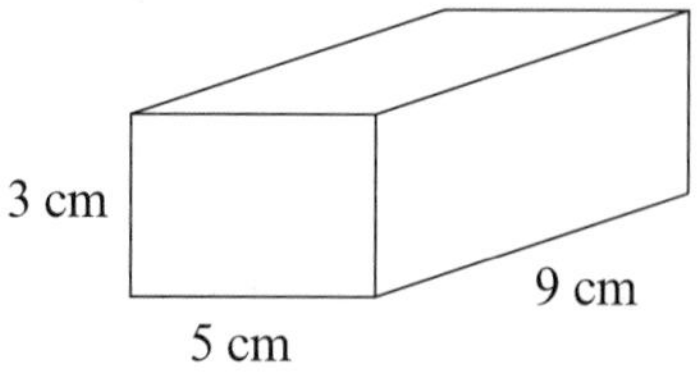


What is the weight of the plastic cuboid?
(3 marks) (NEAB, 1998)

**S4.5** The diagram shows a prism.
The cross-section of the prism is a trapezium.

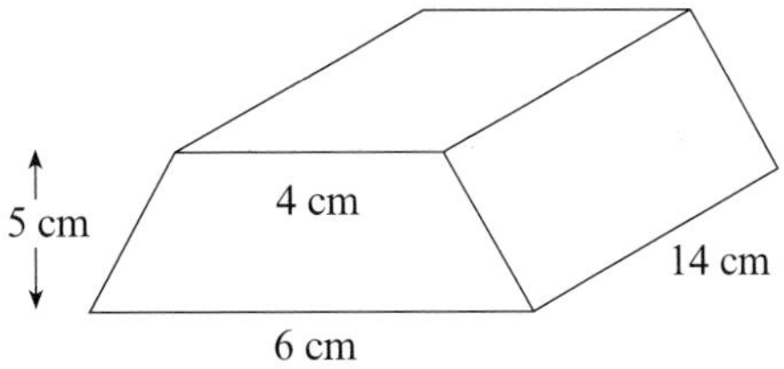


Calculate the volume of the prism.

(4 marks) (NEAB, 2000)

**S4.6** The circle centre $O$ has a radius 10 cm.
The chord $AB$ is 16 cm long.
$X$ is the midpoint of $AB$.
Calculate the area of the segment shaded in the diagram.

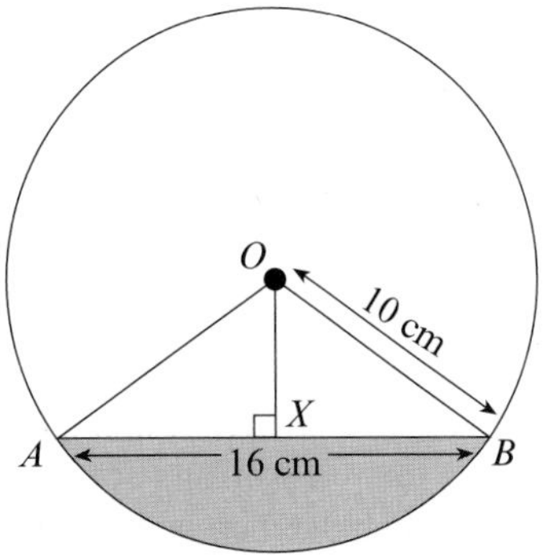


(8 marks) (NEAB, 1998)

**S4.7** The diagram shows a Go Kart track.

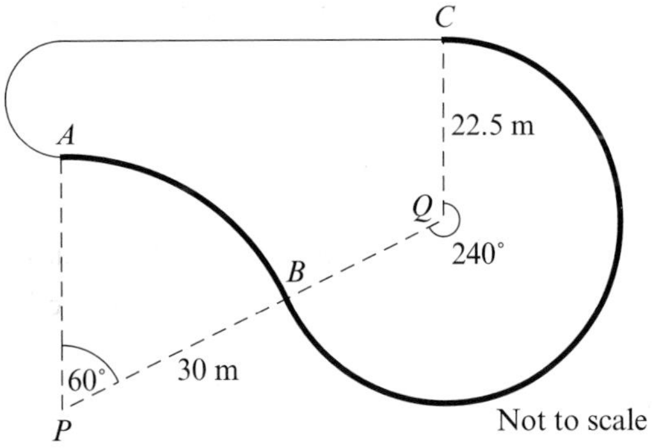


**a** The bend, $AB$, is the arc of a sector with centre $P$, radius 30 m and angle 60°.

The bend, $BC$, is the arc of a sector with centre $Q$, radius 22.5 m and angle 240°.

Calculate, in terms $\pi$, the total distance along the track from $A$ to $B$ to $C$.

Every Saturday Andy and Fay race each other around the track.
If the track is wet, the probability that Andy will win is 0.8.
If the track is dry, the probability that Fay will win is 0.6.
Next Saturday the probability that the track will be wet is 0.2.

**b** Calculate the probability that Andy will win next Saturday's race.

(8 marks) (SEG, 2000)

**S4.8** **a** Calculate the area of a circle of radius 4 cm.
Give your answer in terms of $\pi$.

**b** A square is drawn inside the circle, as shown in the diagram.

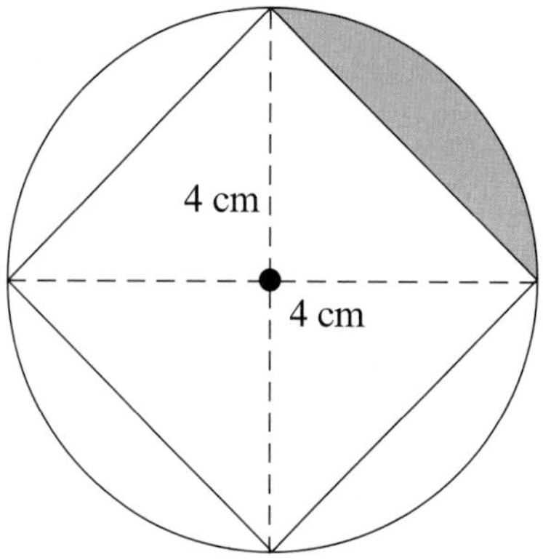


Not drawn accurately

**i** Calculate the area of the square.
Remember to state the units in your answer.

**ii** Calculate the area of the shaded segment.
Give your answer in terms of $\pi$.

(7 marks) (NEAB, 2000)

**S4.9** A running track has 4 lanes.
Each lane is 1 metre wide.
Each straight section is 90 metres long.
The curved sections are semi-circles.
The distance run is measured on the **inside** of each lane.
The total distance around the **inside** of Lane 1 is 400 metres.
The athletes run anti-clockwise in lanes throughout a 400 metre race.

How far in front of the finish line is the starting position in Lane 4 for a 400 metre race?
Show all your working.

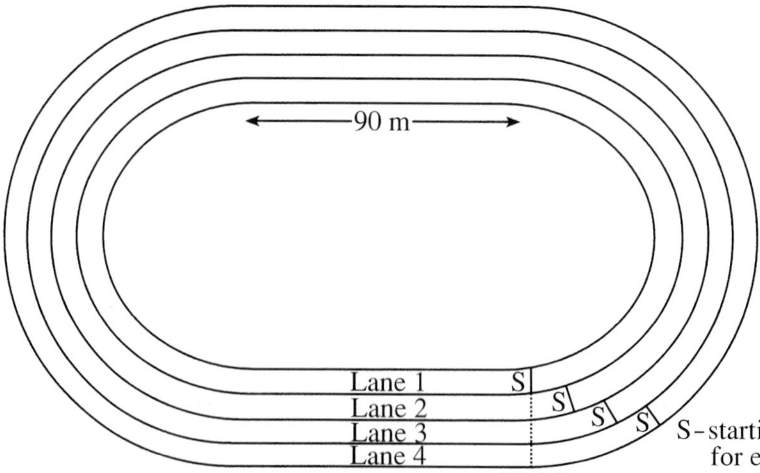


Not drawn to scale

(5 marks) (NEAB, 2000)

**S4.10** The diagram shows a baby's rattle made from a solid cone and a hemisphere.

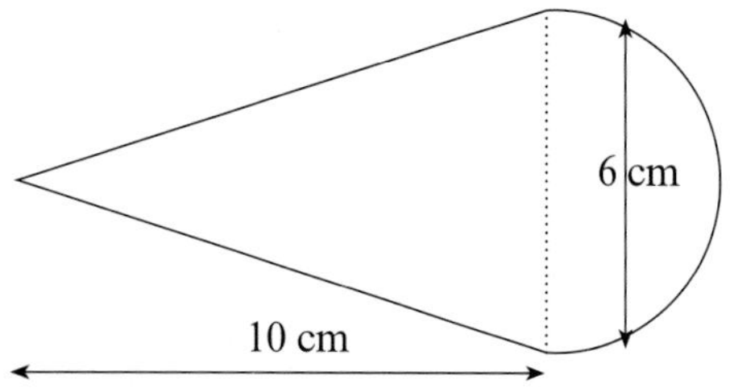


The common diameter is 6 cm.
The perpendicular height of the cone is 10 cm.
Find the total volume of the rattle.

(3 marks) (NEAB, 2000)

**S4.11** A solid cone has a height of 12 cm and a slant height of 13 cm.

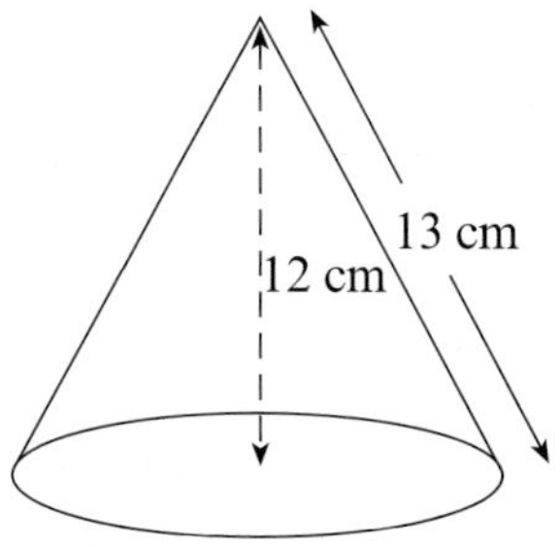


Calculate the **total** surface area of the cone.
Give your answer in terms of $\pi$.

(6 marks) (AQA)

**S4.12** The diagram shows a pepper pot.
The pot consists of a cylinder and a hemisphere.
The cylinder has a diameter of 5 cm and a height of 7 cm.

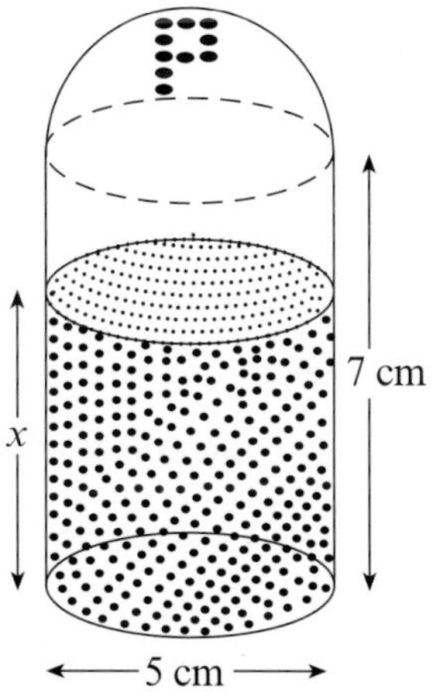


The pepper takes up half the **total** volume of the pot.
Find the depth of pepper in the pot marked $x$ in the diagram.

(5 marks) (NEAB, 2000)

**S4.13** The diagram shows part of the lead framework on a stained glass window.

*AB* and *PQ* are arcs of circles with centre *O*.

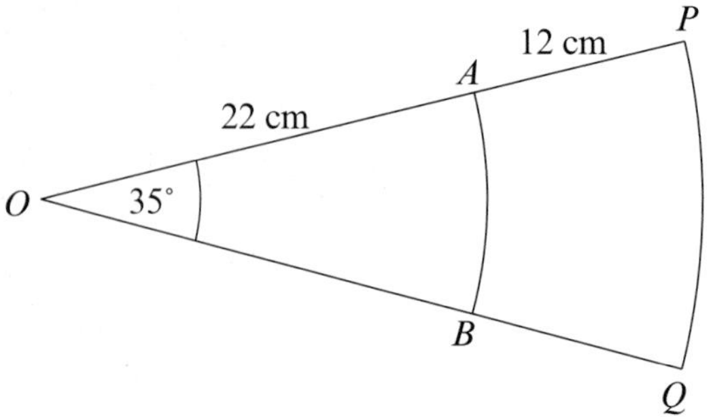


Not to scale

Calculate the total length of the lead.

(4 marks) (SEG, 1998)

**S4.14** A silver earring is in the shape of a solid cone, as shown in the diagram.

The slant height of the cone is 20.5 mm and the diameter of the base is 9 mm.

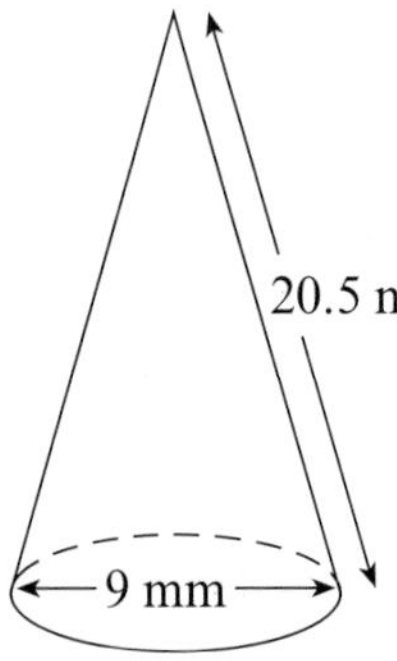


Not drawn accurately

**a** Calculate the volume of the earring.

**b** A similar solid cone of base diameter 13.5 mm is used as a pendant for a necklace.

The pendant is also made from silver.

The weight of one earring is 4.5 grams.

Calculate the weight of the pendant.

(7 marks) (NEAB, 1998)

**S4.15** The diagram shows a window in the shape of a segment, $ABC$, of a circle of radius 48 cm, centre $O$.

The lower edge of the window is 90 cm wide.

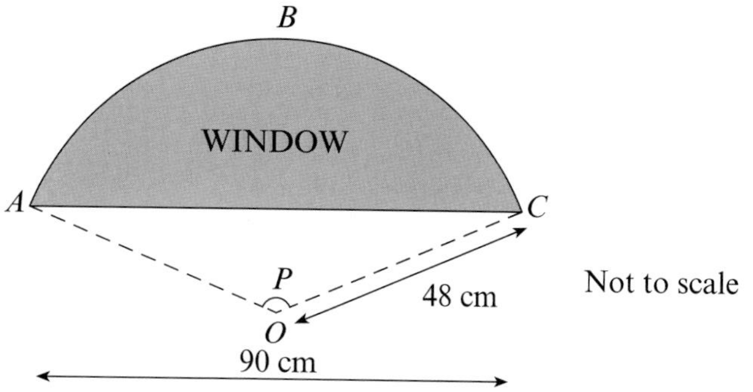


**a** Calculate the size of angle $p$.

**b** Calculate the area of the window.

(8 marks) (SEG, 1999)

**S4.16** $OAB$ is a sector of radius 8.5 centimetres.

The length of the arc, $AB$ is 6.3 centimetres.

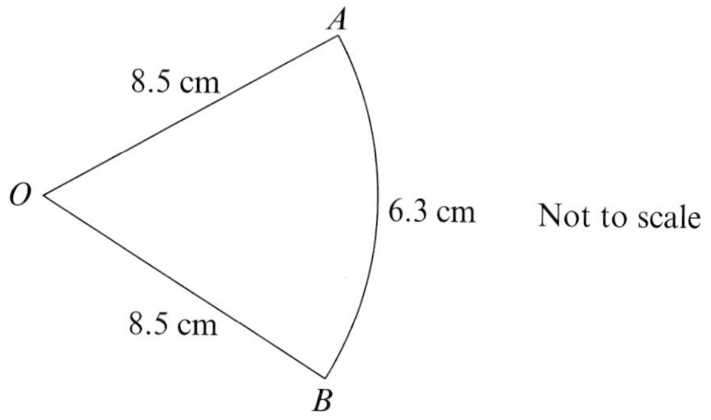


Calculate angle $AOB$.

(3 marks) (SEG, 1999)

**S4.17** $ABCD$ is a rectangle.
$AXY$ is the sector of a circle with centre $A$.
$AB = 80$ cm, $AD = 25$ cm and $DY = 60$ cm.

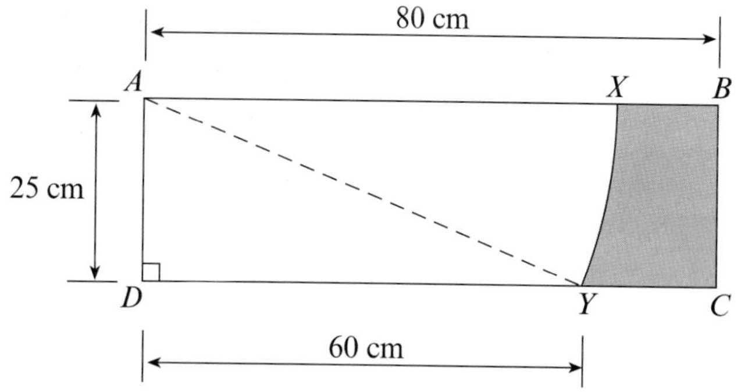


**a** Calculate the length of $AY$, the radius of the sector $AXY$.

**b** Calculate angle $DAY$.

**c** Calculate the shaded area.

(11 marks) (SEG, 1999)

## S5 Construction, loci and scale drawings

**S5.1** The diagram shows a regular tetrahedron.
Each edge is 5 cm long.
Draw an accurate net of the tetrahedron.

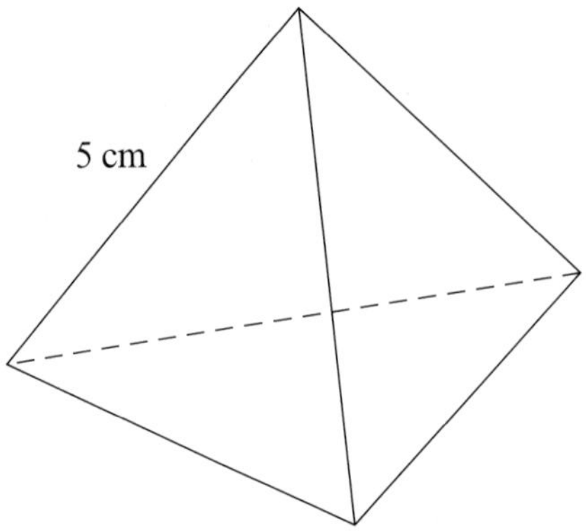


(3 marks) (NEAB, 2000)

**S5.2** The scale diagram shows a plan of a room.
The dimensions of the room are 9 m and 7 m.

Two plug sockets are fitted along the walls.
One is at the point marked *A*. The other is at the point marked *B*.
A third plug socket is to be fitted along a wall.
It must be equidistant from *A* and *B*.

Copy the diagram.
Using ruler and compasses, find the position of the new socket.
Label it *C*.

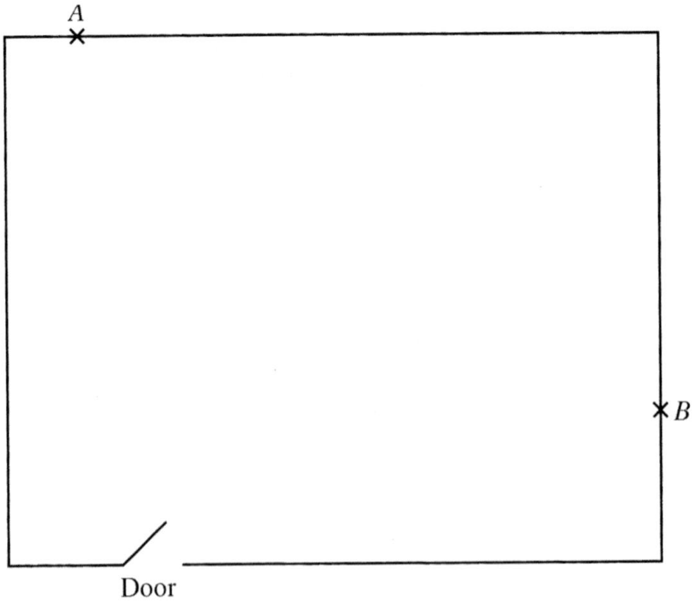


DRAWN TO SCALE

1 cm represents 1 m

(4 marks) (NEAB, 2000)

**S5.4** The circle below has been drawn round the base of a tin.

*PQ* and *RS* are chords of the circle.

Copy the diagram.
Using a ruler and compasses find the centre of the circle.

Label the centre *O*.

You must show all your working.

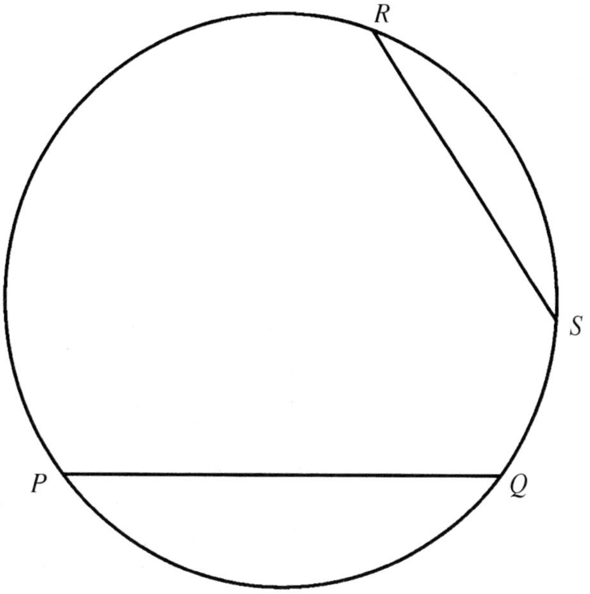


(4 marks) (AQA)

**S5.5** The diagram below shows a sketch of a right-angled triangle.

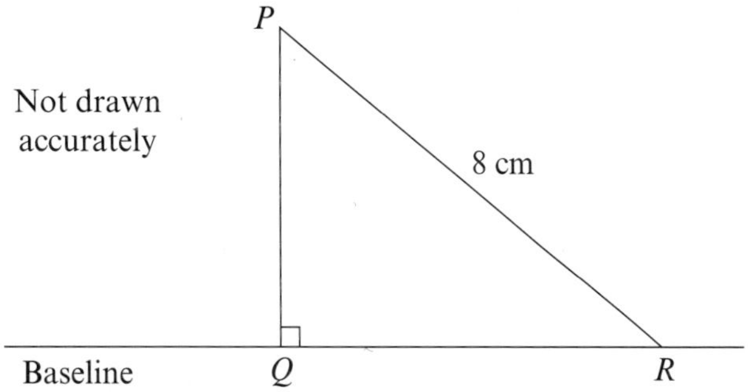


$PR = 8$ cm.

**Using ruler and compasses only**, construct an accurate drawing of the triangle *PQR*.

(4 marks) (NEAB, 2000)

**S5.6** Wendy sketches a toy boat.

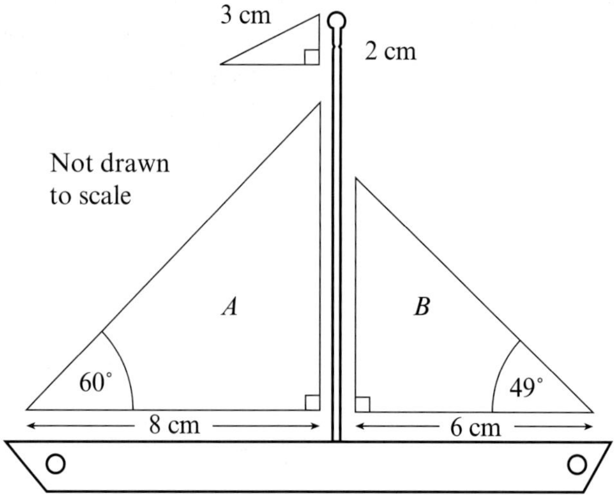


**a** Using ruler and compasses only, draw an accurate diagram of sail *A*.
You must show your construction lines.

**b** This is sail *B*.

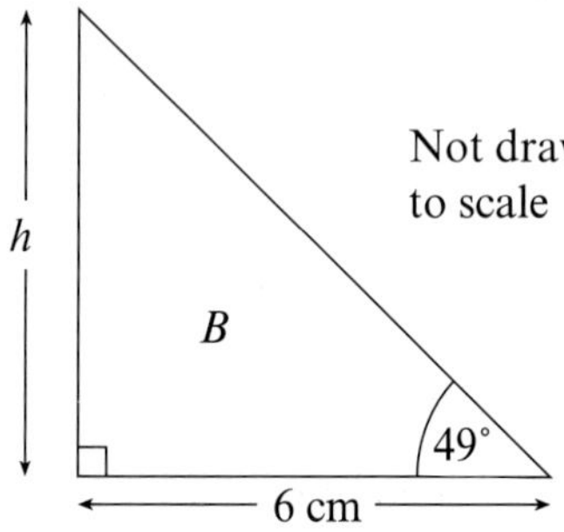


Calculate the height *h* of sail *B*.

**c** This is the flag.

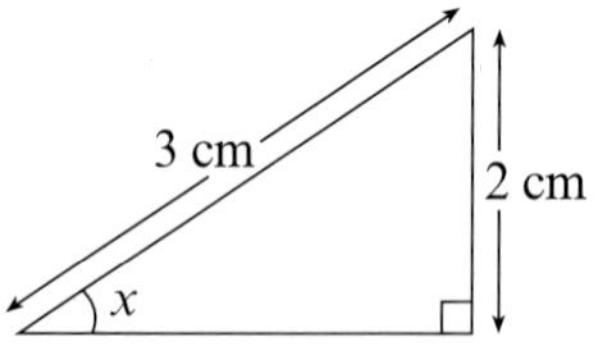


Calculate the angle *x*.

(10 marks) (NEAB, 1999)

**S6**
**Compound measure and graphs**

**S6.1** **a** Water is being poured at a constant rate into a container which has the shape of a prism.

The diagram shows the cross-section of the container.

Sketch a graph to show the height of water in the container as it is being filled.

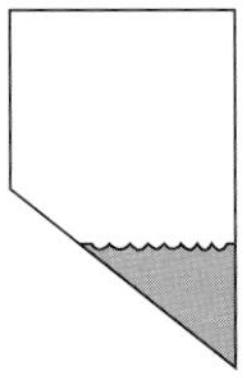

**b** Another view of the container is shown.

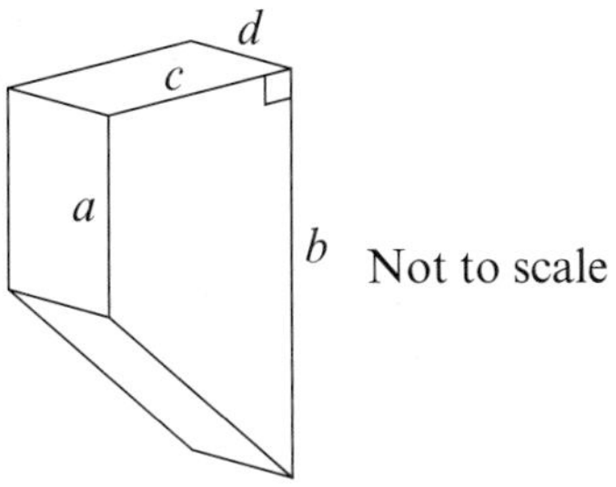


The following formulae represent certain quantities connected with the container.

Which of these formulae represents area?

$$2(a+b) \qquad \tfrac{1}{2}(a+b)c \qquad \tfrac{1}{2}(a+b)cd$$

(3 marks) (SEG, 1998)

**S6.2** The graph shows how the monthly pay of a salesperson depends on sales.

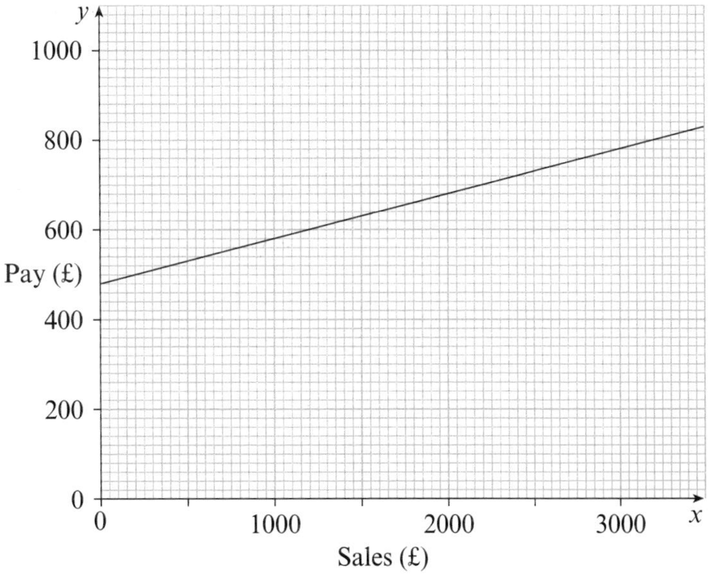


**a** Find the equation of the line in the form $y = ax + b$.

**b** Calculate the pay of a salesperson when sales are £5400.

(5 marks) (SEG, 1999)

**S6.3** In an experiment, different weights are attached to a spring and the length of the spring is measured each time.
The graph shows the results obtained.

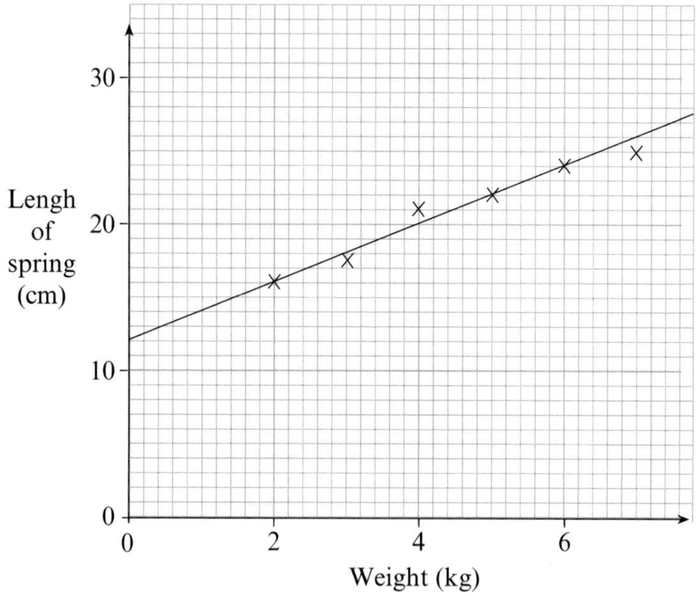


**a** Estimate the length of the spring when no weight is attached.

**b** Calculate the gradient of the line.

**c** Estimate the weight needed to **stretch** the spring by 20 cm.

(5 marks) (SEG, 1999)

**S6.4** Aisha takes part in a race.
The graph shows her distance from the starting line during the race.

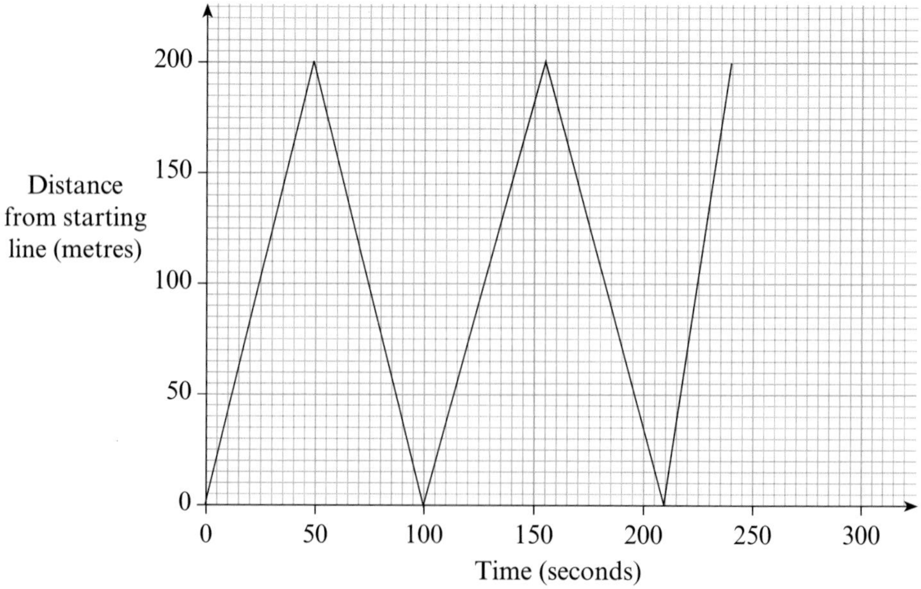


**a** What was her average speed, in metres per second, for the race?

**b** Another runner, Jayne, runs the race at a constant speed of 5 m/s.

On a copy of the diagram draw a graph for Jayne's run.

(4 marks) (SEG, 1998)

**S6.5** The total monthly bill for Ann's mobile phone is made up of two parts: the fixed network charge and the cost of calls.

The graph shows the monthly bill, £$B$, for calls up to a total of 60 minutes.

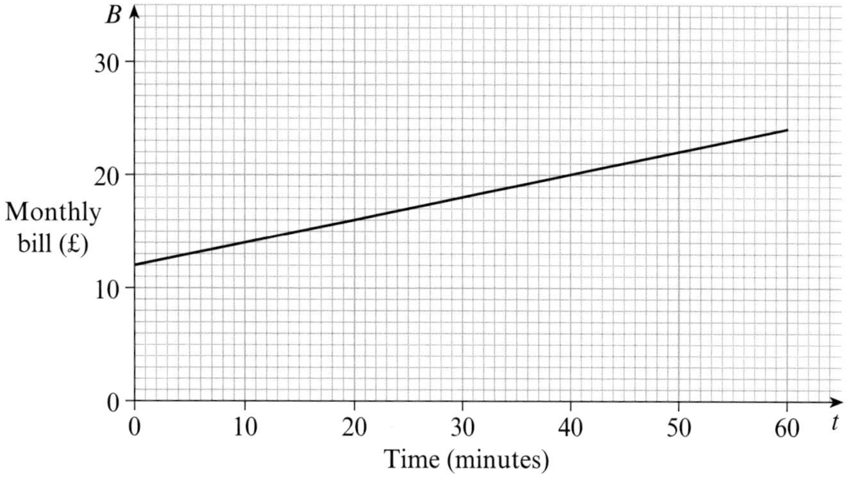


Find the equation of the line.

(4 marks) (AQA)

**S6.6** The graph shows the cost of printing wedding invitation cards.

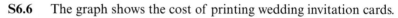

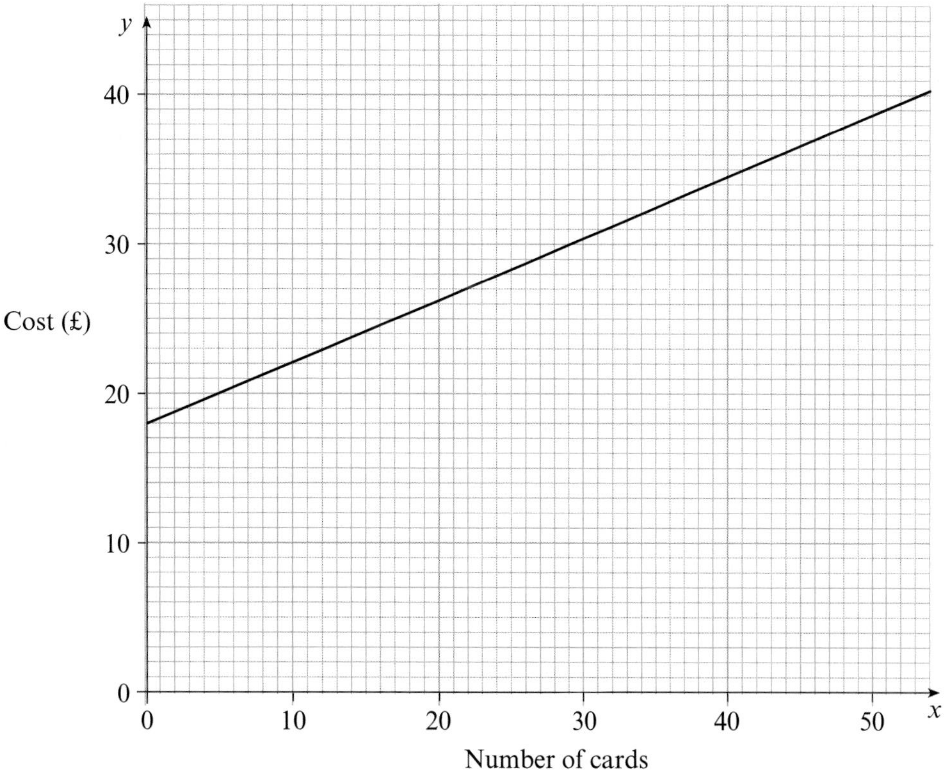


**a** Find the equation of the line in the form $y = mx + c$.

**b** For her wedding Charlotte needs 100 cards to be printed.

How much will they cost?

(5 marks) (SEG, 2000)

## S7
## Pythagoras and trigonometry

**S7.1**

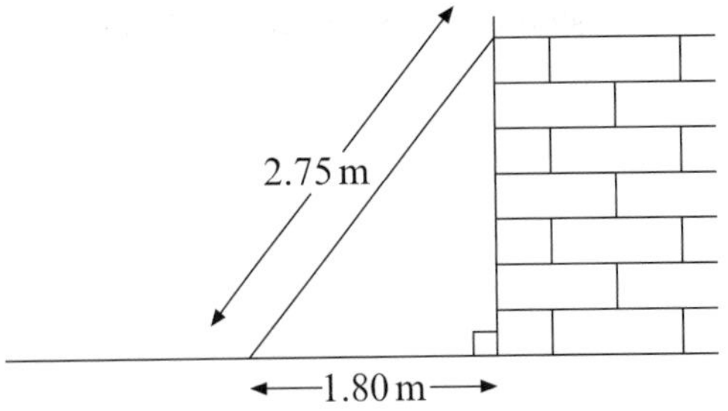


Not drawn to scale

A ladder, 2.75 m long, leans against a wall.
The bottom of the ladder is 1.80 m from the wall, on level ground.

Calculate how far the ladder reaches up the wall.
Give your answer to an appropriate degree of accuracy.

(4 marks) (NEAB, 1998)

**S7.2** Ahmed has this problem to do for homework.

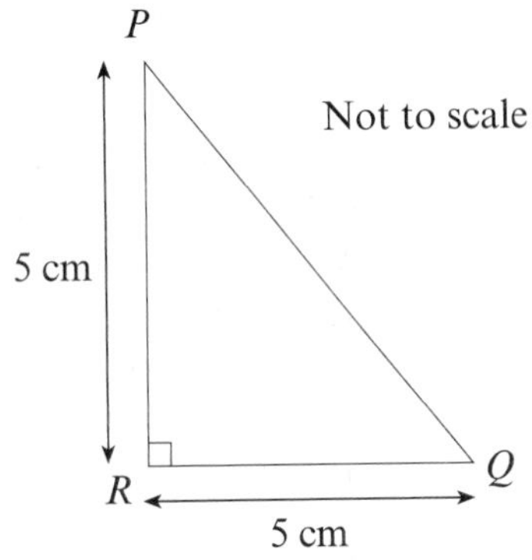


Not to scale

Calculate the length of PQ correct to the nearest cm.

Ahmed's calculator is broken so he thinks he will not be able to do the homework.

His brother says: "You **can** do it, because you only want the answer to the nearest centimetre."

Work out the problem to show that Ahmed's brother is right.

You must show all your working.

(3 marks) (SEG, 2000)

**S7.3** Calculate the length of the line joining the points $A(-3, 2)$ and $B(6, -2)$.

(3 marks) (NEAB, 2000)

**S7.4** An oil rig is 15 kilometres East and 12 kilometres North from Kirrin.

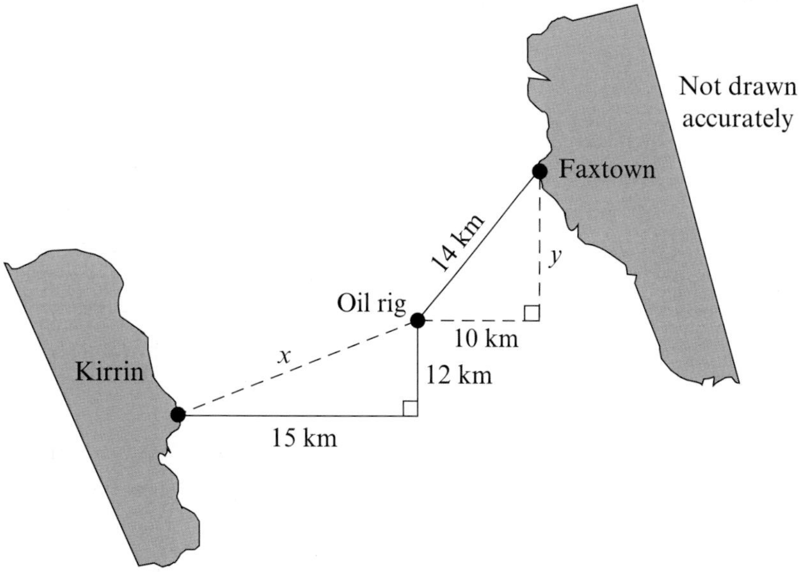


**a** Calculate the direct distance from Kirrin to the oil rig.
(This distance is marked $x$ on the diagram.)

**b** An engineer flew 14 kilometres from Faxtown directly to the oil rig.
The oil rig is 10 kilometres West of Faxtown.
Calculate how far South the oil rig is from Faxtown.
(The distance is marked $y$ on the diagram.)

(6 marks) (NEAB, 1999)

**S7.5** **a** The triangle $ABC$ is shown below.

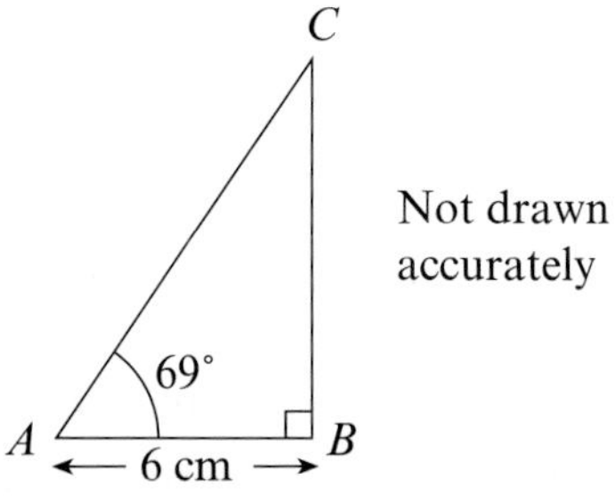


**i** Calculate the length of the side $BC$.

**ii** Calculate the length of the side $AC$.

**b** The triangle $PQR$ is shown below.

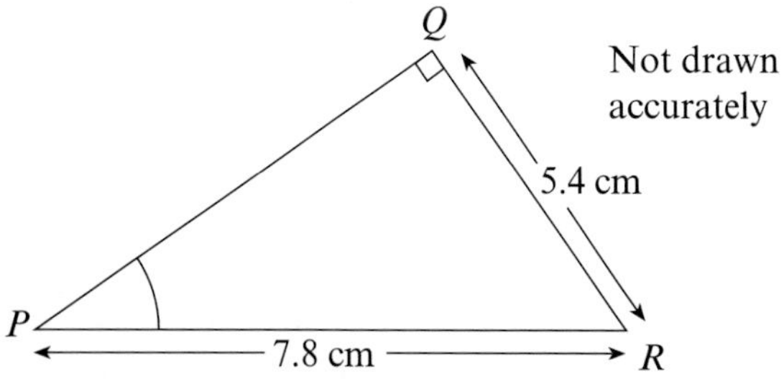


Calculate the size of angle $QPR$.

(9 marks) (NEAB, 1999)

**S7.6** The diagram shows a ramp leading up to the front door of Segville High School.
The ramp is 4500 mm long and it rises at an angle of 6°.

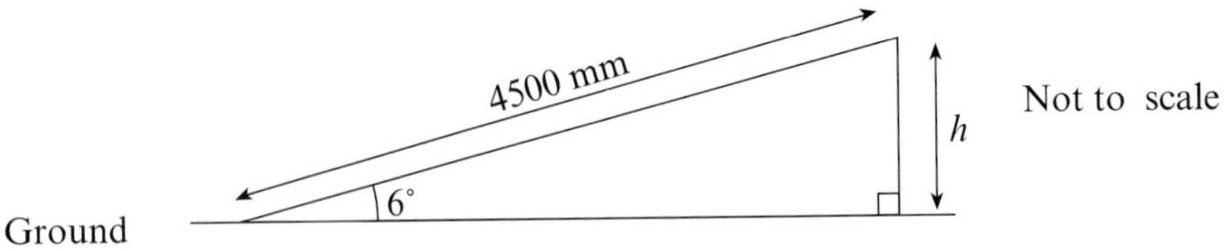


Calculate the height, $h$, of the entrance above ground level.
(3 marks) (SEG, 1999)

**S7.7** The diagram shows a right-angled triangle, $ABC$.

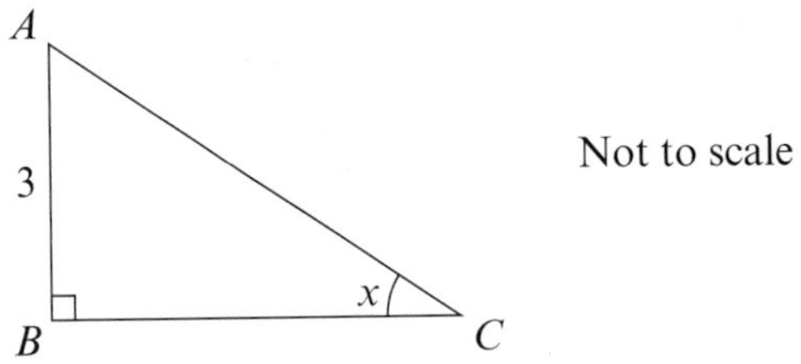


Angle $ABC = 90°$
Tan $x = \frac{3}{4}$.

**a** Calculate sin $x$.

$ABC$ and $PQR$ are similar triangles.

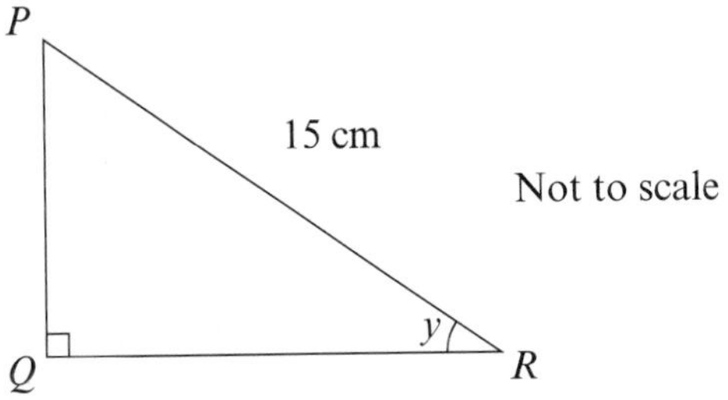


$PQ$ is the shortest side of triangle $PQR$.
Angle $PQR = 90°$ and $PR = 15$ cm.

**b** **i** What is the value of cos $y$?
**ii** What is the length of $PQ$?

(7 marks) (SEG, 2000)

**S7.8** The diagram shows a kite, $ABCD$.

$AE = 12$ cm, $DE = EB = 20$ cm and $BC = 28$ cm.

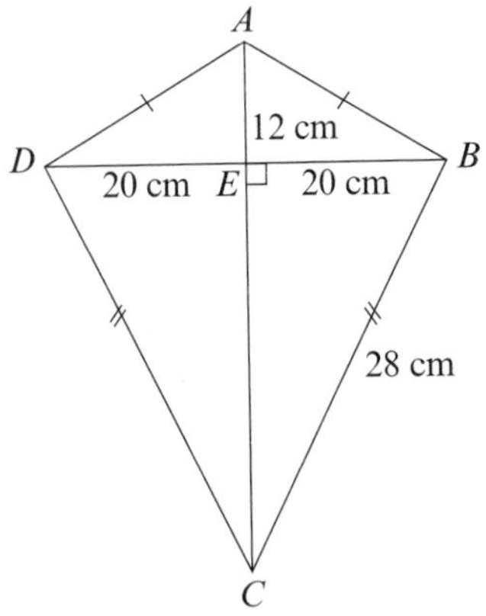


**a** Calculate the size of angle $EBC$.
**b** Calculate the length of $EC$ and hence find the area of the kite.

(8 marks) (SEG, 1998)

**S7.9** In the triangle $ABC$ shown, angle $ABC$ = angle $ACB$.
$BD$ is perpendicular to $AC$ and angle $ABD$ is 45°.

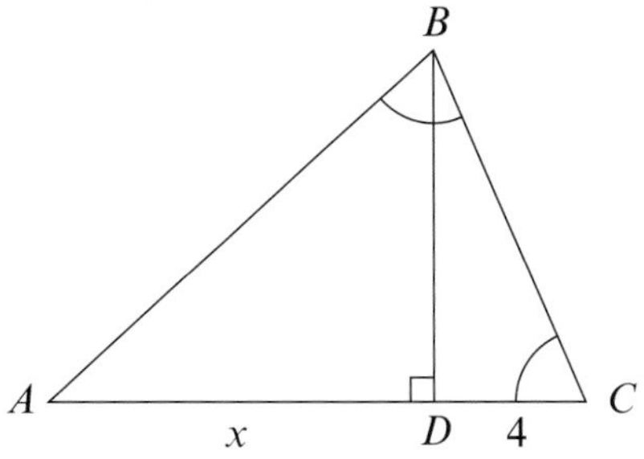


The length of $AD$ is $x$ cm and the length of $CD$ is 4 cm.

**a** Use Pythagoras' theorem in triangle $ABD$ to show that $x^2 - 8x - 16 = 0$.

**b** By solving this equation, find the length of $AC$ correct to one decimal place.

(8 marks) (SEG, 2000)

**S7.10** In triangle $ABC$ the length of $AB$ is 8.3 cm and angle $ABC$ is 20°.
$D$ is a point on $BC$ such that the length of $DC$ is 6.1 cm and angle $ADB$ is 105°.

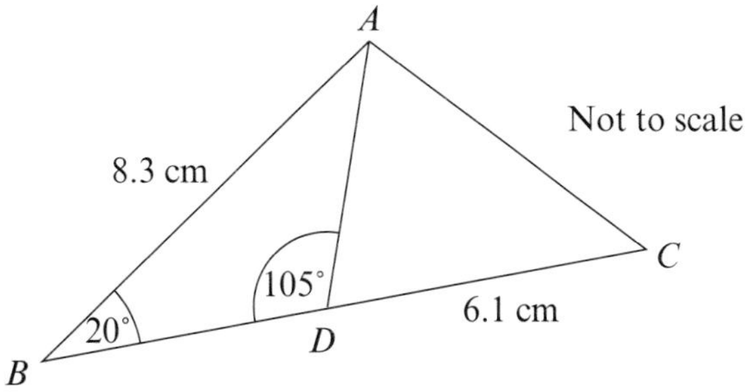


Calculate the length of $AC$.

(6 marks) (SEG, 1999)

**S7.11** A clock has a minute hand that is 10 cm long.
The hour hand is 7 cm long.
Calculate the distance between the ends of the hands when the clock shows 4 o'clock.

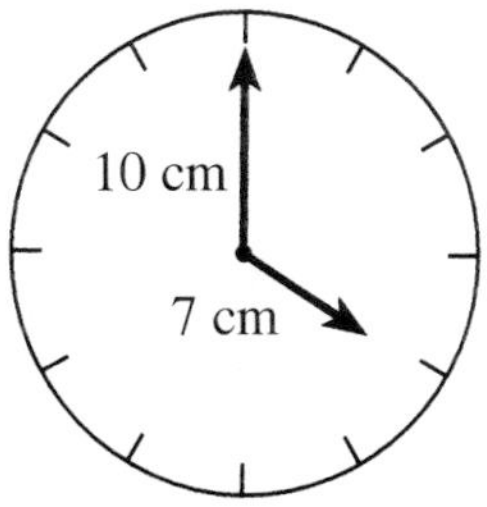


(4 marks) (NEAB, 2000)

**S7.12** A radio mast, $CM$, stands at a corner of a horizontal field, $ABC$.
The angle of elevation of $M$ from $B$ is 5°.

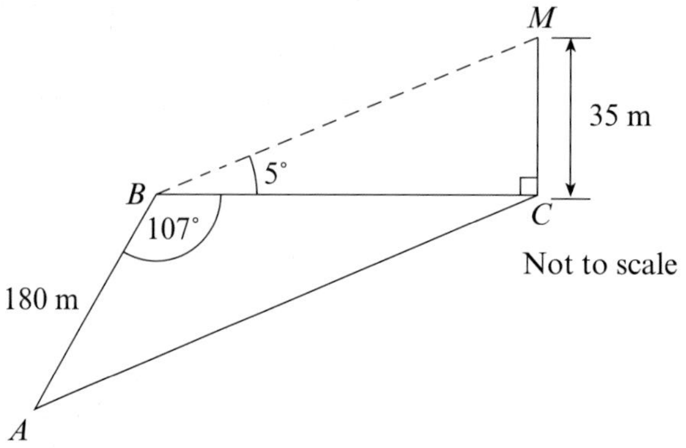


$CM = 35$ m, $AB = 180$ m and angle $ABC = 107°$.

**a** Calculate the length of $AC$.

**b** Calculate the area of $ABC$.

(7 marks) (SEG, 2000)

**S7.13** **a** The sketch shows triangle $ABC$.
$AB = 40$ cm, $AC = 41$ cm and $CB = 9$ cm.

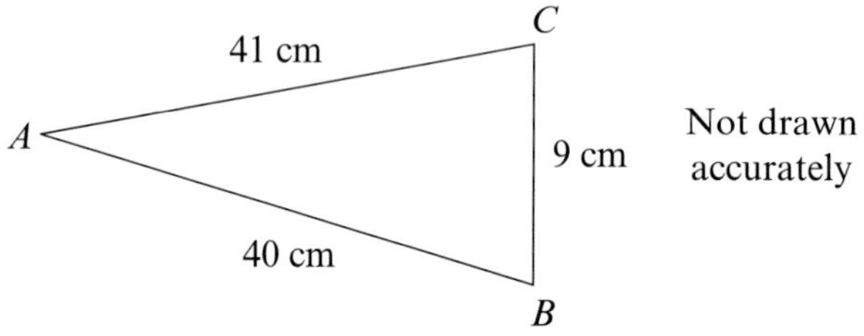


By calculation, show that the triangle $ABC$ is a right-angled triangle.

**b** An open box has internal dimensions 9 cm by 12 cm by 40 cm.

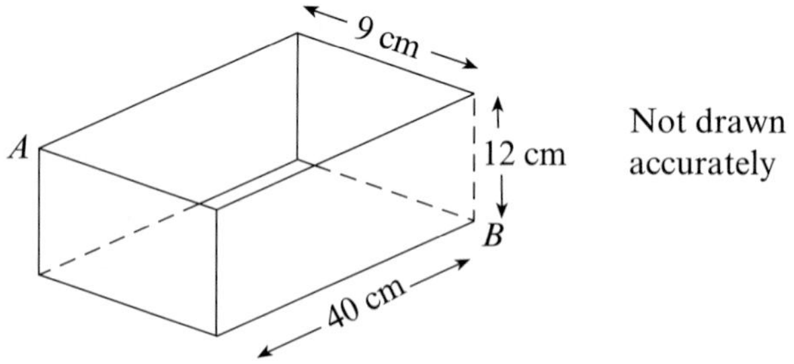


What is the length of the diagonal $AB$?

(5 marks) (NEAB, 2000)

**S7.14**

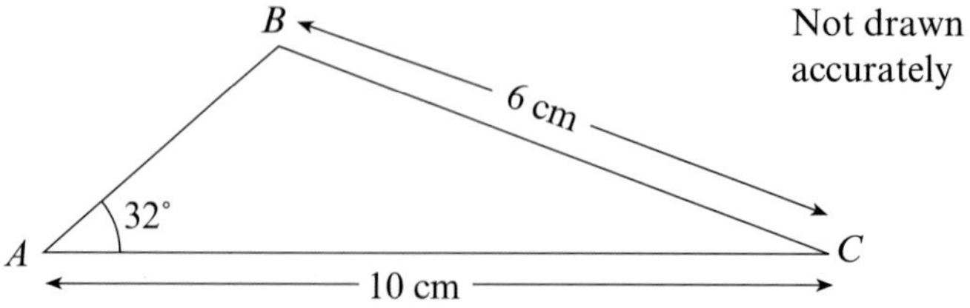


In the triangle $ABC$, angle $ABC$ is **obtuse**.
Angle $BAC = 32°$, $AC = 10$ cm and $BC = 6$ cm.
Calculate the area of triangle $ABC$.

(5 marks) (NEAB, 2000)

**S7.15** In the diagram, the perimeter of triangle $ABC$ is equal to the perimeter of the semicircle, centre $O$, with $AB$ as diameter.

$AC = 12$ cm. $AB = 10$ cm.

Calculate the size of angle $BAC$.

Give your answer to an appropriate degree of accuracy.

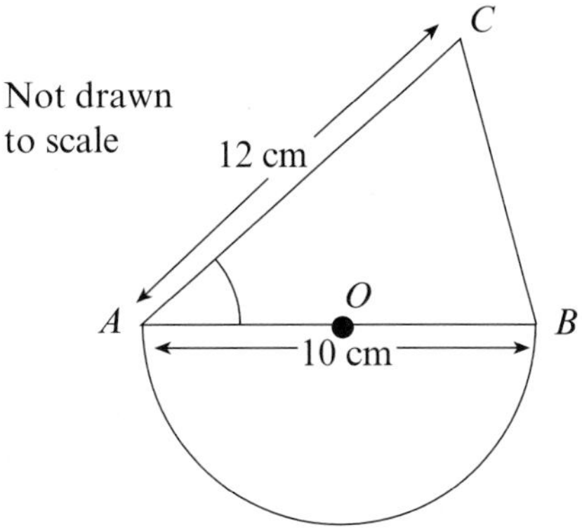


(5 marks) (NEAB, 1999)

**S7.16** In triangle $PQR$, the side $PQ = 7$ cm, the side $QR = 12$ cm and angle $Q = 115°$.

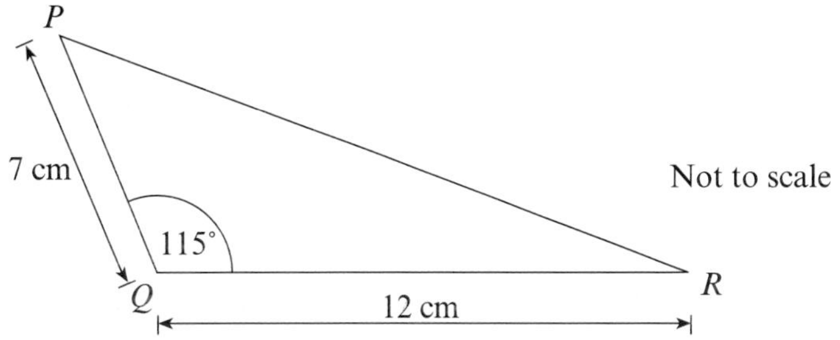


Calculate the area of triangle $PQR$.

(2 marks) (SEG, 1999)

**S7.17** The diagram shows the plan of a garden $PQRS$.

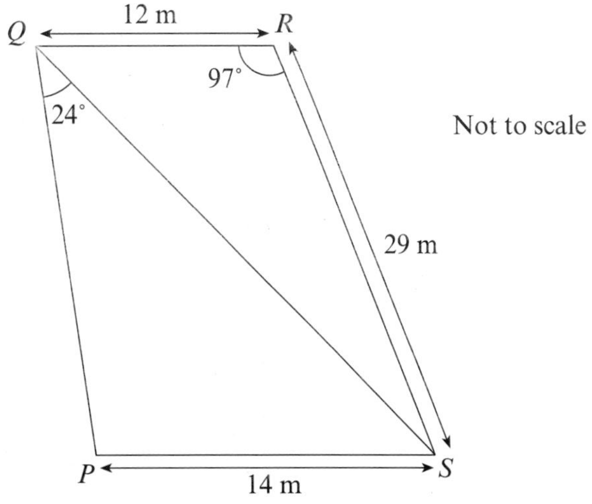


$QPS$ is an obtuse angle.

Calculate angle $QPS$.

(6 marks) (AQA)

**S8**
Transformations

**S8.1** The diagram shows triangles $P$ and $Q$.

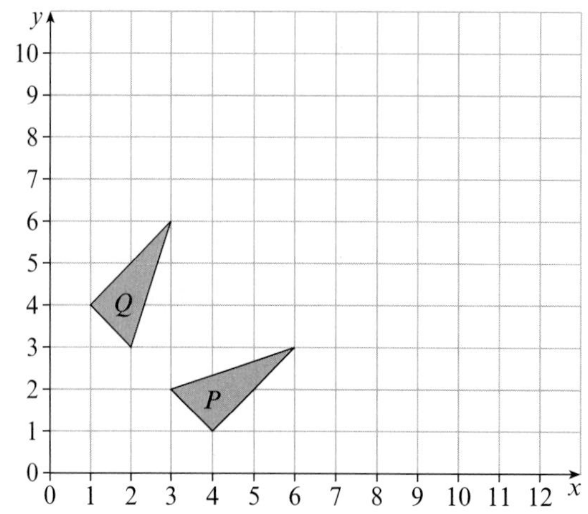


**a** Describe fully the **single** transformation which takes $P$ onto $Q$.

**b** On a copy of the diagram, draw an enlargement of shape $P$ with scale factor 2, centre (3, 2).

(4 marks) (AQA, ••)

**S8.2**

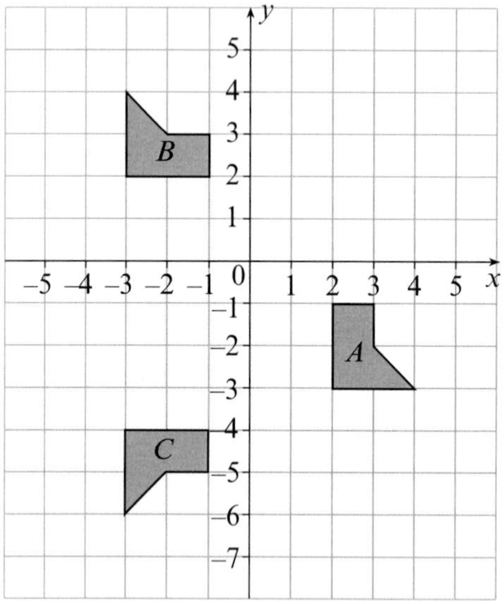


**a** Describe the single transformation that will transform shape $C$ to shape $B$

**b** Describe the single transformation that will transform shape $A$ to shape $C$.

(5 marks) (NEAB, 1999)

**S8.3** Triangles $A$ and $B$ are shown.

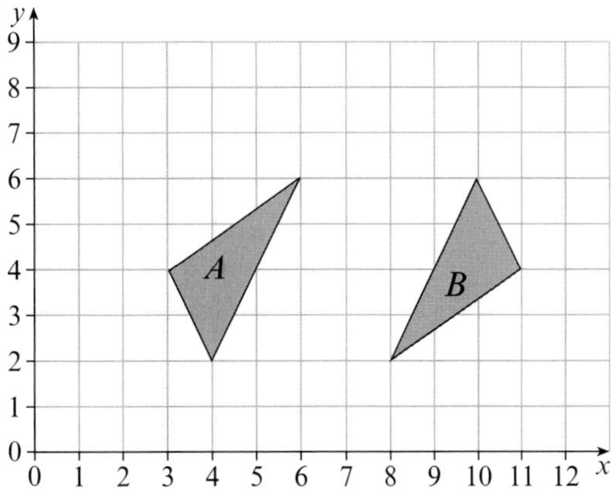


**a** Describe fully the **single** transformation which maps $A$ onto $B$.

**b** $A$ maps onto $C$ by a reflection in the line $y = x$.

Show the position of $C$ on a copy of the diagram.

(5 marks) (SEG, 2000)

**S8.4** The diagram shows three triangles $A$, $B$ and $C$.

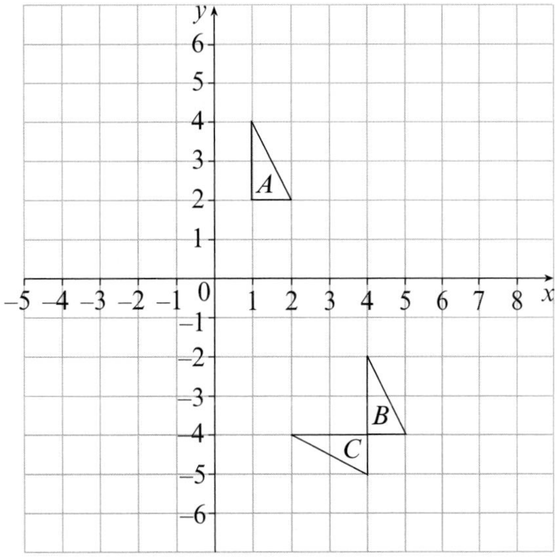


**a** Describe fully the single transformation which maps triangle $A$ onto triangle $B$.

**b** Describe fully the single transformation which maps triangle $B$ onto triangle $C$.

**c** $P$ is a clockwise rotation of 90° with centre (3, –2).
$Q$ is a reflection in the line $y = x - 1$.

Triangle $B$ maps onto triangle $D$ by $P$ followed by $Q$.
Draw triangle $D$ on a copy of the diagram.

(7 marks) (SEG, 2000)

**S8.5** The diagram shows two triangles, *P* and *Q*.

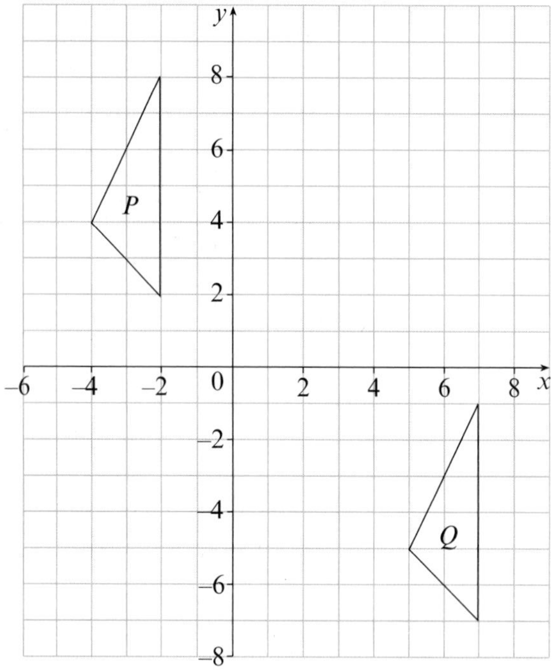


**a** Copy the diagram. Draw the triangle *R* which is an enlargement of triangle *P* with scale factor $-\frac{1}{2}$ and centre (−2, 0).

**b** Describe fully the transformation that takes triangle *R* to triangle *Q*.

(5 marks) (AQA)

**S8.6**

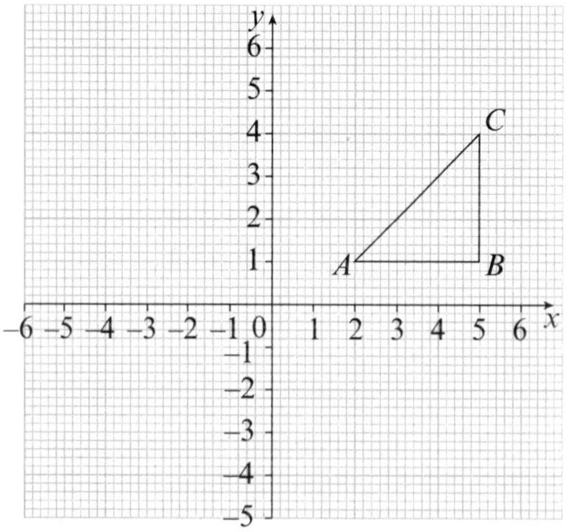


Triangle *ABC* is transformed by a rotation of 90° anti-clockwise about (0, 0) followed by the translation $\begin{pmatrix} 2 \\ -6 \end{pmatrix}$.

**a** Draw the final position of the triangle on a copy of the grid.

**b** Write *A*′ beside the corner of the new triangle that corresponds to *A* on the original triangle.

(4 marks) (SEG, 1999)

**S8.7**

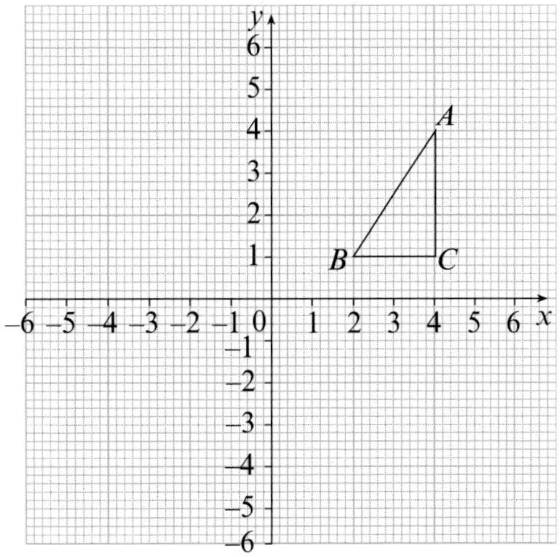


Triangle $ABC$ is transformed by the translation $\begin{pmatrix}-2\\-6\end{pmatrix}$ followed by reflection in the line $y = -x$.

Draw the final position of the triangle on a copy of the grid.

(3 marks) (AQA)

**S8.8** The grid below shows a triangle $ABC$ and a triangle $A'B'C'$.

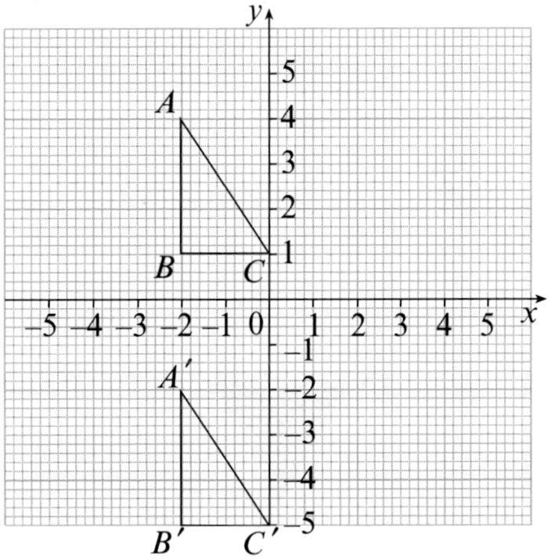


**a** Copy the grid. Draw the triangle $A''B''C''$ which is an enlargement of $ABC$ with a scale factor $-\frac{1}{2}$ with centre (2, 1).

**b** Describe fully the transformation that takes triangle $A''B''C''$ to triangle $A'B'C'$.

(4 marks) (NEAB, 1998)

**S8.9** The triangle, labelled $T$, has vertices (5, 0), (1, 0) and (4, 3).

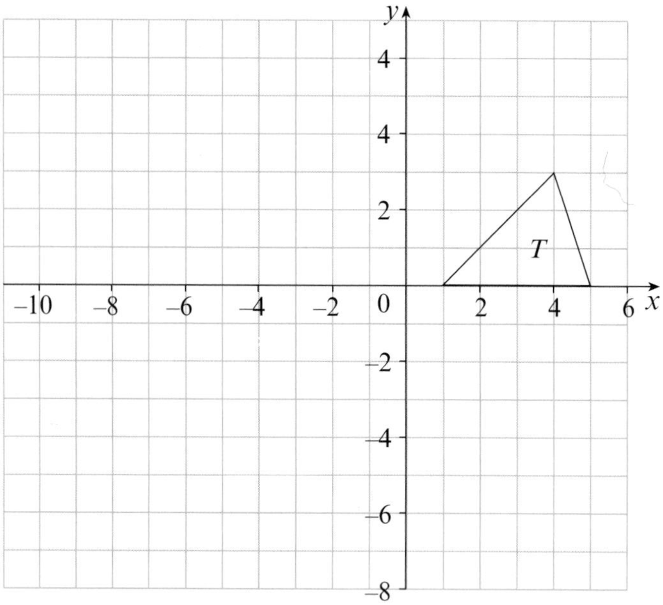


**a** Triangle $T$ is reflected in the line $y = x$.

Copy the grid. Draw the new triangle. Label it $A$.

**b** Triangle $T$ is enlarged by scale factor –2, from centre of enlargement (0, 0).

Draw the new triangle. Label it $B$.

(4 marks) (NEAB, 2000)

**S8.10** The star $ABCDEFGHIJKL$ is made up of 12 equilateral triangles.

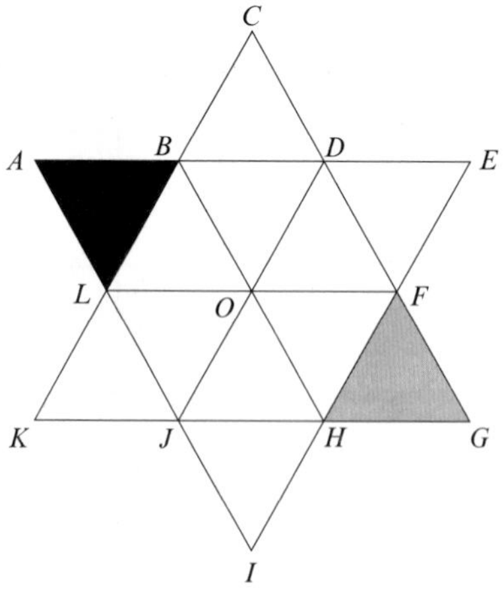


**a** Which triangle will be covered if

**i** triangle $BCD$ is rotated by 60° clockwise about the point $O$

**ii** triangle $BCD$ is enlarged by a scale factor of 2 from the point $C$?

**b** Describe two different **single** transformations that take the black triangle to the grey triangle.

**c** An enlargement takes triangle $CLF$ to triangle $IHJ$.

**i** State the scale factor of the enlargement.

**ii** Mark the centre of enlargement with an X on a copy of the diagram.

(8 marks) (NEAB, 2000)

**S9**
Similarity and congruence

**S9.1** These triangles are congruent.

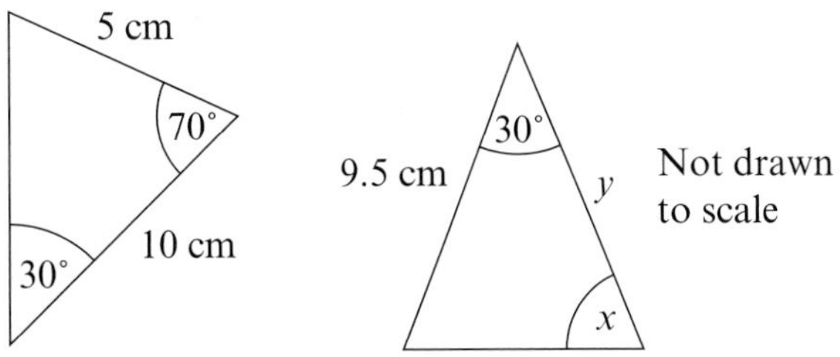


What is the size of

**a** $x$?

**b** $y$?

(2 marks) (NEAB, 1998)

**S9.2** In the diagram $AB$, $PQ$ and $DE$ are parallel.
$AC = CE$.
$AB = 15$ cm. $AC = 9$ cm. $CQ = 8$ cm.

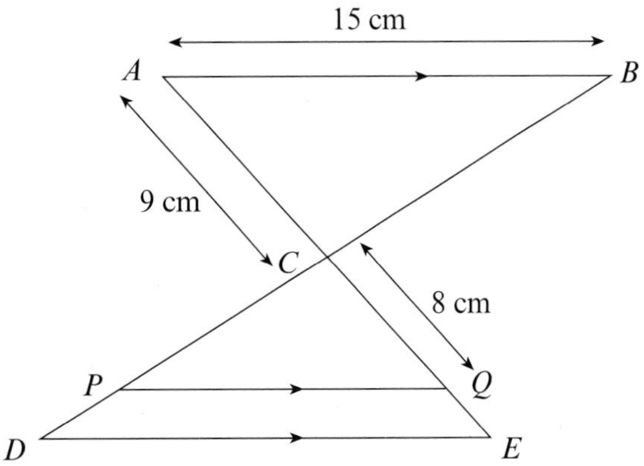


**a** Explain why triangle $ABC$ is congruent to triangle $CDE$.

**b** Calculate the length of $PQ$.

(5 marks) (NEAB, 2000)

**S9.3** At a certain time of day, the shadow of a tree is 11.2 m long. The tree is 9.6 m high.

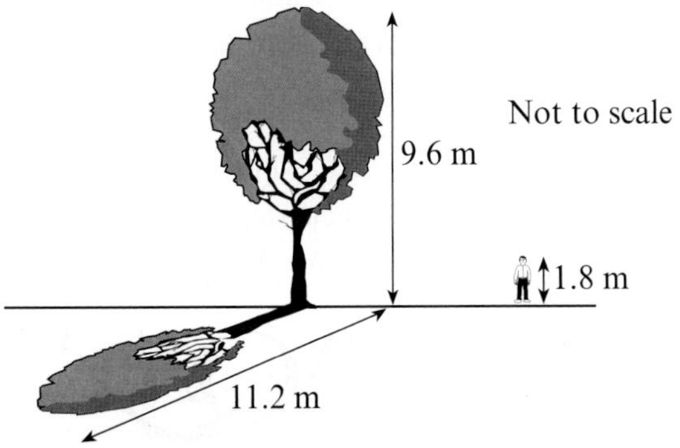


At the same time of day, what is the length of the shadow of a man who is 1.8 m tall?

(3 marks) (SEG, 1999)

**S9.4** $ABC$ is a triangle.
The point $P$ lies on $AB$ and the point $Q$ lies on $AC$.
$PQ$ is parallel to $BC$.
$BC = 12$ cm, $PQ = 8$ cm and $AQ = 6$ cm.

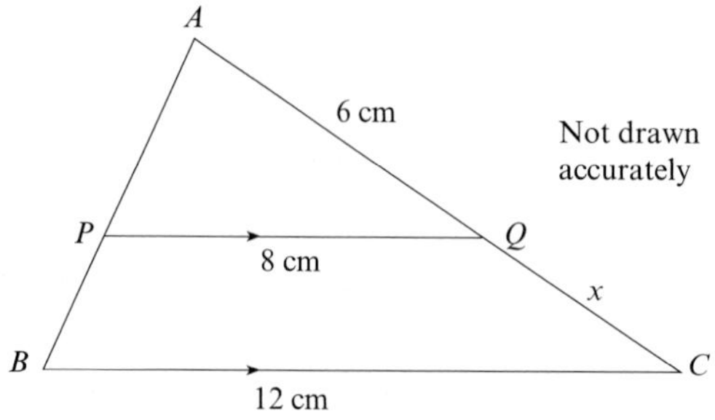


Calculate the length $QC$, marked $x$ on the diagram.
(3 marks) (NEAB, 2000)

**S9.5** Two similar measuring beakers are shown below.

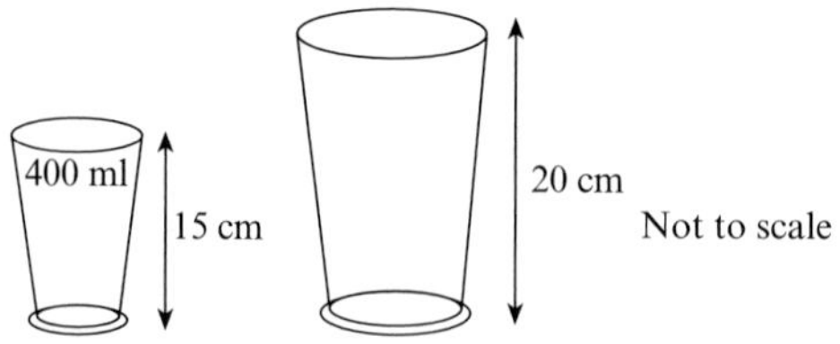


The smaller beaker is 15cm tall and holds 400 ml of liquid.
The larger beaker is 20 cm tall.
How much liquid does the larger beaker hold?
(3 marks) (SEG, 1999)

**S9.6** These cans have the same diameter.

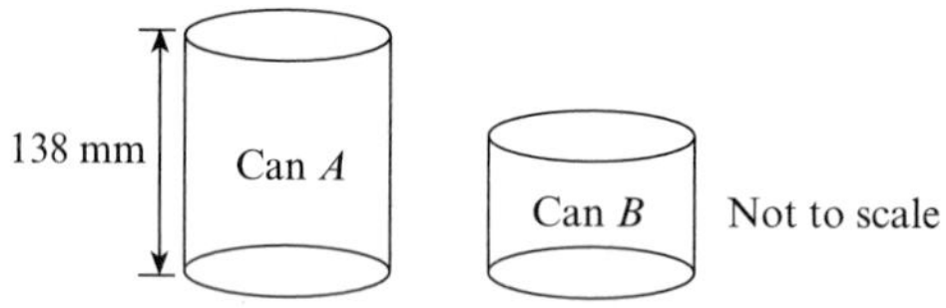


Can $A$ holds 15% more than can $B$ and has a height of 138 mm.
What is the height of can $B$?
(3 marks) (SEG, 1998)

**S9.7** A new AQA sign is to be erected on the Examination Board's offices in Manchester.
The two As in the sign are **similar** in shape.

The cost of each letter A is **proportional to its area**.
The large A costs £250 and is 250 cm high.
The small A is 150 cm high.
How much does the small letter cost?
(3 marks) (NEAB, 2000)

**S9.8** Cone $A$ has a height of 10 cm and a volume of 1280 $cm^3$.

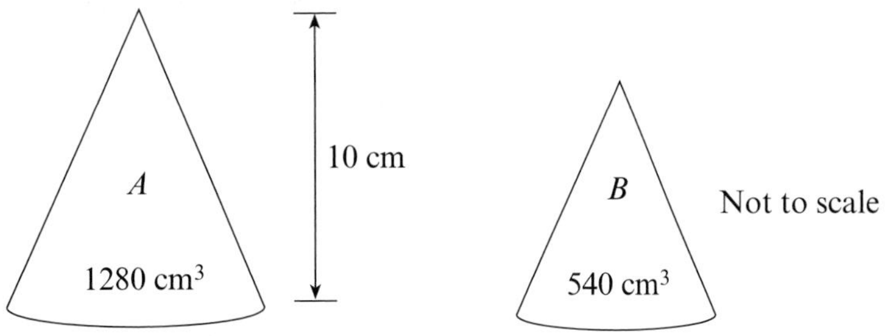


**a** Calculate the radius of cone $A$.

Cone $B$ is similar to cone $A$.
Cone $B$ has a volume of 540 $cm^3$.

**b** Calculate the height of cone $B$.

(6 marks) (SEG, 1999)

**S9.9** These triangles are similar. The lengths are measured in centimetres.

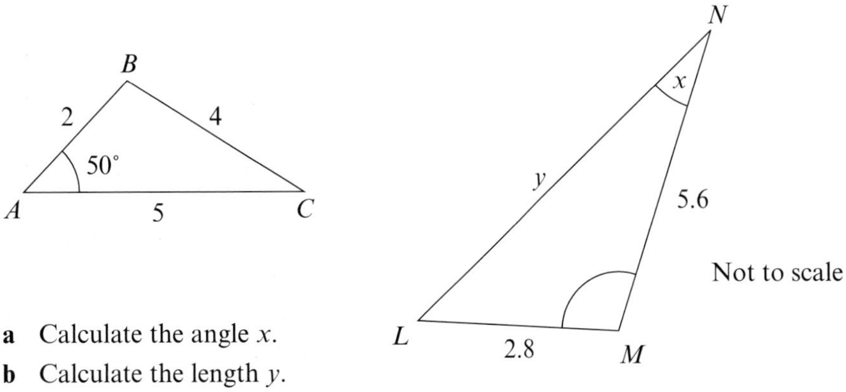


**a** Calculate the angle $x$.

**b** Calculate the length $y$.

**c i** Which of triangles **X**, **Y** and **Z** are **definitely** congruent to triangle $ABC$?

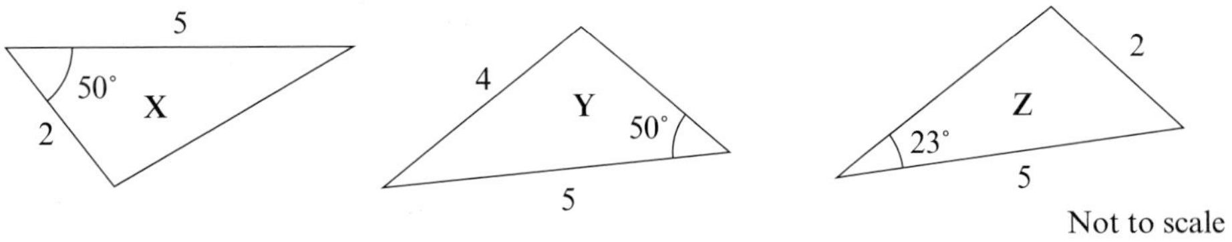


**ii** Explain your answer.

(5 marks) (SEG, 2000)

**S9.10** In the diagram, $A$ and $B$ are the centres of two circles that intersect at $P$ and $Q$.

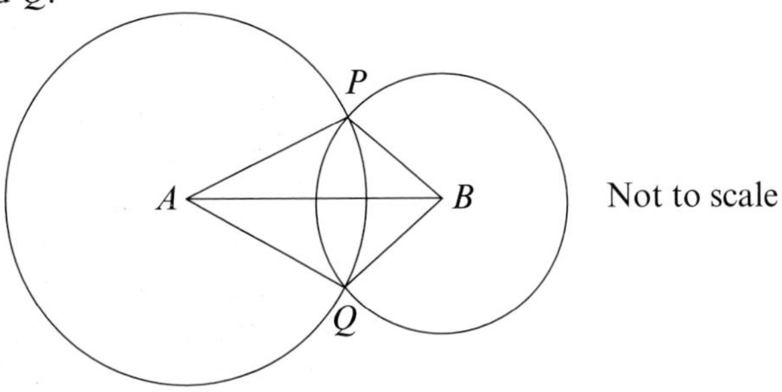


Prove that triangles $APB$ and $AQB$ are congruent.

(4 marks) (AQA)

## S10 Vectors

**S10.1** $OABC$ is a trapezium.

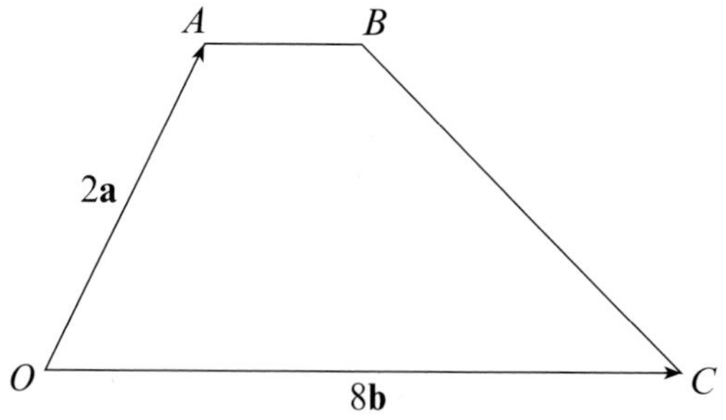


$\overrightarrow{OA} = 2\mathbf{a}$, $\overrightarrow{OC} = 8\mathbf{b}$ and $\overrightarrow{AB} = \frac{1}{4}\overrightarrow{OC}$.

Find in terms of **a** and **b**.

**a** $\overrightarrow{BA}$

**b** $\overrightarrow{BC}$

(3 marks) (SEG, 2000)

**S10.2**

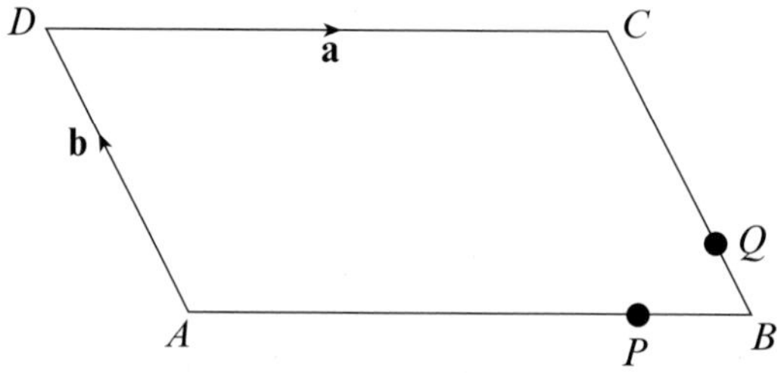


$ABCD$ is a parallelogram. $\overrightarrow{DC} = \mathbf{a}$, $\overrightarrow{AD} = \mathbf{b}$.

$P$ is a point on $AB$ such that $\overrightarrow{AP} = \frac{3}{4}\mathbf{a}$.

$Q$ is a point on $BC$ such that $\overrightarrow{BQ} = \frac{1}{4}\mathbf{b}$.

**a** Find $\overrightarrow{AC}$ in terms of **a** and **b**.

**b** Find $\overrightarrow{PQ}$ in terms of **a** and **b**.

**c** Angle $ABC = 72°$ and angle $ACD = 52°$.
Hence, or otherwise, find the size of angle $PQB$.

(4 marks) (NEAB, 1999)

**S10.3**

$$\overrightarrow{OA} = \begin{pmatrix} 3 \\ 2 \end{pmatrix} \qquad \overrightarrow{OC} = \begin{pmatrix} 3 \\ 6 \end{pmatrix} \qquad \overrightarrow{BC} = \begin{pmatrix} -4 \\ 1 \end{pmatrix}$$

**a** **i** Write down a different column vector which is parallel to $\overrightarrow{BC}$.

**ii** Write down a column vector which is perpendicular to $\overrightarrow{OA}$.

**b** Find $\overrightarrow{AB}$ as a column vector.

(4 marks) (NEAB, 2000)

**S10.4**

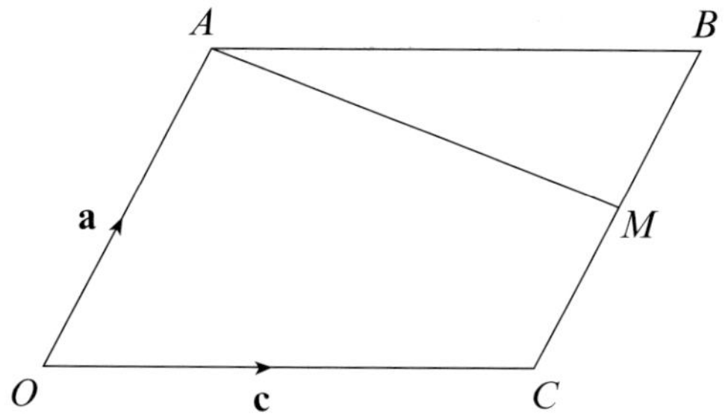


*OABC* is a parallelogram. *M* is the midpoint of *BC*.

$\overrightarrow{OA} = \mathbf{a}$ and $\overrightarrow{OC} = \mathbf{c}$.

**a** Find, in terms of **a** and **c**, expressions for the following vectors.

**i** $\overrightarrow{OB}$

**ii** $\overrightarrow{CM}$

**iii** $\overrightarrow{AM}$

*P* is a point on *AM* such that *AP* is $\frac{2}{3}AM$.

**b** Find, in terms of **a** and **c**, expressions for

**i** $\overrightarrow{AP}$

**ii** $\overrightarrow{OP}$

**c** Describe as fully as possible what your answer to **b ii** tells you about the position of *P*.

(7 marks) (NEAB, 1998)

**S10.5** *OABC* is a quadrilateral.

*M*, *N*, *P* and *Q* are the midpoints of *OA*, *OB*, *AC* and *BC* respectively.

$\overrightarrow{OA} = \mathbf{a}$, $\overrightarrow{OB} = \mathbf{b}$, $\overrightarrow{OC} = \mathbf{c}$.

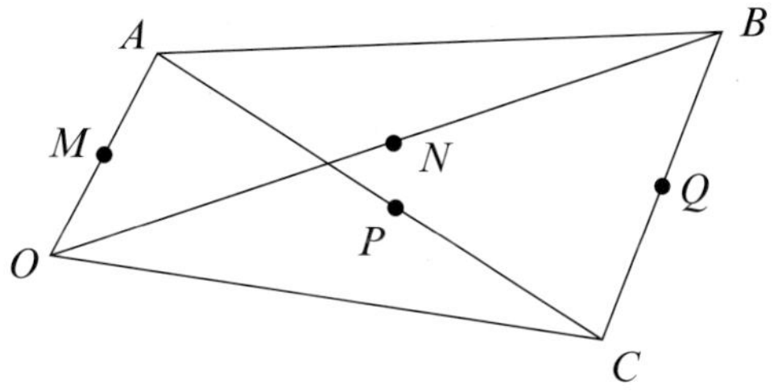


Not drawn accurately

**a** Find, in terms of **a**, **b** and **c**, expressions for

**i** $\overrightarrow{BC}$

**ii** $\overrightarrow{NQ}$

**iii** $\overrightarrow{MP}$

**b** What can you deduce about the quadrilateral *MNQP*? Give a reason for your answer.

(5 marks) (AQA)

**S10.6** $ABCDEF$ is a regular hexagon, centre $O$.

$\overrightarrow{OA} = \mathbf{a}$ and $\overrightarrow{OB} = \mathbf{b}$

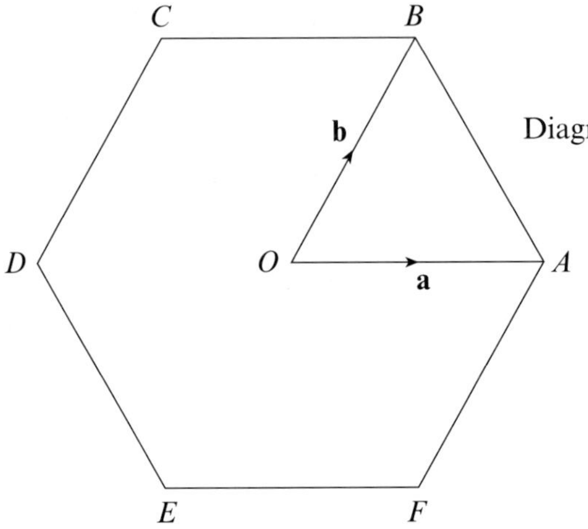


Diagram drawn accurately

**a** Find expressions, in terms of **a** and **b**, for

**i** $\overrightarrow{AB}$

**ii** $\overrightarrow{EC}$

**b** The positions of points $J$, $K$ and $L$ are given by the vectors

$\overrightarrow{OJ} = \mathbf{a} + \mathbf{b}$ $\quad \overrightarrow{OK} = \mathbf{b} - 2\mathbf{a}$ $\quad \overrightarrow{OL} = \mathbf{a} - 2\mathbf{b}$

**i** Draw and label the positions $J$, $K$ and $L$ on a copy of the diagram.

**ii** Find an expression for $\overrightarrow{LK}$, in terms of **a** and **b**.

**iii** What kind of quadrilateral is $ABKL$?
Give a reason for your answer.

(6 marks) (NEAB, 2000)

**S10.7** $OPQR$ is a quadrilateral such that $\overrightarrow{OP} = 4\mathbf{p}$, $\overrightarrow{OQ} = 4\mathbf{q}$ and $\overrightarrow{OR} = 4\mathbf{r}$.

$V$ and $Y$ are points on $OP$ and $OR$ such that $\overrightarrow{OV} = \frac{3}{4}\overrightarrow{OP}$ and $\overrightarrow{OY} = \frac{3}{4}\overrightarrow{OR}$.

$W$ and $X$ are points on $PQ$ and $RQ$ such that $\overrightarrow{PW} = \frac{1}{4}\overrightarrow{PQ}$ and $\overrightarrow{RX} = \frac{1}{3}\overrightarrow{RQ}$.

$VWXY$ is a parallelogram.

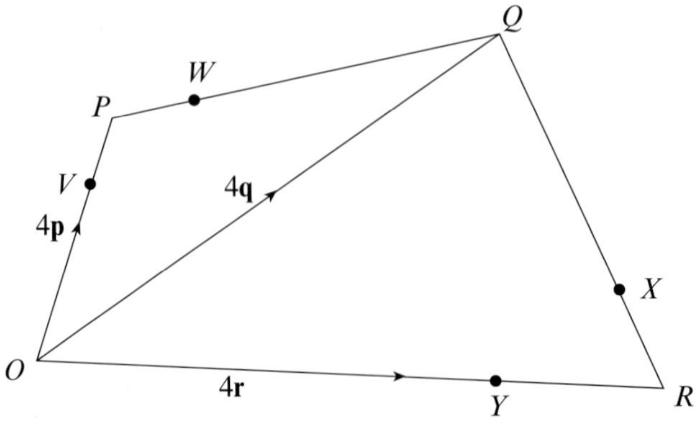


Not drawn accurately

Find expressions in terms of **p**, **q** and **r** for

**a** $\overrightarrow{RQ}$

**b** $\overrightarrow{YX}$

**c** $\overrightarrow{WX}$

(5 marks) (NEAB, 1999)

**S11**
Graphing trigonometric functions

**S11.1** A sketch of the graph of $y = \sin x$ is shown.

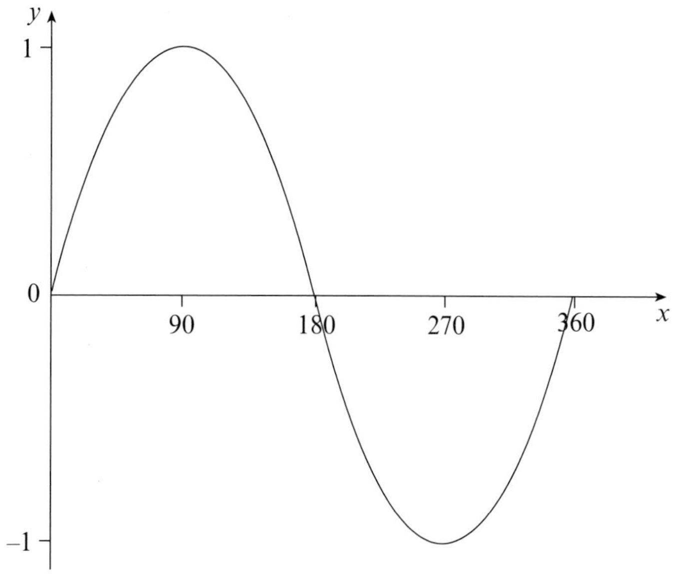


Solve the equation $\sin x = -0.3$ for values of $x$ between 0° and 360°.
(3 marks) (SEG, 1998)

**S11.2** Find two values of $x$, between 0° and 360°, when $\cos x = -0.75$.
(2 marks) (NEAB, 2000)

**S11.3** The diagram shows the graph of $y = \sin x$ for $0° \leqslant x \leqslant 360°$.

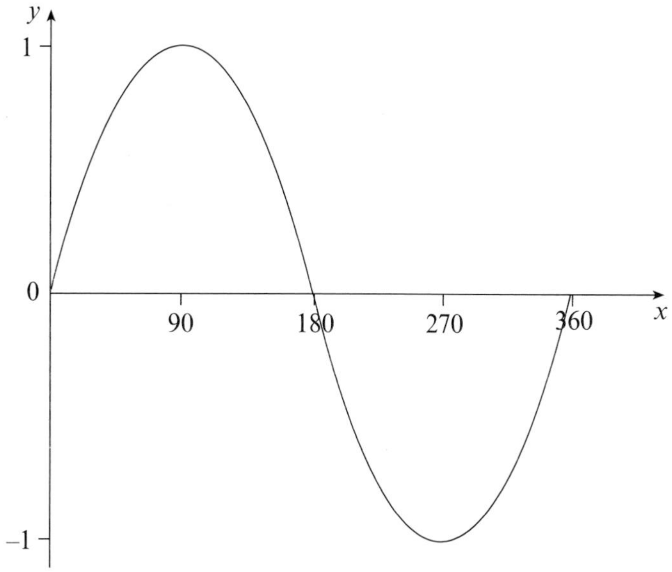


**a** Find a value of $x$, other than 60°, for which $\sin x = \sin 60°$.
**b** Find two values of $x$ for which $\sin x = -\sin 20°$.
(4 marks) (SEG, 2000)

**S11.4** Here is a sketch of the graph of $y = \sin x$ for $o^\circ \leqslant x \leqslant 360^\circ$.

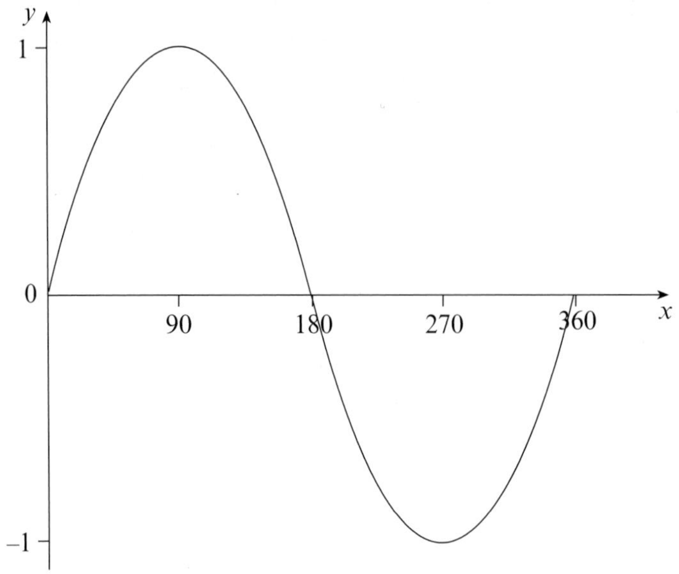


**a** Show on a copy of the sketch the locations of the two solutions of the equation $\sin x = \frac{1}{2}$.

**b** Work out accurately the two solutions of the equation $4 \sin x = -3$ in the range $0^\circ \leqslant x \leqslant 360^\circ$.

(4 marks) (NEAB, 1999)

**S11.5** The graphs of $y = \cos x$ and $y = \sin 2x$ for $0^\circ \leqslant x \leqslant 360^\circ$ are drawn on the grid.

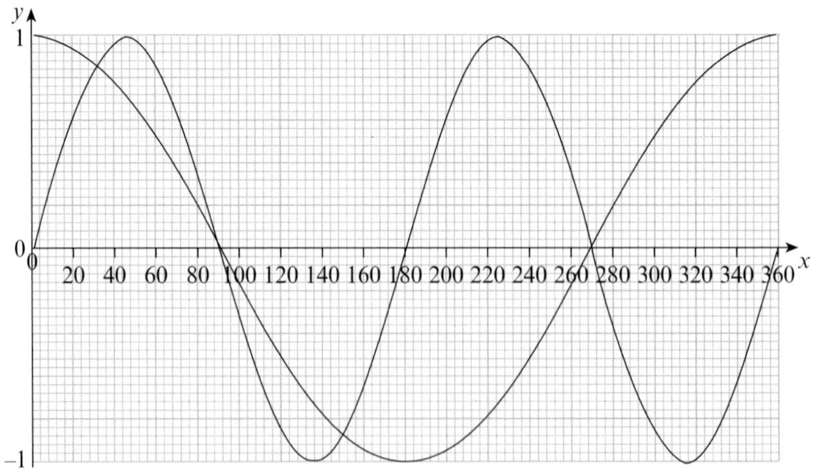


**a** Label each graph on a copy of the grid.

**b** Use the graph to find all the solutions of the equation

$\cos x = \sin 2x$ for $0^\circ \leqslant x \leqslant 360^\circ$.

(4 marks) (NEAB, 2000)

**S11.6** Angles $x$ and $y$ lie between 0° and 360°.

**a** Find the **two** values of $x$ which satisfy the equation $\cos x = 0.5$.

**b** Solve the equation $\sin y = \cos 240°$.

(5 marks) (SEG, 1999)

**S11.7** The diagram shows a sketch of the graph of $y = \cos x$ for $0° \leqslant x \leqslant 360°$.

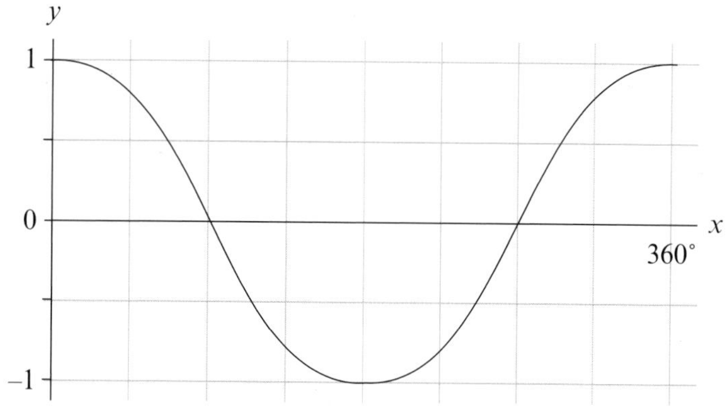


**a** You are given that $\cos 60° = 0.5$.
Write down the other value of $x$ for which $\cos x = 0.5$ for $0° \leqslant x \leqslant 360°$.

**b** Find the solutions of the equation $\cos x = -0.5$ for $0° \leqslant x \leqslant 360°$.

(3 marks) (SEG, 2000)

**S11.8** **a** This is a sketch of the graph $y = \sin x$ for the range $-360° \leqslant x \leqslant 360°$.

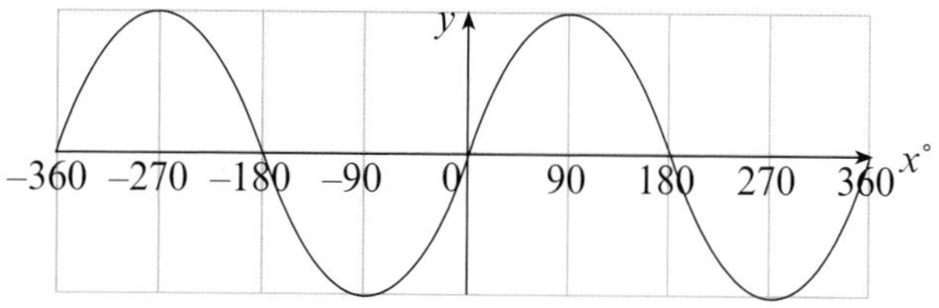


On a copy of the diagram, sketch the graph of $y = -2 \sin x$.

**b** This is a sketch of the graph $y = \cos x$ for the range $-360° \leqslant x \leqslant 360°$.

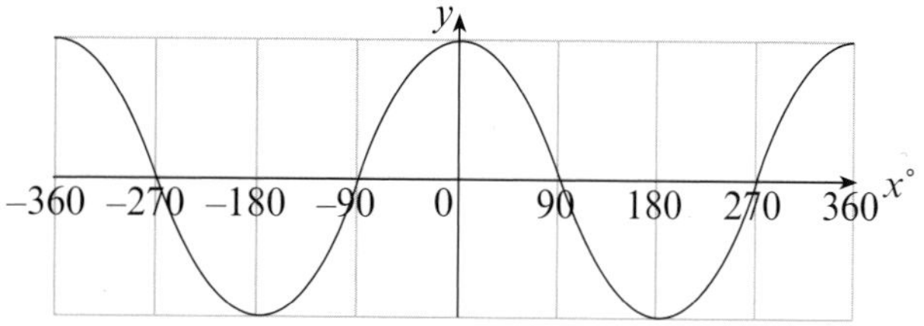


One solution of the equation $\cos x = 0.6$ is $x = 53°$ to the nearest degree.
Find **all** the other solutions to the equation $\cos x = 0.6$ for $x$ in this range.

(5 marks) (SEG, 2000)

# Handling data

**D1**
Probability

**D1.1** Geoff throws a coin 70 times.
He plots the relative frequency of the number of tails after every 10 throws.

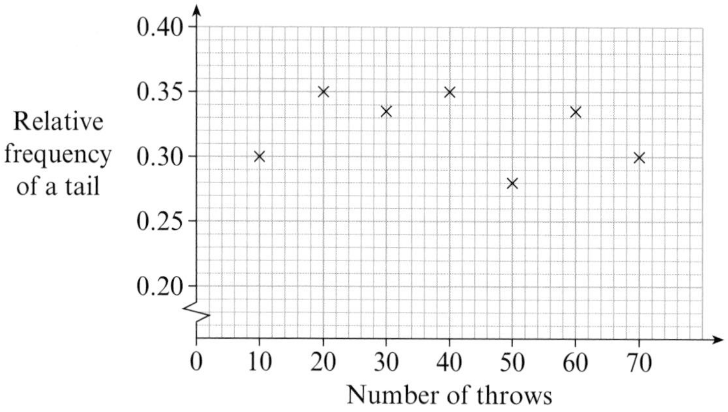


**a** How many tails were obtained in 50 throws?

**b** Use the diagram to estimate the probability of obtaining a tail.

**c** Do you think the coin was biased?
Give a reason for your answer.

(5 marks) (NEAB, 1999)

**D1.2** Ann and Paul each make an ordinary six-sided dice.

**a** Paul throws his dice 12 times.

The number '2' occurs 7 times.

Paul says, 'My dice is no good!'

Explain why Paul may be wrong.

**b** Ann throws her dice 100 times.

The number '2' occurs 19 times.

What is the relative frequency of getting a '2' from Ann's results?

**c** Ann's dice is now thrown 1000 times.

How many times do you expect the number '2' to occur?

(4 marks) (NEAB, 1999)

**D1.3** A bag contains 20 coins.
There are 6 gold coins and the rest are silver.

A coin is taken at random from the bag.
The type of coin is recorded and the coin is then **returned** to the bag.
A second coin is then taken at random from the bag.

**a** The tree diagram show all the ways in which two coins can be taken from the bag.

Write the probabilities on a copy of the diagram.

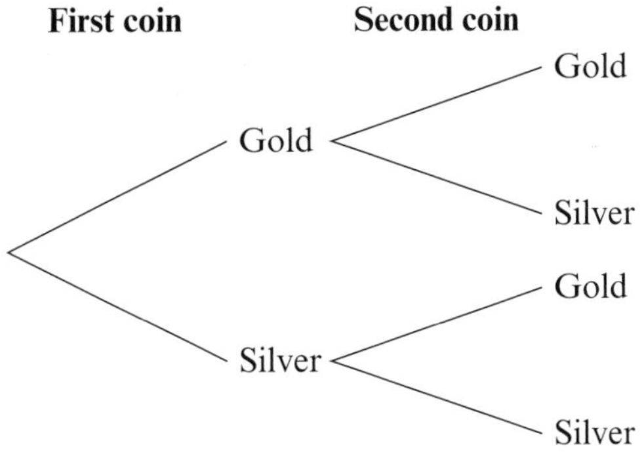


**b** Use your tree diagram to calculate the probability that one coin is gold and one coin is silver.

(5 marks) (SEG, 2000)

**D1.4** One person is to be chosen at random from four men and two women.

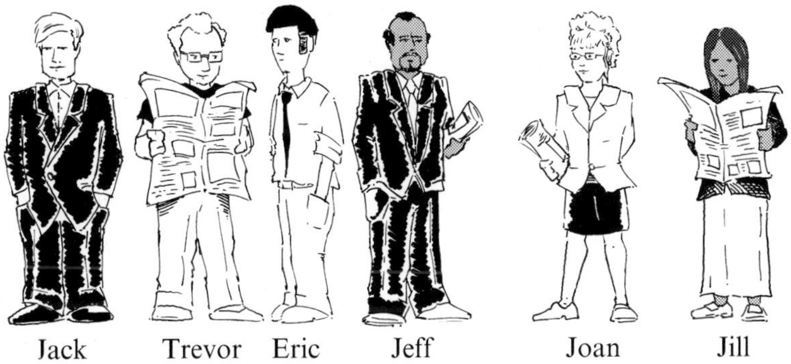


Four events are defined as

Event $J$ Someone with a name beginning with J is chosen.

Event $M$ A man is chosen.

Event $N$ Someone reading a newspaper is chosen.

Event $W$ A woman is chosen.

What is the probability that, if **one person** is chosen at random:

**a** both $J$ and $M$ are true?

**b** both $J$ and $N$ are true?

**c** either $N$ or $W$ is true?

(3 marks) (NEAB, 2000)

**D1.5** The letters of the word C R E A M are written on separate red cards.

The letters of the word C R A C K E R are written on separate blue cards.

All the cards are shuffled.

A red card is taken at random and a blue card is taken at random.

What is the probability of getting a red R and a blue R?

(3 marks) (SEG, 1999)

**D1.6** Jill has a bag containing some black and some white balls.
The probability of taking a black ball from the bag is $\frac{2}{5}$.

**a** Write down a possible pair of values for the number of black and white balls in the bag.

**b** She puts some more black balls in the bag.

The probability of taking a black ball from the bag is now $\frac{2}{3}$.

Using your pair of values from part **a**, find the number of black balls she must have added.

(3 marks) (AQA)

**D1.7** A bag contains ten counters.
Four of the counters are red.
In an experiment, three counters are taken from the bag at random and put in a box.

**a** Calculate the probability that there are **exactly two** red counters in the box.

The same experiment is carried out 600 times.

**b** How many times would you expect there to be **at least two** red counters in the box?

(7 marks) (SEG, 2000)

**D1.8** Two ordinary six-sided dice are thrown.
What is the probability that two fives are thrown?

(2 marks) (NEAB, 2000)

**D1.9** The table shows information about a group of adults.

| | Can drive | Cannot drive |
|---|---|---|
| Male | 32 | 8 |
| Female | 38 | 12 |

**a** A man in the group is chosen at random.

What is the probability that he can drive?

**b** A man in the group is chosen at random and a woman in the group is chosen at random.

What is the probability that both the man and the woman **cannot** drive?

(4 marks) (SEG, 2000)

**D1.10** An office has two photocopiers, $A$ and $B$.

On any one day,

the probability that $A$ is working is 0.8,
the probability that $B$ is working is 0.9.

**a** Calculate the probability that, on any one day, both photocopiers will be working.

**b** Calculate the probability that, on any one day, one of the photocopiers will be working.

(5 marks) (SEG, 2000)

**D1.11** A bag contains packets of crisps of different flavours: 5 chicken, 3 beef and 2 plain.

Abigail takes **three** packets of crisps from the bag at random.

**a** Calculate the probability that she takes

**i** three packets of chicken flavoured crisps

**ii** both packets of plain crisps

**b** Abigail gives, at random, two of her packets to her sister.

Calculate the probability that Abigail's sister receives both packets of plain crisps.

(8 marks) (SEG, 1998)

**D1.12** I have to pay £2 to park in the car park.
The machine which controls the automatic barrier only accepts £1 coins.
Sometimes the machine rejects a coin.
The probability that it rejects any particular coin is 0.1.
If the reject coin is tried again, the probability that it is rejected is 0.4 on each further attempt.

**a** What is the probability that a coin is rejected on the first attempt but accepted on the second attempt?

**b** I have two £1 coins.
If I try each coin no more than twice, what is the probability that I get into the car park?

**c** Write an expression, in terms of $n$, for the probability that a coin is rejected $n$ times.

(9 marks) (NEAB, 1999)

**D1.13** Steve has 6 socks in a drawer,
4 of the socks are black and 2 of the socks are white.

Apart from the colour, the socks are identical.

He takes 2 socks out of the drawer at random.
What is the probability that he gets a pair of socks of the same colour?

(4 marks) (AQA)

**D1.14** A bag contains 4 black buttons and 2 white buttons.

Two buttons are taken from the bag at random.

What is the probability that they are both the same colour?

(4 marks) (AQA)

## D2 Histograms and frequency polygons

**D2.1** A London taxi driver keeps a record of the distance travelled for each of 50 journeys.
The data are summarised in the table below.

| Distance travelled, $d$ (km) | Number of journeys |
|---|---|
| $0 \leqslant d \leqslant 2$ | 12 |
| $2 < d \leqslant 3$ | 11 |
| $3 < d \leqslant 4$ | 10 |
| $4 < d \leqslant 6$ | 10 |
| $6 < d \leqslant 10$ | 6 |
| $10 < d \leqslant 15$ | 1 |

**a** **i** Draw a histogram to illustrate these data.
**ii** Use your histogram to estimate the median distance.

**b** The mean distance is 3.7 km.
Should the taxi driver use the mean or the median to represent the average distance for a journey?
Give a reason for your answer.

(5 marks) (NEAB, 1998)

**D2.2** The students in a school take part in a sponsored silence.
This table shows the 'silent times' of students in Year 7.

| **Silent time** ($t$ minutes) | $0 \leqslant t < 20$ | $20 \leqslant t < 40$ | $40 \leqslant t < 80$ | $80 \leqslant t < 120$ | $120 \leqslant t < 200$ |
|---|---|---|---|---|---|
| **Frequency** | 16 | 20 | 26 | 36 | 8 |

**a** Draw a histogram to represent the data.
This histogram shows the 'silent times' of students in Year 11.

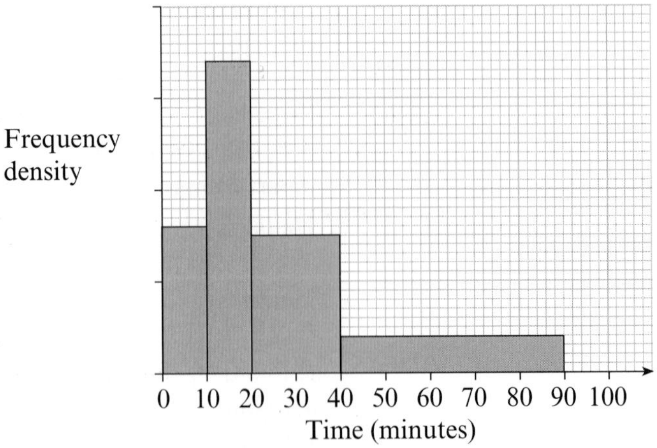


**b** 34 of these students are in the group $10 \leqslant t < 20$.
Two students are chosen at random from Year 11.
Calculate the probability that both of these students are in the group $10 \leqslant t < 20$.

(7 marks) (SEG, 2000)

**D2.3** A leisure company runs two sports clubs, *A* and *B*.
The age distribution of the members of club *A* is shown in this table.

| **Age, $y$ (years)** | $15 \leqslant y < 20$ | $20 \leqslant y < 25$ | $25 \leqslant y < 35$ | $35 \leqslant y < 50$ | $50 \leqslant y < 60$ |
|---|---|---|---|---|---|
| **Frequency** | 13 | 30 | 50 | 45 | 12 |

**a** **i** Show the distribution for club *A* on a histogram.

**ii** Estimate the numbers of club *A* in the age group $40 \leqslant y < 55$.

**b** Club *B* has a total of 250 members.
The distributions of the ages of the male and female members are shown in this table.

| **Age, $y$ (years)** | $y < 20$ | $20 \leqslant y < 50$ | $y \geqslant 50$ |
|---|---|---|---|
| **Male** | 20 | 120 | 30 |
| **Female** | 5 | 60 | 15 |

A survey of the club *B* membership is carried out using a stratified random sample of size 100.

**i** Copy and complete this table to show the number of male and female members of different ages included in the sample.

| **Age, $y$ (years)** | $y < 20$ | $20 \leqslant y < 50$ | $y \geqslant 50$ |
|---|---|---|---|
| **Male** | | | |
| **Female** | | | |

**ii** Give a reason why a stratified random sample should be used rather than a simple random sample.

(9 marks) (SEG, 2000)

**D2.4** The speeds of 100 cars travelling along a road are shown in this table.

| **Speed, $s$ (km/h)** | $20 \leqslant s < 35$ | $35 \leqslant s < 45$ | $45 \leqslant s < 55$ | $55 \leqslant s < 65$ | $65 \leqslant s < 85$ |
|---|---|---|---|---|---|
| **Frequency** | 6 | 19 | 34 | 26 | 15 |

**a** Draw a histogram to show this information.

**b** The speed limit along this road is 48 km/h.

Estimate the number of cars exceeding the speed limit.

(5 marks) (SEG, 1999)

**D2.5** One hundred children were asked to complete a jigsaw puzzle.
The table shows the distribution of the time taken to complete the puzzle by the children.

| **Time** (minutes) | 0– | 4– | 8– | 12– | 16– | 20– | 24– |
|---|---|---|---|---|---|---|---|
| **Frequency** | 0 | 13 | 19 | 30 | 26 | 12 | 0 |

**a** Draw a frequency polygon to show this information.

**b** Calculate an estimate of the mean time taken by these children to complete the puzzle.

(6 marks) (SEG, 2000)

**D2.6** The table shows the speeds of 500 cars travelling on the M1 motorway.
The survey was taken between 8 am and 8.15 am on a weekday.

| Speed, $s$ (mph) | Frequency |
|---|---|
| $20 < s \leqslant 40$ | 52 |
| $40 < s \leqslant 60$ | 102 |
| $60 < s \leqslant 65$ | 174 |
| $65 < s \leqslant 70$ | 106 |
| $70 < s \leqslant 80$ | 42 |
| $80 < s \leqslant 100$ | 24 |

**a** Draw a histogram for the data.

**b** Another survey of 500 cars taken at the same place between 10 am and 10.30 am on the same day gives the histogram below.

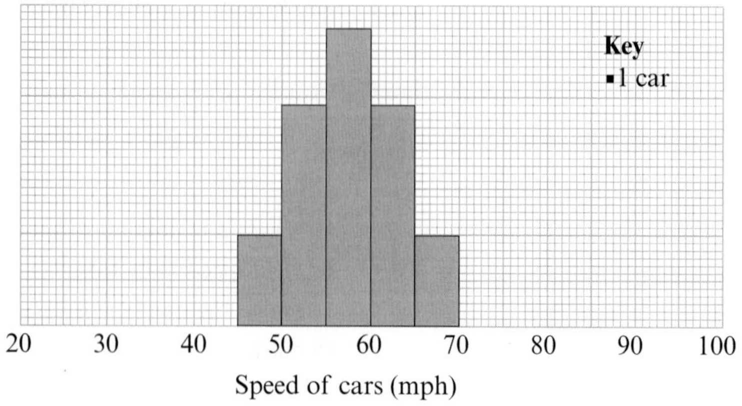


State, with a reason, whether the mean speed has increased or decreased.

(4 marks) (NEAB, 1999)

## D3 Scatter diagrams and correlation

**D3.1** Maizy estimates the lengths of various objects.
This scatter diagram shows her estimates compared with the true lengths.

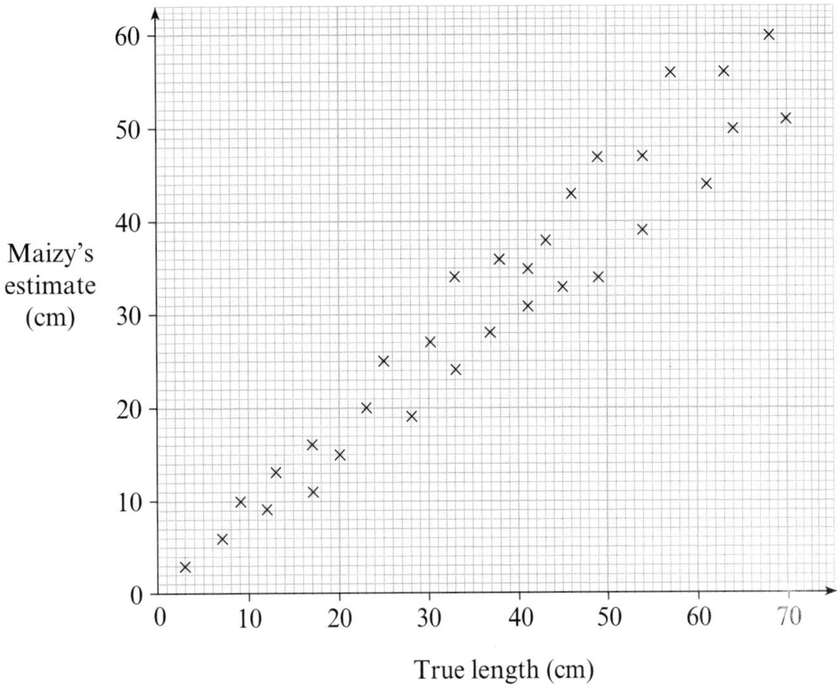


**a** Draw a line of best fit on a copy of the diagram.

**b** Here is part of Maizy's results.

| | | |
|---|---|---|
| True length of object (cm) | *P* | 70 |
| Maizy's estimates length (cm) | 25 | *Q* |

The results labelled *P* and *Q* are missing from the table.
They are not plotted on the scatter diagram.
Use the scatter diagram to estimate the missing results.

**c** Which of your answers to part **b** is likely to be more reliable?

Explain your answer.

(4 marks) (SEG, 2000)

**D3.2** Jane and Sara guessed the weights of various objects. These scatter graphs show their guesses compared with the actual weights.

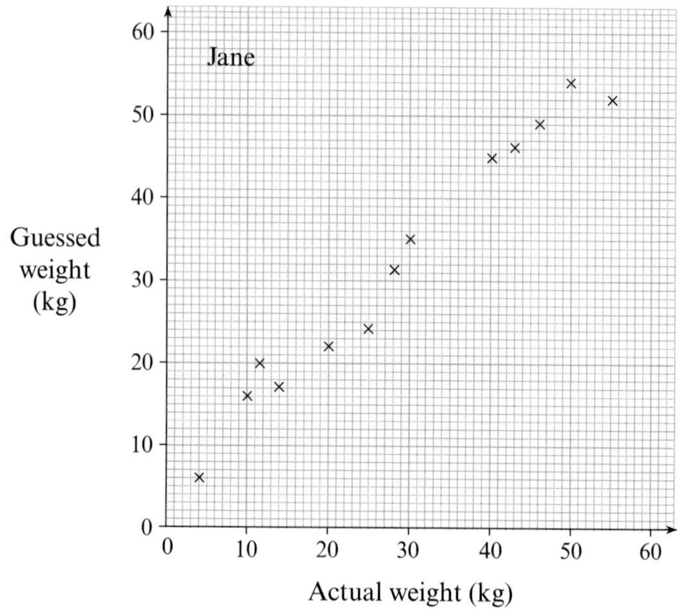


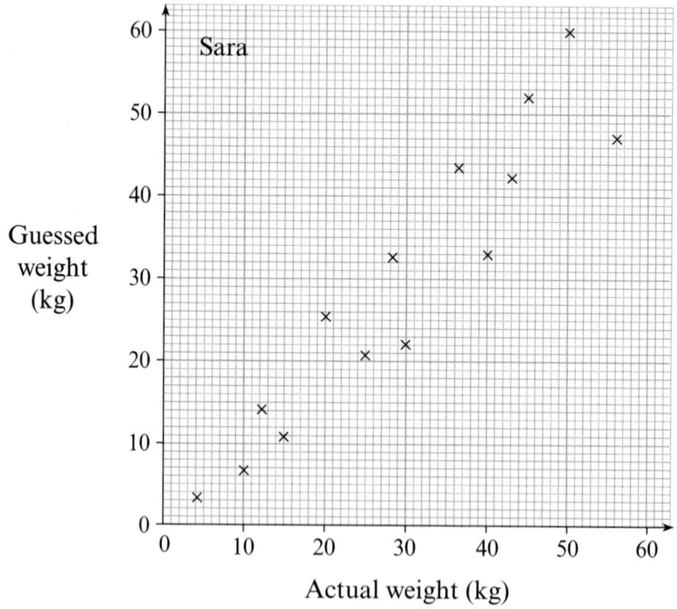


**a** Draw a line of best fit on a copy of Jane's scatter graph.

**b** Comment on the differences between Jane's and Sara's guesses.

**c** One of Sara's guesses on her scatter graph was for an actual weight of 4.5 kg.

Estimate what Jane's guess would have been for this weight.

(4 marks) (SEG, 1999)

**D3.3** The scatter diagram shows the heights of sixteen Year 9 boys and their fathers.

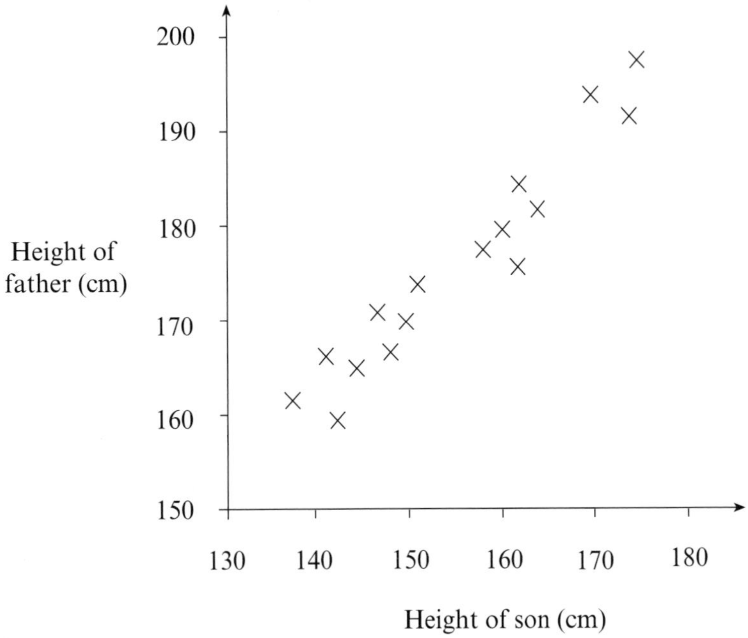


**a** Draw a line of best fit on a copy of the diagram.

**b** Bill, another Year 9 boy, is 155 cm tall.
Use the diagram to estimate the height of Bill's father.
Explain clearly how you obtained your answer.

**c** Look at the distribution of the heights of the fathers.
You want to find an average value for the heights of the fathers in this sample.
Which average, mean, mode or median, would best represent this data?
Give a reason for your choice.

(5 marks) (NEAB, 1998)

**D3.4** The scatter diagram shows the heights of some plants, *d* days after germinating.

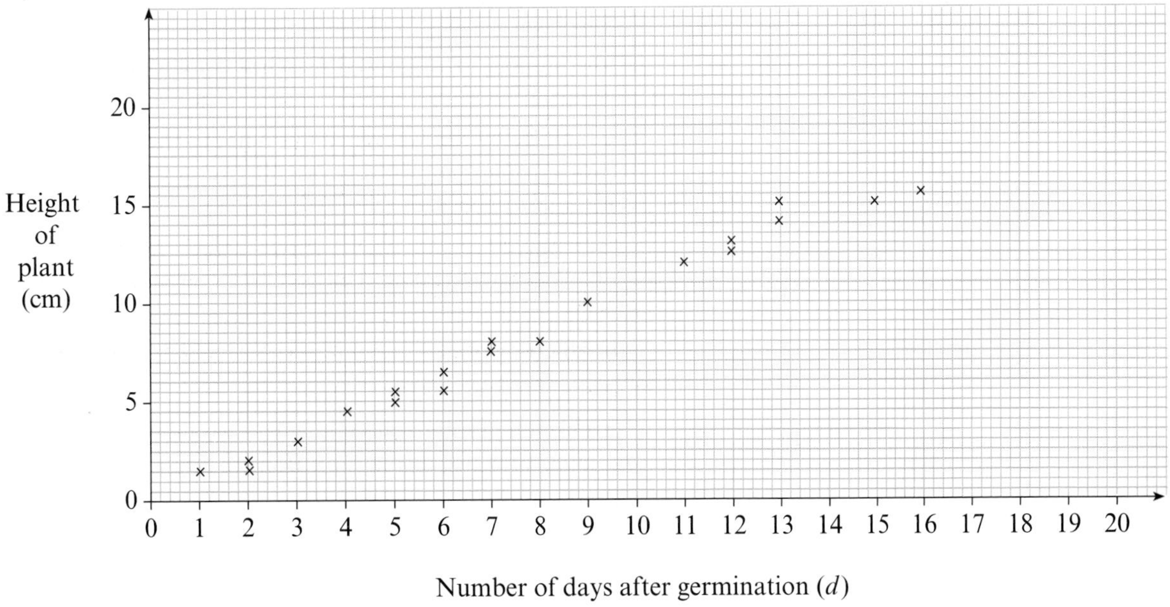

a Draw a line of best fit on a copy of the diagram.

b Use your line of best fit to estimate the height of a plant

i 10 days after germination

ii 20 days after germination

c Which of your two answers in part **b** is likely to be more reliable? Give a reason for your answer.

(4 marks) (SEG, 1998)

**D3.5** Information is collected about the number of organised firework displays and the number of serious injuries on Bonfire Nights over a period of years.

The data are shown in the scatter diagram.

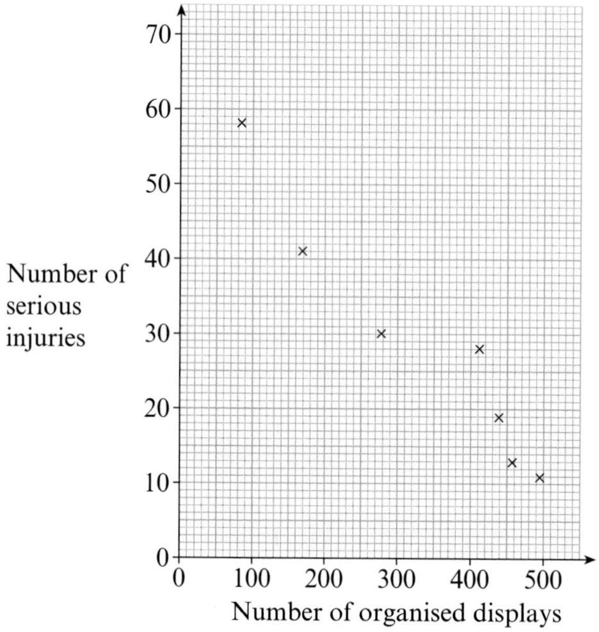


a Draw a line of best fit on a copy of the diagram.

b Use the graph to estimate

i the number of serious injuries if there were 100 organised firework displays

ii the number of organised firework displays if there were 25 serious injuries

c The table below shows information about the number of organised firework displays on Bonfire Nights over a period of years.

| Year | 1987 | 1989 | 1990 | 1991 | 1995 | 1996 | 1997 |
|---|---|---|---|---|---|---|---|
| Number of organised displays | 85 | 169 | 277 | 410 | 432 | 458 | 496 |

Use a sketch graph to show how you would estimate the number of organised firework displays on Bonfire Night in 1993.

(5 marks) (NEAB, 1998)

**D3.6** A hospital carries out a test to compare the reaction times of patients of different ages.
The results are shown as the scatter graph.

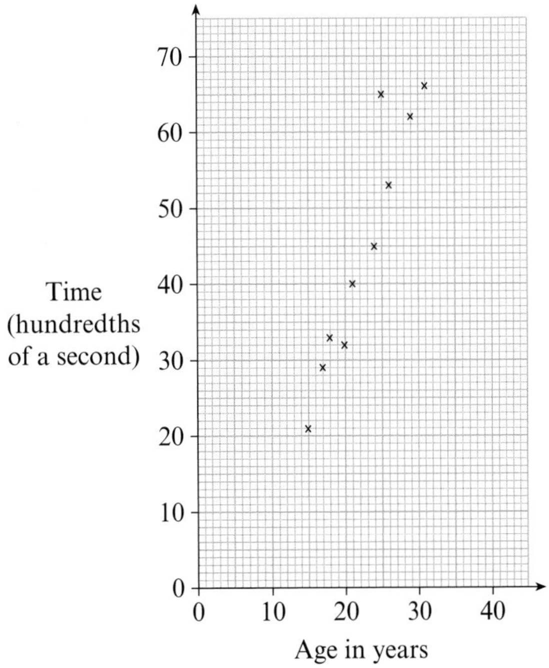


**a** Draw a line of best fit on a copy of the scatter graph.

**b** The hospital is worried about the reaction time of one patient.

**i** How old is the patient?

**ii** Using the line of best fit, what should the reaction time be for this patient?

**c** Hospital records for this reaction test give the following information.

| | 15 year olds | 30 year olds |
|---|---|---|
| | Time (hundredths of a second) | Time (hundredths of a second) |
| Lower quartile | 20 | 61 |
| Median | 22 | 65 |
| Upper quartile | 25 | 76 |

Look at the information for the two age groups.
Write down two comments comparing the two sets of information.

(6 marks) (NEAB, 2000)

**D4**
**Averages and measures of spread**

**D4.1** The table shows the time taken by 40 pupils to do their homework.

| Time, $t$ (minutes) | Number of pupils |
|---|---|
| $0 < t \leqslant 10$ | 2 |
| $10 < t \leqslant 20$ | 3 |
| $20 < t \leqslant 30$ | 12 |
| $30 < t \leqslant 40$ | 11 |
| $40 < t \leqslant 50$ | 8 |
| $50 < t \leqslant 60$ | 4 |

Calculate an estimate for the mean time taken to do the homework.
(4 marks) (NEAB, 1999)

**D4.2** The maximum load for a lift is 1200 kg.
The table shows the distribution of the weights of 22 people waiting for the lift.

| Weight ($w$ kg) | Frequency |
|---|---|
| $30 \leqslant t < 50$ | 8 |
| $50 \leqslant t < 70$ | 10 |
| $70 \leqslant t < 90$ | 4 |

Will the lift be overloaded if all of these people get in?
You **must** show working to support your answer.
(4 marks) (SEG, 2000)

**D4.3** A survey was made of the amount of money spent at a supermarket by 100 shoppers on a Monday. The table shows the results.

| Amount spent $M$ (£) | Number of shoppers |
|---|---|
| $0 \leqslant M < 20$ | 42 |
| $20 \leqslant M < 40$ | 37 |
| $40 \leqslant M < 60$ | 18 |
| $60 \leqslant M < 80$ | 3 |

**a** Calculate an estimate of the mean amount of money spent by these shoppers.

The frequency polygon shows the amount of money spent by 100 shoppers at the supermarket on a Saturday.

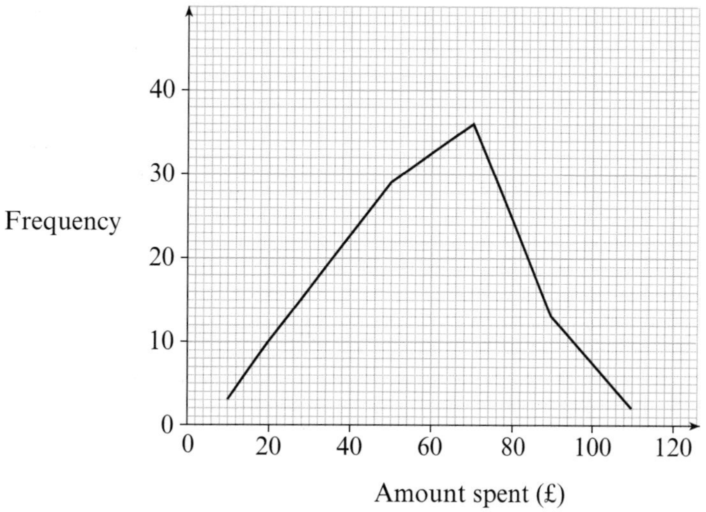


**b** On a copy of the diagram draw a frequency polygon for the amount of money spent by the shoppers on Monday.

**c** Compare and comment on the amount of money spent by shoppers on these two days.

(7 marks) (SEG, 2000)

**D4.4** Farmer Jack has two fields.

In one field, he plants lettuce seeds by hand.
In the other, he plants lettuce seeds using a machine.

A sample of lettuce is taken from each field.
The diameter of each lettuce is measured.
The mean and interquartile range are calculated for each sample.

The results are shown below.

| | Hand | Machine |
|---|---|---|
| Mean diameter | 15.3 cm | 14.7 cm |
| Interquartile range | 3.2 cm | 1.2 cm |

**a** Farmer Jack says, 'It's better to plant by hand.'
Using the data given in the table, give a reason to justify his statement.

**b** His son Peter says, 'No dad, the machine is better.'
Using the data given in the table, give a reason to justify his statement.

(2 marks) (NEAB, 2000)

**D4.5** The table shows the age distribution of females taking part in a marathon.

| Age ($a$ years) | Females |
|---|---|
| $10 \leqslant a < 20$ | 0 |
| $20 \leqslant a < 30$ | 15 |
| $30 \leqslant a < 40$ | 27 |
| $40 \leqslant a < 50$ | 14 |
| $50 \leqslant a < 60$ | 4 |
| $60 \leqslant a < 70$ | 0 |

**a** Calculate an estimate of the mean age of these females.

**b** These frequency polygons show the distribution of the ages of the males and the females taking part in the marathon.

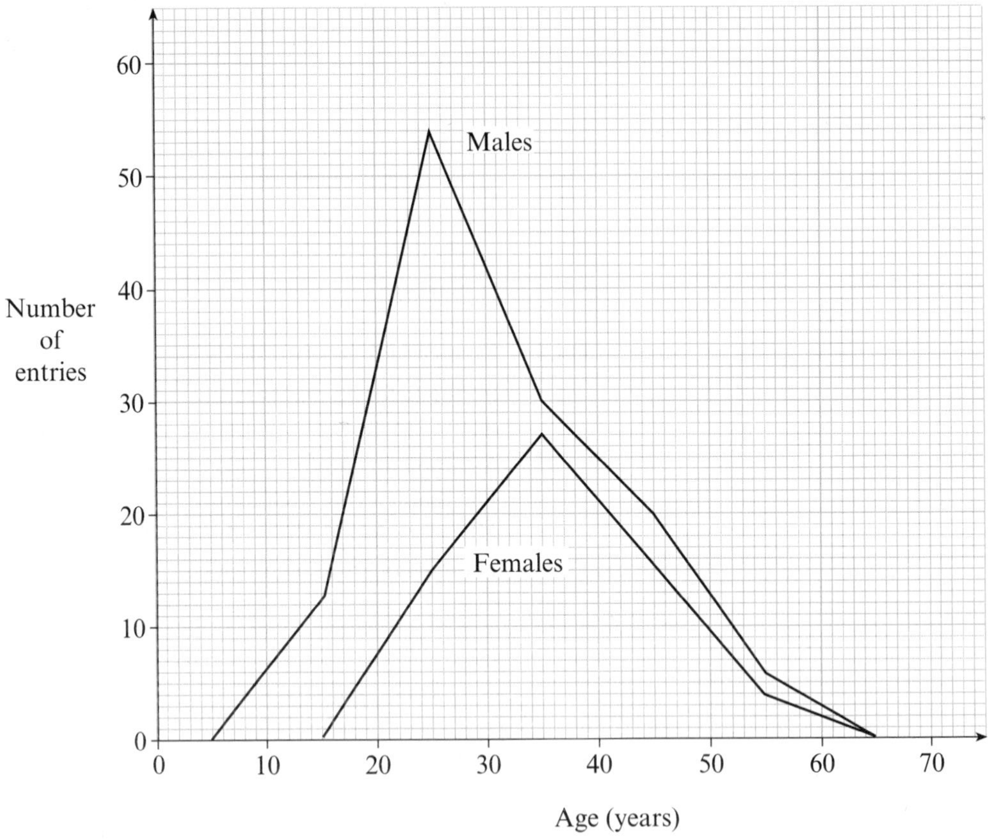


**i** Use the frequency polygons to compare the means and the spreads of the two distributions.

**ii** How many males in the marathon are younger than 30?

(8 marks) (SEG, 1999)

**D5.3** The ages of 500 people attending a concert are given in the table below.

| Age, $A$, years | Number of people | Cumulative frequency |
|---|---|---|
| $0 \leqslant A < 10$ | 20 | |
| $10 \leqslant A < 20$ | 130 | |
| $20 \leqslant A < 30$ | 152 | |
| $30 \leqslant A < 40$ | 92 | |
| $40 \leqslant A < 60$ | 86 | |
| $60 \leqslant A < 80$ | 18 | |
| $80 \leqslant A < 100$ | 2 | |

**a** **i** Copy the table and complete the cumulative frequency column.

**ii** Draw a cumulative frequency diagram.

**b** Use your diagram to estimate

**i** the median age

**ii** the interquartile range

**c** Use your diagram to estimate the percentage of people at the concert who are under the age of 16 years.

(10 marks) (NEAB, 1999)

**D5.4** In a school survey, the heights of 80 boys aged 14 years are recorded. The cumulative frequency graph shows the results.

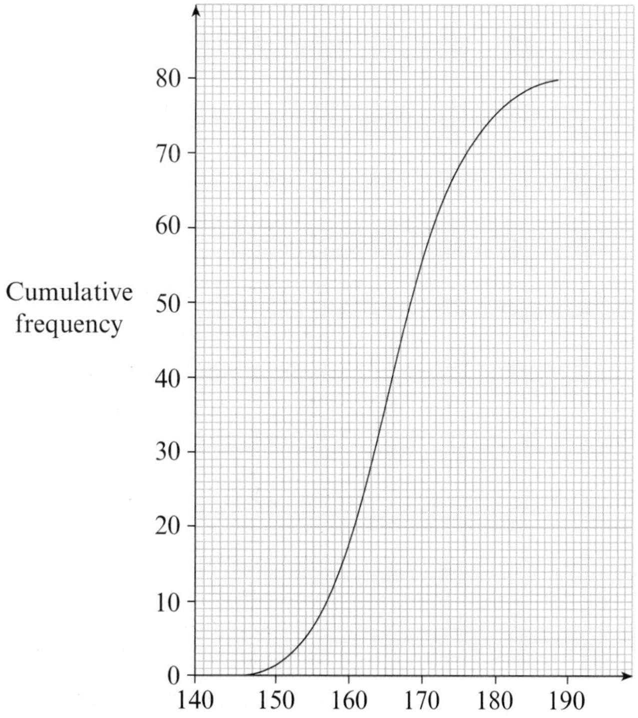


**a** What is the median height of the boys?

**b** What is the interquartile range of the boys' heights?

The heights of 80 girls aged 14 years are also recorded.
For the girls' heights:

| | |
|---|---|
| the median is | 162 cm |
| the lower quartile is | 156 cm |
| the upper quartile is | 168 cm |
| the shortest girl is | 143 cm |
| the tallest girl is | 182 cm |

**c** **i** Use this information to draw, on a copy of the diagram, a suitable cumulative frequency graph for the heights of the girls.

**ii** Compare and comment on the heights of these boys and girls.

(8 marks) (SEG, 1998)

**D5.5** This cumulative frequency diagram shows information about the distances travelled to work each day by a group of adults.

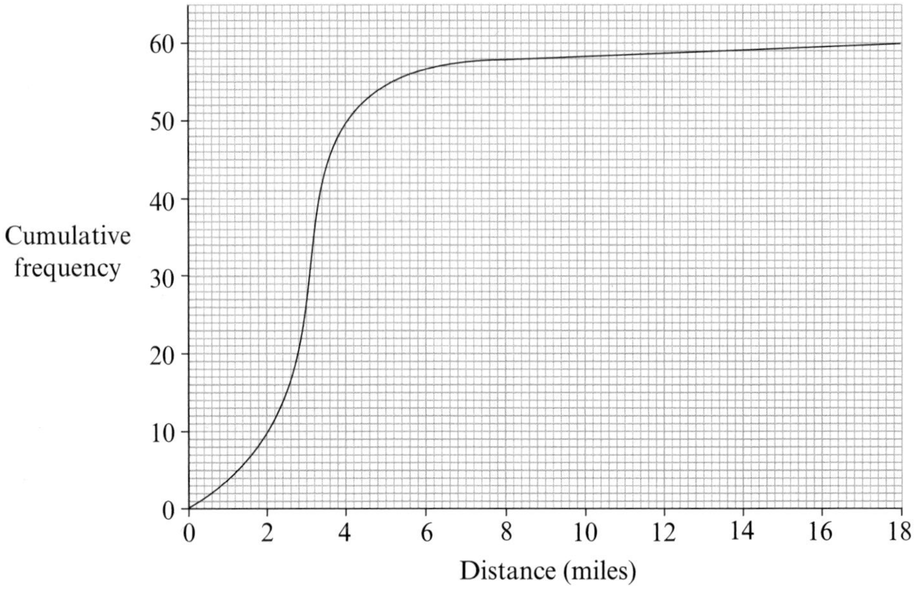


**a** What is the median distance travelled?

**b** Find the interquartile range.
Show clearly how you obtain your answer.

**c** Give a reason why the interquartile range is a better measure of spread than the range for this data.

(4 marks) (SEG, 2000)

## D6 Questionnaires and sampling

**D6.1** Alena and Ben design a questionnaire about the reading habits of pupils in their school.

**a** One of their questions is: 'How often do you read each week?'
Give **one** reason why this question is unsatisfactory.

**b** Ben asks pupils in the school library at lunchtime to complete the questionnaire.
Give **one** reason why this is unsatisfactory.

**c** Alena wants to investigate the reading habits of boys and girls of different ages.

Describe how she could obtain a stratified sample to do this.

(4 marks) (SEG, 1998)

**D6.2** This statement is made on a television programme about health. *'Three in every eight pupils do not take any exercise outside school.'*

**a** Matthew decides to do a survey in his school about the benefits of exercise.
He decides to ask the girls' netball team for their opinions.
Give **two** reasons why this is **not** a suitable sample to take.

**b** This is part of Matthew's questionnaire.

**Question** *Don't you agree that adults who were sportsmen when they were younger suffer more from injuries as they get older?*

**Response** *Tick one box*

☐ *Yes* ☐ *Usually* ☐ *Sometimes* ☐ *Occasionally*

**i** Write down one criticism of Matthew's question.
**ii** Write down one criticism of Matthew's response section.

(4 marks) (NEAB, 2000)

**D6.3** The table shows the number of students in Years 7, 8 and 9 of a school.

| Year | Number of students |
|---|---|
| 7 | 118 |
| 8 | 165 |
| 9 | 142 |

A sample of 100 of these students were asked some questions about homework.
The students were part of a stratified random sample.

How many students in each year group were included in the sample?

(3 marks) (SEG, 1999)

**D6.4** You have been asked by the local library to carry out a survey of the reading habits in the local area.
Comment on the suitability of the following methods for doing the survey.

**a** Choose people from the local telephone directory and ring them up to question them about their reading habits.

**b** Interview people outside the library on a Thursday morning.

**c** Choose 50 men and 50 women in a local shopping centre each day for a week.

(3 marks) (NEAB, 2000)

**D6.5** Two students Lyn and Pat decide to do a survey on school dinners.
Their school has 200 pupils in each of Years 7 to 11.
There are the same number of boys as girls in each year group.

Lyn and Pat each decide to survey 50 pupils.

**a** **i** Lyn visits a different Year 7 form each morning for a week and surveys 5 boys and 5 girls each day.

Comment on this method of sampling.

**ii** Pat gets an alphabetic list of all 1000 pupils in the school and selects every 20th name on the list.
She then surveys these pupils.

Comment on this method of sampling.

**b** Harry also decides to do a survey of 50 pupils.
Explain how he can choose a stratified sample of the school population.

(4 marks) (AQA)

**D6.6** Simone wants to estimate the number of deer in a forest.

She catches 60 deer and puts an orange mark on each of them.

She then releases them back into the forest.

A week later she catches 30 deer and find that 9 of them have an orange mark.

Estimate the number of deer in the forest.

(2 marks) (AQA)

**D6.7** Two companies, $A$ and $B$, are employed to do a survey on sports facilities in a small town with a population of 10 000.
Each company is asked to survey 500 people.

**a** **i** Company $A$ visits the local shopping centre each morning for 10 days and surveys 50 people each time.
Comment on this method of sampling.

**ii** Company $B$ gets the local phone book in which 4000 numbers are listed.
During the evening they ring every 8th name in the phone book and conduct the survey over the phone.
Comment on this method of sampling.

**b** Company $C$ is also asked to survey 500 people.
Explain how they can choose a stratified sample of the population.

(4 marks) (NEAB, 2000)

**Do NOT use a calculator for this exam paper.**

1 Ahmed, Baz and Colin share a bag of 30 Jelly Babies.
Ahmed has $\frac{2}{5}$ of the Jelly Babies.
Baz and Colin share the remaining Jelly Babies in the ratio 5 : 4
What percentage of the bag of Jelly Babies does Baz have? **(4 marks)**

2 **a** Factorise $x^2 + 3x$ **(1 mark)**
**b** Hence, or otherwise, find the exact value of $997^2 + 3 \times 997$ **(2 marks)**

3 Multiply out and simplify the following
**a** $3x^4 \times 2xy \times y^2$ **(2 marks)**
**b** $(2x + 3)(2x - 3)$ **(2 marks)**

4 The formula

$$P = \frac{a^2 + b^2}{2a - 3b}$$

can be used to calculate the value of $P$.
$a = 9.67$ and $b = 4.82$

**a** Write down approximate values for $a$ and $b$ that can be used to estimate the value of $P$. **(2 marks)**
**b** Work out an estimate of $P$ using these approximate values. **(2 marks)**

5 The dimensions of a cone are given.
**a** The following formulae represent certain quantities connected with the cone.

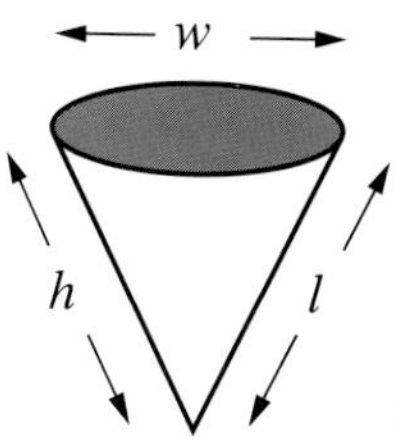


$\pi w \qquad \frac{\pi w^2 h}{12} \qquad 2\pi wl \qquad \frac{\pi w^2}{4}$

Which of these formulae represent volume? **(1 mark)**

**b** Sand is poured into the cone at a constant rate.
Sketch the graph of the depth of sand against time.

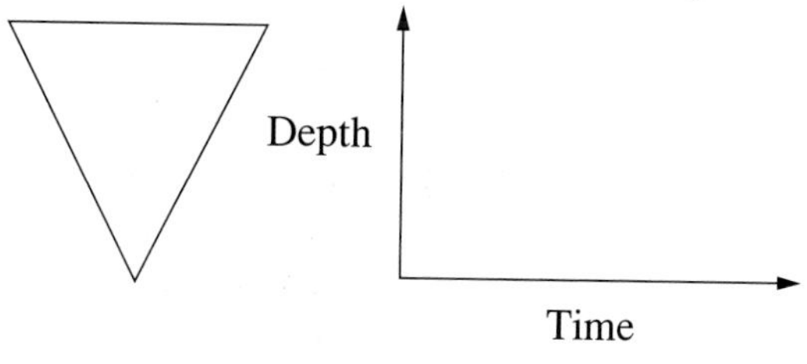


**(1 mark)**

6 **a** Solve the equation $6x + 14 = 4(x + 3)$ **(2 marks)**
**b** Rearrange

$$y = mx + c$$

to make $m$ the subject. **(2 marks)**

7 The map shows an island with three main towns, Alphaville, Betaville and Gammaville.
The map is drawn to a scale of 1 cm represents 15 kilometres.
A radio transmitter is to be installed.

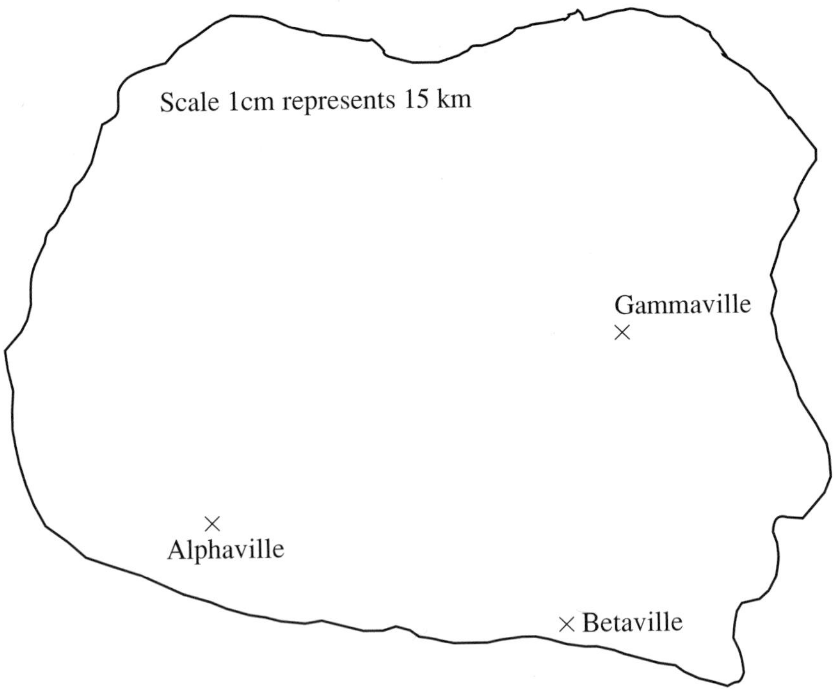


Using tracing paper, make a copy of the map.
The transmitter must be equidistant from Alphaville and Betaville.
The transmitter must be between 50 km and 80 km from Gammaville.
On your copy of the diagram show the possible sites that the transmitter may be drawn.

**(5 marks)**

8 Calculate the distance between the points A(−1, −1) and B(3, 2)

**(3 marks)**

9 **a** Match three of these equations with the graphs shown below.
In each case p is a positive integer.
Equation A: $y = p - x$ Equation B: $y = p - x^2$
Equation C: $y = px^2 + 2$ Equation D: $y = px$

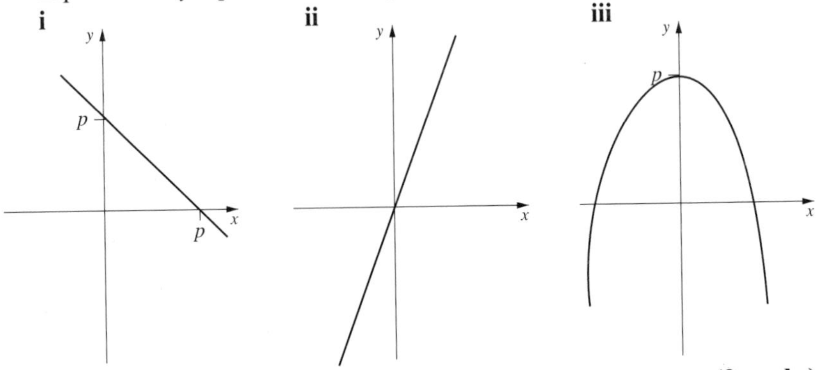


**(3 marks)**

**b** Sketch the graph of the equation you have not chosen.
Mark the values of the points where the graph crosses each axis.

**(2 marks)**

**10** **a** Simplify each of the following expressions
**i** $a^6 \times a^2$ **ii** $a^6 \div a^2$ **iii** $(a^6)^2$ **(3 marks)**

**b** Put a circle round the expression which has the smallest value when $a = 0.2$

$a^6 \times a^2$ $a^6 \div a^2$ $(a^6)^2$

**(1 mark)**

**11** Aimee travels to work by each day.
The probability that the bus is late on any day is 0.3.

**a** **i** Complete the tree diagram to show the possible outcomes for Monday and Tuesday.

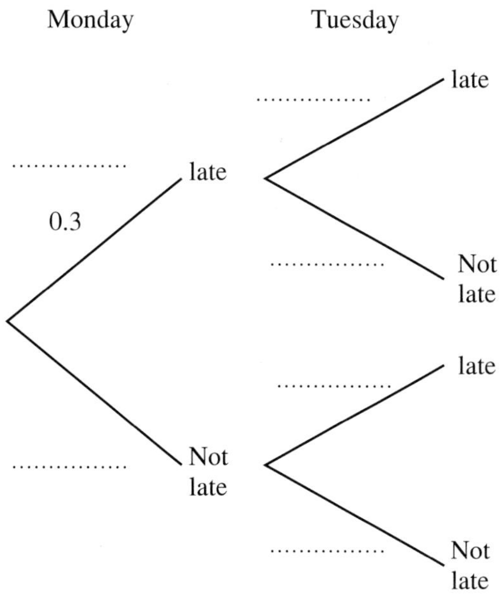


**(2 marks)**

**ii** What is the probability that the bus is not late on both days? **(2 marks)**

**b** What is the probability that the bus is late at least once in a normal week of five days. **(2 marks)**

**12** The table shows the average speed of planets that orbit the sun.

| Planet | Average speed of orbit (km/h) |
|---|---|
| Jupiter | $4.7 \times 10^4$ |
| Mercury | $1.7 \times 10^5$ |
| Neptune | $4.2 \times 10^4$ |
| Pluto | $1.7 \times 10^4$ |
| Saturn | $3.5 \times 10^4$ |
| Uranus | $2.5 \times 10^5$ |

**a** Which planet is travelling the fastest? **(1 mark)**

**b** What is the difference between the average speeds of Neptune and Jupiter?
Give your answer in standard form. **(3 marks)**

13 The energy values of 150 different foods are summarised in the table.

| Energy, $e$ (kcal per 100 g) | Frequency |
|---|---|
| $0 \leqslant e < 50$ | 25 |
| $50 \leqslant e < 100$ | 50 |
| $100 \leqslant e < 150$ | 45 |
| $150 \leqslant e < 200$ | 30 |

Draw a histogram to represent the data. **(3 marks)**

14 Write down the value of

**a** $27^{\frac{1}{3}}$ **(1 mark)**
**b** $5^{-2}$ **(1 mark)**
**c** $16^{\frac{3}{2}}$ **(2 marks)**

15 Mary wants to find out the height of the local church.
She places a 2 metre stick vertically at 100 metres from the base of the church.
She finds that she can line up the top of the stick and the top of the church at a further distance of 10 metres from the church, as shown in the diagram.
Calculate the height of the church.

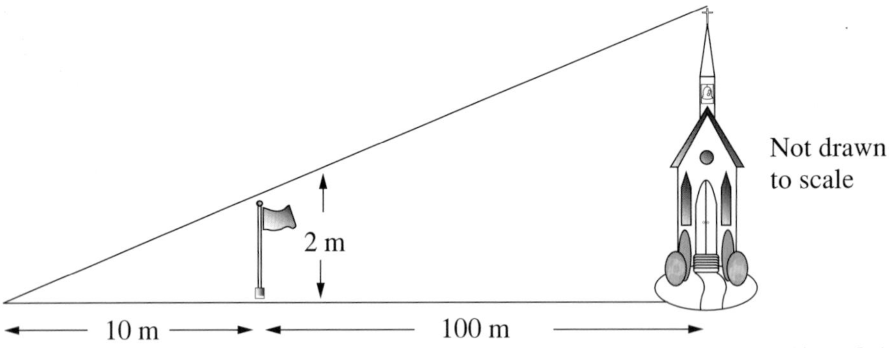


**(4 marks)**

16 A rectangle has sides of length $2x + 5$ and $2x + 1$ cm.
The area of the rectangle is 54 cm$^2$

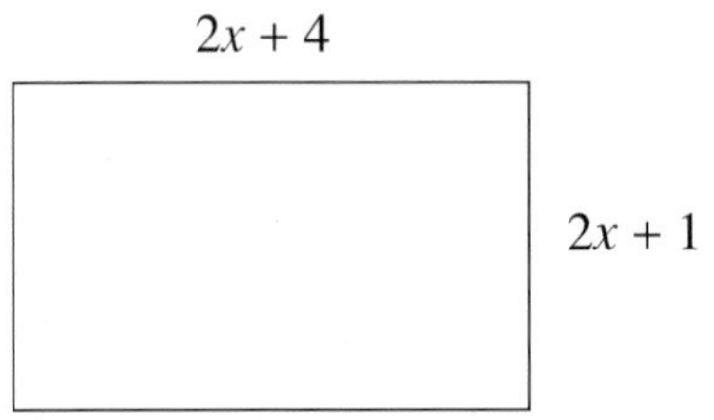


**a** Show that $x$ satisfies the equation

$$2x^2 + 5x - 25 = 0$$

**(3 marks)**

**b i** Solve the equation $2x^2 + 5x - 25 = 0$ **(2 marks)**
**ii** Find the perimeter of the rectangle. **(2 marks)**

**17** **a** $AC$ is a diameter of a circle centre O.
$B$ and $D$ are points on the circumference.
Angle $BAC = 33°$.

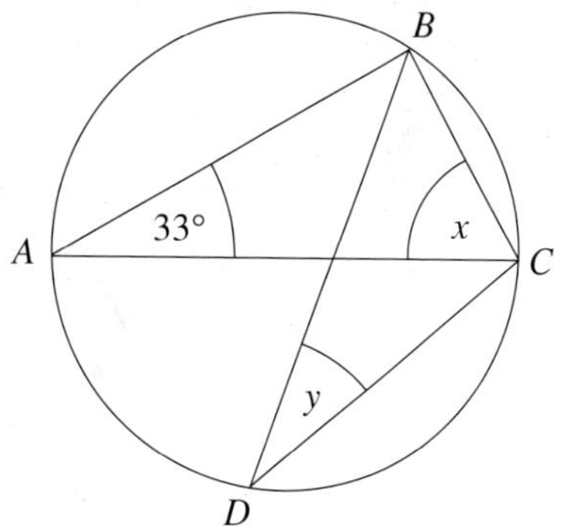


**i** Find angle $BCA$, marked $x$ in the diagram. **(1 mark)**

**ii** Find angle $BDC$, marked $y$ in the diagram **(1 mark)**

**b** $DE$ and $DF$ are tangents to a circle, meeting the circle at $A$ and $C$ respectively.
$B$ is a point on the circumference such that $CB$ is parallel to $DF$.
Angle $FAB = 65°$.
Find angle $BCE$, marked $x$ on the diagram.

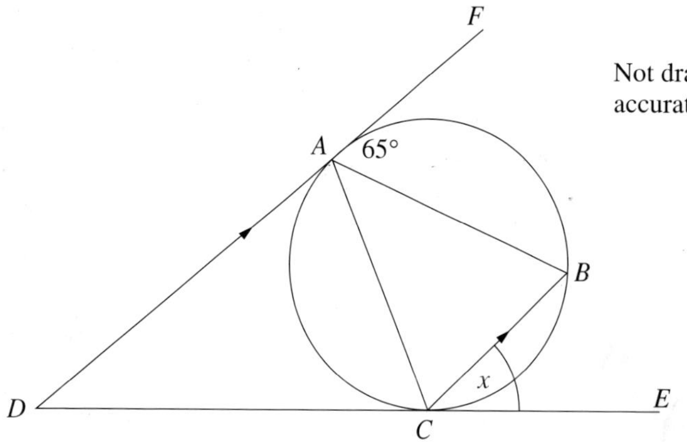


**(4 marks)**

**18** The diagram shows the line joining the origin, O, to the point A(6, 6).

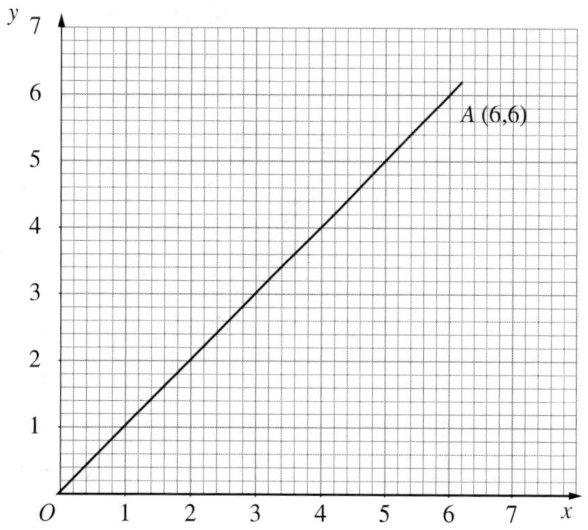


**(3 marks)**

Work out the equation of the line passing through the mid-point of OA and perpendicular to OA.

**19** The table shows the number of students in a school.

| | |
|---|---|
| Boys | 293 |
| Girls | 207 |

A stratified random sample of 150 students were asked some questions about school dinners.
How many boys were included in the sample? **(2 marks)**

**20** A square has a side of $2 + \sqrt{3}$ cm.
What is its area?
Give your answer in the form $a + b\sqrt{3}$ **(2 marks)**

**21** $a$, $b$ and $c$ are connected by the equation

$$\frac{a}{b-a} = \frac{c-a}{a}$$

Rearrange the equation to make $a$ the subject. **(4 marks)**

**22** Prove algebraically that the difference of the squares of any two consecutive integers is an odd number. **(3 marks)**

**23** **a** Sketch the graph of $y = \sin x$ for $0° \leqslant x \leqslant 360°$. **(1 mark)**

**b** One solution of the equation $\sin x = 0.6$ is $x = 37°$, to the nearest degree.
Find the solutions of $\sin x = -0.6$ for $0° \leqslant x \leqslant 360°$ **(2 marks)**

**24** In the triangle $ABC$, $P$ and $Q$ are the midpoints of $AB$ and $AC$ respectively.

$AB = \mathbf{x}$. $AC = \mathbf{y}$.

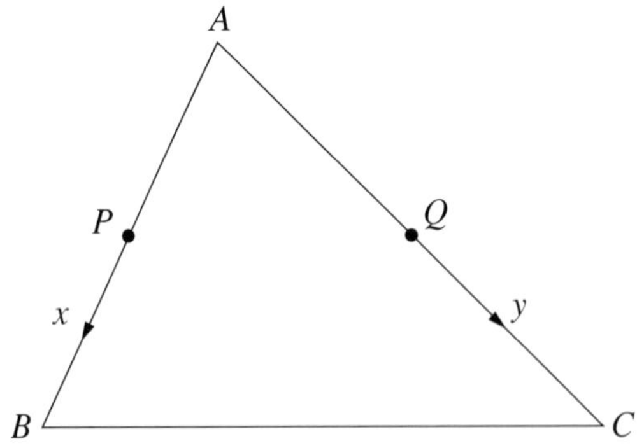


**a** Find in terms of $\mathbf{x}$ and $\mathbf{y}$

**i** $BC$ **(1 mark)**

**ii** $AP$ **(1 mark)**

**iii** $PQ$ **(2 marks)**

**b** What conclusions can you draw about the line segments $AB$ and $ST$? **(2 marks)**

**25** The graph of $y = \dfrac{10}{x} + 4$ is shown.

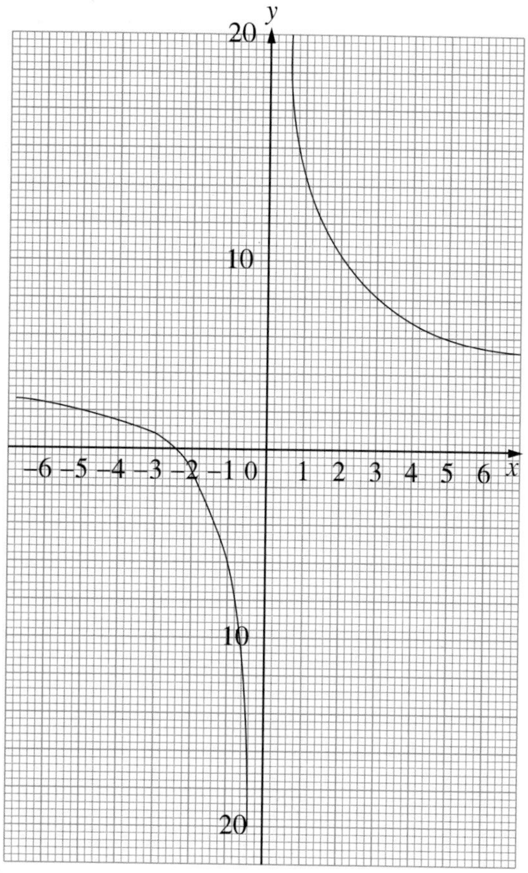


**a** Use the graph to solve the equation

$$\frac{10}{x} + 3 = 0$$

**(2 marks)**

**b** By drawing a suitable line on the graph solve the equation

$$x^2 - 4x - 10 = 0$$

Give your answers correct to 1 d.p. **(3 marks)**

**(Total 100 marks)**

1 In a sale all the prices are reduced by 20%.
I bought a coat for £70 in the sale.
What was the price before the sale? **(3 marks)**

2 Given that $s = ut + \frac{1}{2}at^2$
Find the value of $s$ when $u = 15$, $t = 4$ and $a = -7.5$ **(2 marks)**

3 The sketch graph shows the line $5x + 4y = 14$

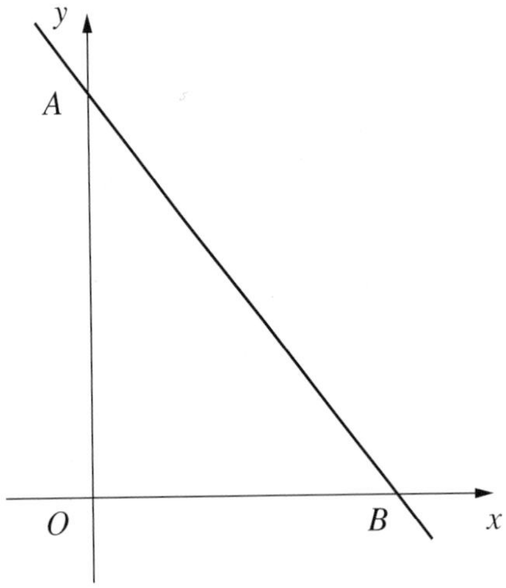


Write down the coordinates of the points $A$ and $B$. **(2 marks)**

4 The diagram shows three quadrilaterals, $P$, $Q$ and $R$.

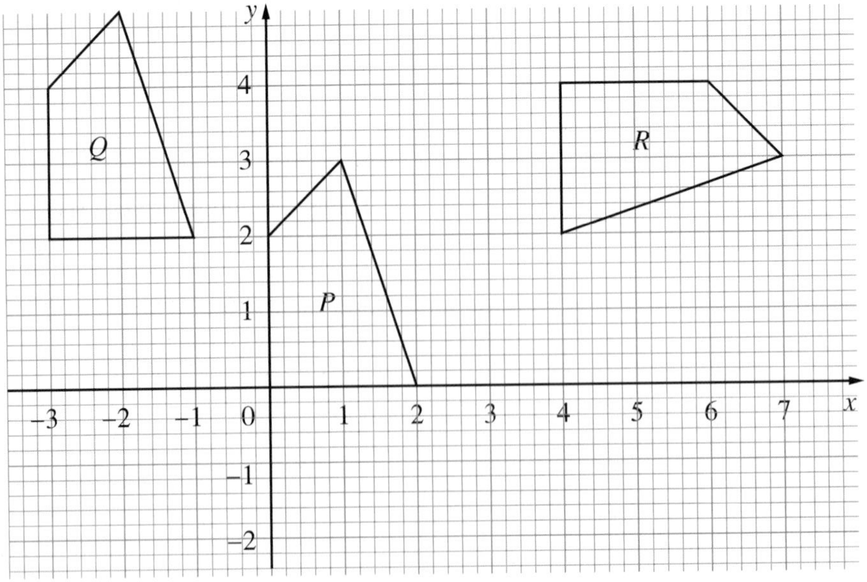


**a** Describe fully the transformation which takes $P$ to $Q$. **(2 marks)**
**b** Describe fully the transformation which takes $P$ to $R$. **(3 marks)**

5 List all the values of $n$, where $n$ is an integer, such that

$$-2 < n - 1 \leqslant 3$$

**(2 marks)**

**6** The diagram shows part of a roof structure.

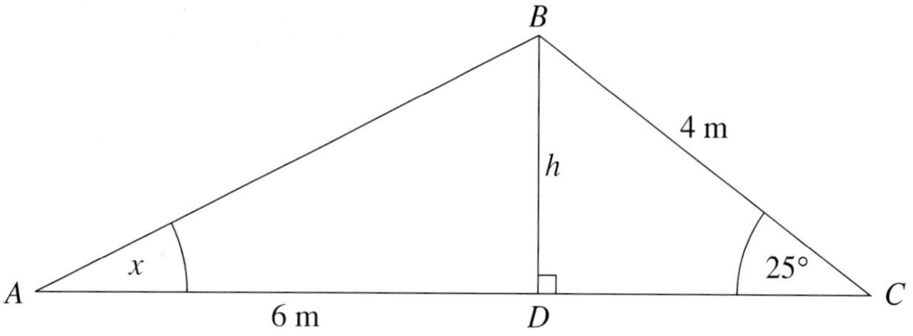


$AD = 6$ m, $BC = 4$ m and angle $BCD = 25°$

**a** Calculate the length of $BD$ (marked $h$ in the diagram). **(3 marks)**
**b** Calculate angle $BAD$ (marked $x$ in the diagram). **(3 marks)**

**7** The rectangle below has a length of $2x + 5$ cm and a width of $3x - 2$ cm.

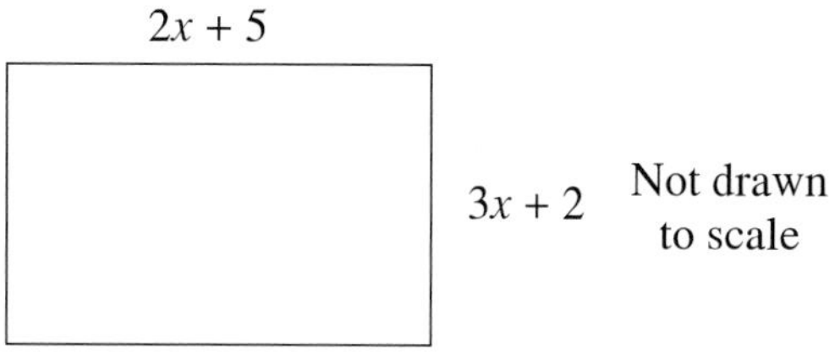


The perimeter of the rectangle is 20 cm.
Find the value of $x$. **(3 marks)**

**8** Solve the following equation

$$\frac{x-3}{2} - \frac{4x+1}{3} = 0$$

**(4 marks)**

**9** At the local Ice Rink the price of an adult ticket is £$x$ and the price of a child's ticket is £$y$.
It costs a family of two adults and three children £20 to go ice-skating.
It costs one adult and four children £18 to go ice-skating.
Use this information to set up and solve a pair of simultaneous equations.
Give the price of each ticket. **(5 marks)**

**10** The table shows the average age of 80 women who take part in a marathon.

| Age ($a$ years) | Frequency |
|---|---|
| $20 \leqslant a < 30$ | 19 |
| $30 \leqslant a < 40$ | 33 |
| $40 \leqslant a < 50$ | 19 |
| $50 \leqslant a < 60$ | 9 |

Calculate an estimate of the mean age of these women. **(4 marks)**

**11** The weights of a sample of 100 apples from an orchard at Holly Farm are shown in the table.

| Weight, $w$ (grams) | $50 \leqslant w < 100$ | $50 \leqslant w < 100$ | $50 \leqslant w < 100$ | $50 \leqslant w < 100$ | $50 \leqslant w < 100$ |
|---|---|---|---|---|---|
| Frequency | 11 | 21 | 32 | 27 | 9 |

**a** Copy and complete this cumulative frequency table. Use it to draw a cumulative frequency diagram for the sample of apples

| Weight, $w$ (grams) | $< 100$ | $< 150$ | $< 200$ | $< 250$ | $< 300$ |
|---|---|---|---|---|---|
| Cumulative Frequency | | | | | |

**(3 marks)**

**b** Estimate the median weight of the apples from Holly Farm. **(1 mark)**

**c** The box-plot below summarises the weights of a sample of 100 apples from Wood Farm.

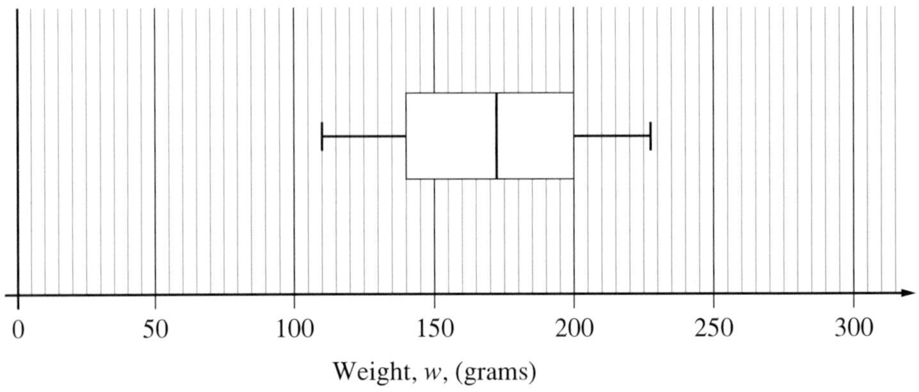


Estimate the interquartile range of the apples from Wood Farm. **(1 mark)**

**d** Compare and comment on the differences in the samples of apples from the two farms. **(2 marks)**

**12** A petri dish contains 2000 bacteria.
The bacteria increase by 12% each day.

**a** How many bacteria will there be after 1 day? **(1 mark)**

**b** How many bacteria will there be after 7 days?
Give your answer to a sensible degree of accuracy. **(4 marks)**

**13** Mr Smith invests £1200 in a unit trust in 1994.
The table shows the value of the investment over the next 6 years.

| Year | **1994** | **1995** | **1996** | **1997** | **1998** | **1999** | **2000** |
|---|---|---|---|---|---|---|---|
| Value £ | 1200 | 1179 | 1296 | 1278 | 1200 | 1410 | 1377 |

The following table shows the 3-year moving average of the value of the investment

| **1995** | **1996** | **1997** | **1998** | **1999** |
|---|---|---|---|---|
| 1225 | 1251 | 1258 | 1296 | |

**a** Calculate the value of the 3-year moving average for 1999. **(1 mark)**

**b** Mr Smith plots the values of the moving averages.

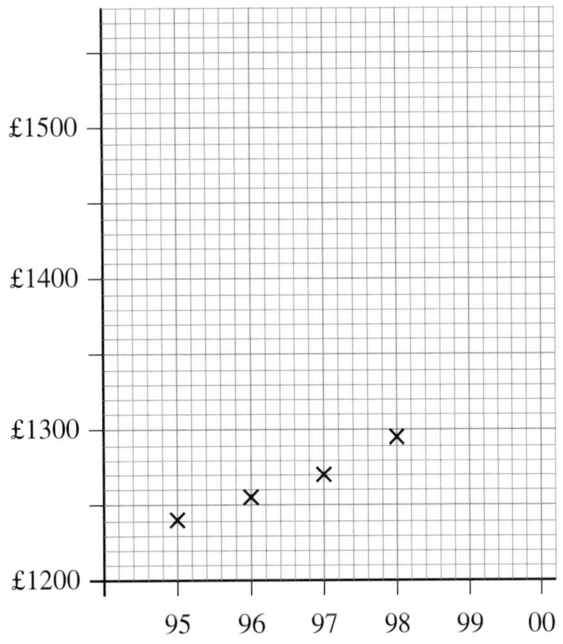


Use the graph to calculate the likely value of the investment in **2001**.
You must show all your working. **(3 marks)**

**14** **a** Write 0.085 as a fraction in its simplest form. **(1 mark)**
**b** Write $0.0\dot{8}\dot{5}$ as a fraction in its simplest form. **(3 marks)**

**15** **a** Factorise $x^2 - 16$ **(1 mark)**
**b** Simplify fully the expression

$$\frac{x^2 - 16}{2x^2 - 7x - 4}$$ **(3 marks)**

**16** The expression $x^2 - 10x + 14$ can be written in the form $(x - a)^2 - b$, where $a$ and $b$ are integers.

**a** Calculate the values of $a$ and $b$. **(3 marks)**
**b** Solve the equation $x^2 - 10x + 14 = 0$, giving your answers to 2 decimal places. **(2 marks)**

**17** A child's toy is in the shape of a cone on the top of a hemisphere, as shown below.
The radius of the hemisphere is 8.5 cm.
The height of the cone is $h$ cm.

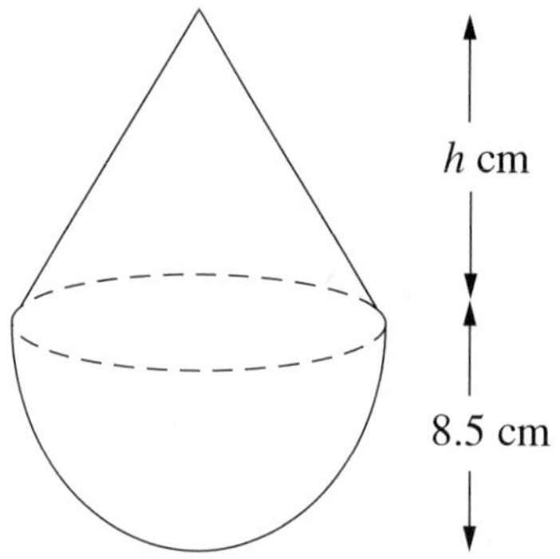


The volume of the toy is 2050 $cm^3$.
Calculate the height, $h$, of the cone. **(5 marks)**

**18** The lengths of the sides of a triangle are 9 cm, 14 cm and 11 cm.
Calculate the smallest angle of this triangle. **(4 marks)**

**19** Rearrange the equation

$$y = \frac{x}{x + 3}$$

to make $x$ the subject **(3 marks)**

**20** To measure the width of a river, Shameela paces two points A and B that are 100 metres apart.
She knows that this distance is accurate to the nearest 5 metres.
At each point she measures the angle between the river bank and a tree on the far bank. These measurements are accurate to the nearest 2°.
Her sketch is shown in the diagram.

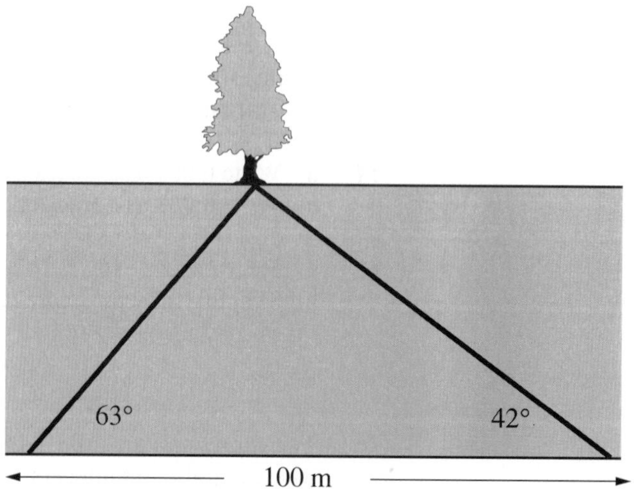


Not drawn to scale

Calculate the maximum possible width of the river. **(6 marks)**

**21** **a** This is the graph of $y = x^2$.

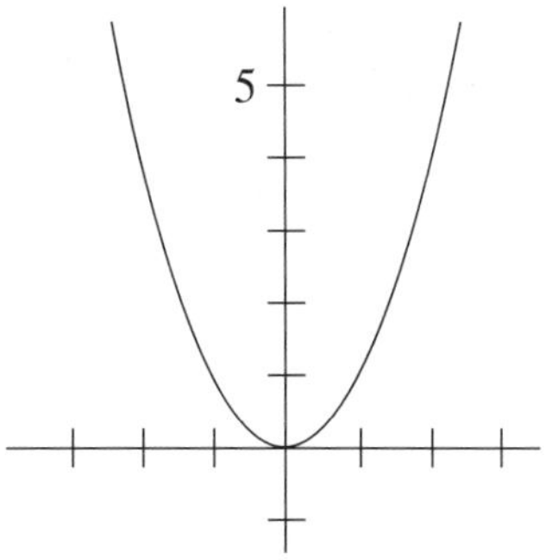


**i** Sketch the graph of $y = (x + 1)^2$ **(1 mark)**
**ii** Sketch the graph of $y = -x^2$ **(1 mark)**
**iii** The point (1, 1) on the graph $y = x^2$ is transformed to the point (1, 3) by a single transformation.
Give the equation of the transformed graph. **(1 mark)**

**b** This is the graph of $y = \sin x$

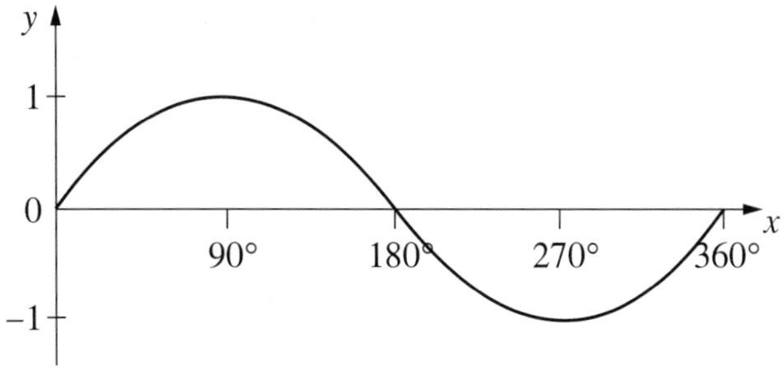


Below are two transformations of $y = \sin x$
Write down the equation of each transformed graph.

**i**

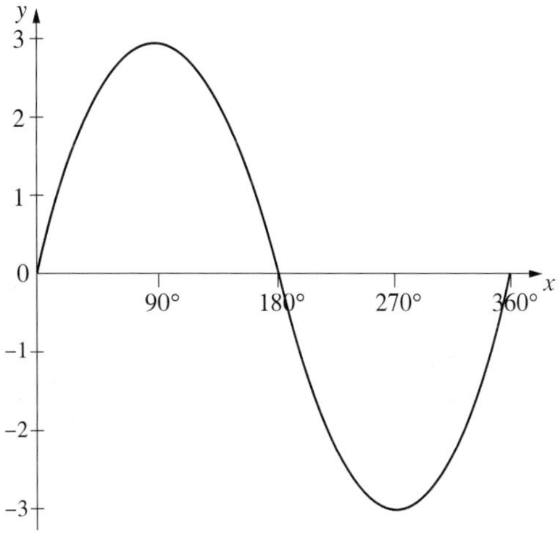


**(1 mark)**

**ii**

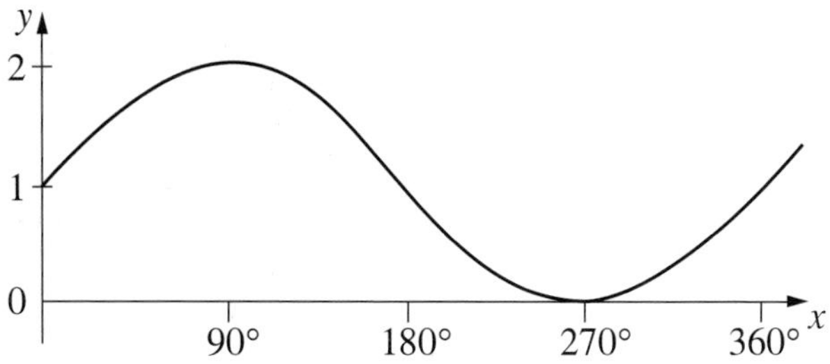


**(1 mark)**

**22** Solve the equation

$$\frac{4}{x+1} + \frac{3}{x-1} = 1$$ **(5 marks)**

**23** A 2-dimensional framework is made from wire.
It is three-quarters of a circle of radius 20 cm with two diameters as shown.

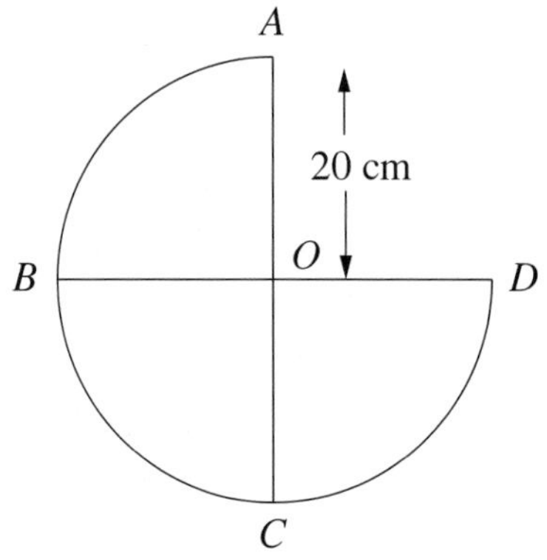


**a** Calculate the total length of the wire in the frame. **(3 marks)**

**b** The frame is now formed into a cone by forming the arc *ABCD* into a circle and joining the radii *OA* and *OD*.

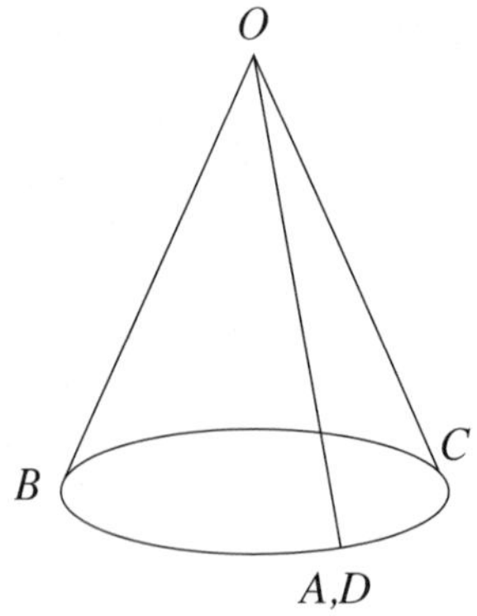


The cone is used as a framework for a lampshade.
Calculate the area of fabric needed to cover the lampshade. **(4 marks)**

**(Total 100 marks)**

## SECTION 1 Fractions

### Starting points

**A2** **a** 14 **b** 15 **c** 24 **d** 4

**A3** **a** 1, 2, 3, 6 **b** 1, 5

**A4** **a** 1 **b** 2 **c** 4 **d** 18

**A5** It has only one factor

**A6** 31, 37

**A7** Any other even numbers have at least 3 factors, i.e. 1, 2 and themselves.

**B1** **a** $2^2 \times 5$ **b** $2^2 \times 3^2$ **c** $5 \times 7 \times 11$ **d** $2^3 \times 3^2 \times 7$

**B2** **a** 2016 **b** 16335

**C1** $\frac{4}{10}$

**C3** **a** $\frac{3}{5}$ **b** $\frac{1}{6}$

**D1** **a** $1\frac{1}{4}$ **b** $2\frac{2}{3}$ **c** $3\frac{1}{3}$ **d** 5

**D2** **a** $\frac{18}{7}$ **b** $\frac{3}{2}$ **c** $\frac{37}{8}$ **d** $\frac{3}{1}$

**E1** **a** $\frac{4}{5}$ **b** $\frac{1}{3}$ **c** $\frac{2}{5}$ **d** $\frac{3}{5}$ **e** $\frac{4}{7}$ **f** $1\frac{3}{5}$ **g** $\frac{5}{7}$ **h** $\frac{5}{13}$ **i** $\frac{5}{16}$ **j** undefined

**E2** **a** $\frac{2}{7}$ **b** $\frac{11}{15}$ **c** $\frac{2}{7}$ **d** $\frac{1}{2}$ **e** $\frac{15}{16}$ **f** $\frac{1}{12}$ **g** $1\frac{1}{4}$ **h** $1\frac{4}{5}$

**E3** $\frac{4}{7}$

### Exercise 1.1

**1** **a** **i** 24 **ii** 20 **b** **i** 180 **ii** 1800

**3** **a**

| Numbers | Product | HCF | LCM |
|---|---|---|---|
| 12, 70 | 840 | 2 | 420 |
| 7, 24 | 168 | 1 | 168 |
| 4, 36 | 144 | 4 | 36 |
| 70, 110 | 7700 | 10 | 770 |

**b** Product ÷ HCF = LCM **c** 72

**4** 8, 160

**5** 31 July

**6** 960 $cm^3$

**7** **a** $2^2 \times 3 \times 7$
**b** 1, 2, 3, 7, $2^2$, $2 \times 3$, $2 \times 7$, $3 \times 7$, $2^2 \times 3$, $2^2 \times 7$, $2 \times 3 \times 7$, $2^2 \times 3 \times 7$
**c** 12
**d** 900 has 27 factors

### Exercise 1.2

**1** **a** $\frac{17}{20}$ **b** $3\frac{3}{4}$ **c** $1\frac{15}{16}$ **d** 4

**2** **a** $1\frac{4}{15}$ **b** $\frac{^-3}{8}$ **c** $\frac{3}{4}$ **d** $4\frac{1}{6}$

**3** **a** $1\frac{7}{9}$ **b** $1\frac{1}{5}$ **c** $\frac{17}{18}$ **d** $5\frac{7}{20}$

**4** $3\frac{1}{3}$

**5** 8

**6** No unit fraction can be greater than $\frac{1}{2}$, so the sum of 2 unit fractions can never be greater than 1.

**8** **a** **i** $\frac{15}{16}$ **ii** $\frac{511}{512}$

**9** $\frac{1}{4}$, $\frac{1}{28}$

**10** $\frac{1}{3}$, $\frac{1}{6}$, $\frac{1}{2}$

**12** $\frac{9273}{18456}$

### Exercise 1.3

**1** L

**2** P and S, Q and U, R and W, T and V

**3** **a** $\frac{1}{(3n)}$ **b** $\frac{8}{x}$ **c** $\frac{(a-b)}{10}$ **d** $\frac{p}{2}$ **e** $\frac{3m}{4n}$ **f** $\frac{y}{8}$ **g** $\frac{z}{15}$ **h** 2

**4** **a** $\frac{x}{5} - \frac{2}{5} = \frac{(x-2)}{5}$ **b** $\frac{2}{m} \times \frac{n}{6} = \frac{n}{(3m)}$ **c** $\frac{y}{x} + \frac{z}{x} = \frac{(y+z)}{x}$ **d** $\frac{6}{n} \div \frac{1}{n^2} = 6n$
**e** $\frac{3t}{10} + \frac{2t}{10} = \frac{t}{2}$ **f** $\frac{2}{(b+3)} \div \frac{1}{(b-2)} = \frac{2(b-2)}{(b+3)}$

**5** A and G, B and C, D and F, E and H

### Exercise 1.4

**1** **a** C **b** $\frac{22}{(3c)}$

**2** E and L, F and G, H and J, I and K

**3** **a** $\frac{3m}{4}$ **b** $\frac{5p}{8}$ **c** $\frac{(5q+2r)}{10}$ **d** $\frac{(2s-st)}{2t}$ **e** $\frac{(4y-5x)}{xy}$ **f** $\frac{17}{2f}$ **g** $\frac{2(p+2)}{(p+3)(p-1)}$
**h** $\frac{(z+12)}{z(z+6)}$

**4** **a** 8 **b** 3 **c** $^-3$ **d** 2

### Exercise 1.5

**1** **b** The numerators are always 3 and the larger denominator is 3 more than the smaller.
**c** $\frac{3}{n} - \frac{3}{n+3} = \frac{9}{n(n+3)} = \frac{3}{n} \times \frac{3}{(n+3)}$

### End points

**A1** **a** 15, 180 **b** 25, 975 **c** 5, 6600 **d** 18, 432
**e** 4, 1512 **f** 1, 28 350 **g** 5, 8000 **h** 1, 12 740

**A2** *a* 21 **b** 630

**A3** 24, 27

**BI** **a** $\frac{5}{9}$ **b** $\frac{5}{18}$ **c** $\frac{7}{10}$

**B2** **a** $4\frac{7}{20}$ **b** $1\frac{3}{4}$ **c** $1\frac{7}{8}$ **d** $4\frac{1}{3}$

**B3** $\frac{20}{81}$

**C1** **a** $\frac{x}{2}$ **b** $\frac{3}{4}$ **c** $3z$ **d** $\frac{(3b-7a)}{ab}$ **e** $\frac{(c+5)}{c(c+1)}$ **f** $\frac{(2d+1)}{(d+6)(d-5)}$

## SECTION 2 Properties of Shapes

### Starting points

**A1** **a** **i** reflex **ii** acute **iii** right angle
**b** **i** equilateral **ii** right-angled
**c** triangles AED and BAD

**B1** **a** 60° **b** 300° **c** 45°

**B2** **a** 35°, 72.5°
**b** The angles in a triangle add up to 180°, so if one angle is more than 90°, the other two must be less than 90°.

**C1** 116°, 116°, 64°, 64°

**D1** 1: trapezium 2: rectangle, square 3: square, rhombus
4: rectangle, square, parallelogram, rhombus

**E1** 1080°

**E2** 142°

**E3** 129°

**E4** **a** 1800° **b** 150°

**F1** **a** A quadrilateral can be split into two triangles by drawing in a diagonal. The angle sum of each triangle is 180°, so the angle sum of the quadrilateral is twice that, 360°.
**b** 900° **c** 12

**F2** **a** 40°
**b** $A\hat{B}C$, + $B\hat{A}C = x$ and $B\hat{C}A = B\hat{C}A = 180° - x°$
so $A\hat{B}C + 180° - x° = x°$
ie $A\hat{B}C = 2x - 180$

### Exercise 2.1

**1** 62°, 120°, 117°

**2** **b** 74°, 53°, 53°, 74°, 53°, 53°

### Exercise 2.2

**1** **a** The corresponding angles are: any two out of *a*, *c* and *e*; any two out of *b*, *d* and *f*; any two out of *g*, *i* and *k*, any two out of *h*, *j* and *l*
**b** The alternate angles are: *b* and *i* or *k*, *d* and *k*, *h* and *c* or *e*, *j* and *e*

**2** 72°, 135°, 45°

**3** **a** **i** The corresponding angles are:
any pair out of ADC, DHG, HLK and LPO;
any pair out of angles EDH, IHL, MLP and QPS;
any pair out of angles ADE, DHI, HLM and LPQ;
any pair out of angles CDH, GHL, KLP and OPS
**ii** The corresponding angles are:
any pair out of angles BDC, DGF, GKJ and KON;
any pair out of angles BDE, DGH, GKL and KOP
any pair out of angles CDG, FGK, JKO and NOR;
any pair out of angles EDG, HGK, LKO and POR

**b** BR and AS are not parallel
**c** The alternate angles are:
angles CDG and DGH, GKL or KOP;
angles FGK and GKL or KOP;
angles JKO and KOP;
angles EDH and DHG, HLK or LPO;
angles IHL and HLK or LPO;
angles MLP and LPO;
angles EDG and DGF, GKJ or KON; angles HGK and GKJ or KON;
angles LKO and KON;
angles CDH and DHI, HLM or LPQ;
angles GHL and HLM or LPQ;
angles KLP and LPQ
**d** **i** 65° **ii** 54° **iii** 61° **iv** 54° **v** 126° **vi** 61°

**4** 42°, 77°, 61°, 72°, 72°, 72°, 68.5°, 68.5°, 68.5°

### Exercise 2.3

**1** **b** **i** stays the same **ii** gets bigger **iii** gets smaller **c** gets smaller
**d** **i** obtuse **ii** right angle **iii** acute

**2** **a** 3 **b** 1, 2, 3

### Exercise 2.4

**1** **a** 56°, 42°, 69°, 104°, 89°

**2** **a** The 12 exterior angles are 12 equal turns which are equivalent to one full turn, 360°.
**b** 30° **c** 150°

**3** 40°, 140°, 70°, 40°

### Exercise 2.5

**1** **a** AMD is isosceles, so angle AMD = 180° – 2 × 22.5° = 135°
**b** **i** 67.5°, 157.5°

**2** **a** **i** 67.5° at A, 67.5° at C, 45° at F
**ii** 112.5° at A, 112.5° at C, 135° at F
**b** 5 **c** 45°, 67.5°, 90°, 112.5°, 135°, 157.5°

### End points

**A1** 42°, 67°, 109°

**B1** **a** **i** C **ii** B and E

**C1** **a** B 120°, 30°, 30°; C 60°, 60°, 60°; D 90°, 90°, 90°, 90°; E 60°, 90°, 120°, 90°
**b** B 60°, 150°, 150°; C 120°, 120°, 120°; D 90°, 90°, 90°, 90°; E 120°, 90°, 60°, 90°

**C2** 45°

## SECTION 3 Algebraic Manipulation

### Starting points

**A1** C

**A2** **a** $13a - 30$ **b** $4 - 5x$ **c** $12x - 29$ **d** $^-5x - 11$ **e** 4

**B1** **a** $\frac{6p + 8}{p(p + 2)}$ **b** $\frac{4}{3}$ **c** $\frac{5}{p(p + 5)}$ **d** $\frac{10}{3}$ **e** $\frac{(5p - 1)}{(p^2 - 1)}$ **f** $\frac{7p - 2}{(3p + 2)(p - 1)}$ **g** $\frac{9p + 23}{20}$
**h** $\frac{5(p + 3)}{4(p - 2)}$ **i** $\frac{8p - 12}{3p + 15}$ **j** $\frac{4(4p + 5)}{3(2p - 1)}$

### Exercise 3.1

**1** **a** $a(a + 5)$, $b(b - 2)$, $c(c - 6)$, $d(12 - d)$, $4(e + 8)$, $7(8 - f)$
**b** $a^2 + 5a$, $b^2 - 2b$, $c^2 - 6c$, $12d - d^2$, $4e + 32$, $56 - 7f$

**2** **a** $5(2p + 2)$, $4(3q - 2)$, $r(2r + 2)$
**b** $10p + 10$, $12q - 8$, $2r^2 + 2r$

**3** $2a$, $b$, $c$, $d + 3$, $e + 2$

**4** **a** $4(x + 3)$ **b** $2(3p - 2)$ **c** $2(a + 4)$ **d** $5(2 - q)$ **e** $t(3t - 4)$
**f** $s(6s + 7)$

### Exercise 3.2

**1** **a** $8b - 10$ **b** $24 - 6x$ **c** $3c - 12$ **d** $11y - 15$ **e** $^-2a$ **f** $5a - 14 + 2b$

**2** A and D

**3** **a** $5a + 5$ **b** $x + 15$ **c** $7a - 6$ **d** $^-4x + 11$ **e** $3x - 23$ **f** $7x + 4$
**g** $x - 9$ **h** $17x + 4y$

### Exercise 3.3

**1** **a** $6ab$ **b** $3pq$ **c** $20xy$ **d** $30pq$ **e** $2x^2$ **f** $a^2b$
**g** $2xy^2$ **h** $6a^2b$ **i** $6p^2q$ **j** $a^5$ **k** $6b^3$ **l** $10bc^3$

**2** A and E, B and F, C and H, G and I

**3** **a** $ab + 4a$ **b** $2mn + 3mp$ **c** $6xy + 4xz$ **d** $ac + c^2$ **e** $p^2 - pq$
**f** $4a^2 - 4ab$ **g** $2ab - 4ac$ **h** $6a^2 + 8ab$ **i** $8pq - 12p^2$ **j** $6p^2q + 4pq^2$
**k** $4x^2y - 8xy^2$ **l** $3x^3y - 3xy^3$

**4** **a** $2x^2 + x$ **b** $5ab + 4a$ **c** $x^2 + x^3 - 2x$ **d** $11a + 2b$
**e** $4p^2 + 6q$ **f** $6a$

**5** C

**6** **a** $9a + 26b$ **b** $10x + 5y$ **c** $^-7x$ **d** $10x^2 + 3xy$
**e** $2a^2b$ **f** $3a^2b + 3ab - 2ab^2 + 2b^2$

**7** **a** $6a^2 + 6ab + 12b^2$ **b** $16x^2y + 20xy^2 - 6xy$
**c** $pq^2 - 4p^2q$ **d** $12m^3 - 2m^2n - 8mn^2$

**8** **a** $\frac{(8x + 5)}{(x + 1)(2x + 1)}$ **b** $\frac{(x - 2)}{(x + 2)(x + 1)}$ **c** $\frac{(3x + 26)}{(x - 3)(x + 4)}$ **d** $\frac{(3x^2 + 10x - 4)}{(x + 2)(x + 4)}$

### Exercise 3.4

**1** **a** $2(x + 7)$ **b** $2(4x - 5y)$ **c** $x(5 + y)$ **d** $p(q + 7)$
**e** $2d(3 + 2e)$ **f** $3a(2b + 3)$ **g** $2a(1 - 4b)$ **h** $3a(a + 4)$
**i** $5y(3x + 4z)$ **j** $ab(a + b)$ **k** $3cd(4c - 5)$ **l** $5y(5x^2 + 3z)$

**3** **a** $b$

**7** **a** $5(x + 2y + 4)$ **b** $3(2a - 3b + 4c)$ **c** $x(2x + 3 + y)$
**d** $7(2x - 4y + 3z)$ **e** $2x(x + 4y + 3)$ **f** $ab(6b + 2a + 5)$

### Thinking ahead to ...

**A** **a** $x^2$, $3x$, $2x$, 6 **b** $x^2 + 5x + 6$

**B** **a** $(p + 3)(2p + 4)$, $(2q + 3)(3q + 5)$ **b** $2p^2 + 10p + 12$, $6q^2 + 19q + 15$

**C** **a** B, C, E, **b** $n^2 - 2n - 8$

### Exercise 3.5

**1** C

**2** D

**3** **a** $a^2 + 4a + 3$ **b** $d^2 + 14d + 45$ **c** $3e^2 + 22e + 7$ **d** $10a^2 + 13a + 4$
**e** $6b^2 + 19b + 15$ **f** $6c^2 + 25c + 14$

**4** **a** $x^2 + 2x + 1$ **b** $x^2 + 4x + 4$ **c** $x^2 + 6x + 9$

**5**

| × | $x$ | $^-4$ |
|---|---|---|
| $x$ | $x^2$ | $^-4x$ |
| $^-2$ | $^-2x$ | 8 |

$(x - 4)(x - 2) = x^2 - 6x + 8$

**6** **a** **i** $(x - 3)(x - 3)$ **ii** $x^2 - 6x + 9$ **b** **i** $(x - 4)(x - 4)$, $x^2 - 8x + 16$
**ii** $(x - 5)(x - 5)$, $x^2 - 10x + 25$
**iii** $(x - y)(x - y)$, $x^2 - 2xy + y^2$

**7** $6x^2 - 13x - 5$

**8** **a** $2z^2 + 3tz + t^2$ **b** $6k^2 + 17km + 5m^2$ **c** $2c^2 + 17cd + 21d^2$
**d** $6p^2 + 13pq - 5q^2$ **e** $5f^2 - 8fg + 3g^2$ **f** $21w^2 - 4vw - v^2$

**9** A and I, B and F, C and H, D and G

**10** **a** $z^2 - t^2$ **b** $4k^2 - m^2$ **c** $4c^2 - 9d^2$
**d** $9p^2 - q^2$ **e** $25f^2 - 9g^2$ **f** $49w^2 - v^2$
The product of the sum and difference between two numbers is equal to the difference between their products.

**11** M and N

**12** **a** **i** $x^2 + 2x + 1$ **ii** $x^3 + 3x^2 + 3x + 1$ **iii** $x^4 + 4x^3 + 6x^2 + 4x + 1$
**b** $x^6 + 6x^5 + 15x^4 + 20x^3 + 15x^2 + 6x + 1$

13 a $a^3 + 3a^2 - 10a$ b $p^3 - 5p^2 - 2p + 24$
c $4q^3 - 22q^2 - 42q$ d $2x^3 + 13x^2 + 13x - 10$

Thinking ahead to ...

A a 6, 6, 6, $(102 \times 103) - (100 \times 105) = 6$
b They are all 6. c $(n + 2)(n + 3) - n(n + 5)$

Exercise 3.6

1 a A 2, 2, 2, ... B 1, 1, 1, ... C 12, 12, 12, ...
b A $(n + 1)(n + 2) - n(n + 3)$ B $(n + 1)^2 - n(n + 2)$
C $(n + 3)(n + 4) - n(n + 7)$
c A 2 B 2 C 12

2 a i 7, 3, 5 ii 6, 6, 20 b i $a + b = c$ ii $ab$

3 a A 4, 2, 0, ... B 7, 6, 5, C $^-3$, $^-2$, $^-1$, ...
b A $(n + 2)(n + 3) - n(n + 7)$ B $(n + 2)(n + 4) - n(n + 7)$
C $n(n + 5) - (n + 2)^2$
c A Each term is 2 less than the previous term.
B Each term is 1 less than the previous term.
C Each term is 1 more than the previous term.

Exercise 3.7

1 a $(x + 1)(x + 7)$ b $(p + 1)(p + 8)$ c $(t + 1)(t + 9)$
d $(m + 2)(m + 6)$ e $(n + 3)^2$ f $(a + 4)(a + 9)$

2 a 8 b 5 c 7 d 11 e $t + 3$ f $p - 8$

3 $(x - 1)(x + 4)$

4 $(x - 4)(x - 2)$

5 a $(p - 1)(p + 3)$ b $(m + 5)(m - 2)$ c $(t + 1)(t - 5)$
d $(r + 4)(r - 7)$ e $(x - 3)(x - 5)$ f $(n - 1)(n - 6)$
g $(t + 3)(t - 4)$ h $(b + 2)(b - 6)$ i $(x + 6)(x - 2)$

6 $(x - 4)(x + 4)$

7 a $(t + 3)(t - 3)$ b $(r + 5)(r - 5)$ c $(4p + 1)(4p - 1)$
d $(3x + 2)(3x - 2)$ e $(x + y)(x - y)$ f $(x + 3p)(x - 3p)$

8 A and C

Exercise 3.8

1 $(2x + 3)(x + 2)$

2 a $(3x - 1)(x - 3)$ b $(4y - 3)(y + 1)$ c $(4p + 1)(p - 3)$
d $(4x + 3)(x - 1)$ e $(3x - 1)(x + 3)$ f $(3a + 1)(a + 3)$
g $(5n - 1)(n - 2)$ h $(3x - 4)(x + 1)$ i $(3x - 1)(x - 1)$

3 Across
2 $(2x + 1)(2x + 1)$ 4 $(x - 1)(x + 1)$ 5 $(x + 1)(3x - 5)$
7 $(2x + 1)(2x + 3)$ 8 $(x - 1)(5x + 1)$ 10 $(x - 1)(x - 4)$
13 $(x + 2)(x - 2)$ 14 $(x - 2)(4x + 7)$
Down
1 $(3x + 1)(3x - 1)$ 2 $(x - 1)^2$ 3 $(2x + 1)(2x - 1)$ 4 $(x - 1)(x - 3)$
5 $(3x + 2)(x + 2)$ 6 $(x + 1)(x - 5)$ 9 $(5x - 4)(x + 1)$
11 $(x + 2)(x - 7)$ 12 $(2x + 3)(2x - 3)$ 13 $(x + 7)(x - 7)$

4 a $a = 36, b = 6$ b $a = 6, b = 3$ c $a = 2, b = 1$ d $a = 4, b = 2$

5 a $3(x - 1)(x - 2)$ b $3(2x + 5)(x + 1)$ c $2(3x - 1)(x + 3)$
d $7(2x + 1)(x - 2)$ e $3(x - 1)^2$ f $4(x + 1)(x - 1)$

6 a $n(n + 1)(n - 1)$
b $n - 1$, $n$ and $n + 1$ are consecutive numbers, so at least one of them must be even, making their product even.

7 a $(x + y)^2$ b $4(x + y)^2$ c $(3x + 4y)(3x - 4y)$

Exercise 3.9

1 a $p + 1, p + 2$
b i $p + p + 1 + p + 2$ ii $3p + 3$ iii $3p + 3 = 3(p + 1)$

2 a i $^-4$ ii 0 iii 6 iv 14 b $(n + 3)(n - 2)$
c The two factors are 5 apart, so one of them must be even, giving an even product.

3 $n^2 - 6n + 9 = (n - 3)^2$ and squares are always positive.

4 a i $y - 8, y - 1, y + 1, y + 8$
ii $(y + 1)(y + 8) - (y - 8)(y - 1)$
iii $(y + 1)(y + 8) - (y - 8)(y - 1) = 18y$
b i $y - n, y - 1, y + 1, y + n$
ii $(y + 1)(y + n) - (y - n)(y - 1)$
iii 320

5 One of the numbers must be a multiple of 2, another a multiple of 3, so the product will be a multiple of 6.

Exercise 3.10

1 a $(a + b)(a - b)$ b i 9600 ii 2200

2 a 3, 5, 7, 9, 11, ...
b The answers are odd numbers.
c $(n + 1)^2 - n^2 = 2n + 1$
d i 2467 ii 246246246245

3 a $(n + 5)^2 - n^2 = 10n + 25$
b i 1255 ii 3045

Exercise 3.11

1 a i $\frac{1}{2}$ ii $\frac{1}{3}$ iii $\frac{1}{4}$ c i $\frac{1}{(n + 1)}$

2 a i $\frac{1}{(n - 1)}$ ii $\frac{(n + 2)}{n}$ iii $\frac{n}{2}$ iv $\frac{2n}{(n - 3)}$
i $\frac{1}{49}$ ii $1\frac{1}{25}$ iii 25 iv $2\frac{6}{47}$

3 a $\frac{(x + 1)(x + 2)}{2}$
b $x + 1$ and $x + 2$ are consecutive integers, so one of them must be even, and so dividing by 2 gives an integer.

End points

A1 a $22b + 16$ b $26b + 14$ c $10ab^2 + 4a^2b$ d $3x^2y - 37xy - 7xy^2$

A2 C

A3 a $a^2 + 9a + 8$ b $b^2 + 8b + 16$ c $5c^2 + 17c + 6$
d $d^2 + d - 2$ e $6e^2 - 5e - 50$ f $3f^2 - 23f + 14$

A4 a $(5 \times 10) - (4 \times 8)$, $(6 \times 11) - (5 \times 9)$, $(7 \times 12) - (6 \times 10)$
b 66 c $(n + 1)(n + 6) - n(n + 4)$
d $(n + 1)(n + 6) - n(n + 4) = 3n + 6 = 3(n + 2)$

B1 a $4(2p + q)$ b $6(a - 2b)$ c $a(a + b)$
d $m(7 + 3n)$ e $2y(4x + 5)$ f $2a(2b + 3a)$
g $xy(3y - 5x)$ h $2pq(2 - 3p)$ i $h(2g^2 - 5h)$

B2 a C and E b B and C c D and F

B3 a $(x + 1)(x + 11)$ b $(x + 2)(x + 7)$ c $(x + 5)(x - 1)$
d $(x + 3)(x - 2)$ e $(x + 4)(x - 5)$ f $(x - 2)(x - 3)$
g $(x + 10)(x - 10)$ h $(2x - 5)(x - 3)$ i $(3x + 4)(3x - 4)$
j $(5x - 1)(x + 1)$ k $(x + 2y)(x - 2y)$ l $(2x - y)^2$

B4 B and D, C and E

B5 a $\frac{(x - 4)}{3}$ b $x > 4$

## SECTION 4 Comparing Data

Starting points

A1 a Discrete b continuous c discrete d discrete

B1 a 15 and 42, 25 b 39.5, 38 c 42.5, 48.8 d 69, 93

C1 a £340 b £340 c £314.89
d The mode and median are reasonable averages to use.
The mean is too low because of one extreme value (£112).

**D1**

| Colour of car | Frequency |
|---|---|
| Red | 5 |
| Blue | 6 |
| Green | 2 |
| White | 3 |
| Silver | 2 |
| Black | 4 |
| Brown | 2 |

**D2** Blue

| Number of children in car | Frequency |
|---|---|
| 0 | 11 |
| 1 | 7 |
| 2 | 4 |
| 3 | 1 |
| 4 | 1 |

**D4** **a** 0 **b** 4

**D5** **a** 1 **b** 0.9

**E1** **a** Red 20.8%, Blue 25%, Green 8.3%, White 12.5%, Silver 8.3%, Black 16.7%, Brown 8.3%
**b** Red 75°, Blue 90°, Green 30°, White 45°, Silver 30°, Black 60°, Brown 30°

**F5** **a** 54.4% **b** 34.8% **c** 10.8%

**F6** **a** 38.8% **b** 33.0% **c** 28.2%

### Thinking ahead to ...

**A** **a** 8 **b** His scores were mainly in the interval 4–7.

### Exercise 4.1

1 **a i** William: 2, 4, 4, 4, 5, 5, 6, 6, 6, 6, 6, 7, 7, 7, 8, 8
Bryony: 4, 4, 5, 5, 5, 5, 6, 6, 6, 6, 7, 7, 7, 9, 9
Daniel: 2, 3, 4, 4, 5, 5, 5, 6, 6, 6, 6, 6, 7, 7, 7, 9, 9, 9, 10, 10
**ii** William's interquartile range = 2.5, Bryony's = 2, Daniel's = 3
**b** Bryony

### Exercise 4.2

1 **a**

| | |
|---|---|
| 4 | 0 6 |
| 3 | 2 3 8 |
| 2 | 0 3 4 6 |
| 2 | 2 2 4 9 |
| 0 | 7 |
| | 1 2 3 4 5 6 |

**b**

| | |
|---|---|
| 4 | 0 |
| 3 | 0 1 2 |
| 2 | 0 2 4 4 4 6 6 |
| 1 | 0 2 5 |
| 0 | |
| | 1 2 3 4 5 6 7 |

**c**

| | |
|---|---|
| 4 | 1 2 4 |
| 3 | 0 0 6 |
| 2 | 4 5 5 |
| 1 | 6 6 8 |
| 0 | 6 6 7 9 |
| | 1 2 3 4 5 6 |

**d**

| | |
|---|---|
| 5 | 6 |
| 4 | 0 0 6 |
| 3 | 0 2 7 8 |
| 2 | 2 4 8 8 8 |
| 1 | 4 8 |
| 0 | |
| | 1 2 3 4 5 6 |

2 **a**

| | |
|---|---|
| 6 | 2 |
| 5 | 0 0 0 1 2 2 3 4 4 6 7 9 |
| 4 | 0 0 0 1 2 4 4 7 |
| 3 | 0 4 8 8 |
| 2 | |
| 1 | |
| 0 | |
| | 1 2 3 4 5 6 7 8 9 10 11 12 |

**b** 50

### Exercise 4.3

1 **b** Supermarket A: 28, Supermarket B: 28
**c** Supermarket A: 48, Supermarket B: 37

2 **a** Group A: 29, Group B: 33.5
**b** Both have a range of 50.

3 **c** A: 42.6, B: 38

### Exercise 4.4

1 17.2

2 **a** 53 **b** 13.6

### Thinking ahead to ...

**A** **b** 1.4, 1.6

### Exercise 4.5

1 **a** 1.55 **b** Round 6

2 **b** Paula 2.7, Zoe 1.9, Matt 2.5

### Exercise 4.6

1 **a** Zeta 61, Carlo 57.5 **b** Zeta 12.4, Carlo 13.3

2 Dean: 5.8, 1.26, Leah: 5.4, 1.19, Jack: 4.8, 1.52

3 **a ii** Eliza 4, Sam 4.5 **iii** Eliza 1, Sam 2.25

### Thinking ahead to ...

1 **a ii** 7 **b ii** 8

### Exercise 4.7

1 Lee 5, Faith 6, Aqib 7, Tegan 4.5

### Exercise 4.10

3 **A** The area of the drops is misleading
**b** The vertical scale does not start at zero

### End points

**A1** **a** 65.5 **b** 11.48

**B1** 60.5, 11.48

**C1** 7.6, 1.49, 1.22

**D1** 7.6, 1.49

**E1** 8, 2.5

**G1** X The vertical scale does not start at zero
Y The areas are misleading

## Section 5 Sequences

### Starting points

**A1** **a** 33, 39 **b** 30, 41

**A2** 31

**A3** **a** 32, 38, 44 **b** 47, 65, 86 **c** 57, 83, 114

**A4** **a** 6 **b** 79

**B1** **b** 46

**c**

| Pattern number ($n$) | Number of matches ($m$) |
|---|---|
| 1 | 6 |
| 2 | 11 |
| 3 | 16 |
| 4 | 21 |
| 5 | 26 |

**d** $m = 5n + 1$

## Exercise 5.1

**1**

| Pattern number of | | Number matches |
|---|---|---|
| 1 | → | 3 |
| 2 | → | 4 |
| 3 | → | 5 |
| 4 | → | 6 |
| 20 | → | 22 |
| $n$ | → | $n + 2$ |

**2** **a**

| Pattern number of | | Number matches |
|---|---|---|
| 1 | → | 4 |
| 2 | → | 8 |
| 3 | → | 12 |
| 4 | → | 16 |
| 5 | → | 20 |
| 6 | → | 24 |

**b** $n \rightarrow 4n$ **c** 400 **d** Pattern 155 **e** Pattern 124

### Thinking ahead to ...

**A** 301

## Exercise 5.2

**1** **a** $m = 2n + 1$ **b** 17

**2** **a** $m = 4n + 1$ **b** 16 **c** 32nd

**3** **a** $m = 6n - 2$ **b** 598

**4** **b** $m = 4n + 2$

**5** **a** $50 \rightarrow 249$, $n \rightarrow 5n - 1$
**b** $2 \rightarrow 22$
$3 \rightarrow 18$
$4 \rightarrow 14$
$11 \rightarrow {}^-14$
**c** $28 \rightarrow {}^-75$
$p \rightarrow 9 - 3p$

### Thinking ahead to ...

**A** **a** 19, 21 **b** 29 **c** 105

## Exercise 5.3

**1** **a** A: $3n + 3$, B: $5n - 4$, C: $10n + 3$, D: $8n - 6$
**b** A: 153, B: 246, C: 503, D: 394

**2** **a** He should have had $3n$ instead of +3.
**b** $3n + 2$

**3** $n + 7$

**4** 1, ⁻2, ⁻5, ⁻8, ⁻11

**5** **a** $22 - 2n$ **b** $32 - 10n$ **c** $1 - 2n$ **d** $15 - 5n$

## Exercise 5.4

**1** **a** 5th number, $40 = 5 \times 8$ **b** 130

**2** **a** 5th triangle number, $15 = \frac{(5 \times 6)}{2}$
**b** 78 **c** $\frac{1}{2}n(n + 1)$

**3** **a** 5th power, $32 = 2 \times 2 \times 2 \times 2 \times 2 = 2^5$
**b** 256 **c** $2^n$

## Exercise 5.5

**1** **c** 204 **d** 80th

**2** **a** $2n + 6$, $2(n + 3)$

## Exercise 5.6

**1** **a** $2^n$ **b** $2^n - 1$ **c** $2^{n-1}$ **d** $2^{n-3}$ **e** $3^{1-n}$ **f** $4^{-n}$

**2** **a** $\frac{n}{(100 - n)}$ **b** $\frac{n^2}{2^{n+1}}$ **c** $\frac{3}{n^2}$ **d** $\frac{(n^2 + 1)}{(2^n + 1)}$

**3** **a** $200 - 4^n$ **b** $4^{5-n}$

**4** **a** 1st number $^-1 = (1 - 1)^2 - 1$
2nd number $0 = (2 - 1)^2 - 1$
3rd number $3 = (3 - 1)^2 - 1$
4th number $8 = (4 - 1)^2 - 1$
5th number $15 = (5 - 1)^2 - 1$
**b** $u_n = (n - 1)^2 - 1$

### End points

**A1** **a** $m = 2n + 3$ **b** 495 matches

**B1** **A** $u_n = 4n + 1$, $u_{20} = 81$
**B** $u_n = n + 5$, $u_{20} = 25$
**C** $u_n = 3n + 6$, $u_{20} = 66$
**D** $u_n = 5n^2$, $u_{20} = 2000$
**E** $u_n = 2 - 5n$, $u_{20} = {}^-98$
**F** $u_n = 8n - 57$, $u_{20} = 103$
**G** $u_n = 12n - 13$, $u_{20} = 227$
**H** $u_n = 9n - 5$, $u_{20} = 175$

**B2** **a** $3^{n+2}$ **b** $2^{n+2}$ **c** $2^n + 2$ **d** $4^n - 3$

**B3** **a** $\frac{2n}{(n^2 + 1)}$ **b** $\frac{(2n + 1)}{3n^2}$

# SKILLS BREAK 1

**1** **a** 90 **b** £24.10 **c** £20.51

**2** **a** regular pentagon, 108° **b** 28p

**3** **b** 1120 **c** 1176

**4** **a** regular hexagon **b** 120° **c** 10 cm

**5** **a** The dimensions of the tiles do not fit exactly into the dimensions of the room, so he should have worked out how many he needs along the length and width and then multiplied.
**b** 198 **c** £332.48

**6** $2\frac{1}{2}$ litres, it is cheapest per litre.

**7** £37.00

**8** £7.45

**9** **a** regular octagon **b** 135° **c** £14

**10** Volume = 18 cm × $\pi$ × radius² = 2500 cm³
So, radius = 6.65 cm and diameter = 13.3 cm = 133 mm

**11** **a** No, if a tub is used to cover 4 m² that is 875 g/m².
**b** increase **c** 5%

**12** 2.2 g

**13** **a** She has used the slant height instead of the height of each triangle.
**b** 442 cm²

**14** **b** Each rhombus is twice the size of each equilateral triangle.
**c** 16.7%

# CONSTRUCTIONS AND LOCI

### Starting points

**A1** **c** 60°

**A2** **a** no **b** yes **c** yes **d** no

**B1** F

**B3** No, the sides may be joined together in a different order, or the angles could be different.

## Exercise 6.1

**2** angle SRQ = 77°

**3** Triangles LMN and PQR look identical.

## Exercise 6.2

**1** **a** ΔEBD **b** ΔDCB **c** ΔCDA
**d** ΔAFD **e** ΔFCB **f** ΔAEC

**2** **a** The angles at P and Q could be different to those at A and B.

**3** **a** congruent by SSS **b** congruent by SAS
**c** not necessarily congruent
**d** congruent by RHS **e** not necessarily congruent
**f** congruent by ASA

Thinking ahead to ...

**A** A, C and H

### Exercise 6.3

**2** Each angle is 60°. When you bisect one of the angles you make an angle of 30°.

### Exercise 6.4

**1** **c** The distances from A and B are equal.

**3** **c** At a single point. **d** yes

**4** **e** They intersect at the centre.

### Exercise 6.5

**1** **a** 2 cm **b** 20 m

**5** **b** 12 km

### Exercise 6.6

**1** **d** A circle with AB as diameter.

**2** **e** 90° **f** It is a right-angled triangle.

**3** angles BCE, BAE, AGH, CGH, AGB, CGB

**4** **a** 28.6 m **b** 20.2 m

### End points

**B1** **e** The gradients are the same.

**D1** **a** congruent by SAS **b** not necessarily congruent
**c** not necessarily congruent **d** congruent by SAS
**e** congruent by ASA **f** congruent by RHS

## SECTION 7 Powers and Surds

### Starting points

**A1** **a** $0.\dot{7}$ **b** $0.1\dot{2}\dot{3}$ **c** $4.5\dot{6}$

**A2** **a** 5.65656 ... **b** 56.5656 ... **c** 565.656 ...

**A3** **a** 0.25 **b** $0.\dot{3}$ **c** $0.\dot{5}\dot{4}$ **d** $0.5\dot{6}$ **e** 0.375

**B1** **a** $3^5$ **b** $4^{-3}$

**B2** **a** 81 **b** $\frac{1}{9}$ **c** 1 **d** $\frac{1}{5}$ **e** 512 **f** 36 **g** $\frac{4}{9}$ **h** $\frac{1}{16}$

**C1** **a** 13 **b** 10 **c** 2.6 **d** 2.1

**C2** **a** 3.5 **b** 10.2 **c** 2.6

**C3** **a** 7 **b** 1.5 **c** $^-10$

**C4** $\frac{1}{5}$ **b** $\frac{7}{10}$ **d** $1\frac{1}{5}$

**C5** Positive and negative numbers have different cubes.

### Exercise 7.1

**1** **a** 2 **b** 1.5 **c** $^-1$

**2** a, d

**3** ***a*** $\frac{3}{4}$ **b** $\frac{2}{3}$ **c** $\frac{1}{5}$

**4** 1.8

**5** **a** 150.0625 **b** 3 **c** $^-91.125$

**6** **a** $6^2$ **b** $2^2$ **c** $5^2$

**7** **a** **i** $x = 4$ **ii** $x = 6$

### Exercise 7.2

**1** **a** $6^5$ **b** $5^4$ **c** $7^8$ **d** 3 **e** $2^{11}$ **f** $5^{-6}$

**2** **a** 5 **b** 0 **c** 6 **d** 10 **e** 5 **f** 5

**3** **a** $x^8$ **b** $x^6$ **c** $x^{10}$ **d** $x^2$ **e** $x^4$ **f** $x^{-2}$

**4** **a** 3 **b** $^-1$ **c** $^-4$

**5** **a** 5 **b** $^-2$ **c** $^-1$

**6** **a** They multiplied the powers instead of adding them.
**b** They divided the powers instead of subtracting.
**c** They subtracted the powers the wrong way round.
**d** They added the powers instead of multiplying them.

**7** **a** $x = 8$ **b** $x = 6$ **c** $x = 1$

**8** **a** $2p$ **b** $\frac{6q^4}{5}$ **c** $2r^4$

Thinking ahead to ...

**A** **a** 5 **b** 9

**B** 2

### Exercise 7.3

**1** a and g, b and h, c and e, d and f

**2** **a** 6 **b** 5 **c** 3 **d** $\frac{1}{7}$ **e** $\frac{1}{2}$ **f** 1

**3** **a** $\frac{2}{3}$ **b** $\frac{1}{3}$

**4** a and e, b and g, c and f, d and h

**5** a and f, b and h, c and d, e and g

**6** a

**7** a 25 b 27 c 9 d 128 e $\frac{1}{8}$ f $\frac{1}{81}$

**8** **a** **i** B, C, D **ii** A, B, C, D, F **iii** B, C, D, E

**9** $p = 243$

**10** **a** $\frac{1}{27}$ **b** 9

### Exercise 7.4

**1** **a** $x = 3$ **b** $y = {}^-1$ **c** $z = 2$ **d** $a = \frac{1}{5}$ **e** $b = \frac{^-1}{3}$
**f** $c = {}^-4$ **g** $p = 2\frac{1}{2}$ **h** $q = 1\frac{1}{3}$ **i** $r = \frac{^-3}{4}$

**2** Across
2 61 3 1024 6 1331 8 27
Down
1 10 2 64 3 16 4 243 5 81 6 17 7 32

Thinking ahead to ...

**A** B and E

### Exercise 7.5

**1** **a** B, D, E

**2** **a** B, C, F, G and H give integer values.

**3** A and H, B and D, C and F, E and J, G and I

**4** **a** $2\sqrt{10}$ **b** $\sqrt{21}$ **c** 3 **d** 8 **e** $3\sqrt{5}$ **f** 45 **g** $11 + 4\sqrt{7}$ **h** $18 + 12\sqrt{2}$
**i** $3 + \sqrt{5} - 3\sqrt{2} - \sqrt{10}$ **j** 2 **k** $33 + 6\sqrt{30}$ **i** 144

**5** **a** $4\sqrt{5}$ m² **b** 36 m² **c** $\frac{(9\sqrt{35})}{2}$ m² **d** $24\sqrt{3}$ m²
**e** $(4\sqrt{15} - 2\sqrt{10} - 6\sqrt{2} + 2\sqrt{3})$ mm²

**6** $\frac{\sqrt{54}}{\sqrt{216}} = \sqrt{(\frac{54}{216})} = \sqrt{\frac{1}{4}} = \frac{1}{2}$

**7** $12\sqrt{2}$ cm

**8** **a** 20 000 000 **b** 2 **c** 3

### End points

**Al** **a** 3 **b** 6 **c** 2

**B1** **a** 8 **b** 6 **c** 3

**B2** **a** $2x^7$ **b** $x^4$ **c** $81x^{12}$

**C1** **a** 10 **b** 16 **c** $\frac{1}{8}$

**C2** **a** $x = \frac{1}{3}$ **b** $x = {}^-2$ **c** $^-\frac{1}{4}$

**D1** **a** $\sqrt{10}$ **b** $4\sqrt{2}$ **c** 63 **d** $5 - 2\sqrt{6}$

**E1** $180\sqrt{6}$

**E2** **i** $\sqrt{2}$ **ii** $\sqrt{3}$

## SECTION 8 Approximation and Errors

### Starting points

**A1** **a** 23.47 **b** 4.16 **c** 142.86 **d** 12.30

**A2** **a** 56.5 **b** 0.3 **c** 6.0 **d** 600.0

**B1** **a** 150 **b** 16 **c** 3400 **d** 6.3 **e** 0.36 **f** 1 100 000

**B2** **a** 30 **b** 200 **c** 0.05 **d** 500 000 **e** 100 **f** 0.000 09

**C1** **a** $8^{10}$ **b** $6^2$ **c** $10^{-2}$ **d** $10^{-2}$ **e** $4^{-10}$ **f** $10^2$ **g** $5^8$ **h** $6^{-12}$

**D1** **a** $3.45 \times 10^5$ **b** $4.17 \times 10^{-3}$ **c** $4.297 \times 10$
**d** $4 \times 10^6$ **e** $2.3 \times 10^{-6}$ **f** $6.413234 \times 10^3$

**D2** **a** 6740 **b** 0.00015 **c** 652.41 **d** 0.000 000 2

### Exercise 8.1
1 2.995 m

2 a 15.65 cm b 15.55 cm

3 a 7 750 000 km$^2$ b 7 650 000 km$^2$ c $7\,650\,000 \leqslant L < 7\,750\,000$

4 $0.16 \leqslant t \leqslant 0.18$

### Exercise 8.2
1 a 755 g b 745 g

2 a 46.2 cm b 1.2 cm

3 a $132 < L < 135$ b 42.89°

4 a 81.64 m$^2$, 81.84 m$^2$ b 38.98 m, 39.02 m

5 62.8875 cm$^2$, 64.5575 cm$^2$

### Exercise 8.3
1 a 6000 b 24 000 c 4 000 000
d 25 000 e 750 f 100 g 250 h 30 000

2 a £18 000 b Both numbers are rounded up

3 100 people per km$^2$

5 a too large b about right c about right d too small
e too small f too large g about right h too small

6 The numbers are a different order of magnitude

### Exercise 8.4
1 a 17784 b 17.784

### Exercise 8.5
1 a $3.6 \times 10^3$ b $1 \times 10^5$ c $5.8 \times 10^5$
d $5.6 \times 10^{-2}$ e $7.6 \times 10^{-4}$ f $6.3 \times 10^{-6}$

2 a 374 600 b 5808 c 34 200 000
d 0.056 75 e 0.000 063 88 f 0.000 000 009 02

3 a $1.38 \times 10^9$ b $1.49 \times 10^5$

### Exercise 8.6
1 875

2 8 minutes 20 seconds

3 a $1.50 \times 10^8$ km, $1.07 \times 105$ km/h b 8808 hours, a year

4 a $5 \times 10^2$, $9.6 \times 10^{-5}$ b 4.8 cm

5 a $7.32 \times 10^{-23}$ g b $1.37 \times 10^{22}$

6 2.6 seconds

7 $6.7 \times 10$ m

8 a $5.3 \times 10^{14}$ m$^2$ b $7.5 \times 10^{-5}$ g

### End points
A1 a 6.75 cm b 9.75 cm c 338.6 cm$^3$, 352.5 cm$^3$ d 208.9 cm$^2$

B1 a 300 000 b 3 000 000 c 0.000 05 d 150

C1 a $3.6 \times 10^5$ b $1 \times 10^5$ c $7.6 \times 10^{-6}$

C2 $6.9 \times 10^0$

C3 12 960 tonnes

## SECTION 9 Solving Equations Algebraically

### Starting points
A1 a $y^2 + 8y + 15$ b $k^2 - k - 12$ c $n^2 - 6n - 27$
d $w^2 - 9w + 20$ e $v^2 - 64$

A2 a $2n^2 - 5n - 12$ b $3w^2 - 22w - 16$ c $12y^2 - 25y + 12$
d $4k^2 - 12k + 9$ e $3v^3 + v^2 - 9v - 3$

A3 a $2y^3 - 2y^2 - 12y$ b $3x^3 + 6x^2 + 3x$ c $x^3 + 3x^2 - 13x - 15$
d $4n^3 - 4n^2 - 5n + 3$ e $x^3 + 6x^2 + 12x + 8$ f $6k^2 - 2k^3 + 8k - 24$

B1 a $(y + 2)(y + 7)$ b $(k + 5)(k - 8)$ c $(x + 6)(x - 2)$
d $(w + 9)(w - 4)$ e $(x - 3)(x - 8)$ f $(y - 1)(y - 8)$
g $(p + 12)(p - 5)$ h $(b + 9)(b - 8)$ i $(x - 12)(x - 4)$
j $(n + 5)(n - 5)$ k $(y + 10)(y - 10)$ l $(a + b)(a - b)$
m $(2y + 6)(2y - 6)$ n $(4x + 6y)(4x - 6y)$ o $(10a + b^2)(10a - b^2)$

C1 a $y = {}^{-}4.5$ b $y = 3.4$ c $x = 2$
d $x = 3$ e $x = 4$

C2 a $x = {}^{-}2$ b $n = \frac{3}{8}$ c $w = 21$ d $k = 2$ e $v = 2.17$

C3 a $y = 8.67$ b $u = 8$ c $k = 10$ d $h = 12$ e $a = 2$

C4 a $w = {}^{-}7$ b $k = 3.5$ c $n = \frac{1}{8}$

D1 9 cm × 15 cm

D2 20

D3 53 fans, 44 fans, 50 fans

### Thinking ahead to ...
A a £2.30 b £6.90 B a £4 b 96p c 28p d 68p

### Exercise 9.1
1 a $x = 6, y = 9$ b $x = 8, y = 1$ c $x = 6, y = 2$
d $x = 2, y = {}^{-}1$ e $x = {}^{-}2, y = 3$ f $x = 3, y = {}^{-}4$

2 $m = {}^{-}1, n = 10$

3 a $v = 2, w = 12$ b $v = {}^{-}1, w = 3$ c $v = 3, w = {}^{-}2$ d $v = 3, w = 2$
e $v = 3, w = {}^{-}1$ f $v = 3, w = {}^{-}2$ g $v = 2, w = {}^{-}1$ h $v = 1, w = 3$
i $v = {}^{-}1, w = 3$

### Exercise 9.2
1 a $a = 10, b = 2$ b $a = 2, b = 5$ c $a = 4, b = {}^{-}3$ d $a = {}^{-}2, b = 8$
e $a = 3, b = 1$ f $a = 6, b = {}^{-}4$ g $a = 2, b = {}^{-}1$ h $a = 4, b = {}^{-}2$
i $a = 3, b = {}^{-}2$ j $a = 2, b = 1$ k $a = 2, b = {}^{-}3$ l $a = {}^{-}1, b = 4$

2 $k = 5, n = 10$

3 $m = 6, n = 1$

4 a $x = 4, y = 2$ b $x = 3, y = 5$ c $x = 2, y = 1$ d $x = {}^{-}1, y = 2$
e $x = 5, y = {}^{-}1$ f $x = 7, y = {}^{-}4$ g $x = {}^{-}3, y = {}^{-}2$ h $x = 3, y = {}^{-}4$
i $x = 2, y = {}^{-}1$ j $x = 3, y = {}^{-}5$ k $x = {}^{-}1, y = 2$ l $x = {}^{-}2, y = 3$

### Exercise 9.3
1 adult £6.50, children £4

2 16 g, 54 g

3 63, 26

4 13, 9

5 123, 157

6 5 and 7 or 4.5 and 6.5

7 4, $^{-}2$

8 16 200, 12 300

9 a 3, 8 b $y = 3x + 8$

10 a 1, 3, $^{-}5$ b $y = x^2 + 3x - 5$

11 a 2, $^{-}3$, 1 b $y = 2x^2 - 3x + 1$

12 a 3, $^{-}1$, $^{-}4$ b $y = 3x^2 - x - 4$

13 8.5 hours, 16.5 hours

14 18, 8

### Exercise 9.4
1 a $^{-}4, {}^{-}1$ b $^{-}6, {}^{-}3$ c $^{-}8, 3$ d $^{-}7, 1$ e $^{-}9, 4$ f $^{-}9, 7$ g $^{-}4, 11$
h $^{-}7, 15$ i $^{-}7, 16$ j 3, 5 k 5, 7 l 5, 6 m 2, 5 n $^{-}28, {}^{-}3$
o $^{-}2, 15$ p $^{-}12, 5$ q 5, 9 r $^{-}15, 3$ s $^{-}15, 14$ t $^{-}19, 21$
u 3, 14 v $^{-}15, 12$ w 7, 8 x $^{-}8, 17$

### Exercise 9.5
1 a 0, $^{-}8$ b 0, 5 c 0, 18 d $^{-}10, 10$ e $^{-}5, 5$ f $^{-}9, 9$

2 a 0, 12 b $^{-}79, 0$ c $-\frac{1}{3}, 0$ d $0, \frac{1}{4}$ e $^{-}13, 13$
f 0, 1 g $0, \frac{2}{5}$ h $-\frac{5}{4}, 0$ i $-\frac{2}{7}, 0$

### Exercise 9.6
1 a $-\frac{1}{4}, \frac{2}{3}$ b $-\frac{1}{3}, \frac{5}{2}$ c $-\frac{1}{2}, \frac{3}{2}$ d $-\frac{1}{5}, \frac{1}{2}$ e $-\frac{4}{3}, \frac{1}{3}$ f $\frac{1}{2}, \frac{2}{3}$ g $-\frac{2}{3}, \frac{1}{3}$ h $-\frac{1}{7}, 1$
i $-\frac{2}{3}, \frac{1}{2}$ j $\frac{1}{5}, 2$ k $^{-}3, \frac{1}{3}$ l $\frac{4}{3}, 4$ m $^{-}2, 3$ n $\frac{2}{3}, 2$ o $-\frac{1}{6}, 2$

2 a $0, \frac{1}{2}$ b 0, 3 c $-\frac{1}{4}, 0$ d $0, \frac{8}{5}$ e $0, \frac{1}{2}$ f $-\frac{1}{5}, \frac{1}{5}$ g $-\frac{1}{4}, \frac{1}{4}$ h 0, 9
i 0, 5 j 0, 1 k $-\frac{1}{3}, \frac{1}{3}$ l $0, \frac{4}{3}$ m 0, 31 n $0, \frac{3}{7}$ o $-\frac{3}{4}, \frac{3}{4}$

## Exercise 9.7

1 **a** −2.35, 0.85 **b** −1.55, 0.80 **c** −4.19, 1.19 **d** −4.30, −0.70
**e** −3.35, −0.15 **f** −1.46, −0.14 **g** 0.23, 1.43 **h** 0.29, 1.71
**i** −0.62, 1.62 **j** −1.85, 1.35 **k** −1.47, 1.81 **l** 0.28, 0.72

2 **a** $\frac{(-1 \pm \sqrt{7})}{2}$ **b** $\frac{(-1 \pm \sqrt{145})}{12}$ **c** $\frac{(1 \pm \sqrt{65})}{8}$ **d** $\frac{(-2 \pm \sqrt{6})}{4}$
**e** $\frac{(-4 \pm \sqrt{11})}{5}$ **f** $\frac{(-9 \pm \sqrt{53})}{14}$ **g** $\frac{-(3 \pm \sqrt{41})}{4}$ **h** $\frac{(5 \pm \sqrt{37})}{2}$ **i** $\frac{3 \pm \sqrt{33}}{2}$

3 **a** −2.77, 1.27 **b** 2.18, 0.15 **c** −1.80, 0.55

## Exercise 9.8

1 **b** −10.48, 0.48

2 −8.36, 0.36

3 $(b^2 - 4ac)$ is negative

4 **a** −12.16, 0.16 **b** −7.12, 1.12 **c** −14.07, 0.07 **d** no solution
**e** no solution **f** −7.16, −0.84

5 **a** $\pm\sqrt{(\frac{13}{2})} - \frac{5}{2}$ **b** $\frac{\pm\sqrt{57} - 7}{2}$ **c** $\frac{\pm\sqrt{5} - 3}{2}$

## Exercise 9.9

1 **a** −4, 9 **b** −5, 12 **c** −7, 9 **d** −9, 3 **e** −4.79, −1.88 **f** −6, 1
**g** −1.27, 6.27 **h** −4.21, 0.71 **i** −0.17, 0.17 **j** −0.2, 0.2 **k** 0, 1.67
**l** −2.91, 0.57 **m** −1.90, 0.40 **n** −1.38, 1.63 **o** −0.1, 0.1

## Exercise 9.10

1 **a** $L(L - 12) = 448$, length = 28 cm, width = 16 cm **b** 32.25 cm

2 18, 24

3 **a** 36 cm (27 cm) **b** 972 cm$^2$

4 37.61°

5 −14.72, 2.72

6 **a** 27 cm, 36 cm **b** 108 cm

7 **a** 8.5 cm **b** 50.14 cm **c** 4.28°

8 50.5, 54.5

9 54 units$^2$

11 **b** $6x^2 + 40x + 48$
**c** $6x^2 + 40x + 48 = 398$, $x = 5$ cm **d** 495 cm$^3$

## End points

**A1** **a** 3, −2 **b** −3, −1

**B1** £4.80, £2.65

**C1** **a** −8, 3 **b** −5, 13 **c** 3, 12 **d** −2, 3

**D1** **a** −6.32, 0.32 **b** 0.35, 8.65 **c** −4.14, −0.36 **d** no solution

**D2** **a** $-4 \pm 2\sqrt{5}$ **b** $3 \pm \sqrt{17}$

**E1** −5.54, 0.54

**E2** $\pm\sqrt{21} - 4$

**F1** 15 cm, 38 cm

# Section 10 Transformations

## Starting points

**A1** **a** $x = 3$ **2b** $y = -1$

**B1** **b i** vertices at (2, 0), (3, 0), (3, 2)
**ii** vertices at (−4, 3), (−2, 3), (−2, 2)

**C1** **a** 90° **b** (0, 1)

**D1** U has vertices at (2, 2), (2, 3), (4, 3);
V has vertices at (−1, −1), (−1, 0), (1, 0)

**E1** **a** vertices at (1, 0), (1, 2), (5, 2)
**b** vertices at (0, 3.5), (0, 4), (1, 4)
**c** vertices at (2, 1.5), (6, 1.5), (6, 3.5)

**E2** **a** vertices at (−1.5, 0.5), (−1.5, −2.5), (−3, −1), (−3, 0.5)
**b** vertices at (4, 0), (4, −4), (2, −2), (2, 0)
**c** vertices at (1, −1), (1, −7), (−2, −4), (−2, −1)

**F1** **a** V, Y, Z

## Thinking ahead to …

**A** **b** B has vertices at (1, 2), (2, 2), (1, 4);
C has vertices at (−2, 1), (0, 1), (0, 2);
D has vertices at (−4, −1), (−2, −1), (−2, −2);
E has vertices at (2, −1), (2, −3), (3, −1);
F has vertices at (−2, 3), (−3, 4), (0, 3)
**c** rotation of 90° about (1, 1)
**d i** rotation of 90° about (0, 1)
**ii** reflection in $y = x - 1$
**iii** rotation of −90° about (−2, −1)

## Exercise 10.1

1 **a** H has vertices at (4, 1), (5, 1), (6, 2), (5, 3), G→H by $\begin{pmatrix} 6 \\ 2 \end{pmatrix}$

2 **a** J has vertices at (3, 1), (4, 0), (3, −1), (2, −1)
K has vertices at (3, 1), (4, 0), (3, −1), (2, −1)
**b** They are the same.

3 1 and 4, 2 and 3

## Exercise 10.2

1 **a** L has vertices at (0, 2), (1, 4), (0, 4), (−1, 3)
M has vertices at (−1, 1), (0, 2), (−2, 3), (−2, 2)
N has vertices at (−1, −4), (−2, −4), (−3, −3), (−2, −2)
**b i** rotation +90° about (1, 1)
**ii** rotation 180° about (1, 2)
**iii** reflection in $y = -x$

## Exercise 10.3

1 **b i** translation $\begin{pmatrix} -2 \\ -2 \end{pmatrix}$ **ii** 180° rotation about (2, 0)
**iii** 180° rotation about (3, 1)

2 **a i** reflection in $x = 5$
**ii** reflection in $y = x$
**iii** rotation 180° about (4, 1)

3 **a** $P_5$ has vertices at (4, 0), (6, 0), (6, −2), (4, −4)
$P_6$ has vertices at (−2, 0), (−2, 2), (0, 4), (0, 0)
$P_7$ has vertices at (0, 2), (0, 4), (2, 6), (2, 2)

## Exercise 10.4

1 **b** B

**c**

| Object | Image | SF | Centre |
|---|---|---|---|
| S | Y | 2 | (4, 8) |
| Y | U | $\frac{1}{2}$ | (8, 8) |
| E | A | 2 | (8, 8) |
| Q | W | −1 | (8, 6) |

**d** They are not in the same orientation.
**e i** Rotate −90° about (8, 8)
**ii** Enlarge SF $\frac{1}{2}$ with centre (8, 12)
**iii** Rotate 180° about (6, 8)

## Exercise 10.5

2 **a** $B_1$ has vertices at (5, 2), (5, 4), (1, 4)
$B_2$ has vertices at (5, 0), (7, 0), (7, 4)
$B_3$ has vertices at (5, 6), (9, 6), (9, 4)
$B_4$ has vertices at (3, 8), (5, 8), (3, 4)
**b i** rotation −90° about (4, 1)
**ii** translation $\begin{pmatrix} 4 \\ 2 \end{pmatrix}$
**iii** rotation +90° about (2, 5)
**c i** enlargement SF 2 with centre (−1, −2)
**ii** enlargement SF 2 with centre (−5, 4)

3 **a** enlargement SF $\frac{1}{2}$ with centre (−1, 0)
**b** enlargement SF $\frac{1}{2}$ with centre (−3, −2)

5 **c i** enlargement SF −2 with centre (0, 1)
**ii** enlargement SF 2.25 with centre (0, 1)
**d i** enlargement SF $-\frac{1}{2}$ with centre (0, 1)
**ii** enlargement SF $\frac{4}{9}$ with centre (0, 1)

Thinking ahead to ...

**A** **b** The transformations have no effect.

Exercise 10.6

**1** **a** translation $\begin{pmatrix} -1 \\ 2 \end{pmatrix}$

**b** rotation +90° about (1, 1)

**c** enlargement SF $\frac{1}{2}$ with centre ($^-$1, 2)

**2** **a** reflection in $y = {}^-x$

**b** the reflection and its inverse are the same

**3** reflection 180° about $(x, y)$; enlargement SF $^-$1, centre $(x, y)$

End points

**A1** **b** K has vertices at ($^-$4, $^-$3), ($^-$1, $^-$3), ($^-$1, $^-$5), ($^-$2, $^-$5), ($^-$2, $^-$4), ($^-$4, $^-$4)

L has vertices at ($^-$3, 1), ($^-$1, 1), ($^-$1, 4), ($^-$2, 4), ($^-$2, 2), ($^-$3, 2)

M has vertices at (1, $^-$1), (2, $^-$1), (2, $^-$2), (4, $^-$2), (4, $^-$3), (1, $^-$3)

**c** **i** rotation 180° about (0, –1)

**ii** rotation 90° anticlockwise about (0, 0)

**iii** translation $\begin{pmatrix} 0 \\ -4 \end{pmatrix}$

**d** **i** translation $\begin{pmatrix} 0 \\ 4 \end{pmatrix}$

**ii** reflection in $y = {}^-3$

**iii** rotation 180° about (1, 0)

**e** **i** rotation 90° clockwise about (1, $^-$1)

**ii** rotation 180° about (0, $^-$3); enlargement, SF – 1, centre (0, $^-$3)

**B1** **a** enlargement SF 4 with centre (3, $^-$1)

**b** rotation +270° about (0, 2)

**c** translation $\begin{pmatrix} 0 \\ 3 \end{pmatrix}$

**B2** reflection in $y = x + 1$

## Skills break 2

**1** **a** 12 **b** 12 **d** 132.7 m$^2$ **e** 1805 cm$^2$ **f** 7.37 m$^3$

**2** **a** 45° **b** 135° **c** 28.97 cm **d** 695.26 cm$^2$

**3** **a** 1110 tons **b** 46 m **c** 9° **d** 2.5° **e** 370 **f** 0.005%

**4** 045°

**5** **a** £7.43 **b** £5.94

**6** **a**

| Vowel | French | rel. freq. | English | rel. freq |
|---|---|---|---|---|
| a | 38 | 0.21 | 40 | 0.24 |
| e | 72 | 0.39 | 49 | 0.29 |
| i | 26 | 0.14 | 35 | 0.21 |
| o | 21 | 0.11 | 36 | 0.21 |
| u | 27 | 0.15 | 10 | 0.06 |

**b** French has longest words.

**7** **a** $\frac{7}{12}$ **b** $\frac{1}{6}$ **c** **i** $\frac{1}{3}$ **ii** $\frac{1}{2}$ **d** loss

**8** **a** £188 **b** £277

## SECTION 11 Using formulas

Starting points

**A1** **a** $d = 5.5$ **b** $y = {}^-14$ **c** $x = 4$ **d** $p = 6$ **e** $x = 1.5454\ldots$

**f** $k = 0.66\ldots$ **g** $n = 18$ **h** $g = {}^-11$

**A2** **a** $c = 7.5$ **b** $x = 0.8$ **c** $x = 12$ **d** $v = 1$

**e** $w = {}^-1$ **f** $a = 19$ **g** $x = 3.4$ **h** $y = {}^-120$

**B1** **a** $3(2bc - 5ax + by)$ **b** $6(3a^2 + 4b^2c)$ **c** $a(3ax - 5y^2 + 2b - c)$

**d** $ax(y - ax + xy - b)$ **e** $3cd(2c - 5b)$ **f** $4ax(2y^2 + 2b - 5a + 4)$

**C1** **a** $x = \sqrt{(4a + 3)}$ **b** $x = \sqrt{\left(\frac{(5c+1)}{3}\right)}$ **c** $x = (3a - 7)^2$ **d** $x = \frac{49}{a}$ **e** $x = \frac{16a^2}{25}$

Thinking ahead to ...

**A** **a** 3 hours 20 minutes **b** 3 lb

Exercise 11.1

**1** **a** $y = \frac{(5x+4)}{3}$ **b** $y = \frac{(8-9x)}{2}$ **c** $y = \frac{(5x-9)}{4}$ **d** $y = \frac{(3x+4)}{5}$

**e** $y = \frac{(2a-4)}{3x}$ **f** $y = \frac{(3a-2x)}{2x}$ **g** $y = \frac{(4x-3)}{6}$ **h** $y = \frac{(5x+6)}{12}$

**i** $y = \frac{(6-5x)}{8}$ **j** $y = \frac{(x+4)}{2}$ **k** $y = \frac{(1-5a^2)}{2}$ **l** $y = \frac{(x^2+6)}{12}$

**m** $y = \frac{(5a+9x)}{10}$ **n** $y = \frac{(6-x^2)}{8}$ **o** $y = \frac{(6x-5)}{8x}$

**2** **a** £10.25 **b** **i** $n = \frac{(c-50)}{15}$ **ii** 129 **c** 63

**3** **a** 900° **b** **i** $n = \frac{S}{180} + 2$ **ii** 20

**d** Putting 600° into the formula does not give a whole number of sides.

**4** **a** **i** 55.924 m **ii** No, closer to 1.5%.

**b** **i** $t = \frac{(x-w)}{wk}$ **ii** 30.5 °C

**5** **a** 3.428 m$^2$ **b** **i** $h = \frac{(S-2\pi r^2)}{2\pi r}$ **ii** 17 cm

Exercise 11.2

**1** **a** 17.6 lb **b** $k = \frac{5p}{11}$ **c** 1.4 kg

**2** **a** $h = \frac{2A}{b}$ **b** 80 cm

**3** **a** 20 °C **b** $F = \frac{(9C+160)}{5}$ **c** 75.2 °F

**4** **a** $y = \frac{4z}{3}$ **b** $x = \frac{3m}{x}$ **c** $y = 2d - 4$ **d** $y = \frac{7k+4h}{4}$

**5** **a** 36.14 km/h **b** $d = \frac{5st}{18}$ **c** $t = \frac{18d}{5s}$

**6** **a** $^-$24 °C **b** $h = 300(G - T)$ **c** 15 000 m

**7** 16.20

**8** **a** $w = \frac{5}{6x}$ **b** $w = \frac{3}{(4 \sin 32°)}$ **c** $w = 3(5a + 4b)$ **d** $w = \frac{a}{t} + b$ **e** $w = \frac{a^2}{b}$

**f** $w = \frac{(c^2-b)}{4ab}$ **g** $w = \frac{(3x-12)}{8}$ **h** $\frac{(3x+2)}{2}$ **i** $w = \frac{(\tan 50° - 3)}{6}$ **j** $w = \frac{(2a+15)}{5}$

**k** $w = 10a - 4$ **l** $w = \frac{(41+2)}{3}$ **m** $w = \frac{6a-5}{2}$ **n** $w = \frac{2a}{3}$ **o** $w = \frac{4ax}{15}$

Exercise 11.3

**1** **a** $w = \frac{(S-2ab)}{(2b+2a)}$ **b** $a = \frac{(S-2bw)}{(2b+2w)}$ **c** 3.6

**2** **a** $w = \frac{2P}{(4+\pi)}$ **b** 50 m

**3** **a** **i** $b = \frac{V}{(ac-2ad)}$ **ii** 14.5 cm **b** 12

**4** **a** $h = \frac{3ac}{(4a-y)}$ **b** $h = \frac{4x}{(3-2a)}$ **c** $h = \frac{(2x-5)}{(3x-a)}$ **d** $h = \frac{(4-b^2)}{(a-c)}$ **e** $h = \frac{7}{(3+x)}$

**f** $h = \frac{9}{(a+b)}$ **g** $h = \frac{(2-y)}{(3a+x)}$ **h** $h = \frac{5}{(1-3a)}$ **i** $h = \frac{3}{(5b+3a)}$ **j** $h = \frac{(2+by)}{(1+4a)}$

**k** $h = \frac{2}{(x-y)}$ **l** $h = \frac{1}{(x-6a)}$ **m** $h = \frac{7}{(2b-3a)}$ **n** $h = \frac{20x}{a}$ **o** $h = \frac{3}{(a-4)}$

**p** $h = \frac{25}{18a}$ **q** $h = \frac{(a-6b)}{6x}$

Exercise 11.4

**1** 9.5 seconds

**2** **b** 8

**3** $r = \sqrt{(\frac{3V}{\pi h})}$

**4** **a** $r = \sqrt[3]{\frac{3V}{4\pi}}$ **b** 2.9 cm

**5** $b = s - \frac{A^2}{s(s-a)(s-b)}$

**6** **a** $b = \sqrt{(y^2 - 3a + 4)}$ **b** $b = \sqrt{\left(\frac{(x-3y)}{2a}\right)}$ **c** $b = \sqrt{\frac{(y-1)}{x}}$ **d** $b = \sqrt{(4 + x)}$

**e** $b = \sqrt{(2x + 5)}$ **f** $b = \sqrt[3]{\frac{(2-3x-4a)}{2}}$ **g** $b = \sqrt[3]{(x - 1)}$ **h** $b = \sqrt{\left(\frac{c}{(1-2a)}\right)}$

**i** $b = \frac{(1+a^2)}{3}$ **j** $b = 4a^2 + 2$ **k** $b = \sqrt{\left(\frac{(w^2-3)}{a}\right)}$ **l** $b = \sqrt{\left(\frac{1}{(c^2-1)}\right)}$

**m** $b = \frac{1}{2}\sqrt{4a^2 - x^2y^2}$ **n** $b = \frac{r^2}{\pi^2} - a$ **o** $b = \frac{k^2y}{4\pi^2}$ **p** $b = \frac{(a+c+x)}{2}$

**q** $b = \frac{(c-3)^2-1}{2}$ **r** $b = \frac{(x-2)^2}{4} + x$

Exercise 11.5

**1** 82.5 km

**2** **a** 121.5 cm **b** 64 cm$^3$ **c** $x = 2$

**3** **a** 42, 42, 47, 42, 42, 96 **b** $^-$8, 12, $^-$193, 12, $^-$28, $^-$64

**c** 6, 12, $^-$4, 12, 0, $^-$15 **d** $^-$14, 0, $^-$16, 0, $^-$28, 5

**e** 10.75, 15.75, $^-$3.63, 15.75, 5.75, $^-$9.63

**f** $^-$14.25, 0.75, $^-$29.88, 10.75, $^-$29.25, 4.13

**4** **a** **i** 31.1 **ii** 21.1 **b** **i** 120 **ii** 130 **iii** 17

**5** **a** $1.098 \times 10^{21}$ **b** $r = \sqrt[3]{(\frac{3V}{4\pi})}$

### End points

**A1** **a** $x = \frac{(5+a)}{3}$ **b** $x = \frac{(a-3)}{6}$ **c** $x = \frac{(5-2y)}{6}$ **d** $x = \frac{(3a-b)}{5y}$ **e** $x = \frac{(2-bc)}{4}$

**B1** **a** $v = \frac{4(a-x)}{3}$ **b** $v = \frac{(3a-14)}{10}$ **c** $v = \frac{(2a+3)}{3a^2}$ **d** $v = \frac{(4ax+8y-5)}{2}$ **e** $v = \frac{^-7a}{12}$
**f** $v = \frac{(bc^2-3)}{2c^2}$ **g** $v = \frac{2w}{(3w-x)}$

**C1** **a** $c = \frac{(3-ax^2)}{x}$ **b** $c = \frac{8x}{(y-a)}$ **c** $c = \frac{(1-2a)}{5}$ **d** $c = \frac{(a^2b-3-2a)}{a^2}$
**e** $c = \frac{(3w-2)}{3x^2}$ **f** $c = \frac{(ax-1)}{(a^2+x)}$

**D1** **a** $y = \frac{25}{x}$ **b** $y = (a+x)^2 - 1$ **c** $y = \frac{4}{(2x-1)^2}$ **d** $y = \frac{ab^2}{9}$ **e** $y = 4\ \frac{a^2-25}{4}$
**f** $y = \frac{8}{(b-a)}$ **g** $y = \sqrt{(3-4a-x)}$ **h** $y = \sqrt{\left\{\frac{(5x+8)}{6}\right\}}$ **i** $y = \sqrt{(2ax-3)}$

**E1** **a** 652 cm$^2$ **b** 8.5 cm

**E2** 7.4

## SECTION 12 Ratio and Proportion

### Starting points

**AZ** **a** 3 : 1 **b** 1 : 3

**B1** **a** 12 : 3 **b** 1 : $2\frac{1}{2}$ : 2 **c** 3 : 4, 6 : 8

**B2** 2 : 5, 1 : $2\frac{1}{2}$ and 10 : 25, 1 : 3 : 2 and 3 : 9 : 6, 10 : 20 and $2\frac{1}{2}$ : 5

**B3** **a** 3 : 2 **b** 2 : 3 **c** 2 : 4 : 1 **d** 10 : 1 **e** 4 : 3 : 6

**C1** W and Y, X and Z

**D1** **a** $\frac{5}{8}$, 0.625, 62.5% **b** $\frac{9}{5}$, 1.8, 180% **c** $\frac{7}{12}$, 0.583, 58.3%
**d** $\frac{15}{11}$, 1.36, 136% **e** $\frac{11}{18}$, 0.611, 61.1% **f** $\frac{17}{24}$, 0.708, 70.8%

**E1** £75, £45, £30

**E2** **a** $1\frac{1}{4}$ hours **b** $1\frac{1}{2}$ hours

**F1** **a** It corresponds to angle K **b** angle L **c** **i** JL **ii** FG

**G1** **a** $\frac{1}{2}$ **b** 4 **c** $\frac{1}{3}$ **d** $\frac{3}{4}$

### Exercise 12.1

1 345 g caster sugar, 450 ml water, 75 g cocoa, 1800 ml chilled milk

2 **a** 60 g **b** 75 ml
**c** 300 g white sugar, 100 ml milk, 100 ml water, 50 g butter, 20 g cocoa

3 **a** 7 drops **b** 175 g

### Exercise 12.2

1 **a** 315 ml **b** 525 ml

2 **a** 250 kg **b** **i** 2 : 1 : 9 **ii** 2400 kg

3 **a** 6.3 cm **b** 1 : 5000

### Exercise 12.3

1 **a** 2.5 **b** 12.75 cm **c** **i** 1.21 **ii** 1.21

### Exercise 12.4

1 **a** MNOP
**b** All corresponding angles are equal and all ratios of lengths of corresponding sides are equal.

### Exercise 12.5

1 Should have divided by 1.25

2 **a** 4.56 cm **b** 4.2 cm

3 **a** 1.6 cm **b** angle LJK is equal to angle LMN

4 12.31 cm

### Exercise 12.6

1 **a** They have the same angles **b** **i** AB **ii** CD
**c** **i** 13.3 cm **ii** 10.7 cm

2 **a** **i** 1.5 **ii** 1.5

3 **b** 3.25 cm

4 It would be the other way up

### Exercise 12.7

1 **d** $h = 0.25b$

2 **a** 4 **b** $b = 4h$

3 **b** no, the line isn't straight

4 **a**

| | W | X | Y | Z |
|---|---|---|---|---|
| Height ($h$) | 1 | 1.25 | 1.5 | 1.2 |
| Area ($A$) | 2 | 3.125 | 4.5 | 2.88 |

**c** no

5 **a**

| | W | X | Y | Z |
|---|---|---|---|---|
| $h^2$ | 1 | 1.5625 | 2.25 | 1.44 |
| A | 2 | 3.125 | 4.5 | 2.88 |

**c** yes

7 **a** **ii** no **b** **iii** yes

### Exercise 12.8

1 **a** approx $\sqrt{2}$
**b** The measurements have been rounded **c** $h = \sqrt{2}w$

2 **a** 36 units$^2$ **b** **i** $h = \frac{36}{w}$ **ii** $wh = 36$ **iii** $w = \frac{36}{h}$

3 **a** 25, 49, 81, 100, 196 **b** $g = 0.4d^2$

### Exercise 12.9

1 **a** $m \propto \frac{1}{v}$ **b** 54 **c** 12 **d** $m$ is doubled

2 **a** $s \propto \sqrt{q}$ **b** $s = 7\sqrt{q}$ **c** 42

3 **a** $p = \frac{90}{t^2}$ **b** 10 **c** $p$ is multiplied by 4 **d** $t = \sqrt{(\frac{90}{p})}$ **e** 5

4 **a** $n \propto \frac{1}{d^2}$ **b** 20 000 **c** $n = \frac{20\,000}{d^2}$ **d** **i** 200 **ii** 102

5 **a** $A \propto d^2$ **b** 3.5 **c** $A = 3.5d^2$ **d** 1263.5 cm$^2$

6 **a** $t = 1.6\sqrt{h}$ **b** 11.3 seconds

7 **a** $f \propto \frac{1}{a}$ **b** 42 **c** $f = \frac{42}{a}$ **d** 2.625 mm

8 **a** $r = \frac{4}{f^2}$ **b** 0.015625 **c** 2

9 **a** $m = 3.2w^3$ **b** 86.4 g **c** 1 : 8

### End points

**A1** **a** **i** 375 g **ii** 10 tablespoons **b** 3 eggs

**B1** EFGH, NOPQ

**C1** **a** 2.5 cm **b** 5.6 cm

**D1** **b** 5.4 cm

**E1** **a** $s = 2.25\sqrt{h}$ **b** 22.5 seconds

**E2** **a** $w = \frac{6}{q}$ **b** 0.6 **c** $w$ is halved **d** 0.8

## SECTION 13 Mensuration

### Starting points

**A1** **a** A, C, F
**b** cuboid, tetrahedron, cylinder, pentagonal prism, octagonal prism, hexagonal prism

**B1** **a** 72 cm$^2$ **b** 84 cm$^2$ **c** 16 cm$^2$

**C1** 4.5 m$^2$

**C2** 8.3 cm

**C3** area = $4\pi y^2$, circumference = $4\pi y$

**D1** A 106 cm$^3$, 182 cm$^2$ B 38.9 cm$^3$, 96.12 cm$^2$ C 45.9 cm$^3$, 82.2 cm$^2$

**E1** 81.8 cm$^3$, 113 cm$^2$

**E2** **a** $768\pi$ **b** $320\pi$

### Exercise 13.1

1 **a** 348 cm$^2$ **b** 243 cm$^2$ **c** 418.7 cm$^2$

2 **a** 559 cm$^2$ **b** 1716.5 cm$^2$

3 **a** 183.7 cm$^2$ **b** more **c** 85.4 cm

4 **a** 154 mm × 136 mm **b** 504 mm$^2$ **c** 1380 mm$^2$ **d** 16288 mm$^2$
**e** between $\frac{1}{5}$ and $\frac{1}{4}$

5 **a** 255 mm$^2$ **b** 763 mm$^2$ **c** 21%

6 **a** $\frac{\pi x^2}{4}$ **b** $(\frac{1-\pi}{4})x^2$ **c** $x^2(\frac{\pi}{2} - 1)$

### Exercise 13.2

1 a $9.5\text{ cm}^2$, $12.7\text{ cm}$ b $792.6\text{ m}^2$, $143\text{ m}$

2 a $192.7\pi\text{ cm}^2$, $(22.7\pi + 34)\text{ cm}$ b $36.0\pi\text{ m}^2$, $(7.4\pi + 19.6)\text{ m}$

3 a $395.8\text{ cm}^2$ b $80.0\text{ cm}$

### Exercise 13.3

1 a $240\text{ cm}^3$, $300\text{ cm}^2$ b $343.6\text{ cm}^3$, $328.5\text{ cm}^2$ c $1291\text{ cm}^3$, $746.7\text{ cm}^2$
d $2204\text{ cm}^3$, $939.4\text{ cm}^2$

2 16.5 cm

3 a $27.2\text{ m}^3$ b i 390 bags ii £1996.80

4 a $46.25\text{ m}^2$ b $555\text{ m}^3$ c $435\text{ m}^3$ d 0.435 litres

5 $603\text{ cm}^2$

6 7 cm

7 a $960\text{ cm}^3$ b £840

8 $43\,800\text{ cm}^3$

9 a $30.2\text{ cm}^2$ b $166\text{ cm}^2$ c $578\text{ cm}^2$ d $1630\text{ cm}^2$

10 a $13\,300\text{ m}^3$ b $10\,400\text{ m}^2$

11 a $1290\text{ cm}^3$ b 182 g

12 a $xyz - \pi r^2 x$ b $2(xy + xz + yz - \pi r^2 + \pi rx)$

13 $R = \frac{6V}{S}$

### Exercise 13.4

1 a $300\text{ cm}^3$ b 75 ml c 170 g

2 a B b C c $575\text{ cm}^3$ d 4% e $44\text{ cm}^3$ f 38 g

3 a $21\,700\,000\text{ cm}^3$ b 21 700 litres

4 $217\text{ cm}^3$, $1.07\text{ g/cm}^3$; $649\text{ cm}^3$, $0.952\text{ g/cm}^3$

### Exercise 13.5

1 a B, C b 1 c A

3 a i C, D ii A, B b A 8, B 1, C 5, D 2

### Exercise 13.6

1 a $168\text{ cm}^3$ b $135\text{ cm}^2$ c $186\text{ cm}^3$

2 $0.512\text{ m}^3$

3 a $41.6\text{ cm}^2$ b $166\text{ cm}^3$

4 a B

### Exercise 13.7

1 a $\pi\text{ cm}^2$, $4\pi\text{ cm}^2$ b 4 times c $\frac{1}{6}\pi\text{ cm}^3$, $\frac{4}{3}\pi\text{ cm}^3$ d 8 times

2 a 34 cm b 191 mm

3 $(1.44 \times 10^{10})\pi$

4 24.8 cm

5 a $348\text{ cm}^3$ b $190\text{ cm}^2$

6 a $707\text{ cm}^2$ b $707\text{ cm}^2$

### Exercise 13.8

1 a $4.69\text{ cm}^3$ b $32.1\text{ cm}^2$

2 a $495.2\text{ cm}^3$ b $396.6\text{ cm}^2$

3 a $392.2\text{ cm}^3$, $27\text{ cm}^2$ b $1696.5\text{ cm}^3$, $1017.9\text{ cm}^2$
c $861\text{ cm}^3$, $850\text{ cm}^2$ d $728.2\text{ cm}^3$, $690.8\text{ cm}^2$

4 a $18.6\text{ cm}^3$ b 12.5%

5 a $138\text{ m}^3$ b $1104\text{ cm}^3$

6 51.5 cm

7 a $h - 6$ b $h = 9.6$ cm, height of part cut off = 3.6 cm c $152.4\text{ cm}^3$

#### Thinking ahead to ...

A a $6\text{ cm}^2$, $24\text{ cm}^2$ b $1\text{ cm}^3$, $8\text{ cm}^3$ c i 1 : 4 ii 1 : 8

### Exercise 13.9

1 a $13.1\text{ cm}^3$ b $124\text{ cm}^2$

2 a 1 : 1.27366 b 9.8 cm c 1 : 1.62222 d 1 : 2.06617

3 a 1 : 2.91003 b $760\text{ cm}^3$

4 1 : 15.625

5 1 : 1.65096

6 14 cm

7 *a* $128\,000\text{ cm}^2$ b $7\,800\,000\text{ cm}^3$

8 a 2.5 : 1 b 6.6 cm

9 a They have scaled by height, but as the shapes are 2D, they should have scaled by area.
b Denopt System's £ sign should be 23 mm high.

10 a $64.0\text{ cm}^2$ b $3560\text{ cm}^3$ c $1.78\text{ cm}^2$ d $49.0\text{ cm}^2$ e $39.1\text{ cm}^3$
f $1060\text{ cm}^3$

11 4.76 m

12 a 2 b $\frac{1}{4}$ c 8 d $\frac{1}{8}$ e $\frac{1}{2}$ f 4

### Exercise 13.10

1 a length b volume c area

2 a $\frac{1}{2}lz(x + y)$ b The other expressions are not for volumes.

3 a $\pi(a + b)$ b $\pi ab$

4 a area b length c length d volume e volume f area

5 It is a mixture of volume and area terms.

### End points

A1 $73.9\text{ cm}^2$

B1 4

C1 a $3.85\text{ m}^3$ b $14.0\text{ m}^2$ c 3850 litres

D1 a 26 cm b $720\text{ cm}^3$

E1 a, b, d

## SECTION 14 Probability

### Starting points

A1 a $\frac{1}{4}$ b $\frac{1}{12}$ c $\frac{1}{6}$

A2 a $\frac{5}{12}$ b $\frac{1}{3}$ c $\frac{2}{3}$ d $\frac{1}{6}$ e 1 f 0

A3 $p(\text{A}) = \frac{1}{4}$, $p(\text{C}) = \frac{1}{12}$, $p(\text{D}) = \frac{1}{6}$, $p(\text{A or B or E or F}) = \frac{3}{4}$

A4 a $\frac{1}{3}$ b $\frac{1}{2}$ c $\frac{5}{6}$

B1 $\frac{1}{5}$

B2 $\frac{5}{8}$

C1 a $\frac{3}{16}$ b $\frac{1}{4}$ c $\frac{4}{7}$ d $\frac{1}{12}$

D1 0.23

D2 a 0.77 b 0.16 c 0.55 d 0.44

D3 all of them

E1 $\frac{1}{6}$

E2 $\frac{1}{3}$

E3 b $\frac{1}{4}$

### Exercise 14.1

1 a Emmelle and Angeline, Emmelle and Bruno, Emmelle and Charles, Emmelle and Danielle, Angeline and Bruno, Angeline and Charles, Angeline and Danielle, Bruno and Charles, Bruno and Danielle, Charles and Danielle
c 10

2 a Frederic and Emmelle, Frederic and Angeline, Frederic and Bruno, Frederic and Charles, Frederic and Danielle, Emmelle and Angeline, Emmelle and Bruno, Emmelle and Charles, Emmelle and Danielle, Angeline and Bruno, Angeline and Charles, Angeline and Danielle, Bruno and Charles, Bruno and Danielle, Charles and Danielle
b 15

3 a 3, 6, 10, 15
b The numbers of handshakes are the triangle numbers.
c 21

4 a Sartre Vol I and Pascal, Sartre Vol I and Zola Vol I, Sartre Vol I and Sartre Vol II, Sartre Vol I and Zola Vol II, Pascal and Zola Vol I, Pascal and Sartre Vol II, Pascal and Zola Vol II, Zola Vol I and Sartre Vol II, Zola Vol I and Zola Vol II, Sartre Vol II and Zola Vol II
b $\frac{1}{10}$ c $\frac{1}{5}$ d $\frac{9}{10}$ e $\frac{7}{10}$

5 $\frac{3}{14}$

### Exercise 14.2

1 a $\frac{25}{81}$ b $\frac{5}{18}$ c $\frac{2}{27}$ d $\frac{1}{12}$

2 a $\frac{5}{12}$ b $\frac{4}{11}$ d $\frac{7}{22}$

3 b $\frac{1}{6}$ c $\frac{5}{9}$ d $\frac{5}{6}$

4 a $\frac{1}{10}$ b 0 c $\frac{3}{10}$ d $\frac{3}{5}$

5 a 1 + 4, 4 + 1, 3 + 1 + 1, 1 + 3 + 1, 1 + 1 + 3, 1 + 1 + 1 + 1 + 1,
d i $\frac{1}{252}$ ii $\frac{1}{4}$ iii $\frac{2}{9}$ e no, 48% f £161.71 profit

### Exercise 14.3

1 b i 0.473 ii 0.4332

2 a 0.2809 b 0.4089 c 0.3102

3 0.0164

### Exercise 14.4

1 a $\frac{4}{16}$ b $\frac{7}{16}$ c $\frac{11}{16}$

3 a $\frac{5}{16}$ b $\frac{4}{16}$

4 a $\frac{10}{16}$ b $\frac{8}{16}$ c Owen has a moustache and an earring

5 a $\frac{4}{16}$ b $\frac{4}{16}$ c $\frac{6}{16}$

### Exercise 14.5

1 b No, some criminals have fair hair and a necklace.

2 a $\frac{11}{16}$ b $\frac{1}{4}$ c $\frac{5}{16}$ d $\frac{1}{8}$ e $\frac{1}{4}$

3 a $\frac{7}{16}$ b $\frac{3}{4}$ c $\frac{3}{4}$ d $\frac{9}{16}$

4 a No, some girls could play hockey and basketball.
b Yes, boys and girls don't overlap.
c Yes, you can't be a driver and a passenger.
d No, tall women can drive buses.
e No, some mechanics can dance.

### Exercise 14.6

1 a 0.12 b 0.32 c 0.47 d 0.53 e 0.13

2 a 0.15 b 0.19 c 0.53

3 0.44

4 a a number can't be less than 5 and greater than 7 b 0.55 c 0.45

### Exercise 14.7

1 b i $\frac{19}{44}$ ii $\frac{39}{44}$ iii $\frac{25}{44}$ iv $\frac{5}{44}$
c Some people had English breakfast and cereal.

2 b i $\frac{7}{50}$ ii $\frac{2}{5}$ iii $\frac{7}{10}$ iv $\frac{27}{200}$

### End points

A1 a $\frac{1}{4}$ c i $\frac{7}{26}$ ii $\frac{25}{39}$

B1 b 0.3544

C1 a mutually exclusive b mutually exclusive
c not mutually exclusive d not mutually exclusive

D1 b i $\frac{9}{34}$ ii $\frac{14}{17}$ iii $\frac{3}{34}$

## Section 15 Grouped Data

### Starting points

A1 a 70, 3 b 70.5, 4

A2 70, 3

B1 a 69.85, 2.63

C1 The 2nd round scores are slightly lower on average.
There is more variation in the 1st round scores.

C2 b The 3rd & 4th round scores are slightly higher on average.
There is more variation in the 3rd & 4th round scores.

D1

| Score | 3rd round | 4th round |
|---|---|---|
| 64–66 | 3 | 2 |
| 67–69 | 11 | 11 |
| 70–72 | 17 | 19 |
| 73–75 | 9 | 8 |

D2 b The modal class is 70–72 for both rounds.

D3

| Score | Frequency |
|---|---|
| 64–65 | 1 |
| 66–67 | 2 |
| 68–69 | 11 |
| 70–71 | 14 |
| 72–73 | 6 |
| 74–75 | 6 |

E1 70.4

E2 70.4

### Exercise 15.1

1 a Frequency densities: 0.08, 0.2, 0.26, 0.16, 0.16, 0.03
b The width of the bar for 700–799 is twice the width of the bars for 900–949 and 950–999.

2 Frequency densities: 0.002, 0.01, 0.02, 0.06, 0.04, 0.009.

### Exercise 15.2

1 a A continuous, B discrete, C continuous

2 Times would be improved if they were rounded down, and distances improved if they were rounded up.

3 a 24.5 b All the ages have been rounded down to the nearest year.
c discrete

### Exercise 15.3

1 Frequency densities: 3, 6, 6.3, 6, 1

2 a Frequency densities: 10, 22, 34, 16, 1
b 19.75, 20.25, 20.75, 21.5
c 20.1

3 a Frequency densities: 3.3, 26, 20, 18, 2.7 b 18.7

4 The men throw further on average. There is less variation in the men's throwing.

5 a Frequencies: 5, 3, 5, 12, 11, 3, 6, 2
b i Frequency densities: 350, 1100, 1300, 1600, 1400, 1600, 450, 133
ii Frequency densities: 250, 300, 500, 1200, 1100, 300, 300, 33

6 a Frequencies: 10, 6, 10, 24, 22, 6, 12, 4
b Mid-class values: 0.13, 0.145, 0.155, 0.165, 0.175, 0.185, 0.2, 0.24
c yes

7 a 0.173 s b 0.170 s

8 a true
b The estimated mean for the finals is smaller than the estimated mean for the semi-finals.

### Exercise 15.4

1 a Cumulative frequencies: 0, 8, 18, 31, 39, 47, 50

2 a Cumulative frequencies: 0, 6, 12, 20, 33, 46, 48

### Exercise 15.5

1 a Cumulative frequencies: 0, 6, 12, 28, 35, 48, 52

2 a 77.9 m b 1.2 m

3 The distances were greater on average in the javelin final.
There was more variation in the distance thrown in the javelin final.

4 b 840 points c 180 points

5 a 880 points b 120 points

6 Scores were higher on average on day 2.
There was less variation in the scores on day 2.

7 a Cumulative frequencies: 7, 18, 28, 38, 45, 46
c Use cumulative frequencies 11.5, 23 and 34.5.
d 61 m, 4.5 m

### Exercise 15.6

1 27

2 14

3 a Cumulative frequencies: 1, 8, 22, 28, 30, 31 b 15

### Exercise 15.7

1 Frequency densities: 5, 18, 22.7, 29.5, 31.5, 17, 13.7, 9.4, 5.5, 1.9

3 **a** 36.8, 35.0, 33.8 **b** 13.0, 12.6, 10.8

4 The estimates improve as the sample size increases.

5 **a** no **b** There are more competitors.

### End points

A1 **a** frequency densities: 1.75, 5, 6.5, 6.5, 1.5
**b** frequency densities: 1.25, 4, 8, 6.5, 0.75

B1 63

B2 63.02

C1 **a** **i** 7.95 **ii** 0.3

C2 6.8, 0.32

D1 The men jumped further but the women were more evenly matched.

D2 The distances were very similar.

E1 **b** **i** 15 **ii** 12

## SKILLS BREAK 3

1 **a** $1.2 \times 10^{-4}$ m **b** $2.97 \times 10^{-3}$ $m^3$

2 **a** $1.87 \times 10^{-3}$ $m^3$ **b** 76

3 695 mm

4 **a** 15 **b** 24%

5 **b** 3457 $cm^3$ **d** $h = \frac{c}{\pi r^2} + \frac{r}{3}$

6 no, it is a 6 4% increase

7 **a** 4, 5.5, 6.6, 7, 6.6, 5.5, 4, 2.5, 1.4, 1, 1.4, 2.5, 4
**c** 6.2 and 11.8 seconds

8 Coffee 75p, tea 52p

9 $\frac{3n}{8}$

10 no, it would need 36.4 litres

12 **a** 131.42 $cm^2$ **b** 122.5 $cm^2$

## SECTION 16 Solving Equations Graphically

### Starting points

A1 **a** 4, 3 **b** $\frac{1}{5}$, $^-8$ **c** 0.8, 1.5 **d** $\frac{5}{4}, \frac{7}{4}$
**e** $^-0.25$, 1 **f** 6, 0 **g** 1, 0 **h** 0, 12

A2 $y = 2.5x + 3$

A3 $y = {}^-x - 2$

B1 **a** $y = \frac{3}{5}x + 3$ **b** $y = \frac{^-5}{2x-4}$ **c** $y = \frac{4}{3}x + 1$ **d** $y = \frac{^-3}{2}x + 2$ **e** $y = \frac{7}{2}x - \frac{1}{2}$
**f** $y = \frac{1}{3}x - 1$ **g** $y = {}^-2x + \frac{5}{4}$ **h** $y = \frac{^-5}{4}x + \frac{3}{2}$ **i** $y = 4x - 2$ **j** $y = \frac{9}{2}x + \frac{3}{2}$
**k** $y = -\frac{1}{10}x + \frac{4}{5}$ **l** $y = \frac{3}{5}x - 5$ **m** $y = -\frac{9}{2}x + 6$ **n** $y = -\frac{1}{2}x + \frac{3}{10}$
**o** $y = \frac{^-3}{10}x + 2$ **p** $y = 8x - 24$ **q** $y = 2x - 2$ **r** $y = 3x + \frac{1}{4}$
**s** $y = -\frac{1}{3}x + 3$ **t** $y = \frac{^-3}{4}x + \frac{9}{4}$ **u** $y = \frac{1}{6}x + \frac{1}{2}$ **v** $y = -2x + 12$
**w** $y = \frac{6}{7}$ **x** $y = \frac{4}{3}x - 4$ **y** $y = -\frac{1}{2}x + \frac{3}{4}$ **z** $y = 2x + \frac{24}{5}$

### Exercise 16.1

1 **a** $^-1$, 2 **b** 1, 4 **c** $^-2$, 2 **d** 2, 3 **e** 2, $^-3$ **f** 1, $^-1$

2 $^-1$, 4

3 **c** 2, 1

4 **b** 2.4, 1.8 **c** they do not cross

5 **a** $2y + 9 = x$, $y + x = 63$ **b** **ii** 18, 45

6 **a** $x + y = 500$, $4x + 25 = y$ **b** **ii** 95, 405
**c** **i** $6x + 40 = y$ **ii** 65.7 litres, 434.3 litres

7 **b** 3, 5

8 $y = 0.5x + 3$

### Exercise 16.2

1 **a** $x = {}^-1$ **b** $x = {}^-1$

2 **b** $x = 1.5$ **c** $y = x^2 - 3x - 8$

4 **a** $x = {}^-2, {}^-3$ **b** $x = 0.5, 2$ **c** $x = {}^-2.5, 0$ **d** $x = 3, 1$
**e** $x = {}^-3, {}^-2$ **f** $x = {}^-4, 3$ **g** $x = 0, {}^-7$ **h** $x = 3, 8$ **i** $x = 3.5, 5$

### Exercise 16.3

1 **a** $y = 10$ **b** intersections are solutions **c** $^-7.2$, 2.2

3 **a** Too little of the graph has been drawn **b** 3.4 **c** less than

4 **a** 2, 3 **c** 0.4, 4.6 **d** $^-0.7$, 5.7

5 **a** **i** $^-1$, 5 **ii** $^-2.2$, 6.2 **b** 2, 2

6 **a** **ii** 1, 3

7 **a** **i** $^-5.7$, 0.7 **ii** $^-4.6$, $^-0.4$ **iii** $^-5.7$, 0.7
**b** **i** $^-3.3$, 0.3 **ii** $^-2.6$, $^-0.4$ **iii** $^-3.4$, 0.4
**c** **i** 4.6, $^-0.6$ **ii** 3.7, 0.3 **iii** 4.2, $^-0.2$
**d** **i** $^-1.8$, 0.3 **ii** 0.9, $^-2.4$
**e** **i** 2.5, 0 **ii** 2.9, $^-0.4$

8 $^-4.2$, 0.2

9 0.2, 1.8

11 **b** 1, 5 **c** $^-1.8$, 2.8

12 **b** 0.2, $^-4.2$ **c** 2.6, $^-1.6$

13 **a** $y = x + 8$ **b** $y = 4 - 4x$

14 **b** 1, $^-7$ **c** 1.3, $^-2.3$ **d** $^-0.8$, $^-37$

15 **b** $^-6.3$, 1.3 **c** $^-5.9$, 0.9 **d** $^-6.5$, 1.5 **e** $^-2$, 2 **f** $^-8.7$, 0.7

### Exercise 16.4

1 **a** **ii** $x = 0$, $y = 0$

2 **b** $x = 0$, $y = 3$

3 $x = 0$, $y = {}^-8$

4 **b** $y = 2$

5 **b** **ii** $x = 3$ **ii** $x = 5.3$ **iii** 3.5

7 **b** $^-x$ **c** $y = {}^-x$, $x = 0$

### Exercise 16.5

1 **i** $^-3.1$ **ii** 2.3, 5.2

2 **a** 6.3 **b** 0.4, $^-3.4$

3 **a** 15.3 **b** $^-0.6$, $^-4.4$ **c** 0.8, $^-3.8$ **d** $^-1.3$, $^-4.7$
**e** **i** 4.8 **ii** $^-3.6$, 0.6 **iii** $y = {}^-2x + 7$

4 **a** **i** $^-3.6$ **ii** $^-0.8$, 3.8

5 **a** 2 **b** $x = 3$ is an asymptote to $y = \frac{2x}{3-x}$

### Exercise 16.6

2 **a** 0, $^-2$ **b** 2.8 or $^-0.4$ **c** **ii** 0, 4 **iii** $^-1.1$, 1.8 **iv** 12.4 or $^-4.4$

3 **a** 3.5 **b** $^-1$, 2

4 **a** 10.4 **b** $^-2.6$ **c** 2.5 **d** $^-3.4$, $^-0.3$, 3.6

5 **a** $^-2$, 1, 2 **b** 4.4 **c** $^-1.7$, 0.4, 2.3

6 **b** **i** 4.4 cm, 12.4 cm **ii** 1.8 cm, 9.8 cm **iii** 6.8 cm, 14.8 cm
**c** 45 $cm^2$ **d** 5.5 cm

7 **a** $A = w(2w - 4)$
**c** **i** 5.3 cm, 6.6 cm **ii** 6.7 cm, 9.4 cm **iii** 7.4 cm, 10.8 cm
**d** 40 $cm^2$ **e** 5.8 cm, 7.6 cm

### Exercise 16.7

1 **a** $y = (x + 3)^2 - 11$ **b** $^-11$ **c** (0, –2)

5 **b** the graphs do not intersect

6 $y = x^2 + 6x + 5$, $a = 1$, $b = 6$, $c = 5$

### End points

A1 **a** 2, $^-1$ **b** 3, 1

A2 £5, £3.50

B1 $x = 2$

B2 **a** $^-4$, 2 **b** $^-2$, 5 **c** $^-1.6$, 2.6 **d** $^-2$, 3

B3 **b** $^-1.2$, 1.7 **c** $^-1.7$, 1.7 **d** $^-2$, 0.5

C2 $y = 5$, $x = 0$

C3 $y = 4$, $x = 0$

D1 **a** $^-1.1$ **b** 2.4, 5

D2 **a** 14.25 **b** 0.7, $^-5.7$ **c** 1.2, $^-4.2$

D3 **b** **i** 5.2 cm, 11.2 cm **ii** 3.4 cm, 9.4 cm **c** 62 $cm^2$ **d** 3.6 cm

E5 $y = x^2 + 4x + 8$, $a = 1$, $b = 4$, $c = 8$

## Section 17 Trigonometric graphs

### Starting points

**A1** **a** **i** (2.60, 2.34) **ii** 3.04 $cm^2$ **b** 75.4°

**A2** 110 $cm^2$

**A3** **a** $\frac{4}{5}$ **b** $\frac{3}{5}$

**B1** **a** 0.77 **b** 0.64

**B2** **a** 24° **b** 66°

**B3** **a** 70° **b** 42° **c** 45°

### Exercise 17.1

**1** **a** 0.68 **b** 0.87 **c** 0.77 **d** 2.30 **e** 1.97 **f** 2.5

**2** **a** **i** 3.86 **ii** 3.70 **iii** 3.19 **b** **i** 30° **ii** 20°

### Exercise 17.2

**1** **b** ($^-$17.3, 10)

**2** **b** **i** 22.98 km **ii** 19.28 km **c** ($^-$22.98, 19.28)
**d** **i** B(20, 70°), C(30, 250°), D(10, 320°) **ii** B(6.84, 18.79), C($^-$10.26, $^-$28.19), D(7.66, $^-$6.43)

**3** **a** (13, 225°) and ($^-$9.19, $^-$9.19), (40, 114°) and ($^-$16.27, 36.54), (25, 291°) and (8.96, $^-$23.34)
**b** (17, 169°), (31, 314°)

**4** **a** $r = \sqrt{(x^2 + y^2)}$ **b** $(r\cos\theta, r\sin\theta)$ **c** $(\cos\theta, \sin\theta)$

### Exercise 17.3

**1** 232°

**2** **a** 0.616 **b** $^-$0.788

**4** **a** $^-$0.809 **b** 0.809 **c** $^-$0.809

**5** **a** sin 53° and sin 127°, sin 233° and sin 307°, cos 233° and cos 127°
**b** 0.799, $^-$0.799, $^-$0.602

**6** **a** $90° < \theta < 270°$ **b** $180° < \theta < 360°$

**7** **a** **i** 0, $^-$1 **ii** 0.616, $^-$7.88 **b** **i** $^-$306° **ii** $^-$260° **iii** $^-$160°

**8** **a** cos 450° = cos(450 − 360)° = cos 90° = 0

### Exercise 17.4

**1** **a** 1, $^-$1 **b** 1, $^-$1

**2** 30°, 150°

**3** **a** **i** 40°, 320° **ii** 145°, 215° **iii** 252°, 288°
**b** **i** 64°, 116° **ii** 154°, 206°

**4** **a** 420°, 660° **b** $^-$60°

**5** **a** 450° **b** $^-$270°

### Exercise 17.5

**1** **a** $^-$4.5, $^-$9, $^-$4.5, 4.5, 9, 4.5, $^-$4.5, $^-$9, $^-$4–5 **c** 9 mm
**d** **i** 360 milliseconds **ii** 180 milliseconds **e** 2

**2** **b** yes **c** The maximum heights of the needle are different

### Exercise 17.6

**1** **a** **i** 3, 1 **ii** 2, $^-$2 **iii** 1, $^-$1

**2** **b** **i** $^-$1, $^-$3 **ii** 3, $^-$3 **iii** 1, $^-$1

**3** **a** 1, $^-$1 **b** 5, $^-$5 **c** $^-$3, $^-$5 **d** 2, 0 **e** 0, $^-$4 **f** 5, $^-$1

**5** A: $y = \cos ax$, $a = 2$ B: $y = \sin x + a$, $a = 3$ C: $y = \cos x + a$, $a = {}^-1$
D: $y = a\cos x$, $a = 2$ E: $y = a\sin x$, $a = 4$

**9** A, C

### Exercise 17.7

**2** **a** **i** 45° **ii** 225° **b** 135°, 315°

**4** **a** 50° **b** 230°, 410°

**5** **a** 60°, 240°, 420° **b** $^-$25°, 155°, 335° **c** $^-$32°, 148°, 328°

**6** **a** **i** 57.3 **ii** 114.6 **iii** 573.0 **iv** 5729.6
**b** **i** $^-$57.3 **ii** $^-$114.6 **iii** $^-$573.0 **iv** $^-$5729.6 **d** 630°

### End points

**A1** **a** 70°, 290° **b** 110°, 250° **c** 140°, 320° **d** 208°, 332°

**A2** cos 41° and cos 319°, cos 139° and 221°, cos 229° and cos 131°

**B2** B and D

## Section 18 Inequalities

### Starting point

**A1** **b** 2, $^-$2, 1.634, $4\frac{3}{4}$

**A2** **a** 5, 6, 7, 8, 9 **b** $^-$2, $^-$1, 0, 1, 2, 3 **c** 68, 69, 70
**d** $^-$11, $^-$10, $^-$9, $^-$8, $^-$7, $^-$6, $^-$5, $^-$4

**A3** ${}^-4 < h < 6$

**A4** ${}^-3 \leqslant g \leqslant {}^-1$, ${}^-4 < g < 2$

**A5** ${}^-3 \leqslant f < 7$

**B2** ${}^-2 \leqslant x \leqslant 6$, $3 < y \leqslant 7$

**C1** **a** $x \geqslant 0$, $y \geqslant 0$, $y \geqslant x$ **b** $x > y$, $y \geqslant 3$

**C4** $y \leqslant 8 - x$, $y > \frac{1}{2}x$

### Thinking ahead to ...

**A** **a** true **b** true **c** true **d** true **e** true **f** false **g** false

### Exercise 18.1

**1** **a** $y < 7$ **b** $s \leqslant 8$ **c** $p \geqslant 6$ **d** $b > 3.5$ **e** $a \leqslant {}^-8$ **f** $x > 8.4$ **g** $k \leqslant \frac{14}{3}$
**h** $a \geqslant 7$ **i** $d \leqslant {}^-5$ **j** $a \leqslant {}^-13$

**2** **a** $a \geqslant 3\frac{2}{3}$ **b** $k < {}^-\frac{14}{5}$ **c** $x > {}^-9$ **d** $n \geqslant \frac{7}{2}$ **e** $q < \frac{^-4}{5}$ **f** $k > 2$
**g** $h < \frac{^-1}{3}$ **h** $c < 2.5$ **i** $h < \frac{12}{11}$ **j** $p \geqslant \frac{16}{7}$

**3** **a** ${}^-4 < x < 1$ **b** ${}^-2 < x < 5$ **c** ${}^-1 < x < 15$ **d** ${}^-3 < x < {}^-1$

### Exercise 18.2

**1** **a** $d \geqslant 6$, $d \leqslant {}^-6$ **b** ${}^-\sqrt{17} < g < \sqrt{17}$ **c** ${}^-7 \leqslant a \leqslant 7$ **d** $h < {}^-4$, $h > 4$
**e** $t < {}^-4$, $t > 4$ **f** ${}^-2 \leqslant c < 2$ **g** $c \leqslant {}^-1$, $c \geqslant 1$ **h** $f < {}^-2$, $f > 2$

**2** **a** $^-$6, $^-$5, $^-$4, $^-$3, $^-$2, $^-$1, 0, 1, 2, 3, 4, 5, 6 **b** $^-$4, $^-$3, $^-$2, $^-$1, 0, 1, 2, 3, 4
**c** $^-$3, $^-$2, $^-$1, 0, 1, 2, 3 **d** $^-$2, $^-$1, 0, 1, 2 **e** $^-$2, $^-$1, 0, 1, 2 **f** $^-$1, 0, 1

**3** squared numbers are never negative

**4** c has no solutions

**5** 1, 2, 3

### Exercise 18.3

**1** **a** (19, 1) represents 20 birds and so is not a solution
**b** (16.5, 2.5) doesn't represent whole birds
**d** £65

**2** **b** $v \leqslant 7$, $n \leqslant 8$
**d** 10, 4 Vikings and 6 Neptune, or 5 Vikings and 5 Neptune

**3** **a** **i** $8E + 10C \leqslant 120$ **ii** $280E + 120C \geqslant 1680$
**b** A: $7E + 3C = 42$, B: $4E + 5C = 60$ **c** $2C \geqslant E$ **d** 13

### End points

**A1** **a** $^-$1 **b** $^-$5, 4, $^-$3, $^-$2, $^-$1, 0, 1, 2, 3 **c** 4, 5, 6, 7
**d** $^-$59, $^-$58, $^-$57, $^-$56 **e** $^-$3, $^-$2, $^-$1, 0, 1, 2, 3
**g** $^-$3, $^-$2, $^-$1, 0, 1, 2, 3

**B1** **a** $x \leqslant 6$ **b** $f > 3$ **c** $a \geqslant 12$ **d** $k > {}^-5$ **e** $s \leqslant 4$ **f** $d > 1.5$
**g** $x > 10$, $x < {}^-10$ **h** ${}^-2 \leqslant x \leqslant 2$

**D1** **a** $S > L$ **b** **i** $S + L \leqslant 200$ **ii** $4S + 5L \geqslant 720$ **iii** $3L \geqslant S$ **d** £899

## Section 19 Working with percentages

### Starting points

**A1** **a** 0.75, 75% **b** 0.625, 62.5% **c** 1.6, 160% **d** 1.25, 125%
**e** 1.2, 120% **f** 0.65, 65% **g** 0.25, 25% **h** 3.17, 317%

**B1** **a** 357.5 miles **b** £290 **c** 1587 **d** £116.20 **e** 16.2 m **f** £1.71
**g** 12.5p **h** 0.24 miles

**C1** **a** 65% **b** 46.2% **c** 12.5% **d** 125% **e** 3%

**C2** 14%

**D1** **a** £13.60 **b** 5.6 mm **c** 1540 g **d** 20p **e** **i** 50.7 mm **ii** 5.07 cm

### Thinking ahead to ...

**A** 45 ml, less

**B** $\frac{3}{20}$

**C** 345 ml

### Exercise 19.1

1 a 87% b 506 g

2 a 413 kg b 518.52 km c 5.49 m d £37 800 e £36.30
f 30 186 tonnes g 3 815 000 litres h 0.69 cm

3 £22 471

4 £1191

5 £3 281 000

6 425 litres per minute = $2.235 \times 10^8$ litres in 2005

7 no, the 20% will be calculated each year

### Exercise 19.2

1 a £18.80 b £51.99 c £217.38 d £22.80 e 85p f £432.89
g £307.85 h £28.85 i £10.99 j £221.72 k 40p l 13p

2 £34.50

3 £31.90

### Exercise 19.3

1 a £115.32 b £29.78 c £63.82 d £56.00 e £158.94 f £12.33
g £42.54 h £680.83 i £84.25

2 a £57 b £11.40

3 £24

4 500 g

5 £55.70

6 £137 515

7 a 70 875 b 25 515 c 16 : 9

8 a £20 384 b $p = \frac{3.15}{4.5}s$

9 147 428 to the nearest person

10 $c = \frac{p}{1.175}$

### Exercise 19.4

1 a £324 b £67.20 c £1350 d £10 200

2 a £5950 b £9450

3 £66 440 000

4 4.6%

### Exercise 19.5

1 a £103.27 b £175.71 c £1217.62 d £1529.58 e £1030.32
f £7015.63 g £197.32 h £2.23 i £14 428.07 j £56.95

2 a £1947.46 b £6947.46

3 9

4 £22 841

5 a 2034 (less than 2 birds left) b 2005

6 $A(1 + \frac{r}{100})^5$

### Exercise 19.6

1 £5108.77

2 £22 930.76

3 Celine

4 £47 098.84

5 a 4.7% b 5.9%

6 1494

### Exercise 19.7

1 £19.64

2 a £1184.60 b £24.68

3 a £267.39 b £9171.84

4 a £38786.04 b £66.19

### End points

A1 556.8 km

A2 426 ml

A3 £52.45

A4 39 203 breakdowns

B1 £998.74

B2 £569.91

C1 £2.27M

C2 380 766

D1 £74.03

D2 4.5%

E1 £12 836.47

F1 a £23.20 b £656.89

## SECTION 20 Circle theorems

### Starting points

A1 b The distances are the same. c yes d both 30°
e both right angles f yes h both the same i yes j yes

A2 b angle AOB = 120°

A3 b OA = OB = OC = 3.6 cm

### Exercise 20.1

1 a 116°, 122° b 94° c 90°, 70° d 107°

2 a angles ADC and AEC b angles CAD and CED
c 58° d 41° e 41°

3 a 74° b 37° c 90° d 53° e 37°

4 a angle EFD b 284° c 38° d 52°

5 a 41° b 49°
c If a tangent (FC) and a chord (EF) meet, the angle between them (EFC) is always equal to the angle in the opposite segment (FGE).
d 61°

6 a 18.5 cm b 8.3 cm c 55.5 cm²

7 a i $10a$ ii $90 - 5a$ b i $180 + 10a + 27 + 43 = 360$ ii $a = 11°$
c $\sqrt{(64 + b^2)}$ d 125°

8 a i $a$ ii $b$ iii $2a$ b $180 - a - b$ c $180 - a$ d $90 - \frac{a}{2} - b$ e $\frac{a}{2}$

### Exercise 20.2

1 Opposite angles are each 90° and so add up to 180°.

2 Not unless it is a rectangle. One pair of opposite angles will sum to more than 180°, the other pair to less than 180°

3 square

4 $v + x = 180° = 12 \times 15°$

6 102°

7 a i 60° ii 120° b 120°
c No-opposite angles not supplementary. Also O not on circumference.

8 a i 67.5° ii 112.5° iii 127° b 106°

9 a i $\frac{1}{2}v$ ii $90 - \frac{1}{2}v$ iii $45 + \frac{1}{2}v$

10 a 28° b 69°

### End points

A1 a 54° b 61°

A2 a angle EBD b angle CAB

B1 a 126° b 54°

C1 62°

D1 54°

D2 72°

D3 36°

E1 The four vertices of the quadrilateral lie on the circumference of the same circle.
The angles at opposite vertices are supplementary.

E2 a $3k$ must be opposite $6k$, $4k$ must be opposite $5k$.
b $9k = 180$, $k = 20°$

## Skills break 4

3 £295 800

4 a 57 691 m³ b 493 256 tonnes

5 a 14 knots b 3 hours 19 minutes c $1\frac{1}{2}$ hours

6 a $C=\frac{5(F-32)}{9}$ b $^-2.2$°C c 55%

9 a 0.61 b 0.43 c 0.25

10 38%

11 70 000 tons

13 a 14° b 28.5 feet c 8.8 fet

14 a 2 000 000 b 2 180 000

15 11 people

## Section 21 Trigonometry in 2-D and 3-D

### Starting points

A1 a i 060° ii 300° iii 300° b 205° c 20 km

B1 b 36.1 km

C1 56°

C2 a 26 km b 15 km

D1 a 30°, 150° b 101.5° c 78.5° d 45°

E1 b 104°

E2 b 20°, 80°

F1 27°

F2 5.1 mm

### Exercise 21.1

1 a 4.9 $cm^2$ b 5.2 $cm^2$ c 5.1 $cm^2$ d 1.85 $cm^2$

2 a 6.0 $cm^2$ b i 30°, 150° c 90°

3 a 46°, 134° b 15 $cm^2$

4 a 3.35 cm, 3.45 cm b 3.6 $cm^2$, 3.9 $cm^2$

### Exercise 21.2

1 a D

### Exercise 21.3

1 b 5.3 cm

2 a 4.2 cm b 5.0 cm c 3.6 cm

3 a 8 cm, 7.5 cm

4 11.6 $cm^2$

5 a 48.5° b 3.2 cm

### Exercise 21.4

1 a 29.4° b 47.5° c 33.8°

2 a 119.5° b 139.2 $cm^2$

3 a 0.57 b 34.8°, 145.2°

4 b i 74.6°, 105.4° ii 51.7°, 128.3°

### Exercise 21.5

1 a i $d^2=e^2+f^2-2ef\cos D$ ii $f^2=e^2+d^2-2ed\cos F$
b ii and iii

### Exercise 21.6

2 94.8°, 6.8 cm, 4.7 cm

3 95.1°

4 26.7 $cm^2$

5 a i 108° ii 36° iii 72°
b 6.5 cm c 16.9 cm d 12.3 $cm^2$ e 8

6 a 11.3 cm b 39.9 cm, 62.9 $cm^2$ c 43.8 cm

7 a 73.5 $cm^2$ b 30.9 cm

8 192 $cm^2$

9 a $\frac{1}{2}nr^2\sin(\frac{360}{n})$ b i 3.09 ii 3.13 iii 3.14 iii 3.14
The answers are tending to $\pi$.

10 9 cm

### Exercise 21.7

1 a 036° b 127.6 km c 051°

2 a 323° b i 39.88 km ii 159°
c i 14.4 km ii 19.2 km d 7.00 km

3 17.5 km

4 a 4573 m b 032°

5 8.016 km

6 15.81 nautical miles

### Exercise 21.8

1 a 142° b i 43 m ii 30° iii 30°

2 a Jameston and Palter
b 11.4 m

### Exercise 21.9

1 a 68.6 m b 3.7°

2 a 13 m b 14.3 m c 25°

3 a 10.7 $cm^3$ b 45° c 5.7 cm d 60°
e 13.9 $cm^2$ f 2.3 cm g 35°

### End points

A1 a 8.2 cm b 8.1 cm

A2 a 82° b 16 cm

B1 a 66.8° b 12.3 cm

C1 a 69.9° b 61.8°

C2 a 5.12 cm b 103.6°

D1 2.21 km

D2 72.9 nautical miles

E1 281.5 m

E2 75°

F1 110 cm

## Section 22 Processing data

### Starting points

D1 a Q b R c P

### Exercise 22.1

1 a not clear, needs specific times

2 a The question seems to expect the answer 'yes'.

3 a It has more than one meaning

4 a 1: Leading
2: Not clear
3: Ambiguous
4: There is no option for between £1.50 and £2.50.

### Exercise 22.3

1

| | Yr 7 | Yr 8 | Yr 9 | Yr 10 | Yr 11 | Total |
|---|---|---|---|---|---|---|
| Boys | 14 | 15 | 17 | 15 | 15 | 76 |
| Girls | 16 | 16 | 15 | 13 | 14 | 74 |
| Total | 30 | 31 | 32 | 28 | 29 | 150 |

2 75 competitors, 45 support staff

3 a Competitors: 45 male, 30 female. Support staff: 32 male, 13 female.

4 a 58.6 million b 417236, 14505, 43515, 24744

### Exercise 22.4

1 c Negative

2 c Positive

### Exercise 22.5

1 Moderate negative correlation

2 Strong negative correlation

3 Weak positive correlation

### Exercise 22.6

**1** **a** 50 miles **b** 80 miles

**2** **a** £75 000 **b** £85 000

**3** **a** 15 miles **b** £110 000

**5** **b** Moderate positive correlation **c** **i** 272 days **ii** 3.8 kg

**6** No, pregnancy doesn't usually last much longer than 9 months.

### End points

**A1** 1: not clear, 2: ambiguous, 3: leading

**C1** 846, 825, 1220, 7109

**D1** **d** Strong negative correlation

**E1** **a** 33 mpg **b** 2 litres

**E2** **a** 18 mpg **b** 4.8 litres

## SECTION 23 Graphs involving compound measures

### Starting points

**A1** 41 minutes

**A2** 47 mph

**A3** 2 minutes 26 seconds

**A4** a 31.3 miles per hour **b** 17 m/s

**B1** 5200 $cm^2$

**B2** 375 square units

**C1** A: Ravi, B: Sally, C: Charlie, D: Joe

**C2** A and B

**C3** B and C

**C4** 9 : 45 am

**C5** **a** 80 miles **b** 135 miles **c** 100 miles **d** 120 miles

**C6** 1 hour 30 minutes

**C7** C, 70 miles

**C8** 100 miles

**C9** **a** 11 : 30 am, 3 : 05 pm **b** same direction, then opposite

**C10** **a** 9.3 mph

**C11** 30 mph

**C12** Faster in the morning (steeper gradient)

**C13** **a** A **b** 12 noon

### Exercise 23.1

**1** Edinburgh

**2** **a** 6.35 km/h **b** bicycle or similar

**3** **a** 11 : 30 am **b** 35.6 km/h

**4** Julie got faster, Simon slowed down

**5** **a** 23 km/h **b** 23 km/h

**6** Julie was travelling most quickly, then Ayesha, then Simon.

**7** **a** 1 : 30 pm **b** 10 : 30 am

**8** **a** 66 km **b** 92 km

**9** 11 : 45 am

**10** 64 km

### Exercise 23.2

**1** **a** **ii** 11.1 km/h **b** **ii** 3 : 16 pm **iii** 2 : 44 pm

**2** **a** **ii** 9 mph **iii** 9.2 mph
**b** **i** 10 miles **iii** 12.8 mph
**iv** Alice was buying cakes as Emily passed
**c** **ii** 20 mph

### Exercise 23.3

**1** The gradient is steeper

**2** **a** stationary **b** constant speed

### End points

**A1** **a** 2.5 miles **b** 26 minutes **c** twice **d** 10.3 mph
**e** **i** 20 : 40 and 21 : 08 **ii** steepest part
**f** 50 minutes **g** 60 mph **h** **i** 21 : 26 **ii** 15.5 miles

**B1** **a** 2 s, 4.5 s, 8.5 s, 15.3 s **b** 0 s, 6.5 s **c** 0 s, 16 s **d** decelerating

## SECTION 24 Vectors

### Starting points

**B1** **a** $\begin{pmatrix}5\\-1\end{pmatrix}$ **b** $\begin{pmatrix}1\\-2\end{pmatrix}$ **c** $\begin{pmatrix}^{-}1\\4\end{pmatrix}$

**C1** **a** $2(2x-y)$ **b** $^{-}2(y-2x)$ **c** $^{-}3(x-3y)$ **d** $3(3y-x)$

**C2** **a** $\frac{1}{2}(2x+y)$ **b** $\frac{1}{3}(x-3y)$ **c** $\frac{1}{2}(3x+y)$ **d** $\frac{1}{3}(4x-y)$
**e** $\frac{2}{3}(x+y)$ **f** $\frac{2}{3}(2y-x)$

**D1** **a** 3 : 1 **b** 1 : :3

**D2** **a** $\frac{1}{4}$ **b** $\frac{3}{4}$

### Exercise 24.1

**2** **a** $\begin{pmatrix}^{-}1\\-3\end{pmatrix}$ **b** $\begin{pmatrix}8\\2\end{pmatrix}$ **c** $\begin{pmatrix}^{-}3\\6\end{pmatrix}$ **d** $\begin{pmatrix}^{-}2\\-6\end{pmatrix}$ **e** $\begin{pmatrix}3\\-6\end{pmatrix}$

**3** $\begin{pmatrix}5\\4\end{pmatrix}$

**4** $\begin{pmatrix}5\\4\end{pmatrix}$

**6** $\begin{pmatrix}5\\-1\end{pmatrix}$

### Exercise 24.2

**2** **a** $\mathbf{u}=\mathbf{p}+\mathbf{r}$ **b** $\mathbf{v}=\mathbf{p}+2\mathbf{q}$ **c** $\mathbf{w}=2\mathbf{q}-\mathbf{p}$

### Exercise 24.3

**1** **b** $^{-}(\mathbf{q}-\mathbf{p})={}^{-}\mathbf{q}-{}^{-}\mathbf{p}=\mathbf{p}-\mathbf{q}$

**5** $2\mathbf{p}-\mathbf{q}$, $2\mathbf{q}-\mathbf{p}$, $^{-}3\mathbf{p}-\mathbf{q}$

**6** $2\mathbf{q}-6\mathbf{p}=-2(3\mathbf{p}-\mathbf{q})$

**7** $^{-}2\mathbf{p}-\mathbf{q}$ and $2\mathbf{p}+\mathbf{q}$, $2\mathbf{p}-\mathbf{q}$ and $\mathbf{q}-4\mathbf{p}$, $2\mathbf{p}+4\mathbf{q}$ and $\mathbf{p}+2\mathbf{q}$

**8** **a** $2\mathbf{p}$, $2\mathbf{q}$, $^{-}2\mathbf{p}$, $^{-}2\mathbf{q}$

### Exercise 24.4

**1** **a** $2\mathbf{c}-\mathbf{d}$

**2** **a** **i** $\overrightarrow{ED}$ **ii** $\overrightarrow{EG}$ **iii** $\overrightarrow{HE}$ **iv** $\overrightarrow{EG}$

**3** **b** **ii** $\frac{1}{3}(\mathbf{p}+3\mathbf{q})$

### Exercise 24.5

**2** **a** **i** $\frac{2}{3}(\mathbf{p}+\mathbf{q})$ **ii** $2\mathbf{q}-\mathbf{p}$

**3** **a** **i** $\mathbf{y}-\mathbf{x}$ **ii** $\frac{1}{3}(\mathbf{y}-\mathbf{x})$

### End points

**A1** **a** $\begin{pmatrix}2\\8\end{pmatrix}$ **b** $\begin{pmatrix}^{-}6\\3\end{pmatrix}$ **c** $\begin{pmatrix}3\\3\end{pmatrix}$ **d** $\begin{pmatrix}3\\2\end{pmatrix}$ **e** $\begin{pmatrix}^{-}6\\5\end{pmatrix}$ **f** $\begin{pmatrix}0\\10\end{pmatrix}$

**B1** **a** $\overrightarrow{CB}+\overrightarrow{BA}+\overrightarrow{AD}$ **b** $2\mathbf{p}+\mathbf{q}$

**B2** **b** $\frac{1}{2}(4\mathbf{p}-\mathbf{q})$

**C1** $3\mathbf{r}+\mathbf{s}$ and $6\mathbf{r}+2\mathbf{s}$, $3\mathbf{s}-\mathbf{r}$ and $2\mathbf{r}-6\mathbf{s}$

**D1** **a** **i** $2\mathbf{b}-2\mathbf{a}$ **ii** $\mathbf{b}-\mathbf{c}$

**D2** **a** **i** $\mathbf{d}-2\mathbf{c}$ **ii** $\frac{2}{3}(\mathbf{d}-2\mathbf{c})$ **iii** $2\mathbf{d}-2\mathbf{c}$

## SECTION 25 Transforming graphs

### Starting points

**B1** **a** $(^{-}2, 3)$, 7 **b** (1, 5), (0, 6) **c** $(^{-}3, 0)$, (0, 9) **d** $(^{-}5, ^{-}1)$, (0, 24)
**e** (8, –4), (0, 60)

**B2** $\begin{pmatrix}6\\5\end{pmatrix}$

**C1** **a** **i** 43 **ii** $^{-}2$ **iii** 13.75 **iv** $^{-}5$ **b** 0.75

### Exercise 25.1

**1** **a** A: $y=0.8y^2-2$, B: $y=0.3\times 2^x$, C: $y=1.6x^2$, D: $y=2.5\sqrt{x}$,
E: $y=0.1x^3$, F: $y=\frac{1}{2}x$

**2** 6, 1.8

**3** $\frac{1}{2}$, 4

### Exercise 25.2

**1** **c** approximately $d=4.9t^2+0.06$ **d** 31 m **e** 4.5 s

**2** **b** $C=80r^2+65$ **c** £1685

3 **c** 20 **d** 155 s

4 A, $p = 640\,000$; C, $q = 0.4$

## Exercise 25.3

1 **a** $y = 2x - 2$ **c** $y = {}^{-}\frac{1}{2}x + 3$ **d** reflection in $y = 0$

2 **b** $y = {}^{-}2x + 1$, $y = 3x - 4$, $y = {}^{-}\frac{1}{2}x - 1$ **c** $y = -4x - 3$

3 $y + x = 8$

4 **b** $y = {}^{-}x - 1$, $y = 2x - 3$, $y = {}^{-}\frac{1}{4}x + 1$ **c** $y = 5x - 1$

5 **b** $p - q = 6$

6 $y = 3x + 1 - 3v + w$

7 $y = kx + l - ka + b$

## Exercise 25.4

2 translation $\binom{2}{0}$

3 A

4 **a** **i** $\binom{-3}{2}$

5 It will map to $y = f(x + 4) + 5$

6 **b** **i** reflection in $y = 0$ **ii** yes

7 **b** reflection in $x = 0$ **c** yes

8 $({}^{-}x)^2 = x^2$

9 **a** C **b** A

12 **a** D **b** A **c** B **d** E **e** F **f** C

## Exercise 25.5

2 **b** reflection in $y = 0$

3 **b** translation $\binom{-1}{-2}$ **c** ${}^{-}3$

4 **a** **i** 3 **ii** 0.8 **iii** 0

5 A: $y = k(x + 1) - 2$, B: $y = k(2x)$

## End points

**A1** 8, 1.5

**B1** **b** 3.6, 6 **c** £14.10

**C1** **b** $y = f(x - 1) - 2$

# In focus 1

1 **a** **i** 1 **ii** 14 **b** **i** 5 **ii** 150 **c** **i** 7 **ii** 252 **d** **i** 10 **ii** 100
**e** **i** 18 **ii** 252 **f** **i** 60 **ii** 13 860

2 **a** $\frac{13}{36}$ **b** $1\frac{1}{6}$ **c** $\frac{7}{10}$ **d** $\frac{9}{10}$ **e** $\frac{5}{8}$ **f** $1\frac{17}{18}$ **g** $4\frac{7}{24}$ **h** $\frac{31}{44}$ **i** $1\frac{5}{8}$

3 **a** $\frac{5}{3}$ 1**b** $\frac{5}{4}$ **c** $\frac{3}{2}$ **d** 3 **e** $\frac{76}{15}$ **f** $\frac{4}{5}$ **g** $\frac{125}{18}$ **h** $\frac{3}{50}$

4 **a** $3\frac{2}{3}$ **b** $1\frac{1}{9}$ **c** $\frac{3}{100}$ **d** $\frac{3}{10}$ **e** $5\frac{1}{2}$ **f** $4\frac{1}{2}$

5 **a** $\frac{1}{3}$ **b** $\frac{25}{144}$ **c** $\frac{1}{3}$ **d** $\frac{1}{8}$ **e** $4\frac{1}{2}$ **f** $\frac{4}{7}$

6 **a** $1\frac{1}{7}$ **b** $\frac{15}{19}$ **c** $1\frac{5}{16}$ **d** $\frac{10}{19}$

7 **a** $\frac{1}{2} + \frac{1}{6}$ **b** $\frac{1}{6} + \frac{1}{12}$ **c** $\frac{1}{12} + \frac{1}{36}$ (other answers possible)

8 **a**

| | | |
|---|---|---|
| $\frac{3}{2}$ | $\frac{1}{3}$ | $\frac{7}{6}$ |
| $\frac{2}{3}$ | 1 | $\frac{4}{3}$ |
| $\frac{5}{6}$ | $\frac{5}{3}$ | $\frac{1}{2}$ |

**b** change $2\frac{1}{4}$ to $2\frac{7}{12}$

9 A and D, B and H, C and G, F and I

10 **a** $\frac{4x}{9}$ **b** $\frac{2}{5}$ **c** $\frac{3}{x}$ **d** 21 **e** $\frac{3x}{10}$ **f** $\frac{10}{3x}$ **g** $\frac{(7x + 2y)}{7}$ **h** $\frac{(2x + 12)}{x(x + 4)}$
**i** $\frac{(x - 2y)}{4}$ **j** $\frac{5}{(x - 1)}$ **k** $\frac{(x - 1)}{(x + 2)}$ **l** $\frac{(y + 2x)}{xy}$ **m** $\frac{(5x - 4)}{x(x - 1)}$ **n** $\frac{(y - x^2)}{xy}$ **o** $\frac{(x - 1)}{(x + 4)}$
**p** $\frac{(5x + 7)}{(x + 1)(x + 2)}$ **q** $\frac{5x - 1}{(x + 1)(x - 2)}$ **r** $\frac{6}{(x + 3)}$ **s** $\frac{(x + 4)}{x(x + 2)}$ **t** $\frac{(6x - 3)}{(x - 1)(x + 2)}$
**u** $\frac{(5y + x + 11)}{(x + 1)(y + 2)}$

11 **a** 6 **b** 9

# In focus 2

1 **a** 47°, 133°, 47°, 133° **b** 51.5°, 128.5°, 51.5°, 128.5°
**c** 11°, 11°, 169°, 11°

2 **a** 34°, 73°, 73° **b** isosceles
**c** **i** 34°, 146°, 34°, 146° **ii** 73°, 107°, 73°, 107°
**d** All sides are of equal length, opposite sides are parallel.
**e** Only pairs of opposite sides are equal in length.
**f** **i** 90° **ii** 73° **iii** 73° **iv** 17° **v** 17°

4 73°, 46°, 92°, 120°, 17°, 72°, 198°, 275°, 120°, 60°, 60°, 60°, 36°

# In focus 3

1 **a** $m^2n^2$ **b** $4p^2q$ **c** $18a^2b$ **d** $a^3g^3$ **e** $20p^3$ **f** $28a^3b$

2 **a** $24a - 18b$ **b** $pn - p^2$ **c** $s^2 + st$ **d** $6mn - 6m^2$ **e** $5x^2 + 5xy$
**f** $4u^2 + 3u$ **g** $5pq + 3pr$ **h** $12mn + 21n^2$ **i** $27ab + 9a^2$

3 **a** $2n + 7m$ **b** $21a^2 - a$ **c** $2b$ **d** $8x + 2xy - 2y$ **e** $4p + 3q$
**f** $8p^2 + 2pq + 5q^2$ **g** $4m^2 + 2mn - 6n$ **h** $2x - 2xy$

4 D

5 **a** $16x^2 + 4x$ **b** $30x^2 + 13xy$ **c** $6q - 6p$ **d** $9a + 6b$ **e** $p^2 - 4p + 6$

6 **a** $\frac{(8p + 2)}{(p^2 - 1)}$ **b** $\frac{2(p - 1)}{(p + 4)(p + 2)}$ **c** $\frac{(9p + 1)}{(2p + 1)(3p - 2)}$ **d** $\frac{(p + 11)}{6}$

7 A $2c + 3$ B $3 + 4b$

8 **a** $p(7 + 3q)$ **b** $m(1 + 3n)$ **c** $4q(2p + 1)$ **d** $3y(1 - 4x)$
**e** $2b(b + 5)$ **f** $6b(3a + 4c)$ **g** $xy(x - 2)$ **h** $3ab(3a + 2b)$
**i** $3mn(4m - 3)$

9 **a** $w^2 + 8w + 12$ **b** $y^2 + 14y + 45$ **c** $x^2 - 6x + 9$ **d** $v^2 - 5v + 6$
**e** $b^2 + 2b - 15$ **f** $b^2 - 8b + 15$ **g** $h^2 - 25$ **h** $t^2 - 9$

10 **a** $4w^2 - 9$ **b** $4p^2 + 4p + 1$ **c** $15x^2 + 8x - 12$ **d** $12g^2 - 43g + 35$

11 **a** $x + 3$ and $x + 4$ **b** $x + 2$ and $x + 6$ **c** $x + 1$ and $x + 6$
**d** $x + 2$ and $x + 3$ **e** $x + 1$ and $x + 12$ **f** $x + 2$ and $x + 4$

12 **a** $(x + 2)^2$ **b** $(x + 4)(x + 5)$ **c** $(x + 4)(x - 1)$
**d** $(x + 5)(x - 3)$ **e** $(x + 2)(x - 3)$ **f** $(x - 1)(x - 10)$
**g** $(x + 2)(x - 4)$ **h** $(x - 1)(x - 2)$ **i** $(x - 2)^2$

13 **a** $\frac{3}{(x + 3)}$ **b** $\frac{4}{(x + 6)}$

14 **a** $x - 2$ and $x + 3$ **b** $x - 3$ and $x + 1$ **c** $x - 3$ and $x - 2$
**d** $x - 2$ and $x + 2$ **e** $x + 2$ and $x + 3$ **f** $x - 3$ and $x + 3$

15 **a** $(3x + 4)(x + 3)$ **b** $3(x - 1)(x - 4)$ **c** $(2x - 1)(x - 4)$
**d** $(2x + 1)3(x - 1)$ **e** $3x(x - 2)$ **f** $(3x + 5y)(3x - 5y)$

16 **a** $p = 8$, $q = 4$ **b** $p = 9$, $q = 3$

17 **a** $\frac{(7p - 5)}{(p - 3)(p + 1)}$ **b** $\frac{(4 - 3a + 3b)}{2(a - b)^2}$ **c** $\frac{9a^2p}{4}$ **d** $\frac{6}{(p^2 - 1)}$

18 **a** $(p + q)(p - q)$
**b** **i** 149 is prime.
**ii** $p + q$ must be greater than 1 as $p$ and $q$ are both positive integers, so $p - q = 1$.
**iii** $p = 75$, $q = 74$

19 264 826 482 647

20 $x + 2$

# In focus 4

1 **a** 64, 63.5, 64.25 **b** 6.05, 6.16, 5.74

2 49.25, 5.74

3 80, 7.56

4 **a** Sally 6.17, 2.07; Gavin 6.5, 1.59
**b** Gavin's scores are higher on average. There is also less variation in his scores.

5 **a** Peggy 6, 2, Toby 5.5, 3; Sue 7, 3.5
**b** Sue's scores are highest on average. Peggy's scores are most consistent.

6 **a** 36, 38, 38.5 **b** 3, 3.5, 3

8 The scale does not start at 0 in the first diagram.
Area (or even volume) should have been used to represent sales, not height.

## IN FOCUS 5

1 **b** $a_n = 2n + 1$, $a_n = 4n$, $a_n = 12n - 8$
**c** 201 matches, 400 matches, 1192 matches

2 **a** $20 \longrightarrow 66$, $n \longrightarrow 3n + 6$ **b** $40 \longrightarrow {}^-153$, $t \longrightarrow 7 - 4t$

3 **a** $n \longrightarrow 4n^2$

4 **a** A $3n - 2$, B $6n + 1$, C $4n - 2$, D $5n + 2$, E $14 - 3n$, F $23 - 9n$
**b** A 88, B 181, C 118, D 152, E $^-76$, F $^-247$

5 **a** $n^2 + 1$ **b** $12 - n^2$ **c** $n^2 - 2$ **d** $n^2$ **e** $2n^2 + 4$ **f** $n^2 - 899$

6 **a** A 2205, B 1, C 2.2, D 5.015
**b** A diverges, B approaches 1, C approaches 2, D approaches 5

## IN FOCUS 6

4 **a** not necessarily congruent **b** congruent by AAS
**c** not necessarily congruent **d** congruent by SSS
**e** not necessarily congruent **f** congruent by SAS

## IN FOCUS 7

1 a, c, e

2 **a** 4 **b** 0.6 **c** $^-3$ **d** 1.1

3 **a** 2.5 **b** 1.6 **c** 2.2

4 **a** $\frac{1}{64}$ **b** $\frac{3}{5}$ **c** $\frac{1}{7}$

5 **a** 3.7 **b** 4 **e** 9 **d** $^-5.6 \times 10^{-35}$

6 **a** $4^8$ **b** $3^{-1}$ **c** $2^0$ **d** $2^{12}$ **e** $6^{-14}$ **f** $6^2$ **g** $6^{-2}$ **h** $6^{-14}$

7 **a** 3 **b** $^-2$ **c** $^-1$

8 **a** $y^5$ **b** $y^{30}$ **c** $81y^8$ **d** $12y^7$ **e** $y$ **f** $2y^{-4}$ **g** $3y^6$ **h** $2y^{-2}$ **i** $\frac{11}{10}$

9 A, D and C, E.

10 **a** 7 **b** 9 **e** 3 **d** 7 **e** 1 **f** 5 **g** $\frac{1}{3}$ **h** $^-4$ **i** 2 **j** $\frac{1}{4}$ **k** $\frac{1}{6}$ **l** $\frac{1}{3}$
**m** 49 **n** 64 **o** 4 **p** 32 **q** 64 **r** 125 **s** 8 **t** 81 **u** 243 **v** 81
**w** $\frac{1}{16}$ **x** $\frac{1}{64}$ **y** $\frac{1}{27}$ **z** $\frac{1}{32768}$

11 **a** $x = 2$ **b** $x = {}^-3$ **e** $x = \frac{1}{3}$ **d** $x = 6$ **e** $x = 1$ **f** $x = \frac{-3}{2}$ **g** $x = \frac{2}{3}$
**h** $x = \frac{3}{2}$ **i** $x = 8$ **j** $x = \frac{-1}{2}$ **k** $x = -\frac{2}{3}$ **l** $x = 1024$

12 **a** 6 **b** $\frac{1}{5}$ **c** $\sqrt{14}$ **d** 10 **e** $8 + 2\sqrt{15}$ **f** $9 - 2\sqrt{14}$ **g** $\sqrt{2}$
**h** $21 + 4\sqrt{5}$ **i** $20 - 6\sqrt{11}$

13 **a** $\frac{1}{3}$ **b** $\frac{4}{11}$ **c** $\frac{1}{99}$ **d** $\frac{14}{111}$ **e** $\frac{5}{9}$ **f** $\frac{1}{11}$ **g** $\frac{6}{7}$ **h** $\frac{3}{11}$

14 b, c

## IN FOCUS 8

1 **a** 5.67 **b** 12.7 **c** 10 **d** 2140 **e** 15.78 **f** 93748000 **g** 16 **h** 500

2 **a** 40 000 **b** 200 **c** 240 **d** 250 000 **e** 350 **f** 0.00075 **g** 120 000
**h** 40 000 **i** 280 000 **j** 45 000

3 **a** 1200, 1.4% **b** 20 000, 9.8% **c** 20, 21.5% **d** 400 000, 26.9%
**e** 16, 47.3%

4 **a** 5.55, 5.65 **b** 53.5, 54.5 **c** 685, 695 **d** 56.3715, 56.3725
**e** 64.995, 65.005 **f** 95, 150 **g** 99.5, 105 **h** 159950, 160050
**i** 12.85, 12.95 **j** 15 995 000, 16 005 000 **k** 999.995, 1000.005
**l** 7.00005, 7.00015

5 **a** 31.6 m to 32 m **b** 59.3475 $m^2$ to 60.9375 $m^2$
**c** £623.15 to £700.78

6 20.2 m/s, 24.1 m/s

7 130.9 $m^3$ 143.8 $m^3$

8 **a** $4.35 \times 10^5$ **b** $1.2 \times 10^7$ **c** $3.21 \times 10^{-4}$
**d** $5.426745 \times 10^1$ **e** $6.54 \times 10^8$ **f** $6.54 \times 10^{-2}$

9 **a** 5230 **b** 0.0000015 **c** 0.1094 **d** 0.03

10 **a** $1.58 \times 10^5$ **b** $2.25 \times 10^8$ **c** $1 \times 10^{-7}$ **d** $7.29 \times 10^{-4}$ **e** $6.48 \times 10^{-4}$

## IN FOCUS 9

1 **a** $x = 4.5$ **b** $x = 4$ **c** $x = {}^-2.5$ **d** $x = 2$ **e** $x = 13$ **f** $x = 7.25$
**g** $x = {}^-26$ **h** $x = 46$ **i** $x = 0.75$ **j** $x = 4$ **k** $x = {}^-2.17$

2 17 cm by 26 cm

3 **a** $x^2 - 2x - 15$ **b** $2x^2 - 7x + 3$ **c** $2x - 15x^2 + 8$ **d** $15x - 4x^2 - 9$
**e** $x^2 - 16$ **f** $4x^2 - 9$ **g** $x^3 - 4x^2$ **h** $2x^3 + 6x^2 - 16x$
**i** $3x^4 + 15x^3 + 12x^2 - 3x$ **j** $x^3 + 9x^2 + 11x - 21$
**k** $10x^2 - x^3 - 22x - 15$ **l** $x^3 + 12x^2 + 17x - 60$ **m** $x^3 - 7x - 6$

4 **a** 3, 1 **b** 2, 1 **c** 1, $^-1$ **d** 3, $^-2$ **e** $^-1$, 2 **f** $^-1$, 2
**g** $^-3$, 5 **h** $^-10$, $^-8$ **i** $^-2$, $^-3$ **j** 4, $^-2$

5 $^-4$ and 5

6 37 £5 notes and 18 £10 notes

7 £8.45, £12.95

8 1778, 472

9 165, 213

10 **a** $^-12$, 7 **b** 1, 7 **c** $^-9$, 6 **d** 3, 12 **e** $^-6$, $^-2$ **f** $^-16$, 20
**g** $^-4$, 15 **h** 6, 8 **i** $^-15$, 9 **j** $^-5$, 13 **k** 4, 5 **l** 2, 14 **m** 0, 3
**n** $^-8$, 8 **o** $^-14$, 14 **p** $^-100$, 100 **q** $^-0.5$, 3 **r** $^-5$, 2 **s** $^-1$, $\frac{2}{3}$
**t** $^-\frac{4}{3}$, 1 **u** $^-\frac{1}{2}$, $\frac{2}{3}$ **v** $^-6$, 0.17 **w** $^-2$, 0.25 **x** $^-\frac{2}{3}$, $\frac{1}{2}$

11 **a** 0.70, 4.30 **b** $^-6.32$, 0.32 **c** $^-7.87$, $^-0.13$ **d** $^-0.30$, 3.30
**e** $^-2.79$, 1.79 **f** $^-3.83$, 1.83 **g** $^-0.56$, 3.56 **h** $^-7.12$, 1.12
**i** $^-0.65$, 4.65 **j** $^-1.30$, 2.30 **k** $^-5.45$, $^-0.55$ **l** $^-1.62$, 0.62
**m** $^-2.64$, 1.14 **n** $^-1.18$, 0.85 **o** $^-0.64$, 0.39 **p** $^-1.09$, 0.66

12 **a** $^-7.90$, 1.90 **b** 0.76, 5.24 **c** $^-3.83$, 7.83 **d** $^-5.12$, 3.12

## IN FOCUS 10

2 **a** vertices: ($^-5$, $^-1$), ($^-11$, $^-1$), ($^-11$, $^-3$), ($^-7$, $^-5$), ($^-5$, $^-3$)
**b** vertices: (4.5, 0), (5, $^-0.5$), (5, $^-1$), (4.5, $^-0.5$), (4.5, $^-0.5$)

3 **b i** Rotation 90° anticlockwise about (1, 1)
**ii** Reflection in $x = 5$ **c i** C **ii** F **iii** E
**c i** Rotation $^-90$° about (6, 2)
**ii** Reflection in $y = -x$
**d i** C **ii** F **iii** E
**e** Rotion 180° about (4, $^-1$)

4 **b** Rotation 180° about (1, 0)

**c i** Rotation 180° about (4, 0)

**ii** Rotation 90° clockwise about (3, 0)

**iii** Translation $\begin{pmatrix} 2 \\ 2 \end{pmatrix}$

## IN FOCUS 11

1 **a** $x = \frac{(4-5a)}{3}$ **b** $x = \frac{(5-a)}{2}$ **e** $x = \frac{(3a-b)}{6}$ **d** $x = \frac{(1+2a)}{7}$ **e** $x = \frac{2}{3} - a$
**f** $x = \frac{(5+3a)}{13}$ **g** $x = \frac{(n+3)}{3a^2}$ **h** $x = \frac{(1-5y)}{ay}$ **i** $x = \frac{(2-y^2-ay)}{a}$ **j** $\frac{(1-3ac^2+ac^3)}{bc^2}$
**k** $x = \frac{(2bc+wy)}{w}$ **l** $x = \frac{(ab-2an+cn)}{an}$ **m** $x = \frac{(ab-a^2)}{3b}$ **n** $x = a\frac{(\pi+2)}{\pi y}$
**o** $x = \frac{(5w-1)}{3w}$ **p** $x = \frac{(2a^2-3a-1)}{2a}$ **q** $x = \frac{-4}{a}$ **r** $x = \frac{(6-a)}{7a}$ **s** $x = 2\frac{(h-ab)}{c}$
**t** $x = \frac{w^2}{4}$

2 11 days

3 **a** £260.10 **b i** $t = \frac{(C-3n)}{0.15}$ **ii** 850 miles

4 **a** $n = 3(a+2)$ **b** $n = 10\frac{(1-x)}{3}$ **c** $n = \frac{a}{2}$ **d** $n = \frac{(3x-2)}{4a}$ **e** $n = \frac{(a^2-b)}{6}$
**f** $n = 2\frac{(5x+9)}{3}$ **g** $n = 3 + (\frac{2}{ax})$ **h** $n = \frac{5a}{6}$ **i** $n = \frac{(3-x^2)}{2x^2}$ **j** $n = 3 - (\frac{x^2}{w})$
**k** $\frac{a}{(b\cos 30°)}$ **l** $n = \frac{3x^2}{2a}$ **m** $n = \frac{2(1+2a^2)}{3a^2}$ **n** $n = \frac{(a+1)}{2}$ **o** $n = \frac{bc}{a}$
**p** $n = \frac{10a}{15-b}$ **q** $n = \frac{3}{(\cos 30°)}$ **r** $n = \frac{(x+ab)}{a}$ **s** $n = \frac{(5x^2+3)}{2}$ **t** $n = \frac{(a+1)}{w}$

6 **a** $p = \frac{3}{(x-y)}$ **b** $p = \frac{1}{(t^2-3)}$ **c** $p = \frac{(c+2)}{(3-a)}$ **d** $p = \frac{(a+6)}{(a+4)}$ **e** $p = \frac{10}{(6-a^2)}$
**f** $p = \frac{(ab-3)}{(2-ab)}$ **g** $p = {}^-x$ **h** $p = \frac{(y-ax)}{(a-b^2)}$ **i** $p = \frac{2ay}{(6a-1)}$ **j** $p = \frac{w}{(\sin x + \sin y)}$
**k** $p = \frac{4}{(1-3w)}$ **l** $p = \frac{1}{b(b-a^2)}$ **m** $p = \frac{(1-b)}{(a-2c)}$ **n** $p = \frac{b(3n+2a)}{(a-3n)}$
**o** $p = \frac{1}{(bc-3a)}$ **p** $p = \frac{2}{(1+3a)}$ **q** $p = 2an - \frac{1}{2}$ **r** $p = \frac{(c^2n-9b)}{3ab}$ **s** $p = \frac{5a}{(5y-2)}$
**t** $p = \frac{2}{(1-2a^2)}$

7 $\pi = \frac{S}{(2rh + 2r^2)}$

8 **a** $n = \frac{(C + 50y)}{(1.85 + y)}$

9 **a** $v = (1 + x)^2$ **b** $v = 1 - 9a^2$ **c** $v = \frac{2x}{\sqrt{3}}$ **d** $v = \frac{1}{2}(a - b)^2 + 1\frac{1}{2}$
**e** $v = \frac{(a^2 + 1)^2}{9}$ **f** $v = \frac{(2x - 1)^2}{5}$ **g** $v = 16 + 3a^2$ **h** $v = \frac{1}{(x + 3)^2}$
**i** $v = (y^2 - 1)$ **j** $v = \frac{\sqrt{(ax - 1)}}{\sqrt{2}}$ **k** $v = \frac{2}{\sqrt{(3 + a)}}$ **l** $v = \sqrt{\frac{3x}{y + 1}}$
**m** $v = \sqrt{\frac{ab}{b - a}}$ **n** $v = \sqrt{2}$ **o** $v = \frac{1}{\sqrt{(a^2 + 1)}}$ **p** $v = \frac{\sqrt{(1 + a^2k)}}{\sqrt{(a^2 - k)}}$
**q** $v = \frac{1}{\sqrt{(3(2a - 1))}}$ **r** $v = \sqrt{(x + 2)}$

10 **a** $r = \sqrt{\frac{\pi R^2 - A}{\pi}}$ **b** 5 cm

11 $g = \frac{4\pi^2 a}{t^2}$

## In focus 12

1 **a** 90 g icing sugar, 405 g margarine, 180 g plain chocolate, 405 g flour
**b** 20 g icing sugar, 90 g margarine, 40 g plain chocolate, 90 g flour

2 **a** 450 ml **b** 630 ml

3 **a** 2.9 cm **b** 1 : 3000

4 **a** 0.75 **b** 8.1 cm **c** both 0.704

5 IJKL and EFGH

6 **i** 3.9 cm **ii** 8.8 cm

7 **b** **i** BD **ii** BC **c** 9.8 cm

8 **a** both 1.75 **c** 0.5 cm

9 **a** $z \propto p^2$ **b** 0.6 **c** $z = 0.6p^2$ **d** 15 **e** $p = \sqrt{(\frac{z}{0.6})}$ **f** 12

10 **a** $l = \frac{36}{w^2}$ **b** 5.76 cm **c** 1.2 cm

## In focus 13

1 2, 3

3 **a** 3 **b** 9

4 19.8 cm, 5.3 cm

5 82.5 $cm^2$

6 **a** 73.5 $cm^2$ **b** 35.8 cm

7 **a** 1540 $cm^3$ **b** 603 $cm^2$

8 **a** 209 000 $cm^3$ **b** 18 200 $cm^2$

9 13.7 cm

10 **a** length **b** area **c** area **d** volume **e** length **f** area
**g** none **h** volume **i** none **j** area **k** none **l** volume
**m** area **n** volume **o** area **p** length

## In focus 14

1 **a** $\frac{1}{6}$ **b** $\frac{1}{3}$ **c** $\frac{2}{3}$

2 **a** $\frac{2}{3}$ **b** $\frac{1}{3}$

3 **a** $\frac{1}{18}$ **b** $\frac{5}{36}$ **c** $\frac{5}{18}$ **d** $\frac{1}{36}$ **e** $\frac{31}{36}$

4 **a** 0.61 **b** 0.84 **c** 0.77

5 **a** **i** $\frac{2}{9}$ **ii** $\frac{7}{54}$ **iii** $\frac{10}{27}$ **b** **i** $\frac{1}{12}$ **ii** $\frac{13}{18}$ **iii** $\frac{5}{18}$ **c** $\frac{5}{7}$ **d** **i** $\frac{1}{12}$ **ii** $\frac{25}{216}$

6 **b** **i** 0.16 **ii** 0.33 **iii** 0.33 **iv** 0.25 **c** **i** 0.0099 **ii** 0.052

7 **a** **i** 0.011 **ii** 0.12 **iii** 0.86 **b** **i** 0.171 **ii** 0.0016 **iii** 0.023

8 **b** **i** $\frac{1}{5}$ **ii** $\frac{4}{15}$ **iii** $\frac{8}{15}$ **iv** $\frac{2}{3}$

## In focus 15

1 **a** frequency densities: 3, 9, 6, 5, 2.8, 2.6 **b** 143 minutes
**c** cumulative frequencies: 15, 33, 51, 71, 82, 100
**d** **i** 142 minutes **ii** 9 minutes **e** **i** 9 **ii** 5 **iii** 42

2 **a** frequency densities: 1, 3, 4, 6, 2.25, 2, 0.8 **b** 159 minutes
**c** cumulative frequencies: 5, 14, 22, 34, 43, 53, 60
**d** **i** 157 minutes **ii** 9 minutes **e** **i** 4 **ii** 10 **iii** 35

3 The men's times were lower on average. The spread is the same for both sets of data.

4 **a** **i** frequency densities: 1.4, 5.4, 4, 0.6
**ii** frequency densities: 0.4, 2, 1.4, 0.44
**b** The total frequencies are different. **c** **i** 28 years **ii** 29 years
**d** The field athletes are slightly older on average.

## In focus 16

1 **a** 2,1 **b** 1, $^-1$ **c** 3, 2 **d** 4, 1.5 **e** 2, $^-1$ **f** $^-1$,2.5 **g** 5, 1.5
**h** 4.5, 3 **i** 3.5, 3 **j** 3.5, 3

2 **a** $x = 0$ **b** $x = {}^-2$ **c** $x = 4$ **d** $x = {}^-3$ **e** $x = 6$ **f** $x = {}^-4$

5 **a** **i** $^-4.9$, 1.9 **ii** $^-3.6$, 0.6 **iii** $^-5.9$, 2.9

6 **b** $^-1$, 2.5

7 0.5, 7.5

8 $^-1$, 5

9 **a** $^-3.8$, 0.8 **c** 0.6

11 $y = 8, x = 0$

12 **a** $^-5.7$ **b** 1.4, 7.2

15 **a** 0, 2 **b** 3.1

18 **a** $^-10$ **b** ($^-2$, $^-10$) **c** $^-6$

## In focus 17

1 (2.6, 1.5), ($^-1$, $^-1.7$)

2 **c** **i** $4.5 \leqslant t \leqslant 13.5$ **ii** $3 \leqslant t \leqslant 15$ **d** 4 **e** **i** 4.5 and 13.5 **ii** 6, 12
**f** **ii** 2, $^-6$

3 **a** 60°, 300° **b** 210°, 330° **c** 135°, 315° **d** 45°, 225°
**e** 155°, 205° **f** 205°, 335° **g** 25°, 205° **h** 75°, 105° **i** 105°, 255°
**j** 105°, 285°

4 **a** **i** 54°, 126°, 414° **ii** $^-54°$, 54°, 306°, 414° **iii** 54°, 234°, 414°
**iv** 26°, 154°, 386° **v** 154°, 206° **vi** $^-26°$, 154°, 334°
**b** **i** 118°, 242° **ii** $^-62°$, 242°, 298° **iii** 90°, 450°
**iv** $^-90°$, 90°, 270°, 450°

7 **b** 63°, 243°

8 **a** 6, $^-6$ **b** 7, 5 **c** 8, $^-8$ **d** 1, $^-1$ **e** 9, 3 **f** $^-2$, $^-6$

9 2, 3, $^-3$

10 **a** 45°, 225°, 405° **b** 40°, 50°, 220°, 230°, 400°, 410°
**c** $^-20°$, 20°, 100°, 140°, 220°, 260°, 340°, 380°
**d** $^-45°$, 15°, 75°, 135°, 195°, 255°, 315°, 375°, 435°

## In focus 18

1 4, 71, $^-3.2$, $^-0.33$

2 **a** 4, 5, 6, 7, 8 **b** $^-3$, $^-2$, $^-1$ **c** $^-5$, $^-4$, $^-3$, $^-2$, $^-1$, 0 **d** $^-6$, $^-5$, $^-4$
**e** none **f** 24 **g** $^-14$, $^-15$ **h** 1, 2, 3, 4, 5
**i** $^-4$, $^-3$, $^-2$, $^-1$, 0, 1, 2, 3, 4 **j** all integers except $^-1$, 0, 1
**k** $^-3$, $^-2$, $^-1$, 0, 1, 2, 3 **l** $^-3$, $^-2$, $^-1$, 0, 1, 2, 3

4 The number of waiters has to be an integer, the weights do not.

6 $x \leqslant 2, x + y < 7$

7 **a** 0 **b** 16

8 **a** $x \geqslant 9$ **b** $t < {}^-7$ **c** $p \leqslant 13$ **d** ${}^-11 \leqslant k \leqslant 11$ **e** $w < 2$ **f** $q > 9.3$
**g** $t \geqslant 8$ **h** $h \leqslant 2$ **i** $f < 3$ **j** ${}^-9 < c < 9$ **k** $g \geqslant 4.5$ **l** $x > {}^-3$
**m** $j < 19$ **n** $d \geqslant {}^-2$ **o** $v < 4.5$ **p** $x \leqslant {}^-13$ **q** $s < 0$ **r** $u \leqslant {}^-3$
**s** $j > \frac{12}{11}$ **t** $r > 3, r < {}^-3$ **u** ${}^-\sqrt{2} < x < \sqrt{2}$

9 **a** $28 < xy < 63$ **b** $11 < x + y < 16$ **c** $1 < \frac{y}{x} < 2.25$

10 **a** **i** $16a + 24p \leqslant 432$ **ii** $8a + 10p \leqslant 200$ **iii** $p \geqslant 6$
**b** At least half of the plants must be apple bushes.
**d** 17 apple, 6 pear, £196

## IN FOCUS 19

1 **a** 63.33% **b** 25.45% **c** 283.33% **d** 62%

2 29%

3 **a** 29.5 kg **b** 2268 km **c** 3702.4 miles **d** 1836 yards

4 23.2 million

5 **a** 426.8 ml **b** £34.77 **c** 22.75 mm **d** 0.74 cm

6 9.5 tonnes

7 **a** £218.37 **b** £13.74 **c** £52.71 **d** £25.56 **e** £315.55 **f** £187.99 **g** £159.07 **h** £10.857 **i** £58.74 **j** £1761.33

8 £155.10

9 £100.71

10 £8510.63

11 **a** Price busters, they are cheaper by 25p

12 **a** £405 **b** £4216 **c** £20.40

13 $R = \frac{I}{PT}$

14 7.5 years

15 **a** £296.92 **b** £959.22 **c** £345.44 **d** £24.14 **e** £61700.34 **f** £21.28 **g** £578.67 **h** £4382.25 **i** £124.88 **j** £1372.66

16 **a** £255.31 **b** £11.06 **c** £38.29 **d** £8.50 **e** £63.83 **f** £76.55 **g** £22.97 **h** £110.63 **i** £16.16 **j** £119.14

17 £100.53

18 £3 157 895

## IN FOCUS 20

1 **b** CD is a diameter

2 **b** 68°

3 136°

4 **a** 109° **b** 33°

5 **b** 96°

6 **b** 110°

7 **a** 2.8 cm **b** 2.8 cm

8 **a** 37° **b** 37.0 $cm^2$

9 **b** 75°

## In focus 21

1 **a** 17.3 $cm^2$ **b** 13.7 $cm^2$ **c** 22.3 $cm^2$ **d** 99.76 $cm^2$ **e** 37 $cm^2$ **f** 14.6 $cm^2$

2 6.6cm

3 **a** 5.2 cm **b** 15.5 cm **c** 48° **d** 50° **e** 81°

4 **a** 5.1 cm **b** 16.0 cm **c** 9.8 cm **d** 13.1 cm **e** 25.4 cm

5 **a** 73° **b** 53.1° **c** 72.1° **d** 59.6° **e** 88.9°

6 **a** 7.3 cm **b** 9.2 cm **c** 9.0 cm **d** 4.9 cm **e** 33.4cm

## In focus 22

1 **a** 1 is too vague, 2 is a leading question, 3 is too vague

7 **d** moderate positive correlation

8 **a** £7200 **b** £4900

9 **a** 320 cc **b** 700 cc

10 **a** £8550 **b** 1220 cc

11 24 female, 36 male

## In focus 23

1 **a** 36 mph **b** 72 mph **c** 27 mph **d** 360 mph

2 3200 km

3 **a** 76.5 miles **b** 38.25 miles **c** 17 miles **d** 46.75 miles **e** 123.5 miles **f** 191.25

4 **a** 30 minutes **b** 2 hours **c** 48 minutes **d** 6 hours 40 minutes **e** 1 hour 5 minutes **f** 1 hour 13 minutes **g** 1 hour 12 minutes **h** 24 minutes

5 **a** 4 hours 23 minutes **b** 12 hours 46 minutes **c** 5.27 hours **d** 8.85 hours **e** 86 km/h **f** 36 mph **g** 26 m/s **h** 42 m/s

6 11 mph

7 244 miles

8 2 hours 26 minutes

9 33 mph

10 1 hour 15 minutes

11 400 m

12 **b** 3, 6, 9, 13.6 sec **c** 6 m **d** 7 s **e** 14 s **f i** 2 m/s **ii** 4 m/s

13 **a** 48 mph **b** 85.71 km/h **c** 52.5 mph **d** 83.33 km/h **e** 53.33 mph **f** 30 km/h

14 **a** 12 noon

15 **b** 6:15 pm **c** 67 mph **d** 60 mph

## In focus 24

1 **a** $\binom{^-4}{10}$ **b** $\binom{3}{12}$ **c** $\binom{^-10}{^-12}$ **d** $\binom{6}{10}$ **e** $\binom{^-1}{9}$ **f** $\binom{7}{0}$ **g** $\binom{4}{3}$ **h** $\binom{6}{8}$

2 **a** $\mathbf{v}-\mathbf{u}$ **b** $\mathbf{w}-\mathbf{v}$ **c** $\mathbf{v}-\mathbf{w}$ **d** $\mathbf{u}-\mathbf{v}$ **e** $\mathbf{u}+\mathbf{w}-\mathbf{v}$

3 **a** $\binom{3}{^-1}$, $\binom{4}{2}$, $\binom{^-4}{^-2}$, $\binom{^-3}{1}$, $\binom{2}{7}$

4 **a** $\overrightarrow{PR}$ **b** $\overrightarrow{PR}$ **c** $\overrightarrow{PS}$

5 It is the zero vector.

6 $\mathbf{w}-\mathbf{u}$, $\mathbf{w}-\mathbf{u}$, $\mathbf{w}-\mathbf{v}$

9 $2\mathbf{v}-3\mathbf{u}$ and $4\mathbf{v}-6\mathbf{u}$, $6\mathbf{u}+4\mathbf{v}$ and $^-3\mathbf{u}-2\mathbf{v}$, $6\mathbf{u}-9\mathbf{v}$ and $3\mathbf{v}-2\mathbf{u}$

10 **a** $2\mathbf{b}-2\mathbf{a}$

11 **a i** $2\mathbf{b}-\mathbf{a}$ **ii** $(\frac{1}{3})(2\mathbf{b}-\mathbf{a})$ **c** 2 : 1 **e** It is the centre of the triangle.

12 **a i** $2\mathbf{p}+\mathbf{q}$ **ii** $\mathbf{p}+\frac{1}{2}\mathbf{q}$ **iii** $(\frac{1}{5})(4\mathbf{p}+\mathbf{q})$

## In focus 25

1 **a i** 1 **ii** 11 **iii** $5\frac{1}{2}$ **iv** $\frac{^-1}{8}$ **v** $^-5$ **vi** 3 **b i** 15 **ii** 0 or $^-6$ **iii** 0.8

2 **b** 0.9, 2 **c i** 320 **ii** extrapolated far beyond range of data.

3 **b ii** $^-5$, 50

4 **b** 0.4, $^-2.4$ **c** 7.6

5 **b i** reflection in $x$-axis **ii** reflection in $y$-axis **iii** translation $\binom{0}{3}$ **iv** translation $\binom{^-3}{0}$ **v** translation $\binom{2}{1}$

7 **b** $y = k(x + 2) + 1$

## EXAM QUESTIONS

**N1.1** No, 3.5 yards is about 3.15 m.

**N1.2** 25 km per hour

**N1.3** 125 m

**N1.4** 0832

**N1.5** **a** 4 hours 51 minutes **b** 5 hours

**N2.1** **a** 12 000 000 **b** 50 **c** 0.5

**N2.3** **a** 8.390 996 312 **b** 1.363 234 15

**N2.4** $40 \times 0.03 = 1.2$, so his answer is about ten times too big.

**N2.6** 14%

**N2.7** $\frac{80}{(0.04 \times 5)} = 400$

**N3.1**  a  17  b  60

**N3.2**  a  24  b  144

**N3.3**  a  3  b  $2^3 \times 3^3$

**N3.4**  $\frac{325}{999}$

**N3.5**  a  $3\sqrt{2}$  b  $7\sqrt{2}$

**N3.6**  b i  $\sqrt{3}$  ii  54

**N3.7**  a i  $7 - 4\sqrt{3}$  b  $2 + \sqrt{3}$

**N3.8**  $3 + 2\sqrt{2}$

**N3.9**  a  $11 + 6\sqrt{2}$
    b i  0  ii  The graph cuts the $x$-axis where $x = \sqrt{3} + 2$.

**N4.1**  a  $\frac{2}{3}$  b  16

**N4.2**  a  $1\frac{4}{5}$  b  $5\frac{1}{3}$  e  5

**N4.3**  5

**N4.4**  a  $\frac{13}{20}$  b  £10 250

**N4.5**  £3.75

**N4.6**  £36

**N4.7**  £950

**N4.8**  £8

**N4.9**  93.75%

**N4.10**  78%

**N4.11**  14%

**N4.12**  a  29 000  b  60 973  c  1989

**N4.13**  32

**N4.14**  £2289.80

**N4.15**  The area is decreased by 23%.

**N4.16**  a  2.4 m  b  1.92 m  c  4 bounces

**N4.17**  $\frac{16}{33}$

**N4.18**  $\frac{7}{33}$

**N5.1**  a i  $3.9 \times 10^5$  ii  390 000  b i  $6.7 \times 10^{-3}$  ii  0.0067  c  $2 \times 10^6$

**N5.2**  a  $24 \times 10^3$  b  $8 \times 10^3$  c  $1.6 \times 10^{11}$

**N5.3**  a i  $\frac{1}{4}$  ii  1  iii  10  b  10  c  $10a^2b^4c$

**N5.4**  a  2.15%  b  30.2 million tonnes

**N5.5**  a  $3.604 \times 10$  b  £1070

**N5.6**  43 minutes

**N5.7**  a  4.25  b i  $4 \times 10^4$  ii  $1.6 \times 10^7$

**N5.8**  a  $\frac{1}{9}$  b  $2^5$  c  $a = 6, b = 2$

**N5.9**  a  5  b  5  c  1

**N5.10**  a  $\frac{1}{16}$  b  6  c  9

**N5.11**  1

**N5.12**  a  $2^0$  b  $2^6$

**N5.13**  a  1.06640625, 1.066650391, 1.066665649
    b  tending to 1.066666 ...  c  1.066666 ...

**N5.14**  a  $x^{1/2}$  b  $\frac{1}{8}$

**N5.15**  $x = {}^-4$

**N6.1**  £69.30, £25.20

**N6.2**  a  35  b  40

**N6.3**  40 to 35

**N6.4**  a  37.5%  b  24

**N6.5**  a  4  b  1.44

**N6.6**  $y = 1.5, x = 400$

**N6.7**  4

**A1.1**  a  19.4°F  b  $^-40°$  c  $C = \frac{(5F - 160)}{9}$

**A1.2**  $x = 3.5$ cm

**A1.3**  2.2

**A1.4**  a  $9x + 15$  b  5 hours

**A2.1**  a  $^-1$  b  2.8

**A2.2**  a  2.25  b  $2x^2 + 5x - 3$

**A2.3**  a  0.25  b  $3x^2 - 5x - 2$  c  0, 8

**A2.4**  1.5, $^-1$

**A2.5**  1.4, 2.1

**A2.6**  a  $2x + 3y = 10.75, x + 2y = 6$  b  £3.50, £1.25

**A2.7**  a  $(0, {}^-1)$  b  $\frac{1}{2}$  c  They do not intersect

**A2.8**  a  $2x(x + 2)$  b  $x^2 - 6x - 7$  c  $^-5, 2$  d  2, $^-1$

**A2.9**  a  $^-\frac{11}{5}$  b  $^-\frac{2}{3}, 5$

**A2.10**  a  $^-\frac{1}{4}$  b  $5y(2 - 3y)$  c  $\frac{3}{4}, 3$

**A3.1**  1.3

**A3.2**  3.4

**A3.3**  a  7.6  b ii  9.6 cm

**A3.4**  a  $(2x - 3)(3x + 4)$  b  $^-0.77, 7.77$

**A3.5**  $^-7, 5$

**A3.6**  $^-3.64, 0.14$

**A3.7**  a  $(2x + 3)(x - 5)$  b  $(x + 5y)(x - 5y)$

**A3.8**  b  20 cm

**A3.9**  $^-0.41, 2.41$

**A3.11**  $^-1.5, 2$

**A3.12**  a  25, $^-5$  b  2.76, 7.24  c  $^-5$

**A3.13**  a  5, $^-18$  b  0.76, 9.24

**A3.14**  2, $^-3$

**A3.15**  a  $\frac{(2x - 1)}{(x + 2)}$  b  4, 1

**A3.16**  $\frac{1}{4}$

**A3.17**  b  $(^-4, 3), (^-3, 4)$

**A4.1**  $y = 6 - \frac{3}{2}x$

**A4.2**  $^-0.4, 2.4$

**A4.3**  a  $^-2.65, 2.65$  b  $^-7$

**A4.4**  $^-2.85, 3.85$

**A4.5**  $^-0.5, 6.5$

**A4.6**  1.6, 3

**A4.7**  a  £17  b  0.25  c  $B = 0.25t + 17$

**A4.8**  a  $(6, 5.5)$  b  2

**A4.9**  b  2, $^-5.5$

**A4.10**  b  1.2  c  $^-5.2$

**A5.1**  $x < {}^-2$

**A5.3**  a  4, 5, 6, 7, 8  b  $y \leqslant 1\frac{2}{3}$

**A5.4**  a  $x > 7$  b i  $12x^5$  ii  $9y^6$  c  $2x^2 - 7x + 3$

**A5.5**  a  $5(a - 2)$  b  $x \leqslant 7$

**A5.8**  $x < {}^-\frac{1}{2}$

**A5.10**  a  $x + y = 8$  c  $30x + 24y \geqslant 180$

**A6.1**  a  $11x + 7$  b  $9x^2 - 6x + 4$

**A6.2**  a i  $13x + 1$  ii  $x^2 + 2x - 8$  b i  $(x + 3)(x + 5)$  ii  $^-3, ^-5$

**A6.3**  a  $24abc$  b  $x$

**A6.5**  $4x^2 + 9y^2 + 12xy$

**A6.6**  a  $3x(x - 2)$  b  $3x^2 - 10x - 8$  c  $\frac{(4w - 3)}{5}$

**A6.7**  a  $t^8$  b  $p^4$  c  $a^4$

**A6.8**  a i  $a^{-3}$  ii  $32p^5q^{10}$  b  $y^2 - 14y + 49$  c  3, 5

**A6.9**  a i  $6a^5$  ii  $3x + 13$  b i  $t = \frac{(v - u)}{a}$  ii  3

**A6.10**  $x = \frac{5y}{(1 - y)}$

**A6.11**  a  $p = \frac{(t + 50)}{7}$  b  $x = \frac{y}{(k + 3)}$

**A6.12**  $\frac{x}{(x + 3)}$

**A6.13**  $h = \frac{s}{2\pi r^2 + 2\pi r}$

**A6.14**  $8y^6x^{12}$

**A6.15** **b** 38.7 $cm^3$

**A6.16** **a** 9.26 **b** $V = \sqrt{(PR)}$

**A6.17** **a** $x^2 + a^2$ **b** $\frac{(x-y)}{(x-2)}$

**A6.18** **b** $^{-}0.78$, 1.28

**A6.19** $\frac{6}{(x-2)}$

**A6.20** $c = \frac{2a}{(1+a)}$

**A6.21** **a** $\frac{17a}{30}$ **b** $\frac{1}{3}$ **c** 8, 6

**A7.1** $n^2 + 2n = n \times (n + 2)$

**A7.2** $r^2 - (r-1)^2 = 2r - 1$

**A7.3** **a** $3n - 1$ **b** $n^2 + 1$

**A7.4** **a** **i** $n^2$ **ii** $4n^2$
**b** One number will be odd, the other will be even. Odd × even = even.

**A7.5** **a** $n^2$ **b** $n^2 + 5$

**A7.6** **a** $n^2$ **b** $n^2 + 1$ **c** $n^2 + n + 1$

**A8.1** $\frac{^{-}3}{5}$, 3

**A8.2** **a** **i** C **ii** A **iii** D

**A8.3** $a = 0.17$, $b = 0.6$ **b** 4.04

**A8.4** **a** $y = 1$ **b** $^{-}6.1$, 1.1

**A8.5** **b** $^{-}1$, 5 **d** $y = 4 - x$

**A8.6** **b** **ii** $^{-}1.2$, 4.2

**A8.8** **a** 5 **b** $\frac{1}{2}$ **c** $x = \sqrt[n]{0.2y}$

**A8.10** $p = 0.4$, $q = 3.6$

**A10.1** **a** $x^{-2}, \frac{1}{x}, x^{\frac{1}{2}}, x$ **b** $x, x^{\frac{1}{2}}, \frac{1}{x}, x^{-2}$

**A10.3** **a** 2.2 **b** 2.1

**S1.2** **a** 105° **b** $15^2 + 36^2 = 39^2$

**S1.3** 36 cm, $q = 135°$

**S1.5** **a** 135°

**S1.6** 108°, 72°, 72°

**S1.7** **a** **i** $x°$ **ii** $2x°$ **b** 64°

**S1.8** **a** 56° **b** **i** 93° **ii** 4.4 cm

**S1.9** **a** **i** 36° **ii** 18° **b** **i** parallel **ii** 36°

**S2.1** 32°

**S2.2** **a** 56° **b** 124°

**S2.3** 62°

**S2.4** **a** 50° **b** 130°

**S2.5** **a** **i** 90° **ii** 25° **b** 137°

**S2.6** **a** 38°

**S2.7** **a** 30° **b** 35° **c** 10°

**S2.8** 118°, 42°

**S2.9** **a** **i** 160° **ii** 10° **iii** 40° **iv** 40°

**S2.10** **a** 79° **b** 25°

**S2.11** 15.8 cm

**S2.12** **a** 34° **b** 146°

**S2.13** **a** 26° **b** 26° **c** 13° **d** 77°

**S3.1** **a** B

**S3.2** $2\pi rh$, $\pi r\sqrt{(r^2 + h^2)}$, $\pi r(l + r)$

**S3.3** **a** area **b** none **c** volume

**S3.4** **a** $2(v + 2w + x + y + z)$ **b** $\frac{1}{2}(x + y)w$

**S4.1** 176.7 $cm^2$

**S4.2** 79.2 $m^2$

**S4.3** 16.2 litres

**S4.4** 128.25 g

**S4.5** 350 $cm^3$

**S4.6** 44.7 $cm^2$

**S4.7** **a** $487.5\pi$ m **b** 0.48

**S4.8** **a** $16\pi$ $cm^2$ **b** 3i 32 $cm^2$ **ii** $4\pi - 8$

**S4.9** 18.86 m

**S4.10** 150.8 $cm^3$

**S4.11** $90\pi$

**S4.12** 4.3 cm

**S4.13** 102.2 cm

**S4.14** **a** 424.1 $mm^3$ **b** 15.2 g

**S4.15** **a** 139.3° **b** 2049 $cm^2$

**S4.16** 42.4°

**S4.17** **a** 65 mm **b** 67.4° **c** 416 $cm^2$

**S5.6** **b** 6.9 cm **c** 41.8°

**S6.1** **b** $\frac{1}{2}(a + b)c$

**S6.2** **a** $y = 0.1x + 480$ **b** £1020

**S6.3** **a** 12 cm **b** 2 **c** 10 kg

**S6.4** **a** 4.2 m/s

**S6.5** $B = 0.2t + 12$

**S6.6** **a** $y = 0.4x + 18$ **b** £58

**S7.1** 2.1 m

**S7.3** 9.8 units

**S7.4** **a** 19.2 km **b** 9.8 km

**S7.5** **a** **i** 15.6 cm **ii** 16.7 cm **b** 43.8°

**S7.6** 470.4 mm

**S7.7** **a** $\frac{3}{5}$ **b** **i** $\frac{4}{5}$ **ii** 9 cm

**S7.8** **a** 44.4° **b** 631.9 cm

**S7.9** **b** 13.7 cm

**S7.10** 6.05 cm

**S7.11** 12.9 cm

**S7.12** **a** 484.3 m **b** 34 430 $m^2$

**S7.13** **b** 42.7 cm

**S7.14** 15 $cm^2$

**S7.15** 16°

**S7.16** 38 $cm^2$

**S7.17** 108°

**S8.1** **a** reflection in $y = x$ **b** vertices at (3, 2), (5, 0), (9, 4)

**S8.2** **a** reflection in $y = {}^{-}1$ **b** rotation of 90° about ($^{-}1$, $^{-}1$)

**S8.3** **a** rotation of 180° about (7, 4) **b** vertices at (4, 3), (2, 4), (6, 6)

**S8.4** **a** translation $\binom{3}{-6}$ **b** reflection in $y = {}^{-}x$
**c** vertices at (2, $^{-}2$), (2, $^{-}4$), (3, $^{-}2$)

**S8.5** **a** vertices at ($^{-}2$, $^{-}1$), ($^{-}1$, $^{-}2$), ($^{-}2$, $^{-}4$)
**b** enlargement with scale factor $^{-}2$ and centre (1, $^{-}3$)

**S8.7** vertices at (5, 0), (5, $^{-}2$), (2, $^{-}2$)

**S8.8** **a** vertices at (3, 1), (4, 1), (4, $\frac{^{-}1}{2}$)
**b** enlargement with scale factor $^{-}2$ and centre (2, $^{-}1$)

**S8.9** **a** vertices at (1, 0), (5, 0), (4, $^{-}4$)
**b** vertices at ($^{-}10$, 0), ($^{-}8$, $^{-}6$), ($^{-}2$, 0)

**S8.10** **a** **i** triangle *DEF* **ii** triangle *CLF*
**b** reflection in the line *JD*, rotation of 180° about *O*
**c** **i** $\frac{^{-}1}{2}$ **ii** The point is in the centre of triangle *OJH*.

**S9.1** **a** 70° **b** 10 cm

**S9.2** **b** 13.3 cm

**S9.3** 2.1 m

**S9.4** 3 cm

**S9.5** 948 ml

**S9.6** 120 mm

**S9.7** £90

**S9.8** **a** 11 cm **b** 7.5 cm

**S9.9** **a** 22° **b** 7 cm **c** **i** X ii SAS

**S9.10** $PB = BQ$ and $AP = AQ$ (radii), so they are congruent by SSS.

**S10.1** a $^{-}2\mathbf{b}$ b $6\mathbf{b} - 2\mathbf{a}$

**S10.2** a $\mathbf{a} + \mathbf{b}$ b $\frac{1}{4}\mathbf{a} + \frac{1}{4}\mathbf{b}$ c 56°

**S10.3** b $\binom{4}{3}$

**S10.4** a i $\mathbf{a} + \mathbf{c}$ ii $^{-}\frac{1}{2}\mathbf{a}$ iii $\mathbf{c}\ \frac{1}{2}\mathbf{a}$ b $(\frac{1}{3})(2\mathbf{c} - \mathbf{a})$ ii $(\frac{2}{3})(\mathbf{c} + \mathbf{a})$

**S10.5** a i $\mathbf{c} - \mathbf{b}$ ii $\frac{1}{2}\mathbf{c}$ iii $\frac{1}{2}\mathbf{c}$ b It's a parallelogram.

**S10.6** a i $\mathbf{b} - \mathbf{a}$ ii $2\mathbf{b} - \mathbf{a}$ b ii $3(\mathbf{a} - \mathbf{b})$ iii trapezium

**S10.7** a $4(\mathbf{q} - \mathbf{r})$ b $\mathbf{q}$ c $3(\mathbf{r} - \mathbf{p})$

**S11.1** 197°, 343°

**S11.2** 139°, 221°

**S11.3** a 120° b 200°, 340°

**S11.4** a 30°, 150° b $x = 229°, 311°$

**S11.5** b $x = 30°, 90°, 150°, 270°$

**S11.6** a 60°, 300° b 210°, 330°

**S11.7** a 300° b $x = 120°, 240°$

**S11.8** a Turn the graph of $y = \sin x$ upside down, and stretch it so that $^{-}2 \leqslant y \leqslant 2$.
b $x = {}^{-}307°, {}^{-}53°, 307°$

**D1.1** a 14 b 0.33
c Yes, the probability of obtaining a tail would be 0.5 otherwise.

**D1.2** a He has not thrown the dice enough times to check the relative frequency.
b $\frac{19}{100}$ c 190 times

**D1.3** b $\frac{21}{50}$

**D1.4** a $\frac{1}{3}$ b $\frac{1}{6}$ c $\frac{1}{2}$

**D1.5** $\frac{2}{35}$

**D1.6** a any multiple of 2 and the corresponding multiple of 3
b the multiple of 4 that corresponds to your answers to part a

**D1.7** a $\frac{3}{10}$ b 200 times

**D1.8** $\frac{1}{36}$

**D1.9** a $\frac{4}{5}$ b 0.048

**D1.10** a 0.72 b 0.26

**D1.11** a i $\frac{1}{12}$ ii $\frac{1}{15}$ b $\frac{1}{45}$

**D1.12** a 0.06 b 0.8676 c $0.1 \times 0.4^{n-1}$

**D1.13** $\frac{7}{15}$

**D1.14** $\frac{7}{15}$

**D2.1** a i frequency densities: 6, 11, 10, 5, 1.5, 0.2 ii 3.25 km
b The median, it is not affected by extreme values.

**D2.2** a frequency densities: 0.8, 1, 0.65, 0.9, 0.1 b $\frac{17}{150}$

**D2.3** a i frequency densities: 2.6, 6.5, 3, 1.2 ii 36
b male: 8, 48, 12 female: 2, 24, 6
c To make sure all age groups are represented.

**D2.4** a frequency densities: 0.4, 1.9, 3.4, 2.6, 0.75 b 65

**D2.5** a midpoints: 2, 6, 10, 14, 18, 22 b 14.2 minutes

**D2.6** a frequency densities: 2.6, 5.1, 34.8, 21.2, 4.2, 1.2 b decreased

**D3.1** b 29 cm, 59 cm c *P*, the correlation is better here.

**D3.2** b Jane's guesses are more closely correlated to the actual weights.

**D3.3** b 175 cm c median, less affected by extreme values

**D3.4** b i 11 cm ii 21.5 cm
c part i, the fit is not good for higher numbers of days

**D3.5** b i 54 ii 370

**D3.6** b i 25 ii 52 hundredths of a second
c The 30 year olds have higher reaction times on average.
The spread of reaction times is similar for each age group (relative to the median).

**D4.1** 33 minutes

**D4.2** Probably, as the estimated mean weight is 56.4 kg, giving 1240 kg in total.

**D4.3** a £26.40 c The shoppers on Saturday spent more on average.

**D4.4** a The diameters are bigger on average for the lettuces planted by hand.
b There is less variation in the diameters of the lettuces planted by machine.

**D4.5** a 36 years
b i The males are younger on average. The spread is larger in the females ages.
ii 67

**D4.7** a The spread is larger for the boys' results.

**D5.1** b 45 c i 52 years ii 28 years
d The people living in the town are younger on average.
There is also less variation in their ages.

**D5.2** a i 670 hours ii 400 hours
c MOONBEAM, they last longer on average and show less spread.

**D5.3** a i cumulative frequencies: 20, 150, 302, 394, 480, 498, 500
b i 26 years ii 17 years c 16%

**D5.4** a 167 cm b 12 cm
c ii The boys are taller on average. They both have the same spread.

**D5.5** a 3 miles b 1.2 miles
c The range is too large because of extreme (high) values.

**D6.1** a Too vague, 'often' could mean different things to different people.
b People in the library are likely to read often.

**D6.2** a He will not be asking anyone who does not take any exercise.
He will not be asking any boys.
b i It is a leading question. ii There is no 'no' choice.

**D6.3** 28, 39, 33

**D6.4** a Not suitable: people without phones and people whose telephone numbers are not listed in the directory will be left out.
b Not suitable: people should be surveyed on different days and at different times of the day.
c Children and working adults may be under-represented.

**D6.5** a i Years 8 to 11 are not represented.
ii She may end up with more boys than girls or vice versa. The year groups will not be equally represented.
b He should randomly select 5 girls and 5 boys from each year group.

**D6.6** 200

**D6.7** a i The people who visit the shopping centre may not be representative of the people in the town. For example, there may be more women than men or children may be under-represented.
ii People without phones and people whose telephone numbers are not listed in the directory will not be represented.
b They should randomly select the same number of males and females from several consecutive age intervals.

## Practice Non-Calculator Paper

**1** $33\frac{1}{3}\%$

**2** a $x(x + 3)$
b 99 700

**3** a $6x^5y^3$
b $4x^2 - 9$

**4** a $a = 10, b = 5$
b 25

**5** a $\frac{\pi w^2 h}{12}$

**6** a $^{-}1$
b $m = \frac{y - c}{x}$

**8** 5

**9** a i to A ii to D iii to B

**10** a i $a^8$ ii $a^4$ iii $a^{12}$
b $(a^6)^2$

**11** a i 0.7, 0.3, 0.7, 0.3, 0.7 ii 0.49
b 0.83193

**12** a Uranus
b $5 \times 10^3$

**14** **a** 3
**b** $\frac{1}{25}$
**c** 64

**15** 22 m

**16** **b** **i** $x = 2.5, ^-5$ **ii** 30 cm

**17** **a** **i** 57° **ii** 33°
**b** 50°

**18** $y + x = 6$

**19** 88, 89 or 90

**20** $7 + 4\sqrt{3}$

**21** $a = \frac{bc}{b+c}$

**23** **b** 217°, 323°

**24** **a** **i** $y - x$ **ii** $\frac{1}{2}x$ **iii** $\frac{1}{2}y - \frac{1}{2}x$
**b** Parallel, and one is twice the length of the other.

**25** **a** $^-3.3$
**b** $^-1.7$, 5.7

## Practice Calculator Paper

**1** **a** £87.50

**2** 0

**3** A = (0, 3.5) B = (2.8, 0)

**4** **a** translation $\binom{-3}{2}$
**b** rotation, centre (4, 0), 90° clockwise

**5** 0, 1, 2, 3, 4

**6** **a** 1.7 m
**b** 15.7°

**7** 1.4 cm

**8** $^-2.2$

**9** $x = £5.20$, $y = £3.20$

**10** 37.25 years

**11** **b** 178
**c** 60

**12** **a** 2240
**b** 4400, 4000

**13** **a** £1329
**b** £1263

**14** **a** $\frac{17}{200}$
**b** $\frac{17}{198}$

**15** **a** $(x - 4)(x + 4)$
**b** $\frac{x+4}{2x+1}$

**16** **a** $a = 5, b = 11$
**b** 8.32, 1.68

**17** 10 cm

**18** 40°

**19** $x = \frac{3y}{y-1}$

**20** 31.4 m

**21** **a** **iii** $y = x^2 + 2$
**b** **i** $y = 3 \sin x$ **ii** $y = \sin x + 1$

**22** 0 or 7

**23** **a** 174.2 cm
**b** 942.5 $cm^2$